RADIATION EFFECTS IN SEMICONDUCTORS

RADIATION EFFECTS IN SEMICONDUCTORS

Edited by

JAMES W. CORBETT
State University of New York at Albany

GEORGE D. WATKINS
General Electric Research and Development Center

GORDON AND BREACH SCIENCE PUBLISHERS

London   New York   Paris

DEDICATION

To
Lee and Ross Corbett

Preface

This conference is one of a series which began at Gatlinburg (1959) and proceeded, Kyoto (1962), Royaumont (1964), Tokyo (1966) and Santa Fe (1967); the next meeting has been tentatively scheduled for 1972 at the University of Reading, Reading, England.

The conference is devoted to the basic science aspects of radiation-induced defects in semiconductors and to closely related subjects. Broadly speaking these aspects include the mechanisms of defect production, the identification of the defects, the nature of the defects (on the atomic scale) and the influence of these defects upon the macroscopic properties of the materials. There have been remarkable advances in the state of understanding of the field in the time spanned by these conferences. In fact, in many respects we know more about radiation-induced defects in semiconductors than in any other system—metals, insulators and ionic crystals. Still the more we know, the more we find out that we don't know. Fifteen years ago a naive faith in the simplicity of these systems was widely held; this faith has now been replaced by an appreciation of an exciting, fascinating complexity which remains to be completely unraveled.

The subject of the conference is an important area of science in itself. It has had a major impact on other areas of semiconductor science as well. Two examples will suffice to illustrate this. The first is the now well-known sensitivity of the radiation-defects to impurities; there is an interplay between semiconductor materials science and the defect science which continues to reveal new impurities or impurities in new states in the semiconductors. The second example is an exciting part of this conference, namely the advances in the theoretical treatment of vacancy defects in the diamond lattice; these may well represent a breakthrough in the treatment of all of the defects which produce levels deep in the forbidden gap, a major unsolved problem in solid-state physics.

The science treated in the conference is exciting in its own right, but it has the additional incentive of being relevant to a number of areas of technology. An early motivation in the field was the use of semiconductor devices in the vicinity of nuclear reactors; this motivation persists but with diminished expectations due to the sensitivity of such devices to radiation. The space programs added a new dimension by using solar cells and other devices in the belts of trapped particles around the earth and in the solar wind itself. The question of hardening devices against the presence of radiation has then a wide continuing interest. The recent emergence of the field of ion-implantation has broadened the relevance yet again; this field, intimately and almost inextricably related to defect studies, has the exciting promise of providing revolutionary advance in semiconductor technology.

To say that the science is relevant to technology is an understatement. *The technology is still very much science-limited.* There have been a number of striking scientific and technological advances which have been inter-related. The complexity of the processes and the sensitivity of device performance to the details of this complexity unequivocally argue that dramatic advances may still be ahead.

The papers at this meeting and the discussions which follow them attest to the progress which has been made and are an exciting portent of what will yet come. While some areas are relatively mature, others are quite virgin and the mature areas still have major areas to be mapped out.

See you in Reading!

J. W. CORBETT
G. D. WATKINS

Acknowledgements

The conference could not have succeeded without the help of many people. The speakers themselves made an indispensible contribution, but a special commendation should go to the members of the organizing committee who helped find the uniformly excellent roster of review speakers.

The vital financial support was graciously provided by the Air Force Cambridge Research Laboratories, the International Union of Pure and Applied Physics, the Office of Naval Research and the State University of New York at Albany. In this regard we must give our special thanks to Clarence D. Turner, Henry M. DeAngelis and Peter J. Drevinsky (AFCRL), Doran W. Padgett and Charles W. Causey (ONR) and Earl G. Droessler and Harry L. Frisch (SUNY/Albany) for their encouragement and unflagging interest in this endeavour.

We thank Professor John C. Corelli who served as Local Committee Chairman.

The conference had over ninety contributed papers submitted for consideration for presentation, but time limitations dictated that only fifty could be accepted. The program committee wrestled with this problem. Many strong papers could not be accepted; in some cases criteria dominated which were somewhat unrelated to the merits of an individual paper, e.g., the question of balancing the program among the various sub-fields. The program committee shall remain unnamed but not unthanked.

And of course special thanks goes to those at SUNY/Albany who actually carried the brunt of the work. The cheery disposition of Mrs. Virginia Cline was much appreciated by the conference participants and is cherished by us; she performed most ably, supported by Mrs. Diane McElroy, in the organizing and secretarial work before, during and after the conference. We also gratefully acknowledge the help of many who worked hard to help the conference, in particular, Paul Brosious, Intaik Chung, Young-hoon Lee, Dave Long, Oscar Neilson, Margaret St. Peters and Fred Strnisa.

Finally a word about the dedication. We do not presume to dedicate the efforts of all who contributed to this work—but special circumstances impell one of us to dedicate his own efforts as is shown.

Members of the Organizing Committee

P. Baruch, *Ecole Normale*
W. D. Compton, *Ford Motor Company*
J. H. Crawford, Jr., *University of North Carolina*
H. Y. Fan, *Purdue*
R. R. Hasiguti, *University of Tokyo*
A. B. Lidiard, *Atomic Energy Research Establishment, Harwell*
E. W. J. Mitchell, *University of Reading*
S. M. Ryvkin, *University of Leningrad*
V. S. Vavilov, *University of Moscow*
F. L. Vook, *Sandia Laboratories*
G. D. Watkins, *General Electric Research and Development Center*

J. W. Corbett, *SUNY/Albany*, Organizing Secretary

Contents

CONTENTS

IMPORTANT UNANSWERED QUESTIONS—1970†
(Radiation Effects in Semiconductors)

F. L. VOOK

Sandia Laboratories, Albuquerque, New Mexico 87115

This paper was given as the opening address at the 1970 Albany International Conference on Radiation Effects in Semiconductors, and it attempts to establish a general overview of the field by concentrating on recent research developments and important unanswered questions. The continuing importance of impurity-defect interactions, of microscopic defect identification, and of the necessity for more theoretical calculations are emphasized. The rapid development of the field of ion implantation and its close relationship with radiation effects studies are pointed out. It is predicted that research in compound semiconductors will increase rapidly with close beneficial interaction with ion implantation studies.

Three years ago Prof. J. W. Corbett opened the Santa Fe Conference with a paper entitled 'Important Unanswered Questions.'[1] The purpose of such a presumptuous talk at the beginning of a conference is, as he stated, to encourage the Conference participants, individually and collectively, to address themselves to important problems beyond the limits of the work they themselves have done, and to stimulate discussion and a critical examination of the progress that has been made and that is still required. I have also noticed that the opening address at one conference has influenced the direction and the progress of research that is reported at the next conference. Many of the questions raised by Prof. Corbett are being answered today. Let us re-examine some of those questions and the developments that have occurred in the intervening years up to this Albany Conference.

One question we may immediately consider is, 'What do we mean by progress?' By 'progress' I believe we mean the development of definitive experimental data and the simplest ideas that will explain all of the data. The field of radiation effects in semiconductors has rarely suffered from a lack of data; rather it has been constantly in danger of choking on a large quantity of unexplained facts. Fortunately much real progress has been made in recent years in interpreting the data. The greatest rate of progress has been made in several fields which were 'ripe' for development. I will attempt to mention several of those fields in this paper.

Three important and related questions raised by Corbett are, 'What are the impurities?', 'Are those all of the impurities?', and 'Are you sure?'. At

Gatlinburg[2] in 1959 we first realized the important role of oxygen in silicon, and there have been continuing studies of the identity of impurity-defect complexes since that time. For example, the identity of several-oxygen associated centers in silicon will be discussed at this conference.[3,4]

Two other impurities in silicon whose roles were specifically questioned by Corbett are carbon and germanium. It is very gratifying to see the research progress that has been made and that is also being presented at this Conference in studying the roles of carbon and germanium in silicon.[5–7] I would like to review some of the work on carbon and oxygen in silicon that illustrates the strong interrelation of several experiments and the continuing importance of defect-impurity interactions in this ripe area of investigation.

In 1965 Newman and Willis[8] studied oxygen- and carbon-doped silicon by infrared absorption measurements of localized modes. They also electron irradiated their samples at room temperature, but saw no radiation-induced bands relating to carbon-defect complexes. They did, however, show that the carbon content of the oxygen-containing silicon was indicated by the strength of a band at 1104 cm^{-1}, which was a shoulder on the interstitial 9μ oxygen band in oxygen-containing Si. In 1967 at Santa Fe, Moyer and Buschert[9] used precision lattice parameter measurements to show that the presence of carbon decreases the equilibrium Si lattice parameter and also increases the change in lattice parameter upon electron irradiation at room temperature, presumably by interaction of radiation-induced defects with the carbon.

Other infrared work had proceeded concurrently which was not known at first to be related to the

---

† This work supported by U.S. Atomic Energy Commission.

carbon problem. In 1966 Ruth Whan had studied electron- and neutron-irradiated oxygen-containing silicon and found two new bands, at 922- and 932-cm^{-1}, which were formed upon irradiation at low temperature in crucible-grown crystals.[10] Because these bands were found in oxygen-containing silicon but not in floating zone silicon they were assigned to oxygen-defect complexes. However, when Herman Stein and I looked at the formation of the 922- and 932-cm^{-1} bands in 1967, we noticed that these bands appeared very strongly in crystals with a large carbon-associated 1104 cm^{-1} band, and we were able to correlate the formation rate of the 922 and 932 cm^{-1} bands with the carbon content of the silicon as shown in Figures 1 and 2.[11,12] In addition, the irradiation temperature dependence of the formation rate of the 922 and 932 cm^{-1} bands had recently been measured[13] and interpreted to involve the impurity trapping of a single Si interstitial or vacancy. Since the formation of the 922- and 932 cm^{-1} bands did not interfere or compete

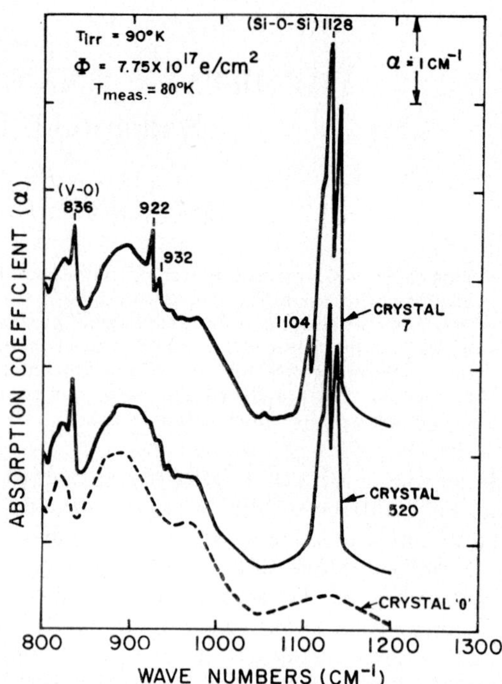

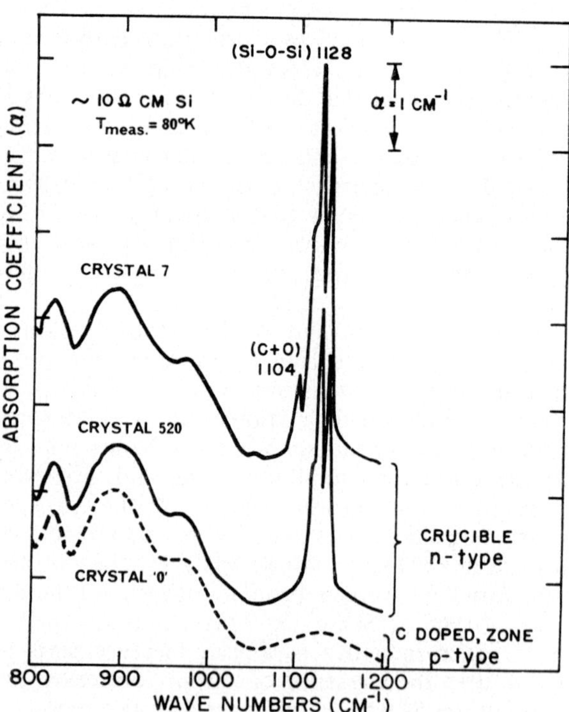

FIG. 1. Absorption coefficient at 80 °K versus wavenumber for a carbon-doped crystal '0', an oxygen-doped crystal 520, and an oxygen plus carbon-doped crystal 7. The data have been displaced along the ordinate to provide separation among the spectra which otherwise would coincide in the 800 to 1000 wavenumber region.

FIG. 2. Absorption coefficients at 80 °K for the crystals of Figure 1 after irradiation at 90 °K with 7.75×10^{17} 2-MeV electrons/cm². Data presentation analogous to that for Figure 1.

with the formation rate of the vacancy-oxygen A center, we concluded that the 922–932 cm^{-1} defect involved the trapping of a single silicon interstitial at a carbon containing center.[11,12] It is interesting to note that Moyer and Buschert rejected Si interstitial trapping and carbon replacement as an explanation for their experiments which were carried out at irradiation temperatures greater than the anneal temperature of the 922–932 cm^{-1} defect.

At this point the investigation was picked up again by Newman and coworkers. They made use of two of the new facts. One was that low temperature irradiations were necessary to observe the 922 and 932 cm^{-1} centers, because the centers annealed at room temperature and could not be seen after a room temperature irradiation. The second useful fact was that irradiations near 100 °K gave a larger formation rate of these centers than irradiation at lower temperatures. These facts have been used again at this conference in the work of Newman and Bean,[5] and by Brelot and Charlemagne[6] in their studies of carbon-containing samples.

Very recently,[14] and also at this conference,[5] Newman and Bean have used isotopic carbon doping of Si to show that a *single* carbon atom is incorporated in the 922–932 cm^{-1} defect. With better sensitivity and resolution they were able to observe the center in crystals that contained only carbon and concluded that the center does *not* involve oxygen. Is the center interstitial carbon? I will leave the development here, but I am eagerly looking forward to the next chapter in this investigation to be presented at this conference.

I think several of the questions about impurities and the warnings they imply are illustrated by the carbon example. What was first thought to be an oxygen-defect complex, was then thought to be a carbon-oxygen-defect complex, and may now be a carbon-defect complex. This change in assignment occurred in spite of the fact that infrared studies are rather selective in studying the defects. How much more difficult must be the interpretation of impurity dependent effects if we have only electrical or other macroscopic measurements to go on and no idea of what the inventory of defects is?

Nevertheless, experiments using carbon- and oxygen-doped silicon do illustrate that deliberately introduced impurities are useful to identify impurity-defect complexes and to study the effect of impurities on the production of defects. There are many papers at this conference that have used deliberately introduced carbon,[6,7] germanium,[7,8] lithium,[15–18] or oxygen[3,4] impurities in silicon to study the impurity dependence of defect production. We need to ask again, 'Are those *all* the impurities?' Since many of the samples contain more than one impurity, we may be studying not only the effect of one impurity, but rather the effects of multiple impurity interactions on defect formation.

The importance of impurities in the interaction with radiation-induced defects must also be considered in other materials. At this conference we will pay particular attention to impurity-related interactions in germanium at low temperature.[19–23] It was first thought that the 35 and 65°K annealing stages did not involve impurities, and then it was shown that the vacancy-oxygen center was formed below nitrogen temperature in germanium.[24] At this conference we have several papers that deal particularly with whether or not and in what manner defects interact with impurities in Ge below liquid nitrogen temperature.[19,21–23] These questions all relate to two more of Jim Corbett's questions which were: 'At what temperature does

long distance migration first take place, and what is moving?'

This raises a most important question, 'How can we identify defects?' The infrared studies indicate the power of this technique, especially when used with isotopic substitution in selectively separating and helping to identify defects. This technique is particularly valuable where little EPR data exist such as for germanium. We all know the success George Watkins has had in identifying defects using EPR measurements.[25] Since Santa Fe, the substitutional germanium-vacancy center in Si[26] and the Zn vacancy in ZnSe[27] have been added to his list of EPR identified defects. He has also identified optical absorption bands corresponding to the Zn vacancy in ZnSe.[27] This gives us a valuable start in the II–VI semiconductors, which he will review later at this conference.[28] Other workers are also using the EPR technique to identify defects. Brower[29] has concluded that the aluminum interstitial in silicon occupies the tetrahedral position, in contrast to the structure of the boron interstitial, which Watkins[25] has shown is nestled asymmetrically between two substitutional Si sites. At this conference new EPR identifications of multiple vacancy centers in silicon will be presented,[3] as well as other EPR studies in silicon and germanium.[16,17,30,31]

The power of recombination luminescence to identify defects will also be demonstrated at this conference, especially when coupled with the stress response.[4] Luminescence studies are not only applicable to silicon[4,15] and the compound semiconductors,[32–34] but improved results for germanium will be presented.[35] Photoconductivity is also being used to provide microscopic information on defects.[36–38] We need more studies of identified defects, particularly in germanium and the compound semiconductors.

Prof. Fan, in his opening address at the Royaumont Conference,[39] raised the question of whether identified defects can account for all the electrically active centers in irradiated Si. At Santa Fe it was concluded,[40] as you will hear from Herman Stein's review paper,[41] that for electron-irradiated *n*-type silicon, the major identified defects are indeed the dominant electrically active defects. In order to come to this satisfying conclusion it was necessary to use the identified defects to determine the production mechanisms of the two major kinds of defects in the defect inventory—the irradiation temperature dependent ITD defects and the irradiation temperature independent ITI defects.[40]

The concept of the 'inventory' of defects, which was developed at Santa Fe, will also be discussed at this conference. For example, Daly and Noffke[31] have used EPR identified defects to determine the inventory of defects in neutron-irradiated silicon.

This brings me to the next question, 'How can the kinetics and energetics of defect motion be used to identify defect annealing mechanisms?' Figure 3 shows the great success that has been obtained in relating many annealing measurements to the EPR identified divacancy and neutral vacancy defects[42] in Si. This graph of the time-temperature relationship of identified defect annealing is extremely useful in correlating macroscopic measurements of annealing in silicon. However, we do not have such a chart for germanium, and we need one. Both intuition and some evidence indicate that the germanium graph is similar to that for silicon with a highly mobile vacancy[24] and a less mobile divacancy.[42] However, in the compound semiconductors a vacancy would have to move on a given sublattice, and I would expect a monovacancy in the compound semiconductors to be less mobile than the AB divacancy. In fact, Watkins has

recently shown that the Zn vacancy in ZnSe has an activation energy of $\sim$1.25 eV,[27,28] which is about the same as that for the divacancy in silicon.

There are several papers on germanium at this conference that use kinetics of annealing of electrical measurements to infer defect annealing mechanisms. One word of caution; kinetic studies are best used either when identified defects are being directly measured or when the results are related to the known kinetics and energetics of identified defects. Where neither of these conditions exists, I think it is very difficult to interpret the macroscopic results.

At this point when we are considering the interpretation of data, I would like to recall another of Corbett's questions, 'Where is the theory?' I do not plan to review the theory since Dr. Stoneham[43] will do that later, but I will merely point out that the number of theoretical papers at this conference is extremely small, and much smaller than at Santa Fe. I believe the two papers to be presented on the electronic structure of the vacancy in silicon[44] and diamond[45] provide extremely valuable new results and show the continuing development of molecular orbital calculations, but I would have expected and hoped for a much greater activity in this area. It seemed that at Santa Fe remarkable progress had been made in the theoretical calculations of defect states,[46–48] and I would have expected this field to have expanded more vigorously.

At Santa Fe, Prof. Seeger and coworkers[49,50] proposed the concept that the structure of intrinsic defects is itself temperature dependent, especially at high temperatures. Because of the large entropy factors for silicon and germanium self diffusion, they suggested that at increasing temperatures interstitials and vacancies in Si and Ge take on a more relaxed or extended character. Seeger and coworkers suggest that self diffusion takes place in Ge and Si via these defects. On the other hand, Kendall and DeVries[51] have recently reviewed diffusion in Si and concluded that Si self diffusion is quite consistent with a divacancy process. In spite of the enormous information available there still is no widely accepted theory of self diffusion in silicon and germanium. Perhaps the analysis and EPR data of the multiple vacancy centers in Si to be presented at this conference[3] will provide us with additional helpful information to quantify ideas about important extended defects in silicon. We should also look for the relationship of such studies to the new field of amorphous semiconductors which will be reviewed by Prof. Cohen.[52]

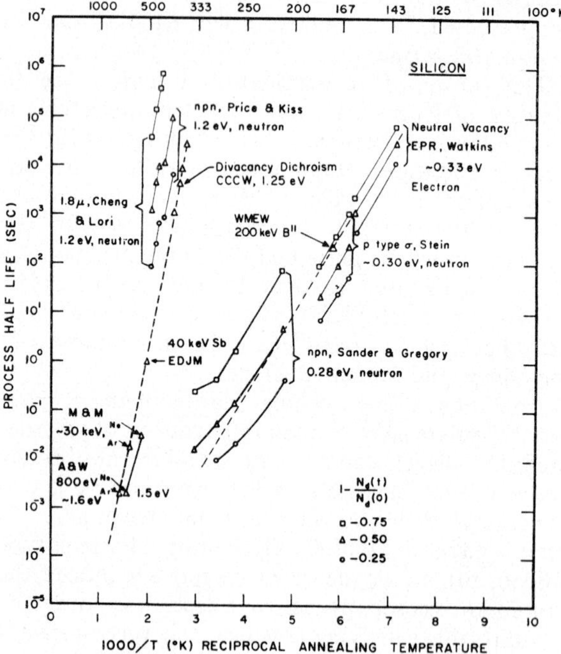

FIG. 3. Time-temperature domain comparison of experimentally observed annealing for electron, neutron, and ion bombardment of Si. The individual experiments are identified n Vook and Stein.[42]

A last question which is implicit in all the other questions is, 'Where are we going?' The large number of papers on germanium at this conference is understandable in terms of the interest and information presented at Santa Fe. The controversial identity and annealing mechanisms of the 35 and 65°K annealing stages were vigorously discussed at Santa Fe and I am sure that will also occur here. What is sorely needed is microscopic identification of the important defects involved, perhaps by EPR or optical measurements.

An area of research which was presented at Santa Fe in which a much greater than anticipated growth has occurred is ion implantation. Recently, the first International Conference on Ion Implantation in Semiconductors was held,[53] and significant progress was reported, especially for silicon. Ion implantation studies are very closely related to radiation effects studies, and I am sure that the review paper by Prof. Mayer[54] will show how radiation damage information and techniques using identified defects in silicon have been used to interpret ion implantation results. Rapid progress has been made by the almost direct transplantation of knowledge and techniques from radiation damage studies to ion implantation studies. It is interesting that the development of the field of ion implantation has rapidly followed the same historical path of radiation effects studies. Room temperature studies of electrical behavior are rapidly being followed by low temperature studies of structural behavior.

Conversely, there are several ways in which ion implantation is being used to study radiation damage effects. For example, the channeling and backscattering techniques developed in conjunction with ion implantation studies have been used in many fruitful studies to be presented at this conference. Other channeling effect lattice location studies[55] have given information which agrees with the EPR results of Watkins[25] on the structure of the boron interstitial in silicon. More results will be obtained as more implantation studies are extended to lower temperature. Furthermore, I am sure that the deliberate introduction of controlled impurities will have a tremendous capability in elucidating impurity-defect studies. In a new direction, ion implantation and radiation damage studies should provide a controlled way of studying the crystalline to amorphous transition in semiconductors, which is of much current interest in solid state physics.

Although radiation damage studies in Si have been very useful in interpreting ion implantation studies, I predict that ion implantation will be more of an equal partner in furthering the understanding of radiation effects in the compound semiconductors. In the closing address[56] at the Santa Fe conference Prof. Crawford asked, 'Where are the compound semiconductors?' Only two of the many papers at the Santa Fe conference dealt with other than the Group IV semiconductors Si, Ge, and diamond. In marked contrast, seventeen of the papers at the Albany conference deal with other than Group IV semiconductors, and *all* the contributed papers in the ion implantation session deal with compound semiconductors. Examples of the increased technological interest in the compound semiconductors are light-emitting diodes made from gallium arsenide, gallium phosphide, and gallium arsenide phosphide, as well as gallium arsenide Gunn effect microwave diodes. Because of great technological motivation as well as academic interest, I predict that at the next radiation effects conference even more papers will deal with compound semiconductors, and many of these will utilize ion implantation. Again what is desperately needed for rapid progress is the microscopic identification of specific defects.

In closing, one final question that I'm happy to say we have answered since the Santa Fe Conference was the first question asked by Jim Corbett 'Where are the Russians?' I'm sure I express the feelings of all of us that we are very happy that our Russian colleagues are able to be represented here in the flesh at the Albany Conference and that we can profit by their contribution both to the program and to the discussion.

REFERENCES

1. J. W. Corbett, *Radiation Effects in Semiconductors*, Ed. F. L. Vook (Plenum Press, New York, 1968), p. 3.
2. The Proceedings of the Gatlinburg Conference on Radiation Effects in Semiconductors was published in its entirety in the August issue of *J. Appl. Physics* (1959).
3. K. L. Brower, this conference.
4. C. E. Jones and W. D. Compton, this conference.
5. R. C. Newman and A. R. Bean, this conference.
6. A. Brelot and J. Charlemagne, this conference.
7. P. Baruch, F. Abel, C. Cohen, M. Bruneaux, D. W. Palmer and H. Pabst, this conference.
8. R. C. Newman and J. B. Willis, *J. Phys. Chem. Solids* **26**, 373 (1965).
9. N. E. Moyer and R. C. Buschert, *Radiation Effects in Semiconductors*, Ed. F. L. Vook (Plenum Press, New York, 1968), p. 444.
10. R. E. Whan, *Appl. Phys. Letters* **8**, 131 (1966); *J. Appl. Phys.* **37**, 3378 (1966).

11. F. L. Vook and H. J. Stein, *Appl. Phys. Letters* **13**, 343 (1968).
12. H. J. Stein and F. L. Vook, *Rad. Effects* **1**, 41 (1969); *Bull. Am. Phys. Soc.* **12**, 346 (1967).
13. R. E. Whan and F. L. Vook, *Phys. Rev.* **153**, 814 (1967).
14. A. R. Bean and R. C. Newman, *Solid State Communications* **8**, 175 (1970); A. R. Bean, R. C. Newman and R. S. Smith, *J. Phys. Chem. Solids* **31**, 739 (1970).
15. E. S. Johnson and W. D. Compton, this conference.
16. B. Goldstein, this conference.
17. J. A. Naber, H. Horiye and B. C. Passenheim, this conference.
18. N. B. Urli, this conference.
19. J. W. MacKay, this conference.
20. V. S. Vavilov, A. V. Spitsyn and M. V. Tchukichev, this conference.
21. J. M. Meese and J. W. MacKay, this conference.
22. R. A. Matula and E. E. Klontz, this conference.
23. J. Bourgoin and F. Mollet, this conference.
24. R. E. Whan, *Appl. Phys. Letters* **6**, 221 (1965); *Phys. Rev.* **140**, A690 (1965).
25. G. D. Watkins, *Radiation Effects in Semiconductors*, Ed. F. L. Vook (Plenum Press, New York, 1968), p. 67; *Radiation Effects on Semiconductor Components* (Journées D'Electronique, Toulouse, France, 1967), Vol. I, A1.
26. G. D. Watkins, *IEEE Trans. on Nuc. Sci.* **NS16**, 13 (1969).
27. G. D. Watkins, *Bull. Am. Phys. Soc.* **14**, 312 (1969), **15**, 290 (1970).
28. G. D. Watkins, this conference.
29. K. L. Brower, *Phys. Rev.* **1**, 1908 (1970).
30. A. Hiraki, this conference.
31. D. Daly and H. E. Noffke, this conference.
32. F. J. Bryant and D. H. J. Totterdell this conference.
33. I. I. Geiczy, A. A. Nesterov and L. S. Smirnov this conference.
34. G. W. Arnold, R. E. Whan, J. K. Maurin and J. A. Borders, this conference.
35. R. J. Spry and J. D. Henes, this conference.
36. C. S. Chen and Corelli, this conference.
37. V. D. Tkachev and M. T. Cappo, this conference.
38. P. Vajda and L. J. Cheng, this conference.
39. H. Y. Fan, *Radiation Damage in Semiconductors*, Ed. P. Baruch (Dunod, Paris, 1965), p. 1.
40. F. L. Vook and H. J. Stein, *Radiation Effects in Semiconductors*, Ed. F. L. Vook (Plenum Press, New York, 1968), p. 99, p. 115.
41. H. J. Stein, this conference.
42. F. L. Vook and H. J. Stein, *Rad. Effects* **2**, 23 (1969); **2**, 139 (1969); and **6**, 11 (1970).
43. A. M. Stoneham, this conference.
44. F. P. Larkins, this conference.
45. R. P. Messmer and G. D. Watkins, this conference.
46. M. Lanoo and A. M. Stoneham, *Radiation Effects in Semiconductors*, Ed. F. L. Vook (Plenum Press, New York, 1968), p. 43.
47. M. Lanoo, G. Leman and J. Friedel, *Radiation Effects in Semiconductors*, Ed. F. L. Vook (Plenum Press, New York, 1968), p. 37; *Phys. Rev.* **164**, 1056 (1967).
48. J. Callaway and A. J. Hughes, *Radiation Effects in Semiconductors*, Ed. F. L. Vook (Plenum Press New York, 1968), p. 27; *Phys. Rev.* **156**, 860 (1967).
49. A. Seeger and K. P. Chik, *Radiation Effects in Semiconductors*, Ed. F. L. Vook (Plenum Press, New York, 1968), p. 53; *Phys. Stat. Sol.* **29**, 455 (1968).
50. G. Schmid, K. P. Chik and A. Seeger, *Radiation Effects in Semiconductors*, Ed. F. L. Vook (Plenum Press, New York, 1968), p. 60.
51. D. L. Kendall and D. B. DeVries, *Semiconductor Silicon*, Ed. R. R. Haberecht (The Electrochemical Society Inc., New York, 1969), p. 358.
52. M. Cohen, this conference.
53. The Proceedings of The International Conference on Ion Implantation in Semiconductors has been published in its entirety in recent issues of *Radiation Effects*.
54. J. W. Mayer, this conference.
55. J. C. North and W. M. Gibson, *Appl. Phys. Letters* **16**, 126 (1970).
56. J. H. Crawford, Jr., *Radiation Effects in Semiconductors*, Ed. F. L. Vook (Plenum Press, New York, 1968), p. 469.

THE THEORY OF DEFECTS IN IRRADIATED SEMICONDUCTORS

A. M. STONEHAM

Theoretical Physics Division, B. 8.9, AERE, Harwell, Didcot, Berkshire, U.K.

Developments in the theory of defects in semiconductors are reviewed, with emphasis on changes since the Santa Fe meeting and on the points which were controversial then. It has become clear that most of the desirable simplifications are invalid for vacancies in diamond and silicon, and probably also for similar systems. In particular one cannot use the one-electron approximation, the rigid lattice approximation, the Jahn-Teller instability with coupling within isolated levels alone, nor some of the conceptually acceptable treatments of the lattice distortion. The successes and weaknesses of the various approaches are reviewed, and some of the potential methods are described.

1. INTRODUCTION

At the last conference in this series Fan[1] gave a lucid summary of the theory up to 1967. I do not intend to repeat his description. Instead I shall concentrate on points which are new and on points which are controversial. My discussion will be confined almost exclusively to isolated vacancy centres in valence crystals such as diamond and silicon, since the majority of theoretical work has concerned these cases. The principles of recent work on the divacancy[2,3] are, of course, similar. There has been no significant progress, to my knowledge, in the understanding of the interstitial, which is the only other elementary intrinsic defect. Nor do theorists seem to have gained sufficient confidence to follow the suggestion at the last conference[4] that the study of vacancy-impurity complexes should be fruitful.

2. ELECTRONIC STRUCTURE IN A RIGID LATTICE

2.1 *Present situation*

I want to outline first some of the calculations of the electronic structure of the vacancy in a rigid (possibly distorted) lattice. The pioneering paper in this field was by Coulson and Kearsley[5] and stimulated almost all subsequent work.

Coulson and Kearsley observed that, in the creation of a neutral vacancy, four bonds were broken. They argued that the observed properties of the defect should be determined mainly by the four electrons on the nearest neighbours which previously participated in the bond, the 'defect electrons'. The energy levels and wavefunctions of this four-electron system were then worked out in detail by standard molecular methods. This approach has been followed by subsequent workers, with changes in detail.[6,7] The advantages of the approach are its rather good treatment of the Coulomb interaction between the defect electrons and the convenient wavefunctions for use in estimates of other observable parameters. Its most conspicuous weakness is the inflexibility of the wavefunction. This inflexibility comes from two restrictions: the assumptions that only the defect electrons are perturbed, and that the defect electrons can be represented by a combination of atomic orbitals determined by symmetry alone. These approximations become better the more compact the centre.

Another approach, initially due to Gourary and Fein,[8] and later extended,[9] also concentrates on the four defect electrons. However variational wavefunctions are used, in a form which greatly simplifies the calculation of the electron-electron interaction. Again the effects of the defect on other electrons are ignored. The important point is that the radial extent of the wavefunction can be varied, and the results confirm that the centre is indeed compact.

These two approaches agree in many respects. The order of the energy levels and the order of magnitude of their separations are similar, although there is some dependence on the one-electron orbitals. The actual order depends on a detailed balance between a number of effects: the electron-electron interaction, which tends to keep spins parallel, the fact that electrons prefer to occupy nodeless (A, rather than T_2 symmetry) one-electron wavefunctions, and certain kinetic energy terms which favour antiparallel spins. These effects have been discussed before[9] and I shall not give details.

Two points emerge from these calculations which caused controversy at the last meeting. The first was the importance of Coulomb correlation, i.e. the need for a proper treatment of the electron-electron interaction rather than a mean field theory of some sort. Thus a 'few electron' model seems necessary, and transitions between 'few electron', rather than one-electron states, considered. The magnitude of the correlation terms—a few per cent of the total energy—is well in line with that for other systems.[10] The importance of these terms does not arise because of particularly large correlation terms. It comes from the fact that the correlation energy varies from state to state and is comparable with the energy separations of the states. The second point to emerge was that several of the states of interest are electronically degenerate, so that a Jahn–Teller instability is likely to be important.

A third approach, also a molecular type of method, has many of the features of the two calculations outlined earlier. Watkins[11] analysed his own very complete resonance data in terms of a one-electron model in which suitable values of the various parameters were chosen to fit experiment, rather than calculated *a priori*. These values could be chosen consistently, and often in the general range expected by direct calculation. Thus his work confirms two of the main features of the other molecular approaches: the spin resonance is apparently dominated by the defect electrons, and the wavefunctions of these electrons are sensibly localised at the vacancy. There can be no real doubt about the localisation, typically 60 per cent at least on the neighbours of the vacancy. Whilst optical spectra may sometimes confuse resonant and bound states, spin resonance does not permit ambiguity. At the same time the results challenge the theoretical conclusion that the one-electron model must be replaced by a few-electron model.

Another one-electron model was used in the careful calculations of Callaway and Hughes[12] and a number of other workers[13,14] This approach followed the method of Koster and Slater[15] which recognises that, given the band structure of the host lattice and a local perturbation, one can exploit the short range of the perturbation. This is particularly attractive conceptually, and one can look at the effect of the defect on all the electrons, not just the defect electrons. The approach, however, has difficulties. It is a one-electron approach, and is not readily extended to examine many-electron effects. The detailed form of the perturbation is hard to estimate. On the one hand the perturbation cannot be found uniquely, even in cases like silicon where only one type of atom occurs and detailed

band structures are available.[16] A symptom of this is that Callaway and Hughes were unable to find bound states for the silicon vacancy without multiplying their potential by an arbitrary correction factor. On the other hand, it is not the potential which is needed directly, but the matrix elements of the potential between Wannier functions. If one asks what perturbation has non-zero matrix elements between Wannier functions at just a few sites, the answer is that it cannot be strongly localised and is also a non-local potential.[13] The most serious objection, however, is that the theory cannot, in principle, give the Hartree–Fock solutions, which are the best possible one-electron solutions, for any but the most weakly-bound states. The reason for the difficulty is easily seen. The perturbation alters the electronic wavefunctions, and these alterations in turn affect the electron-electron interaction. In the Koster–Slater approach one uses an unscreened (or at best a linearly-screened) potential, whereas for deep centres non-linear screening is essential.† The limits of validity (necessary but not sufficient) of such an approach prove to be very similar to those of effective-mass theory: the binding energy must be small compared with the band gap. Moreover, the assertion that the Koster–Slater approach gives energies accurately with respect to band edges is only valid for shallow states, and is misleading otherwise. The method does come into its own in some cases. These include isoelectronic defects[19] and in-band resonances, where the wavefunction is strongly perturbed near the defect but does not fall to zero far from the defect. Two recent abstracts describe the associated problems of phase shifts for scattering from vacancies.[20,21] The various theories are summarised in Table I.

2.2 *One-electron models or few-electron models?*

There are two arguments which suggest that it is possible to use a one-electron model. The first comes from the remarkable success of Watkins in predicting ground states of vacancies in silicon using such a picture. Since the last meeting Larkins[22] has shown that the few-electron model,

---

† Bennemann's calculation[17] goes beyond the linear approximation and does attempt to achieve self-consistency within the one-electron model. His method can, in principle, give solutions equivalent to Hartree–Fock results. It is possible[18] to extend the Koster–Slater method to achieve a degree of self-consistency, but the extended method proves to be just a special and rather inconvenient form of the method of localised orbitals, discussed later.

TABLE I

Theories of the vacancy in the Diamond structure

Model	Authors	Weaknesses
LCAO molecular model	Coulson and Kearsley; Larkins	(1)
Variational molecular model	Gourary and Fein; Stoneham	(1)
A posteriori molecular model	Watkins	(2)
t-matrix model	Bennemann	(2), (3)
Koster–Slater method	Callaway and Hughes; Kilby; Lannoo and Lenglart	(4), (3), (2)

(1) (*a*) Assumption that only defect electrons perturbed;
    (*b*) Inflexible trial wave functions.
(2) One-electron approximation.
(3) Choice of perturbing potential difficult.
(4) Bad for deep defects as poor treatment of electron–electron interaction.

with important correlation terms, also seems to give ground states of the correct multiplicity and symmetry. Thus the one-electron model is *not unique* in its ability to explain the experiments, and one cannot infer the one-electron model from its success in this case. A second argument is the recent attempt[23] to justify the one-electron model by reducing certain electron-electron interaction integrals from vibronic effects. This attempt is in error. The paper constructs many-electron vibronic states as a determinant of one-electron vibronic states. This is wrong: in the corresponding circumstances one should multiply a determinant of one-electron electronic functions by a vibrational function of the appropriate symmetry.

Occasionally one still hears the assertion that if we close our eyes to many-electron effects, they will somehow go away. This is not so. There appears to be no valid reason for believing one-electron models of vacancy centres at present except, with caution, as conceptual tools.

2.3 *Prospects: The method of localised orbitals*

One particularly important method has recently made the passage from molecular to solid state physics. It now seems possible to do a genuine self-consistent Hartree–Fock calculation for defects like the vacancy centres. This is the 'Method of Localised Orbitals', largely due to Adams[24] to Gilbert[25] and to Kunz.[26] The Hartree–Fock method is, of course, a one-electron method, and the one-electron wavefunctions are found by solving self-consistently a set of simultaneous equations for the one-electron energies. The method of localised orbitals takes advantage of the fact that the solution of these equations can be split into two parts. First one solves the problem of self-consistency by finding linear combinations of the one-electron eigenfunctions. It can be arranged that the choice of linear combination is easy to treat and convergence correspondingly rapid. Given these self-consistent combinations it is, of course, trivial to set up a secular equation to give the one-electron eigenvalues and eigenstates. The features of particular value are first that one can examine the effect of the defect on all electrons (both the defect electrons and the rest) in detail, without arbitrary intuitive assumptions: the molecular model can be checked. Secondly, one can treat crystals of indeterminate ionicity, e.g. SiC, the III–V's and the II–VI compounds. Thirdly, since one has accurate one-electron wavefunctions, it is possible to make a good estimate of correlation terms. Finally, the method allows one to retain the conceptual advantages of both the molecular methods (since the local orbitals resemble the atomic orbitals of the molecular model) and the Koster–Slater method (since the relation of the local orbitals to the free-crystal wavefunctions also remains clear). Such conceptual advantages are also shared by a number of approximate methods, for example those based on extended Hückel theory.[27–30]

3. LATTICE DISTORTION NEAR DEFECTS

Properties sensitive to distortion include the order and detailed positions of the electronic energy levels, as well as formation and migration energies. Calculations of the distortion near vacancies have been based on two assumptions.[31–38] First, the distortion results from local electronic reorganisation whose main effects can be represented by appropriate local forces on the nearest neighbours. Secondly, the harmonic approximation† is used to calculate the lattice response to these effective forces.

† In some cases the Gruneisen parameter is used in fixing the interatomic potential, suggesting an anharmonic treatment. The genuine anharmonic terms are, however, grossly oversimplified and apparently not too important. The Jahn–Teller effect, especially with accidental degeneracy, adds a very special sort of anharmonicity; this rather different phenomenon can be treated within the methods we describe.

3.1 *Recent developments*

Recent progress has refined the calculation of both the effective forces and the lattice response. The effective forces have been improved[7,38] by calculation from detailed molecular models with the use of better wavefunctions and with the inclusion of small terms which were arbitrarily dropped in earlier work. In particular one must be careful in including the promotion energy. In the present state-of-the-art there seems to be no justification whatever in using models which assume simple pairwise bonding of the defect electrons calculated using a Morse potential derived from data at a very different internuclear separation.

The improvements in the lattice response stem from two sources. The first is the availability of 'valence-force' potentials[39–41] for describing the interatomic interaction. The second is the improvement of computer methods for finding the equilibrium configuration of a lattice under external forces. The valence-force potential describes interatomic forces in ways familiar to chemists—bond-bending, bond-stretching and similar terms. It is, in fact, a mixture of two-, three- and four-body interactions adjusted to fit phonon dispersion curves. The fits for diamond and silicon are remarkably good, giving better fits than the shell model with far fewer parameters. Although one cannot derive unique interatomic force constants from dispersion data alone[42] this model is manifestly better than others now available.

Given these interatomic and local forces one can calculate the linear response of the lattice, and hence the contributions of distortion to a number of observables. In particular one obtains the Jahn–Teller effect, the relaxation contribution to the formation energy, and the volume change on formation. One striking result is that the local strains produced by external uniaxial stresses are profoundly affected by the presence of the vacancy (Table II). It is grossly inaccurate to assume a

TABLE II

The response to external uniaxial stress of the displacements of the neighbours of a vacancy. The table gives the ratios of these displacements for the crystal containing a defect to the displacements for the perfect crystal.

	Diamond	Silicon
Totally symmetric mode (*A*)	3.2 ± 0.2	10.6 ± 1.0
Tetragonal mode (*E*)	1.5 ± 0.1	1.4 ± 0.1
Trigonal mode (*T*)	1.9 ± 0.1	1.8 ± 0.1

perfect lattice in converting results from stress data into energy charges per unit local strain. Another remarkable feature is that the effective frequencies are much lower than those derived from other methods (Table III). Correspondingly, the relaxation contribution to the formation energy may be

TABLE III

Effective frequencies for different models of the lattice dynamics. Units are 10^{14} rad/sec.

	Larkins and Stoneham[38]	Lidiard and Stoneham[31]	Friedel, Lannoo and Leman[33]
Diamond:			
A mode	0.95	2.02	2.50
E mode	1.13	2.02	2.50
T mode	0.98	1.43	2.50
Raman frequency	2.49	2.50	2.50
Silicon:			
A mode	0.18	0.825	
E mode	0.21	0.825	Not
T mode	0.18	0.584	used
Raman frequency	0.985	0.970	

very large. Indeed, in silicon (but not diamond) the totally symmetric relaxation is so great that the model, as it stands, gives a negative formation energy. The explanation is very probably that there is local rebonding of electrons other than the defect electrons and also the terms in the electronic energy which are of higher order than linear in the displacements become important. The linear terms giving the local forces appear to be reasonable—the volume change per defect is sensible—and the forces are never much larger than the corresponding Morse potential values. Moreover, if we use more primitive lattice dynamic models (e.g. (34)) the formation energy has the right sign and general magnitude. What is very obvious is that it is meaningless to compare experiment with theoretical estimates which ignore lattice relaxation. Papers which omit relaxation include (43) and (44); their results should not be compared directly with experiment.

3.2 *Future developments*

The methods in the last section all make use of the concept of interatomic forces. It is, in fact, possible to calculate formation energies, including lattice relaxation, without using this feature at all. The two points of importance are first, that one wants to know the total energy (electron-electron,

electron-ion, and ion-ion terms) as a function of lattice configuration Q, and second that in certain special but useful cases this energy $E_{TOT}(Q)$ factors into a combination of two types of term: those which depend on the structure but not on the ions, and those which depend on the properties of the ions, not on the structure. For example, if a weak pseudopotential is used[45]

$$E_{TOT}(Q) \sim \sum_q \Gamma_q |S_q|^2:$$

$$S_q = \sum_{\substack{r_I \\ \text{ions}}} \exp(iq \cdot r_I).$$

More general expressions are possible in which the electron-ion interaction need not be weak. The distortion and corresponding energy changes are found as follows. The perfect crystal energy is $E_{TOT}(Q_0)$. When a defect is introduced into some unrelaxed configuration Q_I the energy becomes $E_{TOT}(Q_I)$. The total energy of the defect crystal can be expanded in terms of $[Q - Q_I]$ for small displacements, and the relaxation energy and distortion can be found by methods like those of Kanzaki.[46] The approach has the advantage that there is no assumption of two-body central forces; terms involving many ions are automatically included. The disadvantages are that there may be difficulty in deciding on an electron-ion interaction, and that it is hard to estimate the energy of an inhomogeneous electron system accurately. In several respects the method is an extension of that of Bennemann[43] to include distortion consistently.

3.3 Symmetries of distortions

One of the novel features of Watkin's interpretation[11] of the negative vacancy in silicon was that the Jahn–Teller effect lead to mixed trigonal and tetragonal distortions. His explanation in terms of successive Jahn–Teller effects is appealing, but is hard to relate immediately to the general result[47,48] that for any isolated level in cubic symmetry mixed distortions are never stable. Nor is it obvious how to take this intuitive model over to more complex systems.

This point has now been resolved.[49] Whilst *isolated* levels can show *either* trigonal *or* tetragonal distortions, the Jahn–Teller coupling *between* the non-degenerate levels can lead to mixed symmetry distortions when there is near degeneracy. This inter-level coupling is expected to be important quite often for defects in valence crystals, where near degeneracies occur rather frequently, and

Elkin and Watkins[50] have observed other examples. General results have also been derived which allow discussion of the same phenomena in the more complex many-electron formulation.

4. PROSPECTS AND CONCLUSIONS

One area certain to develop is the prediction of properties which can be measured directly. Several calculations[51,52] have discussed the optical line shape for the vacancy in diamond, for comparison with the GR1 band. Further work on formation energies is also likely, with (sooner or later) an extension to energies of motion. The theory of diffusion and of the reorientation of anisotropic defects has recently been undergoing a profound change[53–56] in that genuine quantum theories have been developed. It is clear that application of these theories will alter some of our views on atomic motion, and may possibly help the understanding of mysteries like interstitial motion in silicon.

Another area of likely development is the study of other crystals than silicon and diamond. Recent work includes discussions of SiC[57] and ZnSe.[58]

The conclusions are depressing in part, since many of the assumptions one would like to make break down. These include:

(A) The one-electron approximation;
(B) The Jahn–Teller effect with only coupling within isolated levels;
(C) The rigid (undistorted) lattice;
(D) The distorted lattice in which electronic energies are linear in the near neighbour displacements.

Against this background there have been some encouraging features. Apart from the benefits of understanding the difficulties, it now seems possible to eliminate many of the approximations and intuitive elements in the theory. Moreover, theorists have been calculating more quantities which have actually been observed, and the interaction with experiments which has resulted can only be fruitful.

REFERENCES

1. H. Y. Fan, *Radiation Effects in Semiconductors*, Ed. F. L. Vook (Plenum Press, New York, 1968), p. 17.
2. J. Callaway and A. J. Hughes, *Radiation Effects in Semiconductors*, Ed. F. L. Vook (Plenum Press, New York, 1968), p. 27.
3. C. A. Coulson and F. P. Larkins, *J. Phys. Chem. Sol.* **30**, 1963 (1969).

4. J. W. Corbett, *Radiation Effects in Semiconductors*, Ed. F. L. Vook (Plenum Press, New York, 1968), p. 3.

5. C. A. Coulson and M. J. Kearsley, *Proc. Roy. Soc.* **A241**, 433 (1957).

6. T. Yamaguchi, *J. Phys. Soc. Japan*, **17**, 1359 (1962).

7. F. P. Larkins, *D. Phil. Thesis*, Oxford, 1969, and to be published.

8. B. S. Gourary and A. E. Fein, *J. Appl. Phys.*, **33** (Suppl.), 331 (1962).

9. A. M. Stoneham, *Proc. Phys. Soc.*, **88**, 135 (1966).

10. See, for example, the review by R. Pauncz in *Physical Chemistry: an Advanced Treatise* III, Ed. H. Eyring, D. Henderson and W. Jost (Academic Press, 1969).

11. G. D. Watkins, Proceedings of the Symposium *Radiation Damage in Semiconductors* p. 97, (Dunod, Paris, 1965); *Radiation Effects in Semiconductors*, Ed. F. L. Vook (Plenum Press, New York, 1968), p. 67.

12. J. Callaway and A. E. Hughes, *Phys. Rev.*, **156**, 860 (1967).

13. G. E. Kilby, *Proc. Phys. Soc.*, **90**, 181 (1967).

14. M. Lannoo and P. Lenglert, *J. Phys. Chem. Sol.*, **30**, 2409 (1969). The assumptions in this paper appear to be very similar to those made by P. Goosens and P. Phariseau, *Physica*, **32**, 1713 and 1724 (1966) although the analysis is totally different.

15. G. F. Koster and J. C. Slater, *Phys. Rev.*, **95**, 1167 (1954).

16. See, for example, the comments of F. Herman and J. C. Phillips at the 1966 Kyoto Conference on the Physics of Semiconductors.

17. K. Bennemann, *Properties of Vacancies and Interstitials*, Ed. A. D. Franklin (N.B.S. Misc. Publ., 287), p. 127 (1966).

18. A. M. Stoneham, unpublished work (1969).

19. R. A. Faulkner, *Phys. Rev.*, **185**, 991 (1968).

20. J. Callaway and A. E. Hughes, *Bull. Am. Phys. Soc.*, **14**, 440 (1969).

21. R. A. Tawil, *Bull. Am. Phys. Soc.*, **15**, 258 (1970).

22. F. P. Larkins, this conference, and to be published.

23. W. E. Hagston, *J. Phys. C.*, 3, 791 (1970).

24. W. H. Adams, *J. Chem. Phys.*, **34**, 89 (1961); *J. Chem. Phys.*, **37**, 2009 (1962).

25. T. L. Gilbert, in *Molecular Orbitals in Chemistry, Physics and Biology*, Ed. P. O. Lowdin (Academic Press, 1964); and in *Sigma Molecular Orbital Theory*, Ed. O. Sinanoglu and K. B. Wibury (Benjamin, 1969).

26. A. B. Kunz, *Phys. Stat. Sol.*, **36**, 301 (1969).

27. R. P. Messmer and G. D. Watkins, this conference.

28. E. B. Moore and C. M. Carlson, *Sol. St. Comm.*, **4**, 47 (1965).

29. E. B. Moore, *Properties of Vacancies and Interstitials*, Ed. A. D. Franklin (N.B.S. Misc. Publ., 287), p. 31 (1966).

30. Particularly good general discussions of the Extended Hückel method are given by R. Hoffmann, *J. Chem. Phys.*, **39**, 1397 (1963); G. Blyholder and C. A. Coulson, *Theor. Chim. Acta. (Berl.)*, **10**, 316 (1968). Most discussions of Hückel theory concentrate solely on the π-orbitals of planar conjugated and aromatic systems. Hoffmann's article considers systems with both σ- and π-orbitals. He concludes that the theory predicts well the relative σ, π importance and the charge distribution, that the three-dimensional shapes are predicted adequately, but that the method performs 'miserably' for spectral predictions. Blyholder and Coulson's paper largely confirms these points, and verifies the rather low accuracy of the method, particularly for total (as opposed to ionisation) energies. The second article by Gilbert in our Reference 25 discusses the reasons for the success of Hückel theory, and some of the limits of its validity. Finally, C. A. Coulson, L. J. Schaad and L. Burnelle, p. 27 of *Proceedings of the 1957 Conference on Carbon* (Pergamon Press), given an informative discussion of the method, and others, as applied to the sequence of structures from Benzene to graphite.

31. R. A. Swalin, *J. Phys. Chem. Sol.*, **18**, 290 (1961).

32. A. Scholz and A. Seeger, *Radiation Damage in Semiconductors* (Dunod, Paris, 1965), p. 315.

33. J. Friedel, M. Lanoo and G. Leman, *Phys. Rev.*, **164**, 1056 (1967).

34. A. B. Lidiard and A. M. Stoneham, *Science and Technology of Industrial Diamonds*, **1**, 1 (Industrial Diamond Information Bureau, 1967).

35. G. Schmid, K. P. Chik and A. Seeger, *Radiation Effects in Semiconductors*, Ed. F. L. Vook (Plenum Press, New York, 1968), p. 60.

36. R. R. Hasiguti, *Lattice Defects in Semiconductors*, Ed. R. R. Hasiguti (Univ. Tokyo Press, 1968), p. 131.

37. A. Seeger and M. L. Swanson, *Lattice in Semiconductors*, Ed. R. R. Hasiguti (Univ. Tokyo Press, 1968), p. 93.

38. F. P. Larkins and A. M. Stoneham, AERE Reports T.P., 386, 409, 410 (1970) to be published.

39. H. L. McMurry, A. W. Solbrig, Jr., J. K. Boyter and C. Noble, *J. Phys. Chem. Sol.*, **28**, 2359 (1967).

40. A. W. Solbrig, Jr., *Bull. Am. Phys. Soc.*, **15**, 102 (1970) and private communication from H. L. McMurry (1968).

41. B. D. Singh and B. Dayal, *Phys. Stat. Sol.*, **38**, 141 (1970).

42. J. C. Slater, *Rev. Mod. Phys.*, **30**, 197 (1958).

43. K. H. Bennemann, *Phys. Rev.*, **137**, A1497 (1965).

44. C. J. Hwang and L. A. K. Watt, *Phys. Rev.*, **171**, 958 (1968).

45. P. S. Ho, *Bull. Am. Phys. Soc.*, **14**, 410 (1969); also A. M. Stoneham and M. Lannoo, unpublished work, 1968.

46. H. Kanzaki, *J. Phys. Chem. Sol.*, **2**, 24 (1957).

47. U. Öpik and M. H. L. Pryce, *Proc. Roy. Soc.*, **A238**, 425 (1957).

48. P. Wysling and K. A. Müller, *Phys. Rev.*, **173**, 327 (1968).

49. A. M. Stoneham and M. Lannoo, *J. Phys. Chem. Sol.*, **30**, 1769 (1969).

50. E. L. Elkin and G. D. Watkins, *Phys. Rev.*, **174**, 881 (1968).

51. M. Lanoo and A. M. Stoneham, *J. Phys. Chem. Sol.*, **29**, 1987 (1968).

52. J. Ritter, *Sol. St. Comm.*, **8**, 773 (1970).

53. J. A. Sussman, *J. Phys. Chem. Sol.*, **28**, 1643 (1967).

54. R. Silsbee, *J. Phys. Chem. Sol.*, **28**, 2525 (1967).

55. R. Pirc, B. Žecš and P. Gosar, *J. Phys. Chem. Sol.*, **27**, 1219 (1966).

56. C. P. Flynn and A. M. Stoneham, *Phys. Rev.*, **B1**, 3966 (1970).

57. F. P. Larkins and A. M. Stoneham, *J. Phys. C.*, **L112** (1970).

58. G. D. Watkins, *Bull. Am. Phys. Soc.*, **14**, 312 (1969).

DISCUSSION

Question (UNKNOWN) Theory calculates the energy levels of the lattice vacancy in diamond in various charge states, but experiment gives quite other quantities—i.e., migration energies for interstitials in germanium and silicon, activation energies for self diffusion in silicon and germanium. In irradiation damage in both silicon and germanium there is evidence that some defect (probably an interstitial) migrates with an activation energy which is of order 0.01 eV. What is the defect?

Answer (STONEHAM, LARKINS and MESSMER) We don't know. We suggest the name lepre-con (the leprechaun is notoriously mobile.)

ELECTRONIC STATES OF THE NEUTRAL VACANCY IN THE DIAMOND LIKE SOLIDS

D. ROUHANI
Faculté des Sciences de Toulouse, France
AND
M. LANNOO AND P. LENGLART
ISEN, Lille, France

1. INTRODUCTION

The electronic states of the neutral vacancy are studied by using the Green's operator method combined with the tight binding approximation. This model is intermediate between the molecular approximation used by Coulson and Kearsley (1957), Yamagushi (1962) and the method of Callaway (1967) which uses a formalism of Wannier function and a pseude-potential.

The vacancy is described as a perturbative potential V. A dynamical state $|\psi\rangle$ of the perturbed crystal is solution of the diffusion equation

$$|\psi\rangle = |\phi\rangle + G V |\psi\rangle$$

when $G(E) = \lim 1/E - H + i\epsilon$ with H the Hamiltonian of the perfect crystal and where $|\phi\rangle$ in a dynamical state of energy E for the perfect crystal. Two cases are to be looked at—the energy E falls into the band gap; the proceeding equation is reduced to

$$|\psi\rangle = G V |\psi\rangle$$

and defines the bound states of the vacancy.

The energy E falls into the valance or the conduction band, we define a variation of the state density $\delta n(E)$

$$\delta n(E) = \frac{1}{\pi} \frac{d}{dE} \eta(E)$$

where $\eta(E)$ in a phase shift operator

$$\eta(E) = \text{Arg dét} (1 - G V)$$

Two bound states are expected and it is shown that 4 places are then missing in the valance band. Therefore 4 electrons are placed in the bound states as in the molecular model.

2. RESULTS

(a) A first study of the vacancy in diamond (Lannoo–Lenglart, 1969) taking into account only one resonance integral between the nearest neighbors gives us the results of the Figure 1.

The A_1 state and the 3 states T_2 are practically degenerate. The degeneracy is removed if we take into account the interaction between the second neighbors as a perturbation. The main results of this model are

the existence of bound states;

their strong localization;

the formation energy of the vacancy is about 9 eV when the parameters of the band structures are determined to fit the cohesive energy per atom to the experimental value 7.5 eV.

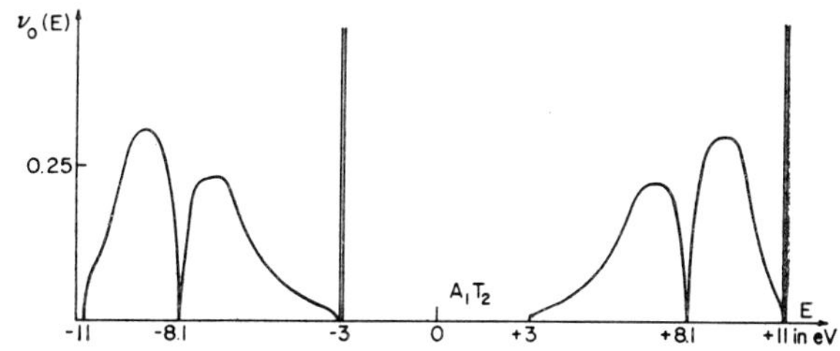

FIG. 1. Density of states of diamond, including the bound states for the vacancy.

(b) For the diamond, a more elaborate model (Djafari–Rouhani, Lannoo, Lenglart) was then studied. The resonance integrals are taken into account till the second neighbors (Figure 2) and we used the band structure calculation of Cohan, Pugh and Tredgold (1963). The results are given by the Figures 3 and 4 for two sets of band structure parameters.

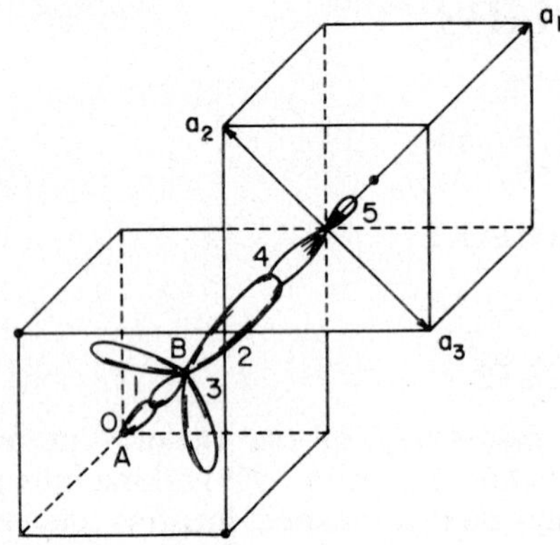

FIG. 2. Positions of atoms and their orbitals which occur in the definitions of the interaction parameters of the diamond crystal.

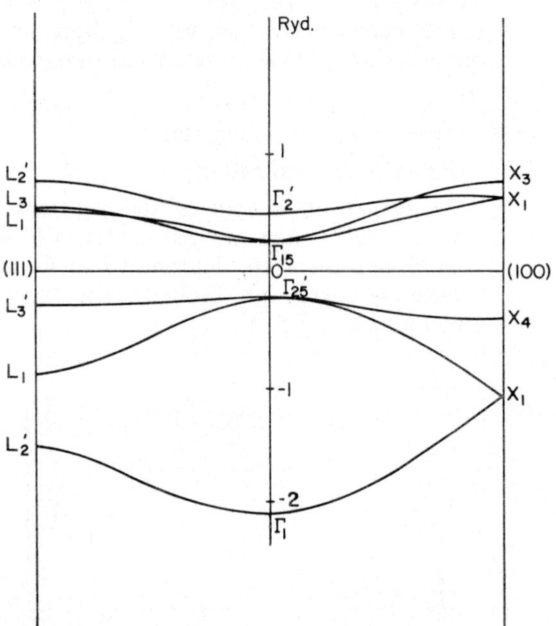

FIG. 3. Band structure of C, case *a* parameters.

As we can see in Figures 5 and 6 there is only a small change in the position of the bound states and the formation energy is 4,6 eV and 5,1 eV.

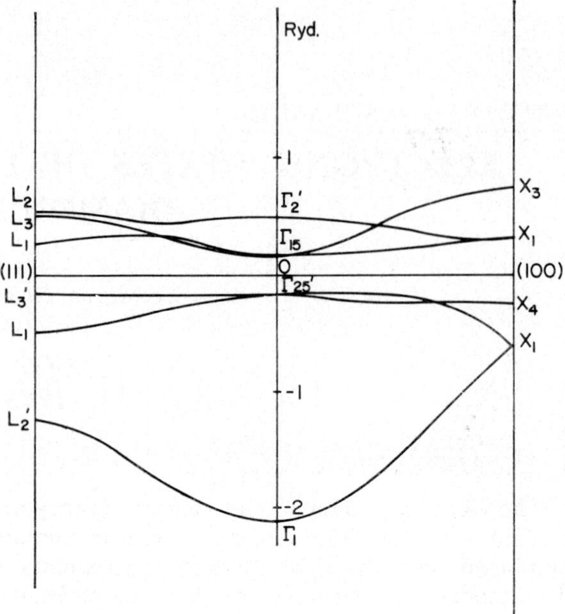

FIG. 4. Band structure of C, case *b* parameters.

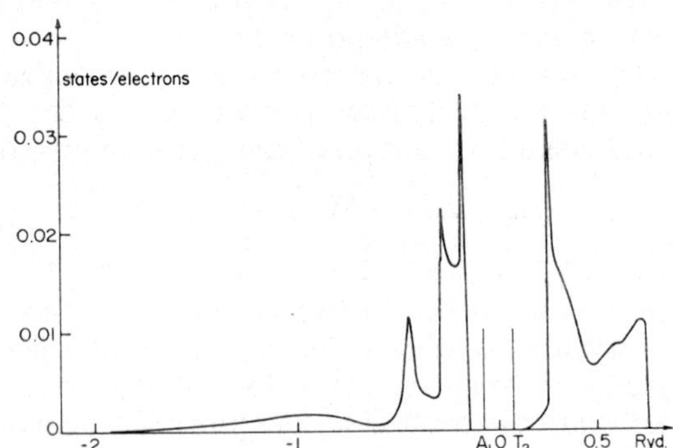

FIG. 5. Density of states of diamond, including bound states; case *a* parameters.

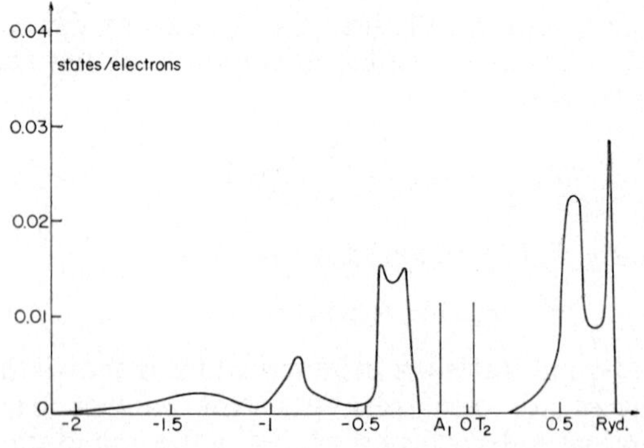

FIG. 6. (Same as 5 except case '*b*').

(c) The preceding model cannot be used for the band structure of silicon; so we try to take

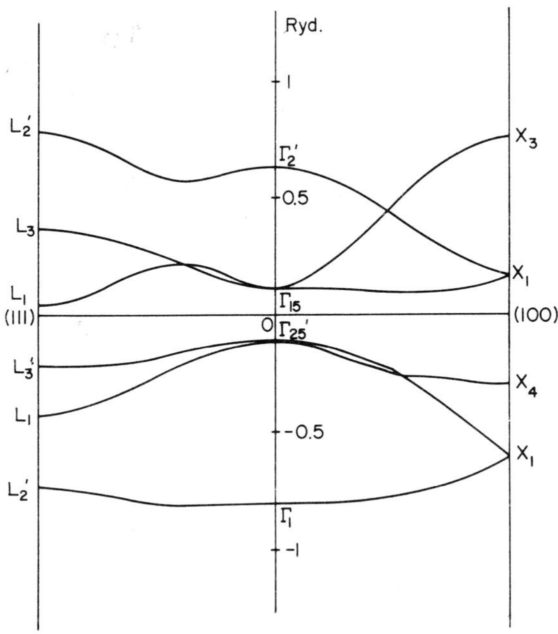

FIG. 7. Band structure of Si.

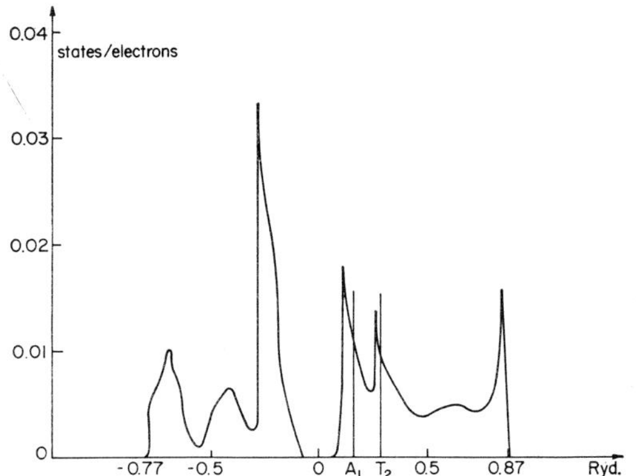

FIG. 8. Density of states of Si including resonant states.

into account an overlap integral between the nearest neighbors. This gives us a better band structure which can be roughly applied to silicon. For Silicon the band structure is fitted to the results of Woodruff (1956), Bassani (1957) and Callaway (1958). In this calculation there is no more bound state (Figures 7 and 8). To describe the state of the vacancy in Si we have to include the relaxation at the beginning of the calculation.

REFERENCES

1. C. A. Coulson and M. J. Kearsley, *Proc. Roy. Soc.*, **A241**, 433 (1957).
2. T. Yamaguchi, *J. Phys. Soc. Japan*, **17**, 1359 (1962).
3. J. Callaway and A. J. Hugues, *Phys. Rev.*, **b56**, 3, 860 (1967).
4. M. Lannoo and P. Lenglart, *J. Phys. Chem. Solids*, **30**, 2409 (1969).
5. M. Djafari-Rouhani, M. Lannoo and P. Lenglart, *J. de Phys.*, **31**, 597 (1970).
6. N. V. Cohan, D. Pugh and R. H. Tredgold, *Proc. Phys. Soc.*, **82**, 65 (1963).
7. T. O. Woodruff, *Phys. Rev.*, **103**, 1159 (1956).
8. F. Bassani, *Phys. Rev.*, **108**, 263 (1957).
9. J. Callaway, *Sol. State Phys.* **7**, 99 (1958).

DISCUSSION

Question (MITCHELL) It is very interesting that you find a transition energy of about 1.8 eV, similar to Coulson and Kearsley defect molecule (modified) calculation and to the GR I band (1.673 eV). As a check on the calculation where did you find the minimum of the conduction band? From your slides it looked like $k = D$ whereas experimentally it is near the zone boundary.

Answer (LENGLART) For the carbon diamond, the calculations have been made in two cases: neglecting the interactions between the bonding and antibonding orbitals; taking into account these interactions. The minimum of the conduction band has been found at a point of the Δ axis, different for the two calculations, but never near the zone boundary.

It has been shown that the energy of the localized states is not very sensitive to the details of the band calculation.

ELECTRONIC STRUCTURE OF THE ISOLATED VACANCY IN SILICON

F. P. LARKINS†

Mathematical Institute, Oxford

and

Theoretical Physics Division, A.E.R.E., Harwell, England.

The many-electron molecular orbital method predicts a ground state for the various charged vacancy centres in silicon which has a spin multiplicity in agreement with the electron spin resonance results. Symmetric relaxation and Jahn-Teller contributions have been included in first order. The quantitative features of the model are very different to those implied by the one-electron model.

1. INTRODUCTION

The defect molecule approach developed by Coulson and Kearsley[1] has been used to calculate the electronic structure of the isolated vacancy in silicon. The single positive (V^+), neutral (V^0), single negative (V^-), and double negatively charged (V^{2-}), states have been investigated. The initial calculation assumes that the vacancy is undistorted and that the atoms neighbouring the defect are sp^3 hybridized. The vacancy electrons are considered to be localized within the defect, but exchange terms between these electrons and the other bonded valence electrons are included. A modified first order linear response theory is used to calculate the symmetric relaxation and Jahn–Teller corrections to the lowest energy levels which result from electron-lattice coupling. The results are compared with the known experimental findings and with the characteristics of the one-electron model.

2. RESULTS FOR UNDISTORTED MODEL

Atomic Hartree–Fock functions determined by Clementi[2] are used to approximate the 3s and 3p orbitals on the silicon atoms in the crystal. Two values for the one-centre coulomb integral, Q, were used. Using the analytic wave functions, $Q = 11.45$ eV, while the semi-empirical estimate determined as in Reference 1 is 8.10 eV. The predicted energy level schemes for the undistorted V^+ and V^- centres are shown in Figures 1 and 2 respectively.

The energies of the symmetry states before configuration interaction are more compact using the smaller value for Q. In some cases symmetry

† Now at Chemistry Dept. Monash University, Victoria, Australia.

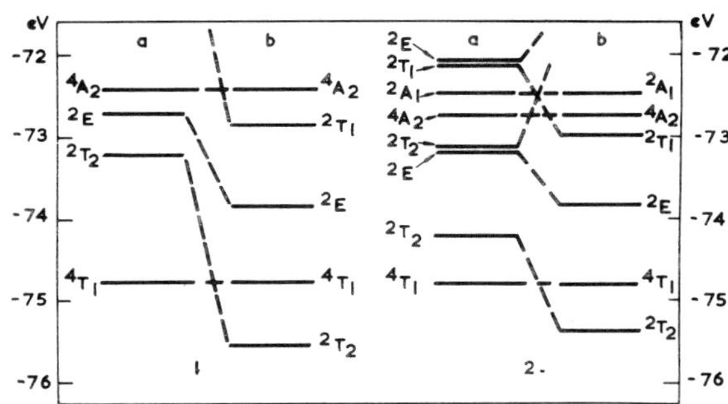

FIG. 1. Electronic structure of the positively charged undistorted vacancy in silicon. a. before configuration interaction; b. after configuration interaction.

   1. $Q = 11.45$ eV   2. $Q = 8.10$ eV

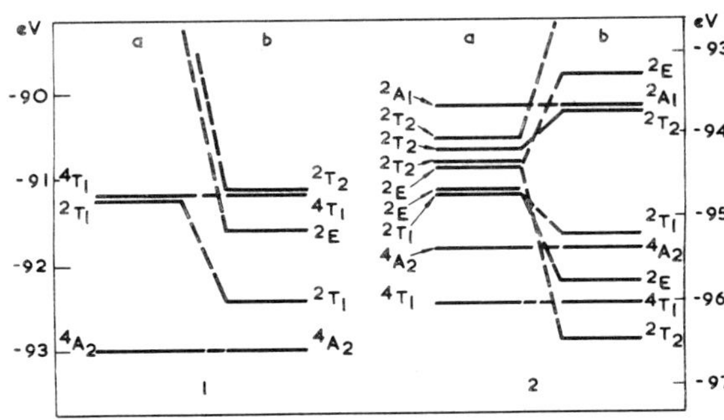

FIG. 2. Electronic structure of the single negatively charged undistorted vacancy in silicon. a. before configuration interaction; b. after configuration interaction.

   1. $Q = 11.45$ eV   2. $Q = 8.10$ eV

states from 'higher' configurations are almost degenerate with like symmetry states from 'lower' configurations before mixing. For example, in Figure 2 the $^2T_2$ symmetry states from the $a_1^2t_2^3$, $a_1t_2^4$ and t_2^5 configurations are separated by only 0.30 eV. If electron-electron interaction terms are neglected the one-electron calculation predicts that these configurations are separated from one another by 2.02 eV. Also with the simple one-electron model all symmetry states belonging to a particular configuration are degenerate, whereas the present many-electron calculations shows that they may be separated by several electron volts. Configuration interaction is even more important in determining the order of the lowest levels than in the diamond calculation.[1]

After configuration interaction the $^2T_2$ level is predicted as the lowest level of the V^+ centre in both cases. For the V^- centre the $^2T_2$ level is lowest when $Q = 8.10$ eV, but the $^4A_2$ level is calculated to be lowest with the alternative value of Q. The lowest $^1E_1$, $^3T_1$ and $^5A_2$ levels of the V^0 centre are close together in both studies with the $^1A_1$ and $^1T_2$ levels approximately 1 eV above these levels for $Q = 8.10$ eV. The V^{2-} centre calculation predicts that the $^3T_1$ level is lowest with $Q = 11.45$ eV and the $^1A_1$ level lowest for $Q = 8.10$ eV.

3. SYMMETRIC RELAXATION AND JAHN–TELLER DISTORTION

The relative displacements of the four nearest neighbours to the vacancy may be described in terms of six normal co-ordinates of A_1, E and T_2 symmetry.[3] All electronic levels couple with the symmetric mode, whereas only degenerate levels couple with the E and T_2 distortional modes. For a given electronic level the total energy change $\Delta E(Q_\alpha)$, within the harmonic approximation, corresponding to a static equilibrium distortion Q_α of the vacancy system, is given by

$$\Delta E(Q_\alpha) = -F_\alpha^2/(2M\omega_\alpha^2)$$

F_α is the appropriate symmetrized force determined within the rigid-atom approximation using the configuration interaction wave functions. ω_α is the linear response frequency for the α mode. The linear response frequencies determined by a dynamic relaxation procedure are much smaller than the Raman frequency[3,4] and lead to unreasonably large distortions when used with the calculated forces. This suggests that non-linear terms may be important. In order to avoid this much more

difficult and unreliable calculation with the wavefunctions at present available we have chosen values for the frequencies based on a sudden approximation. These frequencies lead to more realistic distortions.

For a given centre the symmetric force terms were not constant for all electronic levels. In all cases an inward relaxation of the atoms neighbouring the defect was predicted with a total energy lowering of 1 to 3 eV. However, with the exception of the V^0 centre for $Q = 8.10$ eV, the relative ordering of the lowest level was unchanged when the symmetric relaxation corrections were included. For the V^0 centre the $^1A_1$ level becomes lowest 0.23 eV below the $^1T_2$ level. The inward relaxation is inevitable because the model constrains the atoms neighbouring the vacancy by requiring that they be sp^3 hybridized.

The Jahn–Teller splittings for most of the electronic levels of the V^0, V^- and V^{2-} centres considered are predicted to be very small. In most cases the splittings are less than 0.1 eV. For the V^+ centre the first-order treatment appears to be unsatisfactory. No change in the order of the lowest levels of any centre results from including the Jahn–Teller corrections.

4. COMPARISON WITH EXPERIMENT

The corrected ground states for the various centres are shown in Table I. The electron spin resonance results obtained by Watkins[5] are also included. As in the diamond study the semi-empirical case would appear to give better agreement with experiment.

TABLE I

Predicted ground states for the various charge states of the single vacancy in silicon

Centre	$Q = 11.45$ eV	$Q = 8.10$ eV	Experiment
V^+	$^2T_2$ $\langle 111 \rangle$	$^2T_2$ $\langle 111 \rangle$	$S = \frac{1}{2}$, $\langle 100 \rangle$
V^0	$^3T_1$, $^5A_2$ or ^1E	$^1A_1$ or $^1T_2$	$S = 0$
V^-	$^4A_2$	$^2T_2$ $\langle 111 \rangle$	$S = \frac{1}{2}$, $\langle 110 \rangle$
V^{2-}	$^3T_1$ $\langle 100 \rangle$	$^1A_1$	$S = 0$

The results with $Q = 8.10$ eV suggest that inter-level coupling may be possible in the V^0 ($^1A_1 + {}^1T_2$ coupling) and the V^- ($^2T_2 + {}^2E$ coupling) centres. This latter coupling would qualitatively explain the observed mixed distortion for the V^- centre. Watkins[6] suggests that the Jahn–Teller energies are very much larger than predicted by these studied. The analysis of the stress splitting results assumes that the atoms neighbouring the vacancy move under the applied stress exactly as do the perfect lattice atoms. Detailed consideration of this problem indicates that this is unrealistic and may lead to an overestimate of the Jahn–Teller effect.[3] However in view of the large symmetric distortion higher order terms should be included in a more detailed Jahn–Teller calculation.

5. CONCLUSION

The many-electron defect molecule method predicts the correct spin multiplicities for the ground states of the various charged centres in silicon using a modified treatment. A first-order linear response theory predicts that the symmetric relaxation energies are in general very much larger than the Jahn–Teller energies. Configurational mixing and electron-electron interactions are essential features of the model. The qualitative features of this calculation are very different from those implied by the one-electron model.

ACKNOWLEDGEMENTS

I would like to thank Professor C. A. Coulson and Drs. A. M. Stoneham and A. B. Lidiard for helpful discussions. The award of the Victorian Rhodes Scholarship by the Rhodes Trust and a Senior Scholarship from Wadham College, Oxford, are gratefully acknowledged.

REFERENCES

1. C. A. Coulson and M. J. Kearsley, *Proc. Roy. Soc.*, **A241**, 433 (1957).
2. E. Clementi, *Suppl. IBM J. Res. Dev.*, **9**, 2 (1965).
3. F. P. Larkins and A. M. Stoneham, to be published.
4. A. M. Stoneham (Review paper at this Conference).
5. G. D. Watkins, *Proc. Symp. on Radiation Damage in Semiconductors* p. 97. (1965 Dunod, Paris,),
6. G. D. Watkins, private communication.

DISCUSSION

Question (KOEHLER) The theoretical prediction for the axis of the defect does not agree with experiment for the V^+ (positive vacancy) or for the V^- (negative single charged vacancy). Hence your geometrical predictions do not give one confidence in the model.

Answer (LARKINS) I do not think you should draw that conclusion. The reason for not predicting the correct Jahn–Teller distortions may be explained. Firstly, we have only calculated the linear Jahn–Teller coupling constants; however, if the symmetric distortion is really so large, then higher-order coupling terms between the symmetric and distortion modes will be important and could alter the symmetry of the defect. Secondly, the $\langle 110 \rangle$ distortion of the V^- centre cannot be explained by intra-level coupling effects alone, but must involve inter-coupling as demonstrated by Stoneham and Lannoo. [*J. Phys. Chem. Solids*, **30**, 1769 (1969)]. Coupling between the $^2T_2$ and ^2E electronic levels could produce the correct distortion as suggested in this paper.

Question (MITCHELL) Your results are very sensitive to the value of the one-centre coulomb integral, Q. Could you say how you chose the modified value which you have used?

Answer (LARKINS) The method for determining the semiempirical value of Q is explained in the appendix of the paper by Coulson and Kearsley.

Question (FISCHER) Does your calculation give the ground state energies relative to the band edge?

Answer (LARKINS) No, the aim of such a calculation is to try to determine the symmetry and multiplicity of the ground state of the defect and also possibly the lowest excited states reliably, but not to place them in the band gap.

Question (WATKINS) In the calculations presented have you used the force constants associated with the 'soft' modes or are these those of the restricted model where only nearest neighbors are allowed to move?

Answer (LARKINS) The work reported in this paper has used the force constants deduced from the restricted model. However, the order of the lowest levels and the relative magnitudes of the symmetric and Jahn–Teller distortions are essentially independent of which set is chosen in most of the cases investigated.

AN LCAO–MO TREATMENT OF THE VACANCY IN DIAMOND

R. P. MESSMER AND G. D. WATKINS

General Electric Research and Development Center, Schenectady, New York

A simple molecular orbital treatment which has previously been applied successfully to the substitutional nitrogen donor and boron acceptor in diamond is here applied to the vacancy in diamond. Localized levels associated with the vacancy are found in the gap and an outward relaxation of the four neighbors surrounding the vacancy is predicted. Jahn–Teller effects are found to give stabilization energies of $\sim$0.5 eV. A critical comparison to the 'defect molecule' approach of Coulson and Kearsley and Yamaguchi is included.

1. INTRODUCTION

In previous theoretical treatments, two rather distinct approaches have been adopted in considering the electronic structure of the vacancy in diamond. The first has been to consider a highly localized defect comprising only the four atoms neighboring the vacancy. This has been treated either as a 'defect molecule' made up of 'sp^3 dangling bonds' from each of the four neighbors,[1–5] or in a simple point-ion approximation.[6,7] The second approach has been to start from the perfect lattice band states and by introducing a perturbation due to the missing atom attempt to construct localized defect states.[8,9]

In this paper we describe a new approach, which is somewhat intermediate between these two points of view and as such embodies many of the advantages of both. This approach has been quite successfully applied to the cases of substitutional nitrogen and boron in diamond and has been described elsewhere.[10] The basis of the method is to approximate the solid by a large cluster of atoms which forms a 'large molecule' and then to apply the techniques of molecular orbital theory to obtain the wave functions and energy levels of this system. As we shall see this has the advantage of locating approximately the defect levels with respect to the conduction and valence band edges as does the second approach described above. In addition however it has the advantage of the first approach of producing wave functions for the defect levels which have a clear physical interpretation in terms of the atoms neighboring the vacancy. Another advantage of this method is that it describes the elastic restoring forces of the system, thus avoiding their *ad hoc* introduction as is typical in the first approach.[4,5,11] In applying this 'large molecule' method to the present problem we have chosen the

simplest representation of the LCAO–MO theory— namely, the Extended Hückel Theory.[12] We have previously outlined the basic approximations of the EHT.[10]

2. THE PERFECT CRYSTAL

We adopt as the 'large molecule' the 35 carbon atom cluster as shown in Figure 1. This represents our model of the perfect crystal. The energies of

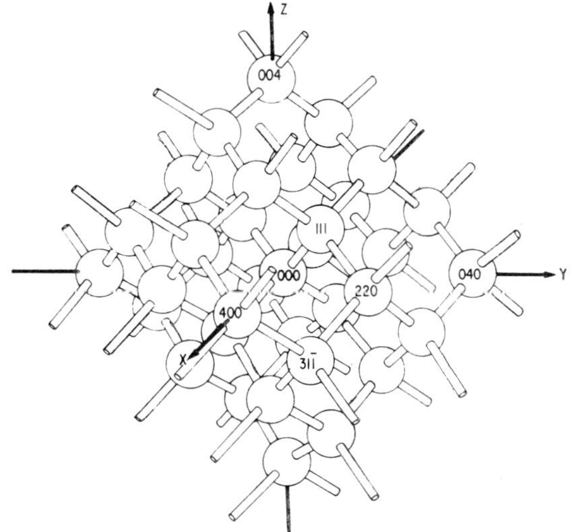

FIG. 1. The 35-atom cluster used for the LCAO–MO calculations.

the 140 MO's which result from the EHT calculation on this model are shown in Figure 2(a). It is apparent from the calculated one-electron energy levels that the model contains enough atoms that the 'band structure' associated with the solid is beginning to emerge. The width of the valence

"

band in the figure is 20.7 eV, which agrees well with the calculated band structure value of 22.5 eV.[13] The band gap in the figure is 9.5 eV, which is to be compared with the experimentally observed value of 5.5 eV.[14] As the size of the cluster is increased, however, the calculated band gap decreases, suggesting that the agreement here is also satisfactory.

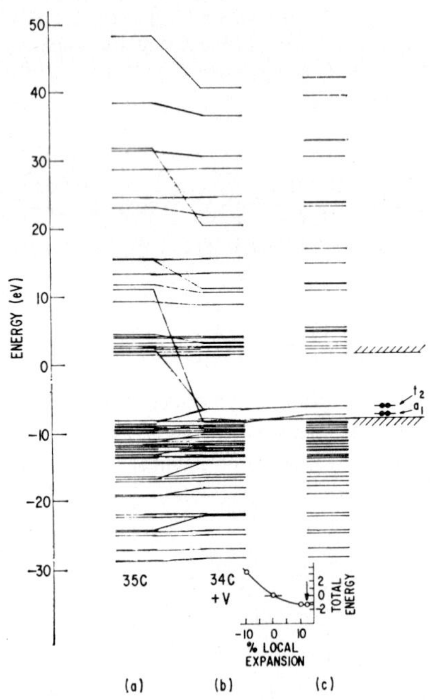

FIG. 2. (a) The energies of the one-electron MO's for the 35-carbon atom 'diamond'. (b) The energies with central atom removed (vacancy), no lattice relaxation. (c) Effect of symmetric relaxation. The total energy has a minimum when the amplitude of the normalized symmetric mode is approx. 13% of the nearest neighbor distance.

In the Extended Hückel Theory, the total energy of the system is given by $E_{\mathrm{TOT}} = \sum_i n_i E_i$, where n_i and E_i are the occupation number and the one-electron energy of the ith MO, respectively. Since this method describes elastic restoring forces,[15] approximate elastic constants may be determined by distorting the lattice and recalculating the total energy.[16] From the 35 carbon atoms of our model we calculate $C_{11} = 8.58$, $C_{12} = 2.67$, and $C_{44} = 5.00$ (10^{12} dynes/cm²), which compare remarkably well with the experimental values for diamond of 10.76, 1.25 (2.75) and 5.76 respectively.[17] We therefore conclude that the elastic forces contained in the model are also realistic.

3. NEUTRAL VACANCY

If the central atom of the 35 carbon cluster is removed and the calculation repeated, the results shown in Figure 2(b) are obtained. Comparing the energy levels with and without the center atom, an electronic level is identified in the forbidden gap close to the valence band edge which we associate with the vacancy. This level is triply degenerate (t_2 symmetry) and in the neutral charge state is occupied by two electrons.[18] The associated wave functions are found to be highly localized, with 54 per cent of the wave function centered on the four nearest atoms to the vacancy and 40 per cent spread over the twelve next nearest.

3.1 *Lattice relaxations*

Since the calculated ground state is degenerate, Jahn–Teller effects must obviously be considered. First, however, let us consider the symmetric relaxation of the four neighboring atoms around the vacancy.

3.1a *Symmetric relaxation*

When the symmetric 'breathing mode' of the four nearest neighbors is investigated, the results shown in Figure 2(c) are obtained. The model predicts an *outward* symmetric relaxation of these atoms by ~13 per cent in contrast to previous estimates based on the localized 'defect-molecule' model.[4,5,11] In addition a new electronic level appears in the gap below the t_2 level; the new level is nondegenerate (a_1). The wave function for this state is also highly localized, with 30 per cent of the wave function on the four nearest atoms and 52 per cent on the next shell.

3.1b *Jahn–Teller distortions*

Distortions of E and T$_2$ symmetry of the atoms surrounding the vacancy can lift the degeneracy of the t_2 orbital and thereby lower the total energy of the system. The total energy, expressed in terms of a particular normal mode distortion coordinate Q_i, must have the form

$$E = V_i Q_i + \tfrac{1}{2} k_i Q_i^2 + \ldots$$

A single calculation with a small normal mode distortion Q_i serves to determine both the Jahn–Teller coupling coefficient V_i (from the splitting of the t_2 level) and the force constant k_i describing the elastic restoring forces (from the change in total energy) for each mode i. The Jahn–Teller energy is then given by

$$E_{JT}^i = - V_i^2/2k_i.$$

The results of a series of such calculations are summarized in Table I. Here distortions both

TABLE I
Jahn–Teller distortions

	No expansion		13% Expansion	
	V(eV/Å)	E_{JT}(eV)	V(eV/Å)	E_{JT}(eV)
Local tetragonal	− 3.88	0.28	− 1.68	0.048
Local trigonal	− 0.30	0.003	+ 1.91	0.28
Additional bulk tetragonal	− 2.32		− 1.92	
Additional bulk trigonal	− 3.24		− 2.15	

with and without the symmetric relaxation are included. For the case of no symmetric expansion and local distortions only ('local' means that only displacements of the four nearest neighbors are considered), we see that a tetragonal distortion is favored. The Jahn–Teller stabilization for the tetragonal mode is 0.28 eV while the trigonal value is only 0.003 eV. Note, however, the surprising result that, starting from the relaxed equilibrium expansion position, the exact opposite situation prevails. Here the trigonal is the more stable, and again by a large margin. Note that even the sign of the trigonal coupling coefficient has changed between the two cases.

The 'bulk' distortions were performed to obtain a rough estimate of the extent of coupling to atom motion other than that of the four nearest neighbors. (Here, 'bulk' means that the whole cluster is strained uniformly by an amplitude equal to that of the corresponding local mode.) As can be seen in the table, the additional contributions to the coupling from these bulk distortions are of the same magnitude as the local coupling constants. This suggests that distortions of the neighboring shells of atoms also give significant contributions to the stabilization and that the resulting Jahn–Teller energies are probably $\sim$0.5 eV.

4. DISCUSSION

4.1 General

We consider the results of this new approach very promising. A localized level has been predicted to exist in the forbidden gap. It is close to the valence band edge which is consistent with observations in silicon[20] that the vacancy has a deep donor state but that its primary role is as an acceptor. In addition the wave function is consistent with the silicon results. The predicted 54 per cent on the nearest atoms is very close to the $\sim$60 per cent estimated from Si[29] hyperfine studies for both the single positive and negative states of the vacancy.[21,22]

Our model predicts an outward relaxation of the nearest atoms. We note, however, from Figure 2 that the localized defect levels have *increased* in energy while the total energy has been lowered. This means that had we used the same approximation as Lannoo[4,11] or Larkins,[5] i.e., attaching the 'defect-molecule' to the rest of the lattice by *equilibrium* force constants, we too would have predicted a collapse. We believe our calculations give the more realistic result and that these atoms are not under zero 'elastic' forces at their normal lattice positions. Instead they feel a strong outward pull resulting from the tendency to form planar sp^2 bonds with their other three neighbors (The driving force is the rather considerable $s \rightarrow p$ promotion energy.)

Our Jahn–Teller calculations reveal a complex, nonlinear coupling to the distortion modes of the neighboring atoms which, in turn, have both local and extended character. Because of this, the exact nature of the lowest energy configuration or the nature of the ground state energy surface in this multi-dimensional configuration space is not clear. Jahn–Teller energies of $\sim$0.5 eV are indicated, however.

Experimentally, the Jahn–Teller distortion for the *silicon* vacancy has been found to be a tetragonal one.[20,21] Study of our calculated wave functions under a tetragonal distortion reveals another interesting feature. We find that the orbitals are tilted $\sim$5° toward the distortion axis from the expected $\langle 111 \rangle$ atom-vacancy direction. This is interesting because experimentally a 7.5° tilt has been previously deduced from Si[29] hyperfine interaction studies for V^+ in silicon.[20] At the time, the origin of the tilt was a mystery. We believe our calculations demonstrate the source of this tilt and that the tilt is not a measure of the magnitude of the distortion at all. The top of the valence band is found to be pure p and the symmetry orbitals that will mix with the defect wave function are correspondingly p orbitals pointing along the distortion direction. These wave functions, of course, display a 54.7° tilt. We suggest therefore that the tilt observed in the localized wave functions results from the proximity of the defect level to the valence band. Consistent with this, V^- in silicon, which is further removed from the band edge has negligible tilt.[20] Indeed the tilt observed in EPR may serve as a sensitive measure of the proximity of a level to the valence band edge.

4.2 *Comparison to the Coulson–Kearsley and Yamaguchi (CKY) model*

Our treatment is a one electron MO treatment and as such does not include correlation effects between electrons. We know on the other hand from the CKY calculations[1,2] that such effects— leading to strong 'configuration interaction'— appear to be very important for the vacancy. Viewed in proper perspective, then, our calculations really serve to give a better one electron wave function from which a configuration interaction calculation should still be performed.

Short of such a detailed calculation, however, it is instructive to compare the two approaches. In our calculation, the $a_1 - t_2$ separation is $\sim$1.3 eV, see Figure 2. In the CKY calculations, on the other hand, this separation was found to be $\sim$6–10 eV. This suggests that when placed in the larger matrix, the more extended character of the wave function (54 per cent on the nearest neighbors vs 100 per cent) serves to greatly reduce the coulomb interactions which give rise to this crystal field splitting. (A direct test of this is afforded by performing an EHT calculation for the localized sp^3 four atom model. The result there for the $a_1 - t_2$ separation is 5.5 eV indicating a reduction by a factor of four in the more extended state.)

Our results tend to indicate therefore that the various coulomb and exchange interactions have been greatly over-estimated in the CKY calculations by virtue of restricting the wave functions to be sp^3 and on the four atoms only. It is interesting to note, however, that the Jahn–Teller couplings, calculated for our extended wave functions remain comparable to those estimated on the CKY model.[4,5,11]

This fact may be pointing the way out of the dilemma that has plagued us in the past in trying to reconcile the experimental results in silicon with the CKY theoretical calculations. In silicon, where the most detailed experimental results are available, electron–electron interactions are concluded to be small and Jahn–Teller interactions appear large.[23]

The CKY calculations however have indicated large electron–electron interactions and considerably smaller Jahn–Teller effects. Our results indicate that using more realistic wave functions which are not so highly localized, tends to reverse the relative importance of electron–electron and Jahn–Teller interactions from that which is predicted by the CKY calculations. Whether or not the order will actually be reversed for the diamond vacancy awaits a proper configuration interaction treatment. Regardless of the outcome of that calculation however, this trend may still be relevant in the case of silicon.

5. CONCLUSIONS

We have described a possible approach to the problem of the deep defect levels of the vacancy in diamond using molecular orbital theory on a large cluster surrounding the vacancy. Some of the advantages of such an approach are:

1. Wave functions of localized states are generated in a form that can be directly related to spectroscopic experimental data (EPR hyperfine constants, etc.).

2. If the cluster is large enough, the electronic level positions with respect to the band edges of the solid can be estimated.

3. Elastic forces are implicitly included in the model allowing relaxations around the defect to be explored. (This feature also makes possible the calculation of activation energies of motion and diffusion of defects, the stability of various configurations, etc. Such studies are currently in progress.)

4. Many instructional features are provided, such as the importance of non-linear effects in considering distortions, insight into the 'tilting' of the 'dangling orbitals', estimates of the percentage of the total wave function which is on the four nearest neighbors, etc.

The method is however, in its present form, strictly an approximate one-electron theory and as such can give no direct information about the importance of electronic correlation. Our calculated functions, however, probably provide a much better starting point for a configuration interaction treatment than does the ansatz of the 'defect-molecule' model, and thus has the potential for giving information about the importance of electronic correlation.

REFERENCES

1. C. A. Coulson and M. J. Kearsley, *Proc. Roy. Soc.* (London), **A241**, 433 (1957).
2. T. Yamaguchi, *Radiation Damage in Semiconductors* (Dunod, Paris, 1965), p. 323.
3. A. B. Lidiard and A. M. Stoneham, *Science and Technology of Industrial Diamonds*, Vol. 1, p. 1, Industrial Diamond Information Bureau, London (1967).
4. J. Friedel, M. Lannoo and G. Leman, *Phys. Rev.*, **164**, 1056 (1967).
5. F. Larkins, Thesis, Oxford University, 1969 (unpublished).
6. B. S. Gourary and A. E. Fein, *J. Appl. Phys.*, **33**, Supp. 1, 331 (1962).
7. A. M. Stoneham, *Proc. Phys. Soc.* (*London*), **88**, 135 (1966).
8. K. H. Bennemann, *Phys. Rev.*, **137**, A1497 (1965).
9. J. Callaway and A. J. Hughes, *Phys. Rev.*, **156**, 860 (1967).
10. G. D. Watkins and R. P. Messmer, *Phys. Rev. Letters* **25**, 656 (1970); *Proceedings of the Tenth International Conference on the Physics of Semiconductors*, Eds. S. P. Keller, J. C. Hensel, and F. Stern (U.S. Atomic Energy Commission, Wash., D.C., 1970), p. 623.
11. M. Lannoo and A. M. Stoneham, *J. Phys. Chem. Solids*, **29**, 1987 (1968).
12. R. S. Mulliken, *J. Chem. Phys.*, **46**, 497, 675 (1949); R. Hoffmann, *J. Chem. Phys.*, **39**, 1397 (1963), and later papers.
13. F. Herman, in *Physics of Semiconductors*, Edited by Dunod (Dunod, Paris, 1964), p. 3.
14. C. D. Clark, P. J. Dean and P. V. Harris, *Proc. Roy. Soc.*, **A277**, 312 (1964).
15. Although neither the nuclear–nuclear nor electron–electron interactions are explicitly introduced, it can be shown that for the EHT method these two neglected terms approximately cancel each other when the charge is evenly distributed over the system. [See J. Goodisman, *J. Am. Chem. Soc.*, **91**, 6552 (1969).]
16. The valence band in our model contains 36 more orbitals than the conduction band due to 'surface states' (or 'dangling' orbitals on the surface atoms, see Figure 1). In our model the energies of these orbitals are in the valence band and as such penetrate well into the bulk. Since it is not possible therefore to distinguish them, the total energy is determined by filling *all* orbitals in the valence band. This must be done in order to properly fill all of the bulk orbitals. In doing this, surface orbitals are also being filled, but we argue that they should not contribute significantly to the elastic forces.
17. H. J. McSkimin and W. L. Bond, *Phys. Rev.*, **105**, 116 (1957). The value for $C_{12} = 2.75$ is from Markham (unpublished); reported in *Physical Properties of Diamond*, R. Berman, Ed. (Oxford Univ. Press, London, 1965), p. 415.
18. In the singly positive or negative charge state, there would be one or three electrons in this orbital, respectively. Although the charge state has not been introduced explicitly in the calculation, the EHT approximation is best when the charge is equally distributed over all atoms (see References 15 and 19). This means that our wave functions and electronic level positions reflect most accurately the neutral charge state. However, they do serve as a guide, of course, for the other charge states as well.
19. G. Blyholder and C. A. Coulson, *Theoret. Chim. Acta* (*Berl.*), **10**, 316 (1968).
20. G. D. Watkins, *Radiation Damage in Semiconductors* (Dunod, Paris, 1965), p. 97.
21. G. D. Watkins, *J. Phys. Soc. Japan*, **18**, Suppl. 11, 22 (1963).
22. In this paper, we will compare our results to the vacancy in silicon, about which the most experimental information is available. An EPR center has been reported by Baldwin [J. A. Baldwin, *Phys. Rev. Letters*, **10**, 220 (1963)] which he attributes to the isolated vacancy in diamond. Our results are also consistent with the inferred properties of that center.
23. G. D. Watkins, *Radiation Effects in Semiconductors*, Ed. F. L. Vook (Plenum Press, New York, 1968), p. 67.

DISCUSSION

Question (STONEHAM) When you observed that the $A - T_2$ splitting was larger in the Coulson–Kearsley model than in yours, did I understand you to suggest that this meant electron correlation would be less in your model? This does not follow directly, since the lower configuration separation should lead to a larger configuration mixing, other features being constant.

Answer (MESSMER) Our point is that the 'other features' are not constant. We suggest that the electron–electron interactions have also been overestimated in the CKY treatment and in rough proportion to the overestimate in the $A - T_2$ splitting (see text).

Question (LARKINS) How do your results depend upon the size of the cluster, in view of the work by Coulson, Schaad, and Burnelle [Proceedings of the 1957 Conference on Carbon (Pergamon Press, 1958), p. 27] on the change in electronic energy levels with the size of the cluster for the graphite layer?

Answer (MESSMER) Our results appear consistent with theirs in that they too find that a cluster of $\sim$35 atoms (our size) is large enough to begin to reproduce the important features of the band structure.

Question (LARKINS) If I understand you correctly, when the configuration of the pseudo-crystal is changed by displacing atoms from their perfect lattice sites, only an electronic energy calculation is performed. Would you comment on the reason for not explicitly including nuclear–nuclear interactions?

Answer (MESSMER) That is correct. In the EHT method neither nuclear–nuclear nor electron–electron interactions are introduced explicitly. There is a justification for this and it is discussed briefly in reference 15 of the text.

Question (STONEHAM) Have you compared your level positions with those of Benneman [Conference on Calculations of the Properties of Vacancies and Interstitials, Shenandoah National Park, Va., May 1–5, 1966 (NBS Miscellaneous Publ. 287, 1967), p. 127]? He uses a t-matrix approach which attempts to be self-consistent, and finds A and T one-electron levels in the band gap.

Answer (WATKINS) Benneman lists four energy states in the gap and it has not been clear to us from his abbreviated account what these states refer to. If his first and second states are indeed A and T_2, as you suggest, then the agreement with our results is quite good. This is rather interesting considering the very different starting points in these two calculations.

Question (STEIN) Have you performed any calculations with more than one vacancy per 35 atom cell which could give an estimate of the average expansion of Si for large-vacancy concentrations?

Answer (MESSMER) No such calculations have been attempted.

Comment (WATKINS) Nor have we yet made any estimates for the symmetric relaxation of atoms beyond the nearest neighbors of the single vacancy.

INVESTIGATION OF POINT DEFECTS IN SILICON AND GERMANIUM BY NON-IRRADIATION TECHNIQUES

A. SEEGER

Max-Planck-Institut für Metallforschung, Institut für Physik, Stuttgart, Germany

The paper reviews the information on vacancies and interstitials from such techniques as diffusion studies, precipitation from supersaturated solid solutions, quenching from high temperatures, and plastic deformation. Self- and impurity diffusion are considered in some detail and the evidence is presented according to which the high-temperature self-diffusion occurs via extended interstitials in silicon and via extended vacancies in germanium. From the diffusion coefficients of group-V impurities and from the precipitation of substitutional nickel impurities the self-diffusion coefficient for a vacancy self-diffusion mechanism is derived and found to be lower than that from high-temperature self-diffusion experiments. This constitutes evidence for a change-over in the self-diffusion mechanism in silicon at about 900 °C to a low-temperature vacancy mechanism with lower activation energy and lower preexponential factor than the interstitial mechanism. It is proposed that self-interstitials in silicon might have a rather high migration energy, $E_{1I}{}^M$; tentatively $E_{1I}{}^M \simeq 0.8$ eV and a donor level at 0.40 eV above the valence band are attributed to the self-interstitials in silicon.

1. INTRODUCTION

The present paper intends to review briefly the evidence on point defects in silicon and germanium that may be obtained from techniques other than those involving radiation damage. These include studies in which the concentration of the point defects to be investigated is essentially the equilibrium concentration as well as approaches in which we are dealing (as in radiation damage) with non-equilibrium concentrations. What connects the various fields is the fact that in general the configuration of a point defect is independent of the way in which it was generated. This means that information on a given defect may be obtained by combining results from different techniques. This approach has been proven to be very powerful in metals. We shall see that in silicon and germanium the number of useful techniques is more limited, but that nevertheless important insights can be gained in that way.

The main techniques are the following:

(A) Defects in equilibrium
  1. Determinations of equilibrium concentrations.
  2. Transport of matter (self-diffusion, diffusion of foreign atoms, electro-transport, etc.).

(B) Defects not in equilibrium
  1. Irradiation (the main subject of this conference).
  2. Plastic deformation.
  3. Quenching from high temperatures.

(C) Precipitation from supersaturated solid solutions [this technique is intermediate between (A) and (B), since the specimen has to be quenched in order to obtain a supersaturated solution, say of copper in germanium, but the intrinsic point defects involved (in the example quoted: monovacancies) are nevertheless essentially in thermal equilibrium].

Of the methods mentioned (other than irradiation) diffusion is by far the most informative. It is fortunate that on account of the technological importance of diffusion in semiconductors a relatively large body of experimental data is available. Since one deals with defect concentrations essentially equal to the equilibrium concentrations, specimen characterization and impurities are much less important than in other approaches. In the present paper, we shall therefore concentrate on diffusion and related studies. Fairly extensive reviews of the field have recently appeared.[1,2] We shall therefore be content with a summary of the main results obtained and devote somewhat more space to recent findings and controversial issues.

2. SELF-DIFFUSION IN SILICON AND GERMANIUM

The available radioisotopes are not too well suited for measurements of self-diffusion coefficients. As a consequence, the tracer self-diffusion data on both Si and Ge extend only over a rather limited temperature range. In the range investigated, the tracer self-diffusion coefficients D^T may be represented, within the accuracy of the experiments, by an Arrhenius law

$$D^{\mathrm{T}} = D_0^{\mathrm{T}}\exp(-Q^{\mathrm{SD}}/kT) \qquad (1)$$

with temperature independent preexponential factors D_0^{T} and self-diffusion activation energies Q^{SD} (k = Boltzmann's constant, T = absolute temperature). The most important results are summarized in Table I.

TABLE I

Self-diffusion data on silicon and germanium

	Si	Ge
Temperature range of measurements (°C)	1100–1400	731–928
$D^{\mathrm{T}}(T_m)$ [extrapolated to the melting temperature T_m] (cm^2 s^{-1})	6.3×10^{-12}*	$6.1 . 10^{-12}$
D_0^{T}(cm^2 s^{-1})	9000[3]	10.8[4]
Q^{SD}(eV)	5.1[3]	2.99[4]

* This number has been given wrongly in Ref. 1.

If compared with typical data on metals, the following features of the experimental results appear noteworthy:

(1) The self-diffusion coefficients at the melting temperatures are several orders of magnitude smaller than they are in metals.

(2) The preexponential factor D_0^{SD} is by about a factor 10^2 larger in Si than in Ge.

(3) The preexponential factors D_0^{SD} are in both Ge and Si much larger than in metals.

(4) The ratios of the self-diffusion coefficients of metals to those of Si and Ge (on a reduced T/T_m temperature scale) increase with decreasing temperature; the activation energies of self-diffusion in Si and Ge are much higher than in metals with comparable melting temperatures.

In attempting to interpret these results, it may be safely assumed that the mechanisms of self-diffusion in Si and Ge are defect mechanisms, as has been found to be true in the majority of crystals investigated. The measured self-diffusion coefficients are so low that it appears unlikely that the *defect* mechanism with the largest self-diffusion coefficients would give a contribution that is still smaller. We shall therefore not consider direct exchange and ring mechanisms any further.

From the point of view of understanding self-diffusion in terms of defect mechanisms the most striking features are the large preexponential factors. The magnitudes of these factors are mainly determined by the entropies of formation

(S^{F}) and migration (S^{M}) of the defects involved. Taking the case of a monovacancy mechanism as an example, we have the following relationship:

$$D_0^{\mathrm{T}} = \frac{a^2}{8}\,f_{1\mathrm{V}}\,\nu_{1\mathrm{V}}^0\,\exp\!\left(\frac{S_{1\mathrm{V}}^{\mathrm{F}} + S_{1\mathrm{V}}^{\mathrm{M}}}{k}\right) \qquad (2)$$

Here a is the length of the cube edge, $f_{1\mathrm{V}} = 1/2$ the correlation factor for tracer self-diffusion, and $\nu_{1\mathrm{V}}^0$ an attempt frequency, the order of magnitude of which may be estimated from the lattice parameter, the atomic mass, and the activation energy for migration, $E_{1\mathrm{V}}^{\mathrm{M}}$. Inserting the appropriate numbers leads for Si to $S_{1\mathrm{V}}^{\mathrm{F}} + S_{1\mathrm{V}}^{\mathrm{M}} \approx 15k$ and for Ge to $S_{1\mathrm{V}}^{\mathrm{F}} + S_{1\mathrm{V}}^{\mathrm{M}} \approx 10\,k$. These values are definitely larger than those for metals, and give rise to serious doubts whether the picture of a monovacancy mechanism familiar from metals is applicable to silicon and germanium.

Essentially three different proposals have been made to solve this 'entropy dilemma.'

(A) Seeger and coworkers[1,5] have put forward the view that such large entropies could only be understood if 'extended defects' were involved. An 'ordinary' vacancy with a small relaxation of the surrounding atoms could not have a formation entropy (which is mainly related to the change in the lattice vibration spectrum) much larger than k and also the entropy of migration could not exceed a few k. If, however, the displacements of the neighbouring atoms are not small compared with the lattice parameter and the 'vacancy' is spread out over a large number of atomic sites, many configurations with similar energy and entropy values are possible. Since a diffusion experiment does not distinguish between these configurations (with the 'centre of gravity' of the vacancy located approximately on a given atomic site) one will observe a rather high entropy of self-diffusion.

(B) For the particular case of silicon, Peart[6] and Goshtagore[7] suggested that divacancies might account for the large entropy factor. It is indeed true that on account of the possibility to take up several orientations, there is an additional entropy contribution to the equilibrium concentration of divacancies. However, in the particular case of the diamond structure, this contributes to D_0^{T} only a factor of 2 and is thus insignificant from the point of view of resolving the entropy dilemma. In order to account for the observed entropies, the divacancy would still have to have an anomalously large entropy of formation, i.e., we are led back to the 'extended defect' concept discussed above. A further, very specific, objection to the importance of the divacancy mechanism comes from the fact that in the diamond structure the divacancy migration energy E_{2V}^{M} exceeds the monovacancy migration energy E_{1V}^{M} (since divacancy migration involves the separation of a vacancy pair to at least third nearest neighbour sites). Since in a defect mechanism the activation energy of self-diffusion is the sum of the defect formation and migration energies,

$$Q^{SD} = E^{F} + E^{M}, \qquad (3)$$

this means that, unless very extreme assumptions about the preexponential factors are made, the divacancy mechanism could only dominate if the divacancy formation energy were less than the monovacancy energy, to wit,

$$E_{2V}^{F} < E_{1V}^{F} \qquad (4)$$

or, in terms of the divacancy binding energy E_{2V}^{B}, if

$$2E_{1V}^{F} - E_{2V}^{F} = E_{2V}^{B} > E_{1V}^{F}. \qquad (5)$$

However, the conclusion that the divacancy binding energy exceeds the monovacancy formation energy is very unlikely to be correct.

In spite of the preceding objections[1] to the divacancy mechanism the Goshtagore proposal was endorsed by Kendall and DeVries,[2] who attributed the self-diffusion coefficient of silicon as listed in Table I entirely to the divacancy mechanism (with $E_{2V}^{F} = 3.4$ eV and $E_{2V}^{B} = 1.59$ eV) and suggested that the self-diffusion coefficient for the monovacancy mechanism is given by

$$D_{1V}^{T} = 6.9 \cdot 10^{-5} \exp(-2.86 \, \text{eV}/kT) \, \text{cm}^{2} \, \text{s}^{-1}. \qquad (6)$$

Here the unlikely conclusion (5) is avoided by making the preexponential factor for the monovacancy mechanism extremely small. In the proposal of Kendall and DeVries, the preexponential factors for monovacancies and divacancies differ thus by a factor of 10^{8}, of which at best a factor of about 10 could be accounted for by the geometrical differences between the two mechanisms. As far as we can see, the remaining factor of 10^{7} is left unaccounted.

(C) Recently Masters[8] has proposed that the self-diffusion of silicon is due to a dumbbell type vacancy (or, as Masters calls it, a pair of semi-vacancies). This proposal gains again a factor of two in the preexponential factor on account of the different orientations such a dumbbell vacancy may take up, and it avoids the difficulties associated with the magnitudes of E_{2V}^{B} and E_{2V}^{M}. However, it is clear from the preceding discussion that the

discrepancies in the entropies cannot be sufficiently reduced to alleviate the existing difficulties.

We see that all discussions lead invariably back to the entropy dilemma. It appears that it can only be resolved by having defects with large formation entropies. This means that a large number of atoms has to be affected by each defect, i.e., that we have to consider 'extended defects.' These may be either of the vacancy type (as we had assumed in our discussions up to now) or of the interstitial type. The conclusion that vacancies in the diamond structure are extended receives support from the neutron diffuse scattering experiments reported at this conference by E. W. J. Mitchell (see his contribution to the conference). These measurements suggest that the displacements of the atoms surrounding divacancies in neutron irradiated germanium are large and have a complicated pattern. This is indeed what one would expect if the divacancies are built up from strongly relaxed monovacancies.

It appears rather unlikely that the spread-out defects have a lower energy of formation than a more concentrated form with a smaller displacement of neighbouring atoms (e.g., such as those considered in various theoretical treatments).[5,9] The reason why the extended configurations dominate at high temperatures is that their large entropy of formation gives them a lower *free* energy of formation at high temperatures. We expect thus a (presumably gradual) change of the dominant vacancy configurations with temperature and hence a curvature of the Arrhenius plot for the self-diffusion coefficient. Estimates of the expected curvature[1] show that it is too small to be evident from the published tracer measurements, which are of limited accuracy and extend over a limited range of temperature. The fact that Eq. (1) describes the existing data well is not in conflict with the above ideas.

If we assume that the activation energy of self-diffusion varies linearly with temperature according to

$$Q^{SD}(T) = Q^{SD}(T_0) + \alpha k(T - T_0) \qquad (7)$$

and that the associated entropy of self-diffusion

$$S^{SD}(T) = S^{SD}(T_0) + \alpha k \quad \ln(T/T_0) \qquad (8)$$

changes between a reference temperature $T_0 = 0.15 \, T_m$ and $0.9 \, T_m$ by

$$\Delta S^{SD} = S^{SD}(0.9 \, T_m) - S^{SD}(0.15 \, T_m) = 8k, \qquad (9)$$

we obtain for germanium $\alpha = 4.5$ and

$$\Delta Q^{SD} = Q^{SD}(0.9 \, T_m) - Q^{SD}(0.15 \, T_m) = 0.35 \, \text{eV}. \qquad (10)$$

Combining this with the evidence [1,10,11] that the low-temperature migration energy of monovacancies in germanium is $E_{1V}^M \approx 0.2$ eV and with the measured high-temperature value of Q^{SD}, we obtain for the formation energy of 'unextended' monovacancies in germanium

$$E_{1V}^F = 2.4 \text{ eV}, \tag{11}$$

a value which compares not unfavourably with the existing calculations (see Sec. 7.3 of Seeger and Chik [1]). We may therefore conclude that from the point of view of the self-diffusion data there is no objection against a low monovacancy migration energy in Ge.

With the appropriate changes the preceding considerations could be carried over to silicon. However, unless the drastic assumption is made that the 'extension' of the vacancy and the temperature dependence of the monovacancy formation energy is much larger than in germanium, the large D_0^T-value of Si cannot be accounted for in this way. This leads us to consider the alternative possibility for an extended intrinsic point defect, the extended self-interstitial. There are in fact quite a number of arguments [1,5] that suggest that self-interstitials might be more important in Si than in Ge, e.g., the larger contraction on melting, the larger ratio of interatomic distance to atom radius, the larger solubility of interstitial solute atoms.

Estimates for the extension, the entropy of formation, and the energy of formation of extended interstitial atoms can be obtained from data on the melts. They indicate [1] that for Si the formation and migration of such extended interstitials could well account for the high-temperature self-diffusion data, whereas in germanium the mechanism would be unimportant compared with the extended vacancy mechanism for the entire temperature range. In silicon, however, one would expect a change-over from the high-temperature interstitial mechanism to the low-temperature vacancy mechanism. Experimental confirmation of this prediction would, of course, constitute strong support for the picture sketched above.

3. DIFFUSION OF SUBSTITUTIONAL IMPURITIES AND DOPING EFFECTS

A large body of data has been accumulated on the diffusion of substitutional impurities in silicon and germanium, particularly of group-III and group-V elements. [1,2] These act as acceptors or donors, respectively, and they will therefore have different interactions with vacancy-type and interstitial-type defects, which may be expected to have different energy levels. Changing the Fermi level by heavy doping will affect these interactions; it constitutes thus an additional tool for studying mechanisms of self- and impurity diffusion. [1,12]

The main experimental results may be summarized as follows:

(a) *Diffusion coefficients of group-III elements,* D_{III}^{imp}.

At $T = 0.94 \, T_m$ the ratio of D_{III}^{imp} to the tracer self-diffusion coefficient D^T is about 30 to 100 for Si and only about 1 to 10 for Ge.

(b) *Diffusion coefficients of group-V elements,* D_V^{imp}:

At $T = 0.94 \, T_m$ the ratio D_V^{imp}/D^T is only 3–10 for Si but 100 to 200 for germanium.

(c) *Doping effect on self-diffusion*:

In Si both n- and p-doping increases the self-diffusion coefficient, whereas in Ge n-doping gives an increase and p-doping gives a decrease of the self-diffusion coefficient. (The existence of these doping effects is an argument against non-defect mechanisms of self-diffusion.)

(d) *Effect of doping on impurity diffusion*:

In Ge the doping effects have the same sign as for self-diffusion, i.e., n-doping increases the impurity diffusion coefficients and p-doping decreases them. In Si the diffusion coefficient of the group-III element In is enhanced strongly by p-doping and suppressed by n-doping, i.e., the doping effects have the opposite directions of those in Ge. On the group V elements, in particular on phosphorus, the evidence is somewhat conflicting. Radioactive tracer measurements by Millea [13] showed that n-doping increases and p-doping decreases the impurity diffusion coefficient (i.e., the directions of the effects are the same as in Ge), whereas the less direct p-n junction method, which has been applied by various authors [14], indicates a large enhancement of the phosphorus diffusion coefficient at high levels of boron (i.e., p-) doping.

The germanium results may all be satisfactorily explained by a vacancy mechanism for both self- and impurity diffusion, provided the vacancy possesses an acceptor level near the valence band edge. This is in qualitative agreement with the results from quenching and irradiation studies. Quantitative agreement cannot be expected, since the diffusion experiments give information on the high-temperature configurations of the monovacancy, which we have argued to be somewhat different from the low-temperature configuration studied in most other experiments.

The empirical facts on Si[1,2,15] are very different from those on germanium. With the exception of the tracer measurements on group-V elements, the Si results cannot be explained in terms of the electronic energy levels of monovacancies (which are known at low temperatures and are assumed not to be modified *drastically* at high temperatures). A straightforward explanation can be given if the defect responsible for diffusion in Si is amphoteric with an acceptor level in the upper half and a donor level in the lower half of the forbidden gap. According to Blount's model[16] the self-interstitial is a good candidate for such an energy level structure. We may take the impurity diffusion and the doping results as an additional confirmation of the interstitial model for high-temperature diffusion in silicon.

Why is it then that the tracer measurement on group-V impurity diffusion show doping effects which have the same sign as those found in germanium and which may thus be qualitatively explained by a vacancy model? The simplest explanation is that the attraction of the acceptor-type vacancies to the donor impurities (which has been demonstrated at low temperatures by the electron spin resonance measurements on the *E*-centre)[17] is so large that it enhances the impurity diffusion via the vacancy mechanism over and above that via the interstitial mechanism. This view-point is supported by the fact that although the self-diffusion coefficients of Si and Ge are approximately the same near the melting points, the impurity diffusion coefficients for group-V elements are almost two orders of magnitude smaller in Si than in Ge. This is immediately understood if, in the case of Si, the enhancement

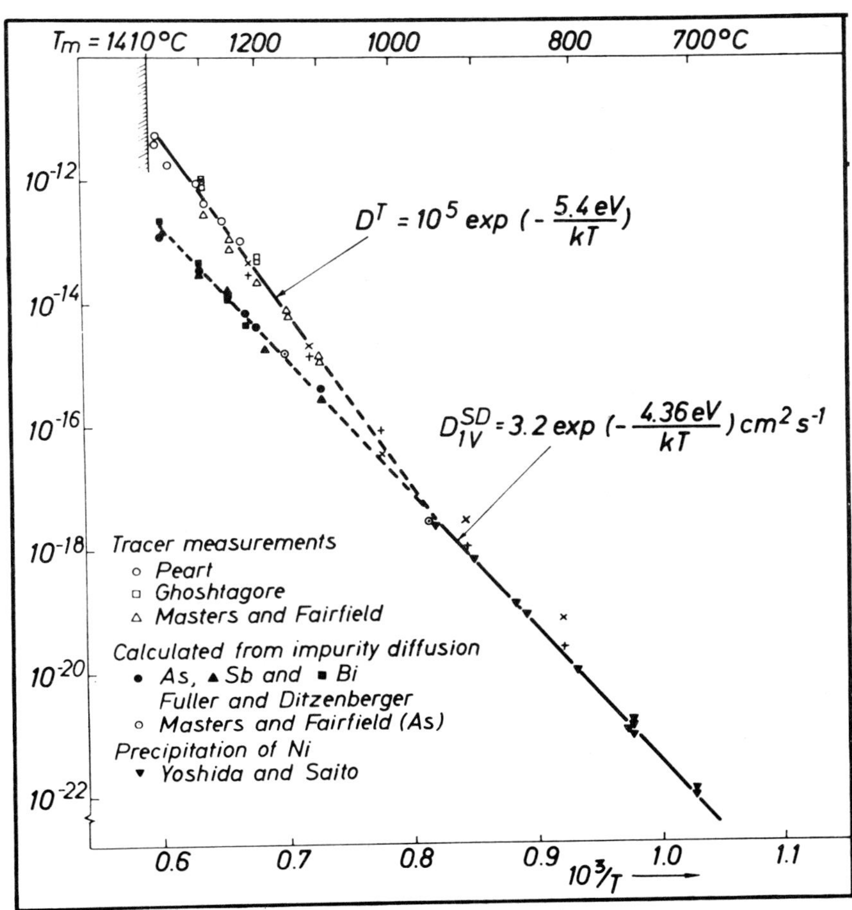

FIG. 1. Comparison of tracer and self-diffusion measurements on silicon with self-diffusion coefficients for a monovacancy mechanism calculated from the diffusion coefficients of group-V impurities and from the time constants of precipitation of nickel from supersaturated solid solutions.

of the impurity diffusion due to the acceptor-donor attraction takes place not from the level of the measured self-diffusion coefficient but from the lower (not directly measurable) level corresponding to the vacancy self-diffusion mechanism.

The preceding idea may be used to deduce the diffusion coefficient for self-diffusion by mono-vacancies, D_{1V}^{SD}, from the diffusion coefficients of group-V elements. Mehrer[18] has calculated and discussed the correlation factor for the diffusion of group-V elements by the monovacancy mechanism, taking into account fully the Coulomb-interaction between the positively charged impurities and the negatively charged vacancies. Figure 1 compares the vacancy self-diffusion coefficients deduced from the measurements of the impurity diffusions coefficients (these themselves are not shown) with the measured self-diffusion coefficients D^T. As we expected on the basis of our model, in Si the calculated vacancy self-diffusion is indeed smaller than the measured self-diffusion coefficient, whereas in germanium the two agree within experimental accuracy. This confirms the view-point that the high-temperature mechanisms of self-diffusion in silicon and germanium are different.

4. PRECIPITATION FROM SUPERSATURATED SOLID SOLUTIONS

In both silicon and germanium a number of metallic elements, e.g., Ni, Cu, Au, dissolve to a measurable extent both substitutionally and inter-stitially. The diffusion rates of the interstitial solutes are many orders of magnitude faster than those of the substitutional atoms, which for their diffusion require the presence and the interchange with intrinsic point defects in the crystals. In these cases the diffusion of the substitutional solute atoms occurs by the so-called dissociative mechanism of Frank and Turnbull[19]: A substitutional atom leaves its lattice site, becomes an interstitial impurity atom and leaves behind a vacancy (reaction constant k_1, in the reverse direction k_2). Then the interstitial impurity and the vacancy migrate separately (vacancy diffusion coefficient D_{1V}). A supersaturation of substitutional im-purities obtained, say, by quenching from high temperatures precipitates out by the migration of the interstitial atoms to nucleation sites. In order to avoid a build-up of a supersaturation of vacancies these have to migrate to sinks, which are usually dislocations (density N_{disl}). There are two possibilities for the rate-determining process:

(i) If $(k_1 + k_2 C_i^{eq}) \gg \beta N_{disl} D_{1V}$ (where C_i^{eq} denotes the equilibrium concentration of solute atoms on interstitial sites and β a geometrical factor of the order of magnitude unity related to the efficiency of the dislocations as sinks for vacancies) the diffusion processes are rate-determining and the time constant for the decay of the supersaturation (which may easily be determined by Hall effect measurements, taking advantage of the acceptor properties of the substitutional impurities) is given by

$$\tau = \frac{k_1 + k_2 C_i^{eq}}{\beta N_{disl} D_{1V} k_1} = \frac{C_s^{eq} + C_{1V}^{eq}}{\beta N_{disl} D_{1V} C_{1V}^{eq}} \qquad (12)$$

or, since the solubility of substitutional atom C_s^{eq} is large compared with the equilibrium concen-tration C_{1V}^{eq} of monovacancies,

$$\tau = \frac{C_s^{eq}}{\beta N_{disl} D_{1V} C_{1V}^{eq}} \qquad (13)$$

Penning[20] realized first that from measurements of τ and known values of β (from theory), N_{disl} (from etch-pit counts, etc.) and C_s^{eq} the so-called macro-scopic diffusion coefficient for a monovacancy mechanism of self-diffusion

$$D_{1V}^{SD} = D_{1V} C_{1V}^{eq} = D_{1V}^T / f_{1V} \qquad (14)$$

may be calculated from (13) for intermediate temperatures and compared with the tracer measure-ments of self-diffusion at high temperatures.

(ii) If $(k_1 + k_2 C_i^{eq}) \ll \beta N_{disl} D_{1V}$ the dissociation reaction described above is rate determining and the time constant is simply given by

$$\tau = 1/k_1. \qquad (15)$$

The change-over from the mechanism (i) to the mechanism (ii) occurs at a temperature at which

$$\beta N_{disl} D_{1V} = k_1 + k_2 C_i^{eq}. \qquad (16)$$

As pointed out by Seeger and Chik[1,21] Eq. (16) and related expressions which we shall not discuss here can be used to determine D_{1V}, in particular the activation energy E_{1V}^M, from the precipitation data. The evaluation of the data of Cu-precipitation in Ge[22] indicated[21] a small value of the activation energy for monovacancy migration, in agreement with the suggestion by Whan.[10,11]

The application of Penning's method[20] for determining D_{1V}^{SD} to Tweet's experimental data[22] leads in the temperature range between 700 °C and about 400 °C to a monovacancy diffusion coefficient in germanium that joins well to the tracer measure-ments at high temperatures. An uncertainty in the experimental value for C_s^{eq} prevents unfortunately a quantitative test of the predicted curvature of the Arrhenius plot (comp. Sec. 2). Nevertheless it can be said that the data are fully compatible with the idea that the high-temperature self-diffusion mechanism in germanium is a monovacancy mechanism.

Figure 1 shows the comparison of the vacancy self-diffusion data for a vacancy mechanism as calculated from the work of Yoshida and Saito[23] on the precipitation of Ni from supersaturated substitutional solute solutions in Si with the tracer measurements of self-diffusion and with the self-diffusion coefficient for a vacancy mechanism as deduced from the data on group-V diffusion. It is clear that while the two sets of data on D_{1V}^{SD} fit well together, this is not so for the directly measured tracer data. Figure 1 supports therefore strongly the view that the high-temperature mechanism of self-diffusion is different from that at intermediate and low temperatures, and that we should expect a break in the temperature dependence of the self-diffusion coefficient at about 900 °C.

Furthermore we may deduce from Figure 1 the self-diffusion coefficient for the monovacancy mechanism in silicon. The activation energy is about 4.4 eV, the preexponential factor about 3 cm²/s. Taking into account that these values are average values over the temperature range from 700 °C to 1400 °C (the accuracy of the analysis does not permit to detect the predicted curvature of the Arrhenius plot of D_{1V}^{SD}), one may say that the preexponential factor agrees rather well with what one would expect for an extended vacancy mechanism. We have thus returned in a satisfactory manner to the starting point of our analysis of self-diffusion in silicon, namely, that the entropies for extended vacancies would be about the same in Si and Ge. Allowing for the temperature dependence of the energy of formation and the energy of migration of monovacancies, one arrives at a monovacancy formation energy in Si of about 4 eV or slightly less. It is gratifying that this value is well below the upper limit of 4.63 eV deduced from the measured cohesive energy by Seeger and Chik,[1] in contrast to what would be obtained from the measured activation energy of self-diffusion, Q^{SD}, by means of the expression $E_{1V}^F = Q^{SD} - E_{1V}^M$.

5. EQUILIBRIUM CONCENTRATIONS OF DEFECTS

The self-diffusion coefficient is (apart from geometrical and correlation factors) the product of the defect diffusion coefficient and the defect concentration. A lower limit for the equilibrium concentration C^{eq} of the defects responsible for self-diffusion may be obtained from the measured self-diffusion coefficient and the assumption that the migration activation energy E^M is zero. One obtains the result that the concentrations C^{eq} at the melting points must be larger than 2.5×10^{-10} for Si and larger than 1.5×10^{-9} for Ge (in the last case use has been made of the fact that the defects are monovacancies).

Upper limits can be obtained if information on the migration energies is taken from elsewhere. Using monovacancy migration energies $E_{1V}^M = 0.33$ eV (for Si) and $E_{1V}^M = 0.2$ eV (for Ge) one obtains for Si

$$C_{1V}^{eq}(T_m) < 1.8 \times 10^{-7} \qquad (17)$$

and for Ge

$$C_{1V}^{eq}(T_m) < 7.4 \times 10^{-8}. \qquad (18)$$

These upper limits are much lower than the concentrations found in metals and make it extremely difficult to detect the presence of point defects in thermal equilibrium. Resistivity experiments on germanium whiskers by Smith and Holland[24] have been interpreted in terms of an equilibrium concentration of monovacancies. Seeger and Chik[11] have obtained a vacancy migration energy of about 0.2 eV from these data. However, it is not clear why the effect is as large as observed. It appears conceivable that in the heavily doped material investigated (resistivity at $25\,°C: 219\,\mu\Omega$ cm) vacancy-dopant interactions have enhanced the vacancy equilibrium concentration considerably above that of pure germanium.

6. QUENCHING FROM HIGH TEMPERATURES

In germanium and silicon, quenching experiments are not nearly as powerful a tool for the study of the defects present in thermal equilibrium at high temperatures as they are in metals. As the estimates of Section 5 show, the equilibrium vacancy concentrations near the melting point are much smaller than in metals and therefore more difficult to detect. The high mobility of monovacancies makes it difficult to quench them. In fact, the only reason one can expect to retain any vacancy type defects at all is that a supersaturation of monovacancies tends to form divacancies, which have a much smaller diffusivity than monovacancies and are rather stable even well above room temperature.

However, a serious competition for the divacancy formation is the formation of vacancy-impurity complexes, which may complicate the issue considerably.

The majority of the quenching experiments on germanium[1] yields acceptor centres, in agreement with the idea that the quenched-in defects are of vacancy type. The quenched-in concentrations are compatible with the estimates of C_{1V}^{eq} given in Sec. 5. The majority of the effective energies of formation observed lie between 1.9 eV and 2.2 eV and are thus somewhat lower than E_{1V}^{F}. At this conference, R. R. Hasiguti and S. Motomiya (discussion of this paper) report that taking into account divacancy formation the experimental results on the quenched-in defects as a function of quenching temperature can be fairly well explained in terms of the energies and entropies used in this paper. The situation for germanium appears therefore quite satisfactory.

Unfortunately, the same cannot be said of the quenching experiments on silicon. The majority of the experiments (see Seeger and Chik[1] and a recent paper by Swanson[25]) find that the quenched-in defects are donors centres. This is in agreement with the idea that the principal defects quenched-in are interstitial atoms. However, the quenched-in concentrations are much higher than one calculates on the assumption that the energy of migration of interstitials is comparable with or smaller than that of monovacancies. The results could be understood if the defects responsible for the high-temperature self-diffusion in silicon had an energy of migration of the order of magnitude of 1 eV or larger. Swanson[25] has observed that the quenched-in donors anneal out near room temperature with an activation energy of migration of (0.81 ± 0.04) eV. Bemski and Dias,[26] Elstner and Kamprath[27], and Swanson[25] all observed that the quenched-in defects in p-Si had an energy level 0.4 eV above the top of the valence band, which agrees quite well with the position of the donor level of the interstitials as deduced by Chik[12] from diffusion measurements.

The same defect appears to have also been observed in experiments on p-silicon. Malovetskaya and coworkers,[28] Vavilov and coworkers,[29] and Stein and Gerreth[30] found strong annealing at or just above room temperature with an activation energy of annealing of 0.9 eV.[31] Recent work by Konozenko and coworkers[32] indicates that both γ- and n-irradiation of high-purity silicon introduce a defect with an energy level of 0.4 eV above the conduction band that anneals out with an activation energy of (0.85 ± 0.1) eV. We propose that this defect is an interstitial atom. The fact that in the earlier work[29] the room temperature annealing was associated with the disappearance of an energy level 0.27 eV above the conduction band is easily understood, since this is one of the energy levels of divacancies. These divacancies may have been annihilated by the interstitials migrating at about room temperature. The experiments discussed here do not distinguish between self-interstitials and impurity interstitials formed by self-interstitial—impurity interactions, but we emphasize that the basic point of our argument was that the self-interstitial silicon should have a migration energy of 0.8 eV or more (according to the quenching data) and a donor level at about 0.4 eV above the conduction band (according to the diffusion data).

So far, self-interstitials have not been observed by electron spin resonance. In Al-doped silicon irradiated at 4 °K, Watkins[33] observed an EPR spectrum which he attributed to interstitial Al++. However, the conclusion that the self-interstitial in p-silicon must have a migration energy $E_{1I}^{M} \sim 0$ is not compelling, since the Al-impurities may have been displaced from their substitutional sites by special interstitials moving along the close-packed directions (e.g., dynamic crowdions). The three-dimensional thermally activated motion of self-interstitials may nevertheless occur at about room temperature.

It is unlikely that a self-interstitial in silicon occupies a position with cubic symmetry. It should therefore be able to take up different crystallographic orientations and to give rise, if present in large enough concentrations, to a relaxation peak. Tan and Berry[34] have observed such a relaxation process (with an activation energy of 0.47 eV and a preexponential time constant $\tau_0 = 4 \times 10^{-15}$ s^{-1}) on Si-bombarded Si single crystals (Figure 2). The τ_0-value and the width of the peak is in agreement with the idea that it is due to the reorientation of a bombardment-induced point defect. Since a large number of self-interstitials have been produced by the treatment, it is not unreasonable to suppose that the relaxation process is due to the reorientation of self-interstitials. A corollary of this is that the interstitial in silicon could be able to rotate without migration, as has been found for self-interstitials in nickel.[35]

It is clearly important and necessary to test the preceding proposal on the nature and the properties of self-interstitials by further experiments.

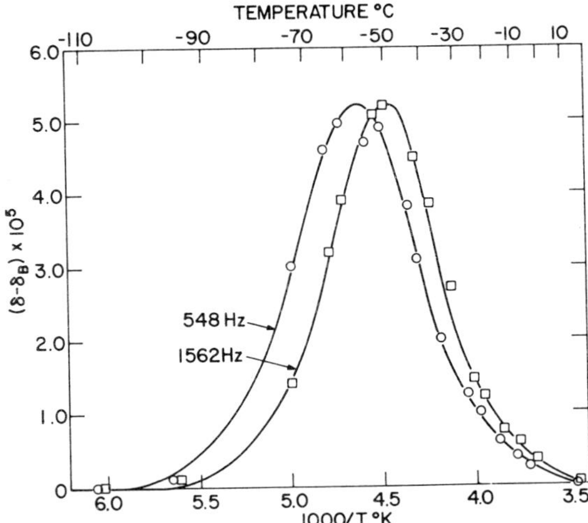

FIG. 2. Internal friction peak produced by room temperature implantation of Si in a Si single crystal to a total dose of $1 \times 10^{16}/cm^2$. Ordinate shows internal friction (expressed as logarithmic decrement), with background damping removed. Measurements in flexural vibration with uniaxial stress along $\langle 110 \rangle$.

7. PLASTIC DEFORMATION

Plastic deformation generates dislocations and point defects, the latter in general with unequal numbers of interstitials and vacancies. This may be an advantage over radiation damage helping to discriminate between interstitials and vacancies. In germanium and silicon, however, plastic deformation is a much less useful tool for the study of point defects than in metals. The reason for this is that because of their low-temperature brittleness, germanium and silicon can only be deformed at fairly high temperatures, at which the majority of the point defects anneal out instantaneously. It was pointed out rather early[36] that on account of the crystallographic nature of the diamond structure point defects should be preferentially created as pairs, i.e., as divacancies and as diinterstitials. It was realized that the divacancy migration and binding energies were presumably large enough to enable the divacancy to be studied in this way. An acceptor level 0.1 eV above the valence band was attributed to the divacancy in germanium.[36,37] It would be interesting to perform similar investigations on silicon where radiation damage studies have yielded much more detailed information on the divacancy than are available on germanium.

ACKNOWLEDGEMENT

The author is grateful to Drs. Berry and Tan for making available and discussing their unpublished results.

REFERENCES

1. A. Seeger and K. P. Chik, *Phys. Stat. Sol.*, **29**, 455 (1968).
2. D. L. Kendall and D. B. DeVries, Semiconductor Silicon ed. by R. R. Haberecht (The Electrochemical Society, Inc., New York, 1969), p. 358.
3. J. M. Fairfield and B. J. Masters, *J. Appl. Phys.*, **38**, 3148 (1967).
4. H. Widmer and G. R. Gunther-Mohr, *Helv. Phys. Acta.*, **34**, 635 (1961).
5. A. Seeger and M. L. Swanson, Lattice Defects in Semiconductors, ed. by R. R. Hasiguti (University of Tokyo Press and The Pennsylvania State University Press, 1968), p. 93.
6. R. F. Peart, *Phys. Stat. Sol.*, **15**, K 119 (1966).
7. R. N. Goshtagore, *Phys. Rev. Letters*, **16**, 890 (1966).
8. B. J. Masters, to be published.
9. A. Seeger and A. Scholz, *Phys. Stat. Sol.*, **3**, 1480 (1963).
10. R. E. Whan, *Appl. Phys. Letters*, **6**, 221 (1965).
11. R. E. Whan, *Phys. Rev.*, **140A**, 690 (1965).
12. K. P. Chik, *Rad. Effects*, **4**, 33 (1970).
13. M. F. Millea, *J. Phys. Chem. Solids*, **27**, 315 (1966).
14. W. H. Drake and A. F. W. Willoughby, Proc. Thomas Graham Memorial Symposium on Diffusion Processes, University of Strathclyde, Sept. 1969.
15. B. J. Masters and J. M. Fairfield, *J. Appl. Phys.*, **40**, 2390 (1969).
16. E. I. Blount, *J. Appl. Phys.*, **30**, 1218 (1959).
17. G. D. Watkins and J. W. Corbett, *Phys. Rev.*, **134A**, 1359 (1964).
18. H. Mehrer, *Z. Naturforschg.* **26a**, 308 (1971).
19. F. C. Frank and D. Turnbull, *Phys. Rev.*, **104**, 617 (1956).
20. P. Penning, *Phys. Rev.*, **110**, 586 (1958).
21. K. P. Chik and A. Seeger, *Helv. Phys. Act.*, **41**, 742 (1968).
22. A. G. Tweet, *Phys. Rev.*, **111**, 67 (1958).
23. M. Yoshida and K. Saito, Japan, *J. Appl. Phys.*, **6**, 573 (1967).
24. R. C. Smith and L. R. Holland, *J. Appl. Phys.*, **37**, 4866 (1966).
25. M. L. Swanson, *Phys. Stat. Sol.*, **33**, 721 (1969).
26. G. Bemski and C. A. Dias, *J. Appl. Phys.*, **35**, 2983 (1964).
27. L. Elstner and W. Kamprath, *Phys. Stat. Sol.*, **22**, 541 (1967).
28. V. M. Malovetskaya, G. N. Galkin and V. S. Vavilov, *Soviet Physics—Solid State*, **4**, 1008 (1962).
29. V. S. Vavilov, G. N. Galkin, V. M. Malovetskaya and A. F. Platnikov, *Soviet Physics—Solid State*, **4**, 1442 (1963).
30. H. J. Stein and R. Gerreth, *J. Appl. Phys.*, **39**, 2890 (1968).
31. H. J. Stein and Frederick L. Vook, Action des Rayonnements sur les Composants a Semiconductor (Journees d'Electronique), Toulouse, 1967, Vol. I, Paper A3.
32. I. D. Konozenko, A. K. Semenyuk and V. I. Khivrich, submitted to this conference.

33. G. D. Watkins, Radiation Damage in Semiconductors (Intern. Conf. Physics of Semiconductors), Ed. P. Baruch, Dunod, Paris, 1965 (p. 97).
34. S. I. Tan and R. S. Berry, to be published.
35. A. Seeger and Fi.-J. Wagner, *Phys. Stat. Sol.*, **9**, 583 (1965).
36. P. Haasen and A. Seeger, Halbleiterprobleme, Vol. IV (ed. by W. Schottky), Vieweg, Braunschweig, 1958, p. 68.
37. A. Seeger, Solid State Physics in Electronics and Tele-communications (ed. by M. Desirant and J. L. Michiels), Academic Press, 1960, Vol. I, p. 61.

DISCUSSION

Question (GOLDSTEIN)   There have been some very high D_0 values reported from self-diffusion studies on compound semiconductors. Have you had any thoughts on this in terms of the entropy considerations you have discussed here?

Answer (SEEGER)   I have looked into the situation for the III-V compounds. The available data appear to be not quite complete enough to perform a complete analysis, but, as you indicate, in some respects the situation appears to be very similar to that found in the group-IV elements. I am therefore of the opinion that similar principles should be applied to the analysis of the diffusion data of III-V compounds.

Question (MASTERS)   In the interests of keeping alive the monovacancy model for diffusion, I would like to point out that the charge state of the vacancy may be very important. Under conditions for which charged vacancies might be expected to predominate over uncharged vacancies, as in the high-concentration diffusion of electrically active impurities in silicon, the experimentally observed activation entropies are very much lower than 10 to $15\,k$.

Answer (SEEGER)   I agree with Dr. Masters that the charge state of the vacancy in silicon may be important for impurity diffusion in silicon. The evidence presented in Figure 1 of my paper suggests indeed that the electrostatic attraction of negatively charged vacancies to positively charged group-V impurities is large enough for the vacancy mechanism to dominate over the interstitial mechanism for the diffusion of these particular impurities. The lower entropy factors observed for group-V impurities are explained in this way, but at the same time the view-point that *self*-diffusion in silicon at high temperatures cannot be due to vacancy diffusion is strengthened.

Question (VOOK)   What are your comments on the suggestion of Kendall and DeVries that a divacancy mechanism can satisfactorily account for self-diffusion in Si? What about the possibility that multiple vacancy centers containing more than two vacancies can account for self-diffusion in Si?

Answer (SEEGER)   The basic feature of the equilibrium concentration of vacancies in silicon is that it must be very low, as discussed in my paper. Since the nth power of the monovacancy concentration enters into the equilibrium concentration of n-fold vacancies, the concentration for n-fold vacancies must decrease very rapidly with n, notwithstanding their high binding energies. I think that for this reason not even divacancies can be important, let alone larger clusters.

ESTIMATION OF THE TRUE FORMATION ENERGY OF A VACANCY IN GERMANIUM

R. R. HASIGUTI AND S. MOTOMIYA

*Department of Metallurgy and Materials Science, Faculty of Engineering,
University of Tokyo, Bunkyo-ku, Tokyo*

In several experiments of quenching of germanium, the formation energy of a vacancy has been found to be about 2 eV.[1,2] This can be an apparent formation energy, which is not the true one. This is considered to be due to the annihilation of vacancies during quenching. If the annihilated vacancy concentration is known, the true formation energy can be estimated.

We made many computer experiments on this problem changing the various parameters concerned. An example is shown in Figure 1. The parameters used in this example are as follows: the binding energy of a divacancy E_B is 0.8 eV, the migration energy of a divacancy E_M^2 is 1.2 eV, the migration energy of a single vacancy E_M is 0.3 eV, the formation energy of a single vacancy E_F is 2.7 eV, the density of dislocations as sinks α is 6×10^3 cm^{-2}, and the entropy of formation of a single vacancy S is $10k$, $9.5k$ or $9k$. The large formation entropy of a vacancy is not unreasonable, because the entropy of activation of diffusion is very large,[3] and the most part of it is considered to come from the formation entropy and not from the migration entropy. Figure 1 shows the concentration of vacancies as a function of quenching temperatures. Solid circles show the equilibrium concentrations at respective quenching temperatures. Open circles show the concentrations of quenched-in vacancies at room temperature, the quenching rate being 500 °C/sec. The broken lines show experimental results (upper line from Reference (1) and lower line from Reference (2)).

The good agreement between the experimental results and the open-circled curves shows that the formation energy of a vacancy can be as large as 2.7 eV. The sum $E_F + E_M = 2.7 + 0.3 = 3.0$ eV agrees with the activation energy of 3.0 eV of self-diffusion in germanium.[4]

The authors with to thank Dr. M. Doyama for his suggestions in making computer programs.

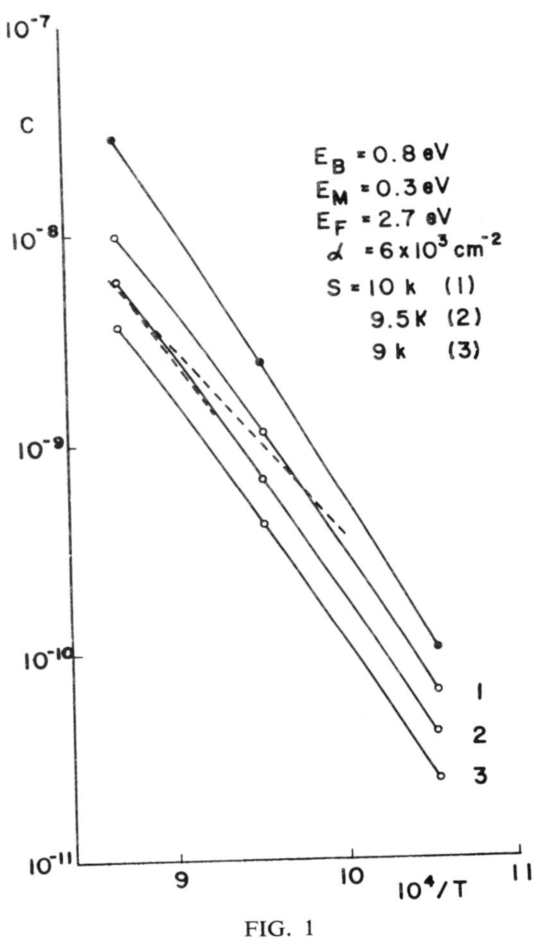

FIG. 1

REFERENCES

1. S. Ishino, F. Nakazawa and R. R. Hasiguti, *J. Phys. Soc. Japan*, **20**, 817 (1965).
2. S. Mayburg, *Phys. Rev.*, **95**, 38 (1954).
3. A. Seeger and K. P. Chik, *Phys. Stat. Sol.*, **29**, 455 (1968).
4. H. Letaw, W. M. Portnoy and L. M. Slifkin, *Phys. Rev.*, **102**, 636 (1956).

REVIEW OF RADIATION EFFECTS IN GERMANIUM†

J. W. MacKAY AND E. E. KLONTZ

Purdue University, Lafayette, Indiana, U.S.A.

The review concentrates on low temperature irradiation and annealing in germanium. Calculations are presented for the charge states to be expected for donor and acceptor centers during irradiation near 10 °K. The calculations are applied to a discussion of the diffusion of an interstitial atom assuming it is mobile in various charge states during irradiation.

The current experimental knowledge is reviewed, as are the various defect models that have been advanced to explain the behavior of germanium in recent years. Suggestions are made of certain experiments that might help in making a rational choice between the possible models.

1. INTRODUCTION

Our understanding of radiation effects in germanium lags far behind that for silicon. The reason is that few of the experiments that are designed to give microscopic information about the structure of defects have been successful in germanium, while substantial results have been obtained for silicon. As a result, defects in germanium are classified according to their connection with some annealing stage of a macroscopic property. Unequivocal identification of a particular defect with a given stage is extremely rare.

Between 4.2 °K and 550 °K there are at least nine annealing stages in *n*-type and *p*-type germanium.[1] In addition there are observations of radiation-induced and light-induced recovery of damage at various temperatures. There is no general agreement on the interpretation of any one of these annealing stages or recovery processes.

Since, at the present time, we lack the means to identify and follow the changes in defect structures with irradiation or temperature changes, only the simplest possible defect arrangements provide any hope for correct interpretation of their identity and characteristics. The simplest conditions are expected to occur when the struck atom is given a minimum of excess energy above that required to produce disorder, and when the temperature is maintained at the lowest feasible value to restrict thermal motion of the defects. For these reasons, most of this review is devoted to low energy electron irradiations at low temperatures.

Even in this restricted field, there is a wealth of phenomena begging for explanation and there is disagreement not only in the interpretation of these

† Work supported by the U.S. Atomic Energy Commission.

phenomena, but also as to the experimental facts.

2. DEFECT CHARGE STATES

Before we proceed to discuss the experimental information and its interpretation, it may be instructive to examine what we expect for the charge states of defects and other centers during irradiation near liquid helium temperature. We will be interested in comparing the expected charge states in lightly doped germanium ($\sim 10^{14}$ chemical impurities per cm³) with degenerate material ($> 10^{17}$ cm⁻³).

In making these estimates we will assume that the temperature is low enough that we can ignore any thermal excitation from states in the forbidden gap to the bands; that the thermal velocity, v, is the same for electrons and holes and that direct excitation of carriers from a center to the band is negligible.

We designate the capture cross section of a center in a given charge state for an electron or a hole by superscripts (0, +, + + etc.) and subscripts (p or n). The superscript indicates the charge state of the center before capture, the subscript indicates whether an electron or hole is to be captured.

The following cross sections are used:

$$\sigma_n^0 = \sigma_p^0 = 10^{-15} \text{ cm}^2$$
$$\sigma_n^- = \sigma_p^+ = 10^{-20}$$
$$\sigma_n^+ = \sigma_p^- = 10^{-12}$$
$$\sigma_n^{++} = \sigma_p^{--} = 5 \times 10^{-12}$$

The value of σ_n^0 was chosen on the basis of measurements of M. Goehring of the Purdue group.[2] He was able to measure the capture cross section of a neutral center for electrons in the band, using pulsed conductivity measurements.

The value of 10^{-15} cm^2 refers to a lattice temperature of about $10\,°$K and is found to be essentially independent of electron temperature, T_e, in the range 50 to $100\,°$K.

The value of σ_n^- at $10\,°$K given above is based on Callcott's measurement[3] of 10^{-18} cm^2 at $T_e = 100\,°$K and a temperature dependence of $\sigma_n^- \propto T_e^{2.5}$.

$\sigma_n^+ = 10^{-13}$ cm^2 at $30\,°$K is an estimate given by Lax.[4] We have assumed a temperature dependence of T^{-2} to obtain a value of 10^{-12} cm^2 at $10\,°$K. σ_n^{++} should be larger than σ_n^+ by a factor of the charge ratio squared.

The rate of generation, G, of electron-hole pairs by an incident electron beam can be estimated from the relation

$$G = \frac{1}{i} \cdot \frac{\mathrm{d}E}{\mathrm{d}x} \cdot \frac{\mathrm{d}\phi}{\mathrm{d}t}$$

where $i \simeq 2$ eV is the average energy per electron-hole pair, $\mathrm{d}E/\mathrm{d}x \simeq 10^7$ eV-cm^{-1} is the energy loss per cm of path by a fast electron in germanium and $\mathrm{d}\phi/\mathrm{d}t$ is the irradiation rate. Irradiation at a rate of 6×10^{10} electrons-cm^{-2}-sec^{-1} (0.01 microamp per cm^2) would result in $G = 3 \times 10^{17}$ cm^{-3} sec^{-1}.

The steady state concentration of excess carriers will be $\simeq G\tau$. Where τ is the excess carrier lifetime. For lightly doped material ($\simeq 10^{14}$ chemical impurities cm^{-3}) we assume that the temperature is so low that the excess carrier concentrations $\Delta n = \Delta p \gg n_0$. Where n_0 is the equilibrium carrier concentration. In this case the value of τ does not enter our calculations. For degenerate material the situation is different. $n_0 \simeq 10^{17}$ cm^{-3} will be much greater than Δn or Δp. We will assume $\tau = 10^{-6}$ sec for degenerate material. Although it is quite likely that τ is actually several orders of magnitude smaller than this, we have chosen to use this value so as not to overstate the differences in probability of finding a center in a positive charge state when degenerate germanium is compared to lightly doped material.

Let us now examine a center that has three possible charge states: 0, -1, -2. If there are N centers (acceptors) of this type per cm^3, we can write the equation for the rate of change of the number N^0 that are in the neutral charge state.

$$\mathrm{d}N^0/\mathrm{d}t = -nv\sigma_n^0 N^0 + pv\sigma_p^- N^- = 0$$

in the steady state. If we take v to be the same for electrons and holes, $N^-/N^0 = n\sigma_n^0/p\sigma_p^-$. A similar calculation for the rate of change of N^{--} gives $N^{--}/N^- = n\sigma_n^-/p\sigma_p^{--}$. The same calculations can be made for a donor center with charge states 0, $+1$, $+2$. As stated above, for lightly doped material, $n = \Delta n = p = \Delta p$ are used to calculate these ratios in lightly doped material. For degenerate material, $n = n_0$, $p = \Delta p = G\tau$.

Table I shows the results of the calculations for two specimens, one lightly doped ($\sim 10^{14}$ Sb atoms per cm^3), the other a degenerate sample with 3×10^{17} Sb per cm^3. $f(0)$ is the fraction of the given type of center that will be found in the neutral charge state (or the fraction of the time a center will have a neutral charge state). The values are given for irradiation rates of 10^{-8} and 10^{-7} amp-cm^{-2}.

TABLE I

Calculated charge state probabilities for donors and acceptors in n-type germanium during irradiation at $T < 10\,°$K. $f(q)$ is the probability of finding a center in the charge state q.

| Irradiation rate | 6×10^{10} cm^{-2} sec^{-1} | | 6×10^{11} cm^{-2} sec^{-1} | |
Impurity concentration	$\sim 10^{14}$	3×10^{17}	$\sim 10^{14}$	3×10^{17}
$f(0)$	1	1	1	1
$f(+1)$	10^{-3}	10^{-9}	10^{-3}	10^{-8}
$f(+2)$	2×10^{-10}	2×10^{-26}	2×10^{-10}	2×10^{-21}
$f(0)$	1	10^{-3}	1	10^{-2}
$f(-1)$	10^{-3}	1	10^{-3}	1
$f(-2)$	2×10^{-10}	2×10^{-3}	2×10^{-10}	2×10^{-4}

We note from the table that the intensity of irradiation does not affect the charge states for lightly doped material. This will be true so long as $\Delta n = \Delta p \gg n_0$. The probability of finding a donor center in the $+1$ charge state is smaller in degenerate material by a factor of 10^5 or 10^6 depending on the value of $\mathrm{d}\phi/\mathrm{d}t$. (These factors would have been larger had we assumed a smaller value of the excess carrier lifetime.) The probability for finding a donor in the $+2$ charge state in degenerate material is smaller by a factor of 10^{14} or 10^{16} than for lightly doped material.

In later sections we will discuss various models that have been proposed to account for the behavior of germanium. One feature of many of these models is that the interstitial atom moves in germanium at low temperatures with an activation energy, $E_m \simeq 0.005$ eV, in some charge state. This implies a diffusion coefficient

$$D = \nu_0 \mathrm{d}^2 \exp(-E_m/kT) \simeq 10^{-6} \text{ cm}^2\text{-sec}^{-1}$$

for an irradiation temperature of 6–8 $°$K.

Table II lists the mean diffusion distance for an

interstitial atom after 100 sec of irradiation at a rate of 6×10^{10} electrons cm^{-2}-sec^{-1}. The diffusion distance, $\bar{X}(q) = \sqrt{2f(q)\,Dt}$, is calculated assuming the data of Table I, and, that in each case, the interstitial is mobile only in the charge state q.

TABLE II

Average diffusion distance, $\bar{X}(q) = \sqrt{2f(q)\,Dt}$, during 100 sec of irradiation at a rate of 6×10^{10} electrons cm^{-2} sec^{-1}, for an interstitial that moves only when its charge state is q.

Impurity concentration	10^{14} cm^{-3}	3×10^{17} cm^{-3}
$\bar{X}(0)$ cm	1.4×10^{-2}	1.4×10^{-2}
$\bar{X}(+1)$	4.5×10^{-4}	1.4×10^{-7}
$\bar{X}(+2)$	2×10^{-7}	2×10^{-15}

After 100 sec of irradiation at the above irradiation rate, the defect concentration should be about 6×10^{12} cm^{-3} for 1.0 MeV electron bombardment. The average distance between vacancies will then be about 5×10^{-5} cm. Comparison of this distance with those in Table II, indicates that any original correlation of an interstitial with its parent vacancy will be lost in lightly doped material if we assume either the 0 or $+1$ charge state to be mobile with an activation energy of 0.005 eV. The $+2$ charge state can experience moderate motion in lightly doped material, but would not move during any reasonable time in degenerate material.

3. *n*-TYPE GERMANIUM

3.1. *Lightly doped*

Germanium doped with 10^{13}–10^{15} group V impurities exhibits a prominent annealing stage, centered at about 65 °K, provided no radiant stimulation is present.[3,5–8] Low energy electron irradiation produces recovery of damage near 4.2 °K and no 65° stage is observed following this radiation annealing.[3,9,10] Similar recovery has also been observed near 4.2 °K when a specimen is subjected to light, filtered so that no band to band excitation of electrons is possible.[6] Other modes of excitation, such as light that can excite electron-hole pairs,[6–8] or X-rays,[11] produce recovery in the range 20–30 °K. The annealing stage that is seen at 35 °K in more heavily doped germanium, is completely absent in lightly doped material.

There are two conflicting reports in these conference proceedings concerning the effect of impurity type on the 65° stage. Meese and MacKay[9] report that As-doped material anneals at a lower temperature than does Sb-doped. Bourgoin and Mollot report that they find no effect of impurity type.[12]

Some properties of the defects produced by electron irradiation of lightly doped *n*-type germanium have been established. Callcott[3] was able to establish that one of the defects involved in the 65 °K annealing stage is a double acceptor and carries a charge of -2 when in thermal equilibrium. He also found that recovery in the 65° stage amounted to more than 90 per cent of the radiation-induced conductivity change for 0.7 MeV irradiation, and that the per cent recovery decreased with increasing bombardment energy to 50 per cent at 4.5 MeV. The large fraction of recovery at lower energies, coupled with the fact that stored energy release has been observed to accompany 65 °K annealing in heavily doped material,[13] leads to the assumption that this annealing is due to mutual annihilation of vacancies and interstitials.

Callcott's results indicate that there are at least two configurations of defects to be accounted for in 1.0 MeV irradiation of lightly doped germanium at low temperature. One configuration is unstable at 65 °K, or at lower temperatures under appropriate radiant stimulation. The other configuration, which increases in importance with increasing bombardment energy, produces as an end result of irradiation and annealing, defects that are stable up to at least 120 °K.

It now appears to be generally agreed that the unstable configuration (65 °K defects) consists of the isolated vacancy and a (sometimes) mobile interstitial atom trapped somewhere in the lattice at some distance from the vacancy, and that the interstitial can be rendered mobile, even at helium temperature, by altering its charge state in the appropriate manner. There the agreement ends and everyone is free to propose his own model.

If we accept the isolated vacancy as one component in the defect models, then it is reasonable to assume that the double acceptor identified by Callcott[3] is the isolated vacancy. This defect disappears in the 65° annealing stage and is also removed by radiation annealing near 4.2 °K. For 0.7 MeV electron irradiation this double acceptor accounts for essentially all the carriers that are removed. Thus we would conclude that if it is the vacancy, then the interstitial, wherever it is, is neutral under equilibrium conditions.

The configuration that results in the more stable defects can be interpreted on the basis of either of two quite reasonable hypotheses. The first[10,14] is that higher energy recoils result in wider separation of vacancy and interstitial and that as the bombardment energy is increased, the fraction of widely separated pairs increases. On subsequent annealing, the more distant interstitials have a lower probability of recombining with the parent vacancy and wander away until they are securely trapped at some other site in the lattice. On this interpretation we have only isolated vacancies and interstitials immediately following irradiation. The second interpretation,[3] is that for bombardment energies that exceed about twice the threshold energy, direct production of divacancies becomes energetically possible. Thus divacancy production should begin at about 0.7 MeV and increase with increasing energy. If annihilation of divacancies by interstitials does not occur at temperatures below 120 °K, then the more stable configuration may consist of divacancies and interstitials.

The way to decide between these two interpretations is obvious once we ask the following question: are all of the acceptors introduced by 4.5 MeV irradiation the same as those produced by 0.7 MeV irradiation? If there are two types, one a vacancy and the other a divacancy, their trapping properties for carriers in the band should be different and should be detectable in the type of experiment described by Callcott.[3] (In retrospect, it is difficult to understand why Callcott and MacKay did not check this point.)

The Purdue group was inspired to ask themselves the above question, after the abstract of the paper by Hyatt and Koehler[10] was made available to them. Very recent results of experiments by M. Goehring show that there are two types of double acceptors present, in nearly equal numbers, following 4.5 MeV irradiation. One of them has the signature of the double acceptor identified by Callcott. *i.e.* A capture cross section for electrons, when it is in its -1 charge state, of 10^{-18} cm^2 for a pulsed field of 40 V-cm^{-1}. The other has a capture cross section of 5×10^{-20} cm^2 under the same conditions. These experiments are still in progress and we do not yet know whether the second kind remains unaltered after 65 °K annealing.

We believe that these results are nearly conclusive proof that higher energy irradiation results in a different type of defect configuration being produced along with the type that is responsible for the 65 °K annealing. The divacancy hypothesis appears to fit the experimental facts.

If the conclusion that divacancies are produced is correct, there may be interesting effects of bombardment energy on the kinetics of the 65 °K annealing stage. For each divacancy that is produced, two interstitials result. If the 65° stage involves only mono-vacancy-interstitial annihilation, the assumption of equal vacancy and interstitial concentrations is incorrect. Irradiation at 4.5 MeV will result in a ratio of interstitials to vacancies of 3. At 1.5 and 1.0 MeV, the corresponding ratios will be 1.6 and 1.3 respectively. These ratios are calculated on the basis of the data of Callcott[3] on the per cent recovery at 65° as a function of energy.

Isothermal annealing studies of the 65° annealing by Zizine[7] showed that the time dependence of the fraction annealed, a, is of the form $a \propto t^{-1/2}$, at long times. This same time dependence has been observed by Meese.[9] It now appears that all of the processes that occur in lightly doped n-type germanium have this same time dependence. Radiation annealing[9,10] is found to produce recovery with $a \propto \phi^{-1/2}$ near 4.2 °K. The optically induced annealing[6] near 4.2° has also been shown to have the $t^{-1/2}$ dependence.[9] The isothermal behavior of the light induced recovery at 20–30 °K has not been reported, but the isochronal annealing looks very similar to the 65° stage, so this process may also show the $t^{-1/2}$ dependence. In addition to these results for lightly doped germanium, Bourgoin and Mollot report that the 35 °K stage, observed in heavily doped germanium, has the same time dependence toward the end of the anneal.

Zizine[7] has pointed out that the $t^{-1/2}$ dependence would be expected for diffusion controlled recombination of correlated vacancy-interstitial pairs. Meese[9] has found that the same time dependence holds if the process is controlled by trapping and re-emission of a very mobile particle by some plentiful trapping centers.

3.2 *Heavily doped*

When degenerate n-type germanium is irradiated by 1 MeV electrons, two annealing stages are observed,[15–19] one at 35 °K, the other at 65 °K. In addition, radiation annealing occurs, near helium temperature, if a damaged sample is irradiated by low energy electrons.[14,15,17,18]

The 65 °K stage is remarkably similar to that in lightly doped material. Callcott[3] reported that, within experimental error, there is no difference in annealing temperature of a specimen doped with 2×10^{14} Sb cm^{-3} on comparing it with a sample containing 2×10^{14} Sb cm^{-3}. Bourgoin and Mollot[12] report a shift of annealing temperature of about 2.5° over a doping range of 4×10^{14} to 6×10^{16} As cm^{-3}.

There are some surprising differences in behavior of the 35 °K annealing stage as observed by different groups. Klontz and MacKay[20] found that the amount of recovery at 35 °K in specimens doped with about 3×10^{17} Sb cm^{-3} is dependent on fluence and previous irradiation and annealing history of the samples. For fluences less than 5×10^{15} electrons cm^{-2} all the recovery was in the 65° stage, no 35 °K stage could be detected. After 10^{16} electrons cm^{-2}, the 35° stage was barely detectable. With larger fluences, the 35° annealing grew in magnitude, becoming equal to the 65 °K stage after about 5×10^{16} electrons cm^{-2}. In another experiment to examine the energy dependence of the annealing stages in germanium,[14] a similar Sb doped specimen was irradiated at different energies in sequence, starting at 0.6 MeV and increasing in steps of 0.1 MeV. At each energy the specimen received sufficient irradiation to change the conductivity by 3 per cent, then an isochronal anneal was made in 5° steps between 20° and 90 °K. In this experiment, no recovery was detectable until the sixth bombardment (1.1 MeV) when it became barely discernible. In none of the anneals did the 35° stage recovery amount to more than 3 per cent of the change produced by irradiation.

Contrast the above described behavior with the results of Penczer and DeAngelis, who performed a similar experiment[17] on the energy dependence of the annealing. They observed 35° annealing immediately following the first irradiation, on samples doped with 10^{18} As cm^{-3}. What is more surprising is that the 35° stage amounted to 30 per cent recovery of the change produced by irradiation while the 65° stage was less than 5 per cent.

Bourgoin and Mollot[19] also report the immediate appearance of the 35° stage in a specimen containing 10^{18} As cm^{-3}, and find that the ratio of recovery at 35° to the recovery at 65° is independent of fluence. In addition, they report that the production rate of defects that anneal in the 35° stage is proportional to impurity concentration. It is not clear in their short report[19] whether the facts stated above apply equally to Sb doped specimens in their experiments. They mention that they have irradiated Sb doped specimens, but quote details only for those containing As. Other experimental details are lacking.

The very different behavior of the 35° annealing stage observed by different groups must have its origin in experimental conditions. It is very tempting to attribute the differences to type of impurity, Sb *vs* As. However, on comparing the experiments of Penczer and DeAngelis to those of Klontz and MacKay, there are several other differences. Most noteworthy: Penczer and DeAngelis used D.C. irradiation, and their specimens were mounted so they were exposed to portions of the cryostat at room temperature. Klontz and MacKay used a pulsed accelerator with a duty cycle of 10^{-4} to achieve the same average irradiation rate, and their specimens were completely shielded from room temperature.

In discussing these experiments we have concentrated our attention on the 35° stage. However it is also true that there are differences in the behavior of the 65° stage. It appears that the essential difference between the experiments of Penczer and DeAngelis and those of Klontz and MacKay is the *distribution* of defects between the 35° and 65° stage. The total recovery in both stages is comparable at every bombardment energy in the experiments of the two groups. However the ratio of 35° recovery to 65° recovery is very different at each energy. Figure 1 sketches the results of the two groups. Note that Penczer and DeAngelis observed both 35° and 65° stages at every energy

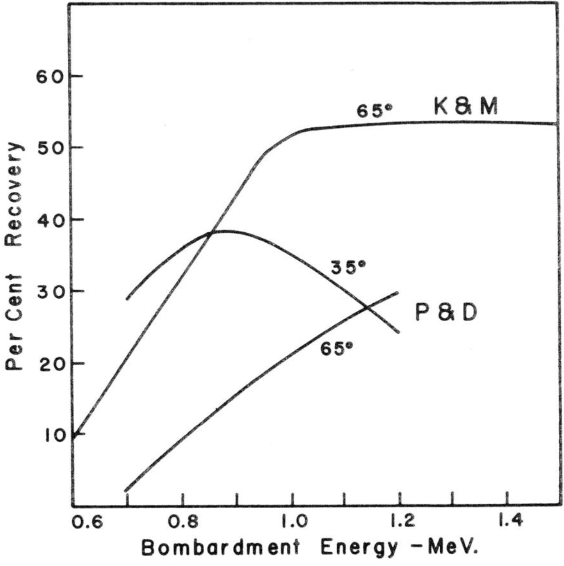

FIG. 1. Comparison of the recovery observed in the 35 and 65 °K annealing stages by Klontz and MacKay (K & M) and by Penczer and DeAngelis (P & D) as a function of bombardment energy.

while Klontz and MacKay saw essentially no recovery in the 35° stage.

3.3 *Oxygen doped*

Whan[21] has studied the infrared absorption of oxygen doped germanium irradiated at 25 °K by 2 MeV electrons. No new bands were observed following irradiation. Isochronal annealing brought out two bands at 719 and 736 cm^{-1}, in the temperature range 58–73 °K. Further annealing caused decay of these bands near 120 °K and a new band grew in simultaneously at 620 cm^{-1}.

Whan convincingly identified the absorption at 620 cm^{-1} to be due to a vacancy-oxygen complex which is the analogue of the Si *A*-center. She was able to show that the 719 cm^{-1} band definitely involved oxygen.

Whan suggested that the simultaneous decay of the 719 cm^{-1} band and growth of the 620 cm^{-1} band implies that the bands are related. On this basis she suggested that the 719 cm^{-1} band is due to an excited state of the vacancy-oxygen complex.

The identification of the 620 cm^{-1} band is definite proof that the vacancy is mobile at 120 °K in germanium. If the 719 cm^{-1} band is an excited state of the vacancy-oxygen complex, then, as Whan suggested, the vacancy is mobile in the temperature range 58–73 °K and must be involved in the 65 °K annealing stage.

In a later paper,[22] Whan casts some doubt on the direct connection between the 719 and 620 cm^{-1} bands. She states in a footnote '... the decay of the complex with the 719 cm^{-1} band is associated with an electronic transition which does not produce the complex with the 620 cm^{-1} band directly or reversibly'. The implication in this paper is that the 719 cm^{-1} band could be due to an interstitial-oxygen complex. This identification would be consistent with vacancy motion at 120 °K, resulting in annihilation of the interstitial with simultaneous formation of the vacancy-oxygen complex.

4. *p*-TYPE GERMANIUM

The behavior of *p*-type germanium is very different from *n*-type.[8,16,23–25] No change is observed following 1.1 MeV electron irradiation at helium temperature. Isochronal annealing results in no changes up to 100 °K. Further annealing finally produces a decrease in hole concentration in the range 120 to 150 °K. No stored energy release

is observed[13] in the range 20 to 90 °K, contrary to results for *n*-type.

The presence of defects can be revealed following irradiation at helium temperature by exposing the specimen to prolonged illumination with light passed through a germanium filter.[23–25,8] Under illumination, the hole concentration decreases with time in a way that can be characterized by two exponentials. One time constant is about 6 hr, the other about 40 hr. These time constants appear to be not strongly dependent on the intensity of the light.

Isochronal annealing, after illumination,[23–25] produces a continuous recovery of hole concentration in the range 40–80 °K, which returns the carrier concentration to the pre-irradiation value. Further annealing, in the range 120–140 °K, produces a decrease in hole concentration as in the specimen that has received no illumination. However, the magnitude of this decrease is dependent on the time of illumination. After 100 hr of illumination the 120–140° stage is completely absent, while 20 hr of illumination has little effect on its magnitude.

There are indications[23–25] that the annealing in the 40–80 °K range is composed of two poorly resolved stages, one centered at about 50 °K, the other at about 65 °K. These stages appear to be correlated with the two time constants that are found when the specimen is subjected to filtered light. For short (10 to 20 hr) illumination, most of the recovery occurs below 60 °K. Longer illumination enhances the recovery above 60 °K and also suppresses the stage at 120 to 140 °K.

Radiation annealing appears to be a major factor in *p*-type germanium. If a specimen that has been irradiated and exposed to about 100 hr of filtered light, is given a further short irradiation, the carrier concentration is returned to the pre-bombardment value. If the specimen is again exposed to filtered light, only a very small decrease in hole concentration is observed and the changes on annealing up to 140° are also suppressed.

The effect of unfiltered light is dramatic. This type of illumination removes all traces of defects, whether it is used following filtered light, or immediately after irradiation.

4.5 MeV irradiation[23] produces damage that has all the characteristics described above but, in addition, there is a decrease in carrier concentration during irradiation ($dp/d\phi = 0.4$ cm^{-1}). The defects that are responsible for this decrease do not appear to take part in the changes produced by illumination or annealing that are described above. Degenerate and lightly doped material appear to behave in the same way.

5. DEFECT MODELS

5.1 *General remarks*

In our introduction we stated that we would restrict our attention to a temperature range where only the simplest defect configurations should occur. It is evident that either we have not imposed severe enough restrictions, or things are not very simple. Any satisfactory model that is constructed to describe the behavior of germanium up to 80 °K must account for motion of some defect in n-type at four distinct temperatures: 65° in the dark, $\sim$28° with unfiltered light, 4.2° with filtered light or irradiation in lightly doped material, and at 35 °K in degenerate material. In addition it should account for the gross qualitative difference between n-type and p-type.

Most of the models that have been put forward are based on the assumption that the interstitial may have different activation energies for migration in different charge states. Table III attempts to summarize four such models which are also dis-

TABLE III

Summary of defect models based on a moving interstitial: (a) Moving particle (migration energy); (b) Controlling mechanism; (c) End reaction; (d) Activation energy for stage.

Stage		Paris	Illinois	Reading	Purdue
65°	a.	I° (0.15 eV)	I° (0.15 eV)	I^- (0.15 eV)	I°, I^+ (0.005 eV)
	b.	Diffusion controlled	Diffusion controlled	Diffusion controlled	(I Sb) dissoc. and $I^\circ \leftrightarrow I^+$
	c.	$I^\circ + V \to 0$	$I^\circ + V \to 0$	$I^- + V \to 0$	$I^+ + V^- \to 0$
	d.	$E_a = E_m(I^\circ)$	$E_a = E_m(I^\circ)$	$E_a = E_m(I^\circ)$	$E_a = E_i(I^\circ) = 0.15$ eV
20–30° Light	a.	I^+ (0.05 eV)		I° (?)	I°, I^+ (0.005 eV)
	b.	Diffusion controlled		Diffusion controlled	(I Sb) dissoc. and $I^\circ \leftrightarrow I^+$; $V^\circ \leftrightarrow V^-$
	c.	$I^+ + V \to 0$		$I^\circ + V \to 0$	$I^+ + V^- \to 0$
4.2° Radiat. anneal	a.	I^{++} (few meV)	I^+ (0.004 eV)	I^+ (?)	I°, I^+ (0.005 eV)
	b.	Diffusion controlled	Diffusion controlled	Diffusion controlled	(I Sb) dissoc. and $I^\circ - I^+$; $V^\circ - V^-$
	c.	$I^{++} + V \to 0$	$I^+ + V \to 0$	$I^+ + V \to 0$	$I^+ + V^- \to 0$
35°	a.	As_I (0.06 eV)			
	b.	Diffusion controlled			
	c.	$As_I + V \to 0$			
	d.	$E_a = E_m(As_I)$			

cussed below. In addition to the mobile interstitial we will also consider the possibility that the vacancy is mobile at 65 °K as suggested by Whan.[20]

5.2 *Paris*

On the basis of his observation that the isothermal annealing in the 65° stage has a $t^{-1/2}$ dependence Zizine[7] proposed that the stage is due to diffusion controlled recombination of vacancies and interstitials. He suggested that the mobile entity is the neutral interstitial and that the observed activation energy of 0.15 eV is the migration energy of the neutral interstitial. To account for the shift of the annealing down to 20–30° under illumination, he proposed that the migration energy, E_m, is dependent on the charge state of the interstitial. In the $+1$ charge state, produced by illumination $E_m^+ = 0.04$ eV.

Bourgoin and Mollot have extended the model.[12,19] To account for optically induced annealing and radiation annealing near 4.2°, they propose that the $+2$ charge state has an even lower migration energy, presumably of the order of a few meV. In this model the germanium atom is mobile during helium irradiation in only the $+2$ state. To account for the 35 °K stage they propose that the germanium interstitial moves during irradiation in heavily doped material to displace a group V impurity atom thus producing As or Sb interstitials. The impurity interstitial is assumed to have a migration energy in its neutral charge state of $E_m = 0.06$ eV.

The first part of the model is consistent with all the experimental information except for the (disputed) impurity dependence of the 65° stage on impurity type.[9] The assumption that only the $+2$ charge state is mobile during irradiation would preserve the correlation implied in the model (see Table II). However, the assignment of the 35° stage to migration of As interstitials appears to be at odds with the rest of the model and with the results of Bourgoin and Mollot.

Tables I and II indicate that the interstitial will experience great difficulty in moving any distance in degenerate material. If we waive this difficulty, it is very hard to see how the interaction energy between I^{++} and Sb^+ in degenerate material can ever be greater than kT. (The Coulomb repulsion will prevent the interstitial approaching closer than R_0, such that $V(R_0) = kT$.) Watkins has observed[26] an interstitial Al^{++} in p-type silicon which he has suggested is a result of displacement by interstitial Si. However, in this case the potential is attractive, and the two atoms may interact quite strongly. Watkins also finds that the interstitial Al is not mobile at low temperatures, but apparently moves at about 473 °K. The mechanism for production of As interstitials would appear to require a dependence of 35 °K annealing on fluence, as 65° defects (germanium interstitials) are converted to As interstitials. Bourgoin and Mollot report that the ratio of the two annealing stages is independent of fluence.

5.3 *Illinois*

Hyatt and Koehler adopt Zizine's proposal that the neutral interstitial is the moving entity in the 65° stage, and that the annealing is diffusion controlled recombination of correlated pairs. To account for radiation annealing near 4.2 °K and the temperature dependence which they observe for this process, they assign an activation energy $E_m^+ = 0.004$ eV to the $+1$ charge state of the interstitial. They make no assignments to account for motion under other conditions.

As we pointed out in Sect. 2, assignment of motion to the $+1$ charge state should cause loss of correlation between the original pairs. Of course some correlation will exist, even in a completely random distribution, but a very high degree of correlation is required to account for the nearly complete recovery for lower energy irradiations.

5.4 *Reading*

The Reading group[8] propose that the interstitial can assume the charge states -1, 0, $+1$. They attribute the 65° stage to migration of the interstitial in its -1 charge state. Illumination produces the neutral charge state and migration of I^0 accounts for the recovery at 20–30 °K under illumination. Radiation annealing near 4.2 °K is due to migration in the $+1$ charge state.

The only criticism of these assignments is that they are in conflict with Callcott's result[3] that the only acceptor present after low energy irradiation is in a charge state of -2 and the first electron is not thermally ionized at an electron temperature of 100 °K.

5.5 *Purdue*

The Purdue group has proposed[25] that the interstitial is mobile in all charge states at helium temperature and that the neutral interstitial can be captured by the group V impurities. It was suggested that the probability of recombination of vacancy and interstitial is small unless a pair simultaneously have opposite charge states, in which case the Coulomb attraction makes recombination very much more probable. In this model the annealing kinetics are governed by the effects of charge state on trapping and recombination cross sections rather than by a direct effect on the migration energy of the interstitial.

Based on his experiments which show a dependence of the 65° annealing stage on the type of impurity, Meese[9] interprets the 65° stage in the following way. During irradiation the interstitials are trapped on the impurities. As the temperature is raised, interstitials begin to be released from the traps by thermal excitation. However, the trapping cross section of the impurities for neutral interstitials is very much larger than the recombination cross section with vacancies, so the concentration of free interstitials remains in equilibrium with the traps. This model has an approximate solution at long times that results in a $t^{-1/2}$ dependence. In Meese's interpretation the activation energy for the annealing is the ionization energy of the free interstitial. The character of the traps does not affect the long time portion of the annealing, but does influence the initial transient.

Radiation annealing can be accounted for on this model. Interstitials are freed from traps by capture of a hole, but spend most of their free time in the 0 charge state, and are retrapped with large probability by the impurities. Again the free interstitial concentration comes into equilibrium with the traps. The recombination cross section of interstitials and vacancies is very small unless they have opposite charge states. Thus the annealing rate is controlled by $P(+, -)$, the probability that the members of a given pair have opposite charge states. $P(+, -) = f_i(+1)f_v(-1) \simeq 10^{-6}$ for lightly doped material (see Table I) and $P(+, -) = 10^{-9}$ or 10^{-8} depending on irradiation rate, in degenerate material. Thus on this model the radiation annealing should proceed 100 to 1000 times faster in lightly doped material than in degenerate material, depending on what irradiation rate is used in the comparison experiments. Callcott[3] reported that the lightly doped material anneals about 100 times faster.

Table I indicates that for diffusion limited models radiation annealing should be faster by a factor of 10^5 or 10^6, if the $+1$ charge state is considered to be mobile. If the $+2$ state is invoked, the factor is 10^{14} or 10^{16}.

Meese[9] is also able to account for the annealing produced by filtered light near 4.2° and the recovery in the 20 to 30° range.

This model appears to give a reasonable description of the 65° stage and the lower temperature, stimulated recovery processes. A major flaw is the fact that assignment of 0.15 eV as the interstitial ionization energy, leaves no way to describe the 35° annealing stage within the framework of the model. Nor does the model give any obvious reason for the much lower production of damage in p-type.

5.6 Whan model

Whan has suggested that the vacancy may be mobile in the temperature range 58–73 °K and could be the mobile entity in that stage. She does not address herself to the question of where the interstitial may be. But, whether the interstitial is immobile or has moved to some nearby trap, the 65° stage could be diffusion controlled recombination of correlated pairs, with the vacancy having a migration energy of 0.15 eV, presumably in a negative charge state.

Whan[21] implies but does not specifically state that the annealing under illumination at 20–30 °K could also be due to the migration of the vacancy in a lower charge state, possibly neutral. Beyond this Whan does not go (and we may have read more into her paper[21] than she intended).

If we were to combine the idea of a very mobile interstitial that is trapped by impurities, with a

vacancy that is mobile at 65 °K and possibly at 35 °K in a lower charge state, there would be no difficulty in constructing models to fit the other temperature ranges.

6. CONCLUSION

One thing stands out as we read back over this review. That is that we know quite a lot about the behavior of germanium under irradiation and annealing at low temperatures, but we know very little for sure. For example: We know for sure that the vacancy is mobile at 120 °K but it may be mobile at 65 °K. We know for sure that charge states are important, but we don't know which defect's charge state is important.

Outstanding questions to which we need answers are the following:

(1) Does the vacancy move at 65 °K?

(2) Does the interstitial remain self-trapped in the lattice up to 65 °K or does it move to some trap?

(3) What is the connection between the 35° stage and the 65 °K stage? Why is the distribution between stages different when observed by different investigators?

(4) Is there an effect of bombardment energy on the kinetics of the 65° stage?

(5) Why are the defects apparently electrically inactive up to 120 °K in p-type?

(6) What is the important difference between germanium and silicon that produces the inversion of sensitivity of n-type and p-type to low temperature irradiation? *i.e.* Why are n-type silicon and p-type germanium resistant to damage, while p-type silicon and n-type germanium are easily damaged?

Questions (3) and (4) will probably be answered soon by continued experimental effort along current lines.

The question of vacancy motion needs to be settled before we can have any real confidence in any description of the 65° annealing stage. It may yield to further infrared absorption studies of the type that Whan[20,21] has made. However, it may also be profitable to look for the vacancy-group V impurity complex (analogue to the Si E-center). One approach to this would be to look for loss of chemical donor states on annealing at 65 °K and also at 120 °K. This approach may be complicated by the presence of oxygen in lightly doped materials.

The question of interstitial-impurity interaction may be answered indirectly if Meese's experiments are confirmed by others. However, it probably could be answered directly if someone had the courage to investigate the effects of low temperature irradiation on the donor excitation spectra in the far infrared.

The answer to question (5) probably lies in the particular defect structure that survives in p-type. So this question may come down to the universal one of somehow obtaining direct structural information in germanium.

At the moment question (6) is merely one to ponder.

REFERENCES

1. R. R. Hasiguti, *Radiation Damage in Semiconductors* (Dunod, Paris, 1965), p. 259.
2. M. Goehring, private communication.
3. T. A. Callcott and J. W. MacKay, *Phys. Rev.*, **161**, 698 (1967).
4. M. Lax, *J. Phys. Chem. Solids*, **8**, 66 (1959).
5. T. A. Callcott and J. W. MacKay, *Radiation Damage in Semiconductors* (Dunod, Paris, 1965), p. 27.
6. Itsu Arimura and J. W. MacKay, *Radiation Effects in Semiconductors*, Ed. F. L. Vook (Plenum Press, New York, 1968), p. 204.
7. J. Zizine, *Radiation Effects in Semiconductors*, Ed. F. L. Vook (Plenum Press, New York, 1968), p. 186.
8. Z. G. Werner, J. E. Whitehouse, S. Ishino and E. W. J. Mitchell, *IX International Conference on the Physics of Semiconductors—Proceedings*, Moscow, 1968, Vol. II, p. 1070.
9. J. M. Meese and J. W. MacKay, this conference.
10. W. D. Hyatt and J. S. Koehler, this conference.
11. M. Goehring, private communication.
12. J. Bourgoin and F. Mollot, this conference.
13. M. P. Singh and J. W. MacKay, *Phys. Rev.*, **175**, 985 (1968).
14. J. W. MacKay and E. E. Klontz, *Radiation Damage in Semiconductors* (Dunod, Paris, 1965), p. 11.
15. J. W. MacKay, E. E. Klontz and G. W. Gobeli, *Phys. Rev. Letters*, **2**, 146 (1959).
16. J. W. MacKay and E. E. Klontz, *J. Appl. Phys.*, **30**, 1269 (1959).
17. R. E. Penczer and H. M. DeAngelis, *Phys. Rev.*, **171**, 862 (1968).
18. H. M. DeAngelis and R. E. Penczer, *J. Appl. Phys.*, **39**, 5842 (1968).
19. J. Bourgoin and F. Mollot, *Phys. Letters*, **30A**, 264 (1969).
20. E. E. Klontz and J. W. MacKay, *J. Phys. Soc. Japan*, **18**, Supplement III, 216 (1963).
21. R. E. Whan, *Phys. Rev.*, **140**, A690 (1965).
22. R. E. Whan, *Radiation Effects in Semiconductors*, Ed. F. L. Vook (Plenum Press, New York, 1968), p. 195.
23. J. E. Whitehouse, *Phys. Rev.*, **143**, 520 (1966).
24. T. M. Flanagan and E. E. Klontz, *Phys. Rev.*, **167**, 789 (1968).
25. R. A. Matula and E. E. Klontz, this conference.
26. G. D. Watkins, *Radiation Damage in Semiconductors* (Dunod, Paris, 1965), p. 97.

DISCUSSION

Question (BOURGOIN) What value of frequency factor did you consider in your calculation of the interstitial migration range? Is it the same value for different charge states of the interstitial? Why did you take this value?

Answer (MACKAY) We used $\nu_0 \simeq 10^{12}$ sec^{-1} for the estimates in all charge states. We expect a frequency factor of the order of kT/h.

Question (BOURGOIN) You stated that in Ecole Normale (Paris) we say that the interstitial which anneals at 65 °K is the neutral interstitial. We just think that there are three different charge states for the interstitial but we do not state exactly what these charge states are.

Answer (MACKAY) I regret that we have apparently misinterpreted your comments in the summary of your paper for this conference where you speak of the interstitial, the ionized interstitial, and the doubly ionized interstitial.

Question (KOEHLER) Why should the production of divacancies by high energy electron irradiation result in a lower percent recovery in the low-temperature annealing stages (i.e., 65 °K and lower) in n-type germanium?

Answer (MACKAY) Only if the divacancy is more stable than the interstitial—more resistant, let's say, than the single vacancy. I don't know why this should be so, but if it is so it would account for it.

Response (KOEHLER) One possible way to achieve this would result if interstitials cluster. When divacancies are generated two interstitials are produced close together. Perhaps di-interstitials are stable and have a high migration energy.

IMPURITY DEPENDENCE OF THE LOW TEMPERATURE ANNEALING IN *n*-TYPE GERMANIUM†

J. M. MEESE AND J. W. MacKAY

Department of Physics, Purdue University, Lafayette, Indiana, U.S.A.

Lightly doped *n*-type Ge, irradiated at liquid He temperatures with 1 MeV electrons, exhibits a large thermal recovery stage at 50–70 °K. We have found that the rate at which this stage anneals depends on the type of group V impurity used to dope the sample. We propose that impurity complexes are involved in this annealing stage. We have also observed the same impurity dependence when this annealing stage is destroyed by radiation annealing at liquid He temperatures. This suggests that one of the defects produced during irradiation is free to migrate at very low temperatures.

1. INTRODUCTION

The 65 °K annealing stage in irradiated *n*-type Ge has been identified as a vacancy-interstitial annihilation. A stored energy release of $\sim$5 eV per defect accompanies this recovery stage.[1,2] There is also a nearly complete recovery in carrier concentration for lightly irradiated samples.[3,4] These experiments lend support to the above identification. Since the 65 °K stage can be shifted to liquid He temperatures by extrinsic optical stimulation[5] or subthreshold electron irradiation,[3] it is clear that the charge states of the defects must be considered.

The 65 °K annealing stage was originally interpreted by MacKay and Klontz as a recombination of close-pairs of vacancies and interstitials.[3] Zizine has proposed that the annealing is rate limited by the diffusion of interstitials to correlated vacancies in order to explain the observed time dependence of the annealing which is proportional to $t^{-1/2}$ for long annealing times.[6] Neither model considers the possibility that impurity complexes may be playing a role in determining this annealing rate.[7]

In this paper we will discuss recent experiments which suggest that impurity complexes do influence the rate of annealing the 65 °K stage.

2. EXPERIMENTAL

Single crystal Ge samples of dimensions $0.8 \times 0.2 \times 0.01$ cm were irradiated and annealed in pairs. One sample of the pair was As-doped while the other was doped with Sb. The group V impurity concentrations of both samples were matched to

† This work was supported by the U.S. Atomic Energy Commission.

within 10 per cent, using DC Hall and conductivity measurements at 77 °K, and were about 4.7×10^{14} cm^{-3}. The pairs of samples were irradiated at liquid helium temperature with 1.0 MeV electrons. Fluences were of the order of 5×10^{13} cm^{-2}; the beam intensity was about 2×10^{11} electrons/cm²-sec. Damage was monitored by measuring pulse-conductivity at 4.2 °K using the 'hot-carrier' technique described elsewhere by Callcott.[4] An electric field of 100 V/cm and a pulse length of $0.66\,\mu$sec were used in these measurements. Callcott has shown that for pulse fields greater than 60 V/cm, the relative change in conductivity is proportional to the change in carrier concentration. This is then related to the number of defects by the relation

$$\frac{\Delta\sigma}{\sigma_0} = \sum_j \frac{Z_j N_j}{n_0}$$

where N_j is the number of the jth type of defect, Z_j is the charge of this defect in electronic units, n_0 is the exhaustion range carrier concentration measured at 77 °K before irradiation, and σ_0 is the pulse-conductivity before irradiation.

All the annealing experiments were performed with the samples in a dark environment. A 1 mil Al-foil window, cooled to 4.2 °K, allowed electrons to enter the irradiation chamber but excluded all light from the samples. This is necessary because the 65 °K stage can be optically annealed at much lower temperatures.[5]

3. RESULTS

Both isochronal and isothermal annealing experiments were performed on pairs of samples doped with different group V impurities, but in

"

nearly equal concentrations. Samples were taken from several different ingots and the samples' positions in the cryostat were changed from one experiment to the next in order to eliminate several possible sources of systematic error. In all cases, the samples doped with As recovered sooner than those doped with Sb, the total recovery in both samples being about 85 per cent.

The impurity dependence is strongest at the beginning of the anneal. In the isochronal anneals,

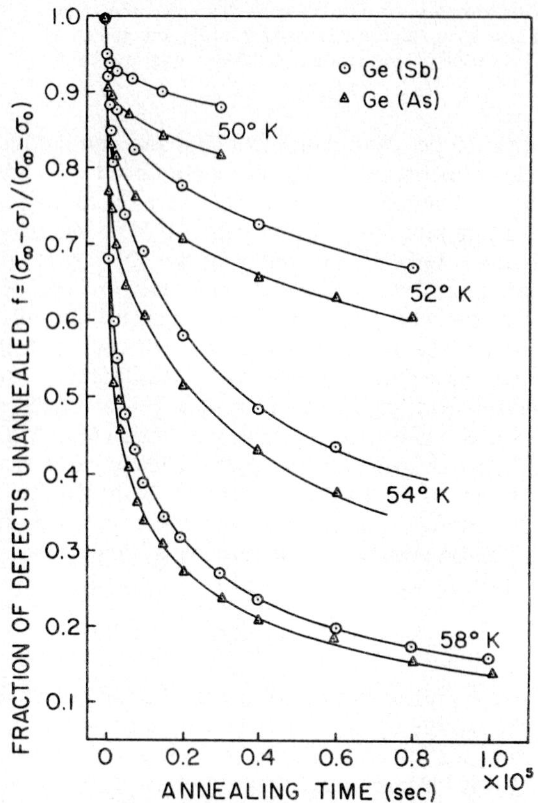

FIG. 1. Impurity dependence in the isothermal annealing of the 65 °K annealing stage.

using temperature pulses of 700 sec, the As-doped samples reached 5 per cent recovery about 5 °K lower in temperature than the Sb-doped samples; at 50 per cent recovery, this difference in temperature was only about 2.5 °K. This same behavior is seen in the isothermal anneals. Figure 1 shows isothermal anneals for As- and Sb-doped samples at several annealing temperatures. The final recovery, σ_∞, was determined by heating the samples to 100 °K for 7 minutes at the end of the

annealing run. The rate of annealing shows the strongest impurity dependence for early annealing times and lower temperatures. For longer times, the annealing curves for the two samples parallel each other indicating that the impurity effect has saturated. The fraction of defects remaining is found to be proportional to $t^{-1/2}$ at longer times in agreement with results reported by Zizine for more heavily doped samples.[1] The activation energy associated with the longer annealing times is 0.155 ± 0.010 eV for both samples. For very short annealing times, we have estimated the activation energy to be about one-third of this value.

Figure 2 shows the result of irradiation at 7 °K with 500 keV electrons on a pair of samples origin-

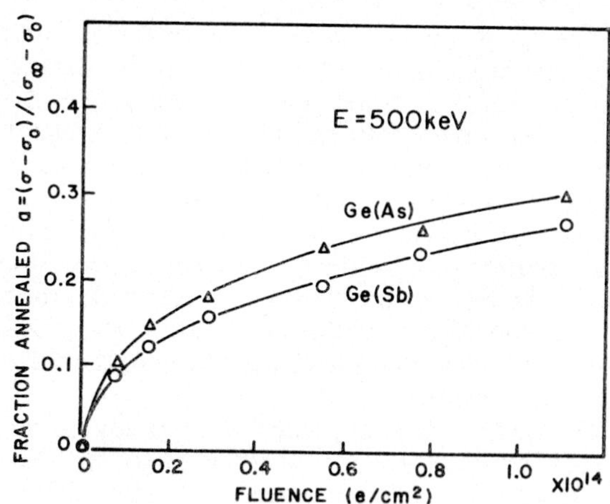

FIG. 2. Impurity dependence in the radiation annealing at 7 °K of the 65 °K stage.

ally damaged at 1 MeV. Both samples radiation anneal, however, the As-doped sample anneals at the faster rate. The fraction of defects remaining is proportional to $\phi^{-1/2}$ over most of the annealing, where ϕ is the fluence of 500 keV electrons.

4. DISCUSSION

Although a recombination of vacancies and interstitials accounts for the high recovery in conductivity and large stored energy releases associated with the 65 °K annealing stage, we see no reasonable way to reconcile the observed impurity dependence with either a close-pair model or a diffusion limited model. In the close-pair model there should be no interaction between the group

V impurities and the vacancy-interstitial pair. In a diffusion limited model, the impurity dependence should occur only after an appreciable fraction of the defects has annealed. Only after longer annealing times can the vacancy interstitial correlation be lost and an interaction occur between the migrating defect and the impurity. However, we observe that the impurity dependence is negligible for very long annealing times.

Figure 3 shows the result of illuminating a sample containing 65 °K defects with light filtered through a thick Ge filter (data by Arimura).[5] We again find that the fraction of defects remaining is proportional to $t^{-1/2}$. The same kinetics are observed for the optical annealing at 4.2 °K, the radiation annealing at 7 °K, and the thermal annealing at 50–80 °K in these lightly doped samples.

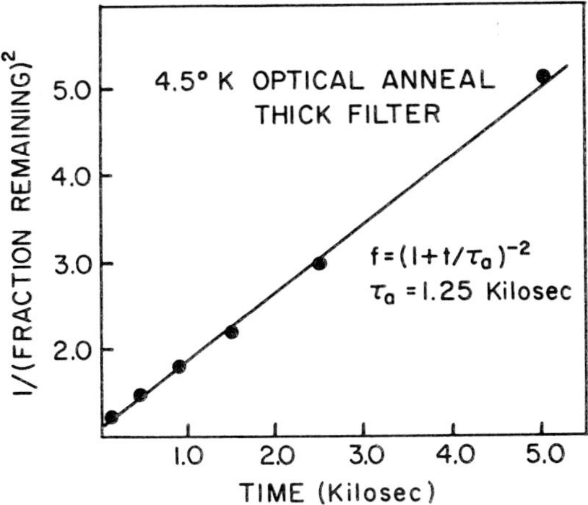

FIG. 3. Annealing kinetics of the 4.5 °K extrinsic optical annealing of the 65 °K stage (data by Arimura).

The impurity dependence in the radiation annealing experiment suggests that one of the defects is migrating over large distances at very low temperatures. Arimura has observed that intrinsic monochromatic light produces no optical annealing at liquid helium temperatures in contrast to the effect of filtered light.[2] Several workers have observed radiation annealing in this same temperature range, however.[3,4] Since both intrinsic light and electron irradiation produce electron-hole pairs, we might expect the annealing rates to be similar for both types of excitation. The sample

temperature for optical stimulation is estimated to be less than 5 °K. However, the electron beam, in the radiation annealing experiments heats the sample to temperatures of 7° to 10 °K. If the interstitial migration energy is of the order of 0.005 eV, then the difference in these two experiments can be explained on the basis of sample temperature alone. This estimate of the interstitial migration energy in Ge agrees well with that made by Watkins for the interstitial in Si.[8]

We propose that the 65 °K annealing stage is initiated by the dissociation of an interstitial-group V impurity complex which is formed during irradiation. If the interstitial and impurity are bound together by sharing an electron as well as by an attractive strain field, then ionization of this electron reduces the stability of the complex. The interstitial is aided in its escape from the impurity strain field by a Coulomb repulsion. Assuming that the vacancy is in a negative charge state, then the probability for annihilation is enhanced when the interstitial is positive and annihilation proceeds. Those interstitials which become neutral before annihilation can again be retrapped at impurities thereby slowing down the annealing rate.

For low levels of interstitial ionization, the above model will produce annealing kinetics which are approximately third-order for longer annealing times. A third-order anneal yields a $t^{-1/2}$ time dependence for the fraction of defects remaining. We associate the 0.15 eV activation energy with the thermal ionization of the free interstitial. The lower activation energy is related to the probability for complex dissociation.

Filtered light at 4.2 °K produces direct excitation of the electron from the complex. However, the vacancy is also partially ionized tending to slow down the annealing rate. In the presence of intrinsic light at 4.2 °K, the complexes can dissociate only by capturing a hole. This is a less likely process since there are equal numbers of excess electrons and holes and the overall probability for annihilation is thereby reduced as observed by Arimura.[5]

Radiation annealing as well as extrinsic light and filtered light optical annealing all show a rapid rise in the annealing rate in the 30–40 °K temperature range.[5,9] We believe this to be due to the increase in conduction band electrons at these temperatures. This increase will reduce the time required for a neutral vacancy to capture an electron. This enhances the recombination probability for all three types of annealing.

REFERENCES

1. J. Zizine, 'Symposium on Radiation Effects in Semi-conductor Components', Toulouse, France, 1967 (*Jour. d'Electron.*, Toulouse, 1969), A23.
2. M. P. Singh and J. W. MacKay, *Phys. Rev.*, **175**, 985 (1968).
3. E. E. Klontz and J. W. MacKay, *J. Phys. Soc. Japan*, **18**, Suppl. III, 216 (1963).
4. T. A. Callcott and J. W. MacKay, *Phys. Rev.*, **161**, 698 (1967).
5. I. Arimura and J. W. MacKay, *Radiation Effects in Semiconductors*, Proc. of the Santa Fe Conf., 1967 (Plenum Press, New York, 1968), p. 204.
6. J. Zizine, *ibid.*, p. 186.
7. J. W. MacKay and E. E. Klontz, *ibid.*, p. 175.
8. G. D. Watkins, *Radiation Damage in Semiconductors*, 7th International Conf. on Semiconductors, Paris-Royaumont, 1964 (Dunod, Paris, 1965), p. 97.
9. J. Naber, Ph.D. thesis, Purdue, 1966, unpublished.

DISCUSSION

Question (HASIGUTI)   When we talk about impurity dependence, we have to distinguish the dependence on impurity concentration from that on impurity species. Do your specimens doped with As and doped with Sb contain the same concentration of dopants?

Answer (MEESE)   Our samples were uncompensated and the concentrations are matched to ± 5 per cent in group V impurity concentration.

Question (MCKEIGHEN)   Zizine plots 'per cent annealed' as being proportional to $t^{-1/2}$ whereas you show 'per cent remaining' as being proportional to $t^{-1/2}$. Are these two results equivalent?

Answer (MEESE)   Yes, except for the intercept. If $f \propto 1/\sqrt{t}$ then $a = 1 - f = 1 - K/\sqrt{t}$ which is linear on a $1/\sqrt{t}$ plot.

Question (BOURGOIN)   Your first slide shows that the difference in annealing rates decreases when the annealing temperature increases. Have you done isothermal annealing at higher temperature, especially at the annealing temperature (65 °K)?

Answer (MEESE)   We have an isothermal anneal at 62 °K which shows no impurity dependence to within experimental error. We emphasize that the impurity dependence is seen only at lower temperatures and earlier rates. This is not a Fermi level shift effect. In the radiation annealing experiment, one can not describe a Fermi level. However, the impurity dependence is still observed.

Response (BOURGOIN)   Since there is no difference between annealing rates at 65 °K your results are identical to mine. I explain such a result in my paper.

Question (MITCHELL)   The ionization energies of As and Sb donors are slightly different. Was the Fermi level in your experiments at the same separation from the defect level in each case?

Answer (MEESE)   We have no way of measuring the position of the Fermi level during a hot carrier measurement. Our model does include, however, a shift in the Fermi level during the anneal.

DEFECTS INDUCED BY ELECTRON IRRADIATION OF p-TYPE GERMANIUM†

R. A. MATULA AND E. E. KLONTZ

Department of Physics, Purdue University, Lafayette, Indiana 47907, U.S.A.

Illumination with filtered light after electron bombardment at ∼8°K reveals defects produced in p-type germanium. During illumination, the carrier concentration decreases as the sum of two exponentials. Identification of the two time constants with two stages of thermal recovery at about 50°K and 65°K indicates the short time constant is associated with the low temperature recovery and the long time constant with the high temperature recovery. Enhancement of the short time component is accomplished by controlling the illumination time, and of the long time component by heat treatment to 80°K prior to exposure by filtered light. Evidence is presented for nearly complete (98 per cent) annealing of defects by white light illumination both before and after exposure to filtered light. This is in contrast to radiation annealing of the exposed defects by the 1 MeV electron beam which anneals only about 80 per cent of the defects and converts the remainder to their neutral state. The rate of introduction of damage increases from very small values for low doses to ∼0.1–0.2 cm⁻¹ for large doses of irradiation. It is believed that the low rate of damage together with the highly mobile defect constituents preclude the existence of isolated vacancies and interstitials.

1. INTRODUCTION

There is no reason to believe that the basic mechanism for the production of defects by high energy (1 MeV) electrons in p-type Ge is any different from that in n-type Ge. The fact that results observed differ so widely for the two materials is evidence that the charge state and mobility of the defects immediately following the electron-atom collision are responsible for those observations. To keep experimental conditions as simple as possible, the irradiations for the present studies were carried out around 8°K, samples were illuminated at about 5°K, and conductivity and carrier concentrations were determined at a reference temperature of 33°K. The relatively high reference temperature was necessary because non-degenerate samples ($p \sim 10^{14}$ or 10^{15} cm⁻³) were used. Simultaneous irradiations and measurements of In-doped and Ga-doped material were made so that differences, if any, due to type of dopant could be observed. Thermal anneals were made by isochronal processes during which the sample was maintained at a given temperature for 300 seconds.

2. BACKGROUND

Results of irradiation of p-type Ge as reported on both degenerate[1–3] and non-degenerate[4] material show similar characteristics. Bombardment pro-

duces very little change in conductivity (Figure 1) corresponding to a carrier removal rate $< 10^{-2}$ cm⁻¹ for non-degenerate material (about 0.4 cm⁻¹ for degenerate material at higher energies[2]). However, a long (∼100 hours) illumination by light filtered through a thick piece of Ge reveals defects produced by the irradiation. Subsequent isochronal annealing produces recovery in conductivity and

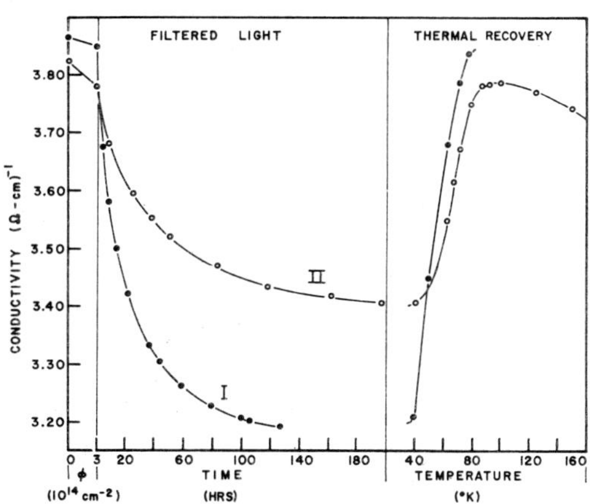

FIG. 1. Defects produced by 1 MeV electrons in p-type Ge are exposed by filtered light illumination. The conductivity recovers by isochronal processes (I). Heat treatment to 80°K prior to filtered light reduces the number of defects, enhances long time component of illumination and high temperature stage of recovery (II).

† Work supported by an AEC contract.

carrier concentration up to 80 °K followed by decreases in those properties up to 150 °K. For non-degenerate samples the recovery at 80 °K is nearly 100 per cent of the initial values. The magnitude of the decrease above 80 °K depends on the duration of the illumination at 5 °K: the longer the illumination, the smaller is the decrease. The decrease in conductivity produced by illumination can be represented by a curve which is the sum of two exponential terms. Ratios of the time constants ($\tau_{\text{long}}/\tau_{\text{short}}$) for the two terms vary somewhat for differing experiments but are generally of the order of 4/1 to 6/1. The short time constant is about 4 to 6 hours. Flanagan[4] has presented evidence that the thermal recovery following illumination consists of two stages which are related to the two components of illumination.

3. CORRESPONDENCE OF ILLUMINATION EFFECTS AND THERMAL RECOVERY

A pair of experiments link the short time constant of illumination with the low temperature stage of anneal. First, two samples were bombarded, illuminated until no further changes in time were observed (110 hours to completion), and then annealed to 80 °K. The temperature at which one-half of the total recovery during the anneal occurred was observed to be 51 °K. The same samples were again bombarded at 8 °K after annealing to room temperature. Following the second bombardment, illumination was carried out for only 5 hours or about 1.3 times the duration of the short time constant. The corresponding half-recovery temperature was then found to have dropped to 41.5 °K. Since the defects which are observed to recover thermally are those which have been exposed by illumination, the conclusion is that the short time constant must be identified with the low temperature recovery stage.

Although in these experiments samples of two different dopants were used, one (indium) of which had 7 times the carrier concentration of the other (gallium), the half-recovery temperatures were within 1.5 °K of each other. This is about the range of the experimental uncertainty which suggests that these low temperature effects are impurity-independent.

4. HEAT TREATMENT

The effect of a heat treatment on the unexposed defects was observed by raising the temperature to

80 °K for 300 seconds immediately after irradiation. The result of this experiment, as indicated in curve II of Figure 1, was that considerably fewer defects were revealed by the filtered light following the heat treatment. Not only was the total number of defects observable by illumination reduced, but the relative numbers as observed in the short and long time constant components was changed. This result is shown in Table I.

TABLE I

Effect of heat treatment at 80 °K on numbers of defects exposed by illumination.

Dopant	Observation	No. of centers (10^{14} cm^{-3})	% Short component	% Long component
Ga	Before heat treatment	0.91	30	70
	After heat treatment	0.25	21	79
In	Before heat treatment	0.73	29	71
	After heat treatment	0.44	16	84

Clearly the long time constant component is enhanced at the expense of the short one. When this treatment is followed by annealing to 80 °K, the relation between the long time constant component of the illumination and the high temperature stage of recovery is evidenced by half-recovery temperatures of 63 °K and 66 °K for the Ga- and In-doped samples respectively.

5. ILLUMINATION WITH WHITE LIGHT

In contrast to the effect of filtered light, white light produces a much different result (Figure 2). Illumination of the Ga-doped sample by white light immediately following bombardment produced no changes (<1.6 per cent) in its electrical

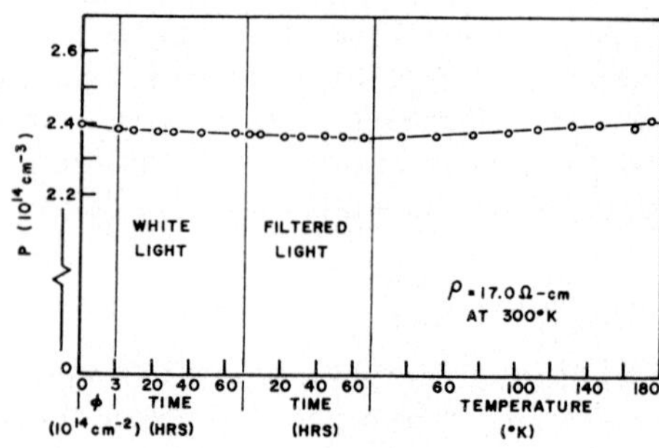

FIG. 2. White light illumination prior to filtered light destroys 98 per cent of defects.

properties. Similar results were observed for the In-doped sample, the two samples having nearly equal ($\sim 2 \times 10^{14}$ cm^{-3}) carrier concentrations at room temperature. Following white light illumination by illumination with filtered light for 65 hours, and then by thermal annealing to 175 °K, still produced no changes. Thus it appears that white light can induce annealing.

This is seen even more graphically when white light is used after filtered light has exposed the defects (Figure 3). Recovery to pre-bombardment

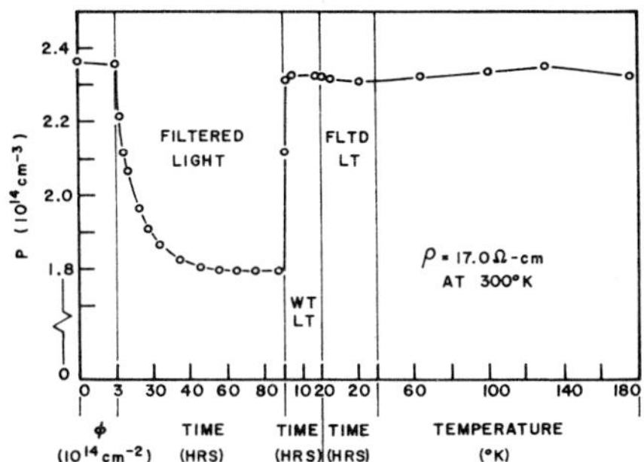

FIG. 3. White light illumination after exposure also destroys defects introduced.

values is rapid and complete. The rate of recovery is exponential with a time constant of about 16 minutes. In this respect there are some differences between the two samples. Subsequent illumination with filtered light followed by a thermal anneal to 175 °K again produces no further changes. The transformation stage above 80 °K[4] has been eliminated by the white light. Absence of this is further evidence of the connection between defects formed at He temperatures and the transformation process above 80 °K.

White light is extremely effective in removing radiation induced defects, at least equivalent to annealing to room temperature. In experiments where irradiation is followed by filtered light, isochronal annealing, and then warming to room temperature, a small number of residual defects remain in the sample. Series of irradiations using the same samples then each start with slightly smaller carrier concentrations. The initial concentration as a function of total irradiation dose received by the sample is plotted in Figure 4A where this decrease is clearly noted. A second series in which irradiations were followed by filtered light illumination, then white light, was made without warming the samples above 33 °K for the entire series. The initial concentration (Figure 4B)

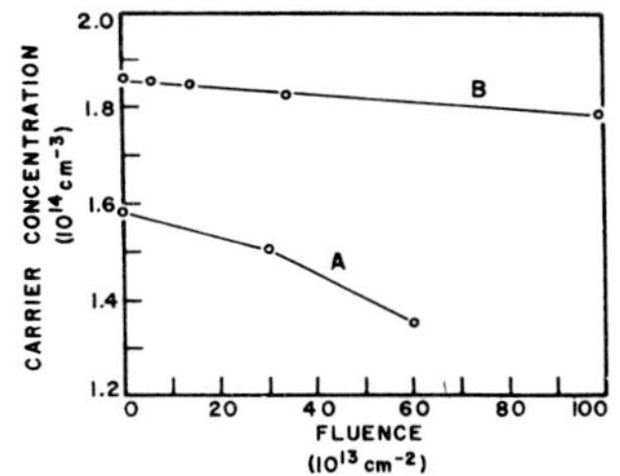

FIG. 4. Effect of residual defects on initial concentration for series of irradiations. (A) Each point represents the initial carrier concentration after sample has received the dose indicated for that point, filtered light, isochronal annealing, and warming to room temperature. (B) Each point represents the initial carrier concentration after sample has received the dose indicated and white light illumination without a temperature increase above 33 °K.

varied similarly in this case to that in which the annealing went to room temperature. The smaller slope to the curve B probably indicates white light is even more effective in removing defects than is the room temperature anneal.

The annealing properties of white light allowed studies to be made of the rate of defect production without raising the temperature above 33 °K. Designating the decrease in carrier concentration by filtered light as Δp after a radiation dose Φ, the effective removal rate $\Delta p/\Phi$, taken as a measure of the rate at which defects are introduced, varies with the dose as shown in Figure 5. The effective removal rate increases from small values for low Φ and reaches a maximum for Φ about 2×10^{14} electrons/cm². Prolonged irradiation tends to

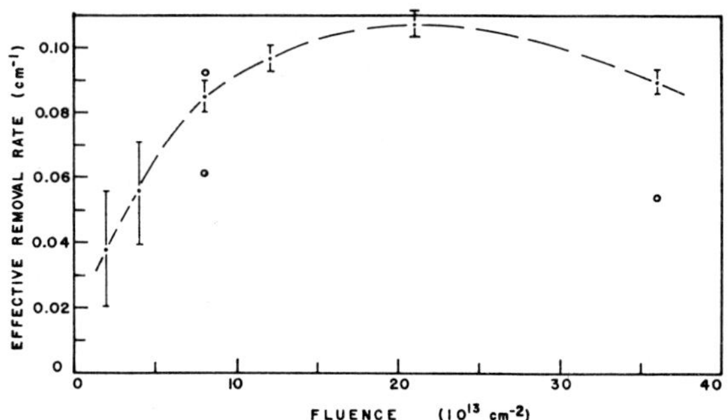

FIG. 5. Effective removal rate variation with dose of irradiation. Open circles are repeated points. For each point, sample was returned to essentially its initial conditions using white light illumination. Prolonged irradiations tend to produce a decrease in the removal rate as noted by the two lower values for the repeated experiments.

cause the removal rate to decrease. Past history, perhaps the accumulated defects, must play a role in determining the net damage rate. Some impurity dependence is noted by the fact that the In- and Ga-doped materials have consistently different effective removal rates although they both qualitatively behave as shown in Figure 5. In any case the removal rate is noted to be a factor of 10 smaller than for n-type material.

6. RADIATION ANNEALING

The 1 MeV electron beam is also capable of removing defects revealed by filtered light, but with somewhat different results from those produced by white light. The rapid recovery induced when 1.8×10^{13} electrons/cm², only 6 per cent of the damage producing dose, are incident on the exposed defects is illustrated in Figure 6. Illumination

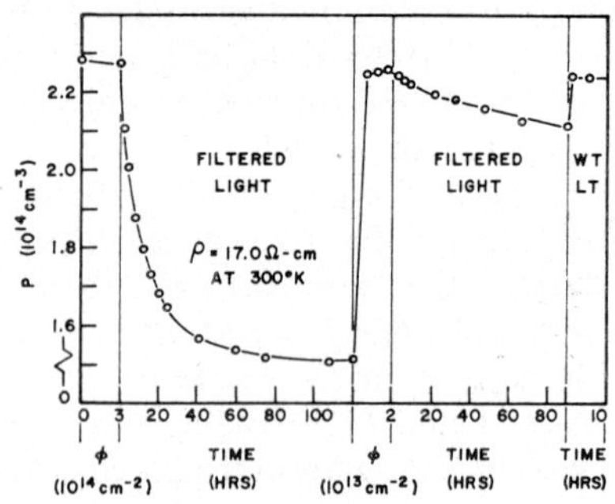

FIG. 6. Radiation annealing of exposed defects removes 80 per cent and converts remainder to neutral state.

again with filtered light indicates the presence of damage centers. Attributing all of the observed defects to the second irradiation would imply a production rate $\Delta p/\Phi$ for the second bombardment which is 4 times that of the first. Such a result suggests that the second irradiation does not actually destroy all of the original defects, but probably reconverts some of the centers back to their neutral state. If it is assumed that the beam creates new defects at the same rate as for the first bombardment, and at the same time either destroys the original centers or converts them to their

neutral state so that they show up under illumination after the second irradiation in the same manner as they originally appeared, it is found that about 80 per cent of the original defects have been destroyed. This compares to ~98 per cent annealing by white light. Thus the process of reconversion by the beam accounts for nearly 10 times as many centers as are converted by white light. Furthermore, the reconversion process enhances the configuration which is responsible for the long time constant component of the filtered light. After radiation-annealing, the long time component represents 90 per cent of the total number of exposed centers compared to 70 per cent for samples not radiation-annealed.

7. SUMMARY AND CONCLUSION

Defects produced by electron irradiation of p-type Ge near liquid He temperature may be observed electrically only after illumination by filtered light. Two time constants of the illumination are closely related to two stages of subsequent thermal recovery and suggest that the defects as formed at the low temperatures consist of two configurations. If the model of the mobile interstitial is accepted together with the low rate of defect production compared to n-type Ge, it is probable that neither of the configurations is represented by the isolated vacancy or interstitial, but rather a more stable arrangement such as a divacancy and an interstitial-impurity or di-interstitial complex. Heat treatment and radiation annealing experiments provide methods for changing the relative concentrations of the two components and in future experiments should help to give more positive identification of the complexes formed.

REFERENCES

1. J. W. MacKay and E. E. Klontz, *J. Appl. Phys.*, **30**, 1269 (1959).
2. J. E. Whitehouse, *Phys. Rev.*, **143**, 520 (1966).
3. Z. G. Werner, J. E. Whitehouse, S. Ishino and E. W. J. Mitchell, *Proceedings of the 9th International Conference on the Physics of Semiconductors*, Moscow, Vol. 2, p. 1070, Nauka (1968), Leningrad.
4. T. M. Flanagan and E. E. Klontz, *Phys. Rev.*, **167**, 789 (1968).

DISCUSSION

Question (MARSH) What is meant by 'filtered light'?

Answer (MATULA) Light from a heated globar element passed through a thick Ge filter. Similarly, by 'white light', we mean light from the heated globar source without going through the Ge filter.

LOW TEMPERATURE ANNEALING OF ELECTRON IRRADIATED GERMANIUM†

W. D. HYATT AND J. S. KOEHLER

Department of Physics and Materials Research Laboratory, University of Illinois, Urbana, Illinois

N-type germanium (4×10^{14} Sb/cm³) has been irradiated with 1.1 MeV electrons at 5 °K. The defects produced have been studied by measuring the voltage-dependent capacitance of a metal-germanium junction at the surface of a germanium sample. These measurements were made at 10 °K and directly gave the fixed charge density near the surface of the sample.

The production and recovery of defects seen near the surface is the same as seen in bulk experiments. A 0.5 MeV electron beam was used to cause radiation annealing of the defects at 5 °K. The fraction recovered during radiation annealing is directly proportional to $\sqrt{t}$. A model based on diffusion-limited recovery theory is used to explain these results. This model is also used to discuss the results of previous experiments. The temperature dependence of the observed recovery at 5 °K gave a defect migration energy of 0.0044 ± 0.0008 eV.

1. INTRODUCTION

Low doped *n*-type germanium ($<10^{15}$ Sb/cm³) that has been lightly irradiated with electrons of less than 1 MeV energy shows almost complete recovery in the 65 °K annealing stage.[1] In an attempt to understand the processes involved in the 65 °K stage Zizine has carried out a careful study of the defect recovery near 65 °K.[2]

It has been observed that these defects can be almost completely annealed at low temperatures (<10 °K) by irradiation with electrons of less than 0.5 MeV energy (radiation annealing),[1,3] or by illumination with broad band light of less than band gap energy.[4] The present experiment is an attempt to gain new information about defect production and annealing in *n*-type germanium by studying the recovery during radiation annealing at low temperatures.

2. MEASUREMNET TECHNIQUE

The differential capacitance technique has been widely used to study impurity atom distributions in semiconductors. This technique requires the use of reverse-biased asymmetrical *p-n* junction, or a similar structure such as a metal-semiconductor junction, located so that a space charge layer is created in the volume of the semiconductor being studied. The conditions under which the charge density of impurity atoms and charged defects in the space-charge layer, or depletion region, can be

obtained from a measurement of the variation of the junction capacitance with applied bias voltage have been discussed by various investigators.[5–8] As the bias voltage (V) is varied, the capacitance (C) of the junction changes allowing a measurement of the fixed charge density (chemical donors and charged defects) $N(x)$ at a distance x from the junction, where

$$N(x) = \frac{2}{\kappa \epsilon_0 q} \left[\frac{dV}{d(1/C^2)} \right]$$

and

$$x = \frac{\kappa \epsilon_0}{C}$$

κ is the dielectric constant and ϵ_0 is the permittivity of free space.

This method has been previously applied to the study of radiation damage in germanium *p-n* junctions by Baruch.[9] In the present experiment silver was evaporated onto a freshly etched germanium surface to form a metal-semiconductor junction. The doping level of the *n*-type germanium used, 4×10^{14} Sb/cc, allowed measurements to be made from 1.5μ to 3.0μ from the surface. This method was used so that changes in the defect distribution due to the possible loss of migrating defects to the surface could be observed.

3. EXPERIMENTAL PROCEDURES AND RESULTS

The defect production irradiations were done at 5 °K with 1.1 MeV electrons with currents of about 2.5×10^{-9} amp/cm². For the radiation annealing,

† This work was supported in part by the U.S. Atomic Energy Commission under Contract AT(11–1)–1198.

the sample was held at a given temperature (5 to 10 °K) and irradiated with a 5×10^{-9} amp/cm² beam of 0.5 MeV electrons for a carefully measured length of time. All measurements were made at 10 °K.

The main experimental results were:

1. Production rates and the amount of thermal annealing were essentially the same as reported for the bulk specimen measurements of Callcott and MacKay.[1] (1.96/cm production rate and 80 per cent recovery for 1.1 MeV irradiation.)

2. The fraction of defect radiation annealed with 0.5 MeV electrons at ~5 °K was directly proportional to $\sqrt{t}$, where t is the 0.5 MeV irradiation time. This proportionality is also a function of temperature and it indicates an apparent migration energy of 0.0044 eV (see Figure 1).

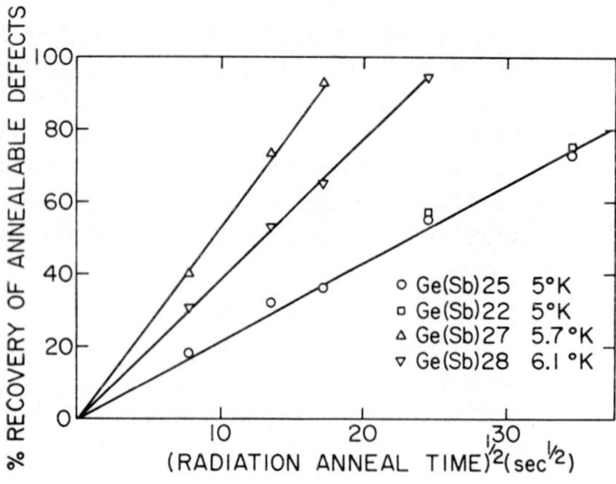

FIG. 1. Recovery as a function of $\sqrt{t}$ for various samples.

4. DISCUSSION

As the recovery near the surface seen here is the same as seen in a bulk experiment, and as a $\sqrt{t}$ dependence is predicted by the diffusion-limited correlated recovery theories of Waite[10] and Simpson and Chaplin,[11] correlated recovery must be considered as a possible explanation for these processes.

Correlated recovery assumes that the interstitial diffuses through the lattice and that interstitial-vacancy annihilation occurs when an interstitial comes within a critical radius r_0 of its own vacancy. Therefore the proposed model is that interstitial-vacancy pairs are formed during irradiation, with the distance between these defects varying due to the distribution in the energy transferred to the

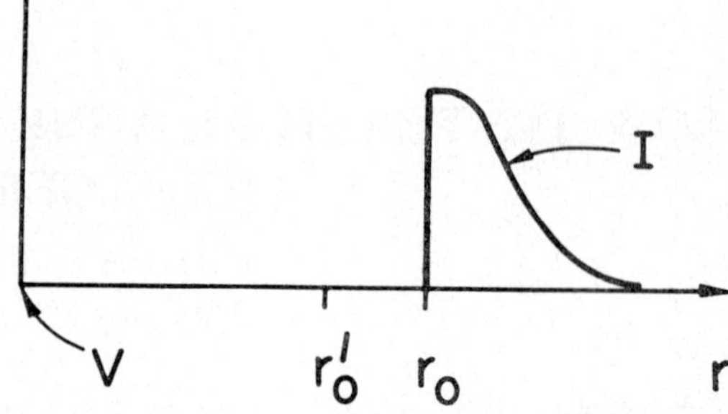

FIG. 2. Proposed model for the distribution of interstitials around their vacancies. No interstitials for $r < r_0$, where r_0 is the critical radius for annihilation for the present experiment, and some distribution of interstitials for $r > r_0$ (r_0' is the critical radius in thermal equilibrium).

interstitials. Interstitials that stop within a distance r_0 from the vacancy will be annihilated during the irradiation. After irradiation the damage will consist of vacancies with their interstitials in a distribution at $r > r_0$, as shown in Figure 2. During the radiation annealing these interstitials are in some non-equilibrium charge state (probably positive).

At 65 °K the interstitials, presumably in their equilibrium charge states (neutral) are able to move with a migration energy of 0.15 eV.[2] The critical radius for annihilation in this case, r_0', is expected to be smaller than in the radiation annealing case, where coulomb attraction of the charged defects is expected to create a larger critical radius of annihilation, r_0, as shown in Figure 2. A calculation for all times can be made on the basis of this

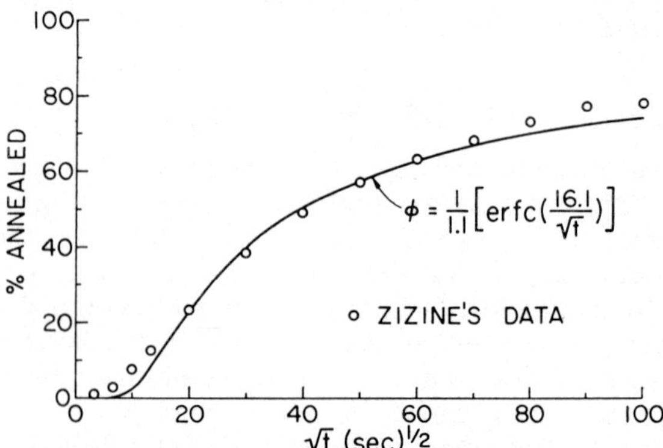

FIG. 3. Zizine's data, compared with calculations that assume a delta function distribution of interstitials at r_0, and a critical radius at r_0', as shown in Figure 2.

assumption, using r_0' as the annihilation radius. This is shown in Figure 3 and compared with Zizine's data.[2] The 'delayed' annealing seen by Zizine at short times is now seen to be characteristic of having a region beyond r_0' that is depleted of interstitials. The long time solution is dependent on $1/\sqrt{t}$ for any distribution.

5. SUMMARY

These studies on n-type germanium $(4 \times 10^{14}$ Sb/cc) show that defects produced by 1.1 MeV electrons at 5 °K are radiation annealed with a 0.5 MeV beam of electrons in such a way that the fraction annealed is proportional to $\sqrt{t}$. This is consistent with a diffusion limited correlated recovery theory and indicates a migration energy of 0.0044 eV. Assumption of different charge states for the interstitial and, therefore, different radii of annihilation for the radiation annealing and thermal annealing cases allows calculation of equilibrium recovery which is consistent with Zizine's work.[2]

REFERENCES

1. T. A. Callott and J. W. MacKay, *Phys. Rev.*, **161,** 698 (1967).
2. J. Zizine, *Radiation Effects in Semiconductors*, Proceedings of the Santa Fe Conference, 1967, Ed. L. Vook (Plenum Press, New York, 1968), p. 186.
3. J. W. MacKay and E. E. Klontz, *J. Appl. Phys.*, **30,** 1269 (1959).
4. I. Arimura and J. W. MacKay, *Radiation Effects in Semiconductors*, Proceedings of the Santa Fe Conference, 1967, Ed. L. Vook (Plenum Press, New York, 1968), p. 204.
5. W. A. Schottky, *Z. Physik*, **118,** 539 (1942).
6. A. M. Goodman, *J. Appl. Phys.*, **34,** 329 (1963).
7. D. P. Kennedy, P. C. Murley and W. Kleinfelder, *IBM J. Res. Develop.*, **12,** 399 (1968).
8. D. P. Kennedy and R. R. O'Brien, *IBM J. Res. Develop.*, **13,** 212 (1969).
9. P. Baruch, *J. Appl. Phys.*, **32,** 653 (1961).
10. T. R. Waite, *Phys. Rev.*, **107,** 471 (1957).
11. H. M. Simpson and R. L. Chaplin, *Phys. Rev.*, **178,** 1166 (1969).

DISCUSSION

Question (MEESE)   Did you see the 35° stage?

Answer (HYATT)   No, we did not see any annealing at 35 °K.

Question (MEESE)   Did you radiation-anneal at lower energies? An ionization model should be energy dependent.

Answer (HYATT)   No, the irradiation annealing was only done at one energy, i.e., at 0.5 MeV.

Question (BOURGOIN)   What do you know about the charge state of the defects in such a depletion layer?

Answer (KOEHLER)   The measurement gives the mobile charge concentration in the conduction band at the measuring temperature and voltage. The specimens are originally n-type and irradiation reduces the mobile carrier concentration. Annealing reverses such changes. Our bias voltage was not large enough to alter the charge state from that usually found in n-type specimens.

Question (BARUCH)   In the differential capacitance measurement, the defect concentration is measured inside the junction depletion region, where the state of charge of the defects can be changed by the shift in quasi-Fermi level. Would this affect your data?

Answer (KOEHLER)   No. In most of the measurements the bias voltage was not sufficiently large to alter the charge state of the defects. In one experiment the bias voltage was made high enough so that levels within about 0.2 eV from the conduction band would have been altered. In this experiment no change in the behavior was seen from that usually found with lower bias voltages.

Question (MACKAY)   Do you observe a change in fixed charge immediately following irradiation at 4.2 °K or do you have to raise the temperature to observe the equilibrium charge state?

Answer (HYATT)   All measurements are made at 10 °K because the bulk resistivity must be kept low in this measurement technique. No change in the fixed charge is seen during the measurement at this temperature.

THE VACANCY-INTERSTITIAL PAIR IN ELECTRON IRRADIATED GERMANIUM

J. BOURGOIN† AND F. MOLLOT

Groupe de Physique des Solides‡ de l'Ecole Normale Superieure, Tour 23, 9 quai Saint-Bernard, Paris 5, France

We have studied the influence of the nature and concentration of donor impurity on the 65 °K and 35 °K stages in electron irradiated *n*-type germanium. Because the nature of the impurity does not influence the 65 °K stage and because this stage is present alone in lightly doped samples, we confirm that it is associated with the annihilation of a vacancy–interstitial pair. Because the 35 °K stage is directly connected to the impurity concentration, and not to the free electron concentration, we conclude that it is associated with the annihilation of a vacancy and inter-stitial–impurity pair. We have shown that the annihilation of the vacancy–interstitial pair occurs through the interstitial mobility at 65 °K, 27 °K and 4.5 °K depending on its charge state, and that the interstitial–impurity mobility occurs at 35 °K.

Our model explains easily the radiation annealing, the behavior of irradiated *p*-type germanium and can be extended to the case of indium antimonide and perhaps of silicon.

1. INTRODUCTION

Depending on doping and irradiation conditions, one stage (at 65 °K) or two stages (at 35 °K and 65 °K) appear in electron irradiated *n*-type germanium.[1,2] Defect creation rate measurements,[3] stored energy measurements[4] and annealing kinetics studies (interpreted as a recombination of correlated pairs through diffusion[5,6]) strongly suggest that the defects associated with the 35 °K and the 65 °K stages (respectively 35 °K and 65 °K defects) are two states of the vacancy-interstitial pair.

In order to identify these two stages and to determine the mobile element in each stage we have studied:

(i) the influence of type and concentration of impurity on the 65 °K stage in lightly doped *n*-type material (impurity concentration lower than 10^{16} cm^{-3});

(ii) the 35 °K stage in highly doped *n*-type material (impurity concentration higher than 10^{16} cm^{-3}).

2. THE 65 °K STAGE IN LIGHTLY DOPED *n*-TYPE GERMANIUM

2.1. *The impurity dependence on the 65 °K stage*

In order to see the influence of the impurity type on the annealing kinetics we have studied[8] two

† Present address: Physics Department, SUNY, Albany, New York.

‡ Laboratoire associe au C.N.R.S.

samples doped respectively with 3.5×10^{14} As cm^{-3} and 3.9×10^{14} Sb cm^{-3} using comparative measure-ments[9] (two samples are placed side by side in the cryostat so that they received the same irradiation dose and the same thermal treatment, and are in the same surrounding radiation field). Isochronal annealing of resistivity for both samples are shown in Figure 1. It can be seen that the annealing temperatures (corresponding to an annealed frac-tion of 0.5) are the same. In addition around this temperature the annealing rates are identical, the difference ΔE between the activation energies being lower than 0.5 meV.

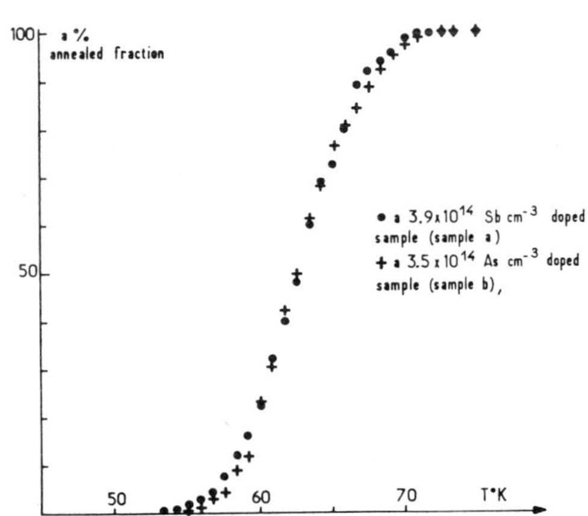

FIG. 1. Annealed fraction of resistivity in an isochronal annealing (4 min °K^{-1}) in the dark of *n*-type germanium electron irradiated at 20 °K with 6.4×10^{14} electrons cm^{-2}.

This difference ΔE, which can be attributed, in a model of diffusion through trapping of interstitials on impurities, to the difference in binding energies of an interstitial on an As or Sb atom, gives a maximum value of $0.7\,kT$ at the annealing temperature for this binding energy E when we estimate a minimum value for $\Delta E/E$: $\Delta E/E \geqslant \Delta r/r = 13$ per cent (Δr is the difference in atomic radius r for an As and Sb atom). Such an interaction is too small to be taken into account at the annealing temperature.

Figure 1 shows that at the beginning of the annealing ($T = 50\,°K$) the annealing rates are somewhat different. This could be explained by the influence of the Fermi level: at this temperature the ionization level of the impurity still plays a role in the Fermi level position. As we shall see in the next paragraph—the ionization energy for As being larger than that for Sb—the As-doped sample must anneal faster than the Sb-doped sample. This is corroborated by the fact that the annealings occurring under illumination are quite identical for both samples (Figure 2).

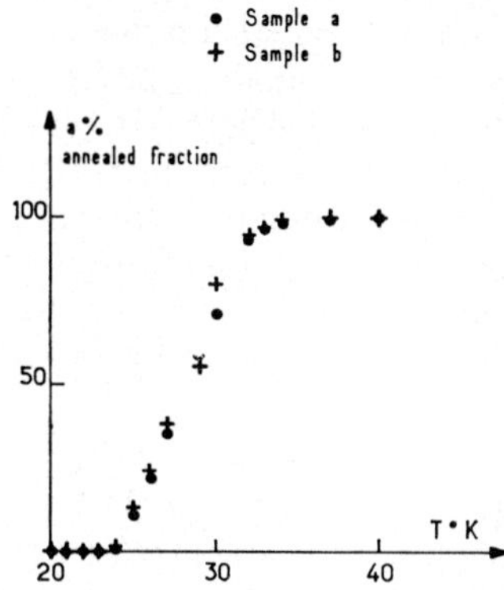

FIG. 2. Annealed fraction of resistivity in an isochronal anneal (4 min °K^{-1}) under illumination (same conditions as for Figure 1).

We believe that there is no interaction between the 65 °K defect and the donor impurities for the following reasons:

(a) As we shall see in the next section, such an interaction gives rise to 35 °K defects.

(b) Since the diffusion coefficient of the interstitial at low temperature is very small, the 65 °K defect creation rate would depend on the average inter-impurity distance, i.e., on the impurity concentration. Instead, it is the same for 10^{14} and 10^{16} cm^{-3} doped samples.

(c) The total recovery at 65 °K would be larger in more highly doped samples. Actually, it is smaller. (In a 10^{13} cm^{-3} doped sample—where the average inter-impurity distance is about 4500 Å— the annealed fraction can be as large as 90 per cent.)

(d) The 65 °K stage appears in a 1.3×10^{14} In cm^{-3} doped sample in which 10^{18} Li cm^{-3} have been diffused (donor concentration before irradiation: 1.5×10^{17} cm^{-3}) and irradiated with 1.9×10^{15} (1.5 MeV) electrons cm^{-2}.

2.2 The electronic excitation mechanism of the 65 °K stage

In lightly doped n-type germanium under illumination, the 65 °K stage is shifted to 27 °K[5] when the defect traps a hole—low injection condition—or even to 4.5 °K when the defect traps two holes—high injection condition (this is the case of a 10^{14} cm^{-3} doped sample below 6 °K: the free electron concentration is very small compared with the electron-hole concentration introduced by illumination[7]).

The activation energy E_2 for the 27 °K reaction depends strongly on light intensity. With an intense light, we found (only at the beginning of the reaction): $E_2 = 70 \pm 15$ meV in a 2.5×10^{14} Sb cm^{-3} doped sample irradiated with 10^{14} electrons cm^{-2}.

The effect of illumination on the 65 °K stage suggests that the 65 °K defect is stabilized by an electronic process. If this defect is stabilized by one or more electrons, the annealing rate, and the annealing temperature, will depend on the proportion of filled defects, hence on the Fermi level.

Therefore, we have studied[8] the influence of the Fermi level E_F on the annealing rate, using comparative measurements of isochronal annealing, with the following samples: 6×10^{16}, 2.2×10^{15}, 4.1×10^{14}, 1.1×10^{13} Sb cm^{-3}. We have observed that the higher the impurity concentration, the higher the annealing temperature T_A. At the annealing temperature, E_F is between 33 and 55 meV below the conduction band, except for the most highly doped sample. This result suggests that the 65 °K defect anneals when E_F goes through a level which we can estimate at $E_1 = E_F + 3\,kT_A$ (between 50 and 70 meV). This has been confirmed (Figure 3) by measurements done on two 5.9×10^{15} As cm^{-3} doped samples, in one of which the carrier concentration (4.3×10^{15} electrons cm^{-3}) having been decreased by copper diffusion.

The dependence of the reaction rate on the free carrier concentration implies that this level is in

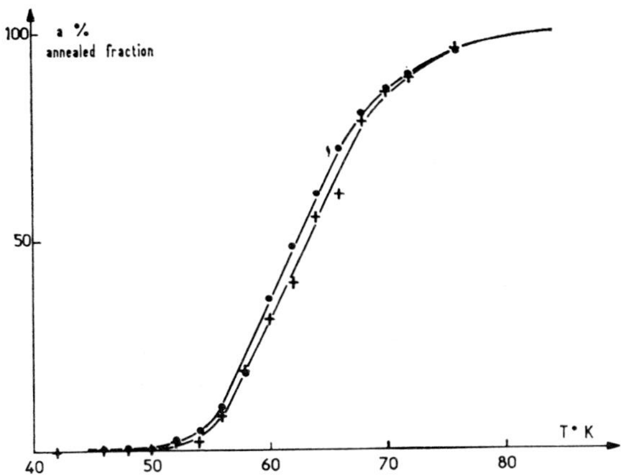

FIG. 3. Annealed fraction of resistivity in an isochronal anneal (1 min 15 s °K⁻¹) of two samples containing 5.9×10^{15} As cm⁻³: ● counterdoped with copper to carrier concentration 4.3×10^{15} cm⁻³, + reference sample with 5.9×10^{15} cm⁻³ carriers.

equilibrium with the conduction band. Since the time for transition between this level and the conduction band is short compared with the time for annealing, the activation energy ($E = 140$ meV) for the process has to be the sum of the electronic excitation energy ($E_1 = 60$ meV) and of the activation energy for the annealing of the ionized defect ($E_2 = 70$ meV), which is approximately the case.

This can be shown considering the equilibrium:

$$N_0 \rightleftarrows N_i + e$$

$$N_i \rightarrow \text{annihilation}$$

between ionized defect N_i and non-ionized defects N_0 for which:

$$\frac{N_0}{N_i} = \frac{n}{N_c} \exp(E_1/kT)$$

Here n is the carrier concentration, and N_c the density of states in the conduction band.

When practically all defects are ionized, the annealing equation (assuming a first order reaction):

$$\frac{dN}{dt} = -KN_i$$

with

$$K = \gamma \exp(-E_2/kT), \quad N = N_0 + N_i$$

gives at the beginning of the reaction an annealed fraction proportional to $\exp[-(E_1 + E_2)/kT]$ when $(n - N)\exp(E_1/kT) \gg N_c$.

Experimentally we found only $E \simeq E_1 + E_2$ since, for an α order reaction and when the proportion β

of ionized defects is not close to 1, the result is $E = aE_1 + bE_2$, a and b depending on α and β.

So the annealing occurs when the Fermi level goes through the level associated with the defect. When the ionization of the defect is induced by illumination—or γ rays—the annealing occurs at 27 °K.

In highly doped n-type germanium the electronic excitation cannot take place. The annealing of the non-ionized defects is thermally induced at 65 °K.

3. THE 35 °K STAGE IN HIGHLY DOPED n-TYPE GERMANIUM

In highly doped n-type germanium, because of donor impurity concentration itself, or of the free electron concentration, or (and) of the irradiation dose (higher in highly doped material than in lightly doped material), a second stage appears at 35 °K. We have shown that the factor influencing the 35 °K defect creation rate is the impurity concentration itself and not the free carrier concentration[6,8]: in a 10^{18} As cm⁻³ doped sample in which the free carrier concentration has been reduced to 1.5×10^{17} cm⁻³ by Ga compensation, the 35 °K defect creation rate is slightly higher than in an 8×10^{17} As cm⁻³ doped sample (Figure 4). This has been corroborated by the study of the influence of the concentration of permanent defects (the permanent defects

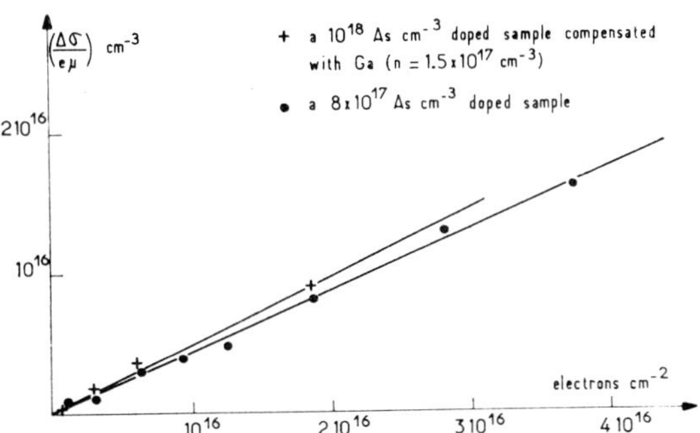

FIG. 4. '35 °K defect' concentration versus irradiation dose in 1.5 MeV electron irradiation at temperature lower than 10 °K of two samples: ● an 8×10^{17} As cm⁻³, + a 10^{18} As cm⁻³ counterdoped with gallium to carrier concentration 1.5×10^{17} cm⁻³.

are the defects which anneal above 80 °K) on the 35 °K defect creation rate: the 35 °K defect creation rate remains constant when the permanent defect concentration increases, that is, when the free electron concentration decreases (Figure 5). We observed that the 35°K defect concentration is linear with the irradiation dose (Figure 6) and that the larger the donor impurity concentration, the larger the 35°K defect creation rate. The 35°K defect creation rate increases with the impurity concentration (Figure 7).

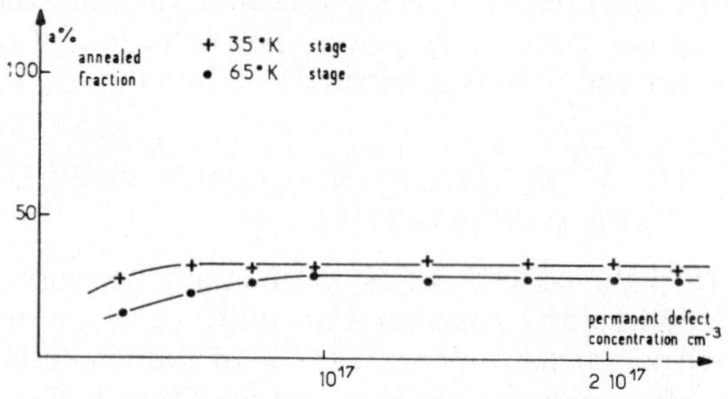

FIG. 5. Annealed fraction of resistivity in 35°K and 65°K stages in a 7×10^{17} As cm^{-3} doped sample irradiated with several doses of 1.5 MeV electrons. After each dose the concentration of permanent defects (annealing above 80°K) increases.

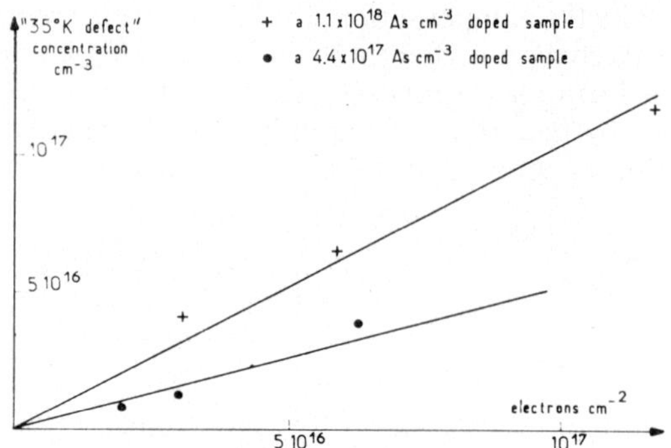

FIG. 6. '35°K defect' concentration versus irradiation dose.

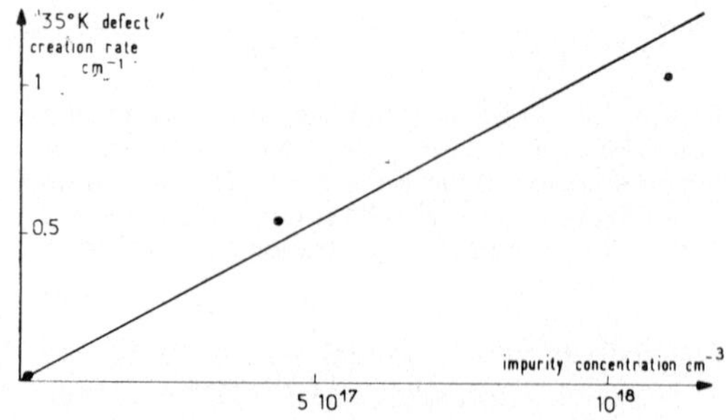

FIG. 7. '35°K defect' creation rate versus As impurity concentration for 1.5 MeV electron irradiation at temperature lower than 10°K.

4. NATURE OF THE 35°K and 65°K STAGES

Because the 65°K stage is present alone in lightly doped *n*-type germanium and because this stage is donor impurity independent, we interpret it as being due to the recombination of vacancy-interstitial pairs.

The 35°K defect is also a vacancy-interstitial pair, an element of which is bound to an impurity atom. This binding cannot be the trapping of an interstitial or a vacancy on an impurity atom (because during annealing such a defect has to go through an intermediate state—the vacancy-interstitial pair—whose elements are stable at least up to 65°K).

The only simple model for the 35°K defect appears to be an exchange between the interstitial germanium atom and a substitutional donor impurity atom (As or Sb) giving rise to an interstitial impurity atom through a mechanism similar to the mechanism proposed by G. D. Watkins[10] to explain the formation of interstitial Al atoms in irradiated Al doped silicon. This defect configuration has been verified: in a lithium doped sample, the 35°K defect creation rate is low compared with the 35°K defect creation rate expected in a sample doped with the same concentration of substitutional impurity atoms. This model is also in agreement with the work of H. M. DeAngelis and R. E. Penczer.[11]

5. THE ANNEALING MECHANISM OF PRIMARY DEFECTS

The 35°K defect identification is interesting, especially because it allows us to determine the mobility of the element involved in every annealing stage. The interstitial mobility under irradiation is necessary for the interpretation of the annealing kinetics of both stages and for the identification of the 35°K defect. After irradiation, at equilibrium, the interstitial is no longer mobile.[8]

The only characteristic of the defect which is different at equilibrium and under irradiation being the charge state of the defect, this implies that the interstitial mobility depends upon its charge state. The interstitial can be already mobile at 4.5°K when it is in a charge state I_0 (high injection condition). When it traps an electron (charge state I_1), its mobility occurs at 27°K (shift of the 65°K stage to 27°K in low injection condition). Then, because of the electronic excitation mechanism, the 65°K stage must be due to the mobility of the interstitial in its equilibrium charge state (I_2). Since the vacancy mobility occurs above 65 °K, the 35 °K stage must be due to the interstitial impurity mobility.

Annealing experiments on irradiated n-type samples doped with copper show that there is a reverse annealing at 90 °K.[12,8] Because this stage is also present in undoped copper n-type samples, it can be attributed to the vacancy mobility (interstitial copper atoms created by irradiation are coming back to substitutional position). This is also in agreement with R. E. Whan's work.[13]

Such a behavior of the vacancy interstitial pair explains easily the radiation annealing (the irradiation induces the interstitial mobility and so the recombination of interstitials with vacancies) and the low defect introduction rate in p-type germanium (the interstitial cannot trap an electron and is always mobile, whatever is the temperature—higher than 4.2 °K—and the pair recombines).

This behavior allows us also to predict the recombination of the pairs in n- and p-type silicon. Contrary to the case of n-type germanium, a low temperature irradiation in n-type silicon usually corresponds to a high injection condition since the free electron concentration is very small and the interstitial, in the I_0 charge state, is mobile. So if, in the temperature range where the interstitial can trap an electron, the interstitial I_1 is already mobile, it is impossible, whatever the irradiation temperature, to observe a stable vacancy-interstitial pair. Then the first stage observed might be due to the vacancy mobility—80 °K, activation energy of 180 meV in n-type silicon (to be compared with 90 °K, 200 meV, in n-type germanium) and 180 °K in p-type silicon (to be compared to the first stage appearing in p-type germanium around 190 °K).

The generality of the behavior of the vacancy-interstitial pair could be tested, studying other semiconductors in which the existence of the pair is observed. In n-type indium-antimonide where stages I and II have been attributed respectively to the annihilation of vacancy-Sb interstitial and vacancy-In interstitial pairs, F. H. Eisen[14] has shown that stage I is shifted towards lower temperature when the free electron concentration decreases in the same way as the 65 °K stage in n-type germanium. On the other hand we observed[15] (Figure 8) that the annealing kinetics of stage I are qualitatively the same as the annealing kinetics of the pairs in n-type germanium. (The order of the reaction varies all along this reaction,

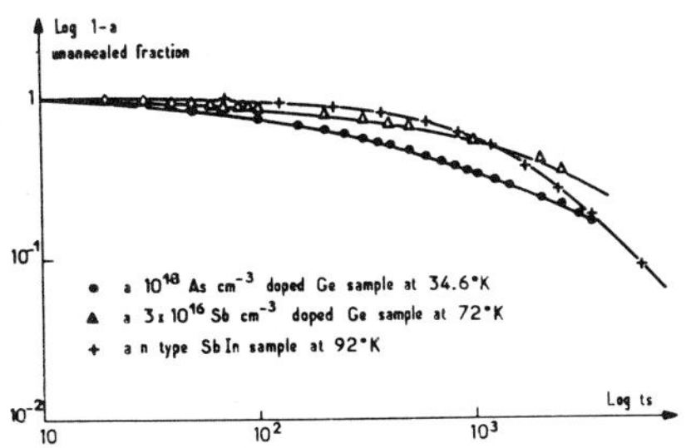

FIG. 8. Comparison of isothermal annealing of stage I in irradiated InSb with low temperature annealing in n-type germanium.

the end of the reaction leading to third order, suggesting a recombination through diffusion, whose apparent behavior is similar to such a reaction.)

6. CONCLUSION

The study of the behavior of vacancy-interstitial pairs is helpful to predict the formation of secondary defects (that is of defects formed by interaction of vacancies and interstitials between them or with different kinds of impurities).

The defects created by the irradiation at a given temperature T are not necessarily the same as those created by irradiation at low temperature and annealing at T since the mobility of the elements of these defects depends on their charge state.

Because we have not to deal with close pairs, a creation rate of secondary defects exponential with T^{-1} (T = irradiation temperature) cannot be explained, as it is in the literature,[16,17] by the difference in barrier heights for the recombination of the pairs stabilized or not by an electron. We think it is rather due to an energy for the formation of a given type of defect: indeed, the 'energetic barrier' is different for different secondary defects (50 meV and 100 meV for the defects associated with the 836 and 956 cm⁻³ bands respectively in electron irradiated n-type silicon[18]). In n-type germanium this can also explain why the recombination rate is larger when the 65 °K stage is shifted to 27 °K than at 65 °K.[5]

To conclude we indicate that the existence of such a barrier for the formation of a given type of defect might depend on the charge state of the elements involved in this defect and so might be different for n- or p-type material. A small difference ΔE_B in barrier height will be the cause of a large difference in the secondary defect creation rates (for instance $\Delta E_B = 20$ meV is sufficient to explain the large ratio $\theta_n/\theta_p = 3.10^{-2}$ of creation rates in n- and p-type silicon at 40 °K).

REFERENCES

1. J. W. MacKay and E. E. Klontz, *J. Appl. Phys.*, **30**, 1269 (1959).
2. J. W. MacKay and E. E. Klontz, *Symposium on Radiation Damage in Solids*, Venice (1962), p. 28.
3. R. A. Calcott and J. W. MacKay, *Phys. Rev.*, **161**, 698 (1967).
4. M. P. Singh and J. W. MacKay, *Phys. Rev.*, **175**, 985 (1968).
5. J. Zizine, *Radiation Effects in Semiconductors*, Ed. F. L. Vook (Plenum Press, New York, 1968), p. 186.
6. J. Bourgoin and F. Mollot, *Phys. Letters*, **30A**, 264 (1969).
7. I. Arimura and J. W. MacKay, *Radiation Effects in Semiconductors*, Ed. F. L. Vook (Plenum Press, New York, 1968), p. 204.
8. J. Bourgoin and F. Mollot, to be published.
9. J. Bourgoin, J. Zizine, S. Squelard and P. Baruch, *Rad. Effects*, **2**, 287 (1970).
10. G. D. Watkins, *Radiation Damage in Semiconductors*, Ed. Dunod, Paris (1964), p. 97.
11. H. M. DeAngelis and R. E. Penczer, *Phys. Rev.*, **39**, 5842 (1968).
12. A. Hiraki, J. W. Cleland and J. H. Crawford, *J. Appl. Phys.*, **38**, 3519 (1967).
13. R. E. Whan, *Phys. Rev.*, **140A**, 690 (1965).
14. F. H. Eisen, *Phys. Rev.*, **123**, 736 (1961).
15. J. Bourgoin, Thesis, Paris University, 1970.
16. F. L. Vook and H. J. Stein, *Radiation Effects in Semiconductors*, Ed. F. L. Vook (Plenum Press, New York, 1968), p. 99.
17. R. E. Whan, *Radiation Effects in Semiconductors*, Ed. F. L. Vook (Plenum Press, New York, 1968), p. 195.
18. R. E. Whan and F. L. Vook, *Phys. Rev.*, **153**, 814 (1967).

DISCUSSION

Question (KOEHLER) Have you measured the frequency factors associated with the 65 °K annealing and the 27 °K annealing?

Answer (BOURGOIN) No, because they depend on the process, which we do not know in detail.

Response (KOEHLER) Hyatt has used $D = D_0 \exp(-E/kT)$ for the 65 °K and the 6 °K annealing. He finds $D_0 = 10^{-15}$ cm²/sec for the 6 °K annealing and $D_0 = 10^{-9}$ cm²/sec for Zizine's 65 °K annealing. The difference is reasonable if in the 6 °K radiation annealing the defect must wait until it is ionized before it can attempt to migrate. Even the value 10^{-9} cm²/sec is very small relative to what one observes in say vacancy diffusion where D_0 is of order 1 cm²/sec.

Reply (BOURGOIN) I think that this is right.

Question (VOOK) You believe, I take it, that any impurity dependence of the annealing of the 65 °K stage depends only on the impurity's effect on the Fermi level. This effect would influence both the measurements of the fraction of defects annealed and the temperature of the beginning of annealing. This annealing would occur for a specific charge state of the primary defects.

Answer (BOURGOIN) We only look at and observe the influence of the Fermi level on the annealing temperature.

Question (MEESE) The measurement techniques are different in our two experiments. D.C. electrical conductivity is subject to impurity scattering effects on mobility and carrier freeze-out, which could partially compensate the impurity dependence. Hot carrier measurements at 100 V/cm are not affected by impurity scattering or freeze-out. We should also consider Group III compensation. In our samples, the Group III concentration was ~1 per cent that of the Group V donor. Do you know the compensation in your samples?

Answer (BOURGOIN) Approximately the same as yours.

THE KINETICS OF THE ANNEALING OF IRRADIATION DAMAGE AT LOW TEMPERATURES IN n-TYPE GERMANIUM

J. E. WHITEHOUSE

J. J. Thomson Physical Laboratory, Whiteknights, Reading, England

The annealing which occurs always near 65K following low-temperature electron-irradiation of n-type germanium is governed by second order (or higher) annealing kinetics. It follows that the experimental activation energy must vary from experiments performed on high resistivity material where the defect concentration is low to those on low resistivity material where the defect concentration is high. A model for the annealing of defects which accounts for this observation and involves the diffusion of ionized interstitials, is suggested.

1. INTRODUCTION

Annealing of irradiation damage in germanium below 78 K was first observed following α-particle irradiation, by Gobell.[1] Subsequent experiments[2-4]; also concerned with the electrical behaviour during isochronal annealing but after electron irradiation, confirmed the presence of a prominent annealing stage in n-type always near 65 K. The feature of the annealing, that it adheres to 65 K for samples containing widely different concentrations of defects might suggest that it is governed by first-order annealing kinetics. The model put forward by MacKay and Klontz,[5] being a development of earlier ideas of Wertheim[6] and themselves,[3] would lead to a first-order process. Experiments employing isothermal[7] and warm-up[8] techniques have now also been used to obtain information regarding the processes occurring near 65 K.

2. ANALYSIS OF EXPERIMENTS

Despite the limited amount of information available in a single isochronal experiment, a more precise test for the order of reaction may be made than that suggested above. Taking first the case of a unimolecular reaction we have

$$-\,\mathrm{d}N/\mathrm{d}t = KN \tag{1}$$

where N is the concentration of defects and K is the rate constant $A \exp -\epsilon/kT$ in which A (the so-called frequency factor) is about 10^{13} sec^{-1} and ϵ is the activation energy. In an isochronal experiment K changes from one measurement to the next. Integration of the above rate equation may therefore only be carried out for a single annealing step with result

$$\ln\,(N_i/N_{i-1}) = -K_i t = -At \exp -\epsilon/kT_i \tag{2}$$

where i refers to any particular annealing step governed by the rate constant $A \exp -\epsilon/kT_i$. N_{i-1} is the number of defects cm^{-3} which remain after the previous annealing step, that is the number cm^{-3} before the ith step. It follows that $\log\log N_{i-1}/N_i$ versus $1/T_i$ will yield a straight line of slope $-\epsilon/2.3\,k$.

For a second order process

$$-\,\mathrm{d}N/\mathrm{d}t = K'N^2 \tag{3}$$

and integration for a particular annealing step yields

$$1/N_i - 1/N_{i-1} = K'_i t = A't \exp -\epsilon/kT_i. \tag{4}$$

A' is the pre-exponential factor appropriate to a second order reaction. Plotting $\log\,(1/N_{i-1} - 1/N_i)$ versus $1/T_i$ gives a straight line again of slope $-\epsilon/2.3\,k$.

Although isochronal annealing does not allow the order to be deduced explicitly, evaluating the data in the two ways suggested above may be helpful as is shown in Figure 1. The particular data used were taken by the author[9] for degenerate antimony doped germanium but identical behaviour is shown in the measurements of Klontz. As will be seen the data is much better explained on the basis of second order kinetics than first order.

Isochronal annealing experiments performed by Callcott[4] in which very small defect concentrations were introduced into lightly doped material do not lend themselves to evaluation in this way because of the small number of data points available. An alternative approach here is to compare the data with the curves to be expected for the annealing of

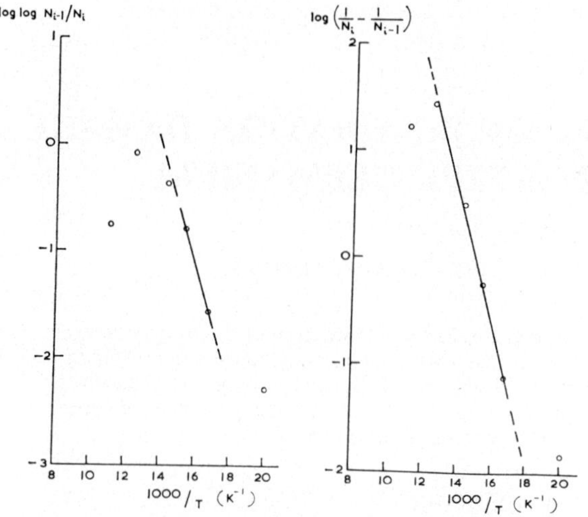

FIG. 1. A comparison of analyses according to first order (left-hand) and second order (right-hand) kinetics of isochronal annealing experiments using degenerate material. (Defect concentration inferred from conductivity data.)

various initial concentrations of defects. Such a family of curves is generated using Eq. 4 in the form

$$N_i = (1/N_{i-1} + A't\exp -\epsilon/2kT_i)^{-1} \qquad (5)$$

and is shown in Figure 2. It is most convenient to take T/ϵ as the variable along the abscissa. The comparison of data to these curves is by no means such a definitive test of the order as that given above but assuming the order one may find the activation energy from the values of $(T/\epsilon)_{1/2}$ and $T_{1/2}$ when the annealing is 50 per cent complete.

Isothermal experiments on the other hand allow

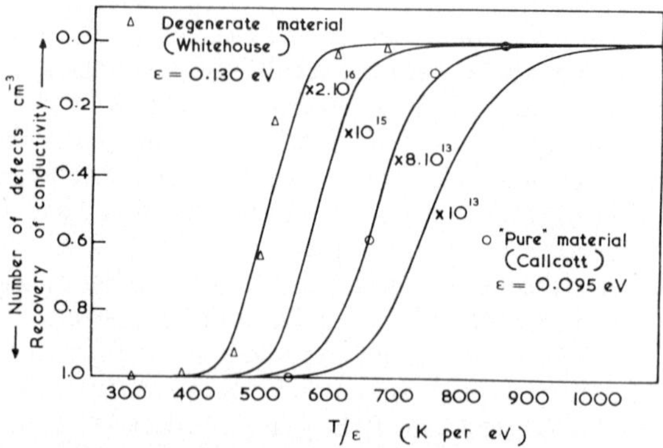

FIG. 2. Annealing curves expected for various initial concentrations of defects and an annealing time of about 500 secs per point. Data points for degenerate and 'pure' material are also shown.

one to determine the order of reaction explicitly and, if a sufficient number of experiments are performed, the activation energy may also be determined. Of course integration of either rate expression to obtain the concentration of defects after any particular time may be performed using the initial conditions ($N = N_0$ at $t = 0$). However, it is convenient at this point to introduce the fraction of defects annealed (a) as a parameter to characterize the annealing. It is a simple matter to show that in terms of this parameter a general rate expression is

$$da/dt = N_0^{(1-n)}K(1-a)^n \qquad (5)$$

where n is the order of reaction and is the slope of the curve $\log da/dt$ versus $\log(1-a)$. Data obtained by Zizine[7] lends itself to interpretation in this way and is shown in Figure 3. It is to be noted that at

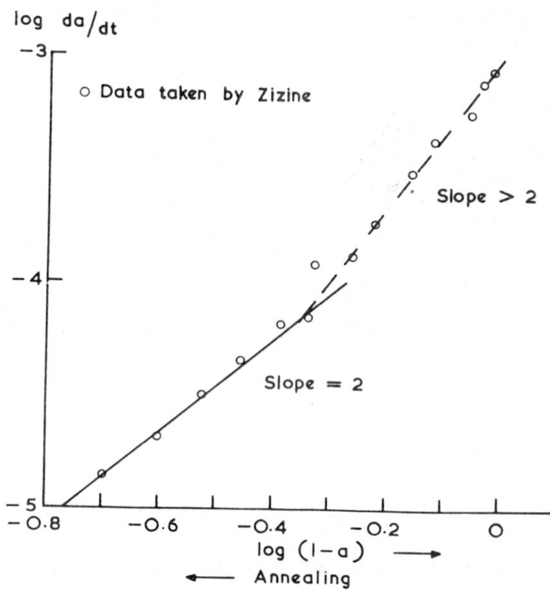

FIG. 3. The analysis of an isothermal annealing experiment performed by Zizine.

long times (about the last 40 per cent of the annealing) the order of reaction is exactly 2. At early times the order is higher than 2 and interpretation along the lines suggested by Waite[10] may be appropriate. Zizine in discussing his own data addressed himself to the behaviour at long times suggesting that this showed the correlated recombination of defects. It would appear that his data at long times may be adequately explained on the basis of second order kinetics.

In independent experiments therefore the order of reaction at 65 K is seen to be 2 or at least much higher than 1. The fact remains that in experiments dealing with widely different defect concentrations the annealing occurs always near 65 K and this is not a feature to be expected for a second order reaction. In some other experiments[8] using very high resistivity material a shift from 65 K has been observed but it does not affect the conclusion based upon the experiments cited.

Phenomenologically one may account for the observations by supposing that the experimental activation energy varies from experiments performed on high resistivity material, where the defect concentration is low,[4] to those performed on low resistivity material, where the defect concentration is high[8] (Figure 2). There is a good basis for expecting that this is indeed the case in that the Fermi energy varies significantly between the materials used and ionization of defects, which will be governed by the Fermi energy, has for sometime been thought to control the annealing.

We shall now show how one may incorporate the Fermi energy into a description of the annealing behaviour.

3. IONIZATION AND ANNEALING OF DEFECTS

The probability of occupancy by an electron of the localized state associated with a radiation-induced defect (which is probably deeper than any state arising from chemical impurity) is the value of the Fermi function at the level. The number occupied per cm³, if there are N cm⁻³ is

$$N^- = N\left[\frac{1}{1 + \frac{1}{2}\exp - (\epsilon_d - \epsilon_F)/kT}\right].$$

ϵ_d is the difference in energy between the localized state and the conduction band edge, ϵ_F is defined in like manner for the Fermi energy, the factor of $\frac{1}{2}$ allows for the two possible spin states of the electron. The number not occupied will be, of course,

$$N^0 = N\left[1 - \frac{1}{1 + \frac{1}{2}\exp - (\epsilon_d - \epsilon_F)/kT}\right].$$

Of more interest when we think of the annealing of defects will be the ratio of the number in one state to the number in the other. From the above this is seen to be the simple expression

$$N^-/N^0 = \frac{1}{2}\exp(\epsilon_d - \epsilon_F)/kT. \tag{6}$$

It may also be regarded as the relative probability of occupancy (the probability that it is filled *rather than empty*). It is clearly in line with the expectation that in say heavily doped *n*-type material where the Fermi energy is high in the gap most defect states will be filled rather than empty whereas in lightly doped material the reverse will be the case.

We now follow similar reasoning to that given by MacKay and Klontz when suggesting their close pair model[5] except that we suppose recombination of interstitials (I) and vacancies (V) to take place by diffusion of interstitials. For simplicity we take the defects to be randomly distributed. This may be the case for defects, initially distributed preferentially, only after some time of annealing. It is suggested that the diffusion proceeds rapidly near 65 K when the interstitial has lost one electron (I⁰). Recombination of vacancies with interstitials which are occupied by electrons (I⁻) is presumed to be very small and is ignored. Symbolically the reactions are

$$\text{I}^- + \text{V} \rightarrow \text{annihilate (negligible)}$$

$$\text{I}^- - e^- \underset{K_2}{\overset{K_1}{\rightleftharpoons}} \text{I}^0 \text{ (quasi-equilibrium)}$$

$$\text{I}^0 + \text{V} \overset{K_3}{\longrightarrow} \text{annihilate}$$

and the important rate expressions are

$$-dN^-/dt = K_1 N^- - K_2 N^0$$

$$-dN^0/dt = K_3 N_V N^0 - K_1 N^- + K_2 N^0$$

where N_V is the concentration of vacancies. Writing $dN^0/dt = 0$ in the expectation that a steady state obtains in the concentration of I⁰ it follows that $-dN^-/dt = K_3 N^0 N_V = (N^0/N^-)(K_3 N_V)N^-$.

The probability per unit time $(K_3 N_V)$ that an interstitial unoccupied by an electron will recombine, is that appropriate to a second order reaction and will include a factor N_V/N_L which allows for the distribution of vacant sites over lattice sites (L). Using Eq. (6) we have

$$-dN^-/dt = N^- \frac{1}{2}\exp - (\epsilon_d - \epsilon_F)/kT.$$

$$S\nu N_V/N_L \exp - \epsilon/kT.$$

S is a configuration number and ν the approximate lattice frequency ($\sim 10^{13}$ sec⁻¹). The activation energy (ϵ) may correspond to the dissociation of ionized interstitials held in complexes, the diffusion of ionized interstitials or the association of interstitials and vacancies, whichever is the rate-limiting step. Taking the number of vacancies to be approximately equal to the number of unionized interstitials

$$-dN^-/dt = A'' \exp - (\epsilon + \epsilon_d - \epsilon_F)/kT(N^-)^2. \quad (7)$$

We see that the experimental activation energy includes both the energy of the localized defect state and the Fermi energy.

4. COMPARISON WITH EXPERIMENTAL RESULTS

The Fermi energy besides being different from one material to another also changes with temperature. However the annealing near 65 K always takes place within a very narrow range of temperatures so that for the moment we take the Fermi energy to remain constant during the annealing. In this way we may make a very simple comparison with experimental results.

From Figure 2 it may readily be shown that annealing would occur at $\sim$65 K for 2.10^{16} defects cm^{-3} if the experimental activation energy were 0.130 eV and for 8.10^{13} defects cm^{-3} if it were 0.095 eV. Such experiments have been performed using material in which the dopant concentration was $\sim 10^{18}$ cm^{-3} (Whitehouse or Klontz) and 4.10^{14} cm^{-3} (Callcott) respectively. The first material is degenerate at very low temperature but the Fermi energy is 0.008 eV below the conduction band edge at 65 K, in the second case ϵ_F is 0.044 eV at 65 K. It follows that in either case $(\epsilon + \epsilon_d)$ is close to 0.14 eV.

The author has shown elsewhere[9] that the temperature dependence of damage production leads to an activation energy of $\sim$0.08 eV if one uses second order kinetics to describe the annealing. It is likely that this corresponds to the annealing of defects maintained in a singly ionized state by the radiation flux, that is corresponds to ϵ in the present discussion. Ishino and Mitchell[8] have suggested that ϵ_d is about 0.06 eV. These values obviously agree very well with the value suggested here for $(\epsilon + \epsilon_d)$.

5. FURTHER SUGGESTIONS AND CONCLUSION

Most germanium used will essentially be in the exhaustion region of its carrier concentration at temperatures as high as 65 K. In such cases one may obtain further insight into the behaviour of

the annealing at 65 K by using the following approximate expression for the Fermi energy

$$\epsilon_F = - kT \ln (N_D/N_C) \quad (8)$$

where

$$N_C = 2(2\pi m^* kT/h^2)^{3/2}$$

arises from the density of states function in the conduction band (the symbols have their usual meaning) and N_D is the number of chemical donors cm^{-3}. Introducing this equation in the form

$$N_D/N_C = \exp - \epsilon_F/kT$$

into Eq. 7 and integrating one obtains the following equation, similar to Eq. 4, for the step by step changes occurring during isochronal annealing

$$1/N_i - 1/N_{i-1} = A't\, N_C/N_D \exp - (\epsilon + \epsilon_d)/kT \quad (9)$$

or

$$(N_i/N_D)^{-1} - (N_{i-1}/N_D)^{-1}$$
$$= A'N_C\, t \exp - (\epsilon + \epsilon_d)/kT. \quad (10)$$

It is a feature of most experiments that N_0/N_D $[i = 0]$ is arranged to be a few percent. That is almost all isochronal experiments have involved studying the removal of about the same fractional change in the original carrier concentration.

The situation simply stated therefore is that as high resistivity material has been used for experiments in which the defect concentration is low and vice versa for high defect concentrations the Fermi energy may operate in such a way to compensate almost exactly for any variation expected in the temperature of annealing due to the different defect concentrations.

Some experiments on very high resistivity material[8] fall outside of this statement. In this case the Fermi energy in the material used was close to the defect localized energy state and the defect concentration was high. The above simple expression for ϵ_F will certainly not suffice. For this reason detailed interpretation of such experiments will be difficult.

REFERENCES

1. G. W. Gobeli, *Phys. Rev.*, **112**, 732 (1958).
2. J. W. MacKay, E. E. Klontz and G. W. Gobeli, *Phys. Rev. Letters*, **2**, 146 (1959).
3. J. W. MacKay and E. E. Klontz, *J. Appl. Phys.*, **30**, 1269 (1959).
4. T. A. Callcott and J. W. MacKay, *Phys. Rev.*, **161**, 698 (1967).
5. J. W. MacKay and E. E. Klontz, *Radiation Damage in Solids* (International Atomic Energy Agency, Vienna, 1963), Vol. III, p. 127.

6. G. K. Wertheim, *J. Appl. Phys.*, **30**, 1166 (1959).
7. J. Zizine, *Radiation Effects in Semiconductors*, Ed. F. L. Vook (Plenum Press, New York, 1968), p. 186.
8. S. Ishino and E. W. J. Mitchell, *Lattice Defects in Semiconductors*, Ed. R. R. Hasiguti (University of Tokyo Press, 1968), p. 185.
9. J. E. Whitehouse, to appear *J. Phys. Chem. Solids*.
10. T. R. Waite, *Phys. Rev.*, **107**, 463 and 471 (1957).

DISCUSSION

Question (BOURGOIN) Your explanation of the influence of the Fermi level on the 65 K annealing stage is similar to mine, but, in your model, the activation energy varies with the free electron concentration. I think it is not right since the activation energy is practically the same in samples doped over the range 10^{14} to 10^{18} cm^{-3}. (Measurements of Zizine between 10^{14} and 10^{16} cm^{-3}, of Bourgoin and Mollot between 10^{17} and 10^{18} cm^{-3}.)

Answer (WHITEHOUSE) For material which is in the exhaustion region of its carrier concentration one may write a simple expression for the Fermi energy. Using this in my annealing equation, the effect of the free carrier density appears in the pre-exponential factor leaving only $(E + E_d)$ in the exponent. It is possible that in your way of determining the activation energy from isothermal experiments you obtain just $(E + E_d)$. I have given a general equation which may also be of some help in interpreting the results for high resistivity material which is not exhausted at 65 K.

Comment (MEESE) We obtain ~ 0.15 eV from isothermal annealing at longer annealing times after the impurity dependence has saturated. For short annealing times we obtain an activation energy of 1/3 this value from isothermal data.

Question (BARUCH) Annealing data by Zizine and more recent data by Bourgoin and Mollot (to be published) follow more closely a 3rd order reaction than a 2nd order, and the 3rd order is, for long times, identical to the $t^{-1/2}$ dependence, characteristic of a diffusion-limited reaction.

Answer (WHITEHOUSE) Those conclusions follow from the time dependence at long times in isothermal experiments. I think it is still something of an open question as to what one may decide from such measurements.

ELECTRON IRRADIATION OF LITHIUM DOPED GERMANIUM AT COLD TEMPERATURES

V. S. VAVILOV

P. N. Lebedev Physics Institute and Moscow State University, Moscow, U.S.S.R.

A. V. SPITSYN

P. N. Lebedev Physics Institute, Moscow, U.S.S.R.

AND

M. V. TCHUKITCHEV

Moscow State University, Moscow, U.S.S.R.

Slices, cut from antimony or arsenic doped n-type germanium ingots, were additionally doped with various lithium impurity concentrations. Samples, suited for Hall effect and minority-carrier diffusion-length measurements, were then irradiated by 1 MeV electrons at 78 or 125 °K and after that in some cases were annealed isochronally. No differences in carrier removal or lifetime damage rates between lithium doped and conventional n-type crystals were found up to 200–230 °K. Hall mobility and lifetime in irradiated lithium doped samples recovered substantially at the annealing temperatures ranges of 240–320 °K and 200–280 °K, respectively, that is at much lower temperatures, than in the crystals without lithium. No $E_c - 0.20$ eV acceptor level radiation defects were observed in the samples with large lithium impurity concentrations.

1. INTRODUCTION

It has been anticipated for a long time, and is now well established in the case of silicon,[1] that certain impurities are able to trap the primary radiation defects and, therefoie, can play an important role in the build-up of secondary defects in semiconductors. Among these impurities the lithium dopant is of particular interest, largely due to its rather peculiar electrical and diffusive behavior in the Ge or Si lattiee. Lithium atoms, diffusing through interstices comparatively fast even at 0 °C and acting at that as shallow level donors, may be easily trapped by various impurities, primarily at charged acceptors and oxygen atoms.[2] With the latter lithium ions give rise to the shallow level donor complexes LiO^+, thus decreasing, as is well known, the A-center introduction rate in silicon.[3] Of course, vacancies are yet other natural sinks for interstitial lithium atoms (and vice versa), so one may expect to find relatively lower concentration of vacancies—both free and bound to impurities—in irradiated lithium doped crystals. In conclusion, it must be mentioned, that by now there is already a wealth of experimental data on irradiations of lithium doped silicon but nothing is known about radiation damage in germanium crystals with the lithium impurity.

2. EXPERIMENTAL PROCEDURE

A number of conventionally doped $(3 \times 10^{14} \mathrm{cm}^{-3} - 3 \times 10^{15}$ cm^{-3} arsenic or antimony)n-type germanium ingots, with dislocations density $\sim 3 \times 10^3$ cm^{-2} and dissolved oxygen content $(3-5) \times 10^{16}$ cm^{-3}, were used. Then into slices, cut from these ingots, various amounts of lithium impurity were diffused from Pb + Li alloy at 350°–400 °C. Samples with lithium (or without it, used for control purposes), suited for either Hall effect or minority carriers diffusion length measurements, were irradiated at 78° or 125 °K by 1 MeV Van-de-Graaf accelerator electrons. Irradiated samples were in some cases annealed isochronally up to 280 or 372 °K, typically for 20 min at each annealing temperature.

3. RESULTS AND DISCUSSION: HALL EFFECT MEASUREMENTS

A more or less standard dc set for Hall effect and conductivity measurements in the temperature range 78–295 °K was employed. Usual bridge-shaped samples were used with current and potential leads soldered to the small area Sn pre-alloyed contacts. The heat treatment of the samples, associated with alloying and soldering, probably induced the lithium precipitation processes (if any) to be nearly completed before irradiation—anyway when referred to the temperature range 280–370 °K —so that they might be expected not to interfere with the radiation damage annealing, studied at the mentioned above temperatures.

The dependence of electron concentration and Hall mobility, both measured at 78 °K, upon the integrated irradiation flux and the annealing temperature, was studied first of all. No noticeable difference both in carrier removal rate and in irradiation-induced mobility degradation behavior between Li doped samples and conventional n-type crystals of the same resistivities, was observed. But concentration, $n/78$ °K, and mobility, $\mu_H/78$ °K, in samples with the lithium dopant responded some-what unusually to the post-irradiation annealing runs. Ordinary, that is, As or Sb doped crystals, irradiated by 1 MeV electrons, are known to exhibit practically no recovery in $n/78$ °K or $\mu_H/78$ °K up to annealing temperature of about 280 °K. On the contrary, we observed a distinct, though not large, reverse recovery of $n/78$ °K (an increase of removed carriers concentration $\Sigma\Delta n/78$ °K) well below room temperature in Ge + As (Sb) + Li samples, irradiated at 78 °K. In fact, when the annealing temperature reached 273 °K, $\Sigma\Delta n/78$ °K was found to exceed by 5–30 per cent the value just after irradiation, $\Sigma\Delta n_{\text{irr}}/78$ °K. At higher annealing temperatures than 280 °K, a usual kind of electron concentration recovery took place, so that by 372 °K the removed carrier concentration, measured at 78 °K, was equal to or only slightly less than the initial value, $\Sigma\Delta n_{\text{irr}}/78$ °K. The strong Hall mobility recovery at temperatures 240°–320 °K is the other annealing peculiarity, characteristic of irradiated Li-doped crystals. The onset of this annealing stage occurs at considerably lower temperatures, than is typical for conventional As- or Sb-doped samples. In Figure 1 is shown how the ratio $(\mu_H/\mu_{H0})/78$ °K varied with the annealing temperature for a sample with 6.5×10^{14} cm^{-3} of donor lithium and irradiated by 3×10^{16} e/cm^2. In some cases, as a rule corres-ponding to the smaller irradiation doses and higher lithium content samples, the Hall mobility, after an annealing run at 270–280 °K, became even greater, than the preirradiation value, μ_{H0}. In Figure 2 are plotted—rather schematically—relative changes in $\mu_H/78$ °K due to the series irradiations at 78 °K and subsequent 30 min anneals at 272 °K (sample with 2.1×10^{15} cm^{-3} of donor lithium). It is worth reminding, that, while the Hall mobility recovers very strongly, the changes in the carriers concentration are quite insignificant. This pheno-

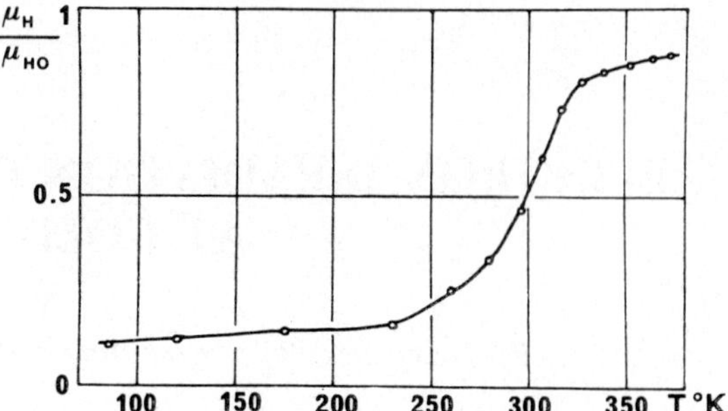

FIG. 1. Dependence of normalized Hall mobility upon the annealing temperature. Ge sample with 6×10^{14} Li/cm^3, 3.5×10^{14} As/cm^3 and irradiated at 78 °K by 3×10^{16} e/cm^2. Measurements at 78 °K.

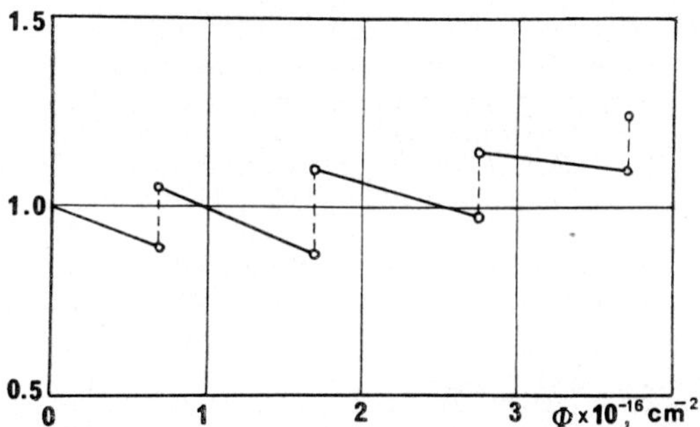

FIG. 2. Behavior of normalized Hall mobility under series of irradiations (solid lines) and subsequent anneals for 30 min at 272 °K (dashed vertical lines). Ge sample with 6×10^{14} As/cm^2 and 2.1×10^{15} Li/cm^3. Measurements at 78 °K.

menon may be explained either by annihilation or by pairing of acceptors (presumably vacancies, free or trapped) and donors (Li interstitial atoms or Li donor complexes). The moving species in both of these suggested processes can possibly be the lithium ions, particularly taking into account the temperatures, at which Hall mobility recovery takes place. In this case, the small reverse annealing of removed carrier concentration may be, probably, due to the Li ions neutralisation at some sinks, associated with radiation defects.

The temperature dependence of electron concen-tration and Hall mobility were also studied in unirradiated and irradiated Li-doped or control samples. As had been expected, no deep energy levels in the upper half of unirradiated Ge (+ Li) crystals forbidden band were observed. Measure-ments were also taken after each of the post-irradiation isochronal anneals; they were extended from 78 °K up to the temperature, some 10° or 15 °K lower, than that of the preceding annealing experiment. The most interesting feature of irradiated Li-doped germanium samples is related to the well known, but still mysterious, $E_c - 0.20$ eV acceptor level radiation defects. They either were not observed at all, which was the case, when the samples with large lithium content had been irradiated, or were found to have smaller concentration, than in the crystals without lithium impurity. For example, concentration of these defects in a sample without lithium impurity and in another one, doped with quite a small amount of donor lithium ($\sim 1 \times 10^{13}$ cm^{-3}), were determined to be 3.9×10^{13} cm^{-3} and 2.2×10^{13} cm^{-3} respectively; crystals came from the same ingot, doped with arsenic impurity (6×10^{14} cm^{-3}), and were irradiated by the same dose of electrons (2×10^{16} e/cm^2). We found no trace of the $E_c - 0.20$ eV level defects in irradiated samples with the lithium donor impurity concentration greater than 3×10^{14} cm^{-3}. Experimental data show, that in rather heavily Li-doped samples the $E_c - 0.20$ eV level radiation defects concentration is negligibly small even at temperatures well below 250 °K, when lithium interstitial atoms (if any) are supposed to become mobile. It looks quite probable, therefore, that Li atoms are liable, prior to irradiation, to pair with some impurity X, thus preventing the radiation-induced primary defects from forming the secondary $E_c - 0.20$ eV level defects with the same impurity atoms. For obvious reasons, the donor As or Sb dopants cannot play the role of the unknown impurity X, in contrast with the recent conclusions by J. W. Cleland and J. H. Crawford.[4] That leaves the following possibilities: (1) X – is a residual acceptor impurity, then the $E_c - 0.20$ eV acceptor level radiation defects are complexes of interstitials with this impurity, which seems somehow not very likely; (2) X is some electrically inactive residual impurity. In this case the $E_c - 0.20$ eV level defects are, most probably, associations of vacancies with this impurity. It is rather tempting to suppose that the impurity in question is dissolved oxygen, LiX being at that the LiO complexes and the $E_c - 0.20$ eV acceptor level radiation defects—A-centers in germanium. But some experimental facts do not seem to fit the latter assumption, or, anyway, are difficult to explain on the basis of it. First, in all the samples used, the lithium donor concentration, including possible LiO^+ complexes, was at least an order of magnitude lower, than the dissolved oxygen content, N_{Ge_2O}. Also, though the $E_c - 0.20$ eV level defect concentration was found to approach, as the integrated flux increased, a certain maximum, $N_{E_c-0.20\,eV}$, characteristic of the given Li-free crystal, and though a correlation was generally noticed between $N_{E_c-0.20\,eV}$ and the value of N_{Ge_2O} in the same sample, but the ratio of these two quantities always remained very small, $(N_{E_c-0.20}/N_{Ge_2O}) = (1-3) \times 10^{-3}$.

Finally, donor defects, which had been supposed to bear some relation to interstitial atoms[5] and had levels $E_c - 0.09$ eV, $E_c - 0.14$ eV and $E_c - 0.24$ eV, were also found in irradiated germanium crystals with lithium. But, contrary to the $E_c - 0.20$ eV level defect behavior, their introduction rate was generally higher in Li-doped samples and these defects appeared to be more stable thermally, than in Li-free crystals. These features may be tentatively attributed to possibly lower concentration of free vacancies in irradiated samples with the lithium impurity, when the electronic-transition-controlled recovery may become vacancies-concentration-limited.

4. RESULTS AND DISCUSSION: MINORITY CARRIERS DIFFUSION LENGTH MEASUREMENTS

Measurements of the 1 MeV-electron-irradiation-induced short circuit current across a surface barrier were employed to obtain data on the minority carrier diffusion length. The method was first described by W. Rosenzweig.[6] Specially fabricated samples were irradiated at 125–130 °K, and then were, as a rule, annealed isochronally.

In Figure 3 the concentration values of irradiation-introduced recombination centers are plotted in arbitrary units against integrated flux for a number of lithium-doped samples. The introduction rate of these centers was found to be practically the same, as was observed in the lithium-free crystals of similar resistivity. Therefore, lifetime damage studies show, that lithium atoms or complexes obviously do not play any important role in the secondary radiation defects build-up at low temperatures. This conclusion is in agreement with the data on carrier removal rate, which were reported in the preceding section.

However, interaction of the lithium impurity with radiation defects becomes evident, as it enhances strongly the lifetime recovery at temperatures higher than 130 °K, but still low enough for no annealing to be observed in ordinary, Li-free,

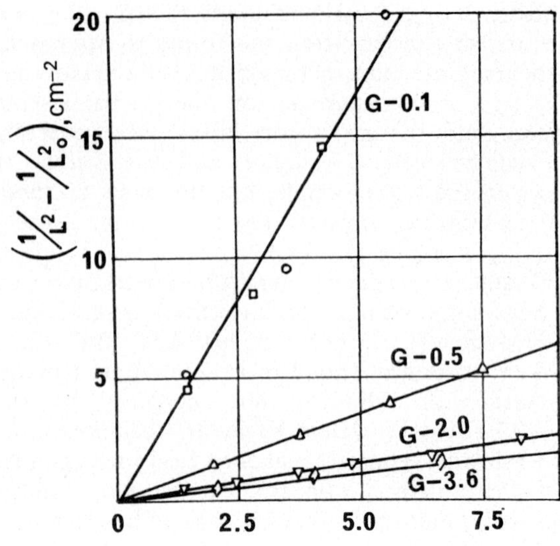

FIG. 3. Introduction of recombination centers by 1 MeV electron irradiation at 130 °K into Ge crystals with various Li concentrations. G–0.1, G–0.5, G–2.0, and G–3.6—samples of resistivities 0.1, 0.5, 2.0 and 3.6 ohm, cm, respectively. Initial ingot with 3×10^{14} As/cm^3 ($\rho \sim 5$ ohm, cm). Concentration of introduced recombination centers (ordinate) in arbitrary units, calculated from the minority carriers diffusion length, L, data.

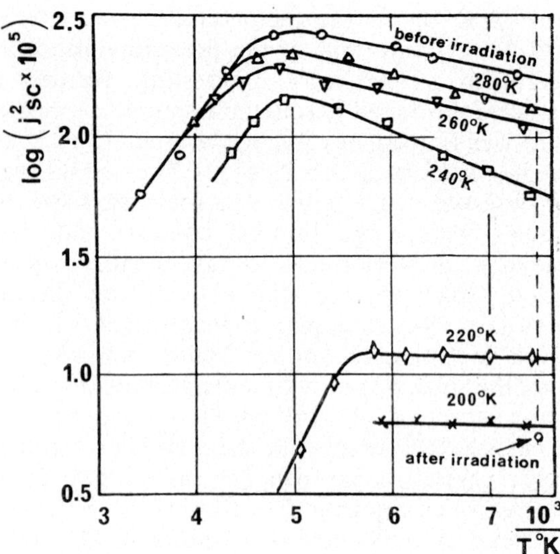

FIG. 4. Reciprocal temperature dependence of minority carriers lifetime (arbitrary units) expressed by the square of short circuit current density, j_{sc}, (ordinate). Ge sample G–0, 1 (see Figure 3), irradiated at 130 °K and then annealed for 20 min at various indicated temperatures.

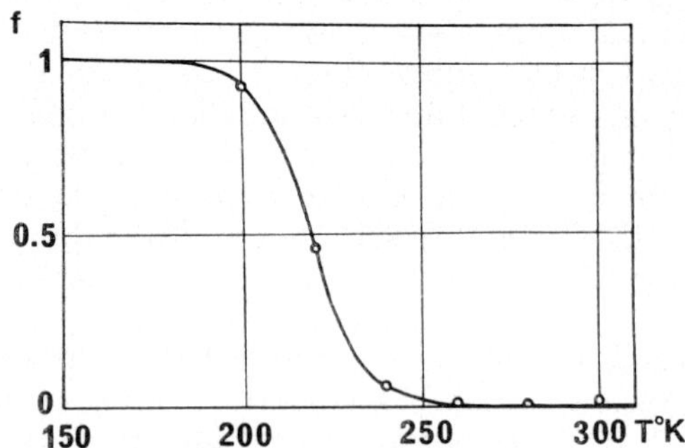

FIG. 5. Annealing temperature dependence for fraction of radiation defects not annealed, f. Data of Figure 4 were used.

crystals. In Figure 4 the values of minority carrier lifetime, in arbitrary units, are plotted as a function of reciprocal absolute temperature for the Li-doped sample G—0, 1 (see Figure 3) before irradiation and after a number of postirradiation isochronal annealing events. It can be seen, that the main recovery occurs within the range 200°–280 °K and by 280 °K the recovery is nearly complete. The data of Figure 4 were used to calculate the unannealed fraction, f, of radiation defects concentration, and the dependence of f on the annealing temperature (Figure 5) was analysed. The activation energy value was obtained to be (0.5 ± 0.1) eV, which is quite close, though maybe just coincidentally, to that of Li diffusion. Characteristic frequency factor was also estimated and found to be $\sim 1 \times 10^8$ sec^{-1}. Finally, it must be mentioned, that the annealing process seems to follow second-order kinetics. If this is the case and if the annealing rate is truly lithium-diffusion-limited, then the mobile lithium concentration must be approximately equal to the value of radiation defects concentration, which is quite small, when lifetime damage is studied. But, at the same time, the Hall mobility recovery, reported earlier in this paper, requires obviously a large amount of diffusing lithium. We failed to explain this disagreement.

5. SUMMARY AND CONCLUSIONS

Experimental data, reported in this paper, show that there are some features of radiation defect introduction and annealing processes, which are characteristic of Li-doped germanium crystals. For example, both Hall mobility and lifetime recovery occur in irradiated germanium with lithium at considerably lower, than usual, temperatures. Another interesting result is, that well known $E_c - 0.20$ eV acceptor level defects are somehow suppressed by the lithium impurity.

Most of the results were tentatively accounted for by lithium diffusion either prior to irradiation or during annealing. Still, additional experimental data are needed to clarify some important points.

ACKNOWLEDGEMENT

The authors would like to thank Mr. S. I. Vintovkyn for his help in carrying out the irradiation experiments.

REFERENCES

1. G. D. Watkins, Paper given at the International Symposium on Radiation Effects in Semiconductor Components, Toulouse, 1967.
2. H. Reiss, C. Fuller and F. Morin, *Bell Syst. Tech. Journ.*, **35**, 535 (1956); R. Weltzin and R. Swalin, *Journ. Phys. Soc. Japan*, **18**, Suppl. III, 136 (1962).
3. I. V. Krjukova and V. S. Vavilov, *Soviet Physics-Semiconductors*, **2**, N9 (1969).
4. J. W. Cleland and J. H. Crawford, Jr., *Phys. Rev. B*, **1**, 713 (1970).
5. A. V. Spitsyn, Paper given at the International Symposium on Radiation Effects in Semiconductor Components, Toulouse, 1967.
6. W. Rosenzweig, *Bell Syst. Tech. Journ.*, **41**, 1575 (1962).

ANNEALING OF POINT DEFECTS IN p-TYPE GERMANIUM AFTER ELECTRON IRRADIATION AT LIQUID NITROGEN TEMPERATURE

H. SAITO, N. FUKUOKA AND Y. TATSUMI

Department of Physics, College of General Education, Osaka University, Toyonaka, Osaka, Japan

Annealing of radiation induced defects in p-type germanium was studied by measuring Hall coefficient and conductivity. The dopant was gallium or indium. It was concluded that the annealing stage between 80° and 140 °K is caused by migration of the vacancy to the sink of an impurity atom. In this stage the vacancy migrates to a substitutional impurity atom and makes an association. The activation energy of the stage was found to be 0.1 eV and it is regarded to be that of the vacancy migration. The model for the annealing stage which occurs in the range 220 to 270 °K is proposed as follows: An interstitial impurity atom migrates to a substitutional impurity atom and makes an association. From the activation energy of the stage, the migration energy of the interstitial impurity atom was concluded to be about 0.4 eV for gallium and 0.7 eV for indium atoms.

1. INTRODUCTION

It has been reported by several authors that irradiated p-type germanium shows four or five annealing stages in the temperature range 80–400 °K according to their dopant impurities. Analyses of some stages have been published.[1-7] However, understanding of the microscopic nature of radiation effects in p-type germanium is much less satisfactory than that in silicon mainly caused by the lack of ESR information for germanium. A more extensive body of experimental work would be necessary to get a better understanding of this material.

The five annealing stages are tentatively nomenclatured as stages I, II, III, IV and V for convenience sake.[4,5] The temperature range for each stage is as follows; stage I (80–140 °K), stage II (150–170 °K), stage III (220–270 °K), stage IV (330–370 °K), stage V (above 380 °K). In this paper the stage I and III are discussed, and annealing models are proposed. The migration energy of the single vacancy in p-type germanium is concluded to be 0.1 eV and migration energies of the interstitial impurity atoms were obtained.

2. EXPERIMENT AND DISCUSSION

Stage I is the reverse annealing stage in the temperature range 80–140 °K after the sample has been irradiated with 1.5 MeV electrons at liquid nitrogen temperature. The activation energy of the annealing has been found to be 0.1 eV regardless of the impurity and concentration (Figure 1).

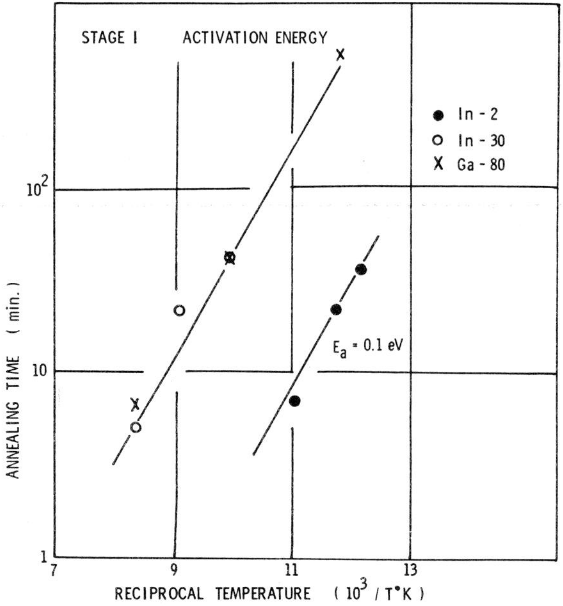

FIG. 1. Time for 50 per cent anneal vs. reciprocal temperature, stage I.

Before discussing the annealing model it would be helpful to discuss the characteristics of the stage.

The change in carrier concentration in this stage can be attributed to two causes. One is the thermal release of electrons from the trap and the other is an annealing of some intrinsic defect. Thermal release of electrons from the trap can be refilled by illumination with visible light and the residual change in carrier concentration is attributed to the annealing of the defect. In this section the intrinsic annealing is discussed.

As discussed in the previous reports, the carrier concentration, conductivity and mobility decreases in the stage.[4,7]

An indium-doped sample was isochronally annealed alternatively in the dark and in light to investigate the effect of charge state on annealing. The result is shown in Figure 2. The plot of

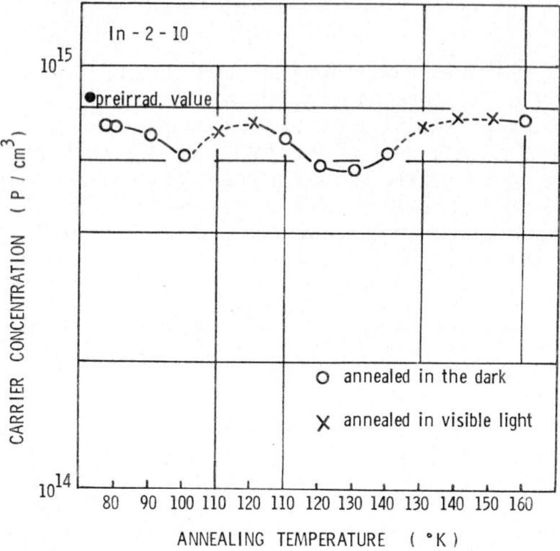

FIG. 2. 20 min isochronal annealing curve of an indium-doped crystal. The circles show the results for annealing in the dark. The crosses show the results for annealing in visible light.

annealing in the dark from 80 to 100°K indicates that the reverse annealing is occurring regularly. When the sample was annealed at 110° and 120°K in light, the carrier concentration increased toward its pre-irradiation value. This observation suggests that stage II is occurring at this temperature. Here it should be remembered that when annealing in the dark, stage II begins at about 150°K. Next, the sample was annealed again in the dark beginning from 110°K, which is lower than the preceding annealing temperature (120° in light). The result is reverse annealing. This means that the defect which anneals in the stage I becomes more stable in light and survived during the annealing at 120°K. An isochronal annealing experiment in light was compared with that in the dark and the result was consistent with that described above. A similar result was obtained for gallium-doped sample. Now it can be concluded that the defect which anneals in stage I becomes more stable when it captures an electron while the defect which anneals in stage II becomes more unstable in light.

To study the effect of impurities on annealing, isochronal annealing studies of a gallium-doped sample and an indium-doped one were made after they were exposed to exactly the same dose. Impurity concentration of the gallium-doped sample was 1.7×10^{14} atoms/cm^3 and that of the indium-doped one was 6.0×10^{14} atoms/cm^3. The gallium-doped sample showed reverse annealing to 140 per cent as measured by the fraction unannealed, while the indium-doped one went up to 220 per cent as shown in Figure 3. Now the question arises as to which is responsible for the difference, the difference in the dopant impurity or the difference in impurity concentration. The amount of reverse annealing for the gallium-doped sample was found to be independent of impurity concentration over a wide range of impurity concentration (5×10^{13}– 5×10^{15} atoms/cm^3). Accordingly the difference should be attributed to some difference in the character of gallium and indium atoms. Hirata *et al.* studied the annealing of electron irradiated silicon and the difference in annealing behaviour was analyzed in terms of the impurity atom size.[8] The covalent radius of the gallium atom is 1.26 Å and that of the indium atom is 1.44 Å, whereas that of the germanium atom is 1.22 Å. Here, it is interesting to refer to the work of Whan.[9] According to her findings, the temperature range where the vacancy begins to move is close to the temperature of stage I. It is reasonable to take vacancy migration into consideration on discussing annealing models. It should be noted here that the indium-doped sample which has larger impurity atom size

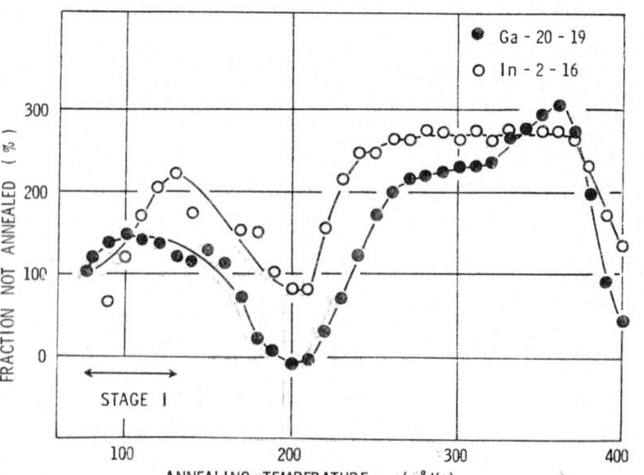

FIG. 3. 20 min isochronal annealing curves of gallium- and indium-doped samples. The unannealed fraction is plotted against temperature.

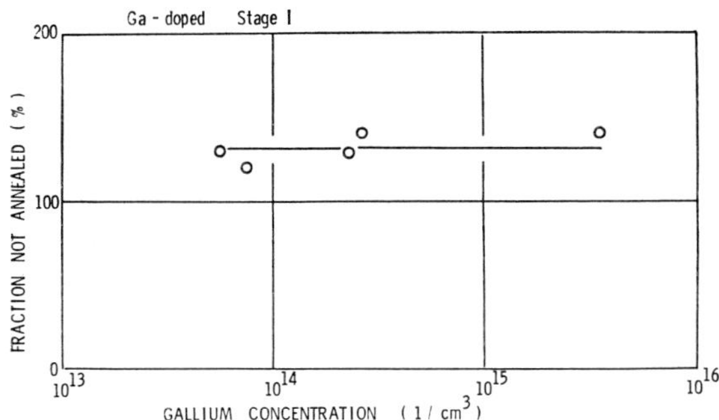

FIG. 4. Amount of anneal in stage I vs. gallium concentration.

than gallium shows larger fractional anneal. It is understandable that if formation of *E*-center-like association occurs in the stage the process is favoured for larger sized impurity from the standpoint of release of lattice strain.

Now, a tentative model for the stage I can be proposed. The defect which migrates in the stage I can be assumed to be the vacancy with a single plus charge. The vacancy migrates in the stage and makes an association with a substitutional impurity atom resulting in an *E*-center-like association. The process is expressed in the following formula:

$$\text{Vacancy}^+ + \text{Impurity (Substitutional)}^-$$

$$\rightarrow \text{Vacancy} \cdot \text{Impurity (Substitutional)}^+ + e^-$$

If this model is adopted, the acceptor loses its ability and this causes intrinsic reverse annealing. The effect of charge state on annealing can be understood by considering Coulomb interaction. When the specimen is irradiated by light, the vacancy with a single plus charge will lose its charge and the Coulomb interaction between the vacancy and ionized acceptor no more exists. This explains that the defect is more stable in light.

The activation energy of the annealing was found to be 0.1 eV from a series of isothermal annealings. This is regarded to be the migration energy of vacancies. The fact that two kinds of indium-doped samples which differ from each other about an order of magnitude in impurity concentration give frequency factors which also differ by about the same order suggests that the impurity atom is the sink (Figure 1, Tables I and II).

Stage III is the annealing stage from 220 to

TABLE I

Activation energies, kinetics and frequency factors for gallium-doped samples.

	Ga-doped Germanium									
Stage	I (80–140 °K)	II (150–170 °K)	III (220–270 °K)			IV (330–370 °K)		V (380°K–)		
Imp. Conc. (atoms/cm³)	8×10^{13} (Ga-80)		2×10^{14} (Ga-20)		2×10^{15} (Ga-1)	8×10^{13} (Ga-80)	2×10^{14} (Ga-20)	8×10^{13} (Ga-80)	2×10^{14} (Ga-20)	2×10^{15} (Ga-1)
Anneal	in dark		in dark	in light	in dark	in dark	in dark	in dark	in dark	in dark
Act. Energy (eV)	0.1		0.38	0.23	0.42	0.85	0.85	1.2	1.2	1.2
Kinetics	1st		1st	1〈〈2	1st	1st	1st	2nd	2nd	2nd
Freq. Factor (1/sec)	2×10^2		5×10^4	4×10	8×10^5	1×10^9	5×10^9	5×10^{12}	2×10^{12}	1×10^{12}

TABLE II

Activation energies, kinetics and frequency factors for indium-doped samples.

	In-doped Germanium						
Stage	I (80–140 °K)		II (150–170 °K)	III (220–270 °K)		IV (330–360 °K)	V (370 °K–)
Imp. Conc. (atoms/cm³)	1×10^{14} (In-30)	8×10^{14} (In-2)	8×10^{14} (In-2)	8×10^{14} (In-2)		—	8×10^{14} (In-2)
Anneal	in dark	in dark	in dark	in dark	in light	—	in dark
Act. Energy (eV)	0.1	0.1	0.35	0.69	0.15	—	1.6
Kinetics	1st	1st	1st	1st	1〈〈2	—	2nd
Freq. Factor (1/sec)	1×10^2	1×10^3	1×10^7	1×10^{12}	3	—	1×10^{18}

270 °K where the traps disappear and a donor level is formed.[4,5] Other characteristics of the annealing stage are as follows: The defect which anneals in the stage is more unstable when it captures an electron.[1,4] The activation energy for stage III was found to be 0.38 eV with traps empty and 0.23 eV with traps full in a gallium-doped sample and 0.69 eV with traps empty and 0.15 eV with traps full in an indium-doped sample. These results are shown in Figure 5 and Tables I and II.

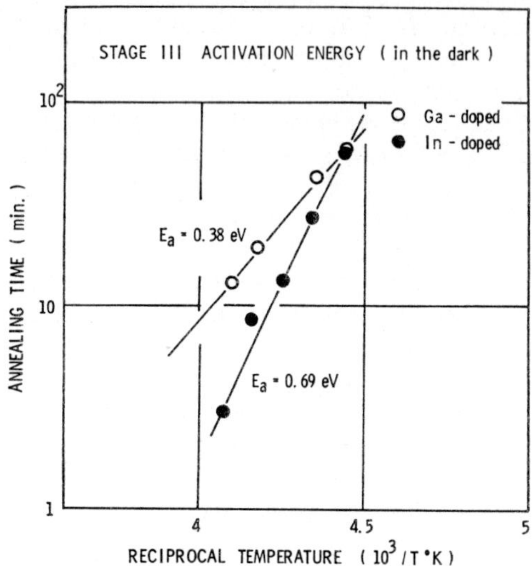

FIG. 5. Time for 50 per cent anneal vs. reciprocal temperature, stage III, in the dark.

The frequency factors were calculated to be 5×10^4/sec for traps empty in the gallium-doped sample. For the indium-doped sample the value is 1×10^{12}/sec. The annealing kinetics was found to be first order for both impurities. The stage is not observed for relatively pure samples.[4]

Now a tentative model for the defect which is annealed in this stage can be considered. The requirements for the model are as follows: (a) The defect which is annealed in the stage must be associated with the impurity atom. (b) It is transformed into a donor which does not exhibit trapping action. (c) It becomes unstable when it captures an electron. (d) Annealing kinetics are first order. (e) The activation energy of the annealing is not dependent upon impurity concentration. The most likely defect which accounts for the trapping property is either vacancy++ or impurity (interstitial)++. If the trap is due to the vacancy, the position of the trapping level and the activation energy of the annealing should be nearly the same, but the observation indicates these values are dependent upon impurity. The existence of the impurity (interstitial) is expected by analogy with defect behavior observed in aluminum-doped silicon and copper-doped germanium.[10,11]

Judging from these considerations, it is deduced that the most probable model for the trapping center is an interstitial impurity atom with double plus charge. If this is the case, what is the sink of the annealing? The number of sinks can be estimated by analysis of the frequency factors, and the results give numbers almost equal to the impurity concentration for two gallium-doped samples which contain 2×10^{14} and 2×10^{15} gallium atoms/cm³ respectively. This suggests that the sink is the impurity atom.

Now, a tentative model for stage III can be proposed. The defect which migrates in the stage is an interstitial impurity atom with a double plus charge. The interstitial gallium atom makes an association with substitutional gallium atom resulting in an association gallium (interstitial) · gallium (substitutional). If this model is adopted, the fact that the stage is strongly dependent on impurity concentration can be understood. The same model is proposed for the indium-doped sample. The results show that the migration energy of the interstitial indium atom is larger than that of the gallium atom. The fact that stage III has first order kinetics is understandable because the concentration of sinks is much greater than that of migrating interstitial impurity atoms. The number of sinks calculated for the indium-doped sample by this simple method does not coincide with the impurity concentration. This may be caused by the fact that the indium atom has a larger atom size than germanium and the entropy factor cannot be neglected, whereas for the case of gallium, its atom size is almost equal to that of germanium and the effect of the entropy factor is small. Experimentally obtained results for the other stages are summarized in Tables I and II.

3. CONCLUSION

Stage I which appears in the temperature range 80–140 °K was interpreted by using the model that the defect which migrates in the stage is the vacancy and that it makes an association with a substitutional impurity atom. The activation energy for the process was found to be 0.1 eV. This activation energy is concluded to be that of the vacancy migration.

The annealing of stage III in the temperature range 220–270 °K was concluded to be caused by the migration of interstitial impurity atom to the sinks of substitutional impurity atoms. If this model is adopted the activation energy for the migration of interstitial impurities is concluded to be about 0.4 eV for gallium atoms and 0.7 eV for indium atoms.

ACKNOWLEDGMENTS

The authors are greatly indebted to Dr. Y. Nakai of the Osaka Laboratory, Japan Atomic Energy Research Institute for his kind help in making irradiation experiments.

REFERENCES

1. W. L. Brown, W. M. Augustyniak and T. R. Waite, *J. Appl. Phys.*, **30**, 1258 (1959).
2. J. E. Whitehouse, *Phys. Rev.*, **143**, 520 (1966).
3. N. Fukuoka, H. Saito, H. Hattori and J. H. Crawford, Jr., *J. Appl. Phys.*, **38**, 4098 (1967).
4. H. Saito, N. Fukuoka, H. Hattori and J. H. Crawford, Jr., *Lattice Defects in Semiconductors*, Ed. R. R. Hasiguti (Univ. of Tokyo Press, Tokyo, 1968), pp. 199–210.
5. H. Saito, N. Fukuoka, H. Hattori and J. H. Crawford, Jr., *Radiation Effects in Semiconductors*, Ed. F. L. Vook (Plenum Press, New York, 1968), pp. 232–237.
6. T. M. Flanagan and E. E. Klontz, *Phys. Rev.*, **167**, 789 (1968).
7. N. Fukuoka, H. Saito and Y. Tatsumi, Japan, *J. Appl. Phys.*, **8**, 1573 (1969).
8. M. Hirata, M. Hirata, H. Saito and J. H. Crawford, Jr., *J. Appl. Phys.*, **38**, 2433 (1967).
9. R. E. Whan, *Phys. Rev.*, **140**, A690 (1965).
10. G. D. Watkins, *Radiation Damage in Semiconductors* (Dunod, Paris, 1965), pp. 97–113.
11. A. Hiraki, J. W. Cleland and J. H. Crawford, Jr., *J. Appl. Phys.*, **38**, 3519 (1967).

DISCUSSION

Question (WOLF) What was the electron fluence?

Answer (SAITO) We used 1.5 MeV electrons from a Van de Graaff and the sample in Figure 3 received 2.5×10^{15} electrons/cm². Specimens of larger impurity concentration received proportionally greater fluence.

Question (WOLF) If the first annealing stage is due to the migration of vacancies to impurity atoms, can it be described by first order of kinetics?

Answer (SAITO) The number of defects which anneal in the stage is much smaller than that of impurity atoms and first order is reasonable.

THE 220 °K DEFECT IN ELECTRON IRRADIATED p-TYPE GERMANIUM

M. SHIMOTOMAI AND R. R. HASIGUTI

Department of Metallurgy and Materials Science, Faculty of Engineering, University of Tokyo, Bunkyo-ku, Tokyo

Lattice defects introduced in p-type nondegenerate germanium by 1.5 MeV electron irradiation at liquid nitrogen temperature was investigated by means of electrical resistivity and Hall coefficient measurements. The annealing behavior of two kinds of defects, which anneal at about 220 °K, was investigated in detail. Each of them has an electron trap. Making use of trap-filling and emptying processes, the 220 °K defects are separated from other defects.

1. INTRODUCTION

The behavior of lattice defects in p-type germanium is not as well investigated as that of n-type germanium. The aim of this paper is to clarify the behavior of lattice defects in electron-irradiated p-type germanium between liquid nitrogen temperature and room temperature.

2. EXPERIMENTAL PROCEDURES

The specimens used are Ga-doped and In-doped non-degenerate p-type germanium. The concentrations of dopants are 7×10^{14} Ga/cc (called 14 Ga specimen), 6×10^{15} Ga/cc (15 Ga), 5×10^{13} In/cc (13 In) and 1.5×10^{15} In/cc (15 In). The 1.5 MeV electrons were used for irradiation at the liquid nitrogen temperature. The resistivity and the Hall coefficient of the specimens were measured in dark as well as after white light illumination.

3. EXPERIMENTAL RESULTS AND DISCUSSION

3.1. *Electron trap*

The illumination of the specimen at 80 °K by germanium-filtered infra-red light after electron irradiation makes the hole concentration and the electrical conductivity decrease. These properties resume their values of the as-irradiated state upon tungsten white light illumination. These processes can be repeated completely reversibly. This shows that irradiation induced defects are in a latent state just after irradiation in that the carrier removal is not apparent because electrons excited from valence band are trapped at the electron trap level of irradiation induced defects, and that these latent defects become apparent by emptying traps by infra-red light illumination.

3.2. *Annealing behavior*

Figure 1 shows a typical result of the 15 minute isochronal annealing behavior of 14 Ga specimen irradiated to the dose of 2×10^{15} electrons/cm². After each isochronal annealing in dark the specimen was cooled to 80 °K, at which temperature the measurements were performed. Then the specimen was illuminated by the white light and the electrical measurements were performed again. The solid circles in Figure 1 show the results of the above isochronal annealing in dark, and the open circles show the results after white light illumination. The triangles in Figure 1 show the difference between the in-dark annealed state and the illuminated or trap-full state. The difference is evidently concerned with the irradiation induced defect with an electron trap, or more directly the triangles in

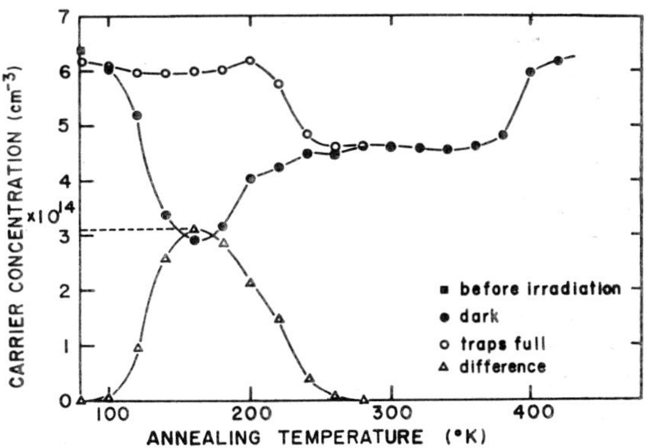

FIG. 1. Isochronal annealing curves of electron irradiated 14 Ga specimen.

Figure 1 show the number of trap-emptied defects.

The annealing results up to 160 °K can be reversibly repeated, which means that the increase of trap-emptied defects between 80 °K and 160 °K is due to the trap emptying process by thermal excitation, and that there occurs no defect annealing in this temperature range. It is considered that the total number of defects concerned is indicated by the number of trap-emptied defects after 160 °K annealing which is shown by the horizontal broken line in Figure 1.

The trap-emptied defects decrease upon annealing above 180 °K and anneal out completely at 280 °K, the annealing stage being centered at 220 °K. On the other hand the hole concentration for trap-full state does not show appreciable change between 80 °K and 160 °K, and after a small peak at 200 °K it decreases parallelly with the trap-emptied defects around 220 °K. Now the above described defect with an electron trap will be called the 220 °K defect in this paper.

It should be emphasized here that the behavior of the 220 °K defect is separated in this paper from the hole concentration change due to other lattice defects by means of trap-filling and emptying processes as described above.

3.3. *Dopant dependence of annealing behavior*

The species of dopant do not affect the annealing behavior. For example, the 15 In specimen, of which the dopant concentration is not very much different from that of the 14 Ga specimen, shows almost the same annealing curves between 80 °K and 420 °K as those of 14 Ga specimen shown in Figure 1. However, the concentration of dopant does affect the annealing behavior in that the temperatures of annealing stages are somewhat shifted depending on the dopant concentration, although the general annealing behavior remains unchanged. For example, the annealing stages of 15 Ga shift somewhat to lower temperatures, and the 220 °K stage of 13 In splits into two substages centered at about 180 °K and 280 °K. This splitting of 220 °K stage means that the 220 °K defect consists of two components, which become more apparent when the dopant concentration is small. These two components will be called the fast and the slow components, because they anneal at lower and higher temperatures, respectively. The annealing of the slow component makes the hole concentration decrease, while the annealing of the fast component does not.

3.4. *Defect introduction rate*

Figure 2 shows the introduction curve of the 220 °K defect as a function of electron dose, the parameters being the dopant species and the

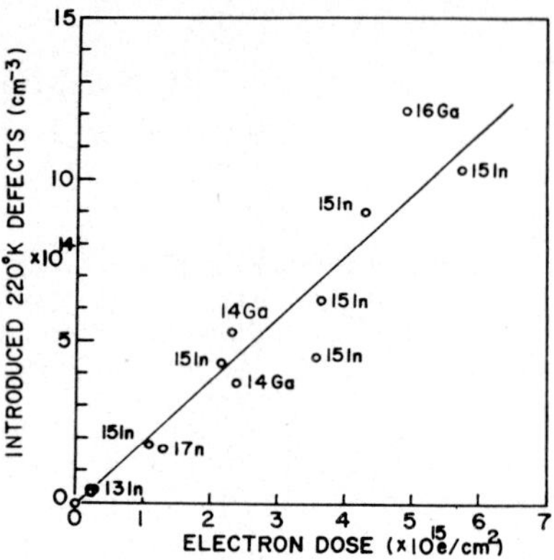

FIG. 2. Introduction rate of 220 °K defects in variously doped specimens.

dopant concentration. The results show that the introduction rate is about 0.2 defects/cm³/(e/cm²) independent of dopant species and concentration.

3.5. *Light-induced annealing*

Figure 3 shows the results of light induced annealing of 15 In specimen performed at three annealing temperatures, 160 °K, 180 °K and 200 °K. The measurement procedure is as follows. After the 15 minute annealing in dark at an annealing temperature, the electrical measurement is performed at 80 °K. Then the annealing is done at the same annealing temperature under white light illumination for 30 minutes. This is followed by

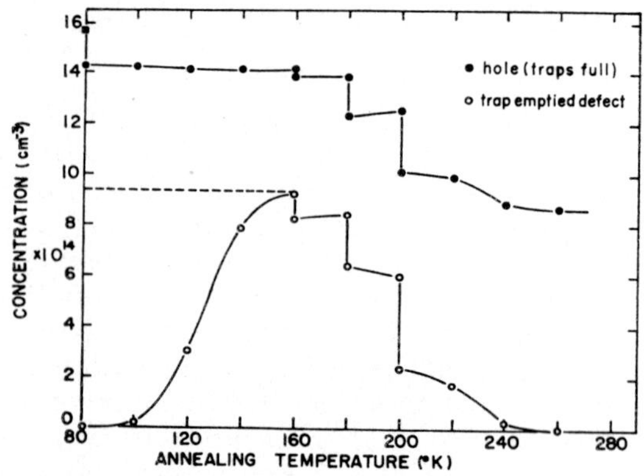

FIG. 3. Light induced annealing of electron irradiated 15 In specimen.

the 15 minute trap-emptying annealing at 160°K and the electrical measurement at 80°K. The vertical lines in Figure 3 show the light induced annealing. The decrease of hole concentration shown by vertical lines connecting solid circles is always smaller than the decrease of 220°K defects expressed by vertical lines connecting open circles. This is understandable, because the 220°K defect consists of the fast and the slow components, and the slow component removes holes while the fast component does not.

3.6. *Energy levels of 220°K defect*

The fast component appears distinguishable only in a specimen with small dopant concentration, and it anneals out at lower temperature as described above. This situation makes the energy level determination difficult for the fast component. We rather eliminated the fast component by means of light induced annealing at 160°K for 30 minutes in order to make accurate determination of energy levels of the slow component.

The energy level of electron trap is determined from the thermal emptying process between 80°K and 160°K. From the isothermal annealings of 14 Ga specimen, E_c–0.21 eV was obtained. This level somewhat changes depending on the concentration of dopant, and about E_c–0.15 eV is estimated for 15 Ga specimen. Similar results were already obtained by other investigators.[1-3] Although in these earlier results the separation technique of subtracting the trap-emptied values from the trap-full values was not used, the results give correct values of energy level by virtue of the nearly constant hole concentration of trap-full state between 80°K and 160°K as shown in Figure 1.

The donor level is determined from the temperature dependence of the Hall coefficient of the trap-emptied state after 220°K annealing, which gives E_v + 0.17 eV.

The energy levels of the fast component are considered to be similar to those of the slow component.

3.7. *Isothermal annealing of 220°K defect*

The isothermal annealing experiment was performed for the 220°K defect at several constant temperatures. The details of the annealing behavior are rather complicated because of their dependence on the charge state of the defect. The detailed results will be published elsewhere, and only the summary will be given here. The annealing kinetics show in principle the first order reaction, which manifests itself somewhat differently according to the charge state and to the mixing of the fast and the slow components. The activation energy and the frequency factor of the annealing of slow component are 0.18 to 0.33 eV and 4.6 to 4.2×10^2 sec^{-1}, respectively, depending on the charge state.

There are some earlier works concerning this annealing stage.[2-4] But these earlier results were obtained from the hole concentration change, which does not necessarily give the true behavior of the 220°K defect.

4. POSSIBLE MODEL OF 220°K DEFECT

The 220°K defect was also observed by Trueblood[5] as Pl center by means of EPR in its trap-full state. The 220°K defect cannot be observed by EPR when the trap is empty.

Now our experimental results show that the 220°K defect does not contain impurities. The possibilities of a vacancy and an interstitial can be ruled out by the existence of two components with similar trap properties, one removing no carrier, the other a carrier on annealing in the same charge state.

As the fast and slow components have similar trapping properties, they must be similar defects to each other. But as the carrier removing property is quite different, they must anneal differently. The defect which satisfies both of the above two requirements is a kind of loosely bound complex which has a potential barrier against its recombination. When it makes recombination annihilation, no carrier removal takes place, but when it dissociates, one of its components would combine with a dopant atom to invalidate its acceptor function, and to remove a carrier. This kind of complex should be a defect similar to a vacancy-interstitial pair in its nature, and yet should be stable up to 220°K.

REFERENCES

1. J. E. Whitehouse, *Phys. Rev.*, **143**, 520 (1966).
2. H. Saito, N. Fukuoka, H. Hattori and J. H. Crawford, Jr., *Lattice Defects in Semiconductors*, Ed. R. R. Hasiguti, University of Tokyo Press, Tokyo and Pennsylvania State University Press, University Park and London, 1968, p. 199.
3. T. M. Flanagan and E. E. Klontz, *Phys. Rev.*, **167**, 789 (1968).
4. W. L. Brown, W. M. Augustyniak and T. R. Waite, *J. Appl. Phys.*, **30**, 1258 (1959).
5. D. Trueblood, *Phys. Rev.*, **161**, 828 (1967).

EPR IN ELECTRON BOMBARDED PHOSPHORUS-DOPED *N*-TYPE GERMANIUM—AN ATTEMPT TO DETECT THE PRESENCE OF BOMBARDMENT-INDUCED INTERSTITIAL PHOSPHORUS

A. HIRAKI†

Denki-Bussei Laboratory, Department of Electrical Engineering, Osaka University, Suita, Osaka, Japan

Effects of electron bombardment on Phosphorus-doped *N*-type Germanium (Ge) were studied by EPR techniques in the hope of detecting the presence of interstitial Phosphorus (P_I) which were ejected from the original substitutional sites through the interaction with the mobile interstitial Ge atoms produced by the bombardment.

Bombardment induced the following two remarkable effects: (1) Extreme broadening of EPR linewidth of the normal substitutional Phosphorus (P_S) donors which indicated that the bombardment produced large internal strains ($\sim 10^{-3}$) at the sites of P_S; (2) Occurrence of new EPR line and study of its *g*-value showed that the corresponding impurities (i) had symmetric wave function with the [001] axis and (ii) were located in the field of stronger internal strain ($\sim 10^{-2}$) than that at P_S. A model of P_I displaced into $\langle 001 \rangle$ directions from ordinary interstitial sites with tetrahedral symmetry due to Jahn-Teller effect was proposed to explain these effects.

1. INTRODUCTION

Hiraki and coworkers[1] studied by mainly electrical measurements the production and annihilation of interstitial impurities in electron bombarded *N*-type Ge. Their production mechanism was the ejection of substitutional impurities (Sb and Cu) out of their original positions through interactions with mobile interstitial Ge atoms produced by the bombardment.

In order to get more microscopic evidence than our previous electrical studies, the present work has been undertaken with EPR techniques.[2]

If in *N*-type Ge the interstitial donor impurities are really introduced by electron bombardment with high yield rates as was pointed out by us,[1,3] their existence might possibly induce at least the following two phenomena.

(1) EPR linewidth of normal donors which survived from the ejection into interstitial sites is broadened due to the internal strains induced by the interstitial donors the reason for this broadening is explained briefly in Section 2.

(2) The presence of interstitial donors might cause new resonance line to occur.

Experiments have been conducted to observe the above mentioned effects (1, 2) on the P-doped Ge because the value of valley-orbit splitting of normal substitutional P-donor (P_S) is appropriate to study

† Present address: Department of Electrical Engineering, California Institute of Technology, Pasadena, Calif., U.S.A.: On temporary leave from Osaka University

(1) from the linebroadening as will be mentioned in Section 2.

2. LINEBROADENING DUE TO BOMBARDMENT-INDUCED INTERNAL STRAIN

Interstitial P(P_I) is expected to induce strain and in addition to this effect, if there is degeneracy in energy of the electronic ground state of P_I, the Jahn-Teller distortion (possibly, displacement of P_I in $\langle 100 \rangle$ directions) occurs at the measuring temperature (1.5 °K) to remove such degeneracy and eventually this leads to strengthen greatly the internal strain through disturbing the regular tetrahedral configuration of host Ge atoms.

The next thing to be mentioned is about the interrelationship between thus induced internal strain and the electronic ground state of P_S whose spin gives rise to the EPR signal. Since by the effective mass approximation the electronic state of normal donors (as P_S) in *N*-type Ge is expressed in terms of Bloch functions associated with the conduction band minima, the four-valley structure in the conduction band leads to a state with 4-fold degeneracy. And the effect of the impurity potential removes this degeneracy to give singlet and triplet states with an energy difference of E_{13} (valley-orbit splitting)—experimentally the ground state was found to be singlet state.

While the internal strain ($\mathscr{S}_i$) at *i*th donor causes the change in the relative valley population among four valleys and induces repopulation from the equi-populated equilibrium state (or isotropic population). This effect consequently gives influence on the electronic ground state of the normal donor. And this influence can be represented by the admixture of the excited triplet state into singlet ground state to introduce a new ground state—the extent of this admixture in ith donor state is proportional to $\sum_u . \mathscr{S}_i / E_{13}$, where $\sum_u$ is the deformation potential constant.

Since this new ground state is originated in the breakup of the equivalent (or isotropic) population in each four valleys, if the electronic g-value in each of the four valleys (or a single valley) is not isotropic the observed g-value of EPR signal from this new ground state might also show to some extent anisotropy and also g-shift from the isotropic value (g_0) which came from the original ground state.

The degree of this anisotropy (or shift) is proportional to $(g_\parallel - g_\perp) . \sum_u . \mathscr{S}_i / E_{13}$, where $(g_\parallel - g_\perp)$ is the measure of the anisotropy of g-value in single valley.[4] And because of the internal strain at each donor (i) being different in strength and direction, random deviation of each g_i from g_0 causes the broadening of linewidth from overall normal donors.

At this point, the very important thing in N-type Ge is that the value of $(g_\parallel - g_\perp)$ which is the main origin of both the linebroadening and the g-shift is 3 orders of magnitude larger than that of Si and this fact guarantees the extremely higher sensitivity of the EPR of normal donors in Ge than in Si to either internal or external strain. And therefore N-type Ge is very ideal material for the study of this kind.

The reason for the choice of P is that the valley-orbit splitting (E_{13}) is adequate in magnitude—if Sb would be chosen instead, due to the value of E_{13} being 10 times smaller than that of P, linebroadening occurs even before the bombardment to make the analysis of bombardment-induced internal strain difficult.

In order to learn if the internal strain is introduced by the bombardment, specimens containing 5×10^{16} and 8×10^{16} P-donors/cm³ with surface planes of (001) and (1$\bar{1}$0) respectively were bombarded by 1.5 MeV electrons at either 0 °C or dry ice temperature with various amounts of total dose. Following the bombardment, EPR measurements at 1.5 °K were made in an X-band cavity capable of uniaxial stress application with magnetic field (H) rotated in each surface plane and their results are reported below.

2.1. *Low-dose bombardment*

As was pointed out by Wilson,[4] linewidths measured with the magnetic field (H) around [100] axis direction can be a very good check of the internal strain. Data on the linewidths ΔH_{msl} as a function of (θ) between H and [100] direction were taken on the specimens with (001) surface which were bombarded with 5×10^{15} and 1×10^{16} electrons/cm². The concentrations of P_I† in these specimens, if produced, were estimated to be about $2 \sim 3 \times 10^{15}$/cm³.

As seen in Figure 1a, just after the bombardment

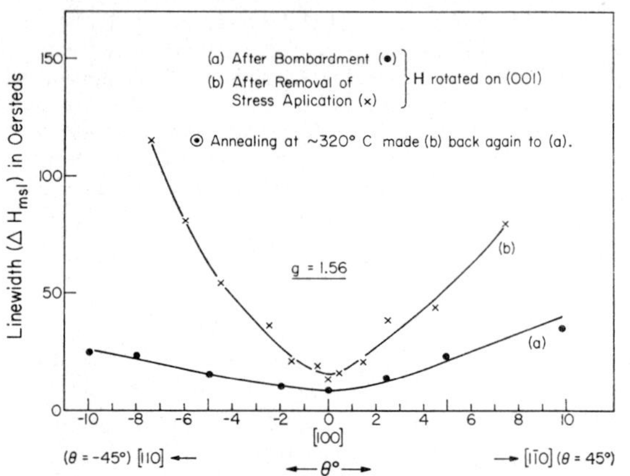

FIG. 1. Effects of low-dose bombardment, uniaxial stress application (T//[110] direction) and annealing upon linewidths (around H//[100]) of normal Phosphorus donors measured with H rotated on (001) surface—deviation of H from [100] is expressed by angle (θ).

linewidths did not show any appreciable broadening effect and they were almost the same as were measured before the bombardment.

But the uniaxial stress (T) application along the [110] direction (T||[110]) brought out very interesting phenomena:

(i) Although before the bombardment the linewidth at H||[1$\bar{1}$0] became narrower by the stress application as was already studied experimentally[5] and theoretically,[6] these lines became broader and broader as the magnitude of the stress increased and finally they disappeared and this line did not appear again even the stress application was removed—

† Since by bombardment vacancies (V) are produced almost the same as P_I in number and these V interact either with P_S to form stable complexes $(V - P_S)$ or with P_I to get annihilated, some fractions of P_I remain unannihilated even at room temperatures[1,3] since these P_I are immobile at this temperature.

g-shifts of lines around $H\|[100]$ under the stress of 1×10^9 dyne/cm², on the other hand, were almost those measured before the bombardments.

(ii) Also after the removal of the stress, the linewidth except at $H\|[100]$† was found to show remarkable broadening (Figure 1b).

Heat treatment around 300 °C for 10 hours almost removed these effects (i, ii) and all linewidths became as narrow as those shown in Figure 1a; at this time uniaxial stress application did not give any after-effects either.

2.2. *High-dose bombardment*

Two specimens with (001) and (1$\bar{1}$0) surfaces were rather heavily bombarded with $1\sim2 \times 10^{17}$ electrons/cm². And this time, remarkable linebroadening was observed in both specimens without any uniaxial stress application which shows that enough internal strains were introduced by the bombardments; in addition a less sensitive response to the external uniaxial stress application than that of specimens bombarded with less total doses indicates the presence of fairly large bombardment-induced internal strain. These results were exhibited in Figures 2b and 3b with the data taken before the bombardments for comparison Figures 2a and 3a.

Of course the line at $H\| (1\bar{1}0)$ disappeared and from this the estimation of the magnitude of the bombardment-induced strain was possible using

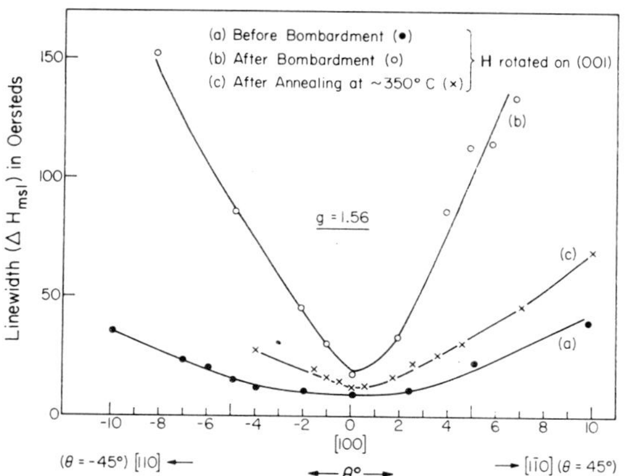

FIG. 2. Effects of high-dose bombardment and annealing upon linewidths (around $H//[100]$) of normal Phosphorus donors measured with H rotated on (001)—deviation of H from [100] is expressed by angle (θ).

† Linewidth at $H//[100]$ is very insensitive to the internal strain due to the valleys locating in $\langle 111 \rangle$ directions.[6]

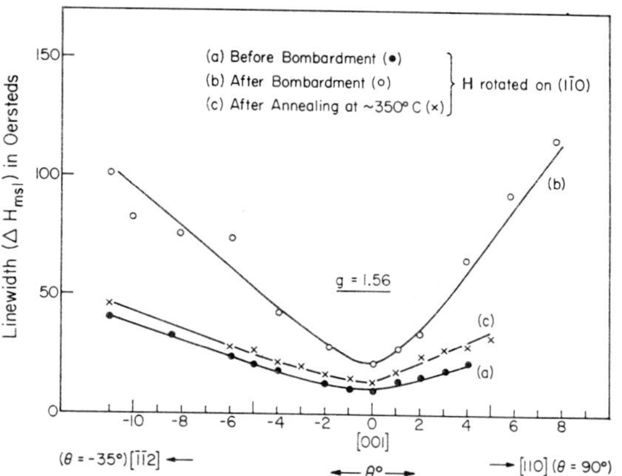

FIG. 3. Effects of high-dose bombardment and annealing upon linewidths (around $H//[001]$) of normal Phosphorus donors measured with H rotated on (1$\bar{1}$0)—deviation of H from [001] is expressed by angle (θ).

the formula[6] derived by Nakeyama and Hasegawa under the assumption that the distribution of the strains is random.

$$\left(\frac{\Delta H_{\mathrm{msl}}}{H}\right)^2 = \frac{64}{15}\left(\frac{g' \cdot \Sigma_u}{g_0 \cdot 3E_{13}}\right)^2 \langle(\Delta\epsilon)^2\rangle \quad \text{(for } H\|[1\bar{1}0]\text{)}$$

where

ΔH_{msl} : linewidth

$\sqrt{\langle(\Delta\epsilon)^2\rangle}$: spatially averaged magnitude of strains

$g' = \frac{1}{3}(g_\| - g_\perp) = -0.35$, $g_0 = \frac{1}{3}(2g_\perp + g_\|) = 1.56$

E_{13} (valley-orbit splitting) $= 2.9 \times 10^{-3}$ eV for P_s in Ge

Σ_u (deformation potential constant) $= 18$ eV for Ge

Taking ΔH_{msl} as 200 Oersted which was the widest observable linewidth limit of our measurements, $\sqrt{\langle(\Delta\epsilon)^2\rangle}$ was calculated to be a little less than 10^{-3}.

This surprising large value of bombardment-induced strain is, in itself, very interesting—this fact was only detected in *N*-type Ge due to the reason stated in the beginning of Section 2—and this strain is equivalent in strength to that induced by the presence of $\sim10^{10}$ dislocations/cm².

As shown in Figures 2c and 3c, annealing of these specimens at around 300 °C also reduced linewidths almost to those measured before the bombardments, suggesting that the origins of the broadening mentioned in Sections 2.1. and 2.2. are

the same and they are not something like dislocations, since dislocations in Ge become mobile
only above 650°C.

3. A NEW RESONANCE LINE

In the measurement where magnetic field (H)
was rotated in ($1\bar{1}0$) plane, besides the resonance
line from normal donors (P_s), a new line was
observed. This line also disappeared due to the
annealing around 300 °C.

As seen in Figure 4, some representative features
of the new resonance are:

(i) Sensitive g-shift to the rotating angle (θ) of
H and these shifts being symmetric with [001] (or
$\langle 100 \rangle$) direction.

(ii) g-values of both new and normal resonances
being almost equal ($g = 1.56$) in the direction of
$H\|$ [001] under the condition of no external stress
application.

(iii) Insensitive or almost no response to the
large external stress ($T\|$ [111]: 9×10^8 dyne/cm² in
comparison with sharp response of normal P-donor
(P_s) resonance. This response was the same as
studied by Morigaki[5] in unbombarded specimens.

The above mentioned features are qualitatively
explained through the model that this new resonance originates in interstitial P(P_I) and these P_I
undergo displacements in [001] (or $\langle 100 \rangle$)
directions due to the Jahn-Teller effect and in this
case P_I is naturally expected to be exposed to very
strong strain field, so that the new line is almost

unaffected by the strong external stress application.

But in order to propose the precise model about
this new resonance, there are several points yet to
be pursued experimentally. One of them is to
know the spread of the wave function of P_I which
also concerns to its energy-level depth and investigations of this kind are now being done by
Morigaki and coworkers and Nishida and coworkers.

The reduction of linewidths of the normal P_s
resonance by annealing at around 300 °C stated in
Section 2 is consistent with the disappearance of
the new line by the same heat-treatment if the
origins of these two phenomena are the same.

4. DISCUSSION

As already mentioned in Sections 2 and 3, there
is no doubt that electron bombardment induces a
large internal strain ($\sim 10^{-3}$) at the sites of normal
donors (P_s) and also much larger ones (probably
$\sim 10^{-2}$) at the impurities which gave rise to the
new resonance.

Then if we adopt the model described in Section 3
that the origin of the new resonance is the displaced
(into $\langle 001 \rangle$ directions) P_I and assume these P_I to
be also the origin of the observed internal strains,
besides the [001] symmetry of g-value in new
resonance, a qualitative understanding becomes
possible concerning the reason why the magnitude
of the internal strain at P_I is larger than that at P_s.

One more experimental support of above mentioned possibility is, as mentioned in Section 2.1,
that in the specimens bombarded with lower total
doses the linebroadening was caused by the uniaxial
stress application in [110] direction. And from the
consideration of the atomic configurations in the
host lattice structure of Ge, the uniaxial compressive
stress in this direction enhances the P_I to displace
into [00$\bar{1}$] direction from the position of tetrahedral
symmetry. In the sense of 'Collective Strain Splitting of Acceptor State in Si' studied by Koonce,[7]
if there is a critical concentration of P_I to perform
their displacements collectively into $\langle 001 \rangle$ (or
[00$\bar{1}$]) directions, this stress application might have
been of some help for the displacements followed
by the relaxation of the surroundings to keep these
P_I clumped; on the other hand, in the specimens
with heavier doses, due to the concentration of P_I
being large enough, they all displace to cause linebroadening of P_s without any assistance from such
external stress application as reported in Section 2–2.

Therefore, the essential thing in the origin of the

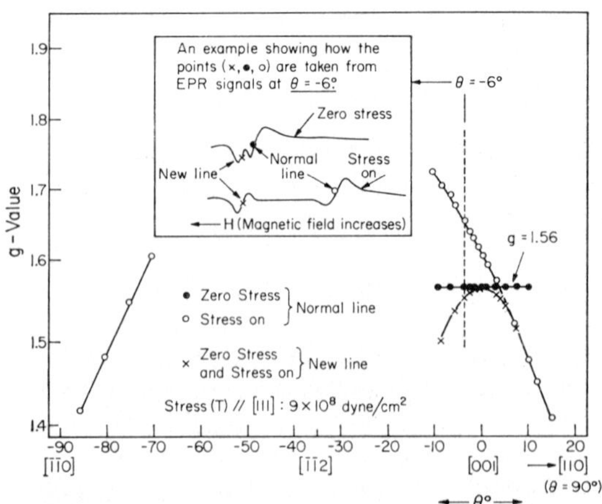

FIG. 4. g-values of both new and normal resonance
lines as a function of angle (θ) between H and
[001] with or without uniaxial stress application
($T//$[111])—H was rotated on ($1\bar{1}0$) surface.

bombardment-induced large internal strain may be the displacements of bombardment-induced P_I from the sites of higher to those of lower symmetry.

The work reported in this paper is at present not complete but the way in which we have been trying to solve the problem may possibly give new information yet to be obtained.

And also there is an interesting possibility to explain the high mobilities of bombardment-induced vacancies in Ge and Si in terms of the internal strain mentioned in this paper.

ACKNOWLEDGMENTS

The author heartily thanks for the precious collaboration and technical guidance of Prof. K. Morigaki of the University of Tokyo, with whom more complete papers on this problem will soon be published. Also much help in the measurements from Miss M. Onda and Mr. S. Toyotomi of the University of Tokyo are appreciated. The stimulating theoretical discussions with Prof. T. Yamaguchi of Tottori University about the interstitial impurities were very helpful. Kind help given to the author in the course of preparation of this paper during his stay in the group of Prof. J. W. Mayer at the California Institute of Technology were also appreciated. Finally the author should express his sincere thanks to Prof. K. Kawabe (head of our Denki-Bussei Laboratory) of Osaka University for his kind and continuous encouragement given for this work.

REFERENCES

1. A. Hiraki, J. W. Cleland and J. H. Crawford, Jr., *Radiation Effects in Semiconductors*, Ed. F. L. Vook (Plenum Press, New York, 1968), p. 224.
2. A. Hiraki and K. Morigaki, *J. Phys. Soc. Japan*, **27**, 1701 (1969).
3. A. Hiraki, *Crystal Lattice Defects*, **1**, 277 (1970).
4. D. K. Wilson, *Phys. Rev.*, **134**, A265 (1964).
5. K. Morigaki and T. Mitsuma, *J. Phys. Soc. Japan*, **20**, 62 (1965).
6. M. Nakayama and H. Hasegawa, *J. Phys. Soc. Japan*, **18**, 229 (1963).
7. C. S. Koonce, *Phys. Rev.*, **134**, A1625 (1964).

DISCUSSION

Question (ALBANY) In a recent work, I believe that Pearlman and coworkers have been suggesting that 4Δ in P-doped and As-doped Ge might be changed by compensation. Is there any possible correlation with this work?

Answer (HIRAKI)   I am afraid that I do not know in detail the work you referred. But since the linebroadening which I mentioned was due to the repopulation effect among valleys caused by the bombardment-induced internal strain, I think there may be no correlation between them.

ONE-VALLEY EPR SPECTRUM IN ELECTRON IRRADIATED Sb-DOPED GERMANIUM

R. R. HASIGUTI, T. NAKANISHI, N. FUNAKOSHI

*Department of Metallurgy and Materials Science, Faculty of Engineering,
University of Tokyo, Bunkyo-ku, Tokyo*

AND

S. TAKAHASHI

National Research Institute for Metals, Meguro-ku, Tokyo

The one-valley EPR spectrum which will be called Ge-T1 spectrum was observed in electron irradiated and annealed (above 200 °C) Sb-doped n-type germanium.

The specimens used were germanium crystals doped with 1.2×10^{16} Sb/cc which were irradiated by the 2.5 MeV electrons to the doses of 5×10^{16} e/cm^2 to 1×10^{18} e/cm^2.

Figure 1 shows the g-shifts of the EPR spectra observed in a crystal irradiated (1×10^{18} e/cm^2) and annealed at 485 °C for 30 min. The g-shifts are expressed by

$$g^2 = g_\parallel^2 \cos^2 \theta + g_\perp^2 \sin^2 \theta$$

where $g_\parallel = 0.82$ and $g_\perp = 1.93$.

These g-shifts are often observed in a $\langle 111 \rangle$ compressed non-irradiated specimen, in which one valley is depressed compared with other three valleys.[1,2] But in the irradiated germanium other origin should be considered, because external or long range internal stress cannot be considered. We propose a divacancy-antimony atom complex which gives rise to T1 spectrum. Here Sb atom is situated at one end of V_2. The Sb donor electron reflects the electronic potential and strain field due to divacancy, and this gives the one-valley T1 spectrum because of the removal of valley degeneracy. The details of this electronic state will be described elsewhere.[3]

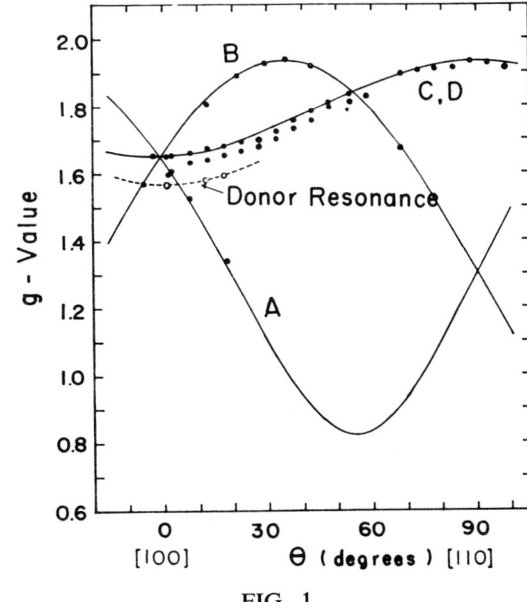

FIG. 1.

REFERENCES

1. R. E. Pontinen and T. M. Sanders, Jr., *Phys. Rev. Letters*, **5**, 311 (1960).
2. T. Mitsuma and K. Morigaki, *J. Phys. Soc. Japan*, **20**, 491 (1965).
3. T. Nakanishi and R. R. Hasiguti, to be published.

MÖSSBAUER-TYPE PHOTOLUMINESCENCE SPECTRA OF IRRADIATED GERMANIUM

R. J. SPRY

U.S. Air Force Materials Laboratory

AND

J. D. HENES

University of Dayton

A Mössbauer-type structure has been observed in the 15 °K spectra of germanium crystals irradiated at room temperature by Co^{60} γ-rays and 700 keV electrons. The spectra consist of an intense band peaking at 0.688 eV accompanied by mainly TA-phonon-assisted lower energy sidebands. The transition producing the luminescence is believed to be the annihilation of an exciton bound to a defect center by forces of shorter range than coulombic attraction. The dependence of the new spectra upon irradiation temperature and certain sample characteristics indicate that the center contains both primary defect(s) and an unknown impurity.

1. INTRODUCTION

Recombination luminescence was first used by Brauenstein to study radiation damage in germanium.[1] He was able to monitor the defect introduction rate through the degradation of the intrinsic band gap transition following electron bombardment, but did not discover any new luminescence bands. While trying to observe luminescence spectra in electron irradiated germanium, Vavilov and coworkers found difficulty in distinguishing radiation-induced bands in the spectral region of 0.53 eV from those caused by recombination at dislocations.[2,3] Broad, relatively featureless post-irradiation bands with intensity maxima at 0.59 eV have also been found in germanium.[4,5]

In the other covalent semiconductors, luminescence has proven more effective in gaining insight into the production and structure of radiation damage defects. Following the initial report of a broad, radiation-induced band in silicon,[4] high resolution measurements revealed a particular type of spectrum consisting of a sharp, intense zero-phonon band and less intense phonon-assisted sidebands.[6,7] Additional spectra found in irradiated silicon,[7–11] and luminescence bands produced by radiation defects and chemical impurities in diamond[12] also demonstrated this same Mössbauer-type structure. The analogy to the Mössbauer effect derives from the fact that similarly shaped spectra may be observed in the recoilless gamma ray emission of radioactive atoms contained within a crystalline lattice. Because it has the same crystal structure and covalent bonding as carbon and silicon, germanium might be expected to demonstrate Mössbauer-type luminescence spectra following the formation of radiation defects. This paper expands the scope of the initial discovery of these spectra in electron- and gamma-irradiated germanium.[13]

2. EXPERIMENTAL PROCEDURE

Photoluminescence spectra were obtained by the method developed by Haynes for his study of band-gap transitions in silicon and germanium.[11] Greater-than-band-gap light was focused upon the front surface of a sample mounted on the cold finger of a vacuum dewar, creating electron-hole pairs within a distance from the surface approximately equal to the reciprocal absorption coefficient. Recombination luminescence emerging from the back side of the sample was analyzed by a grating spectrometer, detected by a cooled lead sulfide photo-resistor, and synchronously amplified by a lock-in voltmeter tuned to the frequency of a chopper placed in the exciting light beam. This system could detect photons with energies as low as 0.35 eV. The sample temperature was obtained using a gold (0.02 per cent iron) vs. chromel thermocouple soldered on the back surface of the sample platelet.

The germanium samples used in this work were obtained from nine different boules, five grown by Semi Elements, Inc. and four by the General Electric Co. They included both *N*- and *P*-type single crystals having carrier concentrations ranging from 3×10^{11} cm^{-3} to 4×10^{14} cm^{-3}. The impurity content of the crystals was obtained by mass spectrographic analyses. The sample platelets were cut to a thickness of about 0.050 inch, usually perpendicular to the 1–1–1 growth direction. They were etched in CP-4A prior to irradiation and re-etched lightly subsequent to each luminescence experiment.

The samples were bombarded by Co^{60} γ-rays at room temperature, and by 700 keV electrons from a Van de Graaff generator at liquid helium, liquid nitrogen, and room temperatures.

3. EXPERIMENTAL RESULTS

Typical spectra of an unirradiated sample, similar to those found by Haynes and coworkers[14] are illustrated in Figure 1. Structure in the liquid

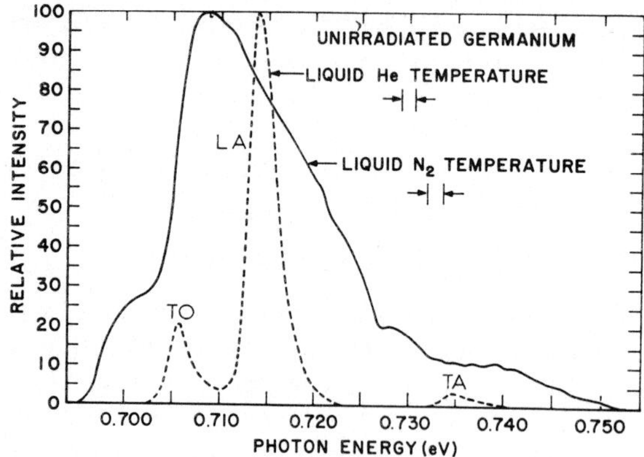

FIG. 1. Luminescence spectra of an unirradiated germanium crystal. Resolution 48 Å for the solid line and 32 Å for the dashed line.

nitrogen temperature curve is completely resolved at liquid helium temperature. The three distinct bands are produced by free exciton recombination, accompanied by emission of transverse acoustic, longitudinal acoustic, and transverse optical phonons. The shape of the longitudinal acoustic phonon band may be described by a Boltzman energy distribution produced by the translational energy of the free excitons. However, this band is broader (2.50 kT) than predicted by the distribution function (1.80 kT) because of lattice scattering of the excitons.[15] The sample temperature during excitation was measured indirectly from the shape of this band as well as directly by the thermocouple. In fact, this indirect method seemed to be a more

reliable procedure, because it provided the temperature in the sample region in which radiative recombination occurred. The sample temperatures obtained from the LA band halfwidth were consistent with those obtained from thermocouple measurements, about 15 °K for liquid helium and 80 °K for liquid nitrogen cooling.

Spectra are shown in Figure 2 which are typically found in certain samples after a room temperature

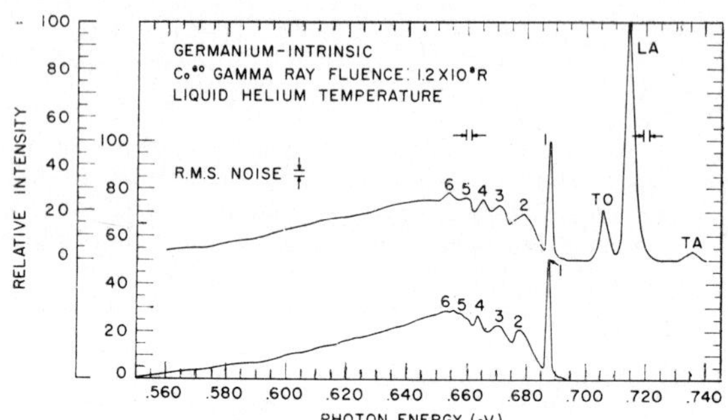

FIG. 2. Luminescence spectra of an irradiated germanium crystal. Resolution 48 Å. Upper curve, 14 days after irradiation; lower curve, 18 days after irradiation.

bombardment of 10^8R of Co^{60} γ-rays. The post-irradiation bands were also found after 700 keV electron bombardment, except that the band gap luminescence usually disappeared for an electron fluence of about 10^{17} cm^{-2}. The minimum detectable electron fluence was approximately 3×10^{15} cm^{-2} for 700 keV electrons. No other spectra were seen above 0.35 eV, and none were seen when the photoluminescence measurements were performed at liquid nitrogen temperature, following room temperature irradiation. The disappearance of the band gap luminescence with time suggests that new defect centers were formed during room temperature storage which were effective recombination centers. However, they may not necessarily have been the centers producing the new luminescence bands, because a quantitative study was not made of the growth of their absolute intensity with elapsed time after irradiation. In addition to the intense band no. 1 peaking at 0.688 eV, there are five other bands in the lower energy portion of the spectra. Band no. 1 is resolution broadened in Figure 2, but by narrowing the spectrometer splits, the halfwidth was measured to be <0.0006 eV.

No new spectra were seen following 700 keV electron bombardment at liquid helium or liquid nitrogen temperatures. The band gap luminescence disappeared for the fluence value of 10^{17} cm^{-2} used in these experiments, but returned to its full pre-irradiation intensity upon bringing the sample to room temperature, and then repeating the 15 °K and 80 °K photoluminescence measurements.

4. DISCUSSION

The first major question to be decided in interpreting the data is the nature of the transition producing the post-irradiation bands, for which four possible models are shown in Figure 3. For

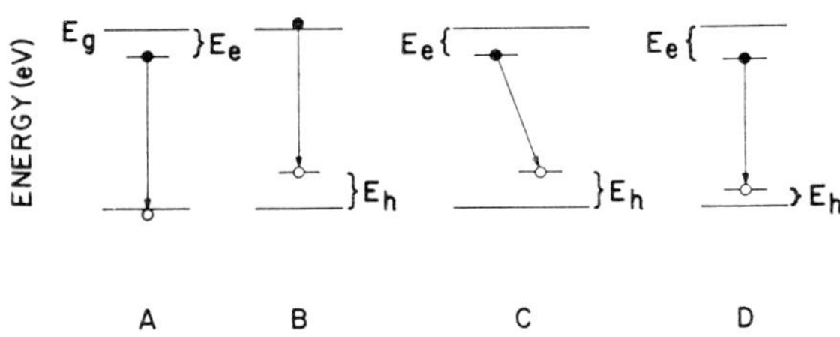

FIG. 3. Schematic representation of possible radiative recombination processes. A: bound electron-free hole; B: free electron-bound hole; C: electron and hole bound at different centers; D: bound electron-bound hole.

processes of the types labeled A and B, consisting of recombination of one free and one bound carrier, the transition rate may be written as

$$R = Nn\sigma v \qquad (1)$$

where N is the density of bound carriers, n is the density of free carriers, σ is the recombination cross section, and v is the velocity of the free carriers. This recombination rate has a functional dependence upon energy of

$$R \sim E\, e^{-E/kT} \qquad (2)$$

which has a halfwidth of 2.45 kT, or 0.0031 eV at 15 °K. Because this is five times larger than the maximum value of the halfwidth of band no. 1, the free-to-bound transitions A and B must be excluded as possibilities. This conclusion is supported by the recent data of Jones, who unambiguously determined from high resolution photoluminescence measurements that the halfwidths of the most intense bands of similar spectra in irradiated silicon are much less than kT over a wide temperature range.[16]

Process C, recombination of distant electron-hole pairs, is known to require donor and acceptor concentrations of 10^{15} cm^{-3} to be detectable, even when ionization energies are only 0.01 eV.[17,18] However, carrier removal rates for room temperature Co^{60} γ-irradiated germanium indicate that the total defect density for the largest fluence of 1.2×10^8R used in the present work was no larger than 7.0×10^{13} cm^{-3}.[19] In addition, contributions from different pairs are not resolved in the spectra of donor-acceptor recombination in germanium, producing a single band having a halfwidth of 0.004 eV, or 7 times larger than the maximum value for band no. 1.[17] For these reasons, pair recombination must also be excluded as a possible explanation of the spectra.

The spectra found here are best explained by bound exciton recombination, illustrated in Figure 3-D. In this case, band no. 1 corresponds to a zero-phonon transition producing a halfwidth much less than kT, while the smaller bands correspond to phonon-assisted transitions. This interpretation of the spectra identifies it as an optical analog of the Mössbauer effect.[20] The energy separations between band no. 1 and the smaller bands are listed in Table I, along with the identification of participating phonons whose energies would correspond

TABLE I

Energy location of radiation-induced luminescence bands of Figure 2. TA: transverse acoustic phonon; LA: longitudinal acoustic phonon; TO: transverse optical phonon.

Band number N_i	Peak (eV)	Band N_i – Band 1 (eV)	Phonon emitted
1	0.688		zero
2	0.679	0.009	TA
3	0.671	0.017	2 (TA)
4	0.665	0.023	3 (TA)
5	0.660	0.028	LA
6	0.654	0.034	TO

to the tabulated energy differences.[21] There is no a priori way to determine whether the lower energy portion of the spectra corresponds to the excitation of local vibrational modes or the normal modes of the lattice. However, the energies of phonons from the normal lattice spectrum fit the data for both germanium and silicon.[7,16] In particular, because of the small value of the germanium TA phonon energy (0.008 eV), the first three sidebands (2–4) unambiguously correspond to single and multiple TA phonon emission, if only normal lattice vibrations are considered. The identity of band no. 5 is also fairly certain, but band no. 6 can possibly be identified with the excitation of four TA phonons. From a careful comparison of the spectral intensities and relative transition probabilities calculated from a multiphonon configuration-coordinate model, Jones has found that the radiation defects of silicon are also most strongly coupled to the TA modes.[16]

It is informative to compare the relative spectral intensities for silicon and germanium using a probability function calculated within the Debye approximation[20]:

$$\frac{I_0}{I} = \exp\left\{ -S\left[1 + \frac{2\pi^2}{3}\left(\frac{T}{\theta_D}\right)^2 \right] \right\} \qquad (3)$$

for $T \ll \theta_D$, where I_0 is the probability for a zero-phonon transition, I is the total transition probability, S is the average number of phonons emitted, T is the Kelvin temperature, and θ_D is the Debye temperature of the individual lattice phonons. For $T = 15\,°K$ we obtain 0.093 for silicon and 0.068 for germanium, using the Debye temperatures of the TA phonons and the value of $S = 2.3$ obtained by Jones for the silicon defect centers.[16] Therefore, the relative intensities should be nearly the same in these two materials, as was qualitatively observed. Although Eq. (3) does not strictly apply at $T = 80\,°K$, it is useful to obtain an estimate of the probabilities at this temperature, which are 0.012 for silicon and 1.6×10^{-6} for germanium. Evidently, there should be no practical possibility for observing the zero-phonon transition in germanium at $80\,°K$, which might explain the lack of sharp line spectra in the 0.59 eV bands found by other authors.[4,5] Unfortunately, no radiation-induced luminescence was observed in the present work at this temperature, eliminating the chance for comparison with theory. This was probably caused by a decrease in the total transition probability with increasing temperature. However, the Mössbauer-type luminescence spectra of irradiated silicon still displayed a significant zero-phonon component at $80\,°K$.[6–10,16]

Lacking the results of Zeeman effect studies of the luminescence spectra, it is difficult to distinguish among several types of bound exciton transitions which could occur, depending upon the number of carriers bound to the defect center and the charge state of the defect. The first model to consider is that of an exciton bound to an ionized donor (acceptor), which also may be described as a hole (electron) bound to a neutral donor (acceptor). The localization energy of the exciton is given by:

$$E_L = E_g - h\nu - E_x = E_e + E_h - E_x \qquad (4)$$

where E_g is the energy gap, $h\nu$ is the energy of the photon corresponding to the zero-phonon transition, E_x is the binding energy of the free exciton, E_e is the donor ionization energy, and E_h is the energy needed to bind the hole to the neutral donor. E_L is found to be 0.055 eV using the values of 0.746 eV[22] for the energy gap and 0.003 eV[23] for the free exciton binding energy. This is also approximately the value of the donor ionization energy, because in Equation (4), E_h should tend to cancel E_x and both are small. Although E_L is five times larger than donor (acceptor) energies for singly charged coulombic centers in germanium, it is shallow enough (7 per cent of E_g) to be treated approximately by an effective mass model. Hopfield has shown that this type of complex will not be stable unless the hole mass to electron mass ratio is greater than 1.4[24] Because this ratio is about unity in germanium and because these transitions have never been observed experimentally for any donor or acceptor, an exciton bound to an ionized coulombic center is an unlikely candidate for explaining the spectra.

It is also possible to bind an electron and hole to an unoccupied, neutral center. Although there is no coulombic attractive force for binding the first carrier, there are other forces that are known to have this capability, i.e., the forces involved in isoelectronic traps[25] and dipole attraction.[26,27] The second carrier is subsequently bound by the coulombic attraction of the first bound carrier. The exciton localization energy is again given by Eq. (4) and has the same value calculated previously. E_e or E_h may not be negligibly small in comparison to the other, so that E_L would be somewhat larger than single carrier ionization energies measured by other techniques. It should be pointed out, however, that there is no reason that a center (especially a defect or defect-impurity complex) could not possess both coulombic and strong, short range forces, making an exciton bound to an ionized donor or acceptor again a strong contender for the correct model. This type of center is known to produce the optical Mössbauer spectra of diamond.[12] The important result of this discussion is that whatever the charge state of the defect, the localization energy of the exciton is $\geq$ the single carrier ionization energies when only two carriers are bound to an attractive center.

Finally, the model of an exciton bound to a neutral, coulombic donor (acceptor) may be completely excluded after brief consideration. The exciton localization energy may be again computed from the first part of Equation (4), but is now related to the donor ionization energy by

$$E_L = \alpha E_e \qquad (5)$$

where α is a constant for each particular semiconductor material, depending on the ratio of electron and hole masses. Hopfield's theory[24] predicts $\alpha = 0.1$ for both silicon and germanium, which was exactly the value obtained experimentally.[12] This would imply E_e or E_h is 0.55 eV for the present defect center, or nearly 3/4 of the energy gap and beyond the range of any effective mass treatment. As pointed out by Jones,[16] central cell corrections must be made for deep ionization energies which have the effect of making the values of E_e calculated from Eq. (5) even larger and more implausible.

Another major problem is the formulation of a defect model which will account for all of the experimental results. The post-irradiation luminescence bands were found in samples cut from only five of the nine boules surveyed. There was no correlation with the sign of the majority carrier or carrier concentration prior to irradiation, except that the samples having the new bands were not intentionally doped, the largest carrier density being less than 10^{14} cm^{-3}. There was also no correlation between production of new bands and the concentration of impurities determined from mass spectrographic analysis. As a matter of record, the prevalent impurities found in all nine boules in parts per million atomic included: C, O, $\sim$1; Si, 1–8; Na, $\sim$1; K$\sim$0.5; and S, 0.06–0.3.

The disappearance of the band gap luminescence during irradiation at liquid helium and liquid nitrogen temperatures indicates that stable recombination centers were produced which were not the same centers associated with spectra of Figure 2. It should be emphasized that only greater-than-band-gap light was impinging upon the samples during photo-excitation, so that optical bleaching should not have occurred while luminescence spectra were recorded. The return of the band gap luminescence after a room temperature anneal is very likely associated with the complete healing of the lattice, although it is possible that defect centers were still present which were not effective recombination

centers. The temperature dependence of the degradation and renewal of the band gap luminescence is consistent with the existence of well known major annealing stages at 65 °K and 150 °K,[28] while long time decay of the band gap luminescence at room temperature is indicative of the mobility of primary radiation defects and their ability to form new complexes. We believe that a model for the recombination center containing both primary radiation defect(s) and an unknown impurity satisfies all of the above observations. It is very likely that this center possesses an electronic energy level <0.055 eV from a band edge after room temperature irradiation. This could be the level located 0.02 eV above the valence band previously reported in room temperature, γ-irradiated germanium.[29,30]

ACKNOWLEDGMENTS

The authors wish to thank Dr. Robert N. Hall of General Electric for germanium crystals, Dr. Emile D. Pierron of Monsanto for mass spectrographic analyses, Capt. Arthur L. Robinson and Dr. Jon Meese for helpful discussions, and Major C. Neale Elsby, Capt. Richard A. House, II, Mr. Anthony N. Fasano, Lt. Mark E. Mount, and Sgt. John L. Singleton for sample irradiations.

REFERENCES

1. R. Brauenstein, *Bull. Am. Phys. Soc. (II)*, **2**, 157 (1957).
2. V. S. Vavilov, A. A. Gippius, M. M. Gorshkov and B. D. Kopylovskiĭ, *Soviet Phys.-JEPT*, **37**, 15 (1960).
3. A. A. Gippius and V. S. Vavilov, *Soviet Phys.-Solid State*, **7**, 515 (1965).
4. Yu. L. Ivanov and A. V. Yukhnevich, *Soviet Physics-Solid State*, **6**, 2965 (1965).
5. A. B. Gerasimov, B. M. Konovalenko and G. L. Éristavi, *Soviet Phys.-Semiconductors*, **1**, 252 (1967).
6. A. V. Yukhnevich, *Soviet Phys.-Solid State*, **7**, 259 (1965).
7. Robert J. Spry and W. D. Compton, *Phys. Rev.*, **175**, 1010 (1968).
8. A. V. Yukhnevich and V. D. Tkachev, *Soviet Phys.-Solid State*, **7**, 2746 (1966).
9. A. V. Yukhnevich and V. D. Tkachev, *Soviet Phys.-Solid State*, **8**, 1004 (1966).
10. M. V. Bortnik, V. D. Tkachev and A. V. Yukhnevich, *Soviet Physics-Semiconductors*, **1**, 290 (1967).
11. Eric S. Johnson and W. D. Compton, *Bull. Am. Phys. Soc. (II)*, **15**, 52 (1970).
12. P. J. Dean, 'Lattices of the Diamond Type', in *Luminescence of Inorganic Solids* (Academic Press, New York, 1966), Chap. 2, pp. 119–203.
13. Robert J. Spry and Jacque D. Henes, *Bull. Am. Phys. Soc. (II)*, **15**, 398 (1970).
14. J. R. Haynes, M. Lax and W. F. Flood *J. Phys. Chem Solids*, **8**, 392 (1959).

15. J. R. Haynes and N. G. Nilsson, 'The Direct Radiative Transitions in Germanium and Their Use in the Analysis of Lifetime', in *Radiative Recombination in Semiconductors* (Academic Press, New York, 1964), pp. 21–31.

16. Colin E. Jones, Ph.D. Thesis, University of Illinois (1970).

17. V. P. Dobrego, S. M. Ryvkin and I. S. Shlimak, *Soviet Phys.-Solid State*, **8**, 1689 (1967).

18. V. P. Dobrego and I. S. Shlimak, *Soviet Phys.-Semiconductors*, **1**, 1231 (1968).

19. P. I. Baranskii, I. D. Konozenko and A. K. Semenyuk, *Soviet Phys.-Semiconductors*, **1**, 958 (1968).

20. D. B. Fitchen, 'Zero-Phonon Transitions', in *Physics of Color Centers*, ed. W. Beall Fowler (Academic Press, New York, 1968), pp. 293–350.

21. R. T. Payne, *Phys. Rev.*, **139**, A570 (1965).

22. R. A. Smith, *Semiconductors* (Cambridge University Press, London, 1961), p. 352.

23. B. Lax, 'Cyclotron Resonance and Magneto-Optical Effects in Semiconductors', in *Semiconductors* (Academic Press, New York, 1963), p. 335.

24. J. J. Hopfield, 'The Quantum Chemistry of Bound Exciton Complexes', in *Physics of Semiconductors* (Academic Press, New York, 1964), pp. 725–735.

25. P. J. Dean, *Journal of Luminescence*, **1, 2**, 398 (1970).

26. Yu. A. Kurskii, *Soviet Phys.-Solid State*, **6**, 1162 (1964).

27. Yu. A. Kurskii, *ibid.*, p. 1795.

28. James W. Corbett, *Electron Radiation Damage in Semiconductors and Metals* (Academic Press, New York, 1966), pp. 95–143.

29. N. A. Vitovskii, T. V. Mashovets, S. M. Ryvkin and V. P. Sondaevskii, *Soviet Phys.-Solid State*, **3**, 727 (1961).

30. N. A. Vitovskii, B. M. Konovalenko, T. V. Mashovets, S. M. Ryvkin and I. D. Yaroshetskii, *Soviet Phys.-Solid State*, **5**, 1338 (1964).

DISCUSSION

Question (MITCHELL) What was your value of S in equation (3)?

Answer (SPRY) We borrowed the value of $S = 2.3$ obtained by Jones for both sets of luminescence bands found in irradiated Czochralski-grown silicon in order to obtain an estimate of the temperature dependence of the relative intensity of the zero phonon band.

Question (MITCHELL) But have you calculated your own value of S from your data?

Answer (SPRY) No, because we have not yet measured the relative areas under the zero-phonon band and the phonon-assisted sidebands.

Question (WATKINS) Why do you believe that there is an unknown impurity associated with the defect producing the new post-irradiation spectra?

Answer (SPRY) These spectra were obtained from samples cut from only five of nine boules used in this work, but there was no correlation with the pre-irradiation sign of majority carrier, carrier concentration, or impurity content from mass spectrographic analysis.

Question (WATKINS) Was there any correlation with the sources of your boules?

Answer (SPRY) Samples from three of five boules from Semi-Elements, Inc. demonstrated the new spectra, but there was no correlation with any characteristic which could be provided by this company. Samples from one boule grown in a nitrogen atmosphere and one boule grown in a helium atmosphere by Dr. R. N. Hall of General Electric always demonstrated the new spectra after room temperature irradiation, but samples from two boules grown in a hydrogen atmosphere did not.

ON THE DONOR-VACANCY TYPE COMPLEXES IN GERMANIUM

V. V. EMTZEV, T. V. MASHOVETS, M. MAXIMOV AND N. A. VITOVSKII

A. F. Ioffe Physico-Technical Institute of the Academy of Sciences of the U.S.S.R., Leningrad

Very unusual energy spectrum and defect production kinetics have been observed under ^{60}Co-γ-irradiation and heat-treatment of high-pure germanium. The results can be explained in terms of vacancy-donor and divacancy-donor complexes. The upper limit of the vacancy migration energy in germanium has been estimated as 0.3 eV.

1. INTRODUCTION

Very unusual energy spectrum and defect production kinetics have been observed under ^{60}Co-γ-irradiation and heat-treatment† of high-pure germanium. The results can be explained in terms of vacancy-donor and divacancy-donor complexes.[1-4] The upper limit of the vacancy migration energy in germanium has been estimated as 0.3 eV.

2. EXPERIMENTAL RESULTS

2.1. *Initial material*

To analyse the nature and concentrations of the impurities in the initial material the Hall-coefficient temperature dependence, the photo-voltaic-effect spectral distribution, the infra-red absorption and the lithium drift data were used.

It appeared, that the concentration of the impurities inserting the deep energy levels into the forbidden gap did not exceed 10^{11} cm^{-3}. The oxygen concentration was $(3-5).10^{16}$ cm^{-3} mainly in the form of GeO$_4$. The dislocation density was $\sim 10^3$ cm^{-2}.

Careful comparison of the experimental Hall data with the results given by the computer showed that the total donor concentration was $(1-2).10^{13}$ cm^{-3} and that they were compensated by the group III acceptors up to 20–80 per cent.

2.2. *Energy spectra*

The main characteristic feature of the spectra observed was that no energy levels appeared in the forbidden gap under gamma-irradiation or heat treatment which could account for the compensation.

In spite of this fact exact compensation took place

---

† The specimens were covered with gold before heat treatment to prevent impurity diffusion into the volume. Heating was followed by quenching in oil. The layer, containing gold, was removed before the measurements.

and the $n-p$ conversion was observed. There were no energy levels in the upper half of the forbidden gap at all. In the lower half the energy levels $E_V + 0.01$ eV and $E_V + 0.09$ eV appeared. These levels began to appear only after the compensation had been completed and that is the reason why these levels cannot be regarded as the compensating levels. It seems to be for the first time, that the electron concentration decreased but no energy levels in the forbidden gap could be responsible for it.

The typical temperature dependence of the hole concentration in the $n-p$-converted material and the corresponding energy scheme are given in Figure 1. They are very similar after gamma-irradiation and heat treatment.

Two concentrations can be found precisely from

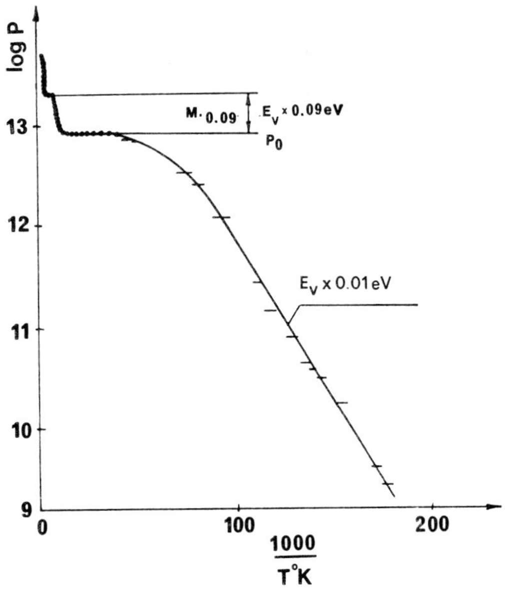

FIG. 1. Temperature dependence of the hole concentration for the irradiated material and the defect energy spectrum.

the curves of the type shown in Figure 1: n—the electron concentration in the impurity exhaustion temperature region (before complete compensation) and $M_{0.09}$—the concentration of the levels $E_V + 0.09$ eV (after $n - p$-conversion). The total concentration of the level $E_V + 0.01$ eV could be only approximately estimated. For this reason further description will concern with only n and $M_{0.09}$. The experimental values of these concentrations will be compared with the calculated ones given by a computer.

2.3. *Creation kinetics of radiation defects*

The irradiation was carried out at two different temperatures (77 °K and 282 °K) and two different intensities of the irradiation were used: $I = 1.0 \times 10^{12}$ cm^{-2} sec^{-1} and $I = 1.4 \times 10^{11}$ cm^{-2} sec^{-1}.

The dose dependences of n and $M_{0.09}$ are shown in Figure 2. The main features of the dependences are:

1. n decreases linearly from the beginning of irradiation till the moment of the exact compensation (φ_1).

2. There is an interval of the doses ($\varphi_1 - \varphi_2$) in which no changes can be noticed, the material remaining 'intrinsic'.

3. After heavier doses ($\varphi > \varphi_2$) the energy levels $E_V + 0.09$ eV arise.

Their rise is superlinear immediately after φ_2, then it becomes nearly linear and at last stops.

4. The maximum value of the $dM_{0.09}/d\varphi$ is about 0.3–0.5 of the $dn/d\varphi$.

5. The value of $dn/d\varphi$ does not depend on either the intensity of the irradiation I or the temperature T. The value of $dM_{0.09}/d\varphi$ does not depend on I but increases as the temperature rises.

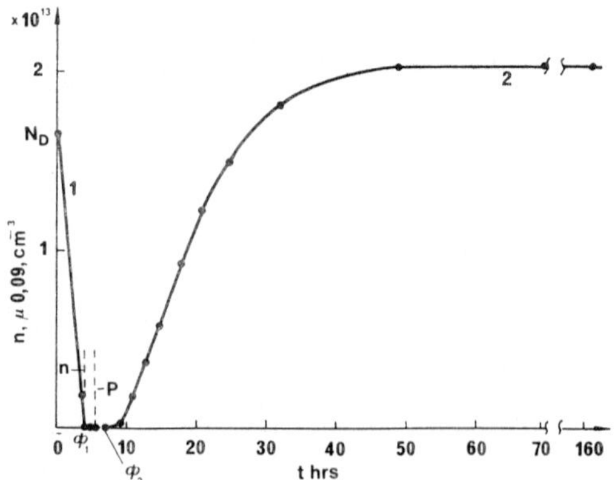

FIG. 2. Dose dependences of the concentrations: 1-n, 2-$M_{0.09}$.

6. The saturation value of $M_{0.09}$ does not depend on either I or T.

2.4. *Creation kinetics of thermal defects*

Figures 3 and 4 show the dependences of n and $M_{0.09}$ on the duration of the heat treatment at two different temperatures (660 °C and 870 °C).

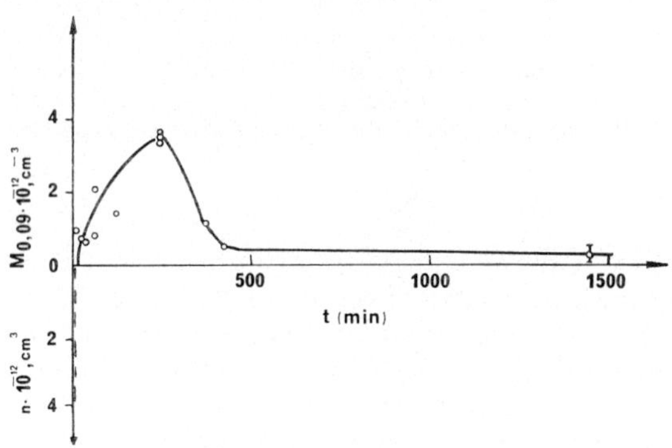

FIG. 3. Thermal defect creation kinetics at 660 °C.

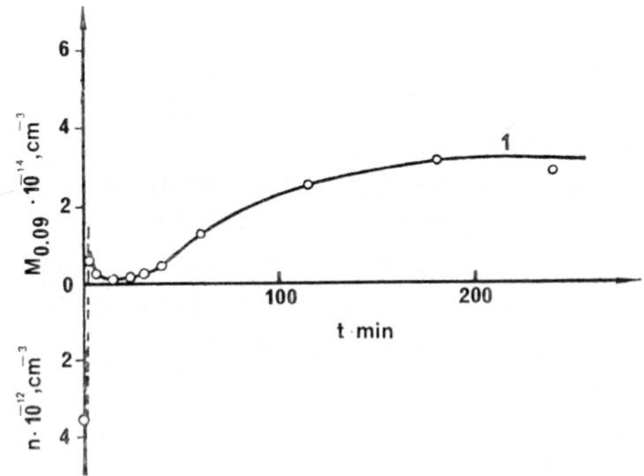

FIG. 4. Thermal defect creation kinetics at 870 °C.

The kinetics of defect production is rather complicated. The process starts with a very rapid decrease of the electron concentration. After the compensation process has been completed the $M_{0.09}$ concentration begins to rise. Then the $M_{0.09}$ concentration passes through a maximum and decreases down to unmeasurable values, the exact compensation recovering. If the temperature is sufficiently high (Figure 4), the $M_{0.09}$ concentration increases after the minimum once again and reaches some limit, which depends on the quenching temperature.

2.5. *Temperature dependence of stationary defect concentrations*

The temperature dependence of the stationary concentration of the defects arising during the initial stages of heat treatment (until $n-p$-conversion took place) and of the $M_{0.09}$ concentration are given in Figures 5 and 6 respectively. The activation energies of the formation process are: for the first type defects—2.2 eV, for the second type ($M_{0.09}$)—3.1 eV.

3. DISCUSSION

The similarity of the energy spectrum of the defects produced by gamma-irradiation and heat

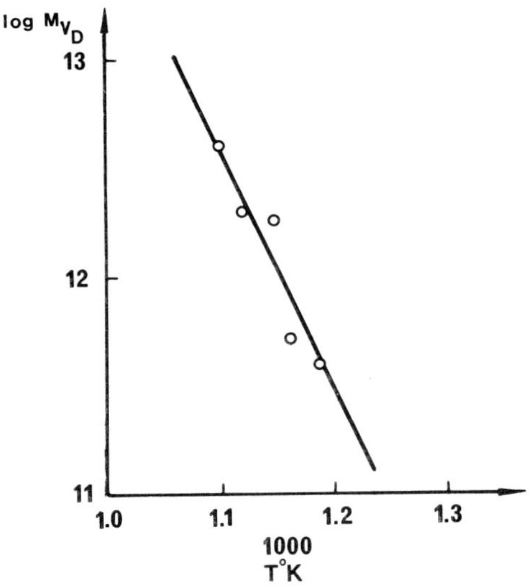

FIG. 5. Temperature dependence of the stationary VD-complexes concentration.

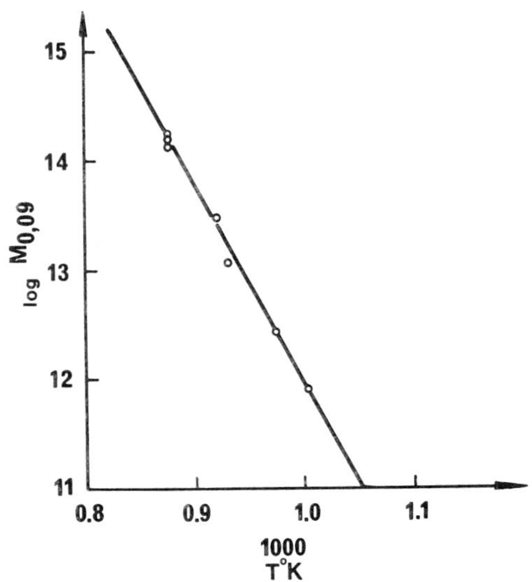

FIG. 6. Temperature dependence of the stationary $M_{0.09}$ concentration.

treatment gives opportunity to develop a general model of the defect production processes in the investigated material.

As the interstitials and vacancies in germanium are mobile at low temperatures,[5] the generating Frenkel pairs mainly recombine. Some of them (a few per cent at room temperature)[6] split however and the vacancies, which become free, can be associated with some impurities or other defects and give rise to come complexes, stable at room temperature.

As it was mentioned above, no energy levels can account for the compensation and we cannot but conclude that it is the disappearance of the donor electrical activity, that causes the compensation.[6] This may result from the formation of the vacancy-donor (VD) complexes.

It was assumed that the centres possessing the $E_V + 0.09$ eV levels can be one of the three following types:

1. Complexes: Vacancy + an unknown impurity X-(VX). The X-impurity can be of any type-oxygen, group III acceptors (A) etc.
2. Free divacancies (VV).
3. Complexes: Group V donor + two vacancies (in either VVD or VDV configuration).

3.1. *Gamma-irradiation*

The equations for the kinetics of the concentrations of free electrons n, free donors M_D, free vacancies M_V, complexes M_{VD} and either complexes M_{VX} or divacancies M_{VV} or complexes M_{VVD} (or M_{VDV}) were written. These equations were combined in the three respective systems.

The solutions of the systems for the cases of complexes VX and divacancies given by a computer for the parameters which give the best approximation to the experiment are shown in Figures 7 and 8.

The experimental values of n and $M_{0.09}$ are given by the points in both the figures. Figure 7 shows that the calculations give reasonable approximation of the experiment to the following items:

1. The dose dependences of n (both calculated and experimental) coincide (curve 1).
2. The calculated ratio of the complexes VX production rate to the electron removal rate (see curves 2 and 1) is nearly equal to the experimental value (see the points).
3. The calculated time at which the curve 2 saturates (the concentration M_{VX} reaches its limit) is not far from the experimental one.

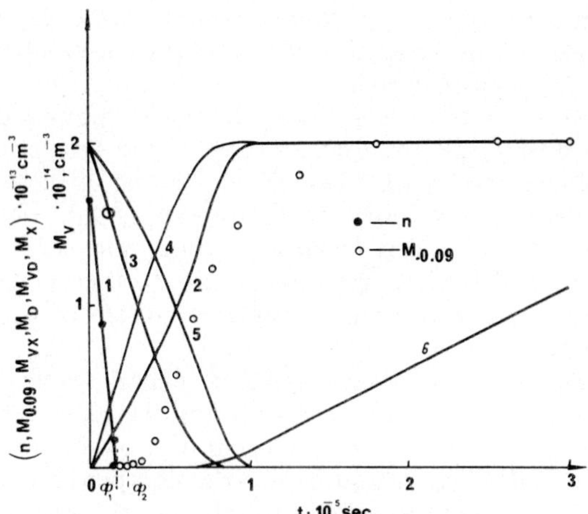

FIG. 7. Concentrations 1-n, 2-M_{VX}, 3-M_D, 4-M_{VD}, 5-M_X, 6-M_V, calculated as the function of the irradiation time under the assumption that VX-complexes are formed. The points—experimental values of n and $M_{0.09}$. $1 = 10^{12}\ cm^{-2}\ sec^{-1}$; $M_{D0} = 2.10^{13}\ cm^{-3}$; $N_A = 3.5 \times 10^{12}\ cm^{-3}$; $M_{X0} = 2 \times 10^{13}\ cm^{-3}$; $\beta_X = 0.4$; $\beta_V = 0$.

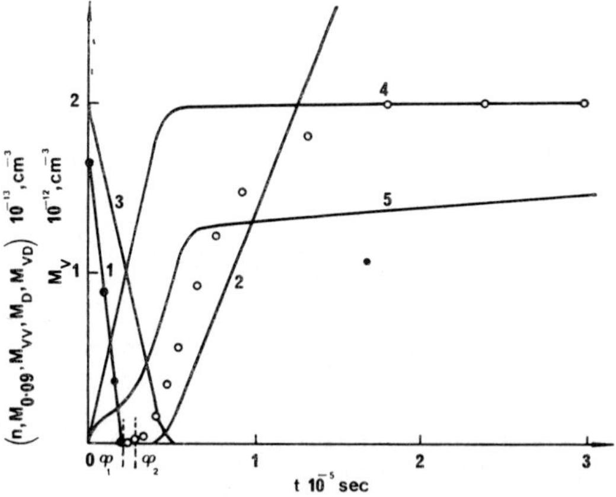

FIG. 8. Concentrations 1-n, 2-M_{VV}, 3-M_D, 4-M_{VD}, 5-M_V, calculated as the functions of the irradiation time under the assumption that divacancies are formed. The points—experimental values of n and $M_{0.09}$. $1 = 10^{12}\ cm^{-2}\ sec^{-1}$; $M_{D0} = 2.10^{13}\ cm^{-3}$; $N_A = 3.5 \times 10^{12}\ cm^{-3}$; $\beta_V = 0.8$; $\beta_X = 0$.

4. The limit of $M_{0.09}$ in this model is equal to the concentration M_X and must not depend on either irradiation intensity or temperature. This was observed in the experiments.

Nevertheless the calculated and experimental results are not in sufficiently good agreement: the calculations predict a marked increase of the concentration $M_{0.09}$ immediately after the beginning of irradiation. In the experiment the $M_{0.09}$ concentration becomes apparent only after the dose φ_2 and rises very slowly at the beginning. This disagreement can be demolished by changing the values of the parameters in the equations, but in this case the agreement with respect to the items 2 and 3 cannot be obtained.

Figure 8 shows that:

1. The calculated and experimental dose dependences of n—coincide.

2. The divacancy concentration reaches measurable values only a long time after the beginning of irradiation (after the dose φ_2).

3. The calculated ratio of the divacancies production rate to the electron removal rate (curves 2 and 1) is equal to the experimental value (see the points).

The disagreement between this model and the experimental results is that the model predicts an infinite increase of the divacancy concentration even if their dissociation is taken into account. This disagreement cannot be removed by the changing parameter values. The limitation of the divacancy concentration takes place if free vacancies can disappear (e.g. on surfaces or dislocations) but under such circumstances the limit of $M_{0.09}$ must strongly depend on irradiation intensity and temperature which is not the case, as the experiment shows.

From what was said above one can conclude that the observed energy levels $E_V + 0.09$ eV can belong neither to the VX-complexes nor to the free divacancies. Nevertheless both models possess some positive features and it seems reasonable to try to combine them.

The following combined model was considered.

1. At the first stages of the irradiation the dominant process is the VD-complex formation.

2. These VD-complex can accept new free vacancies thus becoming the complexes: group V donor + 2 vacancies (VVD or VDV).

The concentration of these complexes started increasing markedly only after the essential part of the donors had turned into VD-complexes, especially after $n-p$-conversion had taken place and kept increasing until practically all the donors turned into VVD or VDV complexes. The total concentration of VVD and VDV complexes cannot exceed the initial donor concentration M_{D_0} and does not depend on either irradiation intensity or temperature.

The system of the Eqs. for this model is:

$$1. \quad \frac{dM_V}{dt} = \sigma N_{ge} I - \alpha M_V (\beta_D M_D + \beta_{VD} M_{VD}) + C M_{VD}$$

$$2. \quad \frac{dM_D}{dt} = -\alpha \beta_D M_V M_D$$

$$3. \quad \frac{dM_{VD}}{dt} = \alpha \beta_D M_V M_D - \alpha \beta_{VD} M_V M_{VD} + C M_{VVD}$$

$$4. \quad \frac{dM_{VVD}}{dt} = \alpha \beta_{VD} M_V M_{VD} - C M_{VVD}$$

$$5. \quad h = M_D - N_A - 2 M_{VVD} - \frac{M_{VD}}{1 + \gamma \exp\left(-\dfrac{\Delta E_{VD} + \mu}{kT}\right)}$$

$$(1)$$

Here σ—effective cross section for the generation of free vacancies, $\sigma = 1 \times 10^{-26}$ cm^{-2}, as in reference (6). $N_{ge} = 4.45 \times 10^{22}$ cm^{-3}—germanium concentration, N_A—group III acceptors concentration, I—irradiation intensity (1.0×10^{12} cm^{-2} sec^{-1} or 1.4×10^{11} cm^{-2} sec^{-1}. C—the time constant for the divacancy dissociation, was set to be 1.5×10^{-6} sec^{-1} (the experimental value for the annealing of $M_{0.09}$ at room temperature), α—the coefficient which is proportional to probability of a vacancy to meet some other centre—it is the volume which a vacancy traverses per second,[6] $\alpha \approx 10^{-16}$ cm^3 sec^{-1} (see below), β_D, β_{VD}—the probabilities for a vacancy to be trapped when it meets a donor or a VD-complex respectively. β_D was set to be unity,[6] β_{VD} was altered to achieve the coincidence of the calculated and the experimental results. μ—the Fermi level. $\Delta E_{VD} = 0.35$ eV—the activation energy of the E-centre type acceptor levels (by the analogy to Si it was assumed that they can exist in the middle of the forbidden gap, though they were not found). γ—degeneration factor which is unknown for so deep levels and was set to be unity.

The initial conditions are:

$$M_V = M_{VD} = M_{VVD} = 0; \quad M_D = M_{D_0} \quad \text{if } t = 0.$$

The experimental values of M_{D_0} and N_A were put into the Eqs.

The β_{VD} parameter was altered to reach the coincidence of the calculated curve and the experimental points. The results are shown in Figures 9 and 10 for the temperatures 282 °K and 77 °K respectively. One can see that the best coincidence occurs if $\beta_{VD} = 0.1$ at 282 °K and $\beta_{VD} \approx 0.001$ at 77 °K. This leads one to a conclusion that the threshold energy for a vacancy to be trapped by VD-complex (or the second vacancy to be trapped by a donor D) is equal to 0.05 eV, assuming that it is some threshold that limits the value of β_{VD}.

Now we shall consider the parameter α, which

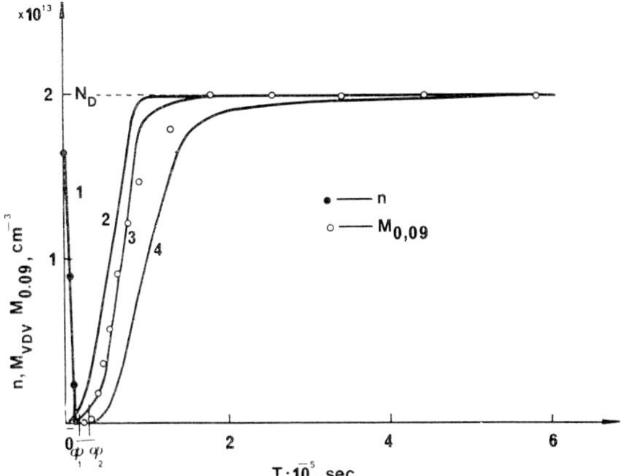

FIG. 9. Calculated dose dependences of the concentrations: 1-n, 2-4-M_{VVD}. β_{VD} for curves 2-1.0; 3-0.1; 4-0.01. $T = 282$ °K, $1 = 10^{12}$ cm^{-2} sec^{-1}; $M_{D_0} = 2 \times 10^{13}$ cm^{-3}; $N_A = 3.5 \times 10^{12}$ cm^{-3}. The points—experimental values of n, $M_{0.09}$.

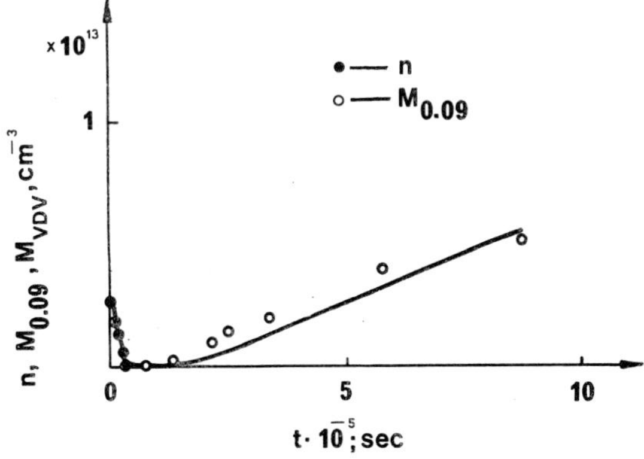

FIG. 10. Calculated dose dependences of the concentrations: 1-n, 2-$M_{0.09}$. $T = 77$ °K, 1-1.4×10^{11} cm^{-2} sec^{-1}; $M_{D_0} = 8 \times 10^{12}$ cm^{-3}; $N_A = 5.5 \times 10^{12}$ cm^{-3}; $\beta_{VD} = 1 \times 10^{-3}$. The points—experimental values of n and $M_{0.09}$.

was not altered but was estimated. The parameter α is a characteristic of the vacancy movement speed, or to be more exact, of the volume which is traversed by a vacancy per unit time.

The value of α can be estimated either as $\alpha \simeq a^2\bar{v}$ or as $\alpha \simeq a^3 v \exp(-E_V^M/kT)$, where a—is the lattice constant, $\bar{v}$—average vacancy speed, v—lattice frequency. Both approximations give α of the order of magnitude

$$(10^{-9} \div 10^{-10}) \exp\left(-\frac{E_V^M}{kT}\right)$$

Experimental dose dependences were compared with the calculated ones for different values α. The results are shown in Figure 11. It can be seen, that if $\alpha > 10^{-16}\,\mathrm{cm^3\,sec^{-1}}$, all the calculated dependences are in good enough agreement with the experiment. The conditions, which must be satisfied to

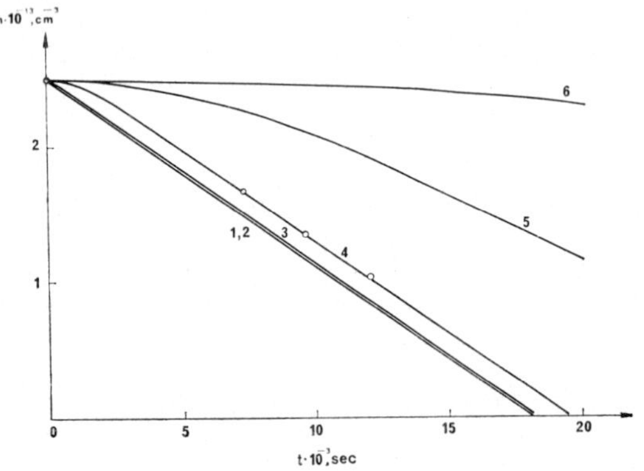

FIG. 11. Calculated dose dependence of the electron concentration for different values of parameter α:

1-$\alpha = 10^{-12}\,\mathrm{cm^3\,sec^{-1}}$ 4-$10^{-16}\,\mathrm{cm^3\,sec^{-1}}$
2-$10^{-14}\,\mathrm{cm^3\,sec^{-1}}$ 5-$10^{-17}\,\mathrm{cm^3\,sec^{-1}}$
3-$10^{-15}\,\mathrm{cm^3\,sec^{-1}}$ 6-$10^{-18}\,\mathrm{cm^3\,sec^{-1}}$
The points—experimental values of n.

observe the linear changes in n with the dose increasing were regarded in reference (6). It can be shown that if these conditions are satisfied, n does not depend on α. Figure 11 shows, that if $\alpha > 10^{-16}\,\mathrm{cm^3\,sec^{-1}}$ the linearity of n and the independence of $dn/d\varphi$ on α occur. It gives $\alpha = 10^{-16}\,\mathrm{cm^3\,sec^{-1}}$ as the lower limit of α and consequently the upper limit for E_V^M is $E_V^M \leqslant 0.3$ eV, which corresponds to that given in reference (7).

3.2. *Heat treatment*

As the energy spectra of the defects arising under γ-irradiation and heat treatment are the same, one can suppose that these defects are of the same nature.

Now we shall consider some temperature regions, in which different processes are the most important.

570°–640 °C

Only VD-complexes are produced. The equilibrium concentration M_{VD} is less than N_D and the n-type conductivity persists. The activation energy for VD-complex formation, found from the temperature dependence of their equilibrium concentration, is $E_{VD}^F = 2.2$ eV; $E_{VD}^F = E_V^F + E_V^M$. If $E_V^F = 2.0$ eV[8] and, therefore $E_V^M \simeq 0.2$ eV, which coincides with the value given in reference (7).

660°–700 °C

The equilibrium vacancy concentration increase leads to the $n - p$ conversion and to the creation of the VDV or VVD complexes ($M_{0.09}$ grows). The maximum in $M_{0.09}$ is a striking feature of the kinetics. The extraction of the complexes from the specimen may cause $M_{0.09}$ to diminish. This assumption was confirmed by a special experiment, which showed, that the specimens, heated for a long time could by no means be turned into the n-type-material again.

700°–870 °C

Two great difficulties arise in accounting for the defect creation processes at these temperatures (see Figure 4).

(a) $M_{0.09}$ increases after the minimum.
(b) The $M_{0.09}$ values in both maxima are greater than the donor concentration.

It means that the defects forming at these temperatures cannot be VVD or VDV complexes, though the energy levels of the former are the same as those of the latter.

The following speculative considerations were used with respect to the mentioned difficulties.

(a) The group $\overline{V}$ atoms can be regarded as a catalizer of the vacancy coagulation process. As soon as two vacancies unite with the group $\overline{V}$ atom, they form a divacancy. This divacancy can liberate and migrate over the crystal, the group $\overline{V}$ atom becoming free and able to accept new single vacancies.

(b) It is necessary to assume that the energy spectrum of a free divacancy is essentially the same as that of a complex $-$ donor $+$ two vacancies.

Under these assumptions the activation energy taken from the temperature dependence of the stationary $M_{0.09}$ concentrations (3.1 eV) can be regarded as a divacancy formation energy E_{VV}^F. This energy is equal to $E_{VV}^F = 2E_V^F - E_{VV}^B$, where E_{VV}^B is divacancy binding energy. If $E_V^F = 2.0$ eV, $E_{VV}^B = 0.9$ eV. This value is in good accordance with the activation energy for the $VDV \rightarrow VD + V$ process which was found in reference (6) (0.96 eV) and with the E_{VV}^B value given in reference (8).

ACKNOWLEDGEMENTS

The authors are very much indebted to Prof. S. M. Ryvkin for his kind attention and interest to the work and to Mrs. N. A. Gunko for her help in computer calculations.

REFERENCES

1. G. D. Watkins and J. W. Corbett, *Phys. Rev.*, **134,** A1359 (1964).
2. G. D. Watkins, *Radiation Effects in Semiconductors*, Proc. of the Santa Fe Conf. (Plenum Press, New York, 1968), p. 67.
3. J. C. Pigg and J. H. Crawford, Jr., *Phys. Rev.*, **135,** A1141 (1964).
4. R. R. Hasiguti, K. Tanaka and S. Takahashi, *Radiation Effects in Semiconductors*, Proc. of the Santa Fe Conf. (Plenum Press, New York, 1968), p. 89.
5. J. W. MacKay and E. E. Klontz, *Radiation Effects in Semiconductors*, Proc. of the Santa Fe Conf. (Plenum Press, New York, 1968), p. 175.
6. N. A. Vitovskii, M. Maximov and T. V. Mashovets, *Sov. Phys. Semicond.*, **v 4,** N. 6 (1970); **v 4,** N. 11 (1970).
7. R. E. Whan, *Phys. Rev.*, **140,** A690 (1965).
8. A. Seeger and K. P. Chik, *Phys. Stat. Sol.*, **29,** 455 (1958).

RECOMBINATION PARAMETERS IN LOW-RESISTIVITY GAMMA-IRRADIATED n-TYPE GERMANIUM†

J. R. SROUR AND O. L. CURTIS, JR.

Northrop Corporate Laboratories, Hawthorne, California

Studies of recombination in ~0.2 Ω-cm As- and Sb-doped Co^{60} γ-irradiated Ge which yield energy levels and the temperature dependence of the electron and hole capture probabilities are reported. For Sb-doped material at $323°K$, the recombination center energy level position (neglecting statistical weight) was found to be 0.361 ± 0.005 eV above the valence band with a possible slight temperature dependence corresponding roughly to one-half the variation of band gap with temperature. The capture probability ratio at this same temperature was 740. For the As-doped case, two different levels appear to dominate the recombination process in annealed and unannealed low resistivity material. The energy level positions relative to the valence band (neglecting statistical weight) are 0.327 ± 0.005 eV and 0.37 ± 0.01 eV at room temperature for the annealed and unannealed samples, respectively. The corresponding capture probability ratios are 650 and 810. As in the case of Sb-doping, the energy level appears to shift with temperature at about one-half the rate of the shift in band gap energy.

1. INTRODUCTION

Recombination in n-type Ge irradiated by Co^{60} γ-rays has been studied extensively.[1-7] However, little has been reported on low resistivity ($\lesssim 0.5$ Ω-cm) material. Such studies will be shown to be particularly informative. A primary goal of this work was to investigate in some detail capture probability temperature dependences and to obtain reliable information about the capture probability ratio, c_p/c_n. Coupled with data from higher resistivity samples, this should yield accurate positions for energy levels associated with the radiation-induced defects. The principal experimental techniques used were measurement of the variation of minority carrier lifetime with temperature (at low excess density) and with excess density (using temperature as a parameter). In the course of the study, improved techniques were developed for utilizing these concurrent measurements in the determination of recombination parameters.

2. EXPERIMENTAL

Measurements of minority carrier lifetime as a function of temperature at low excess density were performed using experimental procedures described previously.[8] The method involves observation of photoconductivity decays. Lifetime as a function of excess density, with temperature as a parameter,

was measured using a modification of a method also described elsewhere.[9] Samples were cut from As-doped and Sb-doped single crystal Ge ingots obtained from Nucleonic Products Co. Ultrasonic soldering techniques were used to prepare ohmic contacts. The sample designation system used in the text, figures, and tables is as follows: NP refers to the manufacturer, Sb and As refer to the dopant, and the numbers give the resistivity in Ω-cm.

After pre-irradiation measurements, irradiations were performed at dry ice temperature in a Co^{60} source (total dose ~5.3×10^6 R). Measurements of lifetime as a function of temperature were made for $1000/T \geqslant 3.3$ $°K^{-1}$ after approximately one hour at room temperature (to achieve defect stability) for both Sb- and As-doped samples (referred to here as the 'unannealed' case). Additional measurements were performed after $\frac{1}{2}$ hour anneals at $72°C$ and $84°C$ ('annealed' case). Lifetime as a function of excess density was measured successively at $1000/T = 3.3$, 3.6, and 3.9 $°K^{-1}$. All excess density measurements were made on samples that were identical to those used in the measurements of lifetime as a function of temperature. After anneals at $72°C$ and $84°C$, further excess density measurements were performed ($1000/T = 2.9$, 3.1, 3.3, 3.6, and 3.9 $°K^{-1}$).

3. RESULTS

Results were analyzed using Hall[10]–Shockley–Read[11] (HSR) recombination statistics. For n-type material (assuming $\Delta n = \Delta p$), the general expression

† Work supported by Harry Diamond Laboratories and the Defense Atomic Support Agency. A more comprehensive account of this work appears in Phys. Rev. **B2**, 4977 (1970).

E

for minority carrier lifetime, τ, due to a single recombination level, can be written as

$$\tau = \tau_0(1 + \Delta n/b)/(1 + \Delta n/n), \qquad (1)$$

where

$$b = [c_p(p + p_1) + c_n(n + n_1)]/(c_p + c_n). \qquad (2)$$

In these expressions, all terms have their usual meaning, and τ_0, the low-excess-density lifetime is given by

$$\tau_0 = [(p + p_1)/c_n Nn] + [(n + n_1)/c_p Nn]. \qquad (3)$$

The quantity b can be conveniently used in the analysis of measurements of lifetime as a function of excess density. A digital computer was, in most cases, used to perform least squares fits to the excess density data. The primary yield of this curve-fitting procedure was values for the parameter b.

In order to separate radiation-induced properties from those present before irradiation, pre-irradiation lifetime was subtracted reciprocally from measured lifetime to yield the radiation-induced component. Figure 1 shows the behavior of

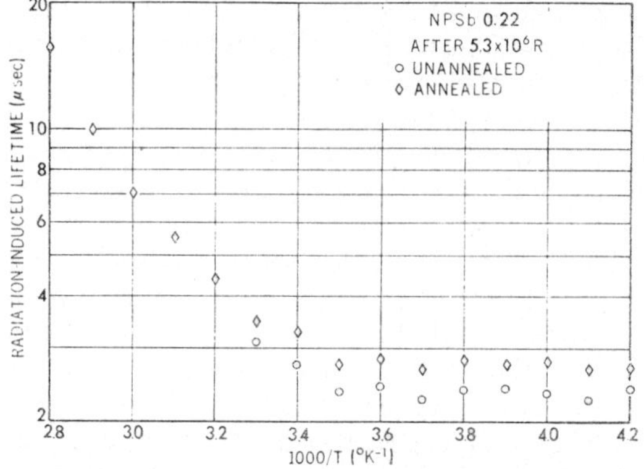

FIG. 1. Radiation-induced lifetime vs. reciprocal temperature for 0.22 Ω-cm Sb-doped Ge (unannealed and annealed cases). The unannealed case refers to samples that were kept at room temperature for $\sim$one hour (to achieve defect stability) before performing measurements at $1000/T \geqslant 3.3\,°K^{-1}$. The annealed case refers to samples that were annealed for 1/2 hour at both $72\,°C$ and $84\,°C$ before performing measurements at $1000/T \geqslant 2.8\,°K^{-1}$.

lifetime as a function of temperature for Sb-doped material. Typical lifetime as a function of excess density curves (annealed case) for this material are

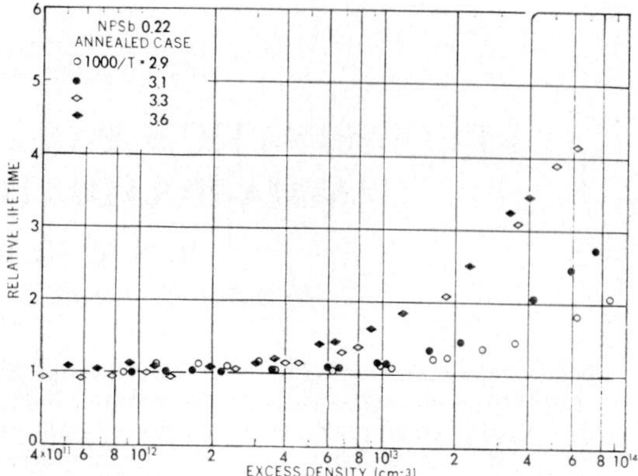

FIG. 2. Relative lifetime vs. excess density for 0.22 Ω-cm Sb-doped Ge with temperature as a parameter (annealed case).

shown in Figure 2 for four temperatures. (For clarity, the $1000/T = 3.9$ curve is omitted because it approximately overlaps the 3.3 and 3.6 curves.) Similar data were obtained for As-doped material. Table I lists b values obtained from fitting curves to the excess density data.

3.1. Sb-*doped material*

Measurements indicate only slight annealing after successive one-half hour anneals at $1000/T = 2.9$ and $2.8\,°K^{-1}$, so the data will be treated simultaneously. Consideration of Eq. (2) and b values obtained from excess density data can yield a conclusion concerning relative magnitudes of the capture probabilities. Since the equilibrium electron concentration, n, is equal to 1.06×10^{16} cm³, we conclude that $c_p \gg c_n$ at all of the temperatures studied.

The temperature dependence data (Figure 1) can be analyzed using Eq. (3), which contains three significant terms: $p_1/c_n Nn$, $1/c_p N$, and $n_1/c_p Nn$. Figure 1 shows that between $1000/T = \sim 3.5$ and $4.2\,°K^{-1}$ lifetime is insensitive to temperature. Two of the terms should exhibit a strong temperature dependence, leading to the tentative conclusion that the $1/c_p N$ term dominates, and thus, that c_p is approximately independent of temperature. The conclusion is tentative because it is necessary to examine the possibility of temperature dependence cancellation in either of these terms. For brevity, we omit the argument but it can be conclusively demonstrated that such a fortuitous cancellation is inconsistent with the present data, and thus that c_p is $\sim$temperature independent.

TABLE I
Values for the parameter b as a function of temperature for Sb- and As-doped Ge.
Values with an asterisk are considered less reliable due to fairly scattered data.

$\dfrac{1000}{T}$ ($^\circ$K^{-1})	Sb-doped (0.22 Ω-cm)		As-doped (0.26 Ω-cm)	
	Unannealed	Annealed	Unannealed	Annealed
2.9		6.59×10^{13}		16.01×10^{13}
3.1		2.74×10^{13}		8.53×10^{13}
3.3	1.65×10^{13}	1.48×10^{13}	1.55×10^{13}	3.49×10^{13}
3.6	1.40×10^{13}	1.04×10^{13}	0.52×10^{13}	1.64×10^{13} *
3.9	1.01×10^{13} *	1.19×10^{13} *	0.33×10^{13}	0.87×10^{13} *

The temperature dependence of c_n can be obtained[12] from the relation

$$c_n = b/Nn\tau_0. \qquad (4)$$

Using the lifetime data of Figure 1 and the b values given in Table I, relative c_n was calculated as a function of temperature and is plotted in Figure 3. Because of the small amount of annealing that occurred (Figure 1), the annealed and unannealed cases are normalized to an equal number of defects N in Figure 3 (for Sb-doped material). For $1000/T \leqslant \sim 3.6$, we can approximate the temperature variation by $c_n = c_0 \exp(-0.07/kT)$, where c_0 is a constant. (This form of the temperature dependence is indicated primarily for comparison purposes, and not necessarily to indicate a physical process.) There is some possibility that the apparent leveling off of c_n at lower temperatures is due to trapping. This might also be the reason for the difference between the b values for the annealed and unannealed cases at low temperatures. The effect of trapping is to increase the observed b values,[13] which would result in an apparently increased c_n.

Detailed analysis of the data of Figures 1 and 2 was performed using the HSR equations and employing the obtained c_n temperature variation and a temperature-invariant c_p. Results of this analysis are given in Table II and Figure 4. Although the slight differences in energy level position obtained could be experimental in origin, they could not arise from a very slight temperature

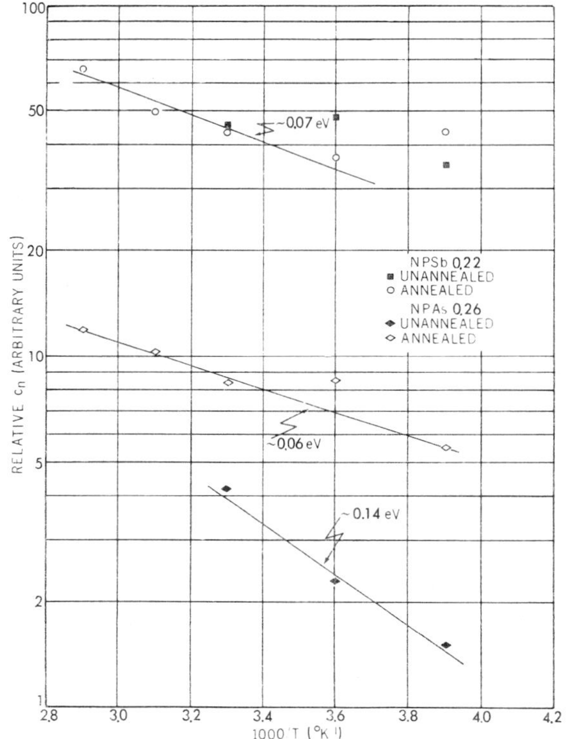

FIG. 3. Relative electron capture probability (c_n) vs. reciprocal temperature for low resistivity Sb- and As-doped Ge. No comparison should be made between any of the curves with regard to relative magnitude. The curves are plotted on the same figure for convenience only.

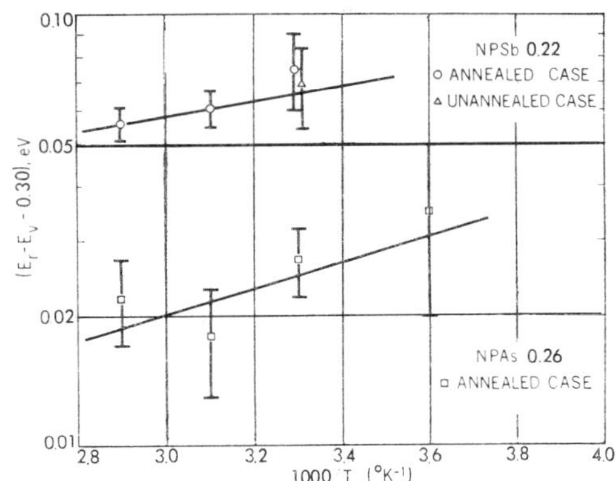

FIG. 4. Recombination center energy level position vs. reciprocal temperature for 0.22 Ω-cm Sb-doped and 0.26 Ω-cm As-doped Ge.

TABLE II

Recombination parameters for As- and Sb-doped material (statistical weight is neglected).

$\dfrac{1000}{T}$ ($°K^{-1}$)	c_p/c_n		$(E_r - E_v)$ eV	
	Unannealed	Annealed	Unannealed	Annealed
As-doped				
2.9	—	510	—	0.322 ± 0.005
3.1	—	540	—	0.318 ± 0.005
3.3	810	650	0.37 ± 0.01	0.327 ± 0.005
3.6	1600	710	—	0.335 ± 0.015
3.9	2670	1060	—	—
Sb-doped				
2.9	—	570	—	0.356 ± 0.005
3.1	—	740	—	0.361 ± 0.005
3.3	850	920	0.369 ± 0.015	0.375 ± 0.015
3.6	760	1020	—	—
3.9	1050	890	—	—

dependence of c_p. An 0.01 eV dependence in c_p would produce errors at the various temperatures of only $\sim$0.002 eV; i.e., if there were such a temperature dependence in c_p, this would raise each apparent energy level value by about 0.002 eV, but would not observably affect the temperature dependence of $E_r - E_v$. However, it is interesting to note that the variation in width of the forbidden gap decreases by roughly twice as great an amount in the same temperature interval.[14] Thus, the actual energy level position could easily be as strong a function of temperature as indicated here. The line indicated through the points of Figure 4 corresponds to one-half the band gap dependence.

In summary, for Sb-doped low-resistivity material, the variation of electron capture probability with temperature was determined and the hole capture probability was found to be nearly independent of temperature. The energy level position (neglecting statistical weight) determined at $1000/T = 3.1\ °K^{-1}$ was found to be 0.361 ± 0.005 eV above the valence band with a possible slight dependence on temperature, corresponding roughly to one-half the variation of band gap with temperature. The capture probability ratio c_p/c_n at this same temperature was 740.

3.2. As-*doped material*

There was approximately no variation of lifetime with temperature for the unannealed case for $1000/T \geqslant \sim 3.5\ °K^{-1}$. Upon re-irradiating to a total dose of 1×10^7 R and annealing for one-half hour at both $1000/T = 2.9$ and 2.8, the lifetime temperature variation changed, exhibiting an increase for $1000/T < \sim 3.8\ °K^{-1}$. Thus, not only do centers anneal, but the physical characteristics of the radiation-induced recombination centers remaining are different than those of the centers which initially dominate.

A consideration of Eq. (2) and the b values, along with the fact that $n = 8.80 \times 10^{15}$ cm^3, again yields the conclusion that $c_p \gg c_n$. Using Eq. (4) in the same manner as for Sb-doped material, the temperature variation of c_n was obtained and is plotted in Figure 3. (No comparison between As-doped and Sb-doped material with regard to relative magnitudes of c_n should be made. They are plotted in the same figure for convenience only.) For the temperature range examined, a relationship of the form $c_n = c_{n0}\exp(-0.14/kT)$ fits the 'unannealed' data reasonably well. For the annealed case, c_n varies as approximately $\exp(-0.06/kT)$, which is approximately the same variation that was obtained for Sb-doped material.

Determination of recombination parameters was made, assuming a temperature-independent c_p, and the results are presented in Table II and Figure 4. In the annealed case, a slight temperature dependence corresponding to one-half the band gap temperature dependence appears reasonable as for Sb-doped material. For the unannealed case, the energy level position at room temperature is determined to be 0.37 ± 0.01 eV above the valence band. For the annealed case, the room temperature result is 0.327 ± 0.005 eV. Thus, the data seem to suggest that the recombination level position changes upon anneal by about 0.04 eV.

In summary, two different recombination levels appear to dominate the recombination process in annealed and unannealed As-doped material. In both cases, the hole capture probability is very nearly temperature independent with the temperature dependence of electron capture probability corresponding to an apparent activation energy of $\sim$0.07 and $\sim$0.14 eV for the annealed and unannealed cases respectively. The energy level positions are (neglecting statistical weight) 0.327 ± 0.005 eV and 0.37 ± 0.01 eV at room temperature for the annealed and unannealed samples respectively. The corresponding capture probability ratios are 650 and 810. As in the case of Sb-doping, the energy level appears to shift with temperature at about one-half the rate of the shift in band gap energy.

4. DISCUSSION

For both As- and Sb-doped material, we have observed that $c_p \gg c_n$, $c_p \neq f(T)$, and that c_n varies as $\sim \exp(-\Delta E/kT)$. (The latter result apparently does not hold for Sb-doped material at the lowest temperature of observation, perhaps due to trapping.) The property $c_p \gg c_n$ implies that the center is negatively charged at equilibrium. Furthermore, the center is probably repulsive to electrons even after capturing a hole, since c_p/c_n is so large. A center that would satisfy these requirements is a double acceptor, doubly negatively charged at equilibrium. The electrostatic force associated with a repulsive center should produce a potential barrier to capture. One might then expect an exponentially temperature dependent c_n for such a center, which would indicate the magnitude of the potential barrier. Since there would be no barrier to hole capture, one should not expect a strongly temperature dependent c_p.

The variation of c_n with temperature observed for both Sb- and As-doped material is qualitatively similar to the temperature variation for slow traps in Si reported by Haynes and Hornbeck.[15] Streetman[4,5] also observed a capture cross section temperature variation of the form $\sigma_n = \sigma_{n0}\exp(-\Delta E/kT)$ for γ-irradiated high resistivity Ge at low temperatures ($< \sim$180 °K). Streetman attributed the lifetime temperature variation at low temperatures to recombination at a center which has a double negative charge at equilibrium (as discussed above), and is also effective for trapping at higher temperatures. Such a doubly-charged double-acceptor site has been discussed by Tyler and Woodbury[16] and Callcott and MacKay.[17] Streetman also examined the effect of increasing majority carrier concentration on recombination in a center that is a trap at higher temperatures, and found the process to be enhanced; i.e., recombina-

tion in this center occurred at increasingly higher temperatures as resistivity decreased.

In the present case, n is approximately one order of magnitude greater than the largest concentration treated by Streetman. Such a large electron concentration can possibly result in enhanced 'trap recombination' to the extent that it is the dominant process even at room temperature and above. Thus, the center that is effective for recombination only at temperatures considerably lower than room temperature in higher resistivity material (Streetman's data) may possibly become the dominant recombination center for low resistivity material for the temperature range examined here. Consideration of previous results for higher resistivity material[7,18] in the light of this possibility is now given.

For Sb-doped material, differences and similarities were found to exist between the results for low and higher resistivity material. The principal difference is the energy level position. For low resistivity material, both the unannealed and annealed cases yielded a position $\sim$0.37 eV above the valence band at room temperature. This position for $\sim$1 and $\sim$5 Ω-cm material was $\sim$0.34 eV at the same temperature. Apparent similarities between low and higher resistivity material include the fact that $c_p \gg c_n$, and the c_n temperature dependence. With regard to c_n, insufficient data exists to obtain its temperature variation for > 1 Ω-cm material.

Two alternate models that can be used to describe the behavior of Sb-doped material exist. The first deduces that the same level is being observed in both low and higher resistivity material. Arguments in support of this model are based upon the premise that trapping influences the higher resistivity results, and not those for low resistivity material. The apparent temperature dependence of energy level positions in higher resistivity material supports this view, as does the c_n variation. The second model assumes that two levels are effective for recombination, one of which may dominate depending on the resistivity of the material being studied. This model arises from the fact that a trap can be converted to a recombination center as the carrier concentration increases, as discussed above. In concept, it seems reasonable to postulate a situation in which one center dominates recombination in high resistivity material and another center, which is effective as a trap in the high resistivity case, dominates recombination at a sufficiently low resistivity. Presumably, there would also be range of resistivities over which both centers were effective for recombination. Many of the results for Sb-doped material support such a model, including the fact that traps present in higher resistivity material are not observed in 0.2 Ω-cm Ge, and that lifetime values tend to be lower than expected from data for higher resistivity samples. If this second model is the correct one, 'trap recombination' dominates in low resistivity Sb-doped material ($\sim$0.2 Ω-cm), and for high resistivity material (considerably greater than $\sim$1 Ω-cm) the other level prevails. In 1 Ω-cm material both centers may be effective for recombination. The present data do not permit a clear choice between the two possibilities given. In fact, the most likely possibility may be one in which both modification of apparent energy level by trapping and enhanced recombination at 'trapping' sites are occurring.

The results for As-doped material seem more readily explainable than the Sb case because of the consistency of the data. The energy level positions obtained in the annealed case for all the higher resistivity results examined agree quite well with that obtained in low resistivity material (0.33 eV at room temperature), indicating that the same center dominates over the entire resistivity range examined. In the unannealed case at low resistivity, a level at $\sim$0.37 eV above the valence band was obtained at room temperature, while for higher resistivity material, the energy level values obtained were intermediate between this value and 0.33 eV. These results again suggest two recombination levels. Thus, we postulate that for As-doped material one level predominates in the annealed case for all resistivities, and that the other level

predominates in the low-resistivity unannealed case ('trap recombination'). For the higher-resistivity unannealed cases examined, both levels are presumably effective for recombination.

REFERENCES

1. O. L. Curtis, Jr. and J. H. Crawford, Jr., *Phys. Rev.*, **124**, 1731 (1961).
2. O. L. Curtis, Jr. and J. H. Crawford, Jr., *Phys. Rev.*, **126**, 1342 (1962).
3. O. L. Curtis, Jr. and J. H. Crawford, Jr., in *Radiation Damage in Semiconductors* (Dunod, Paris, 1965), p. 143.
4. B. G. Streetman, *J. Appl. Phys.*, **37**, 3137 (1966).
5. B. G. Streetman, *J. Appl. Phys.*, **37**, 3145 (1966).
6. O. L. Curtis, Jr. and C. A. Germano, in *Radiation Effects in Semiconductors* (Plenum Press, New York, 1968), p. 331.
7. C. A. Germano and O. L. Curtis, Jr., *IEEE Trans. Nucl. Sci.*, **15**, 34 (December 1968).
8. O. L. Curtis, Jr. and R. C. Wickenhiser, *Proc. IEEE*, **53**, 1224 (1965).
9. C. A. Germano and O. L. Curtis, Jr., *IEEE Trans. Nucl. Sci.*, **13**, 47 (1966).
10. R. N. Hall, *Phys. Rev.*, **87**, 387 (1952).
11. W. Shockley and W. T. Read, Jr., *Phys. Rev.*, **87**, 835 (1952).
12. J. R. Srour and O. L. Curtis, Jr., *J. Appl. Phys.* **41**, 4200 (1970).
13. O. L. Curtis, Jr., *Phys. Rev.*, **172**, 773 (1968).
14. G. G. MacFarlane, T. P. McLean, J. E. Quarrington and V. Roberts, *Phys. Rev.*, **108**, 1377 (1957).
15. J. R. Haynes and J. A. Hornbeck, *Phys. Rev.*, **100**, 606 (1955).
16. W. W. Tyler and H. H. Woodbury, *Phys. Rev.*, **102**, 647 (1956).
17. T. A. Callcott and J. W. MacKay, in *Radiation Damage in Semiconductors* (Dunod, Paris, 1965), p. 27.
18. O. L. Curtis, Jr., R. F. Bass, C. A. Germano and J. R. Srour, *Radiation Effects in Silicon and Germanium*, Harry Diamond Laboratories Report No. 235–4 (AD 685883) (1968).

DISCUSSION

Comment (STREETMAN) It is surprising that a double-negative center near the middle of the band gap should be the most efficient centre present in irradiated material.

Reply (SROUR) Based on your data (Ref. 4), we feel that the center effective for recombination at room temperature in low resistivity material may be the same one which acts as a trap for higher resistivity material. Recombination at this trapping center becomes possible because of the large majority carrier concentration present.

Question (ALBANY) Did you observe other acceptor levels, [such as $E_v - (\sim 200$ mev)] or $E_v - (\sim 100$ mev)] with increasing dose?

Answer (SROUR) We have not performed detailed recombination studies at other doses, but we have observed a shallower level in studies at high injection levels.

Answer (CURTIS) The 0.2 eV level you mention is observed at high excess densities, both in bulk material and in devices. However, at low excess densities we observe only a deeper level, for all resistivities and doses.

Question (ALBANY) Do you mean that the 0.330 eV level is a double acceptor? This might be in agreement with our phonon scattering results in electron-irradiated Sb-doped Ge in the n to p transition region.

Answer (SROUR) Yes, I would be interested in hearing of your results.

PHONON SCATTERING AND INDUCED-ENERGY LEVELS IN ELECTRON-IRRADIATED Sb-DOPED Ge IN THE n TO p-TYPE CONVERSION REGION

G. LAURENCE AND H. J. ALBANY

Centre d'Etudes Nucléaires de Saclay, 91, Gif-sur-Yvette, France

Low-temperature thermal conductivity results on Sb-doped Ge irradiated at $\sim 40\,°C$ by 4 MeV electrons are reported. The change in K upon electron dose Φ was investigated in the n to p-type conversion region for two Sb concentrations 8.4×10^{15} and 4.5×10^{17} cm^{-3}. The thermal conductivity behavior was shown to arise from electron-phonon scattering in relation to antimony levels, irradiation-induced energy levels, and their charge states. The different stages of K variation emphasize the dominant effects of the induced levels determined from Hall effect measurements. The existence of a residual scattering near the n-p transition appears to be a striking feature which might indicate a particular role or structure for the acceptor centres in this region.

1. INTRODUCTION

The low-temperature thermal conductivity K in semiconductors is generally limited by boundary effect, mass-difference scattering, and phonon-phonon interactions. However, recent works have emphasized the influence of electron-phonon scattering on K. This was investigated as a function of doping in several materials, and particularly in Sb-doped Ge[1-3]: in the latter, K was found to decrease with increasing concentration in the studied range $10^{14} - 2 \times 10^{18}$ cm^{-3}. Theoretical studies were carried out especially by Ziman[4] for electrons in a degenerate band, and by Keyes,[5] and Griffin and Carruthers[6] in the case of electrons bound to impurity donors, where the scattering strength depends on the number of neutral donors. Pyle[7] has considered the jump of a carrier from one impurity state to an unoccupied neighbour.

On another hand, the effects of irradiation on the low-temperature thermal conductivity were investigated in different semiconductors. Particularly in Ge,[8-10] the observed variation of K with the flux was ascribed to electron-phonon scattering mechanism.

In this paper we report thermal conductivity results on Sb-doped Ge irradiated at $\sim 40\,°C$ by 4 MeV electrons. In account of the expected sensitivity of the low-temperature thermal conductivity K to energy levels and their charge states, the change in K upon electron dose was studied in the n to p-type conversion region.

2. EXPERIMENTAL RESULTS

The investigations of electron-irradiated Sb-doped Ge were performed for two antimony concentrations, 8.4×10^{15} and 4.5×10^{17} cm^{-3}, corresponding to room-temperature electrical resistivity $\rho_{300°K}$ of 0.25 and 0.011 ohm cm respectively. The supplied single crystals were grown by Czochralski technique. The samples used in thermal conductivity measurements were rectangular parallelepipeds 4×1.5 mm^2, with the long dimension in the $\langle 111 \rangle$ crystal direction. Each sample was submitted to successive electron doses and the integrated fluxes used ranged between $\Phi = 1.5 \times 10^{15}$ and 3.2×10^{18} elect/cm^2. In order to obtain a more uniform production of defects in the volume, the sample was irradiated for equal time on both faces through the 1.5 mm thickness. The thermal conductivity between 5 and $80\,°K$ was measured after each dose, the irradiated sample being characterized after each dose by its room-temperature resistivity $\rho_{300°K}$ and by its type deduced from Seebeck coefficient at $77\,°K$. The experimental conditions have been described elsewhere.[3,9,10]

The variation of $\rho_{300°K}$ with the integrated flux exhibited the well known behavior: with increasing Φ, ρ first increases and then decreases, the maximum corresponding to the conversion of the irradiated material to p-type.

Thermal conductivity results are given for clarification in Figures 1 and 2, which show the variation

G. LAURENCE AND H. J. ALBANY

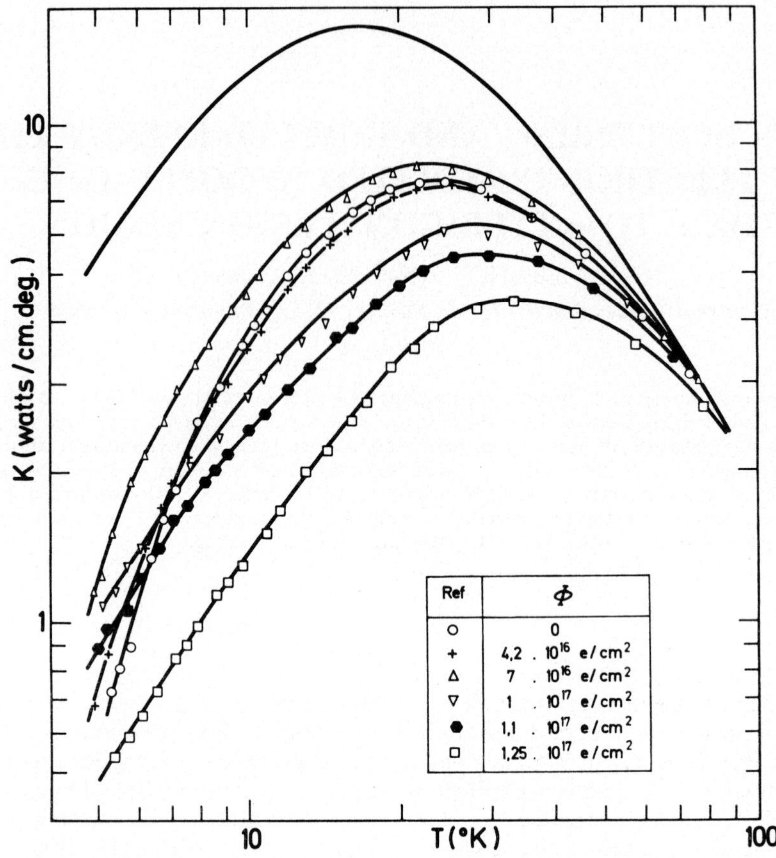

FIG. 1. Variation of the thermal conductivity K as a function of temperature of initially 0.011 ohm cm Sb-doped germanium before and after electron irradiations (sample still n-type). The upper full line curve corresponds to the calculated thermal conductivity using boundary effect, isotope scattering and phonon–phonon scattering.

of K as a function of temperature for initially 0.011 ohm cm sample irradiated with the integrated fluxes indicated. In Figure 1, the irradiated sample was still n-type, and in Figure 2 the irradiated sample was converted to p-type. The upper full line curve corresponds to the theoretical thermal conductivity calculated[2] on the basis of the simple Callaway expression, using for the boundary effect ($\tau^{-1} = V/L$) the Casimir length $L = 0.276$ cm, for isotope scattering ($\tau^{-1} = A\omega^4$) the value $A = 2.4 \times 10^{-44}$ sec^{-3}, and for phonon-phonon interactions ($\tau^{-1} = B\omega^2 T^3$), the value $B = 2.8 \times 10^{-23}$ sec deg^{-3}.

The variations of K are pronounced at temperatures below the peak, while the curves $K(T)$ tend to merge at high temperatures above the peak. In order to emphasize the contribution of the dominant scattering involved in the observed be-

havior, we have plotted in Figure 3 the variation of K at 5.5 °K as a function of $\rho_{300\,°K}$, i.e., increasing flux: ρ_n is the resistivity while the irradiated sample is still n-type, and ρ_p, the resistivity after its conversion p-type.

Similar curves $K(T)$ were obtained for initially 0.25 ohm cm samples. We only report here (Figure 4) the variations of K at 5.5 °K as a function of $\rho_{300\,°K}$ for two samples (points ○○ and △△), irradiated within the integrated flux range $1.5 \times 10^{15} - 2 \times 10^{18}$ elect/cm^2. The behavior of K is similar to that exhibited by 0.011 Ω.cm germanium. It is worth noticing that the calculated value of K at 5.5 °K is $\sim$6 W/cm deg.

The flux dependence of K at low temperatures cannot be explained on the basis of scattering of phonons by strain fields due to induced lattice defects, for this would give rise to a monotonic

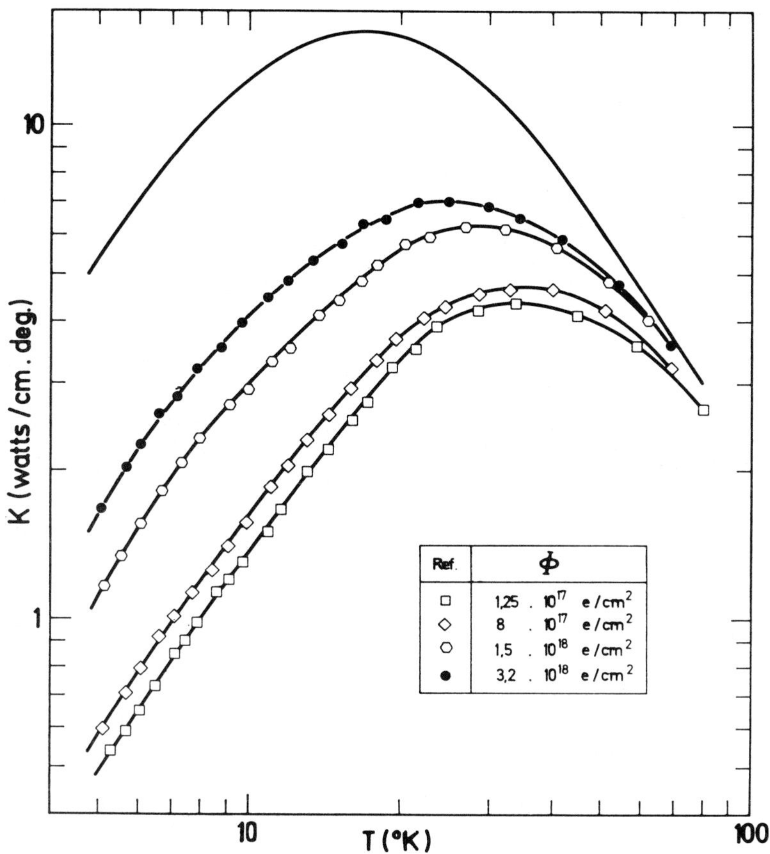

FIG. 2. Variation of the thermal conductivity K as a function of temperature of initially 0.011 ohm cm Sb-doped germanium, after high electron doses (sample converted to p-type). The upper full line curve is the calculated thermal conductivity (see Figure 1).

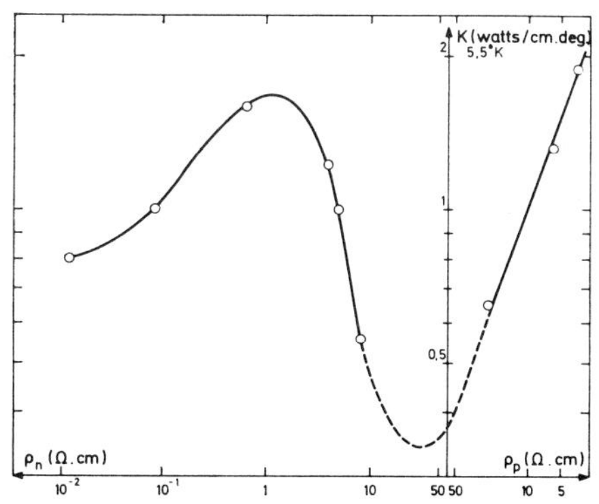

FIG. 3. Variation of the thermal conductivity K at 5.5 °K of initially 0.011 Ω cm Sb-doped germanium as a function of room-temperature electrical resistivity measured after electron irradiations: ρ_n corresponds to the irradiated sample while still n-type, ρ_p, to the sample converted to p-type.

variation of K with Φ. It arises predominantly from electron-phonon scattering associated with irradiation-induced levels.

Hall effect measurements were performed[3] on initially 0.25 ohm cm germanium, submitted to successive electron irradiations in the same experimental conditions. The energy levels obtained with increasing integrated flux are summarized in the table, the irradiated being characterized by its type and its room-temperature resistivity $\rho_{300\,°K}$.

The observed level scheme is in good agreement with that of Konovalenko and coworkers.[11] A similar scheme was obtained by these workers in higher Sb-doped Ge for higher integrated fluxes. The shallow levels $E_V + (\sim 15\,\text{meV})$ observed in their work would correspond in our case to lower resistivity (ρ_p), i.e., to higher integrated fluxes.

By way of illustration, results of Hall effect measurements for initially 0.26 Ω . cm Sb-doped Ge

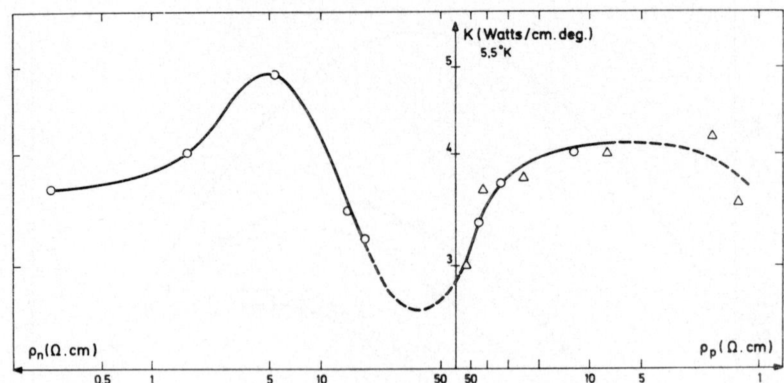

FIG. 4. Variation of the thermal conductivity K at 5.5 °K of initially 0.25 Ω . cm Sb-doped germanium as a function of room-temperature electrical resistivity measured after electron irradiations on two samples ($\bigcirc\bigcirc$, $\triangle\triangle$): ρ_n and ρ_p correspond to the irradiated sample in the n- and p-type regions respectively.

before and after two integrated fluxes corresponding to $\rho_n = 0.9\,\Omega$. cm and $\rho_p = 28\,\Omega$. cm are shown in Figures 5 and 6. The deduced energy levels are given in the table.

predominant effect of one or more of the induced levels.

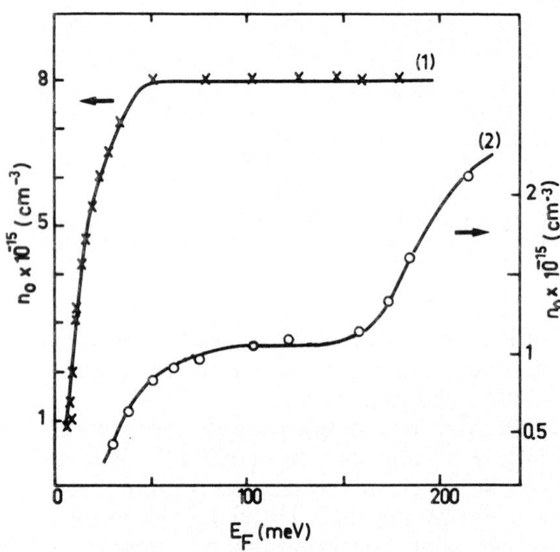

FIG. 5. Electron concentration n_0 as a function of Fermi level E_F for initially 0.26 Ω . cm Sb-doped Ge before (curve 1) and after (curve 2) an electron irradiation ($\rho_n = 0.9$ ohm cm).

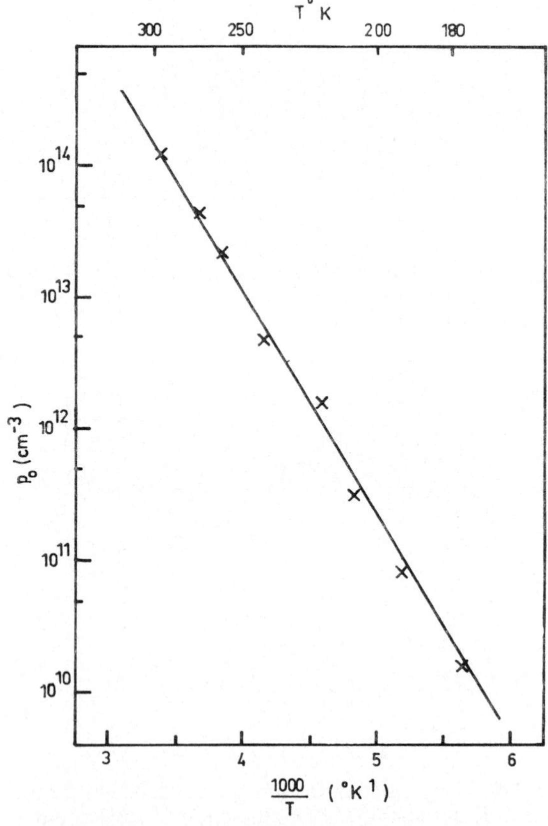

FIG. 6. Hole concentration p_0 as a function of the inverse of temperature $(1/T)$ for initially 0.26 Ω . cm Sb-dope Ge converted to p-type ($\rho_p = 28\,\Omega$. cm) by electron irradiation.

3. DISCUSSION

The above observed K behavior in the n- to p-type conversion region appears to involve electron-phonon scattering in connection with the induced energy levels, and the different intervals of variation of K may be expected to correspond to a

In the n-type region, with increasing Φ, K first exhibits an increase then a decrease reaching values lower than that of the unirradiated material. The observed increase of K for low fluxes arises very likely and dominantly from the decrease of the number of neutral or unionized antimony centers. This is supported by Hall effect results: neutral antimony donor concentration which was initially 8×10^{15} cm^{-3} decreased to 1×10^{15} cm^{-3} after the first irradiation ($\rho_n = 0.9\ \Omega$. cm). This explanation is in agreement with the experimental K results in Sb-doped Ge,[1-3] and is consistent with the Keyes model.

TABLE I

Energy levels vs irradiation use as monitored by room temperature resistivity. The germanium was originally 0.25 ohm-cm n-type.

Type	$\rho_{000°K}$(ohm cm)		Energy levels (meV)
n	0.25 (before irradiation)		$E_c - 10$
n	0.9 (after irradiation)		$E_c - 10$ and $E_c - (\sim 200)$
n	15	,, ,,	$E_c - 200$
p	28	,, ,,	$E_v + 330$
p	12	,, ,,	$E_v + (\sim 200)$
p	2	,, ,,	$E_v + 110$

The following decrease of K in the n-type is not well understood, and the subsequent occurrence of a minimum of K, indicating a residual scattering at the n-p transition is a striking feature. One may assume an additional scattering due to $E_c - (\sim 200$ meV) levels. In the increasing part of K, the contribution of these levels must be considered to be less important than the opposite effect arising from the decrease of the neutral antimony concentration. Tentatively, the decreasing part of K might be ascribed to a dominant influence of the (~ 200 meV) levels. However, this alone cannot explain, after Keyes model, the residual scattering at the n-p transition, for donors and acceptors are expected to be ionized.

Because of the type transition and the observed change in K after the conversion of the material to p-type, one might assume a particular role or a complex structure to the acceptor centers in this transition region which exhibits a significant residual scattering. Thus, the ($E_V + 330$ meV) levels which lie near the middle of the forbidden gap might tentatively be considered to be responsible

for a 'complex' compensation at the n–p transition. It is important to indicate that in a similar study[12] on electron-irradiated n-type GaSb, K exhibits an opposite variation with the flux. At the n–p transition of the material, K approaches the theoretical value in agreement with Keyes model, and is thus consistent with a simple compensation. This emphasizes the more complex nature of the compensation at the n–p transition in electron-irradiated Sb doped Ge. Moreover, Pyle's mechanism might be involved in the above considerations.

In the p-type region, with increasing Φ, K first increases reaching a flat maximum, and then shows at the highest fluxes a tendency to decrease. The increase of K is likely due to a reduction of the mechanism responsible for the residual scattering at the n–p transition. Thus, if this residual scattering is dominantly ascribed to acceptor levels ($E_V + 330$ meV), the increase of K might be related to a modification of these levels, or to a change of their charge state due to the induced 200 and 110 meV acceptor levels. It is possible that the decrease of K which appears at the highest doses arises from a dominant effect of the shallow acceptors levels (~ 15 meV). This decrease is to be related to the results of Vook[8] on germanium irradiated by 2 MeV electrons at $20°K$, and Albany and Vandevyver[1] on germanium irradiated at room-temperature by fast neutrons which are known to introduce shallow acceptors.

REFERENCES

1. J. F. Goff and N. Pearlman, *Phys. Rev.*, **140**, A2151 (1965).
2. H. J. Albany and G. Laurence, *Solid State Commun.*, **7**, 63 (1969).
3. G. Laurence, Thesis (Troisième cycle), Paris, 1969 (unpublished).
4. J. M. Ziman, *Phil. Mag.*, **1**, 191 (1956); **2**, 292 (1957).
5. R. W. Keyes, *Phys. Rev.*, **122**, 1171 (1961).
6. A. Griffin and P. Carruthers, *Phys. Rev.*, **131**, 1976 (1963).
7. I. C. Pyle, *Phil. Mag.*, **6**, 609 (1961).
8. F. L. Vook, *Phys. Rev.*, **138**, A1234 (1965).
9. M. Vandevyver and H. J. Albany, *Physics Letters*, **19**. 376 (1965); H. J. Albany and M. Vandevyver, *J. Appl. Phys.*, **38**, 425 (1967).
10. H. J. Albany and M. Vandevyver, *Phys. Rev.*, **160**, 633 (1967).
11. B. M. Konovalenko, S. M. Ryvkin and I. D. Yoroshetskii, *Fiz. Tverd. Tela*, **5**, 2075 (1963). *Soviet Physics-Solid State*, **5**, 1513 (1964).
12. A. M. Poujade and H. J. Albany, *C. R. Acad. Sci. (Paris)*, **270**, B840 (1970).

DISCUSSION

Question (VOOK) What is the change in thermal conductivity with electron irradiation for GaSb, and how does it compare with Ge?

Answer (ALBANY) The thermal conductivity in electron irradiated GaSb exhibits the opposite behavior than that in irradiated Ge. [See *C. R. Acad. Sci. (Paris)*, **270,** B 840 (1970)[1].

DEFECTS IN SILICON: CONCEPTS AND CORRELATIONS†

H. J. STEIN

Sandia Laboratories, Albuquerque, New Mexico 87115

A review of irradiation produced defects in Si is presented with emphasis on correlations of macroscopic properties with microscopic defects. Such correlation have been successfully used to interpret the electrical properties of irradiated *n*-type Si. Additional correlation studies are needed to obtain a more complete interpretation of the electrical properties of irradiated *p*-type Si. As a further step in this direction, new data are presented on the annealing recovery of the hole concentration in electron-irradiated Al-doped and B-doped Si. The concept of vacancy-rich defect clusters, together with concepts and data on point defects, explain many aspects of neutron radiation damage. The same defects, concepts, and correlations used in studies of electron and neutron produced damage are also useful for interpreting radiation damage produced by ion implantation.

1. INTRODUCTION

Electron and neutron irradiations produce defects in silicon which influence electrical and other macroscopic properties. Moreover, recent efforts to fabricate devices by ion implantation directly involve the lattice disorder produced by the implanted ions.[1] Fortunately, microscopic defect identification[2–5] and general concepts based upon microscopic defect formation and annealing in Si give an understanding of the macroscopic effects. The following general concepts have been especially useful for understanding the effects of irradiation upon the macroscopic properties of Si: (1) direct formation of vacancies and interstitials by high energy particle irradiation at low-temperature, (2) metastable vacancy-interstitial pair formation, (3) charge state dependent stability of metastable pairs and of free vacancy motion, (4) low-temperature motion of vacancies, (5) low-temperature displacement of substitutional impurities, (6) impurity trapping of vacancies and also vacancy-vacancy trapping, and (7) vacancy-rich clustered defect formation by neutron irradiation, proton irradiation, and ion implantation.

Electron paramagnetic resonance (EPR)[2–5] and optical studies[4] have provided microscopic defect identification and specific defect characterization such as annealing behavior, stress response, electronic energy levels and bombarding energy dependence of production. The interpretation of macroscopic measurements in terms of specific defects requires a correlation of these measurements with microscopic defects. Such correlations have

† This work supported by the U.S. Atomic Energy Commission.

worked well for interpreting the macroscopic properties of *n*-type Si,[6] but additional work is needed on correlation studies in *p*-type Si.

2. MICROSCOPIC DEFECTS AND CORRELATION CHARACTERISTICS

2.1. *EPR-Identified defects*

A summary of the characteristics for defects identified in EPR measurements[2–5] on electron-irradiated Si is given in Figure 1. The characteristics include a sketch of the structural model, approximate annealing temperature, and in some cases the annealing activation energies and electronic energy levels. With the exception of the Al

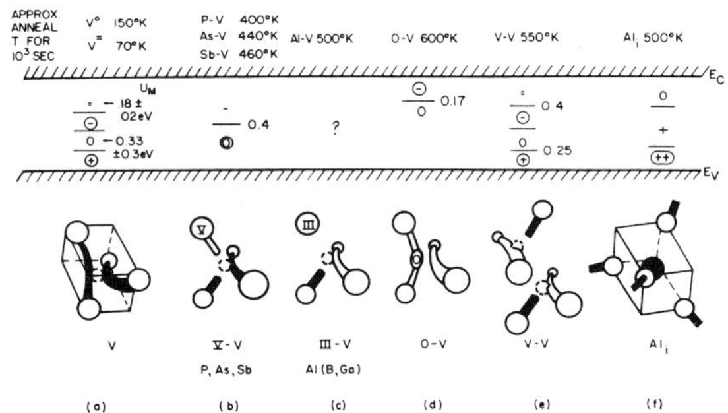

FIG. 1. Summary of irradiation produced defects in Si which have been identified and structurally modelled in EPR studies (References 2–5, 70). Vacancy, Group V-vacancy, Group III-vacancy, oxygen-vacancy, divacancy, and interstitial Al defects are listed. Characteristics of the identified defects such as approximate annealing temperature, energy levels, and annealing activation energy for the vacancy are indicated in addition to a sketch of the structural model.

interstitial, all of the defects incorporate one or more vacancies. Another vacancy-associated defect qualifying for inclusion in such a summary is the Ge-vacancy in Si recently reported by Watkins.[5] Work of Brelot and Charlemagne[7] (this conference) shows that the Ge-V defect is already being used to formulate an additional concept on the atom size dependence for vacancy trapping in Si.

A very important center observed by EPR[8] but not structurally identified, is the K center (G15) in p-type Si. The K center is known to contain oxygen and exhibits a production rate dependence upon bombarding energy[8] which is more like that for the divacancy than for the vacancy.[9] The role of the K center in luminescence and further information on its possible structure are discussed by Jones and Compton[10] (this conference).

Structural models involving multiple vacancies have been proposed by Brower[11] (this conference) for centers labeled P2 through P5, previously observed in heavily neutron-irradiated Si, and also for a new center (S1) observed in both electron and neutron irradiated Si. Thus, the number of defects which have been structurally modeled by EPR studies continues to increase, and other techniques are also being applied[10] in studies of defect structure.

2.2. Correlations of defect characteristics

The certainty of correctly correlating macroscopic properties with a microscopic defect increases with the number of characteristics resolved for a particular defect. For example, there have been no correlations with the A1-vacancy centers because the only identified characteristic is the annealing temperature,[5,12] and it is the same as that for interstitial A1.[5] The correlation method for interpreting electrical properties in terms of microscopic defects has been applied to other vacancy-type defects in Figure 1, and preliminary work on correlating electrical properties with interstitial A1 is discussed in Sec. 4.2.

The correlation method has also been used to relate microscopic defect characteristics measured optically (optical absorption,[13,14] photoconductivity,[15,16] and luminescence[10]) with those measured by EPR.[2-5] These correlations are most complete for the vacancy-oxygen[13] (A center or B1 center) and for the divacancy.[14] There have been no direct correlations of optical properties with free vacancies in Si, however, the annealing growth of vacancy-oxygen[17] and divacancy[18] defects has been attributed to the motion of free

vacancies. Even for electrical measurements, the correlations with the vacancy have been primarily for the neutral charge state of the vacancy. This is because the low temperature ($T < 60\,°K$) necessary to prevent negative vacancy annealing requires degenerate material for conventional electrical measurements. But the production rate of free vacancies is small in n-type material at low temperature,[19] and very long irradiations would be required to obtain a measurable effect in degenerate material.[20] Correlations of specific optical absorption bands and photoconductive stages with Group V-vacancy defects[4] are in the formative stages as are also the correlations with Group III interstitial defects.

2.2A. *Correlations of isochronal annealing, stress response, and electronic energy levels: vacancy-oxygen center*

The characteristics of the vacancy-oxygen center (*A* center) are used in Figures 2 and 3 to illustrate the correlation method which has been repeated many times in defect studies over the past decade. The left side of Figure 2 shows the isochronal

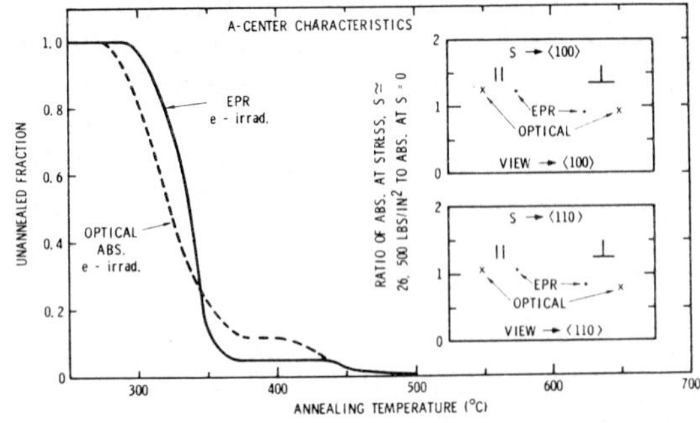

FIG. 2. Correlation of the isochronal annealing (left side) and stress response (right side) for the A center (oxygen-vacancy) as measured by EPR[21] and by optical absorption.[13] S gives the direction of the applied stress, ‖ and ⊥ indicate the E vector parallel and perpendicular to the direction of applied stress.

annealing correlation for the *A* center as observed by EPR[21] and by the localized vibrational mode of substitutional oxygen (829 cm⁻¹).[13] The right side of Figure 2 shows the stress induced alignment measured for the 829 cm⁻¹ band correlated with the alignment predicted using the model of the *A* center deduced from the EPR study. The effect of isotopic oxygen doping[13] upon the vibrational frequency of the localized mode confirmed the role

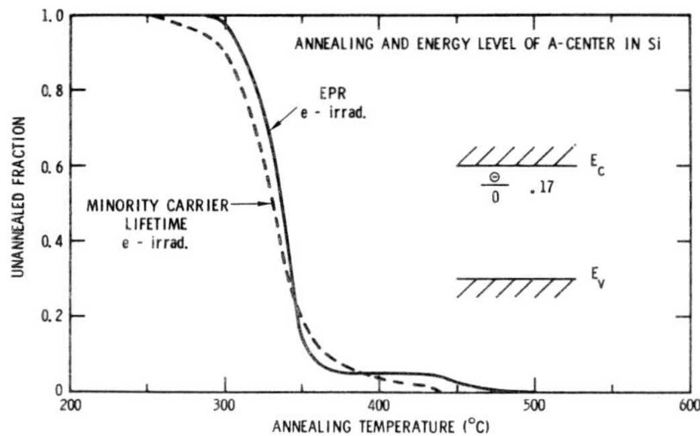

FIG. 3. Isochronal annealing of minority carrier lifetime in *n*-type crucible grown Si[24] correlated with that for the A center measured by EPR.[21] Also a sketch of A center energy level.

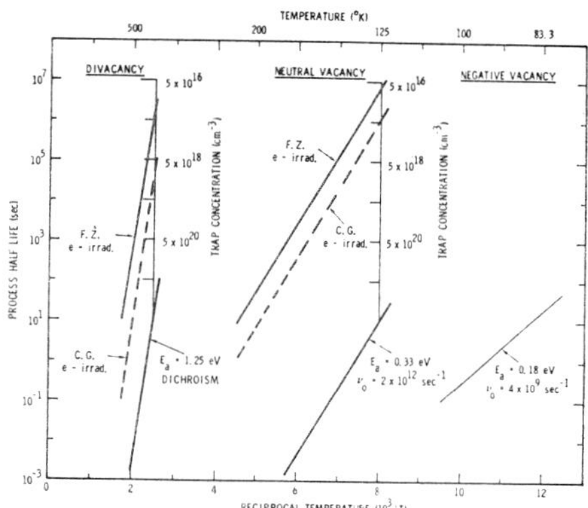

FIG. 4. Annealing half life versus reciprocal temperature for neutral and double negative vacancy and for divacancy defects.[4,14,19] The influence of trap concentration upon the annealing of the neutral vacancy and of the divacancy is illustrated by the difference between vacuum floating zone (F.Z.) and crucible grown (C.G.) Si. The scale for the trap concentration for the divacancy assumes only one jump for reorientation (dichroism). At least two jumps are required and perhaps more on the average. Therefore the trap concentrations given for the divacancy are large by at least a factor of two.

of oxygen in the *A* center. Subsequent optical absorption measurements showed an approximate 1 to 1 correspondence between the loss of the interstitial oxygen and the formation of the *A* center.[22] These experiments lead directly to the concept of impurity trapping of vacancies.

An early correlation[23] of an electrical parameter with a specific defect is shown on the left of Figure 3 where the annealing of minority carrier lifetime[24] degradation by electron irradiation of crucible-grown *n*-type Si is correlated with that for the *A* center as measured by EPR.[21] Similar correlations with the *A* center have been made for carrier concentration measurements[6] by using the characteristic energy level[25] at $E_c - (0.16 + 1.1 \times 10^{-4}T)\,\text{eV}$ (sketched on the right of Figure 3) in addition to the annealing behavior.

2.2B. *Correlations of annealing activation energy, trap concentration, and defect charge state: vacancy and divacancy*

Figure 4 is a plot of the logarithm of the annealing half life versus reciprocal absolute temperature where the slopes give the experimental activation energies for the vacancy[3,4,19] and divacancy defects.[14] All of the results shown in Figure 4 are experimental results or are extrapolated from experimental results except the annealing of the neutral vacancy[19] in the limit of high trap concentration $(5 \times 10^{22}\,\text{cm}^{-3})$, which was obtained by using the experimental activation energy, $E_a = 0.33\,\text{eV}$, and a frequency factor of $2 \times 10^{12}\,\text{sec}^{-1}$. This plot illustrates the charge state dependence of the activation energy for vacancy annealing and also the trap concentration dependence (oxygen concentration dependence in this case) of both

vacancy and divacancy annealing. Activation energies for annealing in macroscopic measurements have been used to infer vacancy[26] and divacancy annealing.[26] The trap concentration dependence and charge state dependence for vacancy annealing have also been applied to studies on *p*-type Si[27] and are further discussed in Sec. 4.1. The trap concentration dependence for divacancy annealing has been used to infer that neutron irradiation[28] and ion implantation produced divacancies[29] anneal primarily in regions of high trap concentrations.

There has been widespread interest in the divacancy. Unlike the vacancy, the Si divacancy is stable at room temperature and is readily observed in optical and electrical measurements. There is also evidence from correlation studies that the divacancy is an important defect in structural measurements,[26] such as anomalous X-ray transmission and particle channeling.[1] The importance of the divacancy relative to other defects increases with incident electron energy, and it is an especially important defect in neutron irradiated[28] and in ion implanted Si.[26,29,30]

Sketches of the divacancy structure, energy levels, and optical absorption bands for the divacancy[14] are shown in Figure 5. The correlation of all these

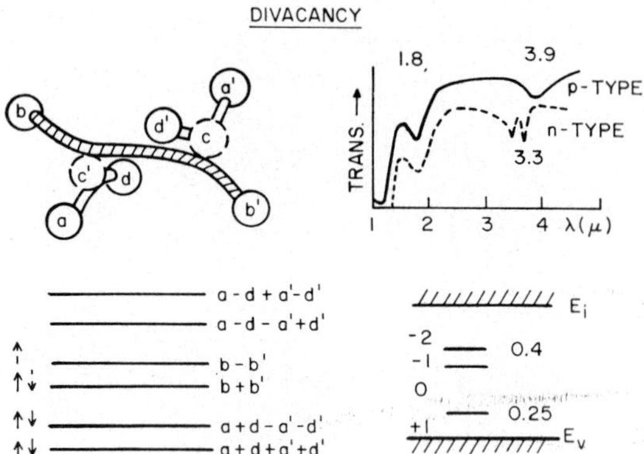

FIG. 5. Sketches of the structural and electronic model for the divacancy and a sketch of optical absorption bands associated with the divacancy.[14]

aspects of the divacancy defect is of current interest. According to the model there is pair-wise bonding of atoms b and b' (or alternately a and a' or d and d'), and it is the filling of the combined b and b' orbitals which largely determines the charge state of the divacancy. The $+1$ and 0 charge states correspond respectively to one and two electrons in the $b + b'$ orbital. The -1 and -2 charge states correspond respectively to one and two electrons in the $b - b'$ orbital. The divacancy has a large effect upon structural measurements because nearest neighbor atoms are pulled off position by $\sim$0.25 Å.[4]

The energy levels (lower right of Figure 5) of the divacancy and their relationship to the optical absorption bands (upper right of Figure 5) have received considerable attention and comprise a part of the work discussed in these proceedings by Tkachev and Lappo;[31] by Chen and Corelli;[32] and by Barnes.[33] The 1.8 μ band is caused by an internal electronic transition which does not lead to photoconductivity.[34] The 3.9 μ band is also caused by a local excitation and has been correlated with the positive charge state of the divacancy.[15] But photoconductivity is observed for the 3.9 μ band and is attributed to emission of a hole from the ground state of the excited defect.[15]

Energy-level correlations with charge states were recently investigated by Cheng and Vajda[35] in a clever experiment using polarized light to preferentially orient the divacancy bonding (i.e. b to b', a to a', or d to d'). Polarized light on an appropriately oriented sample preferentially excites one of the three equivalent bondings; and if the temperature is such that the excited state can reorient but not the unexcited state, then a net alignment of the bonding will be achieved. This was observed for the 1.8 and 3.3 μ bands of the divacancy. Randomization of the 1.8 μ band after alignment does not correlate with the randomization characteristics of either the single negative or single positive charge states of the divacancy measured in EPR studies,[36] and the randomization is therefore ascribed to the neutral charge state of the divacancy. The activation energy to randomize the 3.3 μ band, but not the frequency factor, correlates with the single negative charge state of the divacancy measured in EPR studies. Recent work by Chen and Corelli[32] (this conference) indicates that the sharp peaks near 3.3 μ are probably due to excited states.

An important question about the divacancy has not been answered either directly by EPR and optical studies or indirectly by correlation studies. How do impurities enhance the formation of divacancies? Impurity enhancement of divacancy formation has been noted for oxygen,[14,37] carbon,[38] and boron[28] in Si and the mechanism could be important in considerations of divacancy controlled diffusion.[39] In the case of carbon it was suggested that interstitial Si trapping by carbon resulted in a higher concentration of vacancies and hence in a higher divacancy concentration by vacancy-vacancy trapping.[38] However, it is difficult to apply this interpretation to oxygen because the oxygen is a trap for vacancies and would compete with divacancy formation. Lithium in Si is clearly an exception to the effects of oxygen, carbon and boron because it reduces the 1.8 μ band formation, and additional bands appear at 1.4 and 1.7 μ.[40]

3. ELECTRON IRRADIATION AND MACROSCOPIC MEASUREMENTS: n-TYPE Si

3.1. *Vacancy associated defects*

The characteristics of EPR identified defects summarized in Figure 1 are used as an intial framework for interpreting the electron irradiation produced carrier removal rates and annealing characteristics for n-type Si[6,41] shown in Figure 6. Additional defect characterization from other measurements will be applied to obtain a more complete interpretation of the results shown in Figure 6.

Annealing stages shown in Figure 6 near 400 and 550 °K for the vacuum floating zone (Z) Si

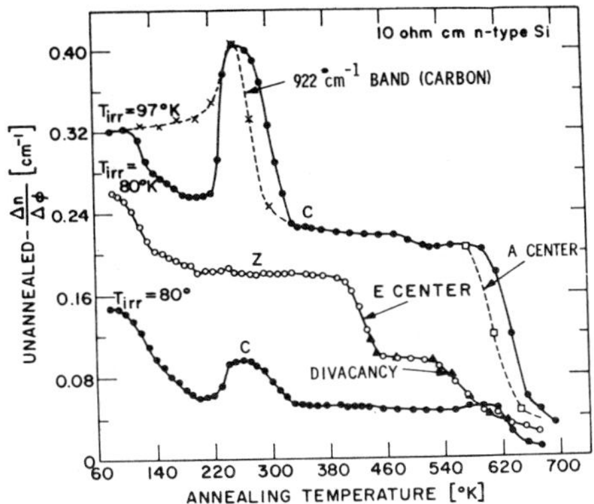

FIG. 6. Isochronal annealing of the carrier removal rate produced by 1.8 MeV electron irradiation of float zone (Z) and crucible-grown (C) n-type Si measured at 80 °K.[41] Divacancy,[14] 922 cm⁻¹ carbon center,[17] E-center[59] and A-center[13] annealing are included for correlation.

correlate with the annealing of the vacancy-phosphorus (E center) and of the divacancy. Likewise, the annealing near 600 °K in the high oxygen content crucible grown Si (C) correlates reasonably well with that for the A center. An electronic energy level at $E_c - 0.18$ eV provided additional evidence[41] for ascribing carrier removal annealing near 600 °K to the A center.[25] The small annealing stage at 500 °K in crucible-grown Si is consistent with divacancy annealing since the divacancy annealing temperature is lowered by a high oxygen concentration as discussed in the previous section.

Figure 6 shows a larger carrier removal rate in float zone than in crucible-grown Si at 80 °K. This can be explained in part by the greater effectiveness of E-center than A-center formation for removing conduction carriers. The formation of an E center converts a positive phosphorus center to a negative E center by the removal of two conduction electrons, whereas A-center formation converts a neutral oxygen center to a negative A center by the removal of only one conduction electron. The difference in the electrical properties for A center and E center formation becomes more pronounced when the measurements are made at room temperature.[42] For example, in 10 ohm-cm Si the Fermi level is below the A-center level but above the E-center level so that the A center does not remove any conduction carriers. There will also be a marked difference in minority carrier lifetime degradation because the hole recombination cross section is much smaller for a neutral A center than for a negative E center.[43] A Fermi level effect combined with an

increasing probability for E-center formation with increasing phosphorus concentration gives a pronounced dependence of the lifetime degradation[43] and carrier removal[42] upon the phosphorus concentration in room temperature studies of electron irradiated Si. Experiments[42,44–46] have shown that phosphorus is 10 to 50 times more effective in vacancy capture than oxygen.

EPR results have also been quite useful for interpreting the carrier removal annealing between 100 and 200 °K. Annealing between 100–200 °K occurs with approximately equal magnitude in both crucible-grown and floating-zone Si at 80 °K and for irradiations at both 80° and 97 °K in crucible grown Si. These defects have been given the name of irradiation temperature independent or ITI defects. According to EPR studies, ∼80 per cent of the divacancies produced by 1.5 MeV electron irradiation of Si at 4 to 20 °K are nonreorientable.[3] Watkins suggested that interstitial Si atoms lock these divacancies, and he noted annealing of the centers near 140 °K.[3] A divacancy-like bombardment energy dependence for production of the irradiation temperature independent (ITI) defects in electrical measurements,[47] and evidence that they are doubly charged,[47] support the association of the ITI defects with the locked divacancy.

Annealing between 100 and 200 °K is, however, more complex than the simple loss of locked divacancies because an irradiation temperature dependent defect growth is observed near 150 °K, particularly in float-zone Si after irradiation at 100 °K.[41,48] Irradiation temperature dependent (ITD) defect production, which is quite apparent in the crucible grown Si results of Figure 6 after annealing to 200 °K, has also been observed in EPR,[2] optical absorption,[6] and in other electrical studies.[6,25,49] A defect production model[6] invoking metastable vacancy-interstitial pairs, in addition to the EPR characterization and structural modeling of defects, has been successfully used to explain the irradiation temperature dependence of the defect production as well as the electrical resistivity dependence of defect production.

3.2. Metastable pair model

The irradiation temperature dependent (ITD) defect formation rate has been modeled (see Figure 7) by postulating metastable vacancy-interstitial pair formation. The pairs are assumed to either recombine or to separate depending upon the charge state of the pair. Results presented in Figure 7 show that the irradiation temperature

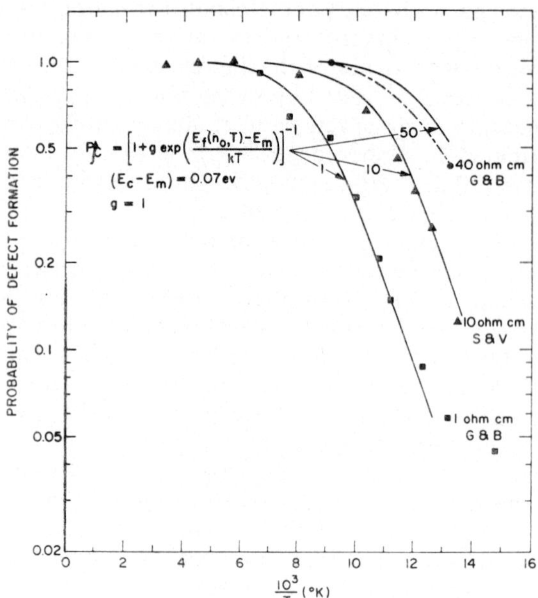

FIG. 7. Probability (P_c) of defect production predicted[6] by the charge-state limited metastable pair model (solid line) compared with electrical measurements of defect production as a function of reciprocal irradiation temperature by Gregory and Barnes (G & B)[49] and by Stein and Vook (S & V).[47] See text for description of terms in probability equation.

dependence of the carrier removal rate[6] (S and V) and the lifetime degradation[49] (G and B) as well as the resistivity effects[6,49] can be explained by this model if the metastable pair energy level E_m is $E_c - 0.07$ eV. The solid lines are the calculated probabilities, P_c, for pair separation.

$$P_c = 1 + g \exp\left[\frac{[E_f(n_0, T) - E_m]}{kT}\right]^{-1}$$

where E_f is the Fermi level which is temperature and carrier concentration dependent, k is Boltzmann's constant, and g is taken equal to unity which is equivalent to assuming that the center has an equal number of ways it can be filled or empty. This model for ITD defect production can also explain the injection stimulated formation[49] of vacancy associated centers in a low temperature irradiation if the metastable pair captures a hole and thereby enhances the probability for vacancy-interstitial separation. ITD defect production occurs for both crucible grown and vacuum floating zone Si and for both phosphorus and As-doped Si.[6]

Irradiation temperature dependent defect production has also been observed in Li-doped Si[50] ($\sim$0.3 ohm-cm) using 1 MeV electron irradiation. Again, the two kinds of defects (ITI and ITD) were formed and the ITD production mechanism was assumed to be metastable pairs. The ITD effect decreases with increasing bombardment energy.[6,47] This is consistent with an expected increase in vacancy-interstitial separation with increasing Si recoil energy, and is inconsistent with the interstitial trapping model[20] proposed to explain the irradiation temperature dependence of defect formation.

3.3. *Interstitial related defects*

A pronounced reverse annealing peak near 250 °K is shown in Figure 6 for crucible grown but not for float zone Si. This reverse annealing does not correlate with any of the defects identified by EPR, therefore, we turn to optical absorption measurements which have also provided identification information on defects.[4]

An interplay between crystal impurity[51,52] and irradiation-produced defect studies[22,53,54] was involved in the recognition that carbon, which is a common contaminant in Si, participates in the defects formed under irradiation. Although optical absorption studies of C-doped Si following room temperature electron irradiation did not show any new bands attributable to irradiation produced defects,[51] X-ray lattice parameter measurements on carbon containing Si[54] did show an effect of electron irradiation at room temperature. The lattice parameter change was smaller, however, than that expected for interstitial carbon formation and was, therefore, attributed to carbon-vacancy formation.

On the other hand low temperature electron irradiation studied on Si showed that two prominent and related optical absorption bands at 922 and 932 cm^{-1} are formed[22,53] when a carbon associated band at 1104 cm^{-1} is present before irradiation.[51] The irradiation temperature dependence for the defect suggested a single irradiation induced defect was involved,[53] but the formation of the 922 and 932 cm^{-1} bands did not reduce the formation rate for the A center.[22] Consequently it was concluded that the formation of the 922–932 cm^{-1} bands involved a single interstitial Si[22,53] atom rather than a vacancy and that carbon was incorporated into the resultant defect.[22] Recent investigations using isotopic carbon doping of Si[38,55] have shown that only one carbon atom is involved in the 922–

932 cm^{-1} absorption bands and that the carbon atom is probably interstitial. The annealing loss of the 922 cm^{-1} carbon band at 300 °K sketched in Figure 6 explains why interstitial carbon is not generally observed after room temperature irradiations.

The annealing observed near 250 °K in the carrier removal measurements is shown in Figure 6 to correspond approximately to that for the 922 cm^{-1} carbon band. The defect responsible for the reverse annealing is clearly an ITD defect as are also the carbon associated localized modes.[53] Therefore interstitial carbon is a likely candidate for explaining the large reverse annealing in the carrier removal data.

3.4. *Other defects*

Multiple impurity atom interactions and impurity atom size dependence of defect behavior in electron irradiated *n*-type Si have also been investigated. For example, lithium is of interest as an interstitial donor impurity in Si and also because it forms complexes with other crystal impurities.[50] (See also Goldstein;[56] and Naber and coworkers[57] (this conference)). Lithium can apparently act as a trap for irradiation produced vacancies and interstitials and then further interact with the irradiation produced centers at room temperature due to its own low 0.65 eV activation energy for motion.[50,56,57]

As noted in Figure 1, the *V*-As and *V*-Sb members of the *E* center family in addition to the *V*-*P* center have been investigated by EPR measurements.[58] The activation energy for annealing of *E*-type centers increases from 0.93 for *V*-*P* to 1.29 eV for *V*-Sb.[58,59] This increase with atom size has been explained as an elastic interaction between the oversized atom and the vacancy.[59] The atom size also influences the pre-exponential frequency factor for annealing, but this effect has not been satisfactorily explained.

In addition to the *E* centers, members of a related family of centers have been observed in electrical measurements.[59] An increase in activation energy with atom size for the related family is twice that for the *E* centers themselves, but an even more rapidly increasing pre-exponential factor lowers the annealing temperature ∼60 °K compared to the corresponding *E*-center. A defect model involving two dopant atoms in substitutional positions near one vacancy[59] has been suggested for the related family. Kimerling and coworkers[60] have recently observed an increase in the magnitude of the

annealing stage for the phosphorous related center with increasing phosphorous concentration. This is consistent with the two-atom defect model.

There is evidence of the vacancy-phosphorous center in recent photoconductivity measurements,[61] and a 1.7 μ absorption band has been associated with electronic population of the vacancy-arsenic center.[4] Other optical absorption bands associated with Group V donors have also been observed[62] between 8 and 9 μ, and these defects anneal below 175 °C. Perhaps the Group V impurity bands between 8 and 9 μ are associated with the two atom defects observed in electrical measurements.[59]

4. ELECTRON IRRADIATION AND MACROSCOPIC MEASUREMENTS: *p*-TYPE Si

4.1. *Vacancy associated defects in p-type* Si

Since vacancy associated defect characteristics were quite important for interpreting electrical measurements on *n*-type Si, we now apply the same procedure to *p*-type Si. Figure 8 shows new

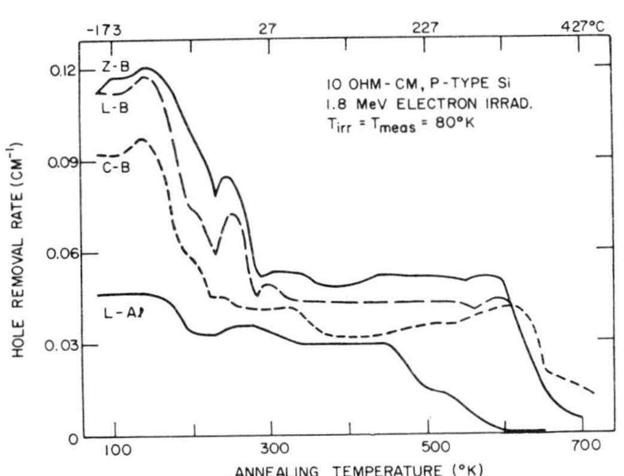

FIG. 8. Isochronal annealing of the hole removal rate produced at 80 °K by 1.8 MeV electron irradiation of B-doped vacuum floating zone (Z-B), crucible-grown (C-B), Lopex (L-B), and for Al-doped Lopex Si (L-Al).

annealing results of the hole removal rate for float-zone (*Z-B*) and crucible-grown (*C-B*) Si doped with boron, and also for Al-doped and for *B*-doped Lopex (*L*-Al, *L-B*) Si after 1.8 MeV electron irradiation at 80 °K. All samples are approximately 10 ohm-cm at room temperature. The hole removal rate for Al-doped Si is smaller than for *B*-doped Si because the Al acceptor level is deeper than that for boron and only partially ionized at the 80 °K measurement temperature.

Recovery of the electrical properties near 175 °K agrees in temperature with the annealing of the neutral vacancy.[19] Injection stimulated annealing for this stage has been observed by Gregory[63] using minority carrier lifetime measurements and by Cheng and Lori[64] using carrier concentration measurements. Such injection-stimulated annealing is consistent with an enhanced annealing of vacancies by altering the charge from neutral to negative. Results for neutral and negative vacancy annealing (see Figure 4) indicate that the negative vacancy, but not the neutral vacancy, would be expected to anneal in 10^2 sec at 80 °K ($10^3/T = 12.5$).

Questions arise, however, regarding the assignment of annealing measured electrically to vacancy annealing. First, how does the vacancy, which will be neutral in $\sim$10 ohm cm Si at 80 °K according to the EPR energy level assignment, exert a large influence upon either the minority carrier lifetime or hole concentration? This could be explained away by assuming vacancy annihilation of some electrically active defect such as an interstitial also introduced by the irradiation. A second objection to ascribing the 175 °K annealing stage to the vacancy has recently been raised by Cheng and Lori.[64] They concluded that the 175 °K annealing stage *cannot* be ascribed to neutral vacancy motion because they did not observe the expected trap concentration dependence when annealing Co^{60} γ-ray or electron irradiated float-zone and crucible-grown p-type Si (500 to 1000 ohm-cm). Our own results for 10 ohm-cm Si shown in Figure 7 indicate more than one stage in the annealing between 140 and 200 °K, and there may in fact be a small shift toward a lower anneal temperature in crucible and Lopex Si than in float-zone Si for the first recovery stage centered near 175 °K. At any rate the precise role of vacancies in the electrical measurements on p-type Si is still an open question.

It may be that metastable pairs are involved in p-type Si as well as in n-type Si. Irradiation temperature dependent defect production has been observed for low energy (300 keV) electron[65] and Co^{60} γ-ray[66] irradiated Si. The slope of the production rate versus $1/T$ is $\sim$0.025 eV rather than the slopes of 0.05 to 0.07 eV observed for n-type Si.[6,49,53,65]

The fact that annealing near 175 °K is prominent following low energy electron[65] and ^{60}Co γ-ray[63,64] irradiation seems to rule out the ITI type defects[6,47,49] observed in n-type Si because these defects have a small probability of formation at low energies. In addition, the ITI defects of n-type

Si are not expected in p-type Si because 'locked divacancies' are not observed in p-type Si.[4] The recovery stage appearing near 210 °K correlates in temperature with the breakup of the Ge-V center in Si,[5] but nothing is known about the concentrations of Ge in these crystals.

A production rate of at least 0.01 cm^{-1} is expected for divacancies[9] by 1.8 MeV electron irradiation of p-type Si, and the electronic energy levels for the divacancy indicate it would remove one hole per divacancy. Only in Al-doped Si, however, is there a recovery of 0.01 cm^{-1} in the hole removal rate near 550 °K where divacancy annealing is expected (see Figure 6). Results for B-doped Si give evidence for an onset of recovery near 550 °K, but this is cut off by a reverse annealing which peaks near 600 °K. Recent studies on Si implanted with boron[67,68] show a loss of substitutional boron at temperatures corresponding to divacancy annealing. Perhaps the divacancy anneals in B-doped Si by interacting with substitutional boron, and the resultant complex anneals between 600 and 700 °K. When the p-type Si contains oxygen the situation is further complicated by K-center formation and its annealing above 600 °K.[69] Correlations of EPR and electrical measurements have indicated that the K-center dominates the minority carrier lifetime[69] and hole removal[8,69] in crucible-grown p-type Si, especially for electron energies greater than 1 MeV. The K center has a level[8,69] at $E_v + 0.3$ eV and anneals in the same temperature range as the A center.[69]

The large difference between defect behavior in B and Al-doped Si is difficult to understand unless it relates to the interstitial position of the displaced dopant atom. The Al interstitial is in the tetrahedral position[70] and does not trap vacancies,[12] but the boron interstitial[3,68] is nested between two substitutional sites and its trapping characteristics for vacancies and divacancies have not been investigated. Interstitial boron trapping of vacancies could be directly determined in low temperature EPR studies by measuring the interstitial boron concentration as the vacancies become mobile.

4.2. *Interstitial associated defects in p-type* Si

It is now known that substitutional carbon,[38,55] boron,[3] or aluminum[2-5,70] impurities in Si are displaced from substitutional sites by irradiation. However, as suggested in the previous sections, it may not be possible to completely separate interstitial and vacancy type defects in macroscopic measurements. A reverse annealing peak in the electrical measurements (Figure 8) on p-type Si is observed in the temperature range where the 922 cm^{-1} carbon peak appears in optical absorption measurements (Figure 6). In p-type Si, however, the peak is more prominent in low-oxygen-content Si than in crucible-grown Si where a high carbon content is expected. Therefore it seems likely that the reverse annealing peak near 240 °K in p-type Si is associated either with interstitial boron or a boron-vacancy complex rather than carbon. The small magnitude for the reverse peak in crucible-grown Si could be explained by a competitive trapping of the vacancies by oxygen in Si if the 240 °K stage is related to a boron-vacancy center. Annealing recovery above the reverse peak in B-doped Si occurs within a 50 °K interval between 240 and 290 °K. A somewhat similar reverse annealing is shown in Figure 8 for Al-doped Si but the recovery extends to 350 °K.

Many studies have shown an annealing recovery in the electrical properties of low oxygen content p-type Si between 300 and 350 °K following irradiation at room temperature.[42,71,72] The magnitude of the stage is larger in Si prepared by the Lopex method than in Si prepared by the vacuum floating zone method. In degenerate Si, recovery near 300 °K has been observed in both crucible grown and float zone material.[20] The electronic energy level associated with the defect annealing between 300 and 350 °K is $\sim E_v + 0.25$ eV,[71,72] and an annealing activation energy of 0.72 ± 0.04 eV has been measured for the defect.[71] This activation energy is quite close to the 0.81 eV[48] measured in crucible-grown n-type Si for annealing between 300 and 350 °K[73] which may be related to interstitial carbon as discussed earlier.

X-ray lattice parameter measurements[52] on Si prepared by the Perfx method indicated that this material has a high degree of perfection. Shown in Figure 9 is a comparison of new annealing results obtained from hole concentration measurements on similarly B-doped Lopex and Perfx Si following 1.8 MeV electron irradiation at 200 °K. Infrared absorption at 607 wavenumbers,[51] indicated $\sim 10^{17}$ carbon atoms per cm^3 for the Lopex crystal and $< 10^{16}$ cm^{-3} for the Perfx crystal. The electrical measurements were performed at 200 °K which gives a high probability for ionization of the boron acceptor centers even in the presence of considerable compensation. The initial hole removal rates are similar for the two materials. A recovery between 250 and 300 °K occurs in both materials

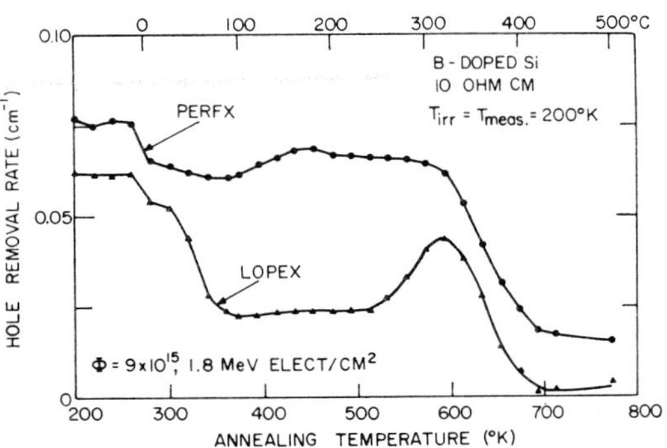

FIG. 9. Isochronal annealing of the hole removal rate produced at 200 °K by 1.8 MeV electron irradiation of B-doped Perfx and Lopex Si.

and, as suggested above, this stage is believed to involve boron. Since the recovery between 300 and 350 °K is much larger in the Lopex than in the Perfx Si it is tempting to ascribe the 300–350 °K annealing stage in Lopex Si to a carbon center. However, similar annealing may be occurring in the Perfx Si but is simply masked by a reverse annealing as other centers are formed. Furthermore, other studies have shown a reverse annealing between 300 and 400 °K in crucible-grown p-type Si where a carbon concentration of the order of 10^{17} cm^{-3} is expected. More work is clearly needed to determine the electrical activity of interstitial carbon and boron in Si.

EPR measurements have supplied considerable information about the interstitial Al defect.[2-5,70] Its formation should have a large effect upon the electrical properties because it not only removes an acceptor center but also compensates the remaining acceptors with two electrons. It is encouraging to observe a recovery stage near 480 °K which correlates in temperature with that for interstitial Al as shown in Figures 8 and 10.

Annealing of the carrier removal for a 0.6 MeV irradiation of Al-doped Si is shown in Figure 10 (solid curve), and the results are compared to annealing of Al$_i$ and Al$_i$-Al$_s$ defects observed by EPR[5] (dashed curve). The irradiation was performed with 0.6 MeV electrons to reduce divacancy formation because the annealing temperatures[5] for the Al$_i$ and the Al$_i$ + Al$_s$ pair are quite close to that for the divacancy. Carrier removal measurements were made at 100 °K to ionize the Al acceptors. The annealing of the hole removal is seen to agree quite well with the annealing of interstitial Al and the interstitial-substitutional Al pairs observed in EPR measurements.[5] These results from Al-doped Si offer a renewed hope for the further utilization of defects identified

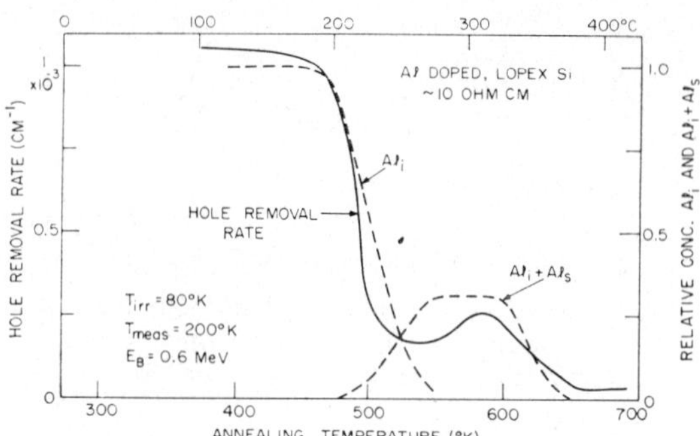

FIG. 10. Isochronal annealing of the hole removal rate produced by 0.6 MeV electron irradiation of Al-doped Lopex Si measured at 200 °K. Interstitial Al (Al_i) and interstitial-substitutional ($Al_i + Al_s$) pair annealing measured by EPR[5] are included for correlation.

in EPR studies to deduce an inventory of electrically active defects in electron irradiated p-type Si.

4.3. *Other defects*

Photoconductivity measurements[74] have shown irradiation formation of centers in crucible-grown Si which: are produced directly at 4 °K, anneal near 520 °K and exhibit slightly different energy levels for B ($E_v + 0.430$ eV) and for Al ($E_v + 0.395$ eV) doping. Because of these characteristics the authors suggested an interstitial dopant defect, but the center has a different symmetry[74] (C_{3v}) than the tetrahedral Al interstitial and therefore does not correlate with EPR identified interstitial Al. If these centers are neutral, as suggested,[74] then they are not expected to be of direct importance in the electrically active defect inventory.

Perhaps the current trend toward defect studies in highly doped and multiply-doped Si will bring new information to bear upon the defect inventory question in p-type Si. Such investigations have already shown that (1) Ge in Al-doped Si is an effective trap for vacancies,[5,7] (2) the presence of carbon,[38] oxygen,[14,37] or boron[28] enhances the formation of divacancies, (3) oxygen enhances the formation rate of interstitial carbon,[38] and (4) room temperature irradiation of heavily Al-doped Si[75] produces Al interstitial-substitutional pairs rather than interstitial Al.

5. NEUTRON IRRADIATION PRODUCED DEFECTS

5.1. *Nature of neutron produced defects*

The observation, by replication electron microscopy, of hillocks ~ 500 Å in diameter on the surface of 10 ohm-cm n-type Si following 14 MeV neutron irradiation and etching (see Figure 11) was

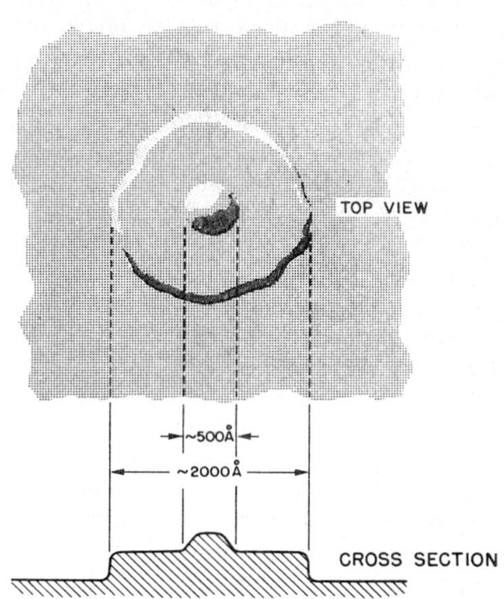

DAMAGE REGION IN NEUTRON-IRRADIATED n-TYPE SILICON AS OBSERVED BY ELECTRON MICROSCOPY AFTER ETCHING (BERTOLOTTI, et. al.).

FIG. 11. Disordered region in 14 MeV neutron-irradiated n-type silicon. Sketched from an electron micrograph by Bertolotti and coworkers.[76] Central 500 Å region is attributed to high resistivity disordered silicon. The surrounding 2000 Å region is attributed to space charge.

the experimental evidence[76] which stimulated investigations on defect clusters in Si. Subsequent infrared measurements,[17] transmission electron microscopy,[77] and electrical measurements[78] led to the currently held concept of vacancy-rich defect cluster formation by neutron irradiation. The central core region of ~ 500 Å diameter sketched in Figure 11 is attributed to a high resistivity, displacement cascade. A surrounding 2000 Å region is attributed to a space charge zone according to a model proposed earlier by Gossick[79] and by Crawford and Cleland.[80] Unfortunately, the obvious extensions of the replication electron microscopy measurements to other resistivities, other neutron energies, and as a function of annealing were never performed.

Current concepts leave the gross features of the defect cluster intact and add information on the nature of the damage and the defects in or near the core regions. Swanson and coworkers[77] (this conference) performed transmission electron microscope studies on neutron irradiated Si and have concluded that the core region remains crystalline rather than converting amorphous. Optical absorption measurements have shown the presence of A centers,[17] divacancies[28,33] and carbon centers[17] in neutron irradiated Si. Annealing growth of the A-center concentration occurs between 225 and 550 °K following neutron irradiation (see Figure 12)

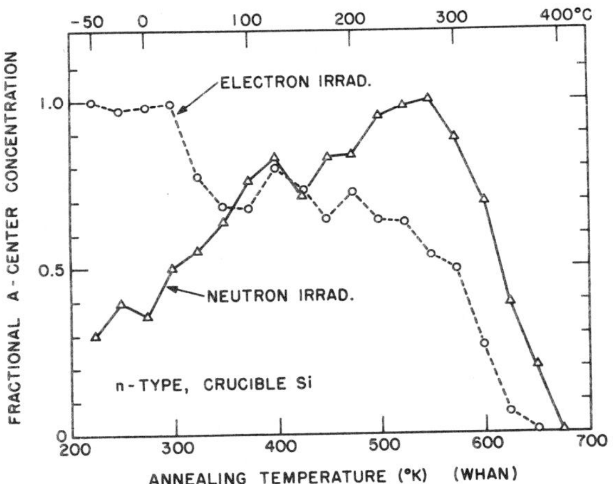

FIG. 12. Optical absorption coefficient associated with vacancy-oxygen (A) centers versus annealing after electron and neutron irradiation. The increase in A-center concentration has been interpreted as vacancy liberation from neutron-produced defect clusters.[17]

compared to an annealing loss of the A-centers for the same temperature range following electron irradiation.[17] Annealing growth similar to that for A centers has also been observed for divacancies following neutron irradiation[18,33] and following ion implantation.[18] Such annealing growth of A centers and of divacancies has been attributed to vacancy motion[17,18,33] in and near the core region. Thus we have the concept of a vacancy-rich defect cluster formed by neutron irradiation and by ion implantation. Subsequent vacancy motion and vacancy-vacancy trapping converts the defect cluster to a divacancy-rich region.[18,33]

The spatial distribution of defects within the core region is also of interest. Using the temperature for the annealing loss of divacancies observed in neutron irradiated Si and the trap concentration dependence for divacancy annealing determined from electron irradiation measurements (illustrated in Figure 4), Cheng and Lori[28] concluded that trap concentrations in or near the defect cluster range between $\sim 10^{17}$ and 10^{21} cm^{-3}. Calculations of Sigmund and coworkers[81] suggest the damage will be concentrated along the Si recoil track with an effective radius slightly smaller than the recoil range. Their calculations also predict an overlapping of any subcascades along the recoil track. Even if the initial vacancy distribution did exhibit some subcascading effects, the presence of vacancy motion below room temperature would tend to remove these effects from room temperature measurements. Therefore it seems reasonable to consider the core region as being continuous along the recoil track. An estimate of the relative number of crystal lattice defects within the damage core and within the space charge region is given in the EPR study by Daly and Noffke[82] (this conference).

5.2. Macroscopic properties

The concepts previously used to interpret the macroscopic properties of electron irradiated Si plus the additional concept of vacancy-rich clustered defect formation have been applied to interpret the macroscopic properties[78] of neutron irradiated Si. Plotted in Figure 13 are unannealed carrier removal

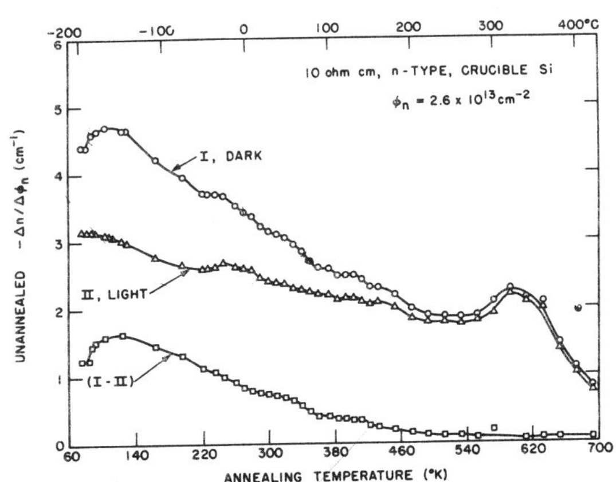

FIG. 13. Unannealed carrier removal per neutron for n-type crucible-grown silicon measured at 76 °K and plotted at the annealing temperature. Convergence of Curve I (measured in dark) and Curve II (measured with light) indicates annealing of light sensitive defects (Curve I–II).[78]

rates[78] versus annealing temperature for approximately-fission-spectrum neutron-irradiated 10 ohm cm n-type Si. Irradiations and measurements were made at 76 °K, and the results are plotted at the annealing temperature. The data for curve I were obtained from carrier concentration measurements in the dark; whereas, the data for curve II were obtained with weak illumination on the sample. The same illumination had an insignificant effect upon electron irradiated samples. Annealing characteristics measured with illumination have many characteristics in common with those observed following electron irradiation. The annealing of the illumination effect is shown by the curve labeled I–II. Except for an initial increase, the illumination effect anneals over a broad temperature range extending to 550 °K. Because this effect is not observed in electron irradiated Si[41] it is attributed to an annealing of defect clusters. The initial reverse annealing is attributed to slow electronic processes associated with the cluster so that equilibrium is not reached in the 20 min annealing periods below 120 °K. Similar illumination dependence and annealing results have been obtained for neutron irradiated p-type Si.[78]

Figure 14 shows a correlation of the annealing loss of the illumination effect and the annealing growth of A centers shown in Figure 12. The close correspondence between the two measurements

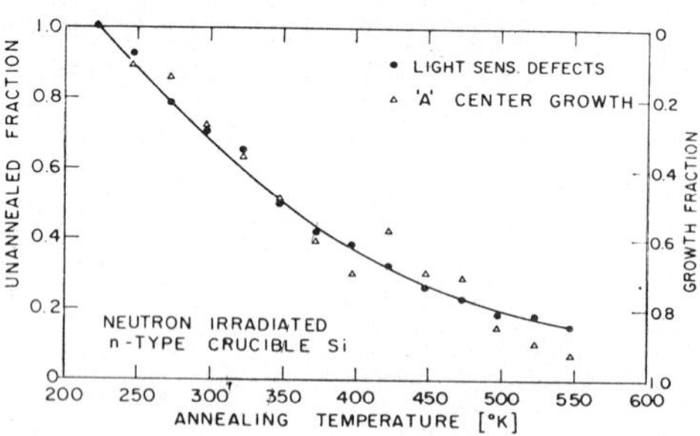

FIG. 14. Comparison of light sensitive defect annealing (Figure 13) and A-center growth (Figure 12) after neutron irradiation. This correlation suggests that the light sensitive defects are the vacancy liberating clusters.

supports the conclusion[78] that the vacancy-rich defect clusters are themselves responsible for the illumination dependence of the electrical measurements.

The formation of defect clusters in Si by neutron irradiation manifests itself in other ways. (1) There is very little irradiation temperature dependence of neutron produced defects between 80 and 100 °K,[42,83] which suggests that metastable interstitial-vacancy pair formation is an unlikely event under neutron irradiation. (2) Crystal impurities have a much smaller effect upon the defect inventory in neutron irradiated than in electron irradiated Si.[42,84,85] This is attributed to the high concentrations of vacancies formed in localized regions and a higher probability for vacancy-vacancy trapping to form divacancies rather than vacancy trapping by impurities in moderately doped Si. (3) The minority carrier lifetime in neutron irradiated Si, even after room temperature annealing, is strongly dependent upon the injection level.[84,86] This effect has been attributed to the potential barriers associated with the defect clusters.[84,86] Defect energy levels within the cluster, corresponding approximately to those for the divacancy, have been used in the modeling of the recombination processes.[86]

5.3. *Other defects*

It is apparent from the annealing data shown in Figures 5, 7, and 13 that a large fraction of the irradiation produced defects anneal below 400 °C, however not all of them anneal below 400 °C.[87] In electrical measurements the remaining defects are important if they significantly compensate the majority carrier concentration or if they affect the minority carrier recombination. For very high fluence irradiations the defects remaining after 400 °C are important in structural measurements of irradiation damage[88,89] as well as in the electrical measurements. These effects are illustrated in Figure 15. Electrical conductivity measurements by Tauke and Faraday[87] on 8 ohm cm n-type Si irradiated with 10^{18} 1 MeV electrons per cm² show annealing between 500 and 600 °C rather than below 400 °C as observed for low fluence irradiations. Two stages of annealing are observed in the structural measurements (anomalous X-ray transmission by Baldwin and Thomas,[88] and electron microscope images by Pankratz and coworkers[89]) following high fluence neutron irradiations. The lower temperature stage (100 to 300 °C) correlates with divacancy annealing but the higher temperature annealing stage extends to 750 °C.

Also included in Figure 15 are disorder annealing results of Mayer and coworkers[90] obtained by Rutherford backscattering measurements on ion implanted Si. The concept of vacancy-rich clustered defect formation has been applied to ion implantation as well as to neutron irradiation

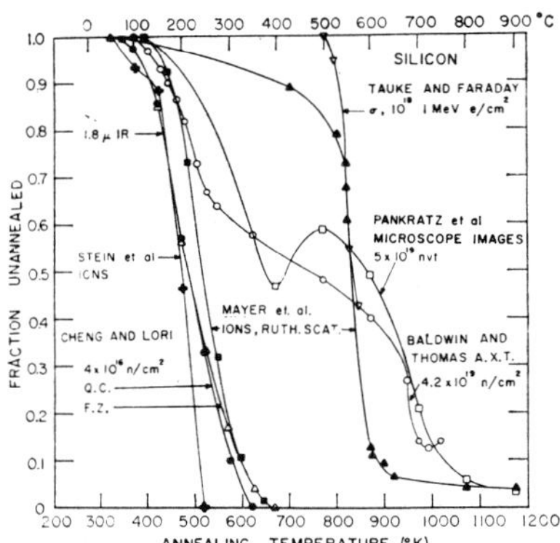

FIG. 15 Comparison of the isochronal annealing of electron, neutron and ion-produced defects in Si: $1.8\,\mu$ infrared absorption band in Si implanted with 5×10^{13} 400 keV oxygen ions/cm^2,[29] and in pulled (Q.C.) and floating zone (F.Z.) Si irradiated with 4×10^{16} n/cm^2;[28] anomalous X-ray transmission (A.X.T.) 4.2×10^{19} n/cm^2;[88] electron microscope cluster images 5×10^{19} nvt;[89] Rutherford scattering measurements[90] of lattice disorder for 40 keV ions in crystalline (1.1×10^{13} Sb/cm^2, 2.0×10^{13} P/cm^2) and in amorphous Si (3×10^{14} Sb/cm^2, 4×10^{14} Ga/cm^2, 2.5×10^{15} As/cm^2); and electrical conductivity annealing in n-type pulled Si irradiated to 10^{18} 1 MeV electrons/cm^2.[87]

damage to explain divacancy formation.[18] However, as the ion fluence is increased, the crystalline structure of the Si is destroyed, and an 'amorphous layer' is formed.[1,90] 'Amorphous layer' is a term applied to describe the condition of the ion damaged layer when crystallographic directions can no longer be detected in Rutherford backscattering measurements.[1,90] Annealing of the disorder produced by an ion fluence less than that to form an amorphous layer correlates in temperature with divacancy annealing. Amorphous layer annealing recovery occurs between 500 and 600 °C, but the 500–600 °C annealing cannot be uniquely ascribed to amorphous layer annealing because it is also an annealing stage in 1 MeV electron irradiated Si which is surely crystalline.

Our current concepts and inventory of identified defects are not sufficient to interpret either the annealing mechanisms or the defects involved in the annealing near 600 °C. Defect annealing in this temperature range is of particular importance in ion implantation because there is sufficient residual damage to strongly affect electrical properties.[91] Electrical measurements on electron irradiated degenerate Si show annealing at temperatures even higher than 600 °C[20] and indicate impurity effects in the high temperature annealing.

The modeling of multiple vacancy defects[11] observed in EPR measurements following high fluence neutron irradiation[92] and ion implantation[93,94] is viewed as a step toward understanding the defects annealing at temperatures >400 °C. There has recently[61] been an observation of an optical absorption band at $8.91\,\mu$ in electron irradiated Si which grows in as the divacancy anneals. The annealing growth and loss characteristics bear a resemblance to that of the P-1 center (odd number of vacancies[11] $\geqslant 3$) observed in EPR studies.[92–94]

Ion implantation combines the effects of both a high concentration of lattice defects and a high impurity concentration. Therefore, studies of both heavily irradiated Si, and irradiation studies on heavily doped and multiply doped materials are of considerably interest for interpreting ion implantation results. Ion implantation is itself important in such defect studies because it provides a controlled method for introducing the dopant impurities.

6. SUMMARY

Concepts such as: (1) low temperature motion of free vacancies, (2) charge state dependent motion of vacancies, (3) impurity trapping of vacancies and vacancy-vacancy trapping, (4) low temperature displacement of substitutional impurities, and (5) trap concentration-dependent annealing of vacancy associated defects, are all well-founded upon microscopic defect identification and characterization. Other concepts such as: (1) metastable vacancy-interstitial pair formation by low energy encounters, and (2) vacancy-rich cluster formation by neutron irradiation or by ion implantation are not directly supported by microscopic defect identification, but they have been very successfully used to explain a wide range of experimental data. These concepts on irradiation produced defects in Si have been quite helpful for organizing information from investigations and for guiding other investigations.

Correlations of stress response and annealing behavior for the A center observed in EPR and optical absorption measurements are used to illustrate the correlation method for studying the same defect in different kinds of measurements. Additional correlations of defect behavior observed in EPR optical absorption, in photoconductivity, and in luminescence measurements are briefly described.

Many macroscopic properties of irradiated Si have been successfully correlated with the characteristics of specific defects. For example, electrical measurements of isochronal annealing have been correlated with the A-center, E-center, K-center, divacancy, and interstitial Al defects. Correlations have given a reasonably detailed understanding of electrically active defect production mechanisms and of the defect inventory for n-type Si, but such a detailed understanding has not yet been achieved for p-type Si. In particular, the role of vacancies in the electrical properties of p-type Si needs clarification, the electrical activity of interstitial carbon needs to be established, and an understanding of the difference in the annealing behavior of Al- and B-doped Si is needed. A need also exists for understanding the mechanisms and defects involved in the high temperature annealing of ion implanted Si.

ACKNOWLEDGEMENT

I would like to thank Dr. F. L. Vook for his valuable comments and suggestions on the manuscript.

REFERENCES

1. J. W. Mayer, L. Ericksson and J. A. Davies, *Ion Implantation in Semiconductors, Silicon and Germanium* (Academic Press, New York, 1970).
2. G. D. Watkins, *Radiation Damage in Semiconductors* (Dunod, Paris, 1965), p. 97.
3. G. D. Watkins, *Radiation Effects on Semiconductor Components* (Journees D'Electronique, Toulouse, France, 1967), Vol. I, paper No. A1.
4. G. D. Watkins, *Radiation Effects in Semiconductors*, Ed. F. L. Vook (Plenum Press, New York, 1968), p. 67.
5. G. D. Watkins, *IEEE Trans. on Nucl. Sci.*, **16**, 13 (1969).
6. F. L. Vook and H. J. Stein, *Radiation Effects in Semiconductors*, Ed. F. L. Vook (Plenum Press, New York, 1968), p. 99.
7. A. Brelot and J. Charlemagne, this conference.
8. N. Almeleh and B. Goldstein, *Phys. Rev.*, **149**, 687 (1966).
9. J. W. Corbett and G. D. Watkins, *Phys. Rev.*, **138**, A555 (1965).
10. C. E. Jones and W. D. Compton, this conference.
11. K. L. Brower, this conference.
12. G. D. Watkins, *Phys. Rev.*, **155**, 802 (1967).
13. J. W. Corbett, G. D. Watkins, R. M. Chrenko and R. S. McDonald, *Phys. Rev.*, **121**, 1015 (1961).
14. L. J. Cheng, J. C. Corelli, J. W. Corbett and G. D. Watkins, *Phys. Rev.*, **152**, 761 (1966).
15. L. J. Cheng, *Radiation Effects in Semiconductors*, Ed. F. L. Vook (Plenum Press, New York, 1968), p. 143.
16. A. H. Kalma and J. C. Corelli, *Radiation Effects in Semiconductors*, Ed. F. L. Vook (Plenum Press, New York, 1968), p. 153; *Phys. Rev.*, **173**, 734 (1968).
17. E. R. Whan, *J. Appl. Phys.*, **37**, 3378 (1966).
18. H. J. Stein, *Appl. Phys. Letters*, **15**, 61 (1969); F. L. Vook and H. J. Stein, *Rad. Effects* **6**, 11 (1970).
19. G. D. Watkins, *J. Physical Soc. of Japan*, **18**, Suppl. II, 22 (1963).
20. L. L. Sivo and E. E. Klontz, *Phys. Rev.*, **178**, 1264 (1969).
21. G. D. Watkins and J. W. Corbett, *Phys. Rev.*, **121**, 1001 (1961).
22. H. J. Stein and F. L. Vook, *Rad. Effects*, **1**, 41 (1969). *Appl. Phys. Letters*, **13**, 343 (1968).
23. J. W. Corbett, G. D. Watkins and R. S. McDonald, *Phys. Rev.*, **135**, A1381 (1964).
24. G. Bemski and W. M. Augustyniak, *Phys. Rev.*, **108**, 645 (1957).
25. G. K. Wertheim, *Phys. Rev.*, **110**, 1272 (1958); *J. Appl. Phys.*, **30**, 1232 (1959); *Phys. Rev.*, **115**, 568 (1959).
26. F. L. Vook and H. J. Stein, *Rad. Effects*, **2**, 23 (1969); **2**, 139 (1969).
27. L. J. Cheng and J. Lori, *Phys. Letters*, **31A**, 281 (1970); *Phys. Rev. B* **1**, 1558 (1970).
28. L. J. Cheng and J. Lori, *Phys. Rev.*, **171**, 856 (1968).
29. H. J. Stein, F. L. Vook and J. A. Borders, *Appl. Phys. Letters*, **14**, 328 (1969).
30. H. J. Stein, F. L. Vook, D. K. Brice, J. A. Borders and S. T. Picraux, *Rad. Effects*, **6**, 19 (1970).
31. V. D. Tkachev and M. T. Lappo, this conference.
32. C. S. Chen and J. C. Corelli, this conference.
33. C. E. Barnes, this conference.
34. H. Y. Fan and A. K. Ramdas, *J. Appl. Phys.*, **30**, 1127 (1959).
35. L. J. Cheng and P. Vajda, *Phys. Rev.*, **186**, 816 (1969).
36. G. D. Watkins and J. W. Corbett, *Phys. Rev.*, **138**, A543 (1965).
37. H. Y. Fan and A. K. Ramdas, *Proc. Int. Conf. of Semiconductor Phys.* (Czech. Acad. Sci., Prague, 1961), p. 309.
38. A. R. Bean, R. C. Newman and R. S. Smith, *J. Phys. Chem. Solids*, **31**, 739 (1970).
39. D. L. Kendall and D. B. DeVries, *Semiconductor Silicon*, Ed. by R. R. Haberecht and E. L. Kern (Electrochemical Society, Inc., New York, 1969), p. 358.
40. R. C. Young, J. W. Westhead and J. C. Corelli, *J. Appl. Phys.*, **40**, 271 (1969).
41. H. J. Stein and F. L. Vook, *Phys. Rev.*, **163**, 790 (1967).
42. H. J. Stein and R. Gereth, *J. Appl. Phys.*, **39**, 2890 (1968).
43. M. Hirata, M. Hirata and H. Saito, *J. Appl. Phys.*, **37**, 1867 (1966).
44. H. Saito and M. Hirata, Japanese *J. Appl. Phys.*, **2**, 678 (1963).
45. E. Sonder and L. C. Templeton, *J. Appl. Phys.*, **34**, 3295 (1963).
46. V. S. Vavilov, E. F. Uvarov and M. V. Chukichev, *Soviet Phys.-Semiconductors*, **3**, 1557 (1970).
47. H. J. Stein and F. L. Vook, *Radiation Effects in Semiconductors*, Ed. F. L. Vook (Plenum Press, New York, 1968), p. 115.
48. E. Matsuura, *Ninth Intl. Conf. on Phys. of Semiconductors I* (Publishing House, Nauka, Leningrad, 1968), p. 126.
49. B. L. Gregory and C. E. Barnes, *Radiation Effects in Semiconductors*, Ed. F. L. Vook (Plenum Press, New York, 1968), p. 124.
50. G. J. Brucker, *Phys. Rev.*, **183**, 712 (1969).
51. R. C. Newman and J. B. Willis. *J. Phys. Chem. Solids*, **26**, 373 (1965).

52. J. A. Baker, T. N. Tucker, N. E. Moyer and R. C. Buschert, *J. Appl. Phys.*, **39**, 4365 (1968).
53. R. E. Whan and F. L. Vook, *Phys. Rev.*, **153**, 814 (1967).
54. N. E. Moyer and R. C. Buschert, *Radiation Effects in Semiconductors*, Ed. F. L. Vook (Plenum Press, New York, 1968), p. 444.
55. R. C. Newman and A. R. Bean, this conference.
56. B. Goldstein, this conference.
57. J. A. Naber, H. Horiye and B. C. Passenheim, this conference.
58. E. L. Elkin and G. D. Watkins, *Phys. Rev.*, **174**, 881 (1968).
59. M. Hirata, M. Hirata and H. Saito, *J. Phys. Soc. of Japan*, **27**, 405 (1969).
60. L. C. Kimerling, H. M. DeAngelis and C. P. Carnes *Phys. Rev. B* **3**, 427 (1971).
61. J. C. Corelli, R. C. Young and C. S. Chen, *IEEE Trans. on Nucl. Sci.*, **17**, 128 (1970).
62. C. S. Chen, J. C. Corelli and G. D. Watkins, *Bull. Am. Phys. Soc.*, **14**, 395 (1969).
63. B. L. Gregory, *J. Appl. Phys.*, **36**, 3765 (1965).
64. L. J. Cheng and J. Lori, *Phys. Rev. B* **1**, 1558 (1970); *Phys. Letters*, **31A**, 281 (1970).
65. R. L. Novak, Ph.D. Thesis, University of Pennsylvania, 1964 (unpublished).
66. H. Djerassi, J. Merlo-Flores and J. Messier, *J. Appl. Phys.*, **37**, 4510 (1966).
67. G. Fladda, K. Bjorkqvist, L. Eriksson and D. Sigurd, *Appl. Phys. Letters*, **16**, 313 (1970).
68. J. C. North and W. M. Gibson, *Appl. Phys. Letters*, **16**, 126 (1970).
69. J. R. Carter, Jr., *J. Phys. Chem. Solids*, **27**, 913 (1966); *IEEE Trans. on Nucl. Sci.*, **13**, 24 (1966).
70. K. L. Brower, *Phys. Rev. B* **1**, 1908 (1970).
71. V. M. Malovetskaya, G. N. Galkin and V. S. Vavilov, *Soviet Physics-Solid State*, **4**, 1008 (1962).
72. V. S. Vavilov, G. N. Galkin, V. M. Malovetskaya and A. F. Plotnikov, *Soviet Physics-Solid State*, **4**, 1442 (1963).
73. Annealing at 300 °K has also been observed for defects quenched in Si. (M. L. Swanson, *Phys. State. Sol.*, **33**, 721 (1969).) The activation energy for the 'quenched in' defects was also found to be 0.81 ± 0.04 eV, but the electronic energy level at $E_v + 0.4$ eV does not correspond to the $E_v + 0.25$ eV observed in the irradiated samples.
74. M. Cherki and A. H. Kalma, *Phys. Rev. B* **1**, 647 (1970).
75. S. D. Devine and R. C. Newman, *J. Phys. Chem. Solids*, **31**, 685 (1970).
76. M. Bertolotti, D. Sette and G. Vitali, *J. Appl. Phys.*, **38**, 2645 (1967).
77. M. L. Swanson, J. R. Parsons and C. W. Hoelke, this conference.
78. H. J. Stein, *Phys. Rev.*, **163**, 801 (1967); *IEEE Trans. on Nucl. Sci.*, **15**, 69 (1968); *J. Appl. Phys.*, **39**, 5283 (1968).
79. B. R. Gossick, *J. Appl. Phys.*, **30**, 1214 (1959).
80. J. H. Crawford and J. W. Cleland, *J. Appl. Phys.*, **30**, 1204 (1959).
81. P. Sigmund, G. P. Scheidler and G. Roth, *Proc. Intl. Conf. on Solid State Phys. with Accelerators*, Ed. A. N. Goland, BNL 50083 (1967), p. 374.
82. D. F. Daly and H. E. Noffke, this conference.
83. K. L. Starostin, L. B. Poretskii, V. N. Mordkovich and Yu. F. Tuturov, *Soviet Physics-Semiconductors*, **1**, 1520 (1968).
84. O. L. Curtis, Jr., *Lattice Defects in Semiconductors*, Ed. R. R. Hasiguti (Pennsylvania State University Press, University Park, 1968), p. 333.
85. R. F. Bass, *IEEE Trans. on Nucl. Sci.*, **14**, 78 (1967).
86. B. L. Gregory, *IEEE Trans. on Nucl. Sci.*, **16**, 53 (1969).
87. R. V. Tauke and B. J. Faraday, *Lattice Defects in Semiconductors*, Ed. R. R. Hasiguti (Pennsylvania State University Press, University Park, 1968), p. 170.
88. T. O. Baldwin and J. E. Thomas, *J. Appl. Phys.*, **39**, 4391 (1968).
89. J. M. Pankratz, J. A. Sprague and M. L. Rudee, *J. Appl. Phys.*, **39**, 101 (1968).
90. J. A. Davies, J. Denhartog, L. Eriksson and J. W. Mayer, *Can. J. Phys.*, **45**, 4053 (1967); J. W. Mayer, L. Eriksson, S. T. Picraux and J. A. Davies, *Can. J. Phys.*, **46**, 663 (1968); L. Eriksson, J. A. Davies and J. W. Mayer, *Radiation Effects in Semiconductors*, Ed. F. L. Vook (Plenum Press, New York, 1968), p. 398.
91. H. J. Stein, *Rad. Effects*, **6**, 175 (1970).
92. Wun Jung and G. S. Newell, *Phys. Rev.*, **132**, 648 (1963).
93. K. L. Brower, F. L. Vook and J. A. Borders, *Appl. Phys. Letters*, **15**, 208 (1969).
94. D. F. Daly and K. A. Pickar, *Appl. Phys. Letters*, **15**, 267 (1969).

DISCUSSION

Question (STREETMAN) If you correct the 80 °K hole removal rate data for the lack of ionization of the Al atoms, do you conclude that the lower removal rate is real for Al-doped compared with *B*-doped material?

Answer (STEIN) No. The defect introduction rates are about the same. There seems to be no advantage to using Al doping in terms of carrier removal rate.

Question (MASTERS) I have a question which pertains to the identification of the divacancy. Do the data for the process of monovacancy association to form divacancies exhibit the expected second-order kinetic dependence upon monovacancy concentration?

Answer (STEIN) I do not believe the kinetics of divacancy formation from monovacancies have been examined in any experiment where both the vacancy and divacancy were explicitly identified. Barnes (this conference) has used the expected second-order dependence of divacancy growth on monovacancy concentration to suggest that an observed annealing loss of a 'band edge' absorption after low temperature neutron irradiation is associated with vacancy-rich clustered defect regions.

Question (NABER) Is anything known about the energy levels of the vacancy and the activation energy of the singly negative vacancy?

Answer (WATKINS) The donor level is believed to be <0.05 eV above the valence band. The position of the other levels is not known. The activation energies of motion indicated in Figure 1 of 0.18 ± 0.02 eV and 0.33 ± 0.03 eV are from annealing studies in *n*- and *p*-type materials respectively where the Fermi level was locked to the corresponding shallow donor or acceptor. From the level structure of Figure 1, these were therefore deduced to be associated with the double negative and neutral charge states, respectively. No information is available about the value for the single negative charge state.

Question (MITCHELL) You showed a slide of some results by Bertolotti indicating a core disordered region of 500°A. However, Bertolotti's technique is an etching technique. The size of the etch pattern can be much greater than that of the underlying defect initiating the etch pattern (compare, for example, etch patterns of dislocations).

Answer (STEIN) The appealing thing about the 500°A dimension observed by Bertolotti and coworkers for the core region is that it coincides approximately with the expected range for Si atoms recoiling from the neutron encounters. I believe we can look forward to more definition of cluster size from other measurements such as transmission electron microscopy (see Swanson and coworkers, this conference) and electron paramagnetic resonance (see Daly and Noffke, this conference).

Question (GIBSON) Could you amplify your comment on the role of the divacancy in the reverse annealing of substitutional boron? Do you have a model of the promotion of substitutional boron to interstitial by a divacancy mechanism?

Answer (STEIN) I do not have a model for the interaction of divacancies with substitutional boron. There is experimental evidence,[28] however, that the presence of boron in Si can enhance divacancy formation, which is also the case for carbon or oxygen.[14,37,38] This may mean that some impurities aid in nucleating divacancy formation. Many divacancies would then be positioned near the impurities and could further interact with the impurities upon annealing.

ELECTRON IRRADIATION EFFECTS IN SILICON AT LIQUID HELIUM TEMPERATURES USING AC HOPPING CONDUCTIVITY†

R. E. McKEIGHEN AND J. S. KOEHLER

Nuclear Engineering Program, Department of Physics, and Materials Research Laboratory, University of Illinois, Urbana, Illinois 61801

Changes in ac hopping conductivity with electron irradiation have been measured in n-type silicon, p-type silicon, and high purity silicon. Irradiations were carried out at both 4.8 and 1.6 degrees kelvin, with measurements made at reference temperatures of 4.2 and 1.3 degrees kelvin, respectively. In the p-type crystals, the changes in ac hopping conductivity depended strongly on the concentration of chemical acceptors, indicating a concentration dependence on impurities of the defect production rate. The production rates at $1.6°K$ were generally very similar to those for a $4.8°K$ irradiation. No significant thermal annealing stages below room temperature were observed in any of the crystals. A small amount of 'reverse annealing', i.e. an increase in ac hopping conductivity was observed below $30°K$. No radiation annealing effects due to a beam of either .500 MeV or .350 MeV electrons were observed, either at $4.5°K$ or at $1.45°K$.

1. INTRODUCTION

The measurement of ac hopping conductivity requires compensating minority impurities (or defects) in a crystal. Conduction takes place by hopping of electrons from occupied to unoccupied localized donor states (for an n-type crystal). The model is that of an electron tunneling from an occupied donor site to an unoccupied site in the presence of the coulomb fields arising from compensating acceptors and ionized donors. M. Pollak and co-workers[1] found experimentally that the ac hopping conductivity (σ_{ac}) is approximately given by:

$$\sigma_{ac} = K\omega^{0.8}N_{min}^{0.85}$$

where K is a constant, ω is the frequency of the ac signal, and N_{min} is the concentration of compensating minority impurities. At the low temperature end (below $\sim 20°K$ for silicon), the conductivity is roughly proportional to the minority impurity concentration and is almost independent of the majority impurity concentration.

2. EXPERIMENTAL

The ac hopping conductivity was determined by measuring the dielectric loss (or dissipation factor) of the samples using a General Radio model 1615-A capacitance bridge with a PAR model HR8 lock-in amplifier used as a null detector. The bridge is capable of 6-figure resolution in capacitance and 4-figure resolution in dissipation factor. Measurements were made at frequencies from 0.1 KHZ to 100 KHZ. Samples were typically 1.0×10^{-2} cm to 1.5×10^{-2} cm thick and had an area from 1.0 to 1.5 cm². Sample electrodes were $\frac{1}{4}$ inch diameter of either evaporated gold or silver. The temperatures of the samples during irradiation ranged from $1.6°K$ to $5.2°K$ and were measured with an Allen–Bradley carbon resistance thermometer mounted next to the samples. Below $4.2°K$ the carbon resistor was calibrated against the vapor pressure above the liquid helium bath. Beam currents ranged from 2.0×10^{-8} to 1.0×10^{-7} amps/cm².

To decide whether the changes observed are to be associated with changes in surface conductivity, four samples with electrodes placed a centimeter apart on the bombarded surface were measured. Two intrinsic, one n-type, and one p-type specimen were used. With various energies from 0.300 to 2.0 MeV and a total fluence up to 1.4×10^{16} e⁻/cm², changes in the surface conductivity were too small to measure. This puts an upper limit on surface conductance of 4.0×10^{-16} (ohm)⁻¹ per square of surface area. Samples typically had only 5 to 10 squares of surface area between front and back electrodes.

3. RESULTS

The changes in σ_{ac} with electron fluence for p-type, aluminum-doped silicon are illustrated in Figures 1 and 2 and for p-type, boron-doped silicon

† This work was supported in part by the U.S. Atomic Energy Commission under Contract AT(11–1)–1198.

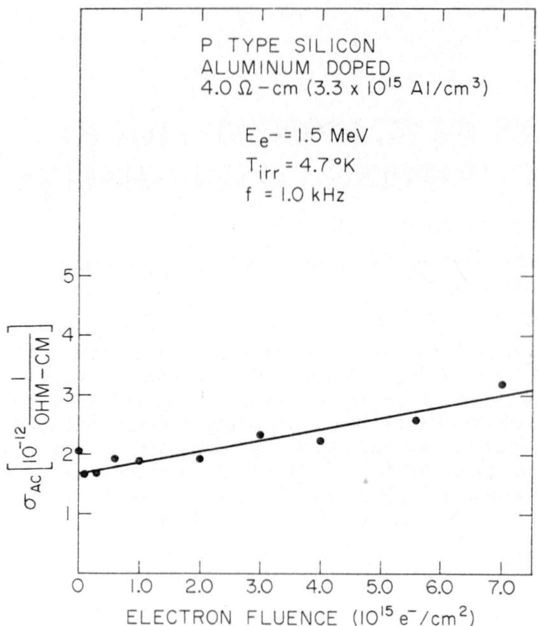

FIG. 1. The ac hopping conductivity, σ_{ac}, at 1.0 kHz vs 1.5 MeV electron fluence for aluminum-doped silicon (4.0 ohm-cm, 3.3×10^{15} Al/cm³). The irradiation temperature was 4.7 °K.

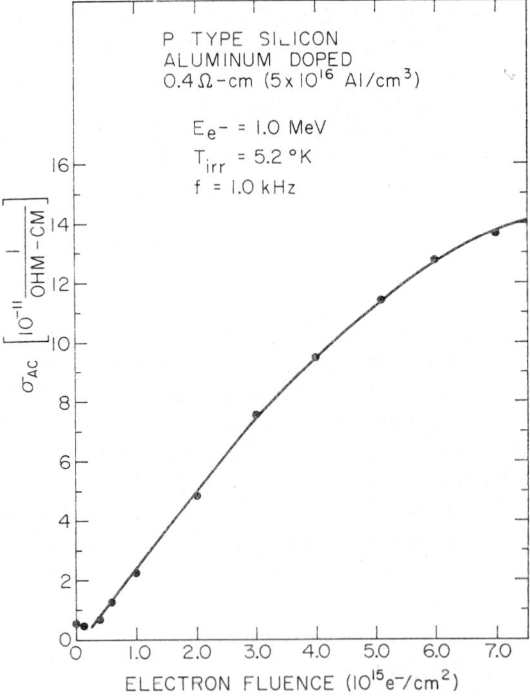

FIG. 2. σ_{ac} at 1.0 kHz vs 1.0 MeV electron fluence for aluminum-doped silicon (0.4 ohm-cm, 5×10^{16} Al/cm³). Irradiation temperature, 5.2 °K.

in Figure 3. The changes are fairly linear with fluence, at least up to 5.0×10^{15} e⁻/cm². The notable thing about the *p*-type samples is that the changes in σ_{ac} for a given electron fluence depend quite strongly on the concentration of impurities in the crystal. The changes in *n*-type, phosphorus doped silicon are shown in Figure 4. The magnitudes of σ_{ac} in the *n*-type sample were much smaller than those typically obtained in *p*-type crystals of similar impurity concentration. In high purity silicon, small initial changes in σ_{ac} were observed, which quickly saturated with electron flux as shown in Figure 5. After irradiation, samples typically

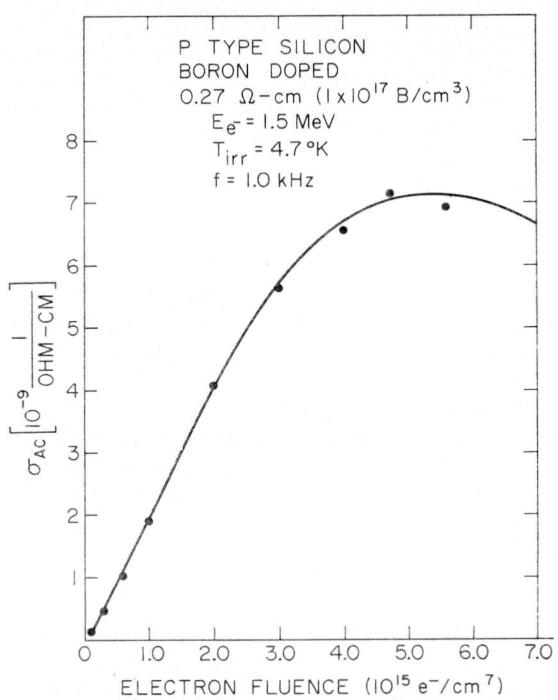

FIG. 3. σ_{ac} at 1.0 kHz vs 1.5 MeV electron fluence for boron-doped silicon (0.27 ohm-cm, 1×10^{17} B/cm³). Irradiation temperature, 4.7 °K.

showed a frequency dependence of the form $\sigma_{ac} \sim \omega^s$, where *s* ranged in value from 0.81 to 0.95. This compares with the value of $s = 0.79$ observed by Pollak and Geballe[1] for crystals compensated with chemical impurities.

Since Watkins'[2] data suggests that interstitials are mobile even at 4.2 °K, irradiations were also performed holding the sample temperature well below 2.0 °K to see if there would be any significant difference in production rates between a 1.6 °K

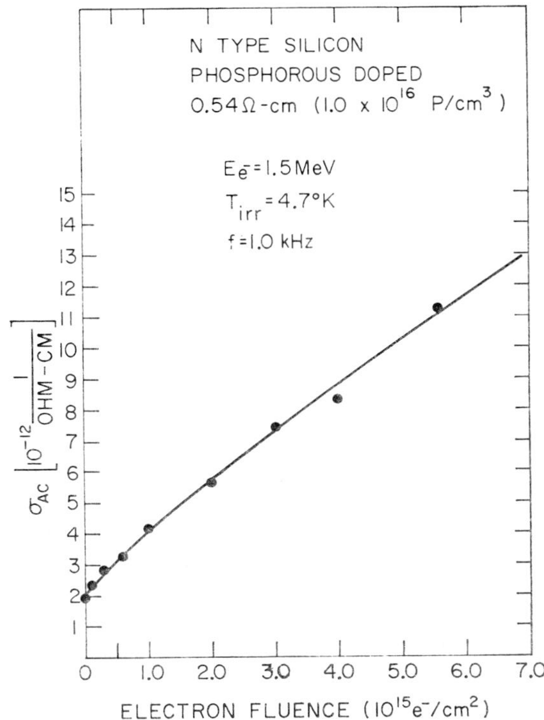

FIG. 4. σ_{ac} at 1.0 kHz vs 1.5 MeV electron fluence for phosphorus-doped silicon (0.54 ohm/cm, 1.0 × 10^{16} P/cm³). Irradiation temperature, 4.7 °K.

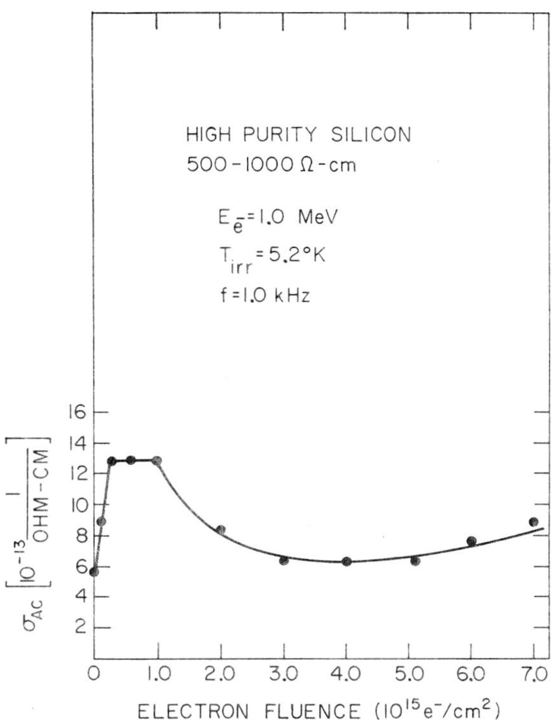

FIG. 5. σ_{ac} at 1.0 kHz vs 1.0 MeV electron fluence for high purity silicon (500–1000 ohm-cm). Irradiation temperature, 5.2 °K.

irradiation and a 5.0 °K irradiation. The results for the various crystals are shown in Figures 6–8. The changes in σ_{ac} for a 1.6 °K irradiation are similar to those observed in a 5.0 °K irradiation. It should also be noted that for an irradiation made at 1.6 °K and measurements made at 1.3 °K, p-type crystals doped with either aluminum or boron of similar concentration yield similar values of σ_{ac} for a given electron fluence. This indicates that the large changes in σ_{ac} observed in a sample doped with 1.0×10^{17} boron/cm³ are due to the concentration dependence and not due to the type of impurity. Measurements were also taken as the samples were allowed to slowly warm up from 1.3° to 4.2 °K. A rise in σ_{ac} from 1.3° to 4.2 °K is expected because the tunneling rate should be temperature dependent. Measurements were then again taken at 1.3 °K to see if any changes ('annealing') took place between 1.3 °K and 4.2 °K (see Figure 9). No significant changes were observed in the p-type specimens but there was a small change in the n-type crystal. Radiation annealing effects due to a 0.500 MeV electron beam following damage introduced by a 1.500 MeV beam were not observed in any of the

samples, either at 5.8 °K, 5.0 °K, or at 1.45 °K. No annealing stages in isochronal (ten minute) anneals up to room temperature following irradiation were observed in any of the samples. A small increase in σ_{ac} on warming to 20°–30 °K was usually observed.

4. DISCUSSION

In p-type silicon, one expects the interstitial to be positively charged and the vacancy to be neutral when the Fermi level is pinned to the acceptor level at these low temperatures. Therefore, the interstitials (silicon or group three impurity) will act as compensating minority impurities in a p-type lattice and one expects σ_{ac} to be proportional to their concentration. Furthermore the vacancies should not contribute to the hopping conductivity in this situation.

In n-type silicon, one expects the interstitial to be neutral (or positive), whereas the vacancy probably has a double negative charge. In this case the vacancies will be acting as the minority impurity and one expects σ_{ac} to be proportional to their

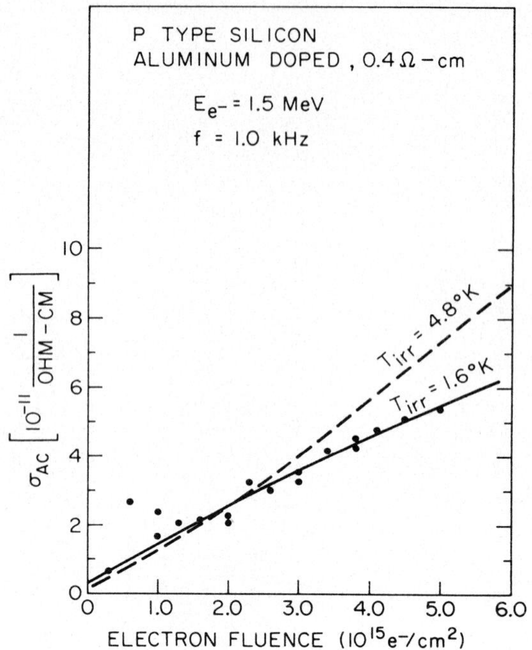

FIG. 6. σ_{ac} at 1.0 kHz vs 1.5 MeV electron fluence at two different irradiation temperatures for aluminum-doped silicon (0.4 ohm-cm).

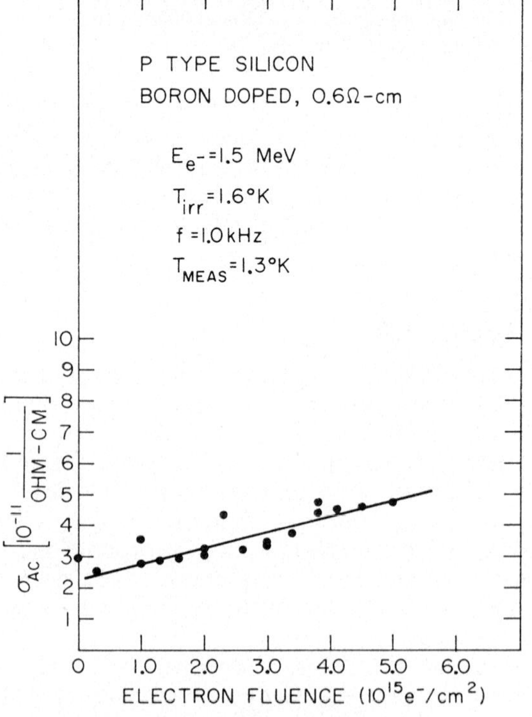

FIG. 7. σ_{ac} at 1.0 kHz measured at 1.3 °K vs 1.5 MeV electron fluence at 1.6 °K. The sample was boron-doped (0.6 ohm-cm).

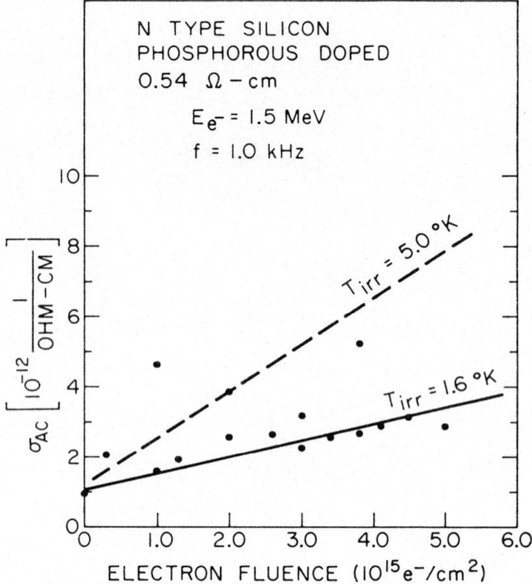

FIG. 8. σ_{ac} at 1.0 kHz vs 1.5 MeV electron fluence at two different irradiation temperatures for phosphorus-doped silicon (0.54 ohm-cm).

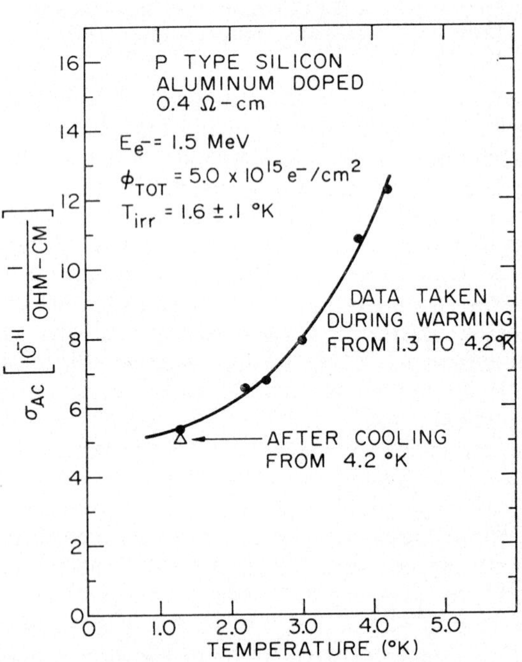

FIG. 9. Temperature dependence of σ_{ac} after 1.6± 0.1 °K irradiation with 1.5 MeV electrons to a fluence of 5.0×10^{15} e/cm². The sample was aluminum-doped silicon (0.4 ohm-cm).

TABLE I

Defect production rates for 1.5 MeV electrons deduced assuming singly charged defects

Crystal type	Dopant	Growth method	Resistivity (ohm-cm)	Dopant concentration (cm^{-3})	Defect production rate (cm^{-1})
P	Aluminum	Float zone	4.0	3.3×10^{15}	0.03
P	Aluminum	Pulled	0.40	5.0×10^{16}	1.8
P	Boron	Float zone	0.60	3.0×10^{16}	1.8
P	Boron	Float zone	0.27	1.0×10^{17}	10.0
N	Phosphorus	Float zone	0.54	1.0×10^{16}	0.02

concentration but independent of the interstitial concentration.

We have estimated the production rate of defects from the changes observed in σ_{ac}, assuming singly charged defects, and the results are summarized in Table I, for an electron energy of 1.50 MeV. These values are only rough estimates, good to ± 50 per cent.

Note that for a crystal doped with $1.0 \times 10^{17}/cm^3$ boron, the production rate is 10.0 cm^{-1}, assuming singly charged defects, or 5.0 cm^{-1}, assuming doubly charged defects. The value of 5.0 cm^{-1} is very close to the theoretically calculated production rate.

In summary the major results are that: (1) there is no significant thermal annealing below room temperature; (2) there is no radiation annealing observed at liquid helium temperatures; (3) the production rates at 1.6 °K are essentially the same as at 5.0 °K (within a factor of two); (4) the production rates seem to depend on impurity concen-

tration in p-type crystals. These results serve to confirm the suggestion of Watkins that the interstitial is mobile at liquid helium temperatures and is either trapped at impurities or interchanges with them in a p-type lattice, creating impurity interstitials which migrate only above room temperature.

The measurement technique employed is also capable of monitoring, for the first time, the small changes produced in non-degenerate n-type silicon during a liquid helium temperature irradiation. The data suggests that during the irradiation most of the interstitial-vacancy pairs recombine in n-type silicon. A few interstitials escape to be trapped at impurities, thus accounting for the small changes observed.

REFERENCES

1. M. Pollak and T. H. Geballe, *Phys. Rev.*, **122**, 1742 (1961), also *Phys. Rev.*, **133**, A564 (1964).
2. G. D. Watkins, in *Radiation Damage in Semiconductors* (Dunod, Paris, 1965), p. 104.

DISCUSSION

Question (COMPTON) Have you observed any changes in the activation energy for low temperature conductivity, the ϵ_3 for a.c. conductivity, with increasing compensation?

Answer (MCKEIGHEN) We have not measured σ_{ac} vs temperature at various stages during the irradiation, as the compensation is being varied by the irradiation. Therefore, we have not plotted σ_{ac} vs 1/T for different values of compensation to see how the activation energy ϵ_3 does vary with irradiation.

Question (SEEGER) If your suggestion is right that the interstitial is very mobile at low temperatures and the only reason you don't see radiation annealing is that interstitials get trapped by impurities, then as soon as your total number of

defects produced exceeds the impurity concentration then the damage production curve should bend over. Have you observed that?

Answer (KOEHLER) The defect production rate does decrease when the concentration of minority impurities approaches the concentration of majority impurities. This occurred consistently in the specimens having various impurity concentrations. It is just what one should expect from the theory of Pollack and Geballe. The negative results obtained in attempts to see radiation annealing were made after considerable irradiation, so that some decrease in the damage production rate had taken place, but no radiation annealing was observed.

Question (ALBANY) The activation energy ϵ_3

associated with hopping conduction is known to change with compensation. Is there any contribution of this effect to the variation observed in the conductivity with the irradiation.

Answer (MCKEIGHEN) Yes, one expects ϵ_3 to vary with compensation and consequently with irradiation. One can think in terms of the expression $\sigma_{ac} = \sigma_0 \exp\left(-\epsilon_3/kT\right)$ where ϵ_3 is being varied by the irradiation. But for defect studies it is more illuminating to consider the expression given in the paper, where for a fixed temperature, $\sigma_{ac} \cong K\omega^{0.80} N_{min}^{0.85}$ where the number of defects (N_{min}) is being increased proportionally to the electron fluence.

Comment (WATKINS) The low damage rate and failure to see annealing in *n*-type silicon suggests that virtually no isolated vacancies are being formed. This is of course consistent with the known strong temperature dependence for damage in the *n*-type material. It would be interesting to use your technique on samples irradiated with higher energy electrons where this temperature dependence disappears and isolated vacancies are known to be formed. Similarly, irradiation with your present electron energies but at higher temperatures ($\sim 77\,^\circ$K) and then cooling to low temperatures for your measurements would be instructive.

STUDY OF ATOMIC DISPLACEMENTS IN He⁺ IRRADIATED SILICON AND A SILICON-GERMANIUM ALLOY, THROUGH RUTHERFORD SCATTERING

P. BARUCH, F. ABEL, C. COHEN, M. BRUNEAUX

Groupe de Physique des Solides de l'Ecole Normale Supérieure†, Tour 23/9, quai Saint-Bernard, Paris 5e, France

AND

D. W. PALMER AND H. PABST

School of Mathematical and Physical Sciences, University of Sussex, Brighton, England‡

The disorder created by He⁺ ions of 275 keV and 1610 keV in Si and in a Si-Ge alloy (0.6 per cent atomic Ge) has been studied at 295 and 85°K through the use of channeling and back scattering of these ions. The disorder is measured through the relative yield of back-scatter for the beam incident in $\langle 111 \rangle$ and $\langle 110 \rangle$ directions. In both crystals, the disorder is not linear with fluence, and shows a tendency to saturate. In many cases the disorder measured along $\langle 110 \rangle$ channels is larger than along $\langle 111 \rangle$ channels, which indicates a prevalence of 'interstitial-type' defects. However, there is no indication of the structure of the defects. Part of the disorder could be attributed to slight displacements of atoms in the strain field of radiation defects. The low temperature irradiations create more disorder than room temperature irradiations, and this excess disorder does not fully anneal at 120°C. In Si(Ge), the number of Si or Ge scatterers is close to the concentration ratio even at high fluences. This shows there is no preferential process by which Ge interacts with Si defects.

1. INTRODUCTION

The use of charged particle channeling and back scattering appears as a selective method for the direct study of interstitial defects in crystals.[1-3] In connection with problems which have arisen in radiation damage and ion implantation studies, we have started a joint program for the study of such defects in silicon and in silicon-germanium alloys. In the latter material, germanium, isoelectronic with silicon, can replace it in any proportion. One should then expect the behavior of the two atoms to be rather similar, except for the size difference. The use of Rutherford back-scattering is here of particular advantage because the large mass difference between Si and Ge allows easy discrimination in the spectra between scattering centers of the two species.

2. EXPERIMENTAL TECHNIQUES

Irradiation and back scattering analysis have been performed at 275 keV, with the Dynagen accelerator of the University of Sussex, and at

1610 keV, with the Van de Graaff at the Ecole Normale Superieure (E.N.S.). The lower energy gives an energy separation (60 keV) between Si and Ge scattering which is not much greater than the energy resolution (20 keV), but the scattering cross section is high. At the higher energy, the large energy separation (382 keV) allows a very precise analysis, although, because of the lower cross-section, a higher dose is needed for the same statistical accuracy.

The samples were mounted on goniometers allowing precise alignment of the crystal directions with the beam. The temperature range covered was 115°K–300°K (Sussex) or 85°K–400°K (E.N.S.). The E.N.S. experimental set-up allowed a fully automated operation, with digital control of angular scanning, dose and data recording. The principles of this system have previously been described,[1] and full details will be published in the near future.[5]

The samples used were *p*-type silicon (pulled, 1–5 Ω-cm, B doped) and *p*-type $Si_{0.94}Ge_{0.06}$ (Semi-Elements; same crystal as used by A. Brelot[6] for optical studies). They were cut perpendicular to a $\langle 111 \rangle$ direction within 1–2°, and then mechanically and chemically polished.

The back scattering yields were measured at 165°

† Laboratoire associe au C.N.R.S. Work supported in part by C.N.R.S. (R.C.P. 157) and D.G.R.S.T.

‡ This research supported in part by the Ministry of Technology.

(E.N.S.) or 150° (Sussex), with surface barrier detectors (resolution 20 keV). The beams were collimated by apertures 1 mm in diameter. The number of counts was integrated over a number of channels corresponding to the same depth for He+ ions scattered either by Si or Ge (0–2000 Å for 1610 keV He+, 200–400 Å for 275 keV He+).

The yield ratios χ have been computed by taking the ratio of counts, within the band specified above, for channeled and unchanneled beams. The difference between χ, for a given irradiation fluence, and χ_0 before irradiation is a rough measurement of lattice disorder. For low disorder, it may be taken as equal to the concentration of displaced atoms.[1]

The radiation damage was created by the same beam as used for analysis, but with the sample being tilted from the main axis, so that the bombardment was carried out in a random direction, as verified by the high yield of back scatter. In the evaluation of damaging fluences, only the doses delivered in unchanneled directions were taken into account, since the channeled irradiations are less efficient (by approximately the factor χ) for damage creation.

3. 275 keV IRRADIATIONS

The variation of χ–χ_0 as a function of fluence has been measured for $\langle 110 \rangle$ and $\langle 111 \rangle$ orientations, for silicon and silicon-germanium at room temperature. The beam current density used for damaging was 1.0×10^{13} He+/cm²/sec [constant to ± 5 per cent for Si or ± 20 per cent Si(Ge)]. The results are plotted in Figures 1 and 2.

In pure Si as in Si (Ge), the damage rate is not constant but is higher for low fluences. There is some evidence of higher damage as observed in $\langle 110 \rangle$ than in $\langle 111 \rangle$. This point will be discussed later. In Si(Ge), the initial damage rate is much higher than in pure Si, but, for higher doses, the slopes are apparently equal. Within experimental limits, it appears that, in Si(Ge), the rate of production of Si and Ge defects in the linear region is the same.

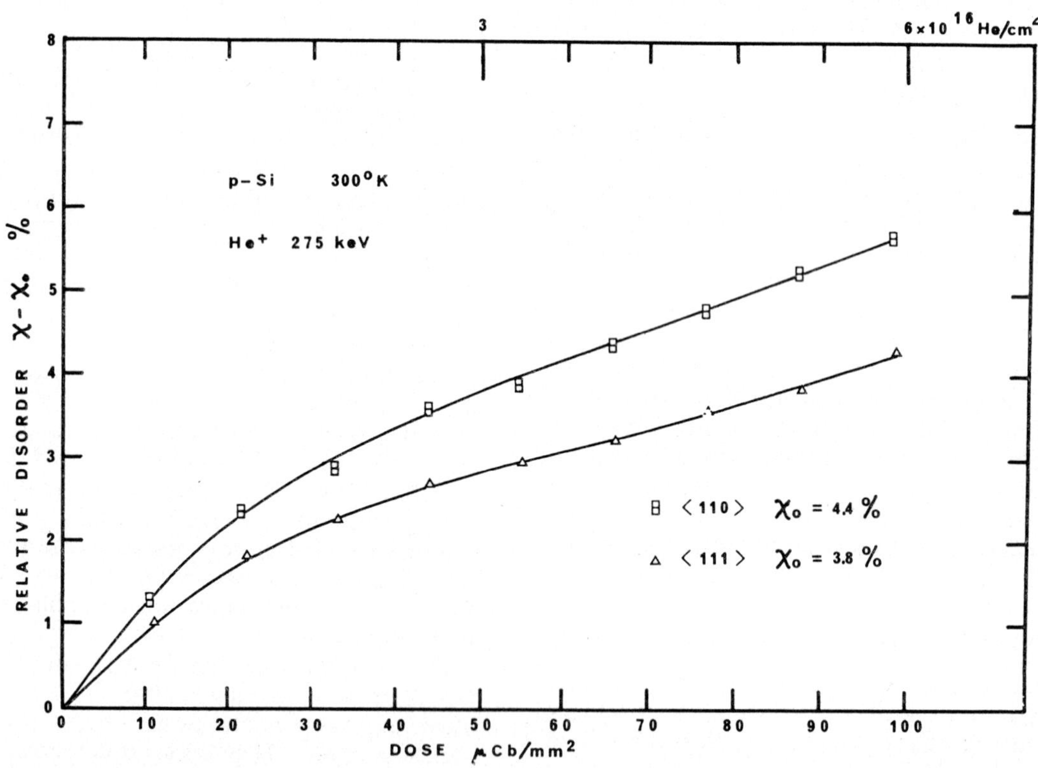

FIG. 1. Variation of the minimum yields in the $\langle 110 \rangle$ and $\langle 111 \rangle$ directions for bombardment of Si (p-type, 1 Ω-cm) by 275 keV He+ ions. For damaging: non-channeling direction, 1.0×10^{13} (± 5 per cent) He+/cm²/sec. For probing: each point 2.4×10^{15} He+/cm² at same dose rate. Damaging and probing both at 300 °K. Values of χ_0 and χ measured for depth approximately 300 Å below surface; statistical standard error in each ± 0.1 per cent.

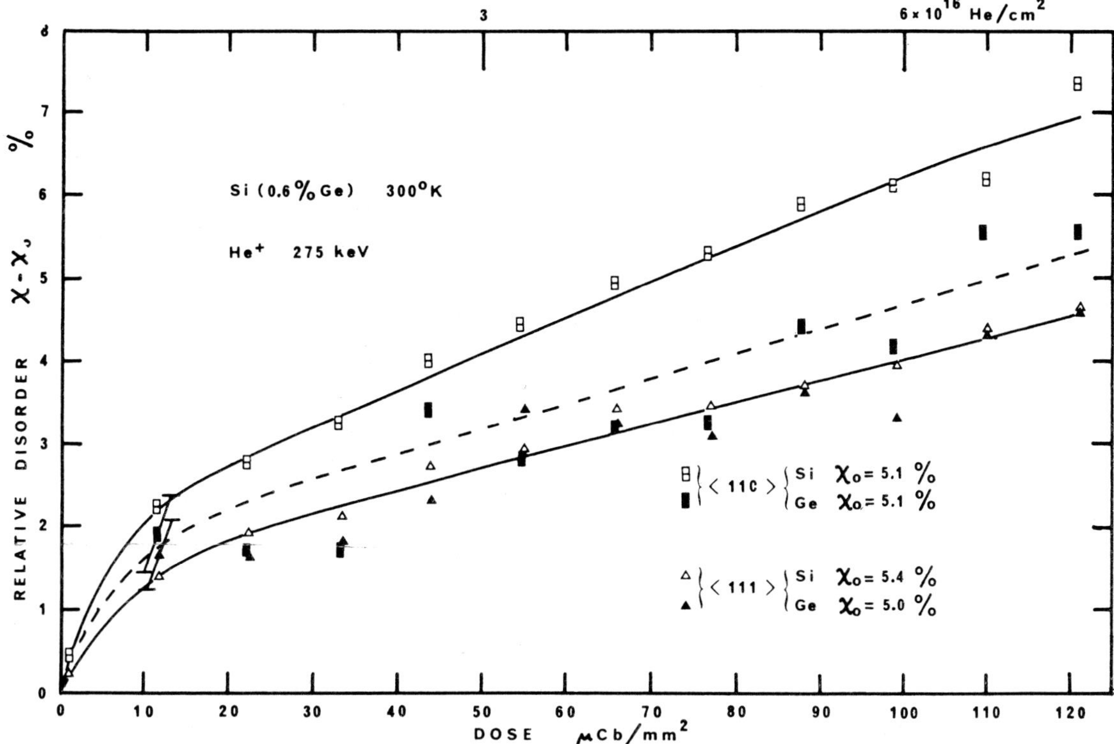

FIG. 2. Variation of the minimum yields in the $\langle 110 \rangle$ and $\langle 111 \rangle$ directions for bombardment of Si (0.6 per cent Ge) by 275 keV He+ ions. Same conditions as for Figure 1 but dose rate constancy was ± 20 per cent, and the statistical error for Ge points was ± 0.5 per cent.

A more detailed study of the damage as a function of fluence and rate for different temperatures is under way and will be published elsewhere, as well as an evaluation of the effect of misalignment on the interpretation of channeling spectra for impurity location.

4. 1610 keV IRRADIATIONS

These irradiations have been performed at room temperature and at liquid nitrogen temperature. The average beam current density was 2×10^{13} He+/cm²/sec for irradiation and probing.

On pure Si (Figure 3), the only data at 295 °K are for a rather heavy dose ($250 \mu C/mm^2 = 1.5 \times 10^{17}$ He+/cm²), which does not allow any estimate of non-linearity. However, it appears that the variation of χ is the same both in $\langle 110 \rangle$ and in $\langle 111 \rangle$, and is less than at the lower energy. The picture changes at liquid nitrogen temperature. Although the data points are scarce, a definite non-linearity can be seen, but not as much as at 275 keV

at room temperature. More important is the very fast increase in χ, leading, for the heaviest dose (1.5×10^{17} He+/cm²), to $\chi - \chi_0 = 14$ per cent for the $\langle 110 \rangle$ direction, and $\chi - \chi_0 = 7.5$ per cent for $\langle 111 \rangle$. The damage rate is then much larger as seen along $\langle 110 \rangle$ than along $\langle 111 \rangle$ channels. After annealing for 40 minutes at 120 °C, a considerable recovery occurs, 47 per cent for $\langle 110 \rangle$ and 62 per cent for $\langle 111 \rangle$.

The Si(Ge) results are plotted in Figure 4, for scattering by Si atoms, and Figure 5, for scattering by Ge atoms. The statistical counting errors are quite large for Ge, and are shown on the graph. However, these are not the only experimental errors, and the reproducibility of different runs, on the same sample but on different points, is quite poor, as shown by the scatter of experimental points (see on Figures 4 and 5, runs 1 and 2 for $\langle 110 \rangle$ axis, at room temperature). The errors may come from beam unstability and divergence from the choice of the reference 'random' spectrum, but also from crystal inhomogeneities, which can be quite large for the Si(Ge) alloy.

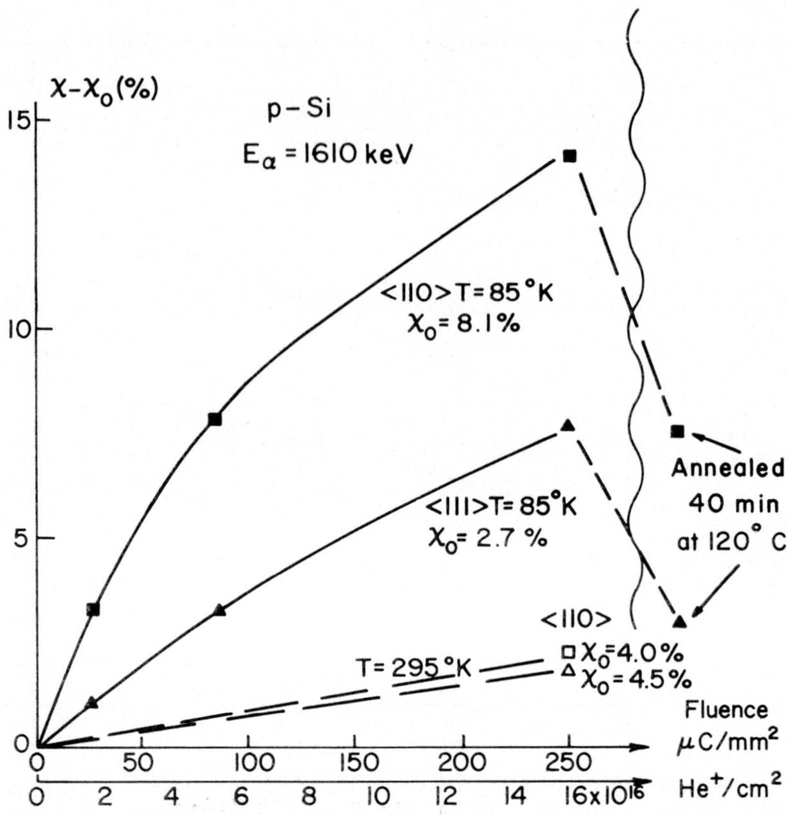

FIG. 3. Variation of the minimum yields in $\langle 110 \rangle$ and $\langle 111 \rangle$ directions for bombardment of Si (*p*-type, 5 Ω-cm) by 1610 keV He$^+$ ions. Beam current density: for damaging, in non-channeling direction, approximately 2×10^{13} He$^+$/cm^2/ sec; for probing, each point 1.3×10^{15} He$^+$/cm^2. Irradiations and probing at indicated temperatures. Values of χ and χ_0 measured for depth 2000 Å; statistical standard error 0.1 per cent–0.2 per cent.

Within these limitations, the most striking feature is the almost exact identity between the Si and Ge data. The other features are the same as found for pure Si: non-linearity vs. fluence, larger damage rate seen in $\langle 110 \rangle$ than in $\langle 111 \rangle$, larger damage rate at low temperatures, and recovery after annealing for 40 minutes at 120 °C (60 per cent on $\langle 111 \rangle$ for Si and Ge, 48 per cent for Si and 30 per cent for Ge on $\langle 110 \rangle$). In all cases, for the same orientation, the damage rate is higher for the alloy than for pure Si, even at room temperature. In no case, Si or Si(Ge), 275 keV or 1610 keV, do the aligned spectra for the damaged samples indicate any completely amorphous layers at the surfaces of the samples or inside the samples.

5. DISCUSSION

5.1. The pure Si results can be compared with similar results obtained in ion-implantation experiments.[2,3,7] The same qualitative features are found: non-linearity of damage vs. fluence and larger damage for low temperature irradiations.

However, the total damage created here is lower, and its structure is certainly different. In the energy range used, the damage collisions are of Coulomb-Rutherford type, yielding a mild damage: for 1.6 MeV He$^+$ ions, the average energy transferred to Si atoms is 228 eV, and the average number of displaced atoms is 5.2 per primary collision (for $E_D = 22$ eV, threshold displacement energy). One should then expect the damage to consist, initially, of isolated point defects (including divacancies or point defect-impurity associations) with a few clusters of high density of damage. With Rutherford cross section for primaries, and hard sphere collisions for secondaries, the calculated rate of creation of defects, at 1.6 MeV, is 1.5×10^4 cm^{-1}, yielding, for the maximum dose used, a defect density of 2.2×10^{21} cm^{-3}, i.e., a 4 per cent concentration

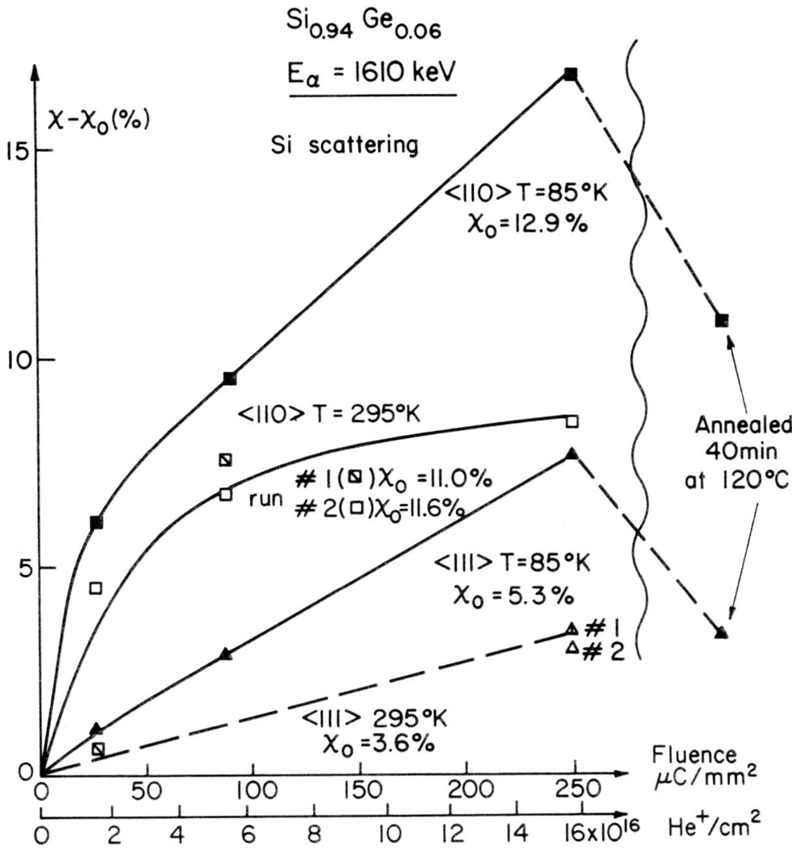

FIG. 4. Variation of the minimum yields in $\langle 110 \rangle$ and $\langle 111 \rangle$ directions for 1610 keV He+ scattered by Si atoms in Si (0.6 per cent Ge). Same conditions as in Figure 3.

of displaced atoms. Such a concentration of point defects seems exceedingly high, and some re-arrangements of point defects should occur.

At low temperature, the maximum value of 14 per cent for $\chi - \chi_0$ in $\langle 110 \rangle$ should not be interpreted as a density of point defects, but of atoms displaced in the strain field of point defects or defect complexes. Since the increase in χ is larger for $\langle 110 \rangle$ than for $\langle 111 \rangle$, this suggests that many of the displaced atoms lie along $\langle 111 \rangle$ rows, by stretching of Si-Si bonds. It is not realistic to say that the damage is due to interstitials, because it is not likely that the interstitial density can reach such a high value, but there is no way of separating the actual interstitial contribution from the total 'interstitial-type' damage.

At room temperature, the damage is much more isotropic both for 1610 keV and 275 keV irradia-tions, reaching 2 per cent and 4 per cent respectively. It might be considered mostly as due to the strain field around point defects and to the point defects themselves, such as divacancies. A maximum di-vacancy concentration of 4×10^{19} cm^{-3}, i.e. 0.1 per cent has been observed after O[7] or Sb and Zn[8] implantation. We might interpret our 2 per cent and 4 per cent disorders as being caused by a large number of relaxed atoms around various point defects.

The variation with energy of the damage rate appears, at first sight, to be consistent with the Coulomb cross-section: the slopes of correspond-ing curves in Figures 1 and 3 (for room temperature) are approximately in the expected ratio of 5.9 to 1. However, we have no indication on the actual shapes of the room temperature curves of Figure 3.

5.2. For Si(Ge), the same general considerations should apply. However, in the 1610 keV experi-ments at both 295 and 85 °K there is a large difference between $\langle 111 \rangle$ and $\langle 110 \rangle$.

The most important point here is the almost

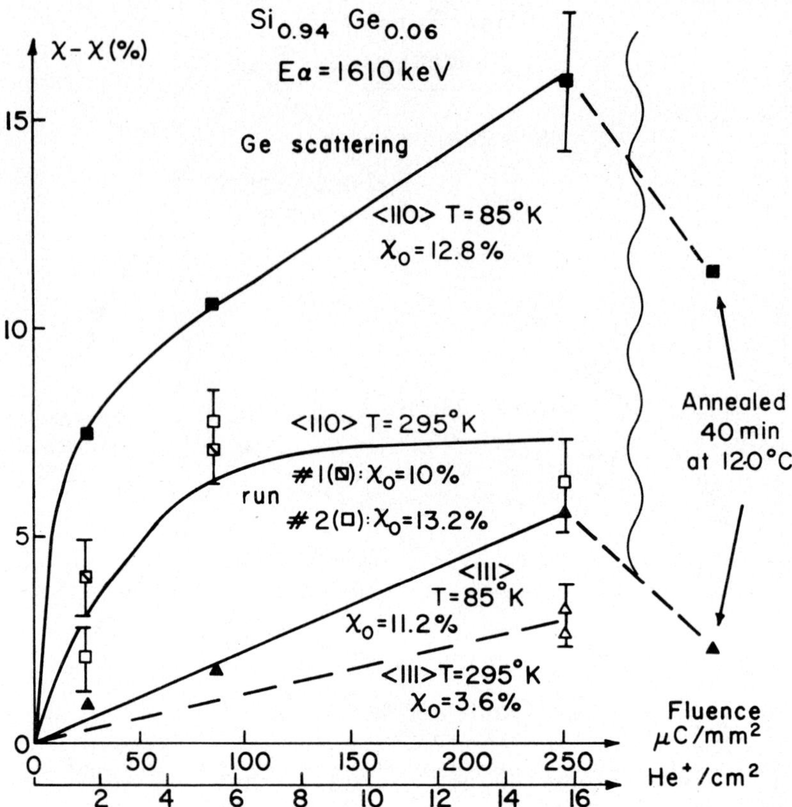

FIG. 5. Variation of the minimum yields in $\langle 110 \rangle$ and $\langle 111 \rangle$ directions for 1610 keV He$^+$ scattered by Ge atoms in Si (0.6 per cent Ge). Same conditions as in Figure 3, but the statistical error is ± 1 per cent; some error bars are shown.

exact equality of damage for Si and Ge in the alloy. The rates of displacement of Si and Ge atoms can be calculated in a way similar to that for pure Si; but now one should take into account mixed cascades, where a primary Ge knock-on can induce a cascade, or where a Ge atom is struck within a cascade. For the 0.6 per cent atomic concentration used, the calculation is straightforward, and yields 0.75 Ge displacement for 100 Si displacements. The theoretical ratio of defect concentrations (Ge displaced/Ge total)/(Si displaced/Si total) should then have the value 0.75/0.6 = 1.25. This difference from unity falls within experimental error, and, for all irradiations we have found that the rates of displacement of Si and Ge are the same. This last result may imply that there is no interaction between Si and Ge defects such as an exchange reaction between Si interstitials and Ge on lattice sites which would increase the number of Ge displacements.

However, in view of the above remarks on defect densities in Si, it appears that most of the excess scattering would come not from defects, but from atoms displaced by the strain field of defects. In this case, the lattice distortions would affect both Ge and Si, and the scattering behavior of the two species would be identical. Evidence for these strains can be found in x-ray observations: Authier and Montenay-Garestier[9] have found such a strain pattern by x-ray topography in proton-irradiated silicon.

6. CONCLUSIONS

The interpretation of back scattering data, even with a mild irradiation is not easy because of the high level of disorder which must be introduced before some experimental accuracy can be achieved The scattering centers are not only lattice defects created by He$^+$ irradiation but also atoms displaced from the regular lattice sites in the strain field of the defects. A further study of the effect of this type of disorder on the scattering yield is necessary. Some of the disorder observed in our experiments can be quoted as 'interstitial type', but it is not yet possible to sort out the contribution of regular interstitials.

The Si(Ge) case shows that germanium atoms in this alloy are displaced by irradiation in the same way as silicon atoms. However, an improvement in sensitivity is needed to obtain data on Ge-vacancy complexes[10] and on the stability of interstitials.

ACKNOWLEDGEMENTS

We thank E. d'Artemare and M. Vidal for Van de Graaff operation, B. Farmery and E. G. Turpin for Dynagen operation, Dr. A. Brelot for supplying the samples of Si(Ge), S. Squelard for sample preparation and Professor M. W. Thompson and Dr. G. Amsel for valuable discussions. One of us (P. Baruch) is grateful to the Faculté des Sciences de Paris and the C.N.R.S. for granting a leave of absence, and to Professor M. W. Thompson for kind hospitality during a stay at the University of Sussex.

REFERENCES

1. E. Bogh, *Can. J. Phys.*, **46,** 653 (1968).
2. S. T. Picraux, et al., *J. Appl. Phys.*, **180,** 873 (1969).
3. F. Eisen, et al., in *Atomic Collision Phenomena in Solids*, Ed. D. W. Palmer, M. W. Thompson and P. D. Townsend (North Holland Publ., 1970), p. 111.
4. F. Abel, G. Amsel, M. Bruneaux and E. d'Artemare, *J. Phys. Chem. Sol.*, **30,** 867 (1970).
5. F. Abel, G. Amsel, M. Bruneaux and C. Cohen, to be published.
6. A. Brelot and J. Charlemagne, this conference (1970).
7. H. J. Stein, F. L. Vook and J. A. Borders, *Appl. Phys. Letters*, **16,** 106 (1970).
8. H. J. Stein, F. L. Vook, D. K. Brice, J. A. Borders and S. T. Picraux, Thousand Oaks Conference on Ion Implantation (1970), to be published.
9. A. Authier and M. T. Montenay-Garestier, in *Radiation Effects in Semiconductors*, Ed. P. Baruch (Dunod, Paris, 1965), p. 79.
10. G. D. Watkins, *IEEE Trans. Nucl. Sci.*, **NS–16,** 13 (1969).

DISCUSSION

Question (MCKEIGHEN) Why did more damage show up when the probing back scattering particle was incident along a $\langle 110 \rangle$ direction as compared to a $\langle 111 \rangle$ direction.

Answer (BARUCH) I refer you to the text where this is covered.

Question (VOOK) Were you bombarding along channeling directions or along a random direction?

Answer (BARUCH) No, we were bombarding in a random direction.

Question (PICRAUX) I would like to comment that your significantly larger initial yields (χ_0) for the $\langle 110 \rangle$ axis than for $\langle 111 \rangle$ axis in the same single for the pre-irradiated crystal in several cases seems quite unusual. The $\langle 110 \rangle$ axis should give the lower initial yield theoretically and I have always observed this to be the case within the same sample for the diamond type lattice. One must be careful when giving to the $\langle 110 \rangle$ orientation to not go to the minor axis which is also approximately 35° from the $\langle 111 \rangle$ axis along the [110] plane but in the opposite ($\langle 100 \rangle$) direction.

Answer (BARUCH) We were indeed surprised with this high value of χ_0 for Si (Ge) in $\langle 110 \rangle$ channeling; we believe it might be associated with grown-in disorder and inhomogeneities in the crystal.

IRRADIATION DAMAGE IN CARBON-DOPED SILICON IRRADIATED AT LOW TEMPERATURES BY 2 MeV ELECTRONS

R. C. NEWMAN AND A. R. BEAN

J. J. Thomson Physical Laboratory, Whiteknights, Reading, England

Irradiation of silicon by 2 MeV electrons at 130 °K leads to the removal of carbon from substitutional sites and the formation of centres with axial symmetry having vibrational modes at 921 and 930 cm^{-1} for ^{12}C; large isotope shifts are found in crystals containing ^{13}C and ^{14}C. This centre is considered to involve interstitial carbon atoms but not oxygen impurities. On annealing such irradiated crystals to room temperature the concentration of these centres is reduced and a new transient centre involving carbon has been detected. Further annealing leads to the formation of the well known 11.56 μm absorption band in pulled crystals and it is shown that this may be correlated with another band at 1115.5 cm^{-1}. Again large isotope effects are found in crystals containing ^{13}C and ^{14}C and this centre is ascribed to a [O-C] complex involving an interstitial carbon atom.

1. INTRODUCTION

Infra-red absorption measurements have shown that the presence of carbon in pulled silicon crystals irradiated at room temperature by 2 MeV electrons has a large effect on the nature of the resulting damage.[1] In particular, the number of divacancy centres formed is greatly increased and the removal of oxygen from unperturbed interstitial sites is correspondingly greater than in carbon free crystals. Many new bands of both vibrational and electronic origin were observed and attributed to [C-O] complexes, and it was inferred that substitutional carbon was ejected into interstitial sites by the replacement mechanism described previously by Watkins[2] for group III impurities.

Subsequently, it was shown[3] that similar irradiations of such crystals at 100 °K led to the formation of sharp local mode absorption bands at 921 and 930 cm^{-1} in samples containing ^{12}C, and corresponding bands at 892 and 904 cm^{-1} when ^{13}C was present. The former bands were also observed in an oxygen-free crystal and they were attributed to an interstitial defect centre involving a carbon atom. In the present paper similar measurements are reported for a pulled silicon crystal containing ^{12}C: ^{14}C in the ratio 2:1 and for other oxygen-free crystals of low carbon content. In addition, the effects of annealing such irradiated crystals up to room temperature are described.

2. EXPERIMENTAL DETAILS

Samples were irradiated with 2 MeV electrons at a flux in the range 10–15 μA cm^{-2} while they were mounted on a cold finger of a liquid nitrogen cryostat. Each sample, which was about 2 mm. in thickness, had a small hole drilled into its side in which was inserted a thermocouple to monitor the temperature during the irradiation and anneals. The measured temperatures depended on the flux used but never exceeded 135 °K and were usually about 120 °K. Infra-red measurements were made at 77 °K by the differential technique[1] immediately after the irradiation and following various annealing treatments.

3. DAMAGE CENTRES PRODUCED BY IRRADIATION

A high resistivity p-type pulled crystal containing about 3.5×10^{17} cm^{-3} of ^{14}C and 7×10^{17} cm^{-3} of ^{12}C was irradiated in stages up to a total fluence of 7×10^{18} electrons cm^{-2} on each side. A strong infra-red absorption band was produced at 834.9 cm^{-1} (A-band), together with two pairs of lines at 921.0 and 930.3 cm^{-1}, and at 866.8 and 881.4 cm^{-1}. For each of these pairs, the line with the lower energy was twice as strong as the other, and the pair of lines of lower energy was weaker than the other by a factor of two, corresponding to the isotopic abundance of ^{14}C:^{12}C. These lines increased in strength with increasing doses of irradiation while the concentration of carbon in unperturbed substitutional sites fell as deduced from the strength of the bands at 607 cm^{-1} (^{12}C) and 573 cm^{-1} (^{14}C). From these observations, and similar measurements made on other samples containing either ^{12}C alone or enriched ^{13}C, it has been possible to allocate an apparent charge of

$\eta \simeq 3.0\,e$ for the damage centre (designated by C(1)) which may be compared with $\eta = 2.6\,e$ for substitutional carbon. It would appear that the centre C(1) has axial symmetry and on the basis of the relative strengths of the two lines it is inferred that the doubly degenerate mode $\omega_\perp$ has a lower energy than the longitudinal mode $\omega_\parallel$. On this assumption, a mean value of ω for this centre for each carbon isotope has been computed and is shown in Table I. The ratios of these values of ω are then found to be in excellent agreement with the corresponding ratios for carbon in substitutional sites C(s), showing that the observed vibrational modes are predominantly due to motion of carbon atoms. However the systematic variations in the values of $(\omega_\parallel - \omega_\perp)$ of 9.3, 12.0 and 14.6 cm^{-1} for ^{12}C, ^{13}C and ^{14}C respectively, are much greater than would be expected for an isolated carbon atom vibrating in a simple static potential well, as this would imply unreasonably large anharmonicity. It therefore seems probable that the carbon atom is coupled preferentially to some other atom which also takes part in the vibrational motion to some extent.

In our previous work,[3] the formation of the centre C(1) was reported for a crystal grown by the floating zone technique which initially contained 6×10^{17} atom cm^{-3} of ^{12}C in substitutional sites. After irradiation to a dose of 1.5×10^{18} electron cm^{-2}, there was a loss of carbon from these sites of about 1.5×10^{17} atom cm^{-3}. It therefore follows that a similar or greater concentration of another impurity would have had to be present if the C(1) centres involve this impurity. This result would seem to exclude oxygen as a possibility since no 9 μm absorption was detected in this sample, implying that the original oxygen concentration was less than 10^{16} atom cm^{-3}. It is therefore suggested that the C(1) centre involves either (1) an interstitial carbon atom in an off-centre tetrahedral site with C_{3v} symmetry or (2) that the substitutional carbon atom traps a silicon interstitial atom. Physically, it may not be sensible to differentiate between these two possibilities, since they may both be regarded as a split interstitial centre, if it is assumed that local lattice relaxation effects occur. It is clear however, that the centre does not consist of an isolated carbon atom occupying a tetrahedral interstitial site and the possibility of some impurity other than oxygen being involved cannot be ruled out.

Our conclusion that the centre is not a [C-O] complex was not in agreement with the work of Stein and Vook.[4] Consequently, we have carried out irradiations of crystals grown by the floating zone technique, but containing lower ^{12}C concentrations of about 10^{17} atom cm^{-3}, which are thus similar to the crystal used by Stein and Vook who

TABLE I

Centre	Carbon isotope	$\omega_\parallel$ (cm^{-1})	$\omega_\perp$ (cm^{-1})	$\omega = \frac{1}{3}(2.\omega_\perp + \omega_\parallel)$ (cm^{-1})	Frequency ratios
	12	607.5		607.5	$\omega^{12}/\omega^{13} = 1.031$
C(s)	13	589.1		589.1	
	14	572.8		572.8	$\omega^{12}/\omega^{14} = 1.061$
	12	930.3	921.0	924.1	$\omega^{12}/\omega^{13} = 1.031$
C(1)	13	904.2	892.2	896.2	
	14	881.4	866.8	871.7	$\omega^{12}/\omega^{14} = 1.060$
Ci C_{3v}?					
	12	858.7	?	858.7	
C(2)	13	834.5†	?		
	14	812.0	?	812.0	$\omega^{12}/\omega^{14} = 1.057$
Ci T_d?					
	12	1115.5	865.2	948.6	$\omega^{12}/\omega^{13} = 1.030$
C(3)	13	1078.3‡	841.8‡	920.6	
[Ci-Oi]	14	1047.0	819.2	895.1	$\omega^{12}/\omega^{14} = 1.060$

† Estimated from an expected ratio of $\omega^{12}/\omega^{13} = 1.03$.
‡ Accuracy may be limited due to presence of other nearby bands.

quoted a carbon concentration of 1.3×10^{17} atom cm^{-3}. An initial irradiation to a fluence of 2×10^{17} electrons cm^{-2} of such a sample definitely led to the formation of detectable absorption lines at 921 and 930 cm^{-1}. However, a subsequent irradiation carried out without an intervening warm up of the sample caused these bands to become so small as to be undetectable; yet a further irradiation caused the centre C(1) to be regenerated. Similar puzzling and apparently anomalous behaviour was found with a second sample cut from the same crystal. It was established that in both of these samples, the concentration of substitutional carbon decreased monotonically with increasing doses of irradiation. It follows that if carbon atoms are ejected into interstitial sites by the first dose of irradiation, they are certainly not returned to substitutional sites by recombination with vacancies during a subsequent irradiation. One possible explanation is that the centre C(1) can trap further silicon interstitials and somehow this prevents the carbon atom from giving rise to any detectable absorption. Finally it should be noted that these latter observations are not in any way at variance with those reported by Stein and Vook.[4]

4. ANNEALING STUDIES

The main effects of annealing irradiated pulled crystals are summarized in Figure 1. During the warm-up the strength of the A-band increased slightly, the $9\,\mu$m oxygen band decreased and the divacancy electronic band at $1.69\,\mu$m increased at first and then decreased in strength. However, of special relevance is the fall in the concentration of C(1) centres starting at about 220 °K and their complete removal at 300 °K. This result is consistent with previous observations,[5] but we now have to consider the fate of the carbon atoms originally in these centres, since they are not returned to substitutional sites.

In both float-zone and pulled crystals a very weak band was observed at 858.7 cm^{-1} after an anneal of about 1 hr at 300 °K; this band then rapidly became less intense for longer periods of annealing at the same temperature. In the pulled sample containing ^{14}C a band with corresponding behaviour was observed at 812.0 cm^{-1} with half the strength of that at 858.7 cm^{-1}. The ratio of these two frequencies is 1.057 and it is therefore inferred that these bands arise from the vibrations of carbon atoms in a new centre labelled C(2) (Table I). The estimated position of the corresponding band for

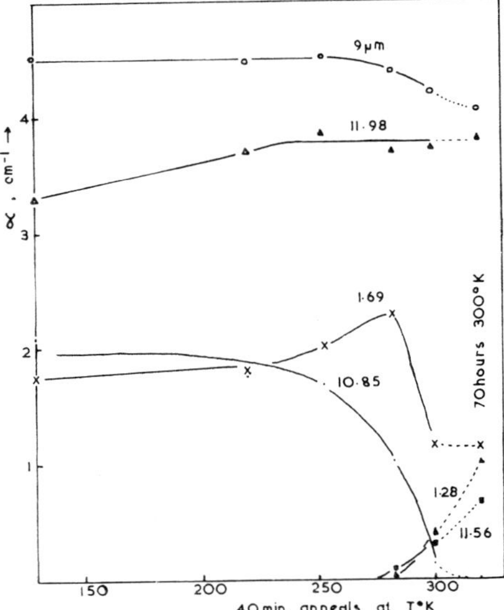

FIG. 1. Peak absorption coefficients of various IR bands following low temperature irradiation, after 40 min anneals at the temperatures indicated.

^{13}C would be 835 cm^{-1} which would preclude its observation in the pulled crystals examined because of the presence of the A-band at the same energy. Since no other associated bands were detected it is possible that this centre may correspond to a carbon interstitial in a site of T_d symmetry. If this is so the transient nature of the band would suggest that such atoms are highly mobile.

During the annealing process the growth of other bands was noted including the $11.56\,\mu$m (865.2 cm^{-1}) band in pulled samples and new structure in the region of $9\,\mu$m as shown in Figure 2 and (Table I). In the sample containing ^{14}C a previously unreported band grew at $12.20\,\mu$ (819.2 cm^{-1}) as shown in Figure 3. This strength of this band was half of that at $11.56\,\mu$m, corresponding to the isotopic abundance of ^{14}C : ^{12}C in the sample, and both bands had identical annealing behaviour; these bands are therefore considered to be isotopic analogues. The situation appears to be more complex than this however, as the two bands discussed above can also be correlated with two bands in the $9\,\mu$m region as shown in Figure 4; the various points correspond to different stages of annealing of several samples. It is therefore thought that these absorption bands again arise from a centre with axial symmetry, designated as C(3), with $\omega_\parallel$ greater than $\omega_\perp$; this latter conclusion is based on the fact that the strength of the lower

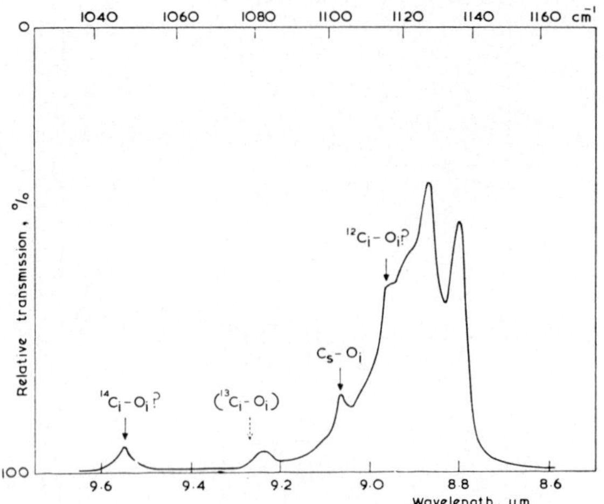

FIG. 2. 80 °K differential absorption spectrum in 9 μm region of a ^{14}C-doped crystal.

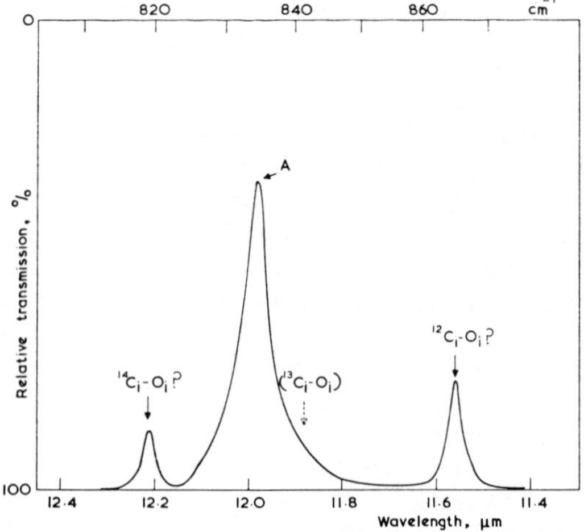

FIG. 3. 80 °K differential absorption spectrum in 12 μm region of the same ^{14}C-doped sample as in Figure 2.

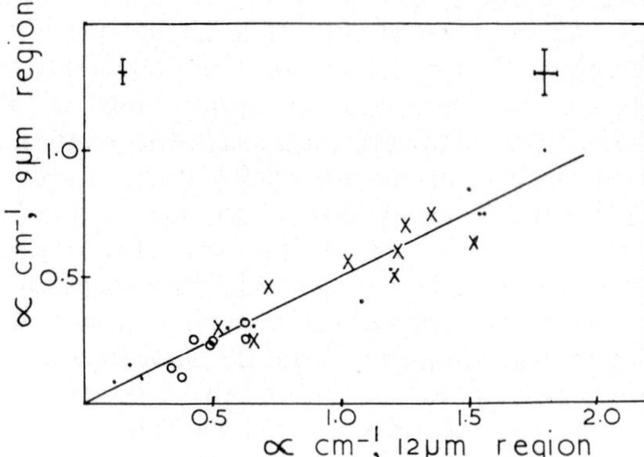

FIG. 4. Correlation of absorption bands in 9 μm and 12 μm regions.

• ^{12}C; 8.96 and 11.56 μm bands measured for a range of samples.

× ^{12}C; 8.96 and 11.56 μm bands measured for a single sample after various irradiation and annealing treatments.

o ^{14}C; 9.55 and 12.20 μm bands measured for a single sample after various irradiation and annealing treatments.

Points for ^{13}C-doped samples are not shown because of the lower accuracy, as explained in the text.

The approximate measurement errors are indicated at the top of the diagram for the extremes of the range.

energy line is approximately double that of the other. Although Figure 2 shows the absorption observed at 77 °K, many of the measurements were made at 4.2 °K so that the line at 1115.5 cm^{-1} could be clearly resolved from those due to interstitial oxygen. In pulled crystals containing enriched ^{13}C, the analogous band at 1078.3 cm^{-1} is close to another band at 1079.6 cm^{-1}, due to another irradiation damage centre, while the line at 841.8 is close to the A-band satellite S_3 reported previously.[1] Assuming the correctness of our interpretation the value of ω^{12}/ω^{14} again has a value

consistent with that found for all the centres involving a carbon atom; this is also true for ω^{12}/ω^{13}, although the accuracy of this ratio is less certain. Centre C(3) was never detected in oxygen-free crystals and it is therefore thought to be due to a carbon atom paired with an interstitial oxygen. In view of our earlier observations on as-grown crystals,[6] concerning the pairing of substitutional carbon with oxygen, it is thought that the centre C(3) must involve carbon in an interstitial site. This conclusion is also supported by the fact that the absolute values of ω for this centre are very similar to those for centre C(1), although the magnitude of $(\omega_{\parallel} - \omega_{\perp})$ is very much greater. This leaves as an unanswered question the energy of the vibrational mode associated with the paired oxygen; attempts to find another correlated mode have not so far been successful.

Finally, it was found that as the divacancy and C(1) centres were reduced in concentration on annealing, the previously reported[1] electronic band at 1.28 μm was formed in oxygen-free crystals and in pulled crystals of high carbon content. It is possible that this band is due to a divacancy centre with a nearby carbon interstitial or to a [C-V] complex; in either case, no associated vibrational absorption has been detected.

5. CONCLUSION

The present results indicate that low temperature irradiation of silicon containing carbon impurities leads to the formation of centres with axial symmetry which almost certainly involve an interstitial carbon atom; it should be noted that the values of ω for C(1) are much higher than those for C(s). On annealing to room temperature the interstitial carbon atoms become mobile, possibly as centre C(2), and then associate with oxygen and probably divacancy centres. The correlation of the C(3) bands shown in Figure 3 would suggest that previous workers should also have detected the band at 8.96 μm. The fact that this has not been reported may indicate that this correlation is accidental, and the possibility cannot be ruled out that the present results are peculiar to the samples examined.

ACKNOWLEDGMENTS

We should like to thank Professor E. W. J. Mitchell for providing the irradiation facilities and one of us A.R.B. wishes to thank the S.R.C. for a research studentship.

REFERENCES

1. A. R. Bean, R. C. Newman and R. S. Smith, *J. Phys. Chem. Solids*, **31**, 739 (1970).
2. G. D. Watkins, in *Radiation Effects in Semiconductors* (Dunod, Paris, 1965), p. 97.
3. A. R. Bean and R. C. Newman, *Solid State Communications*, **8**, 175 (1970).
4. H. J. Stein and F. L. Vook, *Rad. Effects*, **1**, 41 (1969).
5. R. E. Whan and F. L. Vook, *Phys. Rev.*, **153**, 814 (1967).
6. R. C. Newman and R. S. Smith, *J. Phys. Chem. Solids*, **30**, 1493 (1969).

DISCUSSION

Question (STONEHAM)   Your observed isotope effects differ from the simple square-root-mass dependence because the local mode involves the motion of neighboring atoms. Why should these deviations be so similar in all your cases, including the case involving oxygen?

Answer (NEWMAN)   As the frequency of the localized mode increases, the isotope effect should approach a simple square-root-mass dependence. The fact that this is not observed is not understood. For the carbon-oxygen complex, it appears that the centre has axial symmetry. If both atoms are in interstitial sites, then it may be that the carbon and oxygen atoms are not in nearest neighbour sites in which case the perturbation of the carbon modes may arise from the local lattice relaxation associated with the intervening silicon atom.

Question (KOEHLER)   You observe what you believe are absorption bands from interstitial carbon at higher frequency than the absorption from substitutional carbon. This seems remarkable since naively one would expect that substitutional carbon should be more tightly bound than interstitial carbon. How does one understand this?

Answer (NEWMAN)   If a neutral inpurity is inserted into an interstitial site, it might be expected to experience very strong repulsive forces from the neighbouring atoms, which would lead to a high vibrational frequency. For a substitutional carbon impurity we have shown previously that the local force constants are reduced by about 10 per cent, compared with those between pairs of silicon atoms. It must be emphasized however, that the nature of the bonding, or the lack of this, for an interstitial group IV element remains an unanswered question from a theoretical point of view.

Question (LARKINS)   Would you please comment on the relationship between your work and the results and models discussed by K. L. Brower in paper 29?

Answer (NEWMAN)   The defect centers discussed in this paper were ascribed to vacancy clusters with a nearby oxygen impurity. My only comment is that the possibility that a carbon atom may be involved cannot be completely ignored. We have found that in general, pulled silicon crystals contain more carbon than float-zone material.

INFRARED STUDIES OF LOW TEMPERATURE ELECTRON IRRADIATED SILICON CONTAINING GERMANIUM OXYGEN AND CARBON†

A. BRELOT AND J. CHARLEMAGNE‡

Groupe de Physique des Solides de l'Ecole Normale Supérieure§ Tour 23–9, quai Saint-Bernard, Paris 5e, France

Single crystal silicon containing various concentrations of germanium, oxygen, and carbon has been irradiated at 90 °K with 2.8 MeV electrons and the ensuing defects studied by infrared absorption measurements. We find that, in n-type silicon containing 3×10^{20} Ge atoms/cm³, there is no 12μ absorption line (A-center) after a 90 °K electron irradiation. The A-centers appear only after annealing between 200 °K and 280 °K. It is therefore inferred that Ge in silicon acts as an efficient trap for vacancies at low temperatures, the trapped vacancies being released from the Ge-V centers by thermal activation in the 200 °K–280 °K range. These vacancies then migrate through the lattice and are trapped on oxygen atoms. On the other hand we find that the interaction between silicon interstitials and Ge atoms is very low. The 935 cm⁻¹ absorption line is unusually intense in our experiments. Results suggest that the defects ('B-centers') which produce this absorption line are the result of the interaction of the silicon interstitials with oxygen atoms. In the last part of the paper, we study the experimental results concerning the 921 cm⁻¹, 931 cm⁻¹, 959 cm⁻¹ and 966 cm⁻¹ absorption lines. They are found to be associated with defect centers involving carbon atoms and are clearly found to be produced by silicon interstitials. On the other hand, experimental results suggest a fast diffusion of carbon, at 300 °K, in interstitial sites aside from any irradiation effect.

1. INTRODUCTION

The vacancy-oxygen center (A-center) has been studied by Fan and Ramdas,[1] Ramdas and Rao,[2] and Corbett, Watkins and coworkers.[3] Whan (4, 5), and Whan and Vook[6] found several infrared absorption lines between 1000 cm⁻¹ and 800 cm⁻¹ in low temperature irradiated silicon. Subsequently Vook and Stein[7] correlated the 921 cm⁻¹ and 931 cm⁻¹ absorption lines with complex defects involving carbon *and* oxygen. Lastly, Bean and Newman[8] showed these absorption lines come rather from the vibration of carbon than from that of oxygen.

In this paper we shall describe the interaction between radiation defects and neutral impurities and the ensuing associations. We shall use the local mode technique to detect the defects.

As an essential aim of this work, we shall:

(1) study the interaction of radiation defects with Ge impurity atoms;
(2) identify the impurities which are included in the defect complexes producing the various absorption lines which are observed and more particularly we shall separate the roles of carbon and oxygen;
(3) determine if the previous associations come from the interaction of silicon interstitials or of vacancies.

2. EXPERIMENTAL OBSERVATIONS

The properties of the samples used in our experiments are listed in Table I. The concentrations of isolated substitutional carbon and isolated oxygen were estimated from the absorption peaks at 604 cm⁻¹ and 9μ; a differential technique was employed by using a double beam 125 Perkin Elmer spectrophotometer with a pure sample in the reference beam (resolution: 0.7 cm⁻¹ at 604 cm⁻¹).

The irradiations were performed at low temperatures by mounting the samples in a cryostat allowing in situ optical measurements in the 1μ to 50μ range without any further heating. The intensities of the electron beam, produced by a 3 MeV Van de Graaff, extended from 10^{13} to 6×10^{13} e/cm² sec; the detailed irradiation parameters are indicated in Table I.

The sample holder is connected to the cryostat

† Work supported in part by D.G.R.S.T.

‡ This work is a part of the theses being prepared by A. Brelot and J. Charlemagne for the degree of Docteur es Sciences at the Université de Paris.

§ Laboratoire associé au C.N.R.S.

TABLE I

Dopant and its concentration in the samples studied. ϕ is the irradiation fluence. RO and RC are the concentrations of the oxygen and carbon removed by irradiation; A, B, and VV are the intensities of absorption bands associated with A-center, B-center (935 cm^{-1}), and 1.7 μ band, respectively. Measurements were made at 80 °K after 400 °K annealing. A 1 cm^{-1} absorption peak at 1.7 μ is produced by 5×10^{16} divacancies/cm^3 (after Cheng and coworkers[25]).

| Samples | | Dopant concentration $\times 10^{-18}$ | | | | RO | RC | | | |
Type	N°	Ge (a) at/cm³	0 (b) at/cm³	C (c) at/cm³	$\phi \times 10^{-18}$ e/cm²	$\times 10^{-18}$ O/cm³	$\times 10^{-18}$ C/cm³	A cm⁻¹	B cm⁻¹	V-V cm⁻¹
N	1	300	2.5	0.01	1.6	0.7	0.01	2.2	1.2	1
N	1′	300	2.5	0.01	2.6				1	
P	2	350	1.4	0.18	3.3		0.15	2.4		
P	2′	350	1.4	0.18	1.48			0.5		
N	3	0	1.8	0.14	2.3		0.13	1.9	0.6	0.9
N	4	0	<10⁻³	<0.005	1.7	0	0	0	0	0.3

(a) Calculated from density measurements and by using results of Reference (9).
(b) Determined from intensities of the 9 μ absorption line by using the modifications of Pajot[10] on the initial calibration of Kaiser and Keck.[11]
(c) Estimated from the absorption peak at 604 cm^{-1} by using the calibration of Newman and Willis.[12]

by a flexible tube allowing the 'sample in, sample out' technique, the optical windows being always in the light beam. This cryostat allows precise transmission measurements and will be described in more detail elsewhere.[13]

Spectra were recorded at 77 °K (or at 10 °K) immediately after irradiation and then after several annealing stages performed up to 400 °K.

3. RESULTS AND DISCUSSION

3.1. *Heavily* Ge *doped unirradiated silicon*

A concentration of 3×10^{20} Ge/cm^3 is a very high defect concentration in comparison with the one introduced by irradiation. Firstly we shall show that the Ge doped silicon lattice is not very much perturbed in comparison with pure silicon.

In Si-rich Ge–Si alloys, the Ge atoms avoid each other on neighbouring sites; then there is no clustering of Ge atoms in our Ge-doped silicon samples. On the other hand, since Ge and Si both have the same valency, no new electronic states are expected to be introduced by the substitution of a silicon by a Ge atom in the silicon lattice.

We found that the edge absorption is almost identical to the one of pure silicon and there is no evidence of tailing of the densities of states into the forbidden gap due to disorder. The phonon spectrum is only slightly modified in comparison with pure silicon. Aside from a band which appears below the Raman frequency, and which may be

produced by the band mode associated to Ge, no new absorption band appears. We found that the oxygen vibration at 77 °K is not at all perturbed by the possible neighbouring of Ge atoms and therefore we conclude that the oxygen atoms are entirely surrounded by silicon atoms in our samples, in agreement with the conclusion of Hrostowski and Alder.[14] At liquid helium temperature new bands appear close to 9 μ; they have been attributed by B. Pajot[15] to the $Si_3(Si_2O)Si_2$ Ge groups.

3.2. *Vacancy-germanium associations*

Figure 1 shows the absorption spectrum of *n*-type irradiated silicon containing Ge. We see that the intensity of the 12 μ band is hardly measurable which means that only a small number of vacancies, which are mobile at 90 °K under irradiation, are trapped by oxygen atoms. Heating the sample above 200 °K increases the intensity of the 12 μ band. This behaviour is quite different from that which may be observed in *n*-type silicon without Ge, as seen for example in Figure 3.

The vacancies which form *A*-centers between 200 °K and 280 °K can only come by dissociation from trapping centers on which they have been fixed at lower temperatures. We thus deduce that Ge atoms are able to trap the mobile vacancies which are created by irradiation at 90 °K. The Ge–V center begins to release its trapped vacancy at about 200 °K and these mobile vacancies are trapped again on oxygen atoms inducing the 12 μ

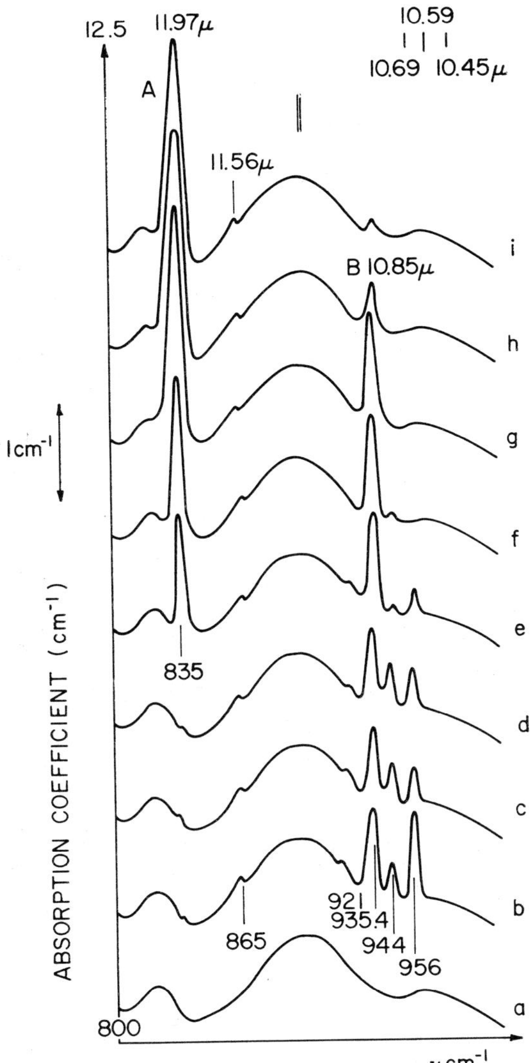

FIG. 1. Infrared absorption spectra of Ge-doped Si (sample 1—see table). Spectra are displaced vertically for clarity. All spectra were recorded at 80 °K. (a) Before irradiation; (b) after 2.8 MeV irradiation at 90 °K, fluence—1.6×10^{18} e/cm², intensity—1.25×10^{13} e/sec-cm²; (c) after three hours in light at 80 °K; remainder after 15 min at (d) 175 °K; (e) 230 °K; (f) 245 °K; (g) 280 °K; (h) 310 °K; (i) 350 °K.

band growth. This result agrees with the recent EPR measurements of Watkins.[16]

When vacancies are released from Ge atoms upon 300 °K annealing, we observe that:

(1) no new band appears except the 12 μ band (see Figure 1);

(2) the number of vacancies which agglomerate to form divacancies is negligible (see Figure 4);

(3) the interstitial-vacancy recombination is negligible because all the interstitials remain trapped on impurities.

Therefore the measurement of the intensity of the 12 μ band after 300 °K annealing gives us the concentration of single vacancies which have been created by the low temperature irradiation. Because the A-center concentration is negligible before annealing we may suppose that all the vacancies are trapped on Ge atoms and we may estimate the concentration of Ge–V centers.

If we assume that the concentration of defect associations is proportional to the product of $P_{V-\text{impurity}}$† times the concentration of impurity atoms, then we can determine the ratio $P_{V-\text{Ge}}/P_{V-0}$ (90 °K). P_{V-0} is calculated from the measured intensity of 12 μ band immediately after irradiation. $P_{V-\text{Ge}}$ is estimated from the measurement of this same band after its complete growth (i.e. after 300 °K annealing). The $P_{V-\text{Ge}}/P_{V-0}$ ratio is equal to about 1 and we conclude that Ge traps vacancies as efficiently as does oxygen because $P_{V-\text{Ge}}$ and P_{V-0} are about the same.

3.3. *Oxygen-silicon interstitial interaction*

We see in Figure 1 that the 956 cm⁻¹, 944 cm⁻¹ and 935 cm⁻¹ bands appear after a 90 °K irradiation. These bands are not produced in floating zone irradiated silicon containing a negligible oxygen concentration (see Table I). There is no correlation between these bands and carbon concentration as we shall see later by comparing Figure 1 and Figures 2–3. The possibility of contamination by an unknown impurity and an ensuing defect association cannot be completely eliminated. The indirect role of oxygen as a vacancy trap which would increase the defect creation rate also cannot be completely ruled out. Nevertheless we think that the considered bands involve oxygen, especially the 935 cm⁻¹ one, since we see that the isolated oxygen band increases when the 935 cm⁻¹ line decreases upon annealing at 320 °K.

On the other hand we shall now show that these bands are produced by the interaction of Si_I with oxygen. In fact, a number of vacancies are trapped on Ge atoms and the intensity of these bands does not increase when vacancies are released by Ge–V centers. From this we conclude

† We shall denote $P_{D-I}(T °K)$ the probability of capture of a defect (vacancy or interstitial) by an impurity, under an irradiation carried out at T °K.

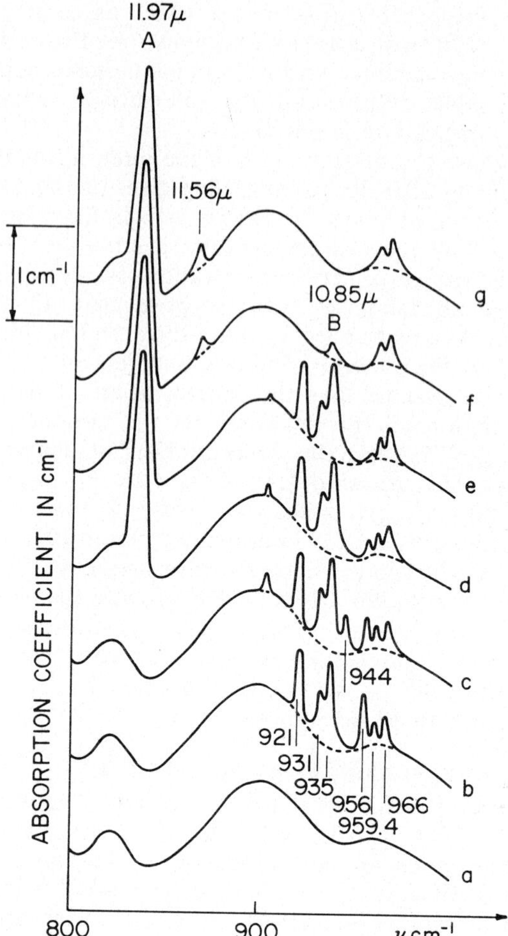

FIG. 2. Infrared absorption spectra (as in Figure 1) of Si doped with Ge and C (sample 2—see table). (a) Before irradiation; (b) after 2 MeV irradiation at 125 °K, fluence—3.3 × 10¹⁸ e/cm², intensity—5.6 × 10¹³ e/sec-cm²; (c) after three hours in light at 80 °K; remainder after annealing at (d) 240 °K; (e) 270 °K; (f) 330 °K; (g) 400 °K.

fact can be very well explained if we remember that the samples 1 and 2 are carbon free while on the contrary there is more than 10^{17} C/cm³ in the other samples. As we shall show later, carbon atoms trap Si_I and so the concentration of Si_I which can interact with O is lower in carbon doped silicon, which reinforces the interpretation of the 935 cm⁻¹ center as arising from Si_I interactions.

3.4. Si_I-substitutional carbon interaction

When we add about 10^{17} carbon/cm³ to the samples we observe after irradiation the following

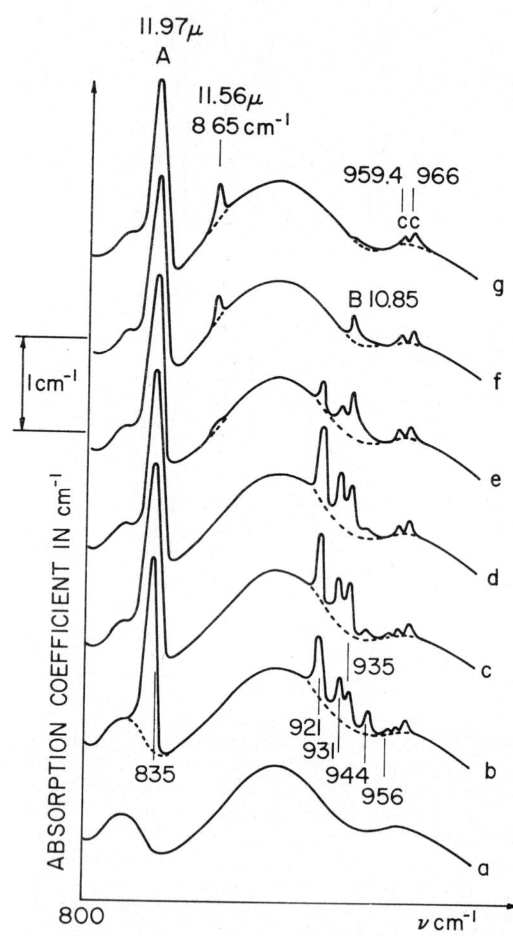

FIG. 3. Infrared absorption spectra (as in Figure 1) of Si doped with C but no Ge (sample 3—see table). The sample was low resistivity n-type containing 5×10^{15} P/cm³ and 10^{17} Li/cm³. (a) Before irradiation; (b) after 2.8 MeV irradiation at 115 °K, fluence—2.3 × 10¹⁸ e/cm², intensity 3.9 × 10¹³ e/sec-cm²; remainder after 15 minutes annealing at (c) 200 °K; (d) 240 °K; (e) 300 °K; (f) 330 °K; (g) 400 °K.

that these bands cannot be associated to a possible V–O center (in a different configuration of an *A*-center). As a conclusion we attribute the 935 cm⁻¹ band to the result of the interaction of mobile Si_I with oxygen.

The intensity of the 956 cm⁻¹ bands depends on electron beam intensity and decreases slowly if the sample is exposed to white light (see Figures 1 and 2).

The 935 cm⁻¹ band is very intense in samples 1 and 2 (see Figure 7) and becomes comparable in magnitude to the 12 μ band (after its growth). This new bands: 921 cm^{-1}, 931 cm^{-1}, 959.4 cm^{-1}, 966 cm^{-1}, so that these bands clearly come from defects involving carbon. The intensities of these bands do not increase when vacancies are released from Ge–V centers. Furthermore, the fact that, when the bands disappear at 300 °K, the isolated carbon concentration does not increase indicates that the centers are not V–C associations. We therefore conclude that the 921 – 931 cm^{-1} center is produced by a defect resulting from the interaction between carbon and Si_I. Our results confirm the indications of Bean and Newman[8] about the 921–931 cm^{-1} bands. They attributed these bands to carbon atoms in interstitial sites by using the fact that a possible V–C center must give a vibration of lower frequency than 604 cm^{-1} (substitutional carbon) and that the A-center band is more intense when the carbon concentration is high.[17]

3.5. Si_I-Ge interaction

Comparing Figures 2 and 3 it is easy to see that Ge atoms do not modify at all the intensity of the 921–931 cm^{-1} bands associated with C-Si_I. We conclude that P_{I-Ge} is very low because the ratio of C and Ge concentration is about 5×10^{-4}. Although the nature and the concentration of defects are quite different, let us note that P. Baruch and coworkers[18] have confirmed this result (after proton irradiation and channeling measurements).

3.6. *Fast diffusion of carbon in interstitial sites*

There are two possibilities to explain the disappearance of the 921–931 cm^{-1} center. We may first suppose that carbon, ejected by Si_I into an interstitial site by Watkins's replacement mechanism,[19] is not strongly bound to the lattice and that it may migrate easily in the lattice by jumping from one interstitial site to another, the silicon atom taking a normal substitutional site. The second possibility is that, when thermal activation dissociates the C-Si_I pair, the Si_I atom is released and migrates afterwards in the lattice, while on the contrary the carbon atom returns to its initial substitutional position. Because we see no increase of substitutional carbon, by monitoring the 604 cm^{-1} band, we conclude that the ejected carbon atoms diffuse interstially at 300 °K in the lattice *independently of any irradiation effect*. This reinforces the hypothesis of Bean, Newman and Smith[17] to explain the complex defects which they observed in room temperature irradiated silicon containing carbon. The migrating carbon atoms

may be trapped by defect centers such as V–V, A-center, etc . . .

3.7. *The 865 cm^{-1} band*

This band was quite recently studied by Cheng and Vajda[20] in neutron irradiated silicon. They observe a saturation of the intensity of this band with the fluence and no dichroism, from which they conclude that this band comes from the A-center vibration in the strain field of an unknown impurity. We could attribute the 865 cm^{-1} band to an A-center perturbed by a carbon atom which has been trapped in its vicinity but the problem seems complicated because:

(1) we do not observe any good correlation between carbon concentration and the intensity of the 865 cm^{-1} band;
(2) there is a 30° shift between the annealing temperature of the 921–931 cm^{-1} bands and the temperature of the appearance of 865 cm^{-1} in the sample 3 (see Figures 3 and 6);
(3) the 865 cm^{-1} band is present in sample 1 (Figure 1) immediately after irradiation and in the case of this sample its intensity does not change at the annealing stage after which the 921–931 cm^{-1} bands have disappeared.

We emphasize the fact that in sample 3, the 865 cm^{-1} band appears at exactly the same temperature at which the B-center (935 cm^{-1}) disappears (see Figure 6). Isotopic shift measurements will allow us to determine the possible role of carbon by using C^{13} and C^{14} doped silicon.

3.8. *Tailing effect near intrinsic absorption*

After irradiation, an intense absorption appears near the frequency corresponding to the gap (see Figure 4a); this absorption decreases slowly with the increasing wavelength. Figure 4c shows that the absorption coefficient at 1 eV decreases progressively with annealing stages. Experiments have shown that this absorption is not uniquely associated to divacancies. The study of this effect requires precise measurement of the reflection coefficient for a sample free of surface contamination. A detailed study of this effect will be published elsewhere.[21]

3.9. *Formation of divacancies from isolated vacancies*

If we suppose that the intensity of the 1.7 μ band gives the V–V concentration, we may conclude (Figure 4b) that when vacancies are released from

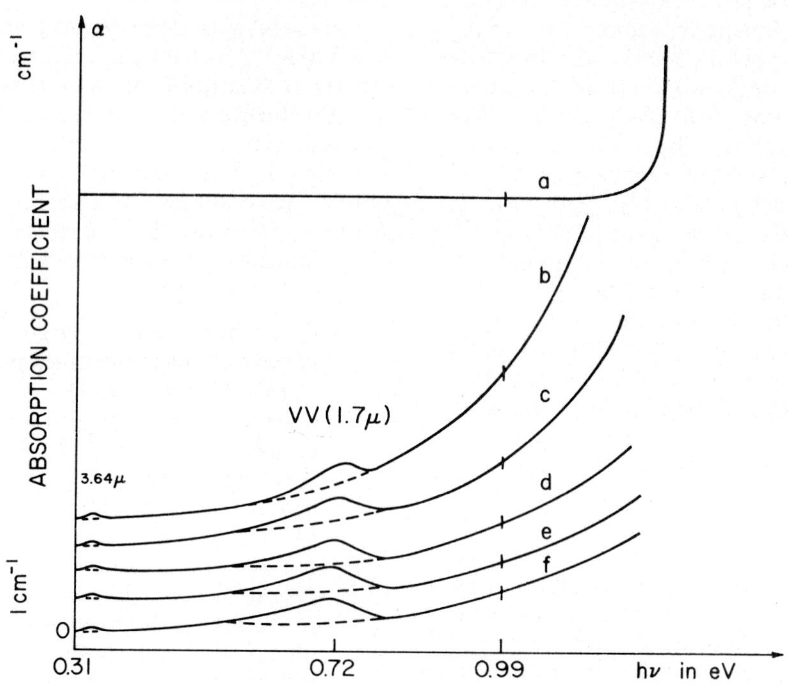

FIG. 4A. Near-edge infrared spectra recorded at 80 °K of silicon doped with Ge (sample 1—see table). 'V-V' means divacancy. The 3.64 μ band appears only under illumination: The reflection coefficient was taken equal to 0.3 in all cases. (a) Before irradiation; (b) after irradiation as in Figure 1; remainder after 15 minutes at (c) 230 °K; (d) 245 °K; (e) 280 °K; 340 °K.

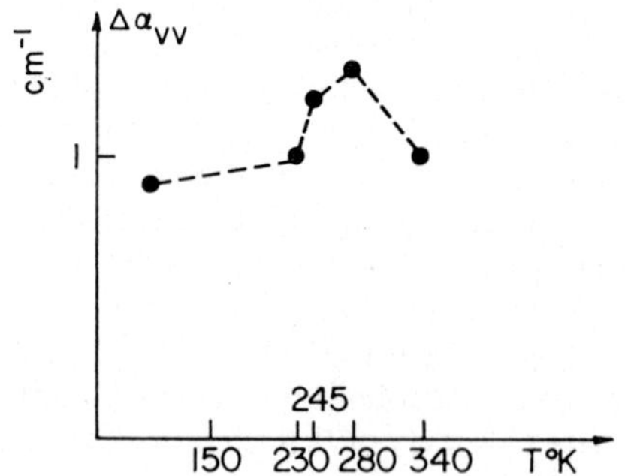

FIG. 4B. Intensity (from Figure 4A) of the 1.7 μ band vs. annealing—the background is subtracted.

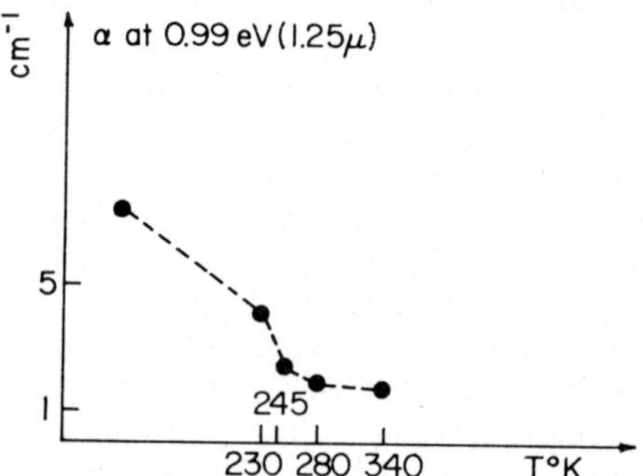

FIG. 4C. Intensity at 1.25 μ vs. annealing.

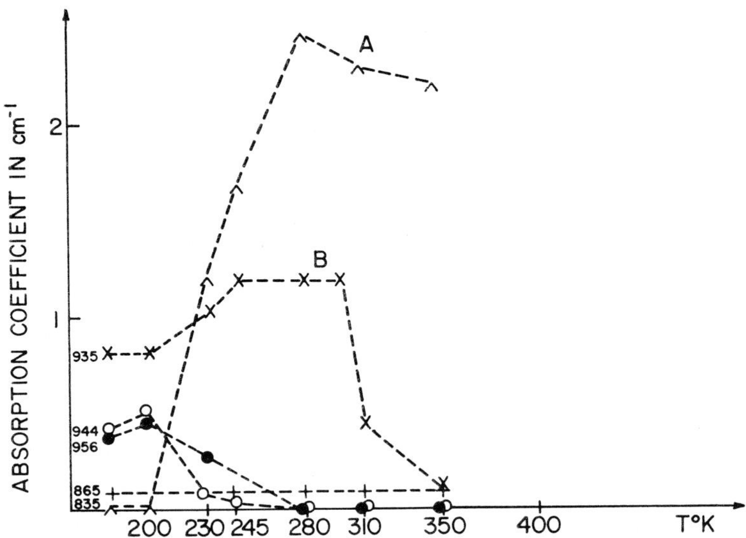

FIG. 5. Annealing curves for the bands in Figure 1.

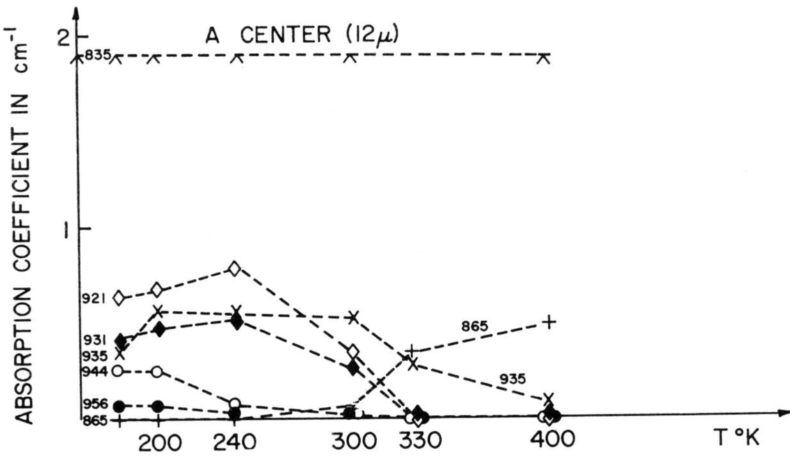

FIG. 6. Annealing curves for the bands in Figure 3.

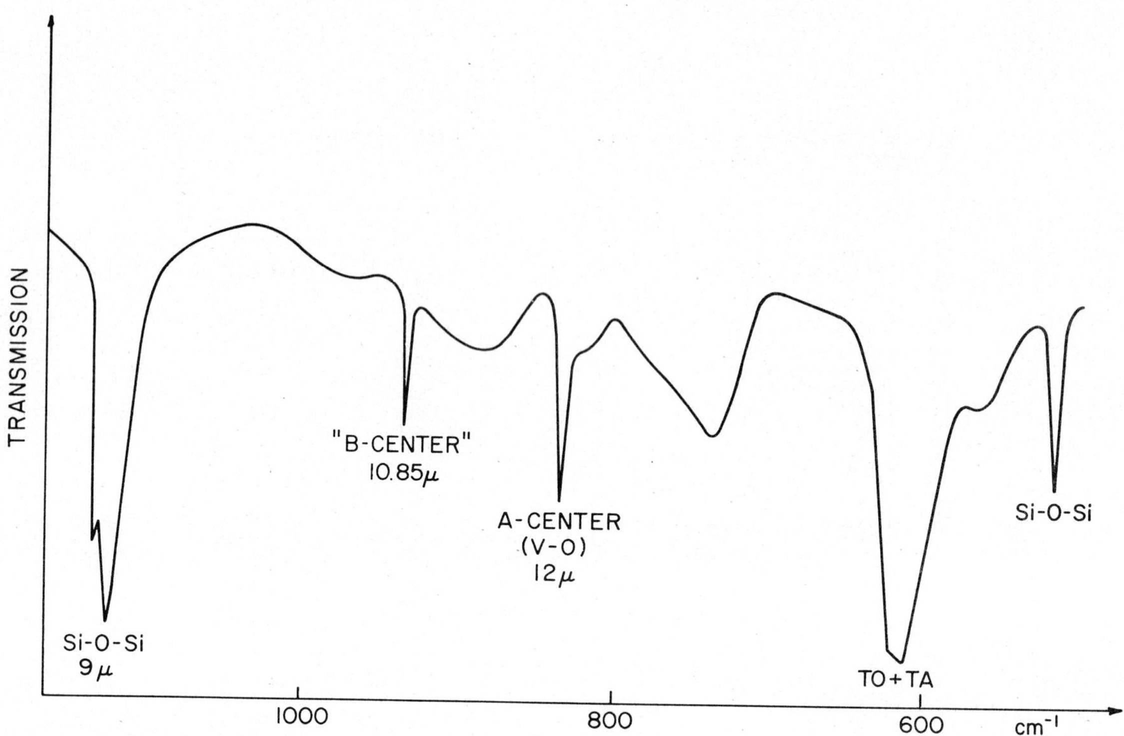

FIG. 7. Transmission spectrum at 80 °K of sample 1′ (see table) after irradiation at 107 °K and annealing to 300 °K.

Ge–V centers by thermal activation, P_{V-V} is the very much lower than P_{V-O}.

3.10. *The 3.62 μ and 1.28 μ bands*

In carbon-doped samples (sample 3; see Table I) a band situated at 1.28 μ appears after annealing at 330 °K. Experimental results suggest that this band may be associated with the center resulting from the interaction of migrating carbon interstitials (or Si_I) with divacancies. This is only a tentative interpretation and further studies are necessary to elucidate this point.

The 3.62 μ band is present after irradiation in samples containing oxygen; it is only observed if the sample is exposed to white light. A more detailed discussion will be published shortly.[22]

4. CONCLUSIONS

We find that in silicon at 100 °K, the capture probabilities of a vacancy either by a Ge atom or by an oxygen atom are almost equal, while the capture probability of an intrinsic interstitial by a Ge atom is very low in comparison with the probability of interaction of interstitials with carbon atoms.

The silicon interstitials can easily be trapped by oxygen and the different defect centers resulting from this trapping produce the observed absorption lines. The concentration of these centers is higher when the concentrations of other possible traps for the interstitials are lower. We have been able to determine the contribution of each impurity —carbon on the one hand and oxygen on the other hand—to these defect-impurity associations.

The use of Ge-doped samples has allowed us to identify vacancy-impurity associations and interstitial-impurity associations. As an example, the study of the interaction of silicon interstitials with carbon atoms has been shown. We believe that this technique, which consists in using samples containing controlled carbon or germanium concentrations, can be easily extended to the study of the interaction of defects with any other kind of impurities.

We wish to emphasize the importance of residual

impurities, such as Ge, O and C, in radiation damage studies. In particular, it is possible that carbon may be playing a leading role in the complex behaviour of Li-doped irradiated silicon.[23] Actually the Li-doped samples which have been generally studied have a low dislocation density and it seems that there is a larger carbon concentration in dislocation free crystals.[24]

On the other hand, it is easy to measure the 921 cm^{-1} absorption band associated with interstitial carbon which suggests the possible use of the infrared absorption measurements of irradiated samples as a method of residual carbon determination in silicon at concentrations lower than 10^{16} atoms/cm^3.

ACKNOWLEDGEMENTS

The authors would like to thank Professor R. Freymann for providing the P.E.M. laboratory facilities.

REFERENCES

1. H. Y. Fan and A. K. Ramdas, Proceedings of the International Conference on the Physics of Semiconductors, Prague, 1960, p. 309.
2. A. K. Ramdas and M. G. Rao, *Phys. Rev.*, **142**, 451 (1966).
3. J. W. Corbett, G. D. Watkins, R. M. Chrenko and R. S. McDonald, *Phys. Rev.*, **121**, 1015 (1961), and *Phys. Rev.*, **135**, A1381 (1964).
4. R. E. Whan, *Appl. Phys. Letters*, **8**, 131 (1966).
5. R. E. Whan, *J.A.P.*, **37**, 3378 (1966).
6. R. E. Whan and F. L. Vook, *Phys. Rev.*, **153**, 814 (1967),
7. F. L. Vook and F. L. Stein, *Appl. Phys. Letters*, **13**, 343 (1968).
8. A. R. Bean and R. C. Newman, *Solid State Comm.*, **8**, 175 (1970).
9. J. P. Dismukes, L. Erstrom and R. J. Paff, *J. Phys. Chem.*, **68**, 3021 (1964).
10. B. Pajot, *Solid State Electronics*, **12**, 923 (1969).
11. W. Kaiser and P. H. Keck, *J.A.P.*, **28**, 882 (1957).
12. R. C. Newman and J. B. Willis, *J. Phys. Chem. Solids*, **26**, 373 (1965).
13. A. Brelot, to be published.
14. H. J. Hrostowski and B. J. Alder, *J. Chem. Phys.*, **33**, 980 (1960).
15. B. Pajot, Thesis, Paris (1969).
16. G. D. Watkins, *IEEE Trans. Nucl. Sci. NS16*, **6**, 13 (1969).
17. A. R. Bean, R. C. Newman and R. S. Smith, *J. Phys. Chem. Solids*, **31**, 739 (1970).
18. P. Baruch, et al., this volume.
19. G. D. Watkins, in *Radiation Damage in Semiconductors* (Dunod, Paris, 1965), p. 97.
20. L. J. Cheng and P. Vajda, *J.A.P.*, **40**, 4679 (1970).
21. A. Brelot and B. Massarani, to be published.
22. A. Brelot, to be published.
23. A. Brelot, et al., to be published.
24. N. Schink, *Solid State Electronics*, **8**, 767 (1965).
25. L. J. Cheng, J. C. Corelli, J. W. Corbett and G. D. Watkins, *Phys. Rev.*, **152**, 7 61 (1966).

DISCUSSION

Question (CROSSMAN) How did you measure the amount of Ge that you had in your sample?

Answer (BRELOT) We used density measurement, which is suitable for these high concentrations.

Question (STEIN) Do you have any information about the possible residual Ge content in Si which has not been intentionally doped with Ge?

Answer (BRELOT) Ge is an heavier atom than silicon; it may therefore be only observed as a broad absorption band associated with the band mode of Ge; so the spectroscopic method is not sensitive and it is difficult to know the concentration of residual Ge atoms when this concentration is low.

NEUTRON IRRADIATIONS OF SILICON STUDIED BY Fe-57 MÖSSBAUER EFFECT†

K. MATSUI

Radiation Laboratory, Faculty of Engineering, University of Tôhoku, Sendai, Japan

R. R. HASIGUTI

Department of Metallurgy, University of Tokyo, Tokyo, Japan

AND

H. ONODERA

Radiation Laboratory, Faculty of Engineering, University of Tôhoku, Sendai, Japan

Fast neutron irradiations of silicon are studied through the cobalt-57 probe incorporated into the lattice.

The cobalt substitutional atoms introduced by fast quenching from around 1100 °C are found to be converted to interstitials by reactor irradiations at 80 °C. The reverse effect in which the irradiations enhance the substitutional atoms also takes place. These conversion effects are accompanied by a marked dependence upon initial Fermi levels. Their efficiencies seem to be proportional to integrated neutron flux.

A strong unstable spectrum best represented by 4 quadrupolar doublets arises before the interstitial-type conversion is observed. These doublets are tentatively assigned to the cobalt interstitials trapped by charged centers. The spectrum dies out with the decay of the induced radiation, suggesting that the pairs become unstable at thermal equilibrium at room temperatures.

1. INTRODUCTION

It is of interest to make use of Mössbauer effect[1] as a tool for microscopic identification of lattice defects.[2] Since defects are hardly found over 10^{-4} in their fractions, the effect has to be studied for parent nuclei embedded in host lattice as probe.

It must be noted in studying the interactions between defects and the probe that the observation is limited to an isomeric state of the daughter. Complexities may arise from radiation-after-effects produced by the disintegration of the parent.[3]

In the case of Co-57/Fe-57 system (14.4 keV M1 transition, $I = \frac{3}{2} - \frac{1}{2}$, $\tau_{\frac{3}{2}} = 9.8 \times 10^{-8}$ sec), a model scheme of electronic states of $3d^n$ series in silicon[4] can be utilized to guess the states of both the parent and the daughter, although discrepancies between the model and experimental observation may exist.

It was proposed by Mössbauer measurements[5,6] that cobalt precipitates and cobalt substitutional atoms give rise to singlet lines at $+0.002$ and at $+0.057$ cm/sec, respectively with respect to 310 stainless steel absorber. Both the singlets are Fermi level independent. The latter was introduced by vacancy liberated from copper simultaneously diffused. 1 MeV electron irradiations up to $10^{18}/cm^2$ did not affect the precipitates.

The present paper deals with an attempt to investigate the interactions between radiation-induced defects and the cobalt precipitates or substitutional atoms in which the cobalt substitutional atoms were introduced by fast quenching from 1000–1200 °C.

2. EXPERIMENTAL PROCEDURE

A small amount of carrier-free $CoCl_2$ solution was smeared on the etched surfaces of single crystals of about $4 \times 1 \times 10$ mm³. The specimens were held in hydrogen, first at 600 °C for 1 hr and then at 900 °C for 1 hr.

Quenching was made by dropping the specimens into ethylene glycol after holding them in argon for 20–60 minutes at quenching temperature. A slower quenching was made for several specimens by sealing them in a quartz capsule to drop into water. The cobalt density in the specimens was estimated to be $10^{14-15}/cm^3$.

Neutron irradiations were carried out in the hydrostatic loop at the Kyoto University Research

TABLE I

Sample characteristics and experimental conditions.
The intensity ratio of the singlet resonance line at $+0.06$ cm/sec to that at $+0.002$ cm/sec is represented by Co(s)/Co(p).

Sample no.	Dopant	$E_f(O)$ (eV)	Quenching treatment	$E_f(Q)$ (eV)	Co(s)/Co(p) (Q)	Neutron dose	$E_f(I)$ (eV)	Co(s)/Co(p) (I)	Fraction† converted	Unstable spectrum
1	P	$E_c - 0.14$	1100 °C	—	0.40	$\phi_{ft}\to\phi_f t = 5.6 \times 10^{17}$ cm^{-2} (40 °C)	—	0.42	$+0.05$	$(-)$
2	B(FZ)	$E_v + 0.4$	× 3 hrs	—	0.44	$\phi_f = 1.6 \times 10^{13}$ cm^{-2} sec^{-1}	—	0.37	-0.16	$(-)$
3	B(FZ)	$+0.28$	(slow)	—	0.48	$\phi_{epi} = 2.4 \times 10^{12}$ cm^{-2} sec^{-1}	—	0.42	-0.13	$(-)$
4	B	$+0.22$		—	0.65	$\phi_t = 3.3 \times 10^{13}$ cm^{-2} sec^{-1}	—	0.72	$+0.11$	$(-)$
5	P	$E_c - 0.14$		—	0.72	$\phi_{ft}\to\phi_f t = 1.4 \times 10^{18}$ cm^{-2} (80 °C)	—	0.29	-0.60	$+ +(100\,\text{hrs})$
6	P(FZ)	-0.31	1150 °C	—	0.70	$\phi_f = 3.9 \times 10^{13}$ cm^{-2} sec^{-1}	—	(0.5)	(-0.6)	$+ +(>100\,\text{hrs})$
7	B(FZ)	$E_v + 0.31$	× 20 min	—	0.60	$\phi_{epi} = 6.0 \times 10^{12}$ cm^{-2} sec^{-1}	—	0.33	-0.45	$+ +(60\,\text{hrs})$
8	B(FZ)	$+0.27$	(fast)	—	0.55	$\phi_t = 8.2 \times 10^{13}$ cm^{-2} sec^{-1}	—	0.52	-0.05	$-(<20\,\text{hrs})$
9	B	$+0.17$		—	0.48		—	0.46	-0.04	$-(<20\,\text{hrs})$
10	P	$E_c - 0.14$	1100 °C	$E_c - 0.38$	0.94		$E_c - 0.53$	0.76	-0.19	$+(<44\,\text{hrs})$
11	P	-0.14	1200	-0.22	1.61		-0.52	1.32	-0.18	$+ +(100\,\text{hrs})$
12	P	-0.22	1100	-0.30	0.84		-0.52	0.80	-0.05	$-(\ll 50\,\text{hrs})$
13	P(FZ)	-0.29	1100	-0.30	0.61	$\phi_{ft}\to\phi_f t = 7.0 \times 10^{17}$ cm^{-2} (80 °C)	-0.54	0.79	$+0.30$	$-(\ll 100\,\text{hrs})$
14	As(FZ)	-0.29	1000	-0.31	0.43	$\phi_f = 3.9 \times 10^{13}$ cm^{-2} sec^{-1}	-0.55	0.78	$+0.58$	$-(\ll 67\,\text{hrs})$
15	P(FZ)	-0.31	950 × 1 hr	-0.43	1.07	$\phi_{epi} = 6.0 \times 10^{12}$ cm^{-2} sec^{-1}	-0.57	0.81	-0.24	$-(\ll 37\,\text{hrs})$
16	P(FZ)	-0.33	1150 (fast)	-0.34	0.74	$\phi_t = 8.2 \times 10^{13}$ cm^{-2} sec^{-1}	-0.55	0.92	$+0.24$	$-(\ll 35\,\text{hrs})$
17	P(FZ)	-0.33	1000	-0.39	0.65		-0.57	0.85	$+0.31$	$-(\ll 50\,\text{hrs})$
18	B(FZ)	$E_v + 0.31$	1100	$E_v + 0.38$	1.23		$E_v + 0.53$	0.90	-0.27	$+(<40\,\text{hrs})$
19	B	$+0.17$	1200	—	0.94		—	0.96	$+0.02$	$+(<67\,\text{hrs})$
20	B	$+0.17$	1050	$+0.27$	0.92		$+0.56$	0.93	$+0.01$	$-(\ll 76\,\text{hrs})$

† defined by $\left[\dfrac{\text{Co(s)/Co(p)}}{(I)} - \dfrac{\text{Co(s)/Co(p)}}{(Q)}\right] \Big/ \left[\dfrac{\text{Co(s)/Co(p)}}{(Q)}\right].$

All measurements were made at $293\left({}^{+2}_{-3}\right)$ °K.

Reactor Institute. The specimens to be irradiated were sealed in quartz tube.

Mössbauer measurements were made using time-mode spectrometers, one fabricated by Elron and the other homemade,[1] coupled to Ar-CO$_2$ (10 per cent) proportional counters. All spectra were measured with respect to 310 stainless steel absorber driven electro-mechanically at room temperatures. The absorption scale indicated is nominal. The positive velocity corresponds to energetically higher states.

Fermi levels were calculated from 4-point-probe resistivity measurements. The sample characteristics and other experimental conditions are summarized in Table I.

3. EXPERIMENTAL RESULTS

The spectra of the quenched specimens agreed with those obtained earlier,[5,6] without the prescribed quenching treatment. The intensity ratio of the singlets $+0.06/+0.002$ cm/sec was enhanced by the faster quenching, as shown in Figures 1(a) and (b). The logarithm of the ratio (cf. Table) showed a tendency to be linear to $1/T_Q$ over 1150 °C to yield activation energies around 2 eV for n-type and 1 eV for p-type specimens, respectively, while no well-defined behaviour was found for lower temperatures.

No essential changes were observed in the spectra by annealings up to 500 °C but for gradual broadening of the singlet at $+0.002$ cm/sec.

The radiation-induced changes in the intensity ratio of the singlets were found to be strongly dependent upon initial Fermi levels $E_f(0)$, as shown in Table I. These are represented by fraction converted defined by

$$f = [\text{Co(s)/Co(p)} - \text{Co(s)/Co(p)}]/\text{Co(s)/Co(p)}$$
$$\text{(irradiated)} \quad \text{(quenched)} \quad \text{(quenched)}$$

where Co(s)/Co(p) = intensity ratio of singlets at $+0.06$ cm/sec to $+0.002$ cm/sec, as justified later.

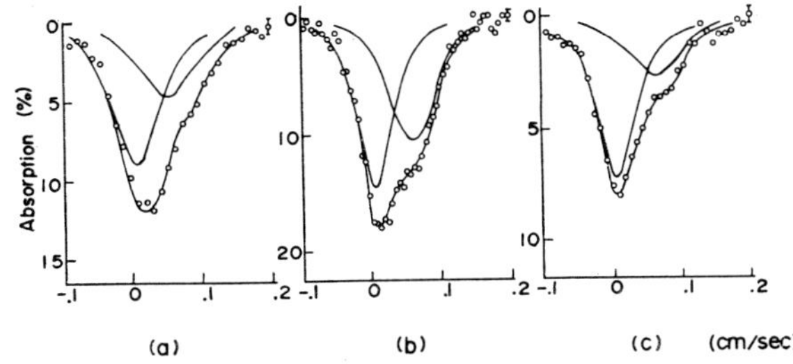

FIG. 1. Effects of quenching rate and neutron irradiation. (a) Slower quenching (specimen-1), (b) faster quenching (specimen-7) and (c) after irradiation by 1.4×10^{18} n/cm² (specimen-7). Absorption scales are nominal.

Figure 2 shows the six different regions to be specified, together with the proportionality of the changes to neutron dose.

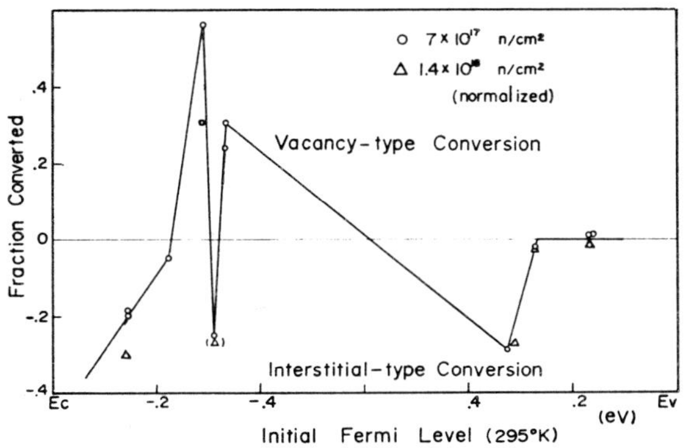

FIG. 2. Fermi level dependence of conversion effects. The fraction converted is normalized for neutron dose 7.0×10^{17}/cm².

The Fermi levels of the quenched specimens $E_f(Q)$ became closer to the middle gap possibly due to the action of quenched-in impurities such as iron-57. The irradiation made block the levels $E_f(I)$ to middle gap.

The strong bremsstrahlung from 1.4 MeV β-decays of silicon-31 (2.62 hrs) produced by thermal neutron capture of silicon-30 (3.09 per cent, 0.11 b.) prevented the Mössbauer measurements of high resolution for several tenth of hours after the offset of the irradiations.

An unstable spectrum arose under such conditions, as shown in Figure 3, which was best represented by 4 sets of quadrupolar doublets of large separations centered at $+0.02$ cm/sec. It was noticed that the spectrum was strong when a

negative value of f was obtained finally. The decay of the spectrum was not of simple nature, as shown in Figure 5 and in Table I.

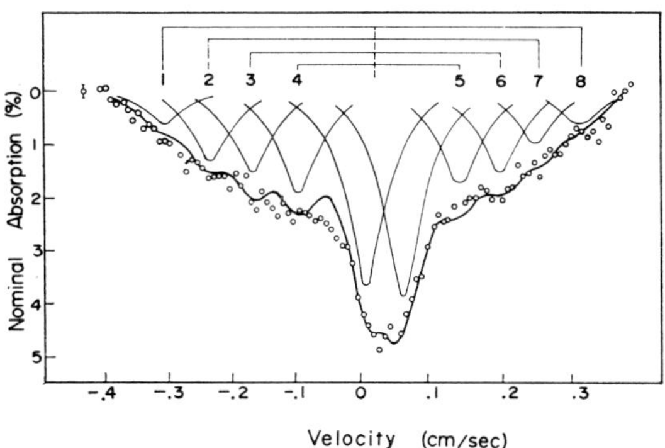

FIG. 3. Unstable spectrum observed after irradiation. Tentatively resolved into 4 quadrupolar doublet lines. Isomer shift = $+0.020$ $\left(^{+0.005}_{-0.010}\right)$ cm/sec. D1 = 0.23, D2 = 0.37, D3 = 0.49 and D4 = 0.60 cm/sec. Specimen-11, $\gamma\|[100]$, 60 hrs after irradiation.

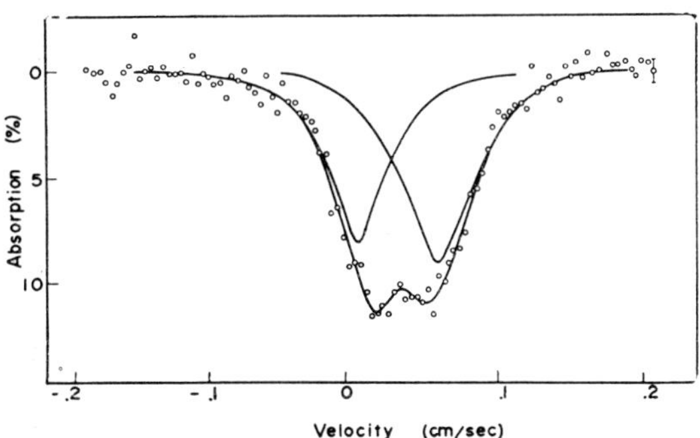

FIG. 4. Unstable spectrum disappeared. Specimen-11, 100 hrs after irradiation.

4. DISCUSSION

The arguments of Norem and Wertheim[5,6] as to the nature of the singlets, Co(p) = $+0.002$ cm/sec and Co(s) = $+0.057$ cm/sec, are considered to be valid.

The absence of magnetic hyperfine splittings[1] in the 'metallic' precipitates can be attributed either to paramagnetic chemical forms, and/or to super-paramagnetism[7] that occurs in finely dispersed (10–100 Å in radii) form. The gaussian line shape of Co(p) is indicative of the latter possibility.

The isomer shift of Co(s), which falls between $3d^4$ and $3d^8$,[8,9] should properly be deduced from molecular orbitals formed between 4 nearest

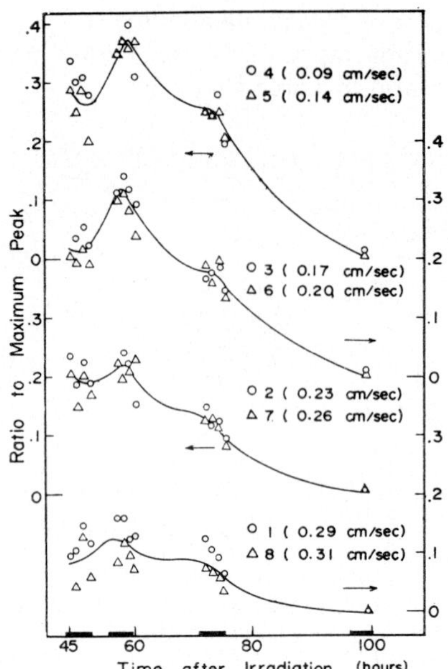

FIG. 5. Decay of unstable spectrum. Specimen-11. The specimen was kept at $-70\,°C$ during the intervals of the measurements.

neighbour silicons, according to the model scheme of Ludwig and Woodbury.[4]

It is expected from the homopolar nature of the bonding that a substantial decreace in the electron density at the nuclear site to give the observed value. The stability of Co(s) experienced in the annealings is in accord with the substitutional model.

The mechanism governing the enhancements of the ratio Co(s)/Co(p) remains controversial. Probably a transition may occur at around $1150\,°C$, such that a dissociative or impurity dependent migration[10] is replaced by conversion of Co(i) by vacancies at higher temperatures. The subject is left to further experiments.

The radiation-induced conversion effects are characterized by fraction converted f as shown in Figure 2.

The principal mechanism which results in $f < 0$ can be interpreted by an interstitial conversion effect similar to those found by Watkins[11,12] for group III acceptors.

Neither migration energies nor charge states of Co(i) are known.[13] Indirect evidence would be Fe(i)$^{+1}$ introduced by quenching having a migration energy 0.78 eV.[14] Ion pair formation with group III acceptors is also known.[4,14] A migration

energy of 0.26 eV is obtained for Co(i)$^{+1}$ following Weiser[15] using ionic radius 0.77 Å estimated by the maximum-electron-density radius of Slater function[16] and repulsive potential constant 0.30 Å.[17]

The Co(i) may safely be assumed to make long range motion at $80\,°C$ under irradiation and at $20\,°C$ under observation.

The precipitation of Co(i) results in $f < 0$. If Co(i) is strongly attracted by vacancy to recombine, $f = 0$ results. A third possibility will be temporary trapping by negatively charged centers, as is the case of Fe(i)$^{+1} - $B(s)$^{-1}$.[14] These processes are illustrated in Figure 6.

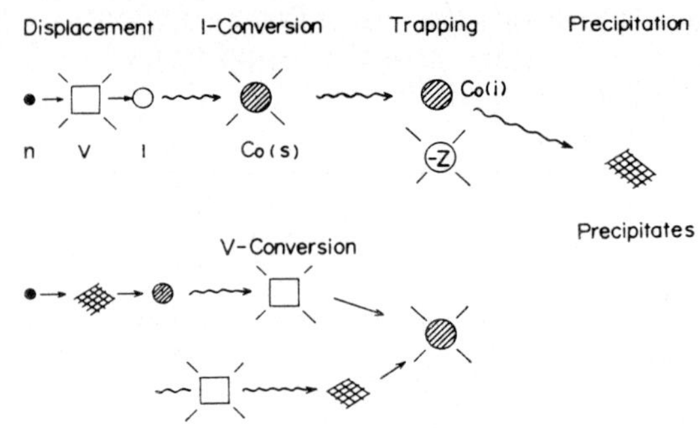

FIG. 6. Some proposed mechanisms of conversion effects.

The unstable spectrum is assigned to Co(i) because of its simultaneous appearance with $f < 0$. It is natural to expect that the Co(i) responsible for the spectrum is a trapped one, since T_{obs} was substantially lower than T_{irr}.

The doublet structure of the spectrum can in fact be related to ion pairs. It is to be noted that the quadrupolar separations are exceptionally large as compaired to those known for Fe-57.[1,9] These 'giant splittings' have recently been found in Co-57 diffused UO$_2$ (max. 0.92 cm/sec),[18] rutile (max. 0.89 cm/sec)[19] and CaF$_2$ (max. 0.70 cm/sec)[20] and interpreted by low spin configurations produced by the axial field of defect,[2] as briefly discussed in Appendix.

It follows from the above that the isomer shift of the interstitial must be $+0.02$ cm/sec, in disagreement with $+0.17$ cm/sec (3d^{7}4s^0)[8,9] expected for localised model of Fe(i)$^{+1}$. The disagreement may be ascribed either to a surviving higher charge state produced by multiple Auger effect[3] to be checked by delayed coincidence technique[21] or to trapped carriers.[6] Still interesting is the fact that a doublet has been reported at $+0.02$ cm/sec with separation 0.08 cm/sec for Fe-57 in a recoil implantation experiment by 36 MeV O^{16} ($+5$) ion beam.[22] The most probable form of the implanted isomer, whose life is limited to 10^{-7} sec, would be interstitial.

It is also interesting that the unstable spectrum decays in an over-all fashion with the internal radiation. The phenomenon can be related to an instability of the pairs at $E_f =$ mid-gap when the trapped carriers disappear.

The details of the pair structures have to be examined by further studies.

The interstitial conversion mechanism leading to $f \leqslant 0$ applies for a proportionality between Co(i) and Co(s) $\times I$ or Co(s) $\times$ neutron dose. Closer examinations on the effect for specimens-1, 5, 10 and 11 (cf. Table) reveal than another mechanism to inducing $f > 0$ is acting in competition. The latter manifests itself explicitly for $E_c - 0.25$ eV $> E_f(0) >$ mid-gap.

The mechanism which results in $f > 0$ is harder to interpret than the previous one. The simplest one will be a vacancy conversion effect, but it must be Fermi level independent by itself. The alternatives would be effective destruction and rearrangement of Co(p) either by energetic knock-on's or by spikes. These are illustrated in Figure 6.

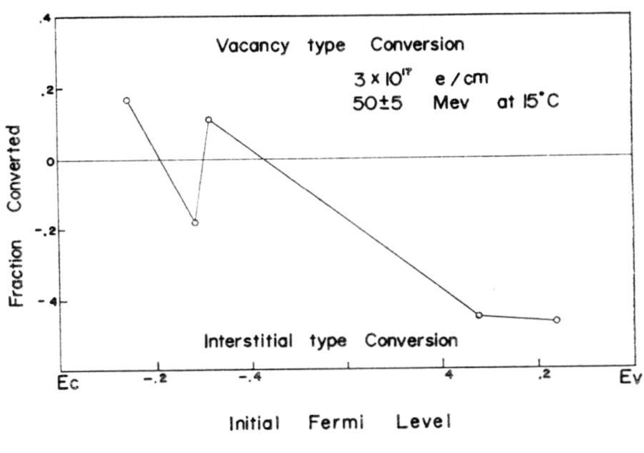

FIGURE 7.

The energy transfer efficiency[23] from silicon knock-on to cobalt atom is so low (5.4 per cent) that the knock-on's capable of the effective destruction must have an energy $1 \gtrsim$ keV. These energetic knock-on's are rare according to multiplication theory[23] and also less effective in displacement due to their large mean-free paths[23] ($\gtrsim 100$ Å) in the cobalt precipitates.

Probably direct evidences of spikes,[24] being effective in heavier materials,[23] would lead to the understanding of the mechanism.

The marked Fermi level dependence of the conversion effects strongly suggests the point defect origin of the processes involved. It is felt, however, too early to figure out its relation to charge states, level schemes, and migration energies of specific defect because of the complexities discussed above.

It is fortunate that more advanced techniques, such as line intensity study[2] and delayed coincidence,[21] can be applied to this problem. It is also interesting to investigate the problem by electron irradiation. No definite results have been obtained by 50 MeV linac irradiation possibly due to its low dose ($\approx 10^{17}$ e/cm²).

5. CONCLUSION

Conversion effects, of interstitial and vacancy type, have been found with marked Fermi level dependence for cobalt substitutionals and precipitates.

The unstable spectrum accompanying the interstitial conversion event has been attributed to cobalt interstitial trapped by negatively charged centers.

Details of the conversion mechanisms and the structures of defects involved are to be clarified by further investigations.

ACKNOWLEDGEMENT

The present work was only performed with the courtesy of Dr. Y. Maeda and Prof. T. Higashimura of the Research Reactor Institute of Kyoto University (Kumatori, Osaka) and at the same time of Dr. F. Sawayanagi of the Research Institute of Physics and Chemistry (Tokyo).

One of the authors (K.M.) was benefited by the support of the Sakkokai Fund (Tokyo).

APPENDIX[2]

The quadrupolar splitting of Fe-57 for transition $I = 3/2 - 1/2$ is given by[1]

$$\Delta = (e^2/2)Q_{3/2}(1 - \alpha^2)(1 - R(s))V_{zz}$$
$$(1 + (1/3)(V_{xx} - V_{yy})^2/V_{zz}^2)^{1/2} \quad (1)$$

where e = electronic charge unit, $Q_{3/2}$ = quadrupole moment of $3/2^-$ state, $1 - \alpha^2$ = covalency factor, $1 - R(s)$ = Sternheimer's shielding factor, and (V_{xx}, V_{yy}, V_{zz}) = components of electric field gradient tensor in diagonal form.

Since $1 - \alpha^2$ and $1 - R(s)$ depend moderately upon electronic structure of each ion, only the factor in (1) dependent on the E.F.G. will be compared in the following, taking $e\langle r^{-3}\rangle = e\int_0^\infty f(r)^2 r^{-3} r^2 \, dr : f(r) = 3d$ radial wave function for each ion as its unit. Also E.F.G. due to lattice ions are neglected, its components being small.[1]

The origin of the 'giant splittings' is attributed to the axial field of a charged defect Ze paired with the iron at a distance $z = R$. The 'unbalanced' field is expected larger than the 'balanced' field of defect-free environment. The defect-free crystal field orbitals are thus rearranged into the axial field orbitals $(d_0, d_{\pm 1}, d_{\pm 2})$ having $V_{zz} = (-4/7, -2/7, +4/7)$; $V_{xx} = V_{yy} = 0$.

$$E_{d_0} - E_{d_{\pm 1}} = (e^2 Z/7)(\langle r^2\rangle/R^3 + (10/3)\langle r^4\rangle/R^5)$$

$$E_{d_{\pm 1}} - E_{d_{\pm 2}} = (e^2 Z/7)(3\langle r^2\rangle/R^3 - (5/3)\langle r^4\rangle/R^5)$$

$$: \langle r^k\rangle = \int_0^\infty f(r)^2 r^{k+2} \, dr \quad (2)$$

exceed even for modest value of R the differences of the term values of free-ion ($\approx 10^4$ cm^{-1}) to supress the interelectronic repulsions acting between electrons to populate these levels. For example, this is expected for Fe(i)$^{+1}$ paired with X^{-2} at $R = 2.25$ Å. At this extreme, the *maximum* value V_{zz} expected from the one-electron orbitals are: $16/7(d^4)$, $14/7(d^5)$, $12/7(d^6)$, $10/7(d^7)$ and $8/7(d^8)$ for Z negative and $-12/7(d^4)$, $-14/7(d^5)$, $-16/7(d^6)$, $-12/7(d^7)$ and $-8/7(d^8)$ for Z positive.

These are called low spin values, as the Hund's spin maximum rule has been assumed to be invalid. The high spin values in defect-free cubic environments are, for d^5 and d^6, 0 and $+4/7$ (max.), which correspond approximately $\Delta = 0$ and 0.3 cm/sec, respectively.[1] The Δ's expected for corresponding low spin cases are 1.1 cm/sec(d^5) and 0.9–1.2 cm/sec(d^6).

The aim of the present discussion is limited to indicate the origin of 'giant splittings'. Quantitative treatments are needed to obtain Δ for specific case.

It is to be noted that the E.F.G. axes are directed along pair axes, which makes the line intensity study of value.

$$I(+)/I(-) = I(-)/I(+) = I(3/2 - 1/2)/I(1/2 - 1/2)$$
$$(V_{zz} > 0) \qquad (V_{zz} < 0)$$
$$= (1/J)\sum_j (3 + 3\cos^2\theta_j)/(5 - 3\cos^2\theta_j) \quad (3)$$

where $I(+)/I(-)$ = intensity ratio of a quadrupolar doublet found at $v(+)$ and $v(-)$ on the velocity scale, J = number of equivalent pair axes, and θ_j = angle between the j-th axis and γ-rays. Both the sign and the direction of the E.F.G. axes hence related informations of the pair can be obtained from (3) by rotating the crystal relative to γ-rays. This may be extended to preferential orientation studies[12] developed for IR and EPR.

For instance, if γ-rays are directed along [100] as was the case for specimen 11 in Fig. 3, the intensity ratio becomes 1.5 for $\langle 100\rangle$ pairs and 1.0 for $\langle 111\rangle$ pairs, respectively. The 4 doublets in Figure 3 appear to arise from $\langle 111\rangle$ pairs within the experimental errors indicated.

It may be added that Goldanskii-Karyagin effect[25] tends to intensify the E.F.G. anisotropy, since the probe nucleus becomes tightly bound as to the emission parallel to pairing axis.[26,20]

It is hoped that these techniques, rather overlooked in this field, may stimulate wider interests.

REFERENCES

1. G. K. Wertheim, *Mössbauer effect, Principles and Applications*, Academic Press, New York and London (1964).
2. K. Matsui, to be published.
3. H. Pollak, *Phys. Status Sol.*, **2**, 270 (1962).
4. G. W. Ludwig and H. H. Woodbury, *Phys. Rev. Letters*, **5**, 96 (1960), ibid., **5**, 98 (1960). *Solid State Physics*, ed's. F. Seitz and D. Turnbull, New York (1962), **13**, 223.
5. G. K. Wertheim, *Proc. 2nd International Conf. on Mössbauer Effect*, p. 294 (comment to M. de Coster, H. Pollak and S. Amerlinckx); ed's. D. M. J. Compton and A. H. Shon. Saclay (1961).
6. P. C. Norem and G. K. Wertheim, *J. Phys. Chem. Solids*, **23**, 1111 (1962).
7. D. W. Collins, J. J. Dehn and L. N. Mulay, *Mössbauer Effect Methodology*, ed. I. J. Gruverman, (1967) New York, 3, 351.
8. L. R. Walker, G. K. Wertheim and V. Jaccarino, *Phys. Rev. Letters*, **6**, 98 (1961).
9. J. Danon, *Applications of Mössbauer Effect in Chemistry and Solid State Physics*, IAEA Report, Vienna (1966), p. 89.
10. A. Seeger and M. L. Swanson, *Lattice Defects in Semiconductors*, ed. R. R. Hasiguti, Tokyo (1968), p. 93.
11. G. D. Watkins, *Disc. Faraday Soc.*, **31**, 86 (1961).
12. J. W. Corbett, *Solid State Physics*, ed's. F. Seitz and D. Turnbull, New York. (1966), suppl. 7.
13. C. B. Collins and R. O. Carlson, *Phys. Rev.*, **108**, 1409 (1957).
14. W. H. Shepherd and J. A. Turner, *J. Phys. Chem. Solids*, **23**, 1697 (1962).
15. K. Weiser, *Phys. Rev.*, **126**, 1427 (1962).
16. J. C. Slater, *Quantum Theory of Atomic Structure*, Wiley, vol. 1, New York (1960).
17. M. F. Millea, *J. Phys. Chem. Solids*, **27**, 315 (1966).

18–20. K. Matsui, E. Yagi and A. Ohkawa, to be published.
21. W. Triftshaüser and P. P. Craig, *Phys. Rev. Letters*, **25**, 1161 (1966), see also Reference 1.
22. G. M. Kalvius, G. D. Sprouse and S. S. Hanna, *Hyperfine Structure and Nuclear Radiations*, ed's. E. Matthias and D. A. Shirley (Amsterdam, 1968), p. 686; see also F. G. Allen, *Bull. Amer. Phys. Soc.*, **9**, 296 (1964) for surface study.
23. L. T. Chadderton, *Radiation Damage in Crystals*, Methuen, London (1965).
24. M. Bertlotti, T. Papa, D. Sette and G. Vitali, ibid. (10), p. 351; see also M. Bertlotti, *Radiation Effects in Semiconductors*, ed. F. L. Vook New York (1968), p. 311.
25. S. V. Karyagin, *Dokl. Akad. Nauk USSR*, **148**, 1102 (1963); see also Reference 1.
26. J. G. Mullen, *Phys. Rev.*, **131**, 1415 (1963).

NOTE ADDED IN PROOF

1. The 50 MeV LINAC electron irradiation gave the preliminary results as shown in Figure 7 to be compared with Figure 2. No unstable spectrum observed.

2. It was newly found that the 'giant quadrupolar lines' are stable at 77 °K. They annihilate very rapidly at 295 °K to leave Co(s) and Co(p) lines, as described in the text.

DISCUSSION

Question (CRAWFORD) How was the Co[57] introduced into the silicon crystals? In the melt?

Answer (HASIGUTI) No, it is diffused in.

Question (BISHAY) I believe you said that Co takes substitutional positions in your silicon; is that easy by your diffusion method?

Answer (HASIGUTI) By quenching from above 1000 °C you get a large percentage of Co substitutional.

Question (HIRAKI) What is the charge state of substitutional Co in Si—I mean, is it a donor or an acceptor?

Answer (HASIGUTI) This is a problem. We do not know definitely. Please refer to Ref. 6.

Question (STREETMAN) Most available evidence indicates that Co has double-acceptor character in Si, with one level near the center of the gap and another level about midway between the valence band and the center of the gap. Did you see effects in your results due to the position of the Fermi level which was reached after Co doping?

Answer (HASIGUTI) We did not especially investigate this. But Table I will explain to you something about it. Also see Ref. 6.

AN EPR STUDY OF FAST NEUTRON RADIATION DAMAGE IN SILICON

D. F. DALY AND H. E. NOFFKE

Bell Telephone Laboratories, Inc., Whippany, New Jersey, U.S.A.

Using electron paramagnetic resonance (EPR) the identity of the point defects produced by fast neutron irradiation of silicon at room temperature has been determined and the concentration of each defect has been measured. Irradiations were performed at an unmoderated fast burst reactor to assure that damage from gamma irradiation could be neglected and that all the observed damage could be attributed to displacements by fast neutrons. Total fast neutron fluence between 1.2×10^{15} n/cm^2 and 7×10^{15} n/cm^2 was used.

The initial rate of removal of the phosphorus donor agrees with the initial carrier removal. However, the production rate for the paramagnetic damage centers is approximately 10 per cent of the carrier removal and less than 1 per cent of the estimated number of displacements per neutron collision.

For samples containing approximately 10^{16} phosphorus donors/cm^3, the neutral donor spectrum is observed simultaneously with the negative divacancy spectrum (Si-G7) in both crucible-grown and float-zone crystals. According to the energy level scheme determined for these spectra in electron irradiated silicon, these spectra cannot appear simultaneously if the sample is in equilibrium and uniformly irradiated. From the observation of these spectra, it is concluded that the damage concentration and hence the depth of the Fermi level is nonuniform on a microscopic scale.

These results are interpreted according to the cluster model for neutron damage. The cluster consists of a core of damaged silicon with the Fermi level at the center of the band gap and a surrounding space charge region. Outside the space charge region, the Fermi level is the same as in undamaged silicon.

It is concluded that the low production rate of the point defects and the non-uniform Fermi level constitute microscopic evidence for the defect cluster model of fast neutron damage in silicon.

1. INTRODUCTION

Fast neutron radiation damage in silicon has been the subject of many experimental and theoretical investigations. However, electron paramagnetic resonance (EPR), which has been used extensively to study the damage centers produced by electron beam irradiation[1,2] has had only limited application to the neutron damage problem.[3-5] This investigation was undertaken to study systematically the identity and the production rate of the various paramagnetic defects produced in silicon by neutron bombardment. Crucible-grown and float-zone silicon were used of both *n*-type and *p*-type conductivity. By determining which microscopic paramagnetic defect centers and how many such centers are produced in various types of silicon crystals by neutron bombardment, the neutron damage process can be better understood.

2. EXPERIMENTAL PROCEDURE

All the neutron irradiations were performed at the White Sands Missile Range Fast Burst Reactor.

The reactor is unmoderated and can be operated in a steady state mode. About three times as many neutrons as γ-ray photons are emitted. As a result, γ-ray displacement damage will be negligible compared to the neutron damage. The samples were irradiated and stored at room temperature and measurements were usually made between two weeks and eight weeks following irradiation. Over this time span, room temperature annealing of the defect concentrations could not be observed. The neutron fluence was measured with sulfur dosimeters. The total fast neutron fluence was determined from the known reactor spectrum. In this paper, all neutron fluences quoted will be total fluence of fast neutrons with energy greater than 10 keV.

All EPR measurements were made after cooling the sample in the dark from room temperature to approximately 35 °K. At this temperature the electrons are all frozen out on donor or acceptor sites or traps. The EPR damage spectra are compared with a sample of silicon powder doped with phosphorus and exhibiting the conduction electron EPR line due to a known number of spins.[6] Under our operating conditions, both the damage spectra

TABLE I

EPR Spectra in neutron irradiated silicon

| | Crucible grown | | Float zone |
Dopant	EPR spectra	Dopant	EPR spectra
$P(1 \times 10^{17})$	donor,†§ Si-G17,§ Si-B1 (vacancy-oxygen),§ Si-G7 (− divacancy)§	$P(6 \times 10^{16})$	donor,†§ Si-G17,§ Si-G7 (− divacancy)§
$P(9 \times 10^{15})$	donor,†‡ Si-G17,‡ Si-B1 (vacancy-oxygen),‡ Si-G7 (− divacancy)§	$P(1 \times 10^{16})$	donor,†‡ Si-G17,‡ Si-G7 (− divacancy),‡
		$P(3 \times 10^{15})$	Si-G7 (− divacancy),‡ Si-G8 (vacancy-phosphorous),§ Si-B2,§ Si-P3§
$P(4 \times 10^{14})$	no spectra‡		
$B(4 \times 10^{14})$	weak spectra‡		
$B(9 \times 10^{15})$	Si-G15,§ Si-P3,§ Si-G11‡	$B(1 \times 10^{16})$	Si-G11,‡ Si-G6‡
		$B(4 \times 10^{16})$	Si-G6‡
$B(2 \times 10^{17})$	Si-G15,‡ Si-G11,‡ Si-G6 (+ divacancy)§	$B(1 \times 10^{17})$	Si-G6 (+ divacancy)§ Si-G11‡

† Donor spectrum is observed before irradiation and decreases with increasing irradiation.
‡ Observed for neutron fluences less than 3.0×10^{15} n/cm^2.
§ Observed for neutron fluences up to 7×10^{15} n/cm.

and the conduction electron spin resonance signals obeyed slow passage conditions. The relative accuracy of concentration measurements on the same defect spectrum is ±20 per cent. The absolute accuracy of these measurements is within a factor of two. Corrections were made for the replicas of the spectrum split off from the main spectrum by the hyperfine interaction with the 4.7 per cent abundant Si^{29} nucleus.

3. RESULTS

The EPR spectra observed in different neutron irradiated samples are summarized in Table I. Practically all the observed defects could be identified with spectra that had previously been studied in electron irradiated silicon.

3.1. *N-Type crucible-grown silicon*

For samples with 10^{16} or more donors/cm^3, a uniform decrease in donor concentration with neutron fluence is observed as shown in Figure 1. This decrease in neutral donor concentration at 30 °K agrees within experimental error with the decrease in free carrier concentration determined at room temperature by a four-point probe resistivity measurement.

The defects increase monotonically with fluence up to some given concentration and then decrease due to deeper damage centers pulling the Fermi level below the paramagnetic state of the defect. This can be observed in Figure 2 for the Si-B1 (vacancy-oxygen center) spectrum. For higher fluences or lower donor concentrations, it would also be observed for the Si-G7 (negative divacancy). The Si-G7 spectrum has been studied as a function of temperature. When the sample was warmed to approximately 60 °K, the Si-G7 spectrum was transformed into another spectrum which is an average over the three different Jahn-Teller distortions of the divacancy. This means that the Si-G7 spectrum observed is 'reorientable' [9] and is not subject to local lattice strains greater than approximately 10^{-3}.

The Si-G17 is only observed for low neutron fluences. This is similar to the behavior in electron irradiated silicon[7] and indicates that a shallow Fermi level is required for the center to be paramagnetic. The simultaneous observation of the donor, Si-B1, and Si-G7 spectra in both 1×10^{17}/cm^3 and 9×10^{15}/cm^3 phosphorus doped crystals will be discussed in the next section. After a neutron fluence of 1.8×10^{15} *n*/cm^2, the 4×10^{14}/cm^3 phosphorus doped sample showed no EPR spectrum if cooled down to liquid helium temperature in the dark. Some weak spectra were observed when the sample was illuminated with white light.

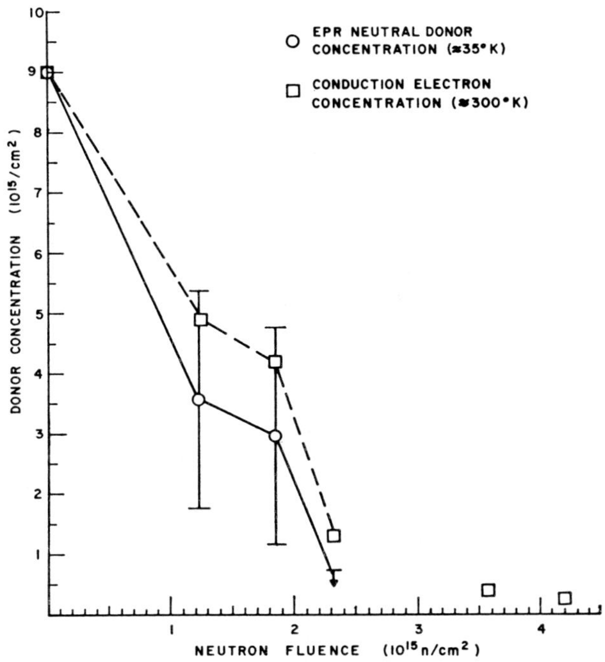

FIG. 1. Neutral donor concentration measured by EPR at approximately 35 °K and free carrier concentration determined by resistivity measurements at room temperature vs fast neutron fluence for an *n*-type crucible-grown silicon sample.

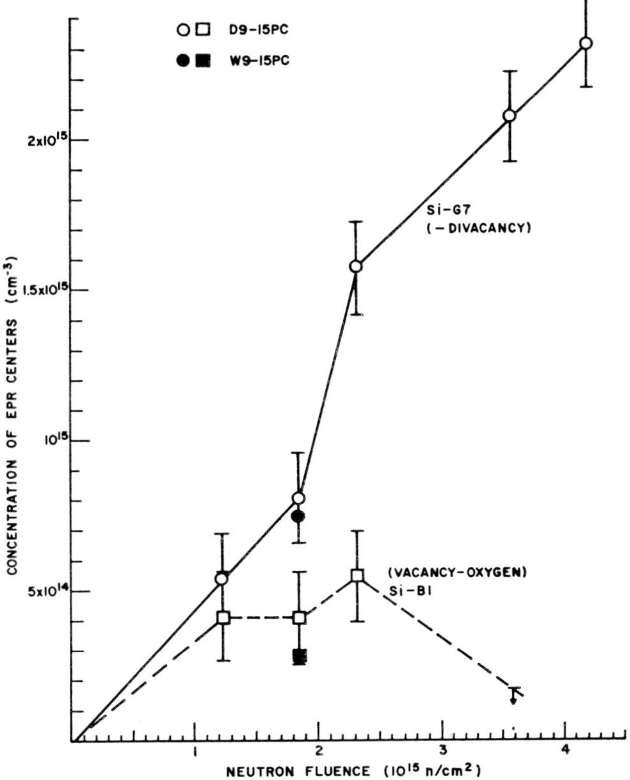

FIG. 2. Paramagnetic defect concentration vs neutron fluence for the same sample used in Figure 1.

3.2. *P-Type crucible-grown silicon*

In the $2 \times 10^{17}/cm^3$ boron doped samples, the Si-G6 (positive divacancy), Si-G15 (K center),[8] and Si-G11 spectra are observed. For a lower boron doping $(9 \times 10^{15}/cm^3)$ and low neutron fluence, the Si-G6 spectrum is not observed although it could merely be so weak that the Si-G15 obscures it. At higher neutron fluence, the Si-P3 spectrum is observed along with the Si-G15. Previously, the Si-P3 had only been observed after much higher neutron fluences.[4] In the $4 \times 10^{14}/cm^3$ boron doped sample, some weak EPR signals corresponding to spin concentrations of $10^{14}/cm^3$ or less were detected. This spectrum could not be identified as one of the known silicon damage spectra and was too weak to be studied in detail.

3.3. *N-Type float-zone silicon*

For the $1 \times 10^{16}/cm^3$ and $6 \times 10^{16}/cm^3$ phosphorus doped float-zone silicon, the spectra are the same as in the crucible-grown samples of similar doping except that the Si-B1 spectrum which requires oxygen is not produced. In the $6 \times 10^{16}/cm^3$ phosphorus-doped samples, the Si-G8 (vacancy-phosphorus center) was not produced even though the Si-G7 (negative divacancy) which requires a

similar Fermi level was observed. However, in the lighter doped sample $(3 \times 10^{15}/cm^3)$ the Si-G8 spectrum did appear after a neutron fluence of $4.3 \times 10^{15}/cm^2$. The Si-G7 (divacancy) spectrum which was observed at fluences less than $3 \times 10^{15}/cm^2$ completely disappeared at the higher fluence. Small signals were observed from the Si-B2[10] spectrum after a neutron fluence of $4.3 \times 10^{15}/cm^2$ and probably from the Si-P3 after a fluence of $7.8 \times 10^{15}/cm^2$. The Si-G8 was also observed up to the latter fluence.

3.4. *P-Type float-zone silicon*

The spectra in this material agree with those observed for *p*-type crucible-grown silicon. For similar neutron fluences, only the Si-G15 which requires oxygen for its formation is not observed. In Figure 3, the concentration of Si-G6 (positive divacancy) versus fluence is shown for different boron doping levels. The production rate of the spectrum is determined from the initial slope of the highest curve. The concentration of these spectra saturate and then decline as the Fermi level moves out of the paramagnetic state. The decline of the concentration is gradual even though the carrier removal rate is ten times the production rate of the

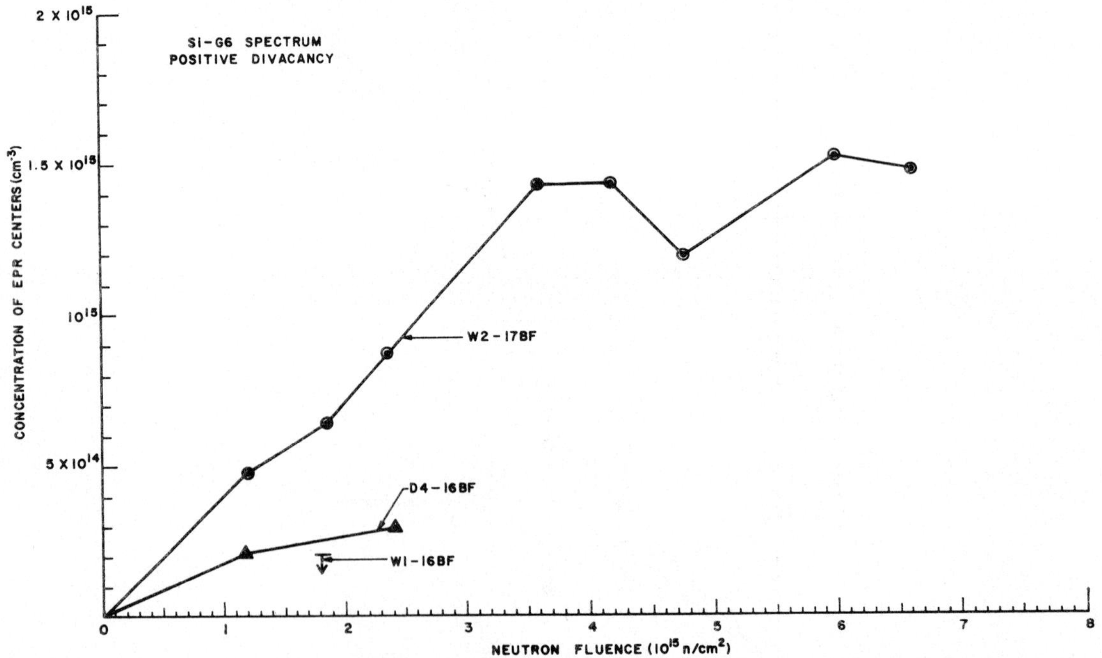

FIG. 3. Concentration of positive divacancies vs neutron fluence for *p*-type float-zone silicon samples with different boron concentrations.

TABLE II

Carriers removed or defects produced per neutron collision $\left(\dfrac{\#/cm^3}{\cdot 165\,\varphi}\right)$

Crucible-grown

Sample	Carrier removal	Donor removal (EPR)	Divacancy (Si-G7)	Vacancy-oxygen (Si-B1)	Other
n-type					
W1-17 PC	85	†	†	†	
D9-15 PC	21	27	2.7	2.0	
W9-15 PC	18	17‡	2.5‡	0.9‡	
p-type					
W9-15 BC	22		<1.0	1.6	0.6 (Si-G11)
					0.5 (Si-P3)‡
W2-17 BC			2.2	0.8	0.2 (Si-G11)

Float-zone

Sample	Carrier removal	Donor removal (EPR)	Divacancy (Di-G7)	Vacancy-phosphorous (Si-G8)	Others
n-type					
W6-16 PX	70	†	†		† (Si-G17)
W1-16 PX	43	38	1.7		1.3 (Si-G17)
D1-16 PF	41	37	2.1		
W3-15 PX	15‡		~3.5	0.6‡	0.4 (Si-B2)‡
p-type					
W1-16 BF	30		0.7		0.1 (Si-G11)
W4-16 BF	72		1.1		
W2-17 BF			2.4		

Additional column headings printed below the table (no data rows on this page):

Crucible-grown				Float-zone		
Carrier removal	Divacancy (Si-G6)	Si-G15	Other	Carrier removal	Divacancy (Si-G6)	Other

† Defect observed. Spin concentration measurement not possible.
‡ Determined from a single point rather than initial slope of the defect production curve.

Notation for samples: M V-YY D G
M   manufacturer (D = Dow, W = Wacker)
V-YY   dopant concentration (V × 10^{YY}/cm³)
D   dopant (P = phosphorous, B = boron)
G   crystal growth method (C = quartz crucible grown, F = float-zone, X = zero dislocation float-zone)

spectrum. The figure shows, as expected, that for lower dopant concentration this decline occurs at lower neutron fluence.

DISCUSSION

In Table II, the number of carriers removed or number of defects produced per neutron collision is listed for the different samples used in this study. The production rate is determined from graphs such as Figures 1, 2 and 3. The number of neutron collisions per cm³ is 0.165 times the neutron fluence.[11,12] It is important to note that the production rate of the damage spectra usually amounts to less than 10 per cent of the carrier removal. Since in most cases, no more than two such spectra are present at any one time, the isolated, paramagnetic damage centers cannot account for as much as one-third of the carrier removal. Also, in Figures 2 and 3, the Si-B1 and Si-G6 concentrations saturate gradually when the Fermi level is pulled out of the paramagnetic state. Since the production rates of these defects are about one-tenth of the carrier removal rate, a uniform distribution of deeper lying traps should produce a decline in spin concentration with approximately ten times the slope of the growth curve. That such a decline is not observed suggests a non-uniform distribution of damage.

In the *n*-type crucible-grown silicon, as mentioned in the previous section, the neutral donor, Si-B1, and Si-G7 (negative divacancy) spectra are observed simultaneously for neutron fluences less than approximately 3×10^{15}/cm³. Likewise, in float-zone silicon, the neutral donor and Si-G7 appear simultaneously. This was observed in samples with phosphorus doping of 10^{16}/cm³ and 10^{17}/cm³. Why this observation was unexpected can be seen in Figure 4 which shows the energy level scheme determined for these defects in EPR studies of electron irradiated silicon. This diagram, adapted from Watkins,[1] shows that these defects are only para-

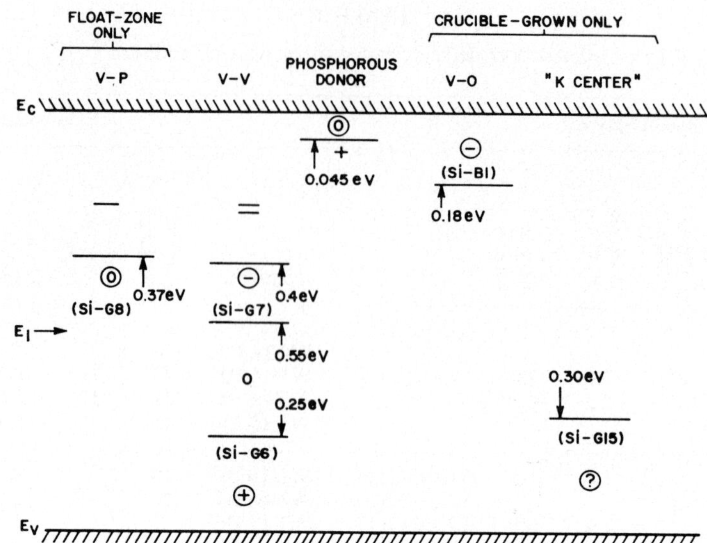

FIG. 4. The energy levels for different charge states of defect centers in silicon. The paramagnetic states are circled and the corresponding EPR spectra listed. (Adapted from Watkins.[11])

magnetic for an appropriate range of Fermi energies, E_F, as follows:

(a) Neutral donor $E_C > E_F > E_C - 0.045$ eV

(b) Si-B1 (vacancy-oxygen)
$$E_C > E_F > E_C - 0.18 \text{ eV}$$

(c) Si-G7 (negative divacancy)
$$E_C - 0.4 \text{ eV} > E_F > E_C - 0.55 \text{ eV}.$$

Since we have cooled the samples slowly in the dark to insure that the electronic states of the defect are in equilibrium, since neutron irradiation is necessarily uniform, and since inhomogeneities in the local dopant concentration would still leave that concentration large compared to the defect concentration (especially in the case of 10^{17}/cm³ doping), we can conclude that the depth of the Fermi level must be nonuniform on a microscopic scale. This would result from an extremely nonuniform distribution of the neutron damage centers.

Both of these observations—(1) low production rate of point defects, and (2) non-uniform Fermi level—can be explained on the basis of the cluster model of neutron damage.

Gossick proposed this cluster model for neutron damage in 1959.[13] The primary and secondary ions displaced by the neutron collision give up most of their energy by atomic displacement along their tracks with a higher concentration at the end of their range. Thus, in a local region, there is a high concentration of displacement defects ($\geqslant 10^{18}$/cm³). It is well known that a uniform distribution of radiation damage to this concentration will reduce the conductivity of silicon to near the intrinsic value, thereby indicating that the Fermi level is near the center of the gap. Hence, it is reasonable to assume that the Fermi level at the high local concentration of the defects is also near the center of the gap while in the surrounding, undamaged material it is at the same level as before the irradiation. This leads to a potential difference between this central region (which we will call the core) and the surrounding undamaged silicon. The result is the damage cluster—a high resistivity core surrounded by a space charge region in which the free carriers have been swept away leaving the donor ions in their positive charge state which is not paramagnetic.

Although the shape of the damage cluster is most likely a tree-like sequence of tracks,[14,15] the essential results can be understood by considering Gossick's model of a damage sphere 300 Å in diameter. This is illustrated in Figure 5. All possible depths of the Fermi level below the conduction band are found within the space charge region. Near the core, the depth is almost $E_g/2$. At the outside of the space charge region the depth is equal to the donor ionization energy. The size of the space charge region will depend on the donor concentration of the crystal and the temperature as shown in Figure 5. Using Gossick's calculations,[13

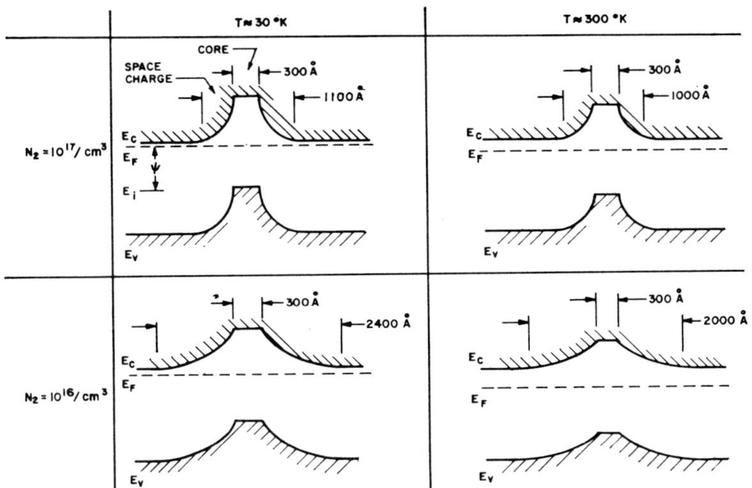

FIG. 5. The energy of the conduction and valence bands surrounding the defect core. The volume of the space charge region is proportional to ψ and inversely proportional to donor concentration.

and assuming one cluster (core plus space charge) per primary collision, the fraction of the crystal volume inside the space charge region can be calculated. This is shown in Figure 6. The results are indicated for Fermi level depths corresponding to room temperature ($\psi \approx 0.3$ eV) and cryogenic temperatures ($\psi \approx 0.6$ eV). In the process of cooling down the sample, the space charge volume can be frozen-in when ψ (the difference between E_F and the center of the gap) has some intermediate value.

For the neutron fluences used in this experiment, it can be seen that samples with donor concentrations between $10^{16}/cm^3$ and $10^{17}/cm^3$ will have an appreciable fraction of the crystal outside the space charge region. These samples can then show both donor EPR spectra from the outside and divacancy (Si-G7) EPR spectra from inside the space charge region. This explains the simultaneous appearance of these two spectra. Also the appearance of the vacancy-phosphorus (Si-G8) spectrum only in a heavily irradiated sample with $3 \times 10^{15}/cm^3$ phosphorus concentration can be explained. If the defects are formed by the capture of vacancies which diffuse away from the core during formation at room temperature, then vacancy-vacancy capture will be more probable near the core where the vacancy concentration is higher. Away from the core, the vacancy concentration will be less than the dopant or oxygen concentration and vacancy-impurity centers will be more probable. This process places the divacancies near the core where the Fermi level is deep enough to put them into the

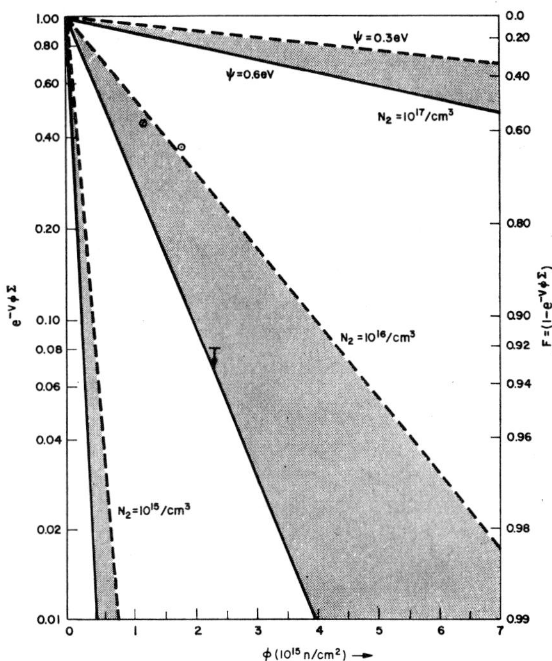

FIG. 6. The fraction of the crystal outside the space charge region, $\exp(-v\phi\Sigma)$ where v is the space charge volume of an individual cluster, ϕ is the neutron fluence, and Σ collision probability (0.165 cm⁻¹). Curves are drawn for the typical room temperature (0.3 eV) and low temperature (0.6 eV) positions of the Fermi level. Points are plotted from the data of Figure 1.

paramagnetic state (Si-G7). Likewise, the vacancy-oxygen (Si-B1) centers will be near the edge or outside the space charge region where the Fermi level is shallow and they can be paramagnetic. The vacancy-phosphorus center will also be located in this outer region where the Fermi level is shallow. However, as shown in Figure 4, in order to be paramagnetic this center requires $E_F < E_C - 0.4$ eV. Hence, it can only be paramagnetic when the space charge region fills the crystal as it will for lightly doped and heavily irradiated samples.

The other anomaly in our results was the fact that the carrier removal was about ten times the concentration of the defects which produce EPR spectra. In the damage cluster model, most of the damage is in the core. A fraction of this damage on the outside of the core compensates the donors in the space charge region and thus produces the carrier removal. These compensating levels are near the center of the band gap and are probably not paramagnetic. Intrinsic irradiated silicon contains very few EPR spectra. Hence, the centers responsible for the EPR spectra are not the ones responsible for most of the carrier removal.

Also, moving the Fermi level out of the paramagnetic state of a center depends on the overlap of adjacent space charge regions. Hence, the concentration of these centers does not decline at the same rate as the carrier removal.

It should also be noted that the expected number of silicon lattice displacements per neutron collision is approximately 1500. The recombination of interstitials and vacancies at room temperature has been estimated to remove 80 per cent to 90 per cent of the displacements leaving 150 to 300 displacements per collision.[16] Although this is about one order of magnitude greater than the carrier removal, it has also been explained by the cluster model. Most of the displacement damage is in the center of the cluster core which does not contribute compensating centers and therefore does not affect the carrier removal.

5. CONCLUSIONS

Although the defects observed by EPR comprise only a small part of the neutron damage, we have been able to use them in this study as microscopic probes of the Fermi level in the neighbourhood of the damage.

The following results are noted for this study.
(1) Paramagnetic defects account for less than one-third of the carrier removal.
(2) Donor removal and carrier removal account for less than 2 per cent of the theoretically expected number of neutron displacements.
(3) The Fermi level is non-uniform on a microscopic scale.

From this evidence we conclude that the damage cluster model gives the best description of neutron damage. The paramagnetic point defects observed in the EPR experiment are formed by diffusion of vacancies away from the damage core into the surrounding space charge region and into the undamaged region of the crystal.

ACKNOWLEDGEMENTS

We wish to acknowledge the assistance of Mr. Leo Flores of the Nuclear Effects Directorate, White Sands Missile Range, who carried out all of the reactor irradiations. The authors benefited from the advice of D. K. Wilson and had useful discussions with R. R. Holmes and J. M. McKenzie.

REFERENCES

1. G. D. Watkins, *Proceedings of the Symposium on Radiation Damage in Semiconductors* (Dunod, Paris, 1965), p. 97 and references therein.
2. J. W. Corbett, *Electron Radiation Damage in Semiconductors and Metals* (Academic Press, New York, 1966). Pages 60–66 contain a table listing most of the EPR spectra observed in irradiated silicon.
3. M. Nisenoff and H. Y. Fan, *Phys. Rev.*, **128**, 1605 (1962).
4. W. Jung and G. S. Newell, *Phys. Rev.*, **132**, 648 (1963).
5. E. G. Wikner and H. Horiye, *IEEE Trans. Nucl. Sci.* NS-13, **6**, 18 (December 1966).
6. This standard sample was supplied by E. A. Gere of Bell Telephone Laboratories, Murray Hill, N.J.
7. G. D. Watkins, J. W. Corbett and R. M. Walker, *J. Appl. Phys.*, **30**, 1198 (1959).
8. N. Almeleh and B. Goldstein, *Phys. Rev.*, **149**, 687 (1966)
9. G. D. Watkins, *Radiation Effects on Semiconductor Components*, Ed. F. Cambou (Journees d'Electronique, Toulouse, France, 1968), Vol. I, paper A1
10. D. F. Daly and K. A. Pickar, *Appl. Phys. Letters*, **15**, 267 (1969).
11. H. J. Stein, *J. Appl. Phys.*, **38**, 204 (1967).
12. F. L. Vook and H. J. Stein, *Rad. Effects*, **2**, 23 (1969).
13. B. R. Gossick, *J. Appl. Phys.*, **30**, 1214 (1959).
14. V. A. J. Van Lint, et al., General Atomic Report February 1967 (unpublished).
15. R. R. Holmes, *IEEE Trans. Nuc. Sci.* NS–17, No. 6 137 (Dec. 1970).
16. L. J. Cheng and J. Lori, *Phys. Rev.*, **171**, 856 (1968).

DISCUSSION

Question (VOOK)   Have you been able to quantitatively obtain the neutron induced defect cluster sizes for both the core and the surrounding space—charge regions from your measurements?

Answer (DALY)   Our measurements, as well as other workers measurements of carrier removal, indicate how much of the crystal is occupied by the space-charge region.   This is indicated in Figure 6.   From this one can determine the space charge volume.   For the dopant concentrations we use, the effect of the core size is negligible.

Question (BROWER)   In the more heavily phosphorus-doped silicon samples, might you not expect to see neutral *E*-centers located in the core of the cluster?

Answer (DALY)   If we accept the estimated core size of 300 Å diameter, for our $10^{17}/cm^3$ doped sample there will only be 1 or 2 phosphorus atoms in each core.   The much larger vacancy concentration right after collision, would make the divacancy formation more probable.   The phosphorus atoms might also be involved in other non-paramagnetic defects, so that the number of Si-G8 spectra formed in the core would be undetectable.

Question (WITTELES)   Have you calculated and measured the E-center contribution to the total carrier-removal-rate as measured on the total bulk sample?

Answer (DALY)   We calculated an E-center formation of 0.6/neutron collision and were able to show this formation to increase as we moved away from the cluster center.   However, we have not measured the *total* contribution of the E-center to the carrier removal rate.

Question (STEIN)   Have you attempted to use the phosphorus donor resonance to estimate the size of the damage core when the donor concentration is large and the space charge region would be small?

Answer (DALY)   The highest donor concentration we have used is $10^{17}/cm^3$.   For this concentration the space charge volume is more than an order of magnitude larger than the damage core volume.   If we went to higher donor concentrations, over-lap between neighboring donor wave functions would alter the donor EPR spectrum.   Then it could not be used to measure donor concentration.

STRUCTURE OF MULTIPLE-VACANCY (OXYGEN) CENTERS IN IRRADIATED SILICON†

K. L. BROWER

Sandia Laboratories, Albuquerque, New Mexico 87115

Analysis of a new paramagnetic spin 1 spectrum, labeled Si-Sl, indicates that it corresponds to the neutral charge state of the Si-B1 center in an excited spin-triplet state. An extension of this analysis to the Si-Pl, -P2, -P3, -P4, and -P5 centers suggest that these centers comprise a family of defects which consist of a string of neighboring vacancies in a single (110) plane. In particular, it appears that the Si-P1 center is an odd ($\geqslant 3$)-vacancy defect; the Si-P2 center is a 2-vacancy, oxygen defect; the Si-P3 center is a 4-vacancy defect; the Si-P4 center is a 3-vacancy, oxygen defect of symmetry lower than C_{2v}; and the Si-P5 center is a 3-vacancy, oxygen defect having C_{2v} symmetry.

1. INTRODUCTION

One of the simplest and one of the most important defects formed in irradiated silicon is the vacancy. Previous electron paramagnetic resonance (EPR) studies have demonstrated the mobility of the vacancy[1] and its tendency to combine with itself to form a divacancy[2] or with an impurity atom to form a vacancy-impurity pair such as the Si-B1 center.[3] These results suggest the possibility that multiple-vacancy-impurity complexes having a more extended structure can also be formed. The EPR studies presented in this paper indicate that such defects do indeed exist.

Recently, in looking at irradiated silicon samples under illumination with EPR, we found a new spectrum, labeled Si-Sl, which is prevalent in *n*- or *p*-type, electron- or neutron-irradiated, crucible-grown silicon. This spectrum corresponds to a defect having a spin of 1 and is of interest because it is similar in many respects to the previously unidentified Si-P1 (Si-N), -P2 (I, I′), -P3 (II, III), -P4 (V, VI), and -P5 (VII, VIII) centers. These paramagnetic centers account for the most prominent of the complex spectra observed at various annealing stages in high-fluence, fast-neutron-irradiated silicon, and they were first observed by Nisenoff and Fan,[4] and Jung and Newell.[5] The Si-P1 and -P3 centers are impurity independent, and the Si-P2, -P4, and -P5 centers are oxygen dependent.[4,5] More recent EPR studies have shown that the Si-P1 and -P3 centers, which are impurity independent, are also the dominant paramagnetic defects in low-fluence, ion-implanted silicon.[6,7]

† This work supported by the U.S. Atomic Energy Commission.

Although some of the characteristics in the structure of these defects were deduced by Nisenoff and Fan,[4] and Jung and Newell,[5] the overall structure of these defects was not established. The general aspects of a more detailed analysis[8] are presented in this paper and suggest that these centers comprise a family of defects which consist of a string of neighboring vacancies (in some cases dependent on oxygen) in a single (110) plane.

2. STRUCTURE OF MULTIPLE-VACANCY (OXYGEN) DEFECTS

2.1. Si-S1 *center*

The spectrum of the Si-S1 center for $H_{\parallel}[001]$ is shown in Figure 1. This spectrum was also observed in some samples of LOPEX and vacuum-float-zone silicon, but its intensity was less than that observed in crucible-grown silicon by at least a factor of 20. This defect is stable at room temperature and anneals out at $\sim$400 °C like the Si-B1 (vacancy-oxygen) and -G15 (oxygen-dependent) centers.[8] The dependence of the Si-S1 center on oxygen suggests that oxygen is incorporated in the structure of this defect. However, the fact that the Si-S1 spectrum is observed in P-, B-, Al-, or Ga-doped silicon indicates that this defect does not incorporate either Group III or V impurities.

A spin Hamiltonian which accurately[8] describes the EPR spectrum of this defect is of the form

$$\mathscr{H} = \beta \mathbf{H} \cdot \mathbf{g} \cdot \mathbf{S} + \mathbf{S} \cdot \mathbf{D} \cdot \mathbf{S}$$

$$+ \sum_i \left(\mathbf{S} \cdot \mathbf{A}_i \cdot \mathbf{I}_i - \frac{\mu_n^i}{\mathbf{I}_i} \beta_n \mathbf{H} \cdot \mathbf{I}_i \right) \tag{1}$$

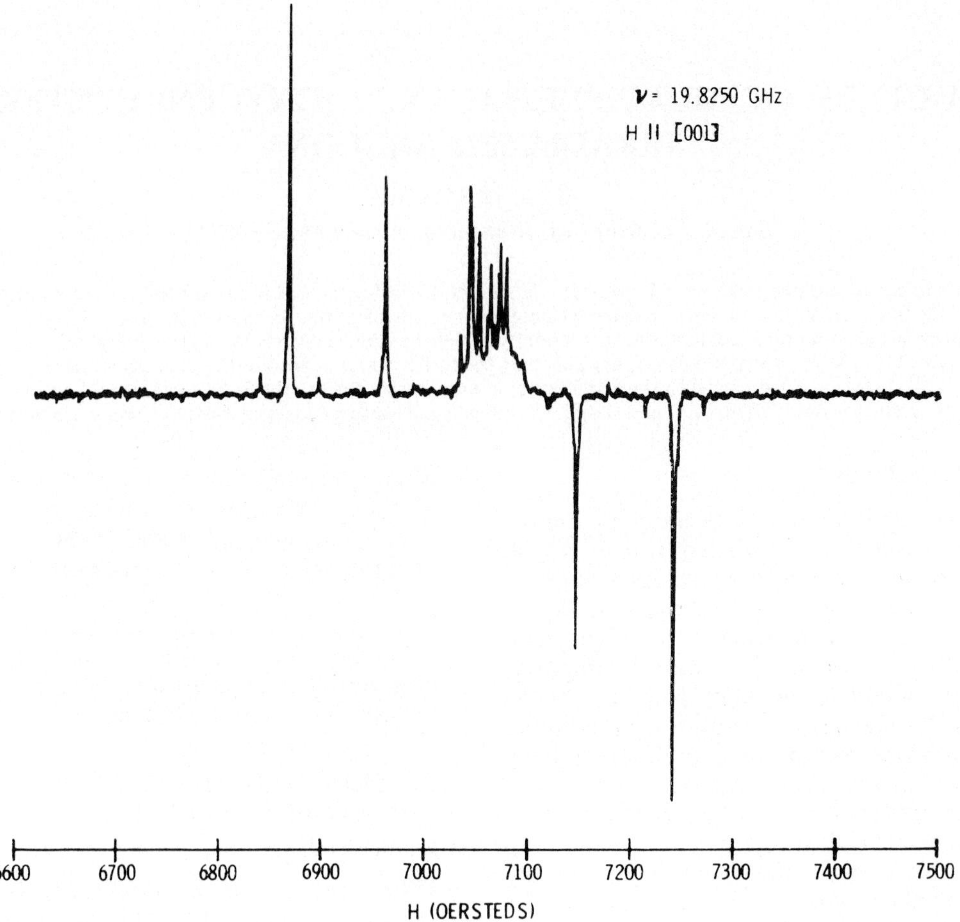

FIG. 1. EPR spectrum observed under illumination in Al-doped, crucible-grown silicon following a 2 MeV electron irradiation of 5×10^{16} e/cm². The lines at $\sim$6870, $\sim$6960, $\sim$7150, and $\sim$7240 Oe constitute the Si–S1 center for $H\|$[001]. Part of the spectrum at $\sim$7060 Oe corresponds to the Si–G15 center. As a result of populating the spin-triplet levels of the Si–S1 center by light, a population inversion is achieved within the spin-triplet levels which is manifested by spin resonance lines of negative intensity (stimulated emission transitions) and positive intensity (stimulated absorption transitions).

where $S = 1$, $I = \frac{1}{2}$ for ^{29}Si which is 4.70 per cent naturally abundant, and $\mathrm{Tr}\,(\mathbf{D}) = 0$. The numerical values for the coupling tensors in this spin Hamiltonian as deduced from the Si-S1 spectrum are tabulated in Table I.

The Si-S1 spectrum has at least two characteristics which place restrictions on the signs of the coupling tensors. The first characteristic in the observed Si-S1 spectrum is the positive or negative intensity of the resonances as illustrated in Figure 1. As a result of populating the spin-triplet levels of the Si-S1 center by light, a population inversion is achieved within the spin-triplet levels which is manifested by spin resonance lines of negative intensity (stimulated emission transitions) or positive intensity (stimulated absorption transitions). The second characteristic of the Si-S1 spectrum that restricts the signs of the coupling tensors is the positions of the forbidden ^{29}Si hyperfine lines in relation to the positions of the allowed ^{29}Si hyperfine lines. An anlysis of these experimental results gives the sign of the **D** tensor unambiguously and indicates that the sign of the **g** tensor is opposite the sign of the **A** tensor. The eigenvalues of Eq. (1) for the two allowed sign combinations of the **g** and **A** tensors are slightly different because the sign of

TABLE I

Coupling tensors for several defects. In the case of the spin 1 centers, we have chosen those coupling tensors which were determined from the K-band spectra. The elements with subscript 33 are parallel to the $[\bar{1}10]$ direction, and the elements with subscripts 11 and 22 are in the $(\bar{1}10)$ plane. The element g_{11}, D_{11}, and A_{11} are rotated off from the $[001]$ direction by ω, φ, and ψ, respectively. The defects in Figures 2 and 4 are defined with respect to the same crystallographic axes as these coupling tensors.

Defect	T °K	Spin	g Tensor $\Delta g \simeq \pm 0.0002$				D Tensor in units of 10^{-4} cm^{-1} $\Delta D \simeq \pm 0.2$				A Tensor in units of 10^{-4} cm^{-1} $\Delta A \simeq \pm 0.5$				
			g_{11}	g_{22}	g_{23}	ω	D_{11}	D_{22}	D_{33}	φ	A_{11}	A_{22}	A_{33}	ψ	Ref.
Si-S1	15	1	2.0058	2.0076	2.0102	90.0	-219.2	$+116.8$	$+102.4$	90.0	-71.8	-41.1	-41.2	54.56	
Si-P1	77	$\frac{1}{2}$	2.0020	2.0104	2.0117	57.5	—	—	—	—	$(-)138.$	$(-)75.$	$(-)75.$	54.7	4
Si-P2	300	1	2.0023_1	2.0098_9	2.0091_2	52.8	$(-)68.60$	$(+)28.69$	$(+)39.91$	81.7					5
Si-P3	300	1	2.0009_6	2.0101_8	2.0099_4	55.6	$(-)15.22$	$(+)\,7.77$	$(+)\,7.46$	96.3	$(-)68.8$	$(-)44.1$	$(-)43.6$	$+53.5$ $+233.5$	5
Si-P4	300	1	2.0042_9	2.0070_5	2.0089_8	93.5	$(-)27.87$	$(+)14.71$	$(+)13.16$	83.9					5
Si-P5	300	1	2.0051_0	2.0072_3	2.0089_9	90.0	$(-)30.60$	$(+)14.76$	$(+)15.84$	90.0					5

the nuclear Zeeman interaction is the same in both cases. A perceptibly better fit of the experimental data to Eq. (1) is achieved for $+\mathbf{g}$ and $-\mathbf{A}$ than for $-\mathbf{g}$ and $+\mathbf{A}$. The model we propose for the Si-S1 center is consistent with the signs of the coupling tensors in Table I.

An analysis of the EPR data indicates that the Si-S1 spectrum corresponds to the neutrally charged Si-B1 center in an excited spin-triplet state. The molecular and electronic structure of this defect is illustrated in Figure 2. The ^{29}Si hyperfine interactions indicate that the two unpaired electrons in this defect are localized mostly in s,p hybrid orbitals directed along the [111] and $[\bar{1}\bar{1}1]$ directions from atoms A and B. This picture is also consistent with the symmetry and anisotropy in the $\mathbf{g}$ tensor expected for two spins in dangling sp^3 hybrid orbitals pointed along [111] and $[\bar{1}\bar{1}1]$ directions from well separated silicon atoms at A and B. Each of these atoms is tetrahedrally bonded to only three nearest neighbor silicon atoms.

Since even- and odd-vacancy (oxygen) defects can be distinguished by the symmetry and anisotropy in their $\mathbf{g}$ and $\mathbf{A}$ tensors, the number of intermediate (oxygen-) vacancies in either an odd- or even-vacancy (oxygen) defect can be estimated from the spin-spin interaction. The magnitude, anisotropy, and sign of the $\mathbf{D}$ tensor suggest that the spin–spin interaction is dominated by the magnetic dipole–dipole interaction between the unpaired electrons. Figure 3 shows the results of a calculation of the principal values of the $\mathbf{D}$ tensor as a function of the separation between two unpaired electrons assuming a magnetic dipole–dipole interaction. In this calculation, we represented the one-electron orbitals for the paramagnetic electrons by 3s,3p hybrid orbitals whose localization and 3s and 3p character agreed with that deduced from the ^{29}Si hyperfine interactions.[8] Only the calculated $\mathbf{D}$ tensors for multiple vacancy structures in which there is no significant overlap in the one-electron orbitals are plotted in Figure 3. A comparison between our calculated $\mathbf{D}$ tensor for various N-vacancy structures and our assignment of the previously measured $\mathbf{D}$ tensors[5] for various spin 1 centers is shown in Figure 3. The smooth progression in the experimental values, as well as our assignment of the Si-P2, -P3, and -P5 centers in terms of particular N-vacancy (oxygen) structures as deduced from the calculated $\mathbf{D}$ tensors, suggests that the Si-S1 center is a 1-vacancy (oxygen) center. Furthermore, stress measurements[8] have revealed that the time-temperature dependence in the alignment of the Si-S1 center is the same within experimental error as that of the neutral Si-B1 center. Consequently, we have concluded that the Si-S1 spectrum corresponds to the neutral charge state of the Si-B1 center in an excited spin-triplet state.[9]

2.2. Si-P5

The similarity in the coupling tensors of the Si-S1 and -P5 centers (see Table I) suggests that the two centers themselves are similar. The symmetry of the $\mathbf{g}$ and $\mathbf{D}$ tensors indicate that the

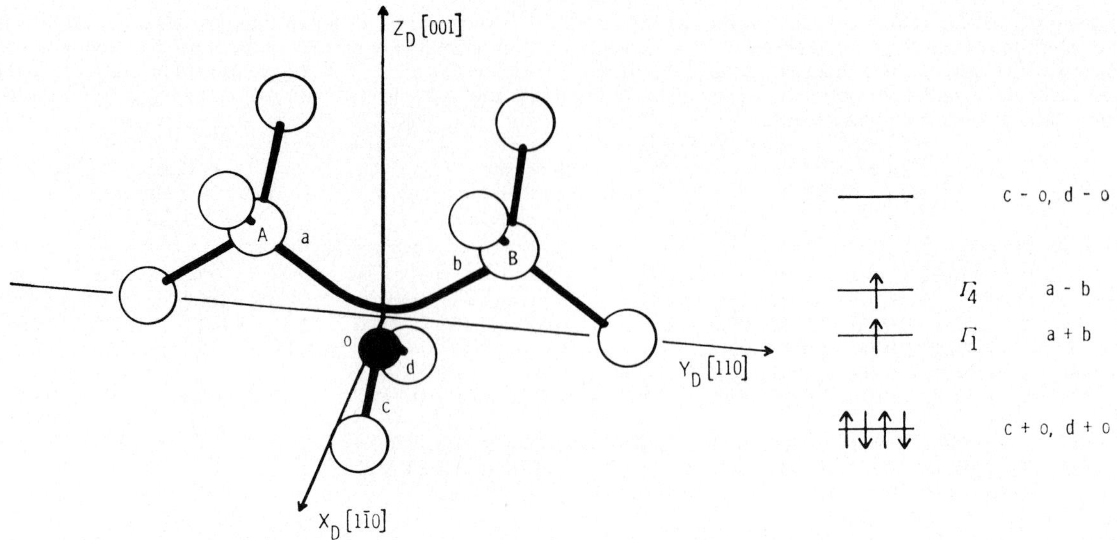

FIG. 2. Molecular and electronic structure of the 1-vacancy, oxygen center. According to Watkins and Corbett (Ref. 3), the oxygen atom, which is represented by the black ball, is bonded between two silicon atoms adjacent the vacancy. Under illumination this defect can be excited into a triplet-spin state as shown and is associated with the Si-S1 spectrum.

Si-P5 center is also an odd-vacancy (oxygen) defect having C_{2v} symmetry. The symmetry and anisotropy in the **g** tensor indicate that the paramagnetic electrons are in s,p hybrid orbitals directed along the [111] and [$\bar{1}\bar{1}$1] directions from the end silicon atoms. However, the magnitude and symmetry of the **D** tensor, assuming a magnetic dipole–dipole interaction, indicate that there are three vacancies in a (110) plane (see Figure 3). The symmetry of the **g** and **D** tensors suggests that the configuration of oxygen atom(s) in this defect is compatible with C_{2v} symmetry.

2.3. Si-P4 center

The coupling tensors for the Si-P4 center differ only slightly from those of the Si-P5 center (see Table I) and suggest that the Si-P4 center is also a 3-vacancy, oxygen defect. However, the symmetry of the **g** and **D** tensors belonging to the Si-P4 center indicates that the symmetry of this defect is lower than C_{2v}. Presumably this break in symmetry is due to a perturbation within or near this defect. For example, this perturbation could be due to an oxygen configuration which does not have C_{2v} symmetry.

2.4. Si-P3 center

Of the spin 1 centers observed by Jung and

Newell,[5] the Si-P3 center was studied by them in the greatest detail. This center is also simpler in the sense that it is impurity independent. The ^{29}Si hyperfine interactions as well as the symmetry and anisotropy in the **g** tensor indicate that this center is an even-vacancy defect.[10] The magnitude of the **D** tensor suggests, assuming a magnetic dipole–dipole interaction, that the Si-P3 center is a 4-vacancy defect. The molecular and electronic structure of a 4-vacancy defect having C_{2h} symmetry is illustrated in Figure 4. With respect to this model, the nuclei at *A* and *B* are accurately equivalent according to the observed ^{29}Si hyperfine interactions which suggests that this defect has C_{2h} symmetry.

2.5. Si-P2 center

The symmetry and anisotropy in the **g** tensor of the Si-P2 center suggest that this defect is an even-vacancy (oxygen) center. The magnitude of the **D** tensor indicates that it is a 2-vacancy (oxygen) defect. Because the ^{29}Si hyperfine interactions have not been measured, we can only say that this defect, including the oxygen configuration, appears to have a symmetry of C_{2h} or lower.

2.6. Si-P1 center

It is interesting to consider the structure of the

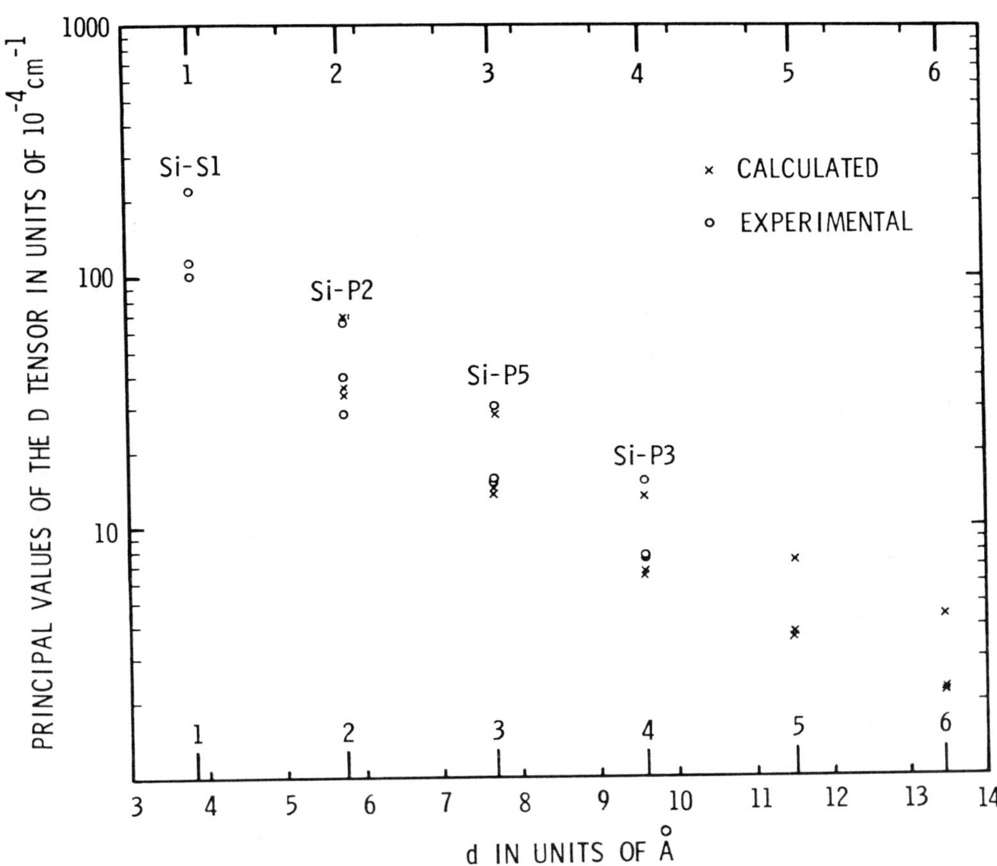

FIG. 3. Calculated and experimental values of $-D_{11}$, D_{22}, and D_{33} as a function of d, the separation in Å between the end silicon atoms of an N-vacancy defect, and N. The particular elements of the calculated $\mathbf{D}$ tensor can be identified in that $-D_{11} > D_{33} > D_{22}$. For 2–, 4–, and 6-vacancy centers, the calculated φ as defined in Table I corresponds to 89.2°, 89.8°, and 90.5°, respectively.

Si-P1 center in terms of the models we have proposed for the other spin 1 centers observed in neutron-irradiated silicon. Since the Si-P1 center has a spin of $\frac{1}{2}$, it is either a negatively or positively charged defect. The ^{29}Si hyperfine interaction indicates that for $T \lesssim 77\,°$K, the paramagnetic electron is localized on only one of the end silicon atoms. For $T \gtrsim 77\,°$K, Nisenoff and Fan[4] observed that the paramagnetic electron was able to hop back and forth between two silicon sites. The relationship between the $\mathbf{g}$ tensors for these pairs of silicon sites is compatible with a model of the Si-P1 center in which there are an odd number of vacancies in a single $(\bar{1}10)$ plane with s,p hybrid orbitals along the [111] and [$\bar{1}\bar{1}1$] directions. In fact as the temperature is raised, the anisotropy in the $\mathbf{g}$ tensor tends to approach that expected for an odd-vacancy defect. Although we are unable to determine the number of vacancies in the Si-P1

center or the existence of other perturbations within or near this defect, the properties of this defect are quite distinct from those of the vacancy or divacancy. Thus the EPR data suggest that the Si-P1 center is an impurity independent defect consisting of an odd number ($\geqslant 3$) of vacancies in a single (110) plane.

3. SUMMARY

Our analyses suggest that the Si-P1, -P2, -P3, -P4, and -P5 centers comprise a family of defects which consist of a string of neighboring vacancies (in some cases dependent on oxygen) in a single (110) plane. One might also include in this family the divacancy as well as the vacancy and 1-vacancy, oxygen centers. The overall structure of these defects is summarized in Table II. In the case of the larger defects particularly, it is impossible to

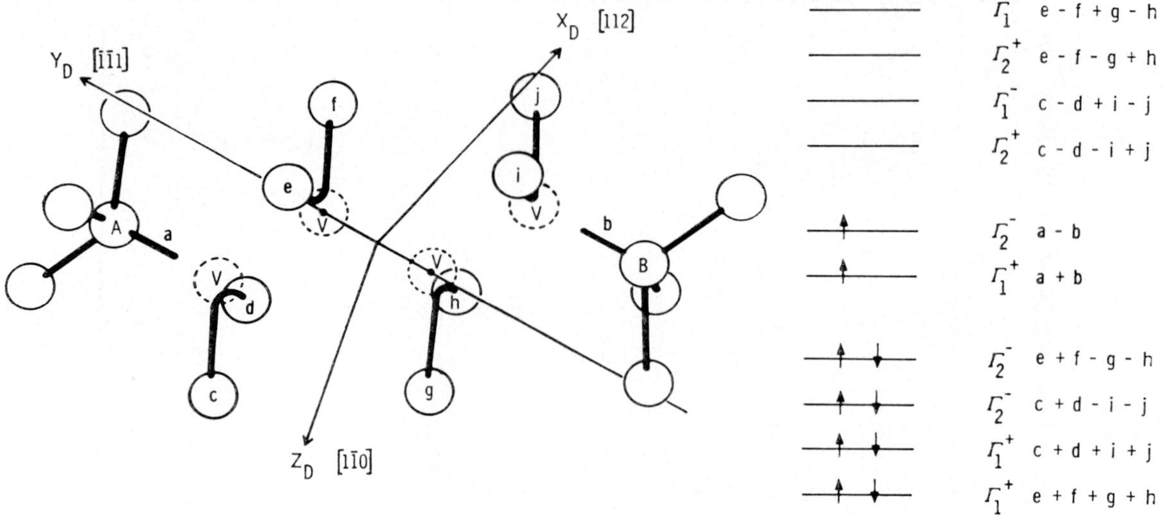

FIG. 4. This model of the molecular and electronic structure of a 4-vacancy having C_{2h} symmetry is consistent with the EPR data associated with the Si-P3 center. The linear combination of orbitals which transform according to the various irreducible representations of the point group C_{2h} were deduced from molecular orbital theory.

pinpoint all of the details in the molecular structure of these defects from the EPR data alone. The models we have proposed correspond to the simplest molecular structures which are consistent with the EPR data.

TABLE II
Tabulation of centers and their overall structure.

Center	Spin	Overall structure
Si-S1	1	1-vacancy, oxygen
Si-P2	1	2-vacancy, oxygen
Si-P4	1	3-vacancy, oxygen
Si-P5	1	3-vacancy, oxygen
Si-P3	1	4-vacancy
Si-P1	$\frac{1}{2}$	odd ($\geqslant 3$)-vacancy

ACKNOWLEDGEMENTS

Discussions with D. K. Brice, H. J. Stein and F. L. Vook on various aspects of this work have been particularly enlightening. The author is also grateful to D. F. Daly who called our attention to a numerical error in our initial calculation of the **D** tensor. The competent assistance of N. D. Wing who carried out the EPR measurements is very much appreciated.

REFERENCES

1. G. D. Watkins, *J. Phys. Soc. Japan*, **18**, Suppl. II, 22 (1963).
2. G. D. Watkins and J. W. Corbett, *Phys. Rev.*, **138**, A543 (1965).
3. G. D. Watkins and J. W. Corbett, *Phys. Rev.*, **121**, 1001 (1961).
4. M. Nisenoff and H. Y. Fan, *Phys. Rev.*, **128**, 1605 (1962).
5. W. Jung and G. S. Newell, *Phys. Rev.*, **132**, 648 (1963).
6. K. L. Brower, F. L. Vook and J. A. Borders, *Appl. Phys. Letters*, **15**, 208 (1969).
7. D. F. Daly and K. A. Pickar, *Appl. Phys. Letters*, **15**, 267 (1969).
8. K. L. Brower, *Phys. Rev.*, submitted for publication.
9. At the Albany Conference, we believed that the multiple vacancy (oxygen) centers contained two more vacancies than assigned to them in this paper. At that time, we did not have the crucial results of our recent stress measurements on the Si-S1 center, and our calculated **D** tensor was two times larger than it should have been.
10. J. W. Corbett. Solid State Physics, Ed., by F. Seitz and D. Turnbull (Academic Press Inc., New York, 1966), Suppl. 7, p. 79.

DISCUSSION

Question (KOEHLER) Would there be any infra-red absorption bands associated with these EPR centers?

Answer (BROWER) There could be but they have not yet been observed. In the case of even-vacancy defects having C_{2h} symmetry, one might expect optical transitions similar to those that have been observed for the divacancy. In the case of odd-vacancy defects having C_{2v} symmetry, symmetry arguments indicate that electronic transitions are allowed between the $\Gamma_1 \leftrightarrow \Gamma_1$, $\Gamma_2 \leftrightarrow \Gamma_2$, and $\Gamma_4 \leftrightarrow \Gamma_4$ states with E_z polarized light; the $\Gamma_1 \leftrightarrow \Gamma_2$ and $\Gamma_3 \leftrightarrow \Gamma_4$ states with E_x polarized light; and the $\Gamma_1 \leftrightarrow \Gamma_4$ and $\Gamma_2 \leftrightarrow \Gamma_3$ states with E_y polarized light.

Comment (WATKINS) As shown in your figures, the molecular orbital energy levels for the multiple vacancy defects are very similar to those for the divacancy. This suggests that optical transitions associated with the defects may also be similar. A word of caution might be in order therefore for those who use the $1.7\,\mu$ band as a monitor of isolated divacancies in experiments where these multiple defects may also be important. Perhaps these multiple vacancy centers also contribute to the observed band. This question could be checked directly in the optical experiments by polarized bleaching studies. The isolated divacancy has a Jahn–Teller distortion which can be aligned by polarized light. The kinetics of the recovery from this alignment can in turn be studied by observing the absorption with polarized light [L. J. Cheng and P. Vajda, *Phys. Rev.*, **186**, 816 (1969)]. The multiple vacancy centers do not have Jahn–Teller effects and cannot 'reorient' and therefore their contribution could be separated out.

Question (DALY) Have you looked for Jahn–Teller reorientation of the spectrum at higher temperature?

Answer (BROWER) We looked for the effects of stress on the Si–S1 spectrum at 15 °K and saw no stress dependence. (See Ref. 9).

Question (HALE) Why were some of the EPR lines inverted?

Answer (BROWER) As a result of populating the excited spin-triplet levels by light, a population inversion is achieved which is manifested by resonance lines of positive intensity (Figure 1) which correspond to stimulated absorption transitions and lines of negative intensity which correspond to stimulated emission transitions. These resonances were observed in the dispersion mode under passage conditions which give absorption-like spectra.

PIEZOSPECTROSCOPIC STUDY (78 °K) OF RADIATION-PRODUCED ABSORPTION BANDS IN Si†

C. S. CHEN AND J. C. CORELLI

Department of Nuclear Engineering and Science, Rensselaer Polytechnic Institute, Troy, New York, U.S.A.

A detailed study has been made on two sharp sub-bands, $3.45\,\mu$ (2900 cm^{-1}) and $3.61\,\mu$ (2765 cm^{-1}) of the $3.3\,\mu$ broad defect absorption band in irradiated Si. Our results indicate that the $3.45\,\mu$ and $3.61\,\mu$ bands are observed in all types of Si (n-, p-type, and intrinsic) subjected to fast neutron reactor irradiation. Spectra were measured at 78 °K under the effect of a variable uniaxial compressive stress with polarized light. A relatively large dichroism is observed for all three stress directions $\langle 100 \rangle$, $\langle 110 \rangle$, and $\langle 111 \rangle$. We also confirm that both sharp bands exhibit splitting and energy shifts under the action of stress. The model of the divacancy defect given by Watkins and Corbett[5] has been utilized to explain the data. The dichroism can be accounted for by the effects of Jahn–Teller alignment which arises from the strong bonding of nearest neighbor atoms of the divacancy. The splittings of the bands are small indicating that the Jahn–Teller distortion is little altered in the ground and excited states. The splittings appear to be most sensitive to stress along the vacancy-vacancy axis and a tentative model for the transitions is given.

1. INTRODUCTION

Radiation-induced infrared absorption bands near $3.3\,\mu$ were first observed and studied by Fan and Ramdas[1,2] in both neutron and electron irradiated Si. Stress-induced dichroism measurements have been reported at both room[2] and low temperature (78 °K).[2] In addition to reporting dichroism for the $3.3\,\mu$ band, Rao and Ramdas[3] also observed splitting of the band at 78 °K under the effect of stress in the $\langle 111 \rangle$ direction. The stress-induced dichroism has been identified as due to electronic reorientation and the defect direction was determined to be $\langle 110 \rangle$.[2,3] From a study of the elevated temperature stress response and annealing behavior of the $3.3\,\mu$ band, Cheng and coworkers[4] have concluded that the $3.3\,\mu$ band is correlated with the -2 charge state of the divacancy defect. However, a detailed study of the fine structure of the $3.3\,\mu$ band which consists of a broad band with three sharp sub-bands at $3.32\,\mu$, $3.45\,\mu$ and at $3.61\,\mu$ has not yet been made. In addition there are no quantitative values for either dichroism or splitting when stress is applied to the sample at low temperature. In previous low temperature stress studies,[2,3] a differential thermal expansion was used to apply the stress and neither the temperature nor the stress value was well defined.

In research to be reported on in this paper, we investigated the $3.3\,\mu$ band produced by fast neutron irradiation in all types of Si samples (n-, p-type, and intrinsic). The temperature dependence of the growth of the broad band and the two more intense sharp sub-bands ($3.45\,\mu$ and $3.61\,\mu$) were measured. We also present detailed measurements at low temperature (78 °K) of the stress-induced dichroism and splitting for the sharp bands. The divacancy model of Watkins and Corbett[5] has been used for an interpretation and discussion of our data. The divacancy is an intrinsic defect consisting of two adjacent vacancies with a D_{3d} point symmetry. Jahn–Teller distortion lowers the symmetry to C_{2h}. The $1.8\,\mu$, $3.3\,\mu$ and $3.9\,\mu$ infrared absorption bands are identified[4] as X–Y electric dipole transitions for which the dipole moment is close to the [110] direction and located in the X–Y plane (see Figure 2[4]). Further details of the divacancy defect properties may be found elsewhere.[4,5] The stress-induced dichroism was interpreted by Cheng and coworkers,[4] as due to Jahn–Teller alignment arising primarily from two pairs of bonding atoms in a model which utilized a linear combination of atomic orbitals (LCAO) to obtain the wave functions of the energy states. The results to be given in this paper offer substantial additional experimental evidence confirming the validity of the divacancy model of Watkins and Corbett.[5]

2. EXPERIMENTAL METHODS

Many of the experimental methods used in this investigation have been described previously,[6] and

† Supported by a Themis contract.

will therefore not be treated here. Samples ($\approx 2 \times 3 \times 20$ mm) were cut from slabs of silicon whose crystallographic orientations were determined from optical or X-ray alignment. The long dimension of the sample was either a [111], [110], or [100] direction and the faces polished to a mirror-like surface through which the infrared light was passed were either [001], or [1$\bar{1}$0] directions. The samples were irradiated to a fluence of 4–6×10^{16} n/cm² with fast neutrons in a reactor at a temperature below 60 °C.

The measurement of the infrared spectra were made at 78 °K using an optical dewar[6] equipped with a concentric rod and tube to allow one to transmit a compressive stress to the sample during the optical measurement. A two-ton hydraulic press was used to transmit pressure to the sample held between the rod and the tube. A Perkin–Elmer model 621 infrared spectrophotometer was used to record the spectra. The optical spectra were measured with varying and known stress applied to the sample.

3. RESULTS AND DISCUSSION

3.1. *General*

Radiation-induced absorption bands at $1.8\,\mu$, $3.3\,\mu$, $3.9\,\mu$ and $5.5\,\mu$ were first observed by Fan and Ramdas[1] in both electron- and fast neutron-irradiated silicon. The $3.3\,\mu$ broad band was observed both at room temperature and low temperature. However, the $3.45\,\mu$ and the $3.61\,\mu$ sharp bands were observed together with the broad band only at low temperature. These sharp absorption bands were suggested to arise from an excitation of a ground state to several excited states in a doubly charged donor-like defect center. The bands were observed only if the Fermi level position was above $E_c - 0.21$ eV.[1] In our study, we observed the $1.8\,\mu$ band and the $3.3\,\mu$ band, with the two sharp bands at $3.45\,\mu$ and $3.61\,\mu$ simultaneously in all types of samples at 78 °K. We did not observe the $3.3\,\mu$ band at room temperature. A simultaneous observation of the $1.8\,\mu$ and $3.3\,\mu$ bands in P-doped Si has been reported recently by Cheng and Vajda.[7] They found that the activation energy for the dichroism recovery of the $3.3\,\mu$ band agrees with that of the Jahn–Teller reorientation activation energy of the singly negatively charged divacancy. According to Fan and Ramdas,[1] the $3.3\,\mu$ band was not observed in p-type samples or n-type samples of high resistivities and thus this led Cheng and coworkers[4] to identify the band as

the doubly negative charged state of the divacancy defect. However, the interpretation of Cheng and coworkers[4] may not be correct as will be shown later.

3.2. *Temperature effect*

The temperature dependence of the intensities of the broad band and the two sharp bands has been measured at $T = 68$ °K, 78 °K, 82 °K, 100 °K, and 133 °K. Both broad and sharp bands are observed only if the temperature is $\leqslant 150$ °K. The intensities of the $3.45\,\mu$ and the $3.61\,\mu$ sharp bands have a different temperature dependence. In addition, the ratio of the integrated absorption in the $3.61\,\mu$ band to the total integrated absorption in the broad band versus T^2 has been plotted and has an exponential rather than a linear temperature dependence. Moreover, we do not observe shifting of the sharp bands versus temperature (68 °K to 133 °K).

The temperature dependence of the ratio of the relative intensity of the $3.45\,\mu$ to the $3.61\,\mu$ band rules out the possibility of a transition from a single ground state to various excited states.

From our observations the $3.3\,\mu$ band does not exhibit the phonon structure given by Fitchen[8] for zero-phonon transitions.

3.3. *Low-temperature* (78 °K) *stress studies*

The application of uniaxial compressive stress at 78 °K produces a large dichroism in three stress directions [100], [110], and [111] for both the $3.45\,\mu$ and $3.61\,\mu$ bands. A typical measurement of the spectra from $3.23\,\mu$ to $3.84\,\mu$ (or from 3100 cm^{-1} to 2600 cm^{-1}) with [111] stress using polarized light is given in Figure 1. The smallest dichroism has been observed in [110] stress, viewing from the [1$\bar{1}$0] direction as shown in Figure 2. A rather small energy shift in [100] stress and a small splitting in the [110] and [111] stress directions have been observed for both bands under the effect of stress up to 3000 kg/cm², as shown in Figure 3. An energy change per unit strain, $dE/d\epsilon$ along the strong bonding atom direction—Z axis in the divacancy model of Watkins and Corbett[4,5]—has been found adequate to describe the stress response of dichroism in EPR divacancy measurements[5] and in the Cheng and coworkers[4] IR divacancy studies. The formulas for dichroism ratios have been calculated and are given by Cheng and coworkers.[4] However, there is no explanation[4] for the splittings of the bands first found by Rao and Ramdas.[3]

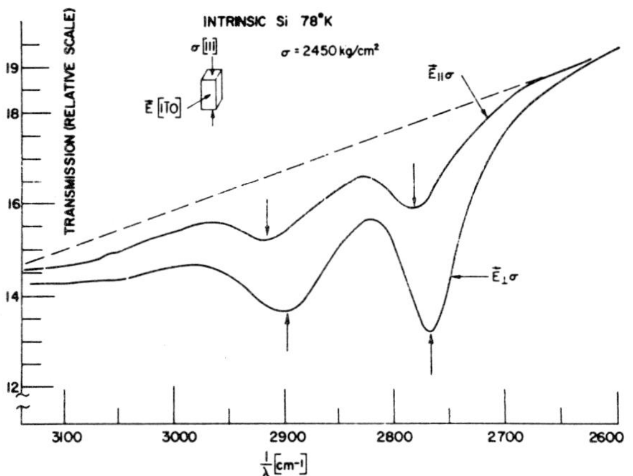

FIG. 1. Dichroism and splitting exhibited by the $3.45\,\mu$ and $3.61\,\mu$ bands in neutron-irradiated intrinsic Si during stress at 78 °K and simultaneous measurement of the spectra. Arrows indicate positions of the minima in the peaks.

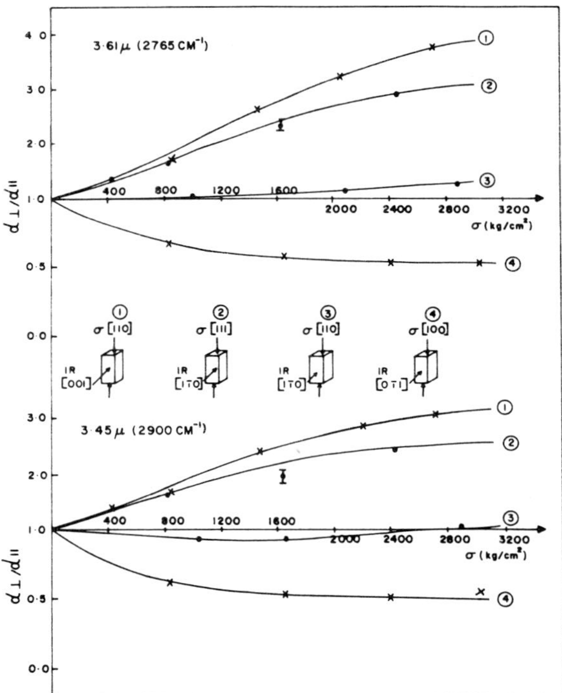

FIG. 2. Dichroism ($\alpha_\perp/\alpha_\parallel$) exhibited by the $3.4\,\mu$ (2900 cm^{-1}) and the $3.61\,\mu$ (2765 cm^{-1}) bands vs. stress for three different stress directions and four different viewing directions for neutron irradiated intrinsic Si. The measurements were made at 78 °K while the stress was on the sample. The solid lines are the calculated values using the parameters given in Table I of text.

The energy of a defect in a strain field can be written as:[4]

$$\sum_{ij} B_{ij}\,\epsilon_{ij} \tag{1}$$

where ϵ_{ij} are the components of the strain tensor and B_{ij}, the components of a second rank 'piezospectroscopic defect tensor'. According to Kaplyanskii,[9] the symmetry of the divacancy defect is classified as a type-I monoclinic center, and in the principal axis system of the defect (Figure 2, Cheng and coworkers[4]), the **B** tensor must have the form[4]

$$\mathbf{B} = \begin{bmatrix} B_1 & B_4 & 0 \\ B_4 & B_2 & 0 \\ 0 & 0 & B_3 \end{bmatrix} \tag{2}$$

In previous EPR studies[5] of the divacancy and also in the IR studies[4] of the $3.9\,\mu$ and $1.8\,\mu$ bands, it was found that one parameter sufficed to describe the stress response. In terms of **B**, the analysis assumed $B_1 = B_2 = B_4 = 0$ and

$$B_3 = M = (dE/d\epsilon)_{\text{Si-Si}}, \tag{3}$$

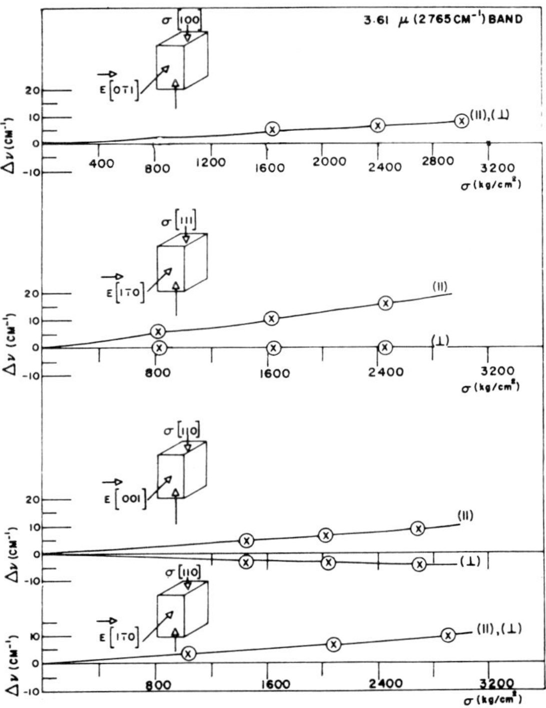

FIG. 3. Frequency splitting and shifts plotted vs. stress for the $3.61\,\mu$ (2765 cm^{-1}) band. The directions of applied stress and incident light are shown on the Figure.

the charge in energy per unit strain along the 'bonding' atom direction associated with the Jahn–Teller distortion. The results of a similar analysis[10] for the data of Figure 1 are shown by the solid lines in the figure. The values of M and θ, which describes the dipole transition moment direction,[4] used in the fit for each line are given in Table I. We

TABLE I

Piezospectroscopic defect tensor components (eV) and transition dipole moment angle θ.

band (μ)	B_1	B_2	B_3	B_4	θ
3.45	0	0	21.4	0	2.1°
3.61	0	0	18.6	0	10.1°

see that the theoretical and the experimental values are in good agreement in all four stress directions (1–4) for both the 3.45 μ and the 3.61 μ bands. We feel that this close agreement constitutes definitive confirmation of the fact that the strong bonding between the nearest Si atoms in the Z direction of the divacancy in the model of Watkins and Corbett[5] dominates the stress response of dichroism. The values of M can be compared to 24 eV for the ($+1$) charge state and 32 eV for the (-1) charge state of the divacancy as determined by EPR.[5]

The splittings of the bands under the action of stress also must be expressable in the form of Eq. (1). Analysis of these results is not as yet complete but certain qualitative features can be pointed out: (1) The magnitudes of the splittings are very small and the components of the corresponding piezospectroscopic tensor are clearly an order of magnitude smaller than those of Table I. From this we must conclude that the Jahn–Teller distortion in the excited state is very similar to that of the ground state so that both states couple almost identically to the stress field. (2) The splittings are very similar to that predicted for a trigonal center.[9] The 'trigonal' axis corresponds to the vacancy-vacancy axis direction of the divacancy.

In terms of the molecular orbital model of Figure 3, Cheng and coworkers,[4] these observations suggest the transition labeled γ for the 3.3 μ band. Such a transition does not disturb the electrons in the Jahn–Teller 'bonding' orbitals explaining (1). It still should reflect sensitivity to distortions along the vacancy-vacancy direction, however, in agreement with (2). As pointed out by Cheng and coworkers,[4] however, this transition

does not exist for the -2 charge state of the divacancy.

3.4. *Nature of optical transitions*

From the preceding discussion we can list the following: (1) the observation of the 3.3 μ band with the 3.45 μ and 3.61 μ sharp bands in all types of Si (p-, n-type, and intrinsic); (2) the temperature dependence of the ratio of the relative intensity between the 3.45 μ and 3.61 μ bands; (3) the sharpness of the two bands compared to the 1.8 μ and 3.9 μ bands in the divacancy defect center; (4) the rather small splitting of the bands under the effect of uniaxial compressive stress, and the different dichroism and splittings between these two sharp bands stress directions which causes two different **B** and θ values for the two, we believe that these observations give evidence contradicting the fact that the 3.3 μ band is due to the (-2) charge state of divacancy defect. Further experimental evidence is needed before a definitive charge state assignment can be made. However, the transitions of these bands are most likely the γ-transitions[4] between the two weak bonding states of the divacancy defect.

4. SUMMARY AND CONCLUSION

We wish to make the following definitive conclusions:

1. The 3.3 μ band with the 3.45 μ and 3.61 μ sharp bands are observed in all types of neutron-irradiated Si (p-, n-type, and intrinsic) at temperature $\lesssim 150\,°$K.
2. The ratio of the relative intensity between the 3.45 μ and 3.61 μ band exhibits a temperature dependence implying that the bands may arise from different charge states of the divacancy defect.
3. Both sharp bands exhibit large dichroism under the effect of uniaxial compressive stress at 78 °K, which arises from the Jahn–Teller alignment in the divacancy model of Watkins and Corbett.[5]
4. Both sharp bands have a rather small energy shift and splittings under the effect of uniaxial compressive stress at 78 °K.

ACKNOWLEDGEMENT

We thank Dr. George D. Watkins of the General Electric Company for giving us the irradiated samples and for his generous offer of time in discussing and interpreting our results.

REFERENCES

1. H. Y. Fan and A. K. Ramdas, *J. Appl. Phys.*, **30**, 1127 (1959).
2. H. Y. Fan and A. K. Ramdas, in *Proceedings of the International Conference on Semiconductors Phys.* 1960 (Czechoslovakian Academy of Science, Prague, 1961), p. 309.
3. M. G. Rao and A. K. Ramdas, *Bull. Am. Phys. Soc.*, **10**, 123 (1965).
4. L. J. Cheng, J. C. Corelli, J. W. Corbett and G. D. Watkins, *Phys. Rev.*, **152**, 761 (1966).
5. G. D. Watkins and J. W. Corbett, *Phys. Rev.*, **138A**, 543 (1965); **138A**, 555 (1965).
6. A. H. Kalma and J. C. Corelli, *Phys. Rev.*, **173**, 734 (1968).
7. L. J. Cheng and P. Vajda, *Phys. Rev.*, **186**, 816 (1969).
8. D. B. Fitchen, *Phys. of Color Centers*, Ed. W. B. Fowler P. 318 (1968), O-Phonon line transition.
9. A. A. Kaplyanskii, *Opt. i. Spektroskopia*, **16**, 602 (1964), English transl. *Opt. Spectry.* (USSR), **16**, 329 (1964).
10. Using the formulas given in reference 4.

DISCUSSION

Question (BRELOT) We do not observe in our experiments the $3.6\,\mu$ absorption band in oxygen-free silicon after 1.7×10^{18} e/cm²–3 MeV irradiation. The intensity of the $1.7\,\mu$ band is $0.3\,\text{cm}^{-1}$. We do observe the $3.6\,\mu$ band in oxygen-doped silicon irradiated with the same fluence. My question is therefore: how can you explain the strong correlation between the intensity of the $3.6\,\mu$ band and the oxygen concentration?

Answer (CORELLI) We see the $3.6\,\mu$ band at 78 °K in 40–50 MeV electron-irradiated (to fluences $\gtrsim 10^{18}$) and fast neutron-irradiated (to fluences $\gtrsim 10^{16}$/cm²). If you do not observe it using 3 MeV electrons it is perhaps due to the fact that you have not given your sample enough dose. In answer to your question, 'Why is more divacancy associated 1.8, $3.6\,\mu$ band produced in 0-containing Si?', I can say that we too find this to be true and believe it may be due to secondary interactions where oxygen atoms receive energy from the incoming particles and go on to 'produce more divacancy defects'.

ABSORPTION MEASUREMENTS IN NEUTRON IRRADIATED SILICON†

C. E. BARNES

Sandia Laboratories, Albuquerque, New Mexico 87115

Silicon samples have been neutron irradiated at 76 °K with fluences sufficient to allow measurement of the 1.7 μ divacancy band at 76 °K. The growth of the divacancy concentration and the recovery of the edge absorption were studied as a function of annealing temperature between 76 °K and 550 °K. Immediately after irradiation the divacancy concentration is about 25 per cent of its maximum value which is attained at 330 °K, the temperature at which the divacancies begin to anneal out. Increases in the 1.7 μ band intensity and recovery of the edge absorption can also be achieved at 76 °K by illuminating the sample with intense sub-bandgap light or white light. The experimental results suggest a neutron-induced cluster model in which the cluster is a vacancy-rich region whose annealing characteristics are controlled by the liberation and motion of vacancies. The injection effects can be explained by analogy to the charge state dependent mobility of the Si vacancy.

1. INTRODUCTION

Recent studies have shown that the divacancy production rate in irradiated silicon depends on several factors. In the case of electron irradiation, the divacancy production rate is greater if the sample contains appreciable amounts of oxygen or carbon.[1,2] For neutron irradiation, the impurity content is not so critical,[3] but the concentration of divacancies present does depend on the irradiation temperature.[4,5] This effect has also been recently demonstrated in ion-implanted Si.[6] Cheng and Lori[3] have shown that there is a large divacancy concentration present following neutron irradiation at 50 °C. Stein[4] determined the divacancy concentration by monitoring the photoconductivity at 3.9 μ as a function of annealing temperature after neutron irradiation at 82 °K. Immediately after irradiation the divacancy concentration was very small but increased markedly upon annealing to room temperature. This result implied that the large divacancy concentration observed by Cheng and Lori[3] was the result, in large part, of cluster reordering. One of the problems associated with the observation of the 3.9 μ photoconductivity band is that the Fermi level must be between the valence band and $E_v + 0.26$ eV.[7] Consequently, it is possible that the divacancy growth curve at low annealing temperatures was distorted because of the necessity for Fermi level recovery.[4]

In contrast with the 3.9 μ photoconductivity band, the 1.8 μ divacancy absorption band is observed when the Fermi level is between approximately $E_v + 0.26$ eV and $E_c - 0.22$ eV.[3,7,8] For the heavy neutron irradiations required to observe defect absorption bands, one can be reasonably sure that the Fermi level lies in this interval immediately after irradiation of silicon having a resistivity greater than approximately 1 ohm-cm. Therefore, in comparison with photoconductivity measurements, the growth of the 1.8 μ band should more accurately reflect divacancy growth at low annealing temperatures in low temperature-irradiated material.

We have employed measurements of the 1.8 μ absorption band and the absorption edge following neutron irradiation at 76 °K and subsequent thermal annealing to 550 °K (277 °C) to obtain additional information about neutron-induced defect clusters in Si. We have also investigated the effect of illumination-induced carrier injection upon the annealing characteristics of the divacancy and band edge absorption to help explain injection stimulated recovery of minority carrier lifetime previously observed at 76 °K in Si neutron irradiated at 76 °K.[9]

2. EXPERIMENTAL PROCEDURE

Chemically polished Si blocks measuring 0.5″ × 0.5″ × 0.2″ were used in this study. Absorption spectra were measured at 76 °K using a Cary-14 spectrophotometer for the range 1.0 μ to 2.5 μ and a Perkin–Elmer model E-1 single beam, grating instrument for wavelengths between 2.5 μ and 4.0 μ. Irradiations were performed in the Sandia Annular Core Pulsed Reactor (ACPR) at 76 °K and also 300 °K. The ACPR has a large sample cavity which allowed the introduction of wide-mouthed

† This work was supported by the U.S. Atomic Energy Commission.

"

Dewars containing liquid nitrogen and the samples. The reactor was operated in the steady-state mode for the irradiations. The neutron fluence for neutron energies greater than 10 keV was determined by multiplying sulfur activation results by 15, the boron-shielded plutonium-to-sulfur ratio for the ACPR.[10]

3. RESULTS

3.1. *Thermal annealing effects*

Absorption spectra illustrating some important aspects of the results are shown in Figure 1. Note the break in the abscissa and that the absorbance increases in the downward direction along the ordinate. The curve for 40 ohm-cm crucible-grown *p*-type Si was taken after annealing to 228 °K. At this point the divacancy band (which is at $1.7\,\mu$ at $76\,°K^{(8)}$), has attained about 75 per cent of its maximum intensity. An annealing temperature of 228 °K was selected because it clearly illustrates a distortion observed in the absorption edge between

$1.05\,\mu$ and $1.20\,\mu$. This 'kink' may be present immediately after irradiation at 76 °K, but it cannot be observed because the edge absorption is so strong. The 'kink' anneals completely between 210 °K and 270 °K. Whether or not it is actually a neutron-induced near-edge absorption band is a difficult question to answer. If so, it is associated with a rather shallow energy level. The 'kink' may also be a peculiarity of the absorption edge associated with a particular amount of damage. However, it is not observed in samples irradiated at room temperature to the same damage level. Additional experimental information will be required before the source of this 'kink' can be clearly defined.

The second curve in the short wavelength range in Figure 1 is for 0.1 ohm-cm *n*-type crucible grown Si after anneal to 330 °K, the point at which the divacancy concentration reaches its maximum for *p*-type Si. Note, however, that the $1.7\,\mu$ band is fairly weak in the *n*-type sample. In contrast with the difference between the divacancy band in *n*- and

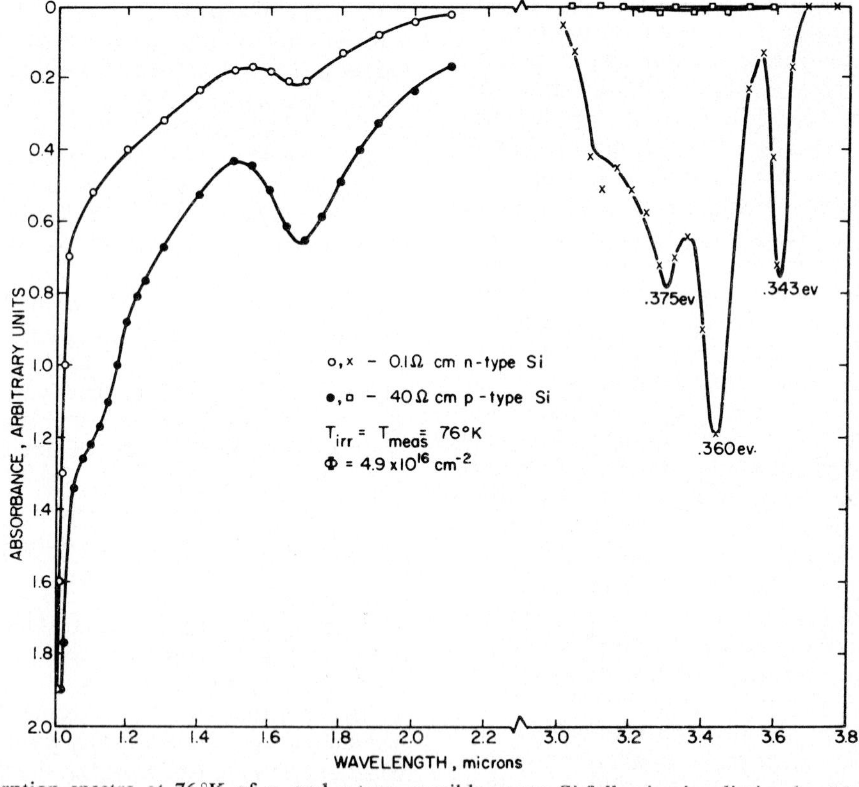

FIG. 1. Absorption spectra at 76 °K of *n*- and *p*-type crucible-grown Si following irradiation by 4.9 (10^{16}) n/cm² at 76 °K. Note the break in the abscissa and that the absorbance increases in the downward direction along the ordinate. The *p*-type sample (●) was annealed to 228 °K prior to the short wavelength measurement and then to 330 °K (□) prior to the long wavelength measurement while the *n*-type sample (o, x) was annealed to 330 °K.

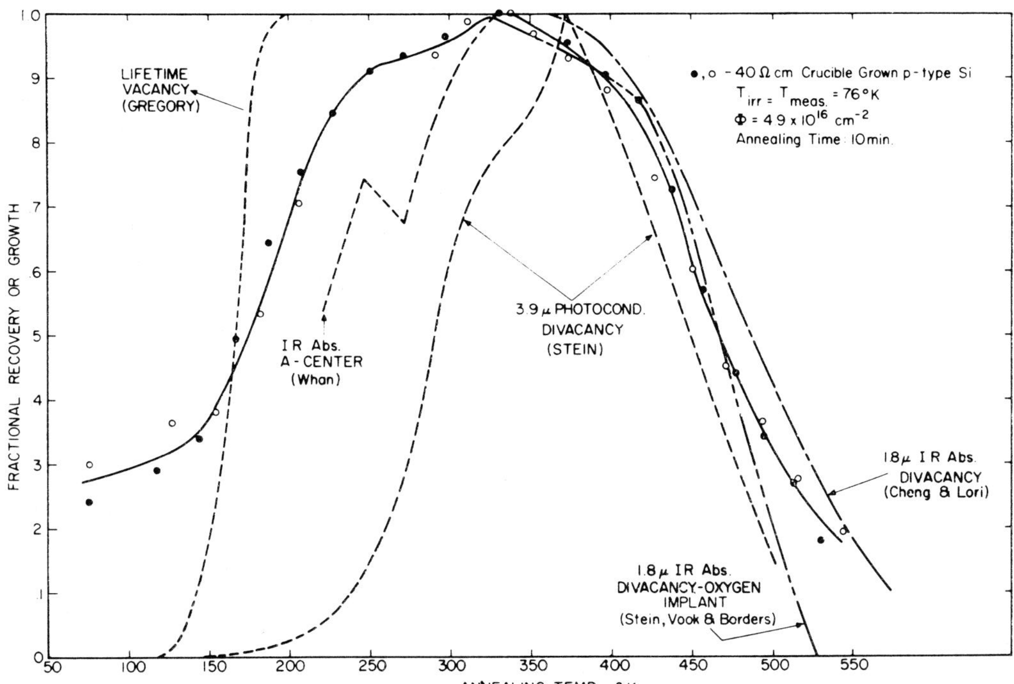

FIG. 2. Growth and anneal of the $1.7\,\mu$ divacancy absorption band for p-type crucible-grown Si irradiated at 76 °K. The annealing time at each temperature was 10 minutes. The following curves are shown for comparison: growth and recovery of $3.9\,\mu$ divacancy photo-conductivity band, after *Stein*[4]; vacancy annealing in γ-irradiated p-type Si based on minority carrier lifetime measurements, after *Gregory*;[12] growth of A-center infrared absorption band in neutron-irradiated Si after *Whan*;[11] annealing of $1.8\,\mu$ divacancy band in oxygen-implanted Si after *Stein, Vook* and *Borders*;[19] and annealing of $1.8\,\mu$ divacancy band in Si neutron irradiated at 50 °C, after *Cheng* and *Lori*.[3]

p-type Si, the absorption edge in the n-type sample has recovered to about the same degree as a p-type sample annealed to 330 °K. Fan and Ramdas[8] have shown that the $3.3\,\mu$ absorption band, due to the double negative charge state of the divacancy, complements the $1.7\,\mu$ band; that is, if E_F is above $E_c - 0.22$ eV absorption due to the divacancy will be observed at $3.3\,\mu$ instead of $1.7\,\mu$. For the low resistivity n-type sample shown in Figure 1, one might expect significant recovery in E_F following anneal to 330 °K for a fluence of 4.9×10^{16} cm^{-2}. Consequently, weak absorption at $1.7\,\mu$ may indicate that most of the divacancies are in the doubly negative charge state, and that absorption at $3.3\,\mu$ should be observed. That this is the case is illustrated on the right side of Figure 1 where an absorption spectrum near $3.3\,\mu$ is shown for 0.1 ohm-cm n-type Si which agrees with that observed by Fan and Ramdas.[8] After annealing to 330 °K, no absorption is observed near $3.3\,\mu$ for the higher resistivity p-type Si.

The complete growth and annealing spectrum of the $1.7\,\mu$ divacancy band for two 40 ohm-cm crucible-grown p-type Si samples is shown in Figure 2. Immediately after irradiation at 76 °K, the divacancy concentration present is approximately 25 per cent of the maximum concentration attained after annealing to 330 °K. In contrast with this result, Stein[4] observed only a very small concentration of divacancies immediately after irradiation at 82 °K. The apparent lack of divacancies in Stein's case can probably be attributed to the fact that the Fermi level was above $E_v + 0.26$ eV immediately after irradiation.

Several types of growth and annealing curves have been included in Figure 2 for purposes of comparison. The knee in our divacancy growth curve occurs at about 140 °K, the same temperature at which vacancy motion becomes significant in γ-irradiated p-type Si according to the minority carrier lifetime annealing curve shown in Figure 2. In contrast with this similarity, divacancy growth as detected by the $3.9\,\mu$ photoconductivity measurements[4] is not significant until much higher

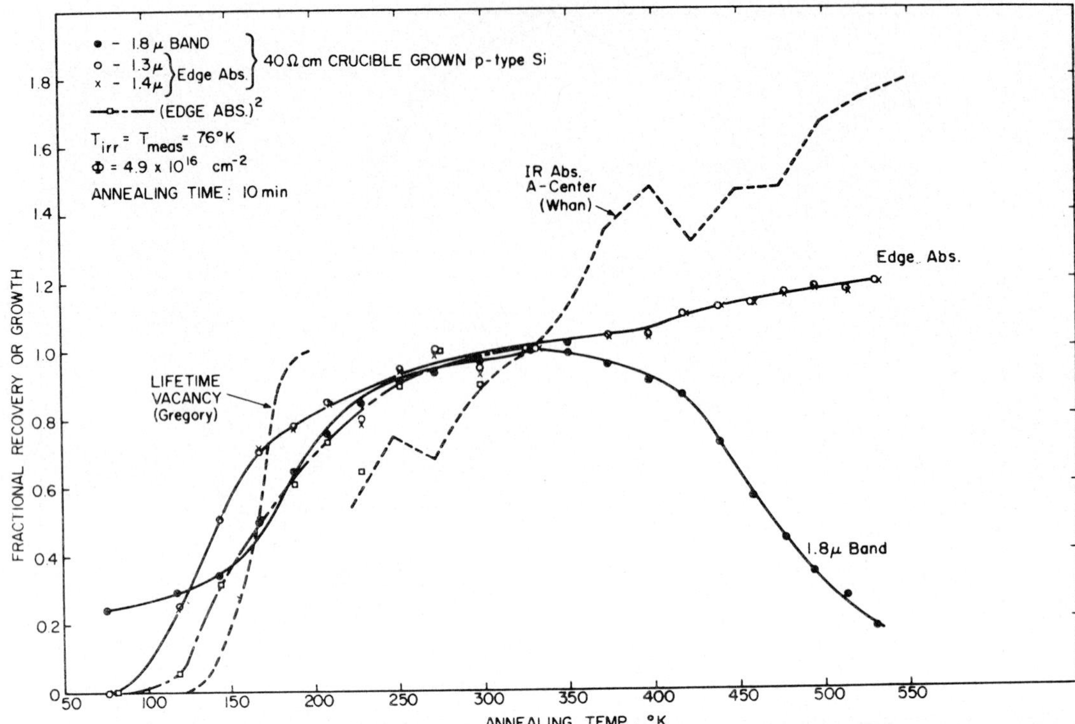

FIG. 3. Comparison of divacancy growth and anneal with edge absorption anneal at $1.3\,\mu$ and $1.4\,\mu$ for the p-type crucible-grown Si irradiated at 76 °K. Note that the square of the edge absorption coincides with the divacancy growth curve between 140 °K and 330 °K. Vacancy anneal and A-center growth curves shown in Figure 1 are included for comparison.

temperatures ($\approx 200\,°K$) are reached, possibly because of the necessity for Fermi level recovery. Intermediate between these two cases is the A-center growth curve.[11] Figure 2 shows that the annealing loss of divacancies for various neutron irradiated and ion-implanted Si samples all agree with each other rather well. Differences in the annealing curves can probably be attributed to variations in annealing times, temperature intervals, and divacancy sink concentrations.

A comparison between the edge absorption recovery at two different wavelengths and the divacancy growth and annealing curve is shown in Figure 3. The edge absorption begins to anneal at about 80 °K so that at low annealing temperatures the edge absorption recovery is more rapid than the divacancy growth. Figure 3 also indicates that the edge absorption annealing starts at temperatures lower than the onset of vacancy motion in p-type Si.[12] At high temperatures, the edge continues to anneal slowly as the divacancies anneal out. Comparison of the three curves shown in the high temperature range indicates that the annealing divacancies are probably a major source of A-centers. This agrees with earlier evidence[5] pointing to oxygen as a divacancy sink.

3.2. *Injection effects*

The various aspects of injection-stimulated $1.7\,\mu$ band growth are shown in Figure 4. In all cases, the samples were immersed in LN_2, and a projection lamp was used as the illumination source. For the p-type sample, injection was first carried out through Ge and Si filters in order to eliminate the possibility that annealing was due to heating from the intense white light. As shown in Figure 4, both $1.7\,\mu$ band growth and edge absorption recovery were induced even with the filters interposed between the light source and the sample. Further growth and recovery was achieved in the p-type sample by $7\frac{1}{2}$ hours of illumination with white light, and finally by a thermal anneal at 330 °K for one hour. The results indicate that injection causes approximately equal changes (≈ 50 per cent) in the $1.7\,\mu$ band growth and the edge absorption recovery.

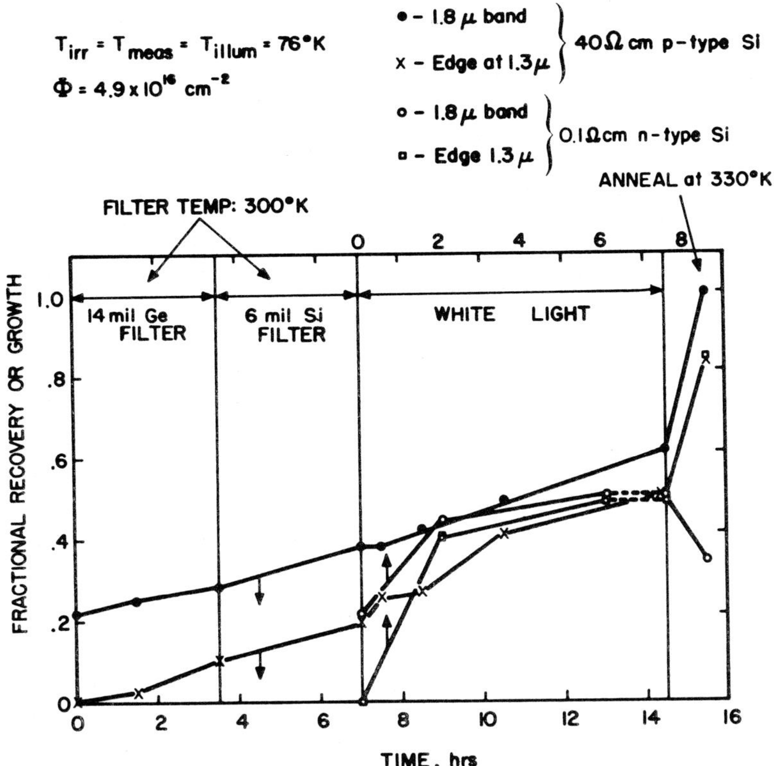

FIG. 4. Characteristics of injection-stimulated 1.8 μ divacancy band growth and edge absorption recovery at 76 °K. The *p*-type sample was illuminated through a Ge filter held at 300 °K and then a Si filter held at 300 °K before being exposed to white light. The time scale for the *n*-type sample is at the top of the graph. Dashed lines merely indicate a transfer of points to the boundary of the white light exposure zone. Note that the last zone is for thermal annealing.

The low resistivity *n*-type sample was subjected to injection only with white light. The injection and thermally stimulated recovery of the edge absorption in this sample was similar to that in the *p*-type sample. However, the divacancy band at 1.7 μ decreased following the thermal anneal for the reasons given in the discussion of Figure 1.

3.3. *Introduction rates*

Divacancy introduction rates have been calculated based on the divacancy concentration per unit peak absorption coefficient calculated by Cheng and Lori.[3] For the 40 ohm-cm *p*-type sample, the introduction rate immediately after irradiation at 76 °K is 0.4 cm⁻¹ while, after anneal to 330 °K (maximum divacancy concentration), it is 1.5 cm⁻¹. The results indicate that the introduction rate for a room temperature irradiation is somewhat higher: 2.3 cm⁻¹. All of these values are less than the introduction rate of 5.7 cm⁻¹ found by Cheng and Lori[3] following irradiation at 50 °C. Such comparisons are somewhat tenuous, however, because of differences in the dosimetry and neutron spectra and also in the methods used to obtain the absorption coefficient.

4. DISCUSSION

Alternative models can be proposed to explain the results depending on whether or not one assumes that the divacancies are present immediately after irradiation at 76 °K. Suppose, for example, that the divacancies are introduced in a metastable charge state for which the 1.7 μ band is not observed; the doubly negative charge state, for example. Then, the growth of the 1.7 μ band following injection or heating would be due to a transfer of the divacancies to a more stable charge state instead of to actual defect annealing and divacancy growth. Wilson[13] used a similar argument to explain injection-stimulated recovery in neutron-irradiated Si diodes. There are a few difficulties with applying this charge transfer model to our case. First, there is very little independent evidence that defects are introduced in a metastable charge state during neutron irradiation to which they cannot be returned following injection. The lack of ability to cycle the defects between charge states makes experimental verification difficult. Second, since heating and injection also result in edge absorption recovery, one must assume for the charge transfer model that the increased edge absorption following irradiation is closely associated with a particular charge state of the divacancy. This further implies that the edge absorption is due to light-induced ionization or excitation of the divacancy. It does not seem plausible that such a specific light absorption process would manifest itself as a strong absorption beginning near the edge and extending continuously into the infrared. Third, photoconductivity results[7] indicate that the energy level for the double negative charge state is at about $E_c - 0.4$ eV while the positive charge state extends from E_v to $E_v + 0.26$ eV.[7] For the metastable charge state model, excitation into or out of these levels must begin at 140 °K, the point at which the 1.7μ band begins to grow. However, this temperature corresponds to a value of $kT \approx 0.012$ eV which is less than 1/20 of 0.26 eV, the level nearest a band edge. Consequently, the thermal energy available at 140 °K does not appear to be sufficient to cause a transfer in divacancy charge state.

A more attractive model is based on the assumption that heating and injection cause defect cluster annealing with the resulting production of divacancies. For this case, the neutron-induced defect cluster is viewed as a vacancy-rich region which begins to reorder upon heating to 140 °K, the temperature corresponding to the onset of isolated vacancy motion in p-type Si.[12] Annealing of the defect cluster occurs at 76 °K under the influence of injection because the cluster is rendered unstable by the capture of electrons created by light absorption. For both types of recovery the liberated vacancies move through the crystal and form divacancies, A-centers[11] and other defects. The strong absorption near the edge is presumably associated with the defect clusters so that as annealing proceeds, either by heating or injection, the clusters shrink in size and the absorption edge recovers toward its pre-irradiation characteristic. A com-

parison of the edge absorption annealing with the 1.7μ band growth curve provides further support for the model if it is assumed that the formation of divacancies is a bimolecular process as one would expect for the annealing model. Then the divacancy growth rate should be proportional to the square of the concentration of liberated vacancies which, in turn is proportional to the edge absorption annealing rate. It follows that the square of the edge absorption annealing curve should coincide with the divacancy growth curve. This does appear to be the case, as shown in Figure 3, with the exception of low annealing temperatures.

Considerable evidence exists in the literature to support the above model. Vacancy motion and capture by oxygen following low temperature neutron irradiation have been demonstrated by Whan[11] as shown in Figures 2 and 3. These measurements provided quite convincing evidence of vacancy liberation and motion because the A-center absorption band which was measured was due to an internal vibrational excitation and, hence, did not depend on Fermi level position. Stein[14,15] has shown that carrier removal annealing occurs between 76 °K and 300 °K in both n- and p-type Si neutron irradiated near 76 °K. In addition, Barnes[9] observed lifetime recovery in the same temperature range in p-type Si. With regard to injection-stimulated annealing, it has been established that the mobility of the vacancy is charge state dependent and that vacancy motion can be induced in p-type Si at low temperatures by injecting electrons into the sample.[12,16] Injection-stimulated recovery of minority carrier lifetime[9] has been observed in p-type Si neutron irradiated at low temperatures. In γ-irradiated p-type Si, Cheng and Lori[17] observed injection-stimulated recovery of the carrier concentration. Injection-enhanced transient annealing of neutron damage has been observed in N/P solar cells irradiated at 300 °K.[18] All of these studies support the annealing model which is based on charge state dependent motion of vacancies away from a vacancy-rich cluster.

5. CONCLUSIONS

The divacancy growth and edge absorption recovery observed during the annealing of Si samples neutron irradiated at 76 °K can be satisfactorily explained by a model in which vacancy liberation and motion are the determining factors. The neutron-induced defect cluster is assumed to be a vacancy-rich region which begins to reorder at about the same temperature $(140\,^\circ K)$ at which isolated neutral vacancy motion becomes significant in p-type Si. The experimental results provide some indication that divacancy formation from liberated vacancies is a bimolecular process as one would expect. Injection-stimulated annealing occurs at $76\,^\circ K$ because the cluster is rendered unstable through the capture of minority carriers in a manner similar to the injection stimulated motion of vacancies in p-type Si.

ACKNOWLEDGEMENT

The author wishes to thank H. J. Stein for many valuable discussions and L. V. Hansen for assistance in performing the experiments. We are also indebted to C. Carter and L. Jennett for sample preparation.

REFERENCES

1. H. Y. Fan and A. K. Ramdas, in *Proc. Int. Conf. of Semiconductor Physical* (Prague), Czeck. Acad. Sci., 309 (1961).
2. A. R. Bean, R. C. Newman and R. S. Smith, *J. Phys. Chem. Solids*, **31**, 739 (1970).
3. L. J. Cheng and J. Lori, *Phys. Rev.*, **171**, 856 (1968).
4. H. J. Stein, *Appl. Phys. Letters*, **15**, 61 (1969).
5. L. J. Cheng and J. Lori, *Appl. Phys. Letters*, **16**, 324 (1970).
6. F. L. Vook and H. J. Stein, 1970 *Int. Conf. on Ion Implantation in Semiconductors*, Thousand Oaks, Calif., May 4–7.
7. A. H. Kalma and J. C. Corelli, *Phys. Rev.*, **173**, 734 (1968).
8. H. Y. Fan and A. K. Ramdas, *J. Appl. Phys.*, **30**, 1127 (1959).
9. C. E. Barnes, *IEEE Trans. Nucl. Sci.*, **NS-16**, 28 (1969).
10. Unfortunately, the irradiations had to be terminated prior to completion of the experiments because of a minor explosion due to the generation of ozone or nitrogen oxides. Approval for more irradiations was not obtained in time for inclusion of additional results in this paper.
11. R. E. Whan, *J. Appl. Phys.*, **37**, 3378 (1966).
12. B. L. Gregory, *J. Appl. Phys.*, **36**, 3765 (1965).
13. D. K. Wilson, *IEEE Trans. Nucl. Sci.*, **NS-15**, 77 (1968).
14. H. J. Stein, *J. Appl. Phys.*, **39**, 5283 (1968).
15. H. J. Stein, *Phys. Rev.*, **163**, 801 (1967).
16. G. D. Watkins, *Symposium on Radiation Effects in Semiconductors*, Toulouse, France (1967).
17. L. J. Cheng and J. Lori, *Phys. Rev.*, **B1**, 1558 (1970).
18. B. L. Gregory and H. H. Sander, *IEEE Trans. Nucl. Sci.*, **NS-14**, 116 (1967).
19. H. J. Stein, F. L. Vook and J. A. Borders, *Appl. Phys. Letters*, **14**, 328 (1969).

DISCUSSION

Question (STEIN) There seems to be some uncertainty about the charge state of the divacancy when the optical absorption bands near 3.3 microns are observed. Would your conclusions be affected if the bands near 3.3 microns were associated with the single negative rather than the double negative charge state of the divacancy.

Answer (BARNES) The conclusions drawn would still be valid since the 1.8 micron band has been attributed to the neutral charge state of the divacancy. Consequently, for n-type material, Fermi level recovery due to annealing would lead to a transfer to the negative charge state with the appearance of the bands near 3.3 microns. The only difference would be that not as much recovery would be required if the 3.3 micron bands were due to the single negative charge state.

INVESTIGATION OF DIVACANCIES IN SILICON

V. D. TKACHEV AND M. T. LAPPO

Semiconductor Physics, Byelorussian University, Minsk, U.S.S.R.

Infrared absorption and photoconductivity caused by divacancies in n- and p-type silicon crystals after fast-neutron irradiation were studied. A differential method was used for measurement of the absorption spectrum in the range of 1.0 to 25 microns in integral and monochromatic light. The isochronal and isothermal annealings of separate absorption bands were carried out. Energy position of neutral divacancy $VV^{\circ} (E_v + 0.31 \text{ eV})$ and double-minus charged state of the divacancy $VV^= (E_c - 0.42 \text{ eV})$ were determined from absorption and photoconductivity spectra. The level of single-minus charged state of the divacancy $VV^- (E_c - 0.56 \text{ eV})$ can be found only from photoconductivity. Analysis of half-width, shape and temperature dependence of absorption bands 0.344 eV, 0.360; 0.375 eV and 0.403 eV ($T = 12 \,^\circ\text{K}$) permits us to show that they are caused by inner transitions in $VV^=$ center. The structure of photoconductivity connected with $VV^=$ was also observed. The possibility of observation of absorption bands associated with $VV^=$ center in p-type crystal is due to that of the center being an effective trap for electrons. The kinetics of electron trapping by $VV^=$ was also studied. The annealing of the complex defect regions occurs as the decay of divacancies and gives the increase of their concentration ($T = 150$–$180\,^\circ\text{C}$).

The annealing of divacancies occurs at temperatures 220–250 °C.†

† Summary of paper accepted but not presented at the conference.

RECOMBINATION LUMINESCENCE IN IRRADIATED SILICON-EFFECTS OF UNIAXIAL STRESS AND TEMPERATURE VARIATIONS†

C. E. JONES AND W. D. COMPTON

Coordinated Science Laboratory and Department of Physics, University of Illinois, Urbana-Champaign, U.S.A.

Luminescence in irradiated silicon consists of a spectral group between 0.80 eV and 1.0 eV which seems to be independent of impurities while a lower energy group between 0.60 eV and 0.80 eV is seen only in pulled crystals. The small halfwidth and temperature dependence of the sharp zero-phonon lines observed in these spectra indicates that the luminescence arises from a bound-to-bound transition. A model is proposed for the transition mechanism. Stress data taken on the 0.79 eV zero-phonon line in pulled crystals can be fit by either a tetragonal [100] defect symmetry or by conduction band splitting effects. It is suggested that the 0.79 eV zero-phonon line and the 0.60 eV to 0.80 eV spectral group arise from the EPR G-15 center. Stress data on a zero-phonon line at 0.97 eV associated with the 0.80 eV to 1.0 eV spectral group can be explained by a trigonal [111] defect. The divacancy is tentatively suggested as responsible for this luminescence spectra.

The characteristic band-to-band luminescence seen in silicon disappears after irradiation with neutrons, γ-rays, or high energy electrons. In its place are found several lower energy spectra. In order to obtain information on the nature of the transitions and the defects involved, an investigation was undertaken of the influence of the measurement temperature and of uniaxial stress upon the luminescence spectra.

The luminescence observed at 34 °K after irradiation with 2.5 meV electrons at 0 °C is shown in Figure 1. The structure between 0.80 eV and 1.0 eV, except for the peak labeled F at 0.94 eV, occurs in the same intensity ratios and seems to be independent of any impurities. The structure between 0.69 eV and 0.80 eV is also seen as a unit but only in pulled crystals which have oxygen as an impurity.[1] The narrow peaks labeled A, B, C, D, E, F, and G are zero-phonon lines and much of the broad structure to the left of the strong lines C at 0.79 eV and G at 0.97 eV can be identified with transitions involving phonons. The first broad peaks which lie approximately 18 meV lower in energy than the zero-phonon lines C and G can be identified with the momentum conserving transverse acoustical phonon assisted transition. Most of the next lower energy broad peaks can be related to transitions involving two transverse acoustical phonons. Weak structure, 58 meV below the C

and G lines, is tentatively identified with momentum conserving transverse optical phonon assisted transitions. The Huang-Rhys factors for both spectral groups is approximately 2.3 with $S_i \simeq 1.5$ for the TA phonon and $S_i \simeq 0.4$ for the tentatively identified TO phonon. The zero-phonon lines C and D form a doublet, the intensity of the high energy component being determined by a Boltzmann factor. The high temperature intensity ratio is 1:1.

There has been some question as to whether the luminescence was due to a bound to bound or bound to free transition.[1,3] The half-width of the strong zero-phonon lines at 0.79 eV and 0.97 eV is given as a function of temperature in Figure 2. The narrow halfwidths and the non-linear dependence of the halfwidths upon temperature show that a free carrier is not involved in the transitions. The broadening can be explained by the interaction of the defect with the lattice phonons. A Debye spectrum for the lattice phonons does not fit the data but the effect of the strongly coupled momentum conserving transverse acoustical phonon, as applied in formalism developed by McCumber,[2] fits the broadening and is shown by the solid line in Figure 2. The lines also shift to lower energy with increased temperature, but the shift follows the band gap shift too closely to determine the direct phonon interaction on this parameter.

These data allow certain conclusions to be drawn about the nature of the transitions. Consider first a neutral defect with a trapping level at a depth a little less than the energy difference between the

† Work supported in part by NASA-JPL under Contract 952383 and in part by the Joint Services Electronics Program (U.S. Army, Navy and Air Force) under Contract DAAB-07-67-C-0199.

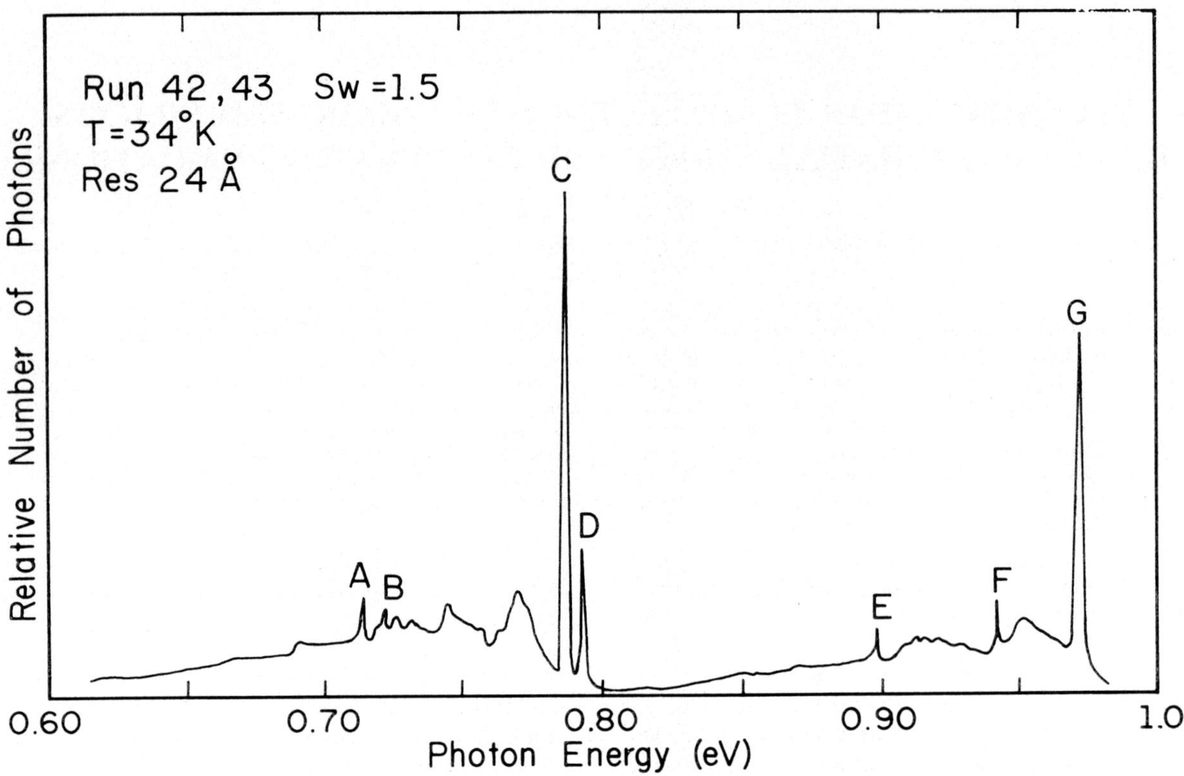

FIG. 1. Luminescence spectrum at 34 °K of 80 ohm-cm *n*-type pulled silicon, phosphorus doped, irradiated with 10^{17} el./cm² at 2.5 meV.

luminescence energy and the 1.16 eV band gap energy. This defect could trap a carrier from the nearest band and then the charge carrier with the opposite charge would be attracted to the defect with a small binding energy. The luminescence would occur when the electron and hole annihilate each other. The bound state luminescence which has been observed in silicon doped with chemical impurities has involved electron-hole pairs or excitons bound to neutral defects. These centers do not have deep trapping levels in the neutral state, and the defect energy is observed to be approximately one tenth the donor or acceptor ionization energy.[4] The observed spectra do not fit this relationship since such a model would require that the donor or acceptor energies would then be larger than the band gap energy. A second model could involve excitons bound to ionized centers. These have not been observed for the chemical dopants in silicon, and theories by Hopfield[5] and Jean-Marc Levy-Leblond[6] suggest that these complexes may not even be stable in silicon.

The dependence of the lower energy luminescence group upon oxygen, irradiation damage, and thermal annealing[9] suggest that the structure between 0.6 eV and 0.8 eV arises from a center studied in spin resonance and called the G-15[7] center or the K-center.[8] This EPR center has a spin of 1/2, a symmetry axis in the $\langle 221 \rangle$ direction, and corresponds to a hole trapped ~ 0.3 eV from the valence band. It is proposed that the luminescence arises when the neutral center first traps a hole into a level at approximately 0.3 eV, and then an electron to this charged center with a binding energy of about 0.06 eV. This is the right order of magnitude for the localization energy of an electron to a hole bound this deeply. The luminescence arises from the recombination of these charges.

The influence of uniaxial stress of about 1.5×10^9 dynes/cm² upon this defect is shown in Figures 3 and 4 for the 0.79 eV peak. The splittings, especially the single line for [111] directed stress, correspond most closely to that expected from a defect with a [100] tetragonal symmetry

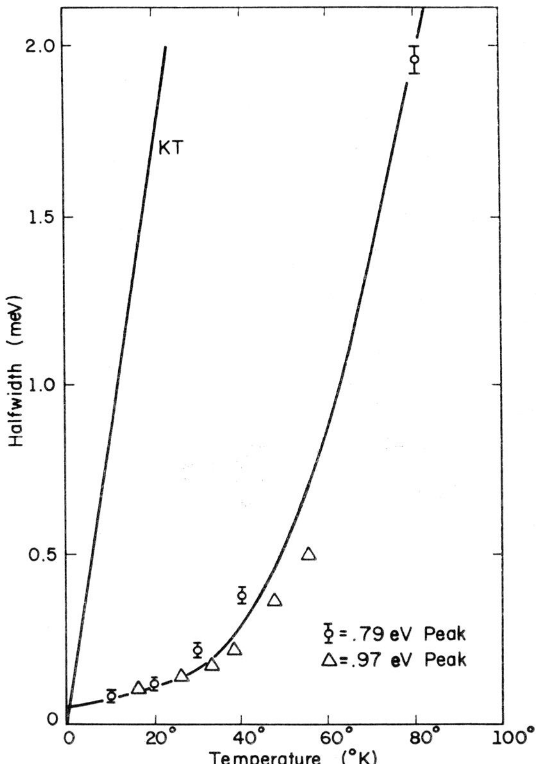

FIG. 2. The halfwidth of the zero-phonon lines at various temperatures. The solid line is the broadening calculated from the 18 meV transverse acoustical phonon interaction in McCumber's formalism.

The correspondence is not complete, however, as can be seen in Figure 4. A second possible explanation for the stress splitting is that the energy of the weakly bound electron splits in the same way as the conduction bands in silicon. This would also give the doublet, singlet, doublet stress splittings observed. The first alternative is considered to be the most likely.

The radiation dependence, the lack of a relationship to an impurity, and the thermal annealing data[9] tend to associate the higher energy structure with the divacancy. The effects of uniaxial stress upon the luminescence are shown in Figure 5. The splitting is that expected from a trigonal $A \rightarrow E$

transition as shown in Figure 6. This is consistent with the assignment of the divacancy as being responsible for this luminescence.

The conclusions from these data are:

1. The luminescence involves a bound-to-bound transition.
2. The luminescence is probably due to a neutral center with a deep trapping level, but the usual models involving an exciton and a charged or neutral donor or acceptor are probably unlikely.
3. A number of facts suggest that the G-15 defect studied with EPR may be responsible for the luminescence associated with the zero-phonon line at 0.79 eV. The stress splitting can either be explained by a [100] tetragonal defect or by band effects.
4. The stress dependence of the defect responsible for the 0.97 eV zero-phonon line appears to have [111] trigonal symmetry and is consistent with an assignment of this spectral group to the divacancy. The stress splitting of this luminescence is not consistent with band splitting effects.

The authors would like to acknowledge helpful discussions with J. J. Hopfield and G. D. Watkins.

REFERENCES

1. R. J. Spry and W. D. Compton, *Phys. Rev.*, **175**, 1010 (1968).
2. G. F. Imbusch, W. M. Yen, A. L. Scharvlow, D. E. McCumber and M. D. Sturge, *Phys. Rev.*, **133**, A1029 (1964).
3. M. V. Bortnik, V. D. Tkachev and A. V. Yukhevich, *Fiz. Tekhn. Pol.*, **1**, 353 (1967) [English transl.: *Soviet Phys.-Semiconductors*, **1**, 290 (1967)].
4. R. J. Haynes, *Phys. Rev. Letters*, **4**, 361 (1960).
5. J. J. Hopfield, *Physics of Semiconductors Proceedings of the Seventh International Conference*, Ed. Dunod (Academic Press, New York, 1964).
6. Jean-Marc Levy-Leblond, *Phys. Rev.*, **178**, 1526 (1969).
7. G. D. Watkins, *Seventh International Conference on the Physics of Semiconductors, Volume Three, Radiation Damage in Semiconductors*, Ed. Dunod (Academic Press, New York, 1964).
8. N. Alemleh and B. Goldstein, *Phys. Rev.*, **149**, 687 (1966).
9. E. Johnson and W. Dale Compton, see paper in this Conference.

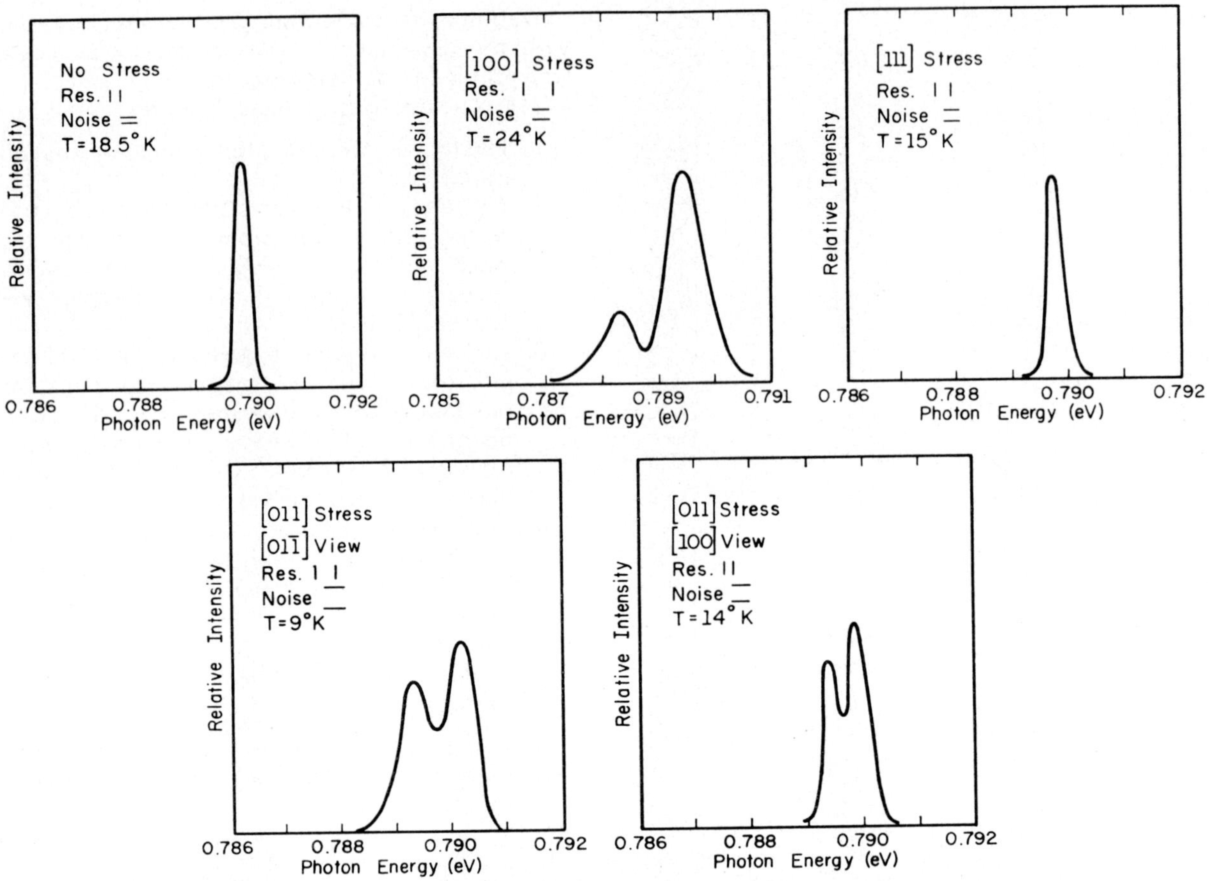

FIG. 3. Stress splittings observed for the 0.79 eV zero-phonon line. The stress is approximately 1×10^9 dynes per cm². The samples are phosphorus doped *n*-type pulled silicon with a resistivity of 80 ohm-cm. They have been irradiated with 10^{17} el./cm² at 2.5 meV at 0 °C. Measurement temperatures are given on each figure.

0.79 eV Peak

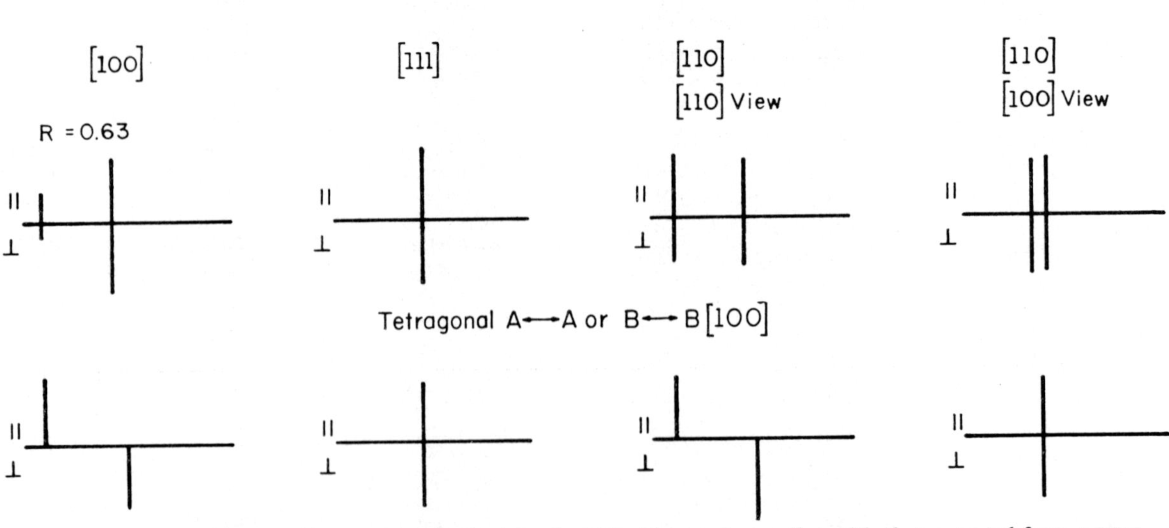

FIG. 4. Comparison of the stress splittings observed for the 0.79 eV zero-phonon line with that expected from a tetragonal [100] defect. The intensity polarized parallel to the stress is represented above the horizontal line and that perpendicular to the stress is represented below. $R = I_\perp / I_\parallel$.

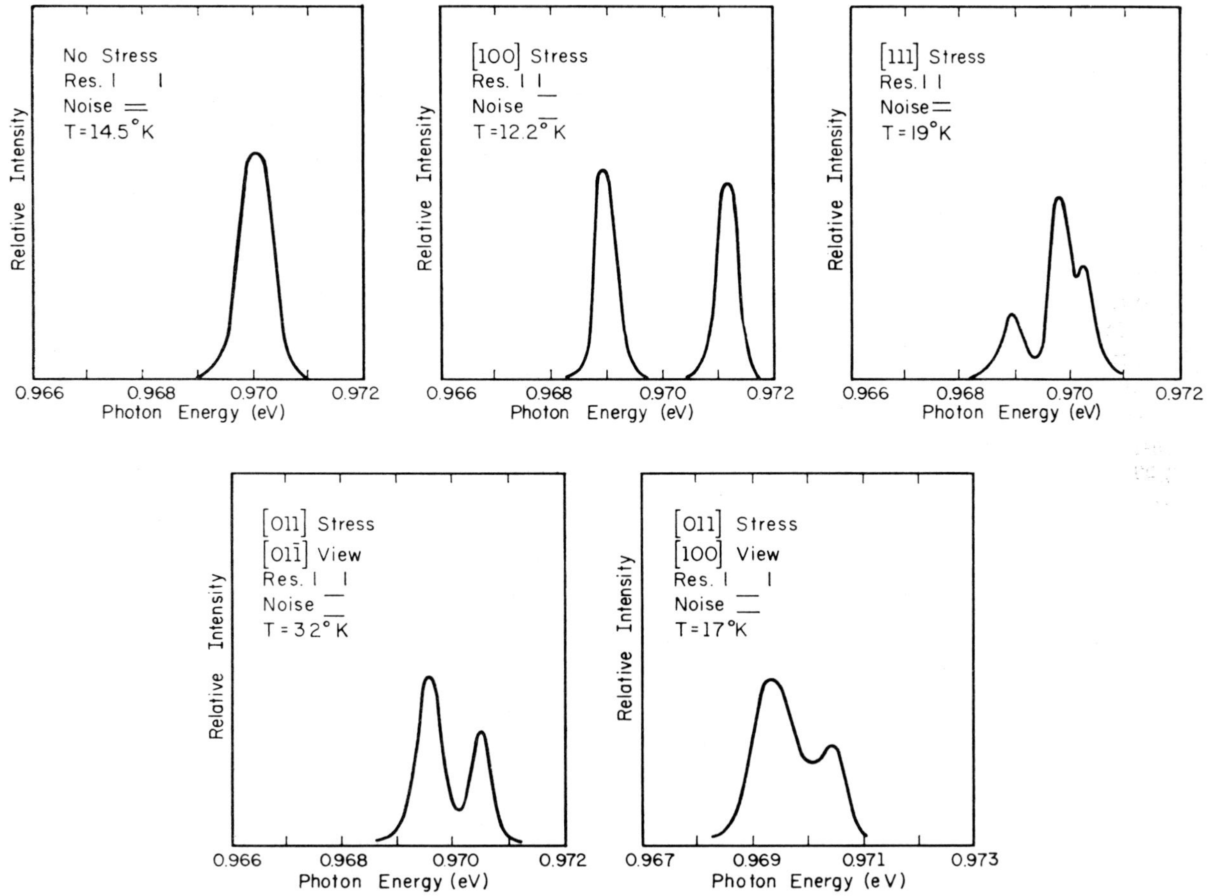

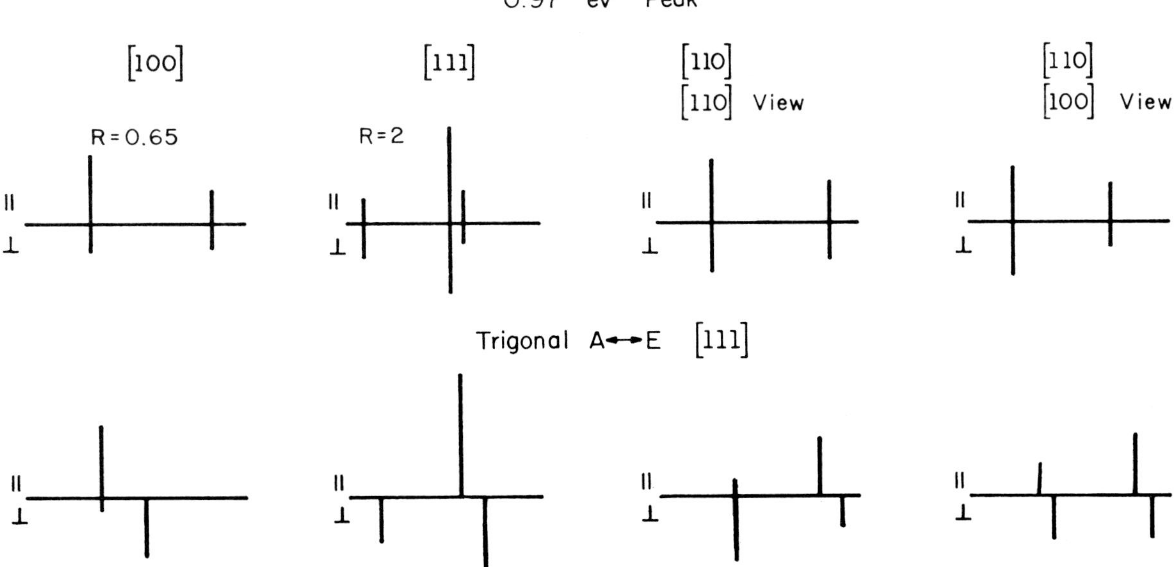

FIG. 5. Stress splittings observed for the 0.97 eV zero-phonon line. The stress is approximately 10^9 dynes/cm². The samples are phosphorus doped n-type pulled silicon with a resistivity of 80 ohm-cm. They have been irradiated with 10^{17} el./cm² at 2.5 meV at 0 °C. Measurement temperatures are given on each figure.

FIG. 6. Comparison of the stress splitting observed for the 0.97 eV zero-phonon line with that expected from a trigonal $A \to E$ [111] defect transition. The intensity polarized parallel to the stress is represented above the horizontal line and that perpendicular to the stress is represented below. $R = I_\perp / I_\parallel$.

DISCUSSION

Question (WATKINS) The identification of the .97 eV luminescence with the divacancy may have some difficulties. First, the divacancy is a double acceptor and at least in *n*-type material one might not expect luminescence from it. (Trapping a hole still leaves it negatively charged and repulsive to electrons for recombination.) Secondly, the significant intensity in the zero-phonon line is somewhat surprising considering the large Jahn-Teller energies known to be associated with the divacancy and the large changes in this energy vs. charge state.

Answer (JONES) The divacancy seems to be the best fit to what we think we know about the 0.97 eV luminescence. If another defect could be found which depends on irradiation, is independent of chemical doping, anneals with the divacancy, has a $\langle 111 \rangle$ symmetry, and has a trapping level in the neutral state of approximately 0.16 eV, it would also fit the data. The creation of neutral defects in both *n* and *p*-type crystals by the strong band gap excitation light is certainly not out of the question. The strong zero-phonon line does not seem unreasonable to me. For electron states near the conduction band the direct recombination with a hole violates conservation of crystal momentum. The spectra are dominated by the higher order transitions involving momentum conserving phonons. As the electron levels get deeper, the conservation requirement gets weaker and the first order zero-phonon transition becomes stronger. The strength of the observed zero-phonon lines relative to the phonon-assisted structure is about what I would expect from defects this deep in the band gap. I am not sure what effect local Jahn–Teller distortions have on this picture.

Comment (DALY) The luminescence decay time of these two spectra has been studied by W. P. Knox and myself. For the 0.97 eV spectrum, the decay time is less than 4 micro-seconds, the response time of our detector. For the 0.79 eV spectrum the decay time is 95 micro-seconds at 20 °K and decreases with increasing temperature. This decrease obeys an activation energy of approximately 11 meV.

RECOMBINATION LUMINESCENCE IN IRRADIATED SILICON— EFFECTS OF THERMAL ANNEALING AND LITHIUM IMPURITY†

E. S. JOHNSON AND W. D. COMPTON

Coordinated Science Laboratory and Department of Physics, University of Illinois, Urbana-Champaign, U.S.A.

Luminescence in irradiated silicon has been used to determine the thermal stability of the defects responsible for the recombination. It was found that the defect responsible for the zero-phonon line at 0.97 eV has an annealing behavior similar to that of the divacancy and the zero-phonon line at 0.79 eV anneals in a manner similar to the G-15 or K-center. Annealing at temperatures up to 500 °C generates other defects whose luminescence is distinct from that seen previously. Addition of lithium to the material produces defects with new characteristic luminescence. Of particular importance is a defect with a level at $E_g - 1.045$ eV.

Irradiation of silicon with high-energy electrons, neutrons or gamma-rays introduces defects which act as efficient recombination centers.[1] The previous paper by Jones and Compton[2] indicates that studies of the luminescence spectra can reveal some of the features of the local symmetry of the defects and something about the nature of the recombination process. In this paper, we will discuss the thermal stability of these defects and the role of the lithium impurity upon this luminescence. In particular, a comparison will be made between the thermal stability of the defects responsible for the luminescence and the stability of defects as measured by other techniques. Some of the properties of several new centers, found only after a high-temperature anneal, will also be described.

Silicon was irradiated at -10 °C with 3 MeV electrons to a fluence of 10^{17} e/cm². Figure 1 shows the spectrum from n-type pulled Si that was stored at room temperature after irradiation. The measurement temperature is between 10 and 25 °K. The spectrum is dominated by zero-phonon lines at 0.97 eV and 0.79 eV, suggesting that defects are involved with energy levels at about $E_g - 0.97$ and $E_g - 0.79$ eV, respectively. Isochronal annealing experiments were carried out in an effort to obtain information on the thermal stability of these defects. The results for the center responsible for the zero-phonon line at 0.97 eV for pulled and float zone silicon are shown in Figure 2. The line is observed to anneal between 250 °C and 300 °C with a substantial difference between pulled and float

† Work supported in part by NASA-JPL under Contract 952383 and in part by the Joint Services Electronics Program (U.S. Army, Navy and Air Force) under Contract DAAB-07-67-C-0199.

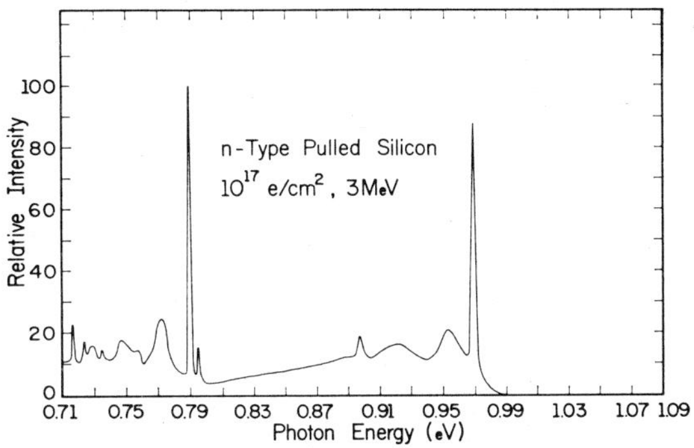

FIG. 1. Luminescence spectrum from 100 ohm-cm n-type pulled silicon, phosphorus doped, irradiated to a fluence of 10^{17} e/cm². Sample was stored at room temperature prior to measurement near liquid helium temperature.

zone material. This behavior compares well with the annealing of the divacancy as observed with the 1.8 μ infrared absorption band[3] and the variation in annealing between the pulled and float zone as observed in EPR.[4] Thus, we believe that luminescence near 0.97 eV arises from recombination at the divacancy and that the responsible energy level is about 0.195 eV from a band edge, presumably the valence band.

There are other defects with energy levels about 0.19 eV from a band edge, e.g., the Si G-1, A center.[5] Since the Si-A center does not anneal between 250 °C and 300 °C, this fact seems to eliminate the Si-A center from consideration as being responsible for this lumines cence. Other defects have annealing behavior that resembles that observed here. Cherki and Kalma, using infrared photoconductivity, have found radiation induced defects which they have identified as interstitial boron and aluminium which anneal near 250 °C.[6] These centers, however, have energy levels at $E_v + 0.430$ eV and $E_v + 0.395$ eV, respectively. The impurity independence of the center associated with the 0.97 eV luminescence and the different annealing temperatures seen for pulled and float zone silicon strongly suggest that the interstitial impurity centers are not responsible for this luminescence.

Similar isochronal annealing studies were carried out on the 0.79 eV zero phonon line. The results are shown in Figure 3. The dashed curves show the annealing of the center associated with the 0.79 eV zero phonon line, seen only in pulled silicon. This annealing behavior compares well with the annealing shown by the solid curve of an oxygen-dependent center, the Si G-15,[7] or the K center.[8,9] The persistence of the 0.79 eV luminescence to high temperature, up to 450 °C, agrees particularly well with the EPR data.

Zero-phonon and phonon-assisted lines associated with new centers are seen after annealing above 300 °C. The spectra depend critically on the type of material examined. The results for *n*-type pulled silicon are shown in Figure 4. New zero-phonon lines at 0.949 eV, 0.925 eV, and 0.767 eV can be seen after 400 °C anneal. The lines at 0.949 and 0.925 eV agree well with the work of Bortnik for *n*-type pulled silicon.[10] The sharp line at 0.767 eV has not been seen previously. The same lines are seen in *p*-type pulled Si along with an additional line at 0.761 eV. There is no agreement with the work of Bortnik for *p*-type pulled silicon. In *p*-type float zone silicon a new center appears following anneal in the temperature range 300 to 450 °C whose luminescence has a zero-phonon line at 1.108 eV. A number of sharp lines are seen in float zone material following a 600 °C anneal. Only the return of the luminescence associated with the recombination of the free exciton, the intrinsic preirradiation luminescence, is seen in pulled crystals above 500 °C.[11] Only intrinsic luminescence has been seen in unirradiated samples annealed to high temperature.

Identification of the centers seen after the high-temperature anneals cannot be done with any great accuracy at this time. Centers seen in pulled crystals but not in float zone apparently depend on oxygen or carbon for their formation. The very shallow center seen only in *p*-type float zone silicon between 300 °C and 450 °C may be boron dependent. Its energy is very close to the boron acceptor

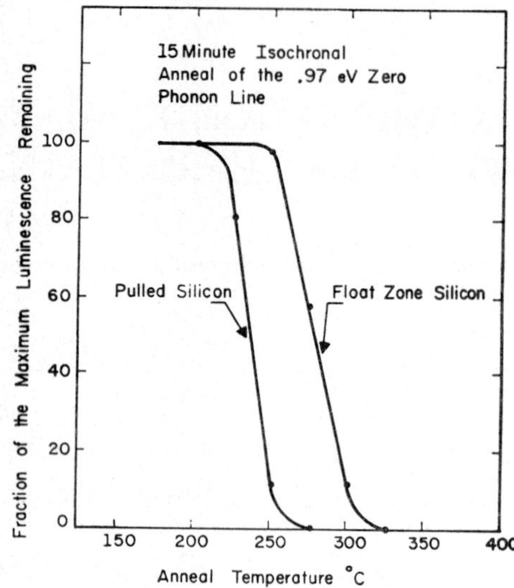

FIG. 2. Isochronal anneal (15 minute) of the 0.97 eV zero-phonon line for electron-irradiated pulled and float zone silicon.

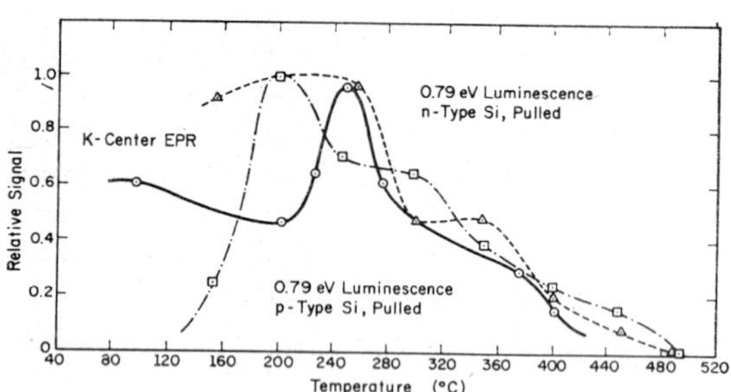

FIG. 3. Isochronal anneal (20 minute) of the 0.79 eV zero phonon line for electron-irradiated pulled silicon. Solid curve gives the isochronal annealing of the K center.

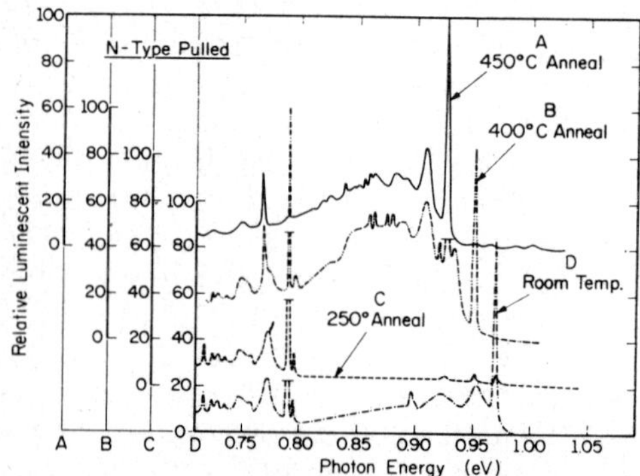

FIG. 4. Luminescence spectra from 100 ohm-cm *n*-type pulled silicon, phosphorus doped, irradiated with 10^{17} e/cm² at 3 MeV, following high-temperature anneal (20 minute).

level of $E_v + 0.045$ eV, but this is very likely coincidental.

The effect of lithium impurity on irradiated silicon has been studied by many investigators.[12] Recent studies indicate that recombination luminescence is capable of giving substantial information on lithium-dependent centers involved in radiative recombination. In particular, diffused lithium greatly changes the luminescence spectra of irradiated silicon. Spectra from lithium-diffused silicon have been taken using samples with electrically active lithium concentrations between 10^{16} and 10^{18} cm³ following electron irradiation with fluences between 10^{17} and 10^{18} e/cm². The spectrum for n- and p-type float zone silicon annealed at room temperature is shown in Figure 5. It is characterized by a strong zero-phonon line associated with a center having a level at $E_g - 1.045$ eV. A spectrum taken with higher optical resolution is shown in Figure 6. Five additional sharp lines of lesser intensity are shown. These additional lines always appear in the spectra of lithium-diffused float zone silicon. The lithium-dependent center associated with the strong line at 1.045 eV is quite stable, annealing between 350 °C and 400 °C. Furthermore, in certain lithium-diffused float zone samples, the 0.97 eV line, seen in all lithium-free samples, is also clearly present.

Lithium-diffused n-type pulled silicon gives a spectrum that is characteristic of both lithium containing material and material that is free of lithium. The spectrum from this material is shown in Figure 7. The intensity of the 1.045 eV line is quite weak.

Lithium-diffused p-type pulled silicon gives a recombination luminescence spectrum following room temperature irradiation that is identical to the spectrum seen from lithium-free silicon. Brief anneals at 50 °C or long periods at room temperature leads to a luminescence spectrum that shows clear evidence of lithium defects which are the same as are found in n-type diffused pulled silicon. Strong lines at 0.785 eV and 1.001 eV and numerous weak lines are seen in lithium-diffused pulled crystals following 300 °C anneal.

The conclusion from these data are:

(1) The annealing of the center responsible for the 0.97 eV line is in good agreement with the annealing behavior of the divacancy.

(2) The annealing of the center responsible for the 0.79 eV line is in good agreement with the annealing behavior of the K-center.

(3) New centers are seen following high-tem-

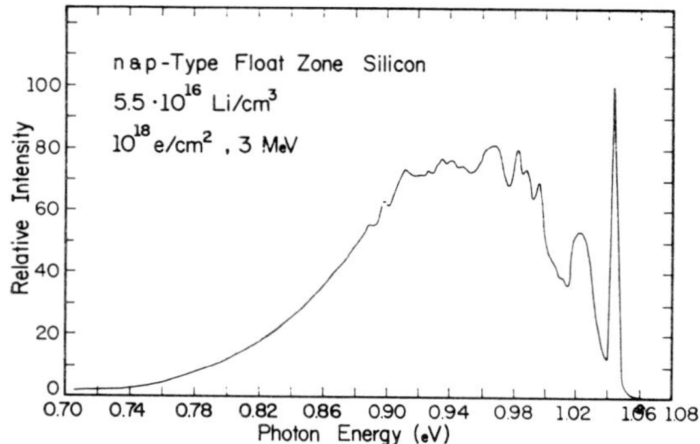

FIG. 5. Luminescence spectrum from lithium-diffused n- and p-type float zone silicon ($5.5 \cdot 10^{16}$ Li/cm₃) irradiated to 10^{18} e/cm² at 3 MeV and stored at room temperature prior to measurement.

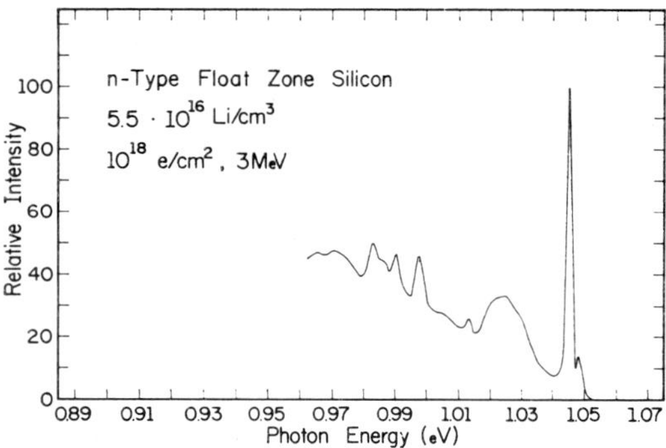

FIG. 6. Luminescence spectrum from lithium-diffused n- and p-type float zone silicon ($5.5 \cdot 10^{16}$ Li/cm³) at high resolution. Five additional lines having half-widths comparable to the main line at 1.045 eV are observed.

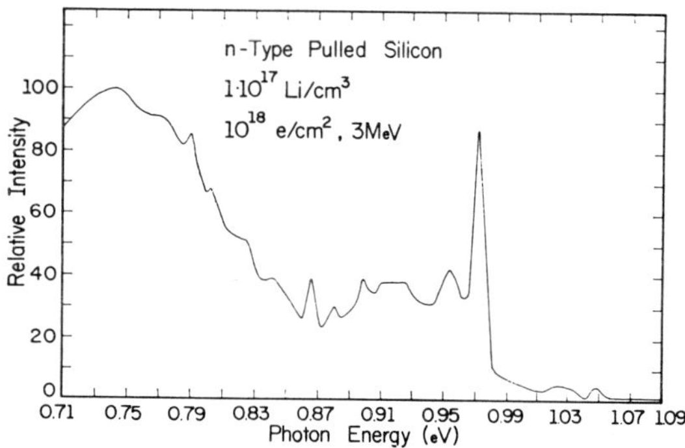

FIG. 7. Luminescence spectrum from n-type pulled silicon containing 10^{17} Li/cm³, irradiated to 10^{18} e/cm² at 3 MeV, and stored at room temperature prior to measurement.

perature anneal which depend upon the initial impurities in the crystals. Four centers are found in pulled silicon, one center is found in float zone silicon.

(4) Recombination luminescence in lithium-diffused silicon differs greatly from that seen in lithium-free silicon. The major recombination center specifically dependent upon lithium for its formation has a level at $E_g - 1.045$ eV. The center is very stable and is independent of other chemical dopants. Centers present in lithium-free silicon are also observed.

REFERENCES

1. R. J. Spry and W. D. Compton, *Phys. Rev.*, **175,** 1010 (1968).
2. C. E. Jones and W. D. Compton, see this meeting.
3. L. J. Cheng, J. C. Corelli, J. W. Corbett and E. D. Watkins, *Phys. Rev.*, **152,** 761 (1966).
4. G. D. Watkins and J. W. Corbett, *Phys. Rev.*, **138,** A543 (1965).
5. J. W. Corbett, E. D. Watkins, R. M. Chrenko and R. S. McDonald, *Phys. Rev.*, **121,** 1015 (1961).
6. M. Cherki and A. H. Kalma, *Phys. Rev. B***1,** 647 (1970).
7. G. D. Watkins, *Seventh International Conference on the Physics of Semiconductors*, Vol. 3, *Radiation Damage in Semiconductors*, Ed. Dunod (Academic Press, New York, 1964).
8. N. Almeleh and B. Goldstein, *Phys. Rev.*, **149,** 687 (1966).
9. B. Goldstein, J. J. Wysocki, N. Almeleh and P. Rappaport, *Analysis of Radiation Damage in Silicon Solar Cells and Annealing or Compensation of Damage by Impurities*, C.F.S.T.I., N67-28927.
10. M. V. Bortnik, V. D. Tkachev and A. V. Yukhnevich, *Soviet Physics-Semiconductors*, **1,** 290 (1967) English translation.
11. J. R. Haynes, *Phys. Rev. Letters* **4,** 361 (1960).
12. B. Goldstein, see this meeting; J. A. Naber, H. Horiye and B. C. Passenheim, see this meeting.

DISCUSSION

Question (BRUCKER) What was the temperature of the irradiation?

Answer (JOHNSON) $-10\,^\circ$C.

Question (ALBANY) Have you observed any Li-clustering during storage at room temperature in your highly Li-doped silicon?

Answer (JOHNSON) Samples stored for $\sim$18 months at room temperature were examined by a four-point probe to measure the electrically-active lithium. Resistivity values obtained were generally within 10 per cent of the pre-storage value. This is about the accuracy of the four-point probe.

Li-DEFECT INTERACTIONS IN ELECTRON-IRRADIATED
n-TYPE SILICON BY EPR MEASUREMENTS†

B. GOLDSTEIN‡

RCA Laboratories, Princeton, New Jersey, U.S.A.

and

Groupe de Physique des Solides,§ Ecole Normale Superieure et Faculté des Sciences, Paris, France

Single crystal silicon, both with and without oxygen, has been diffused with lithium to concentrations $\sim 10^{17}/cm^3$, irradiated with 1 to 1.5 MeV electrons, and the ensuing defects studies by EPR measurements. The presence of oxygen strongly affects the properties of these defects. Measurements have indicated the presence of two new defects which involve Li—one in O-containing material and one in O-free material. The defects are observed in their electron-filled state, and indicate a net electron spin of 1/2. The defect spectra disappear (with time) at room temperature, and can be explained by the formation of other Li-involved defects which lie deeper in the energy bandgap and are not visible by EPR. Electron irradiation at 40 °K followed by annealing at higher temperatures show that both EPR defects described above begin to form at about 200 °K and begin to decrease at about 275 °K—just as does the 250 °K reverse annealing observed generally for *n*-type Si. Based on these data, and the work of others, it is suggested that both defects form as a result of the motion of Si interstitials which produce a (Li-O-interstitial) complex in O-containing Si, and a (Li-interstitial) complex in O-free Si.

1. INTRODUCTION

Recent studies have indicated that the general radiation damage behavior in *n*-type Czochralski grown Si, doped with different Group V donor atoms, is quite similar for bombarding electrons having energies of 1 to 2 MeV. For example, the configuration and symmetries of the defect complexes, their stability at room temperature, their effect on carrier concentration, and their temperature behavior are generally very similar or follow very similar patterns.[1,2] However, the radiation damage properties of *n*-type Si doped with lithium seems to be very different.[3] For example, defect complexes which involve Li are not stable at room temperature,[4] and their effects on minority carrier life-time are different from those when Li is absent.[5] Recent work on Li-doped Si has presented results of optical[6] and electrical[7] measurements. We will present and discuss in this paper some results of electron paramagnetic reson-

ance (EPR) measurements of Li-doped Si after electron irradiation.

2. EXPERIMENTAL RESULTS

2.1. *Apparatus and procedures*

Two types of single crystal Si were used: one was grown by the Czochralski method and contained ~ 5 to $10 \times 10^{17}/cm^3$ oxygen; the other was grown by vacuum float-zone techniques and contained less than $5 \times 10^{13}/cm^3$ oxygen.[8] We expected that oxygen might play a role in the radiation damage behavior because of its pairing with Li and its effect on Li diffusion,[9] and its trapping of primary defects.[10]

The source of electrons was a Van de Graaff accelerator which produced a uniform, mono-energenic beam of electrons. After the irradiations, the samples were refrigerated and stored at liquid nitrogen temperatures until measurements were made. Thus, we will be examining the initial defect-complex situation.

For one group of experiments, involving irradiations at 250 °K, EPR measurements were made with a Varian spectrometer operating at about 9.1 GHz in the absorption mode. The measurements were made at 27 °K. Uniaxial compressional stress up to 140 kg/cm² could be applied to the sample by transmission along a lucite rod to the

† The research reported in this paper was partially supported by Air Force Cambridge Research Laboratories, Office of Aerospace Research, under Contract No. F19628-68-C-0133; by the Delegation Generale de la Recherche Scientifique et Technique, and by the Centre National d'Etudes Spatiales.

‡ Permanent address: RCA Laboratories, Princeton, New Jersey 08540.

§ Laboratoire Associé au Centre National de la Recherche Scientifique.

sample which was pocketed by two teflon cups at the ends of the bar. The application of external stress served to remove the ground state degeneracy of the donor level system produced by isolated interstitial Li,[11] thus enabling us to observe an identifiable EPR absorption.[12]

For a second series of measurements, involving irradiations at about 40 °K followed by annealing studies, a Thompson-Houston EPR spectrometer was used. Its operation and performance were essentially the same as the Varian apparatus. However, a special dewar arrangement had to be constructed to accommodate *in situ* low temperature electron irradiations and EPR measurements. This dewar is shown schematically in Figure 1. It is essentially a quartz tubular extension attached to

a metal cryostat[13] with indium gasket vacuum seals. All EPR measurements in this series were made at liquid H_2 temperatures, 20 °K.

2.2. *Electron paramagnetic resonance measurements*
2.2A. *Irradiation at 250 °K*

In Figure 2 we show the EPR spectrum of Li-doped, O-containing Si bombarded by 1 MeV electrons, under conditions described in the figure. It consists of a 4-line group, a single line due to the (Li-O) donor, and, at the extreme left, one of the doublet lines due to phosphorus—the trace contaminant in this material. Note that the conduction electron resonance, indicated by the arrow, has the same g-value (with our instrumentation) as that of the (Li-O) donor. The appearance of the 4-line

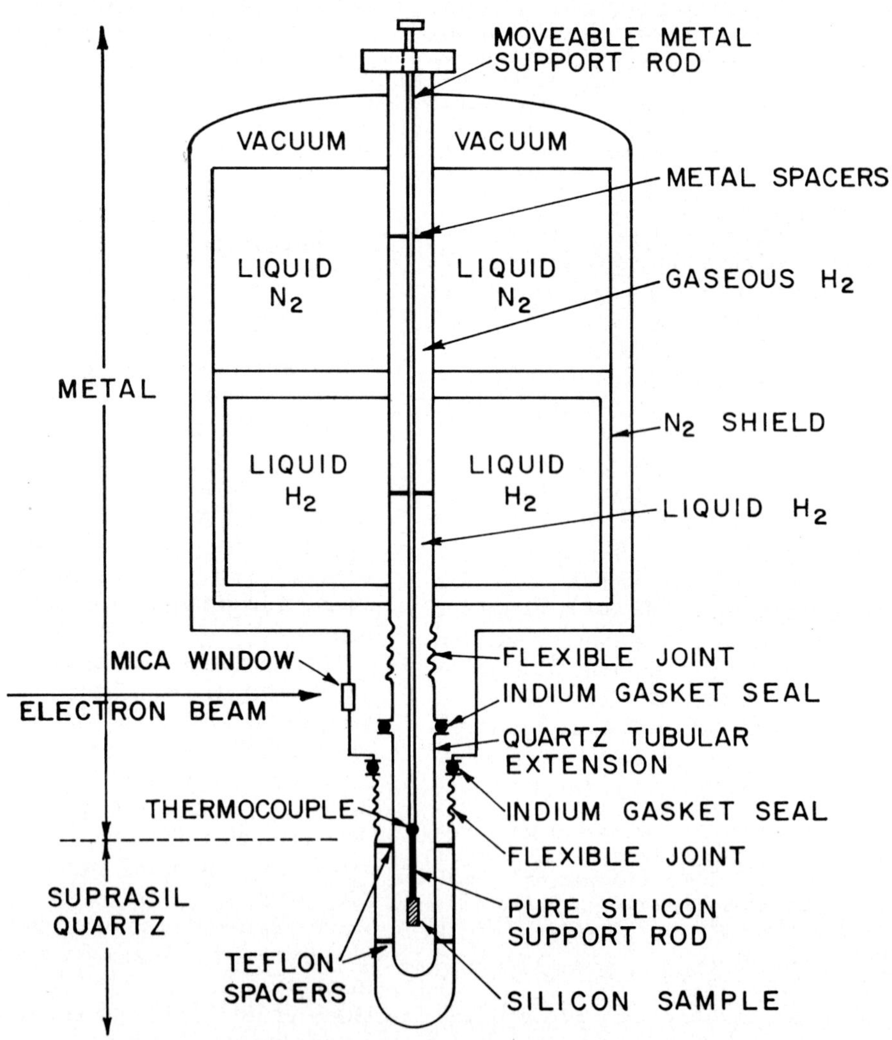

FIG. 1. Schematic representation of metal-and-quartz cryostat.

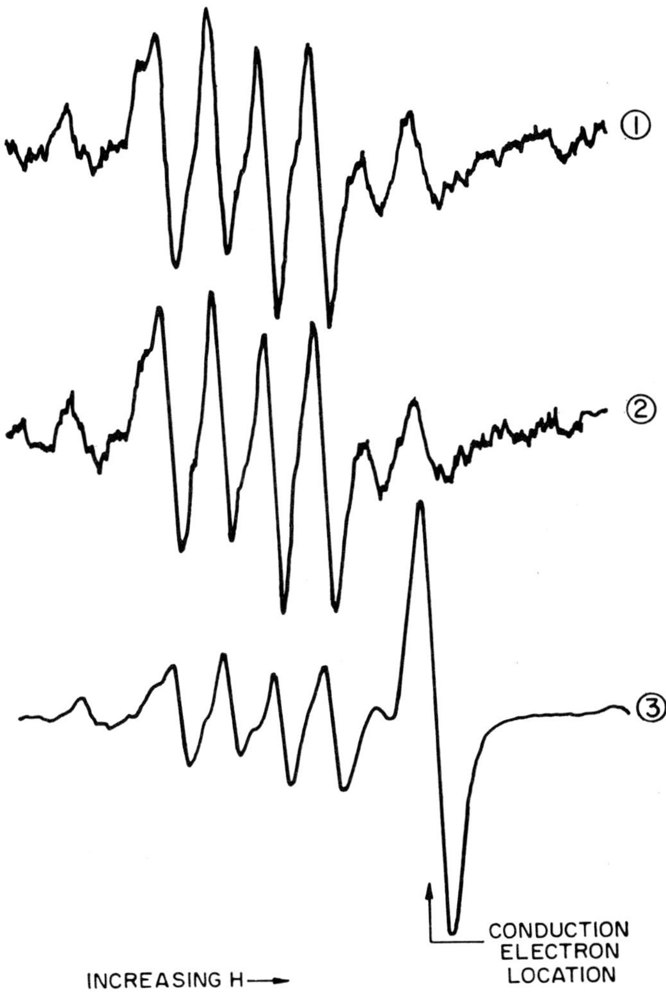

FIG. 2. EPR spectrum of electron-irradiated, O-containing Si showing the $Li^{(7)}$ hyperfine quadruplet and the (Li-O) donor line. Curve 1—zero stress and no light; 2—stress 140 kg/cm² no light; 3—stress as 2 with band-gap light.

group marks it as a hyperfine quadruplet. As the fluence increases, the magnitude of the quadruplet spectrum increases while that of the (Li-O) donor resonance decreases. The 4-line spectrum indicates an interaction with the $Li^{(7)}$ nuclear spin of 3/2. The hyperfine splitting of this spectrum is 3.1 gauss, roughly ten times greater than that reported for the Li shallow donor.[14] This spectrum appears only in Li-doped Si, and only after electron bombardment. We conclude from these data that this spectrum is due to a Li-defect complex, with a net electron spin of 1/2, whose electron is more tightly bound than that of the (Li-O) donor.

Irradiation with bandgap light increases the intensity of the defect resonance and, to a much greater degree, that of the (Li-O) donor resonance. This indicates that since the material was *n*-type, and at least some of the shallow donor states remain (unionized) and can be repopulated,[15] we are seeing the defect in its filled state. Assuming that the bandgap light has essentially saturated the defect levels, an introduction rate of about 0.02 cm⁻¹ can be measured. This is much smaller than the total carrier-removal rate (0.25 cm⁻¹): hence only a fraction of the total defects formed are visible by EPR.

The *g*-value of the 4-line spectrum (at its center) is 2.0046 ± 0.0008. Its angular dependence is shown in Figure 3, where with **H** in the (110) plane we can view all primary crystallographic directions.

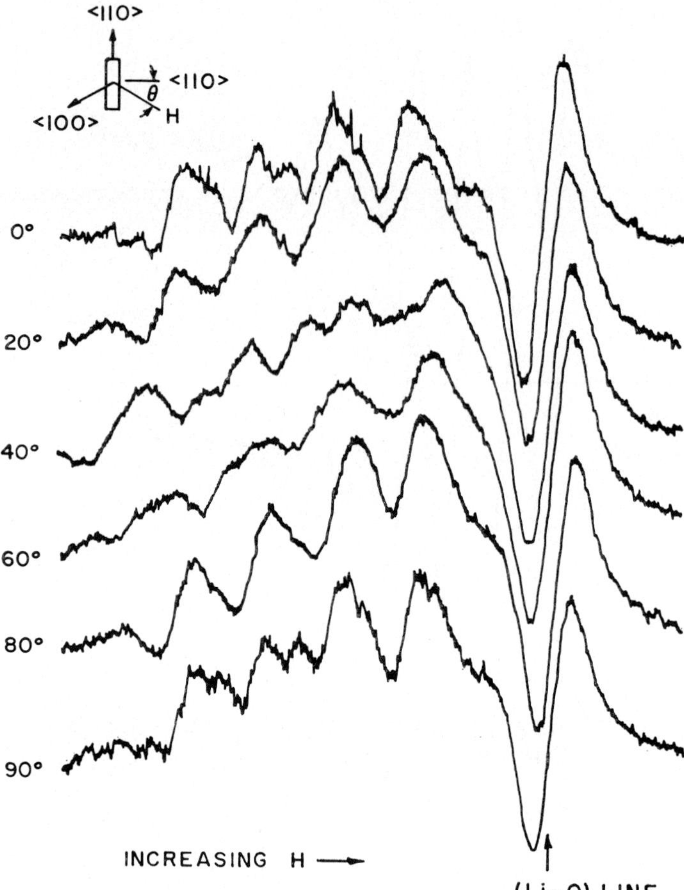

FIG. 3. Angular dependence of EPR spectrum of Figure 2.

The g-tensor is isotropic although the hyperfine interaction is not. Note that there is no indication of axial symmetry so frequently observed for defects and complexes in electron-irradiated Si.[16] We will return to this point later in the discussion. The hyperfine lines seem best resolved in (or near) the $\langle 110 \rangle$ and $\langle 100 \rangle$ directions. In addition, they appear to be doublets in these directions, and may reflect the distribution of the hyperfine axis among three equivalent directions; however, the poor signal-to-noise ratio precludes any firm discussion of these data.

The O-free material shows no trace of the 4-line spectrum after electron irradiation, suggesting that the Li-complex discussed above also involves oxygen. Rather, a single line appears, as shown in Figure 4. It has a g-value of 2.0090 ± 0.0008, is isotropic, and exhibits no measurable structure. Since it appears only in Li-doped material after irradiation, it too, is suggested to be a Li-defect complex. Its introduction rate is ~ 0.01 cm^{-1}, so that again we are seeing only a fraction of the total number of defects formed. It is only slightly affected by light or strain. External stress reduces its peak magnitude, although it does not change its isotropic character. Normally, one might expect the effects of a varying strain field to change the populations of the shifting energy levels involved in the EPR transitions, and so change the appearance of a group of narrow lines, or the shape of a broader line, but such effects were not observable.

As the fluence decreases, the one-line spectrum decreases and the Li donor line increases, as one would expect. At the same time, the observed strain sensitivity of the Li line increases. Note in Figure 4 that the Li line has essentially zero strain sensitivity. As the fluence decreases, this sensitivity increases until at zero fluence, the Li line all but disappears in the absence of externally applied stress. It is as though at high fluences, i.e., in the

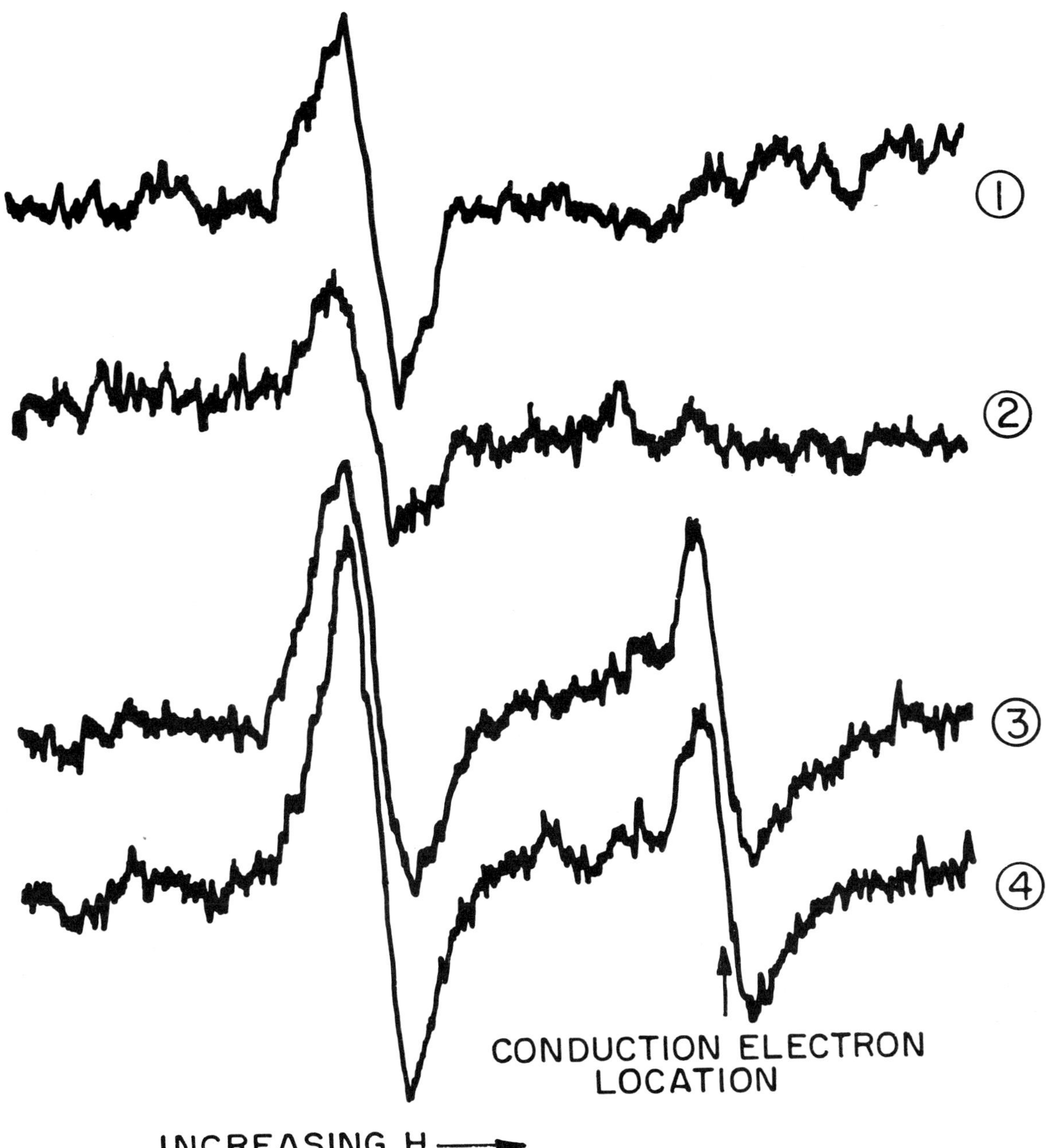

FIG. 4. EPR spectrum of electron-irradiated, O-free Si.   Curve 1—zero stress, no light; 2—stress 140 kg/cm², no light; 3—140 kg/cm², band-gap light; 4—zero stress and band-gap light.

presence of a large number of defects, the Li is already in an appreciable strain field and does not need external stress to lift its ground state's degeneracy.   This may suggest that the reason (Li-O) has a visible EPR absorption is that the neighboring O produces a local region of strain in which the Li finds itself, which is much greater than the random strains ordinarily found in the lattice.[11]

2.2B. *Irradiations at 40 °K*

The purpose here was to produce primary defects at a temperature where neither defect nor

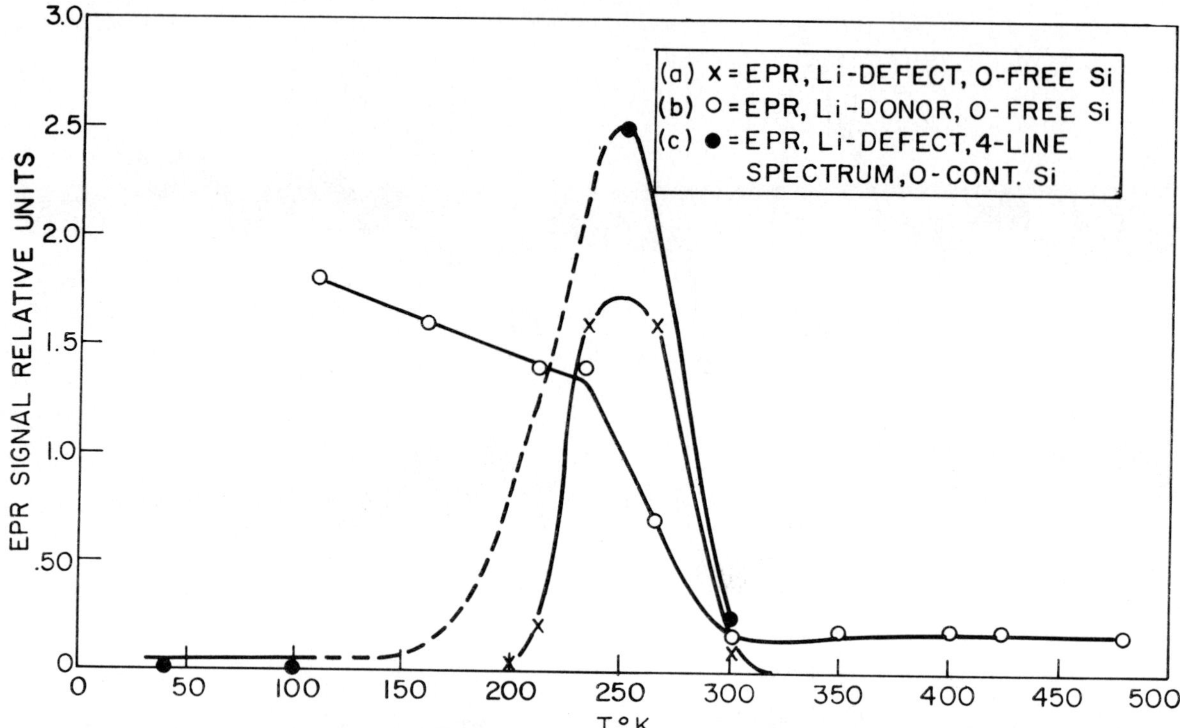

FIG. 5. Temperature dependence of Li EPR lines: (a) Li-defect line in O-free Si. $T_{irr.} = 40\,°K$. (b) Li donor line in O-free Si. $T_{irr.} = 40\,°K$. (c) Li-defect 4 line spectrum in O-containing Si. For (a) and (b) the abscissa is the annealing temperature. For (c) the abscissa is the irradiation temperature. Measurement temperature is 20 °K. The dotted line indicates that for this curve no data points were taken in this region.

impurity was mobile, and then heat the Si to successively higher temperatures before recooling to 20 °K for the EPR measurement. In O-free material, irradiations with 1 MeV electrons at fluences up to $8.4 \times 10^{18}/cm^2$ produced no defect spectra[17] at temperatures up to 475 °K (the limit dictated by the indium vacuum seals, see Figure 1). After irradiation with 1.5 MeV electrons, however, the Li-defect 1-line spectrum, shown originally in Figure 4, begins to appear at about 200 °K. Simultaneously with the growth of this line, a decrease of the Li donor line was observed. At temperature beyond 250 °K, the Li-defect line decreases and finally disappears at about 300 °K. Above 300 °K there is essentially no change in the defect spectra until about 475 °K where a new group of defect lines begin to emerge. However, because of the temperature limitation, these could not be studied in any detail. In Figure 5 we have plotted the temperature behavior of the intensity of both the Li-defect and Li donor lines, where it can be seen that the decrease of the Li donor line is clearly associated with the rise of the Li-defect line.

We next irradiated at 95 °K, hoping to increase the introduction rate of the primary defect(s) which may be involved. We were unsuccessful, however, and even though strong temperature dependent behavior for defect production has been reported for n-type material[1] we did not observe any in this experiment.

In O-containing Si we could not produce the 4-line spectrum under any circumstances when we first irradiated with 1 MeV electrons at a low temperature and then heated the sample. Only by irradiating at higher temperatures could this spectrum be produced. This behavior is also plotted in Figure 5, with the important exception that for the O-containing sample the abscissa is the irradiation temperature.

3. DISCUSSION

Several properties of the radiation damage resulting from 1 to 1.5 MeV electron irradiation of Li-doped Si can be deduced from the experimental data presented here. The role of oxygen is clearly important. Its pairing with Li and with defects, and its effect on Li diffusion are probably all involved. Depending on whether or not O is present, EPR data show that different defects form, carrier-removal data show that total defect introduction rates are different, and the low temperature irradiation data indicate that the formation mechanism of the defect complexes is strongly affected.

The paramagnetic damage centers discussed here are comprised of Li, O and some primary defect(s) in O-containing material, and of Li and some primary defect(s) in O-free material. In addition, temperatures of about 200 °K are required, suggesting that some considerable thermal energy is necessary for their formation. In their observable state, the centers are filled and have an electron spin of 1/2. Their energy levels are almost certainly deeper than the shallow donors. The centers which are observable by EPR are only a fraction of the total number of defects produced.

At room temperature, the EPR spectra and the carrier concentration change with time. These changes are different for O-containing and O-free material. In the O-containing case the damage spectrum, the (Li-O) donor spectrum, and the carrier concentration all decrease with time. This strongly suggests that deeper levels, not visible by EPR, are forming. The fact that bandgap light does not restore, or even enhance, the spectra suggests that the original damage levels and donor levels (seen in Figure 2) have disappeared (or at least been changed) rather than been merely de-populated. One likely mechanism for this is that Li diffuses and complexes with an existing defect center, even one which may already contain one Li atom. Such a mechanism is consistent with Li's mobility at room temperature, and is also in line with previous suggestions for the formation of Li-defect complexes at room temperature.[4]

In O-free material the situation appears to be more complex. Two behaviors have been observed. The first is the disappearance of the EPR damage spectrum with time and, simultaneously, the growth of both the donor spectrum and the carrier concentration. This suggests an actual dissociation of the Li-defect complex, which makes more Li available to resume its role as a shallow donor. The second, observed in some samples, is the simultaneous decrease in the Li-defect EPR, the shallow donor EPR, and the carrier concentration. Here, just as in the O-containing case, the evidence points to the formation of deeper defect levels which removes Li donors in the process.[18] Here too, the diffusion of Li may be important, and the

fact that these changes take place more rapidly than analogous changes in O-containing material is consistent with the more rapid diffusion of Li in O-free material.[9]

Our discussion concludes with an attempt to define the paramagnetic damage centers described here and the way in which they form. It has been previously reported that phosphorus-doped[1] or Li-doped,[7] *n*-type Si which contains O has a reverse annealing stage (in total carrier-removal and reciprocal mobility) starting at about 200 °K, i.e., a defect begins to form at this temperature and the formation does not depend on the dopant. We show this behavior in Figure 6. We show our own data, described earlier, in Figure 5. Examination of these two figures suggests that all these results originate from the same, or closely related, threshold process. Since, at 200 °K none of the impurities is mobile, we suggest that the formation of these defect complexes is due to the fact that some lattice defect has become available and sufficiently mobile to be trapped by an impurity or an impurity-involved complex.

The identification of the defect whose availability and/or mobility allows it to be trapped is a matter of some speculation. The low temperature electron irradiation produces primary defects, i.e., vacancies, divacancies, and interstitials. Among these the weight of existing evidence seems strongly to favor the interstitial. The divacancy itself cannot be the primary defect involved since it is known not to become mobile until about 500 °K.[19,20] Furthermore, it should not be the vacancy, for if it were, we would expect to see the defect complex form at a much lower temperature than 200 °K. On the other hand, some evidence does exist which suggests that in *n*-type Si the interstitial becomes mobile at temperatures of about 140 to 170 °K,[21] and that it can complex with O.[22] These facts tend to point to the interstitial. Finally, consider that the *g*-values of the EPR spectra of both Li-defect complexes are isotropic; specifically they do not have the typical axial symmetry found for defects involving broken or dangling Si bonds (e.g., vacancies). We consider the isotropic character of the *g*-values to be supportive evidence for the interstitial rather than the vacancy being the primary lattice defect involved. Thus, we suggest that the defects we are seeing here are a (Li-O-interstitial) complex in O-containing material, and a (Li-interstitial) complex in O-free material.

If the interstitial can indeed complex both with O[22] and with (Li-O), then this competition for the

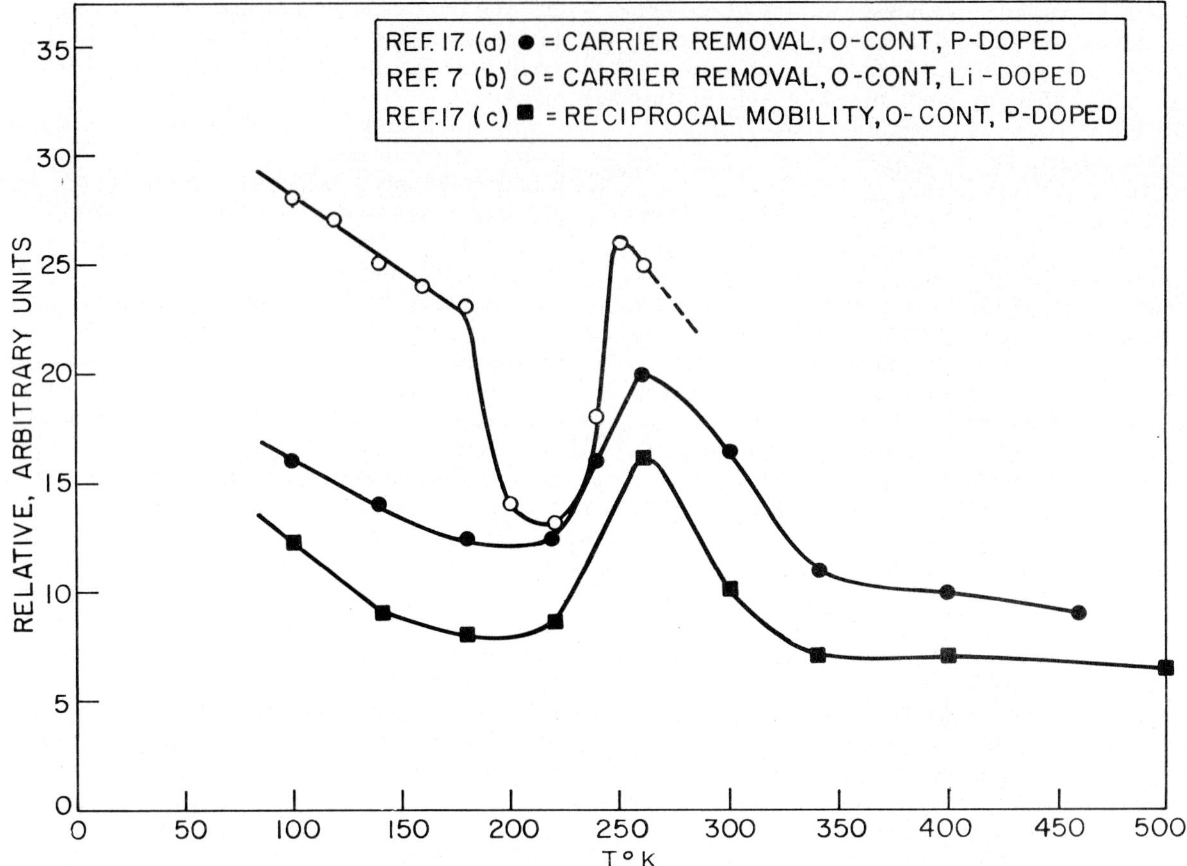

FIG. 6. Annealing curves in terms of unannealed damage vs. temperature. Note that all curves show 'reverse annealing', i.e., defect formation, at about 250 °K.

interstitials' capture could explain why in O-containing material only irradiation at the higher temperature (e.g., $\sim 250\,°K$) enables us to observe the (Li-O-interstitial) complex. If we irradiated at lower temperatures (40 °K) the interstitial would have a chance to be preferentially trapped by O before being trapped by a (Li-O) pair. Note that in O-free material where such competition would not be present, we *can* irradiate at 40 °K and still see the formation of the (Li-interstitial) complex at 200 °K.

[In the light of very recent reports of the behavior of carbon in silicon reported at this meeting (R. C. Newman and H. R. Bean; A. Brelot and J. Charlemagne) it should perhaps be noted that for the float-zone material used in this paper, the carbon concentration, based on localized mode optical measurements, is estimated to be less than $\sim 5 \times 10^{15}/cm^3$. (A. Brelot, private communication.)]

ACKNOWLEDGEMENTS

The author is very pleased to acknowledge the timely assistance of F. Kolondra who operated the Van de Graaff accelerator. He would also like to thank G. Brucker and B. Faughnan for stimulating and helpful discussions, and N. Di Guiseppe for his technical assistance. He would also like to express his deep appreciation to the members of the Groupe de Physique des Solides of the Ecole Normale Superieure, for their professional and personal hospitality—with special words of thanks to P. Baruch, A. Brelot and M. Squelard for their time and helpfulness in many situations.

REFERENCES

1. F. L. Vook and J. L. Stein, *Radiation Effects in Semiconductors: Proceedings of the Santa Fe Conference* (Plenum Press, New York, 1968), p. 99.
2. E. L. Elkin and G. D. Watkins, *Phys. Rev.*, **174**, 881 (1968).

3. B. Goldstein, et al., Final Report, Contract NAS 5-9131 (May 1966).
4. J. Wysocki, et al., *Appl. Phys. Letters*, **9**, 44 (1966).
5. V. Vavilov, *Radiation Damage in Semiconductors*, Dunod, et al., Paris (1965).
6. R. C. Young, et al., *J. Appl. Phys.*, **40**, 271 (1969).
7. G. Brucker, *Phys. Rev.*, **183**, 712 (1969).
8. B. Goldstein, *Phys. Rev. Letters*, **17**, 1 (1966).
9. E. M. Pell, *Phys. Rev.*, **119**, 1222 (1960).
10. G. D. Watkins, et al., *J. Appl. Phys.*, **30**, 1198 (1959).
11. R. L. Aggarwal, et al., *Phys. Rev.*, **138**, A882 (1965).
12. G. D. Watkins, *Bull. Am. Phys. Soc.*, **10**, 303 (1965).
13. J. Bourgoin, et al., *Rad. Effects* (to be published).
14. G. Feher, *Phys. Rev.*, **114**, 1219 (1959).
15. The Fermi level here must be somewhere within $\lesssim kT$ (0.0023 eV) of the shallow donor energy.
16. G. Watkins, *Radiation Damage in Semiconductors*, Proceedings of Conf. at Rayaumont, Dunod, et al., Paris (1965).
17. The overall sensitivity of our instrumentation was such that we should have seen defect introduction rates as low as 10^{-4} cm^{-1}.
18. Hall measurements (of carrier concentration and mobility) have recently been made which indicate that both Li-defect formation and Li-defect dissociation can take place with time at room temperature in Li-doped, electron-irradiated Si (G. Brucker, private communication).
19. G. Watkins and J. Corbett, *Phys. Rev.*, **138**, A543 (1965).
20. L. Cheng, et al., *Phys. Rev.*, **152**, A761 (1966).
21. G. Watkins, *Radiation Effects on Semiconductor Components*, Proc. of the Toulouse Conf., Journees D'Electronique (1967).
22. R. E. Whan, *J. Appl. Phys.*, **37**, 3378 (1966).

DISCUSSION

Question (MCKEIGHEN) How did you determine that the oxygen content in the floating zone samples was only 10^{13} oxygen/cm^3? I understand that it is difficult to determine oxygen content much below 10^{16} oxygen/cm^3.

Answer (GOLDSTEIN) Yes, it is for optical methods. My figure is based on the absence of any EPR absorption in strain-free, Li-doped, float-zone silicon. If oxygen is present (e.g., in pulled material) a very prominent EPR line due to the (Li-O) shallow donor is easily seen whose spin density correlates quite well with the oxygen concentration.

Question (CORELLI) What is the fraction of total defects you are observing in your EPR measurements?

Answer (GOLDSTEIN) The ratio of the 'spin density' of the observed EPR spectra to the total carrier removal is ~ 0.02.

LITHIUM—AN IMPURITY OF INTEREST IN RADIATION EFFECTS OF SILICON†

J. A. NABER, H. HORIYE AND B. C. PASSENHEIM

Gulf Radiation Technology, a Division of Gulf Energy & Environmental Systems, Company, San Diego, California, U.S.A.

The introduction and annealing of defects produced in lithium-diffused float-zone n-type silicon by 30-MeV electrons and fission neutrons are studied. The introduction rate of recombination centers produced by electron irradiation is dependent on lithium concentration and for neutron irradiation is independent of lithium concentration. The introduction rate of Si-B1 centers also depends on the lithium concentration. The annealing of electron- and neutron-produced recombination centers, Si-B1, and Si-G7 centers in lithium-diffused silicon occurs at much lower temperatures than in nondiffused material.

1. INTRODUCTION

The effects of impurities on the production and annealing of radiation-induced defects in silicon have been studied extensively. It is known that impurities do play a significant role in the production and annealing of point defects in electron-irradiated silicon.[1–4] Lithium as an impurity has been of particular interest because of its high mobility, even at temperatures as low as 300 °K. Lithium diffused into n-type silicon exists interstitially and acts electrically as a donor.[5] The initial observations of lithium in radiation effects were made by Vavilov.[6] Further work has been performed by Wysocki,[7] Carter,[8] Bruckner,[9] and Young.[10]

The present work describes minority-carrier lifetime, electron-spin resonance, infrared absorption, and electrical conductivity measurements performed on lithium-diffused bulk silicon after irradiation with 30-MeV electrons and fission neutrons at 295 °K. The experiments were performed on both float-zone and quartz-crucible grown n-type silicon.

2. EXPERIMENTAL TECHNIQUES

The electrical conductivity and minority carrier lifetime measurements were performed by standard methods. The electrical conductivity measurements were performed by the four-probe method. The minority carrier lifetimes were measured by both the photoconductivity decay[11] and steady-state photoconductivity[12] techniques. All lifetime measurements were taken at injection levels of <1 per cent. The cryostat used for the electrical conductivity and minority carrier lifetime measurements permitted these measurements to be made simultaneously on the same sample immediately after irradiation. Annealing experiments were also performed in this cryostat. The electron spin resonance (ESR) measurements were performed by the use of a superheterodyne spectrometer at 9.2 GHz with magnetic field modulation at 100 Hz. These measurements were made at 20 °K. The infrared absorption measurements were performed with a Perkin-Elmer Model 112 double-pass prism spectrometer.

The samples were vacuum float-zone (FZ) and quartz-crucible (QC) grown phosphorus-doped silicon with initial resistivities ranging from 0.4 ohm-cm to 10^4 ohm-cm. The samples were lithium-diffused by the paint-on[7] or lithium-tin[13] bath diffusion techniques. The donor concentrations after lithium diffusion ranged from about 10^{14} to 10^{18} cm^{-3}. Contacts were soldered to the samples with lead-tin solder after gold preforms containing arsenic were alloyed to the silicon by heating for 7 minutes at 450 °C.

Four microsecond wide pulses of 30-MeV electrons from the Gulf Radiation Technology (Rad Tech) linear accelerator (Linac), and fission neutrons from the Rad Tech accelerator pulsed fast assembly (APFA) were used to introduce damage into the samples.

It should be noted that after 30-MeV electron and neutron irradiations at 295 °K, measurements

† This work was performed for the Jet Propulsion Laboratory, California Institute of Technology, sponsored by the National Aeronautics and Space Administration under Contract NAS7-100.

of the minority carrier lifetime were made immediately. The ESR irradiations were performed at both 150 and 295 °K, while the infrared irradiations were performed near 150 °K. The samples used in the ESR and infrared absorption experiments were stored at liquid-nitrogen temperature immediately after irradiation until they were measured.

3. EXPERIMENTAL INVESTIGATIONS

3.1. *Minority-carrier lifetime*

3.1A. *Degradation*

The minority carrier lifetime is inversely proportional to the density of recombination centers, which can be lattice defects produced by particle irradiation. The 30-MeV electron irradiations were performed on lithium-diffused FZ n-type silicon with resistivities between 11 and 0.25 ohm-cm. This corresponds to donor (lithium densities) of 4.5×10^{14} and 2.5×10^{16} cm^{-3}, respectively.

The minority-carrier lifetime degradation constant K, defined by

$$\frac{1}{\tau} = \frac{1}{\tau_0} + K\Phi,$$

where τ_0 is the lifetime prior to irradiation and τ is lifetime after a fluence Φ, was determined.

The degradation constant for samples with lithium density of 1.5×10^{16} cm^{-3} and fluence of 10^{13} e/cm^2 was $(1.8 \pm 0.3) \times 10^{-7}$ cm^2/sec. For lithium density of 4.5×10^{14} cm^{-3} and fluence of 10^{12} e/cm^2, the degradation constant was $(5.0 \pm 2.0) \times 10^{-8}$ cm^2/sec. The inverse lifetime as a function of fluence is linear in both cases.

The degradation constant for samples irradiated with fission neutron fluence of 2×10^{10} n/cm^2 ($E > 10$ keV) and with lithium density of 1.5×10^{15} cm^{-3} was $(6.4 \pm 0.4) \times 10^{-6}$ cm^2/sec. Again, the inverse lifetime as a function of the above fluence was linear.

3.1B. *Annealing*

Isothermal and isochronal annealing were performed on lithium-diffused float-zone n-type silicon after electron and neutron irradiations. The minority-carrier lifetime is used as the damage indicator. For isochronal annealing, the unannealed fraction of the annealable damage can be expressed in terms of the minority-carrier lifetime by

$$\frac{\dfrac{1}{\tau_T} - \dfrac{1}{\tau_f}}{\dfrac{1}{\tau_0} - \dfrac{1}{\tau_f}},$$

where τ_f is the lifetime after complete annealing, τ_0 is the initial lifetime before annealing, and τ_T is lifetime after annealing to temperature T. For isothermal annealing, τ_T is replaced by τ_t where τ_t is the lifetime at any time, t. The isochronal annealing consisted of 5-minute anneals at each temperature. In each case, the samples were annealed until at least 95 per cent of the damage was removed. Figure 1 shows the unannealed fraction of the annealable damage as a function of temperature for an isochronal anneal of an electron-irradiated sample, while Figure 2 shows the isothermal annealing at 295 °K for the same lithium-diffused n-type FZ sample with lithium density of 1.7×10^{16} cm^{-3}. The annealing is first-order. The analysis of both the isochronal and isothermal annealing data yields an activation energy of 0.80 eV ± 0.10 eV. The effective frequency factor for this annealing is 10^{10} sec^{-1}.

A similar set of experiments was performed on electron-irradiated lithium-diffused n-type FZ silicon with a lithium density of 4.5×10^{14} cm^{-3}. The annealing was first-order and yielded an activation energy of 0.75 ± 0.10 eV and an effective frequency factor of 5×10^7 sec^{-1}.

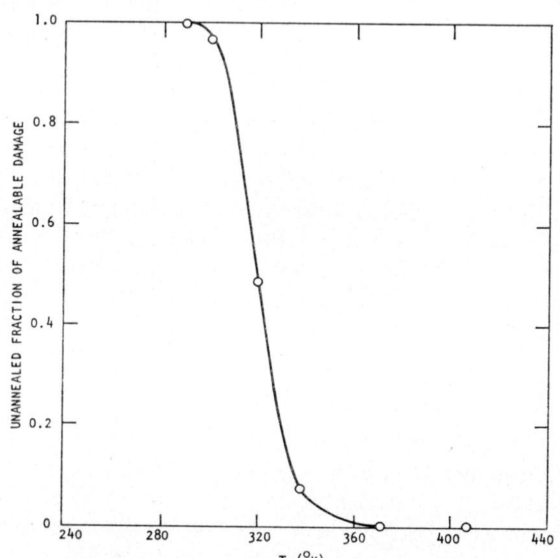

FIG. 1. Isochronal annealing of lithium-diffused FZ n-type silicon after 30–MeV electron irradiation, lithium density of 1.7×10^{16} cm^{-3} and fluence of 10^{13} e/cm^2.

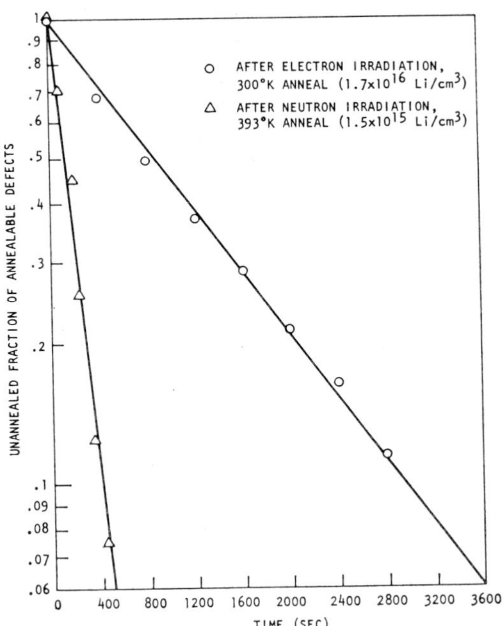

FIG. 2. Unannealed fraction of annealable defects as a function of time, lithium-diffused FZ n-type silicon.

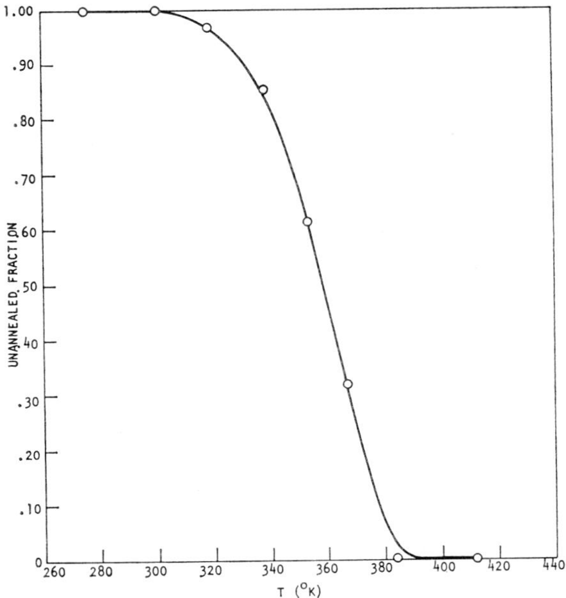

FIG. 3. Unannealed fraction of defects vs. isochronal anneal temperature for neutron-irradiated, lithium-diffused n-type silicon with lithium concentration of 1.5×10^{15} cm^{-3} and neutron fluence 2×10^{10} n/cm^2 ($E > 10$ keV).

The isochronal and isothermal annealing of the neutron-irradiated lithium-diffused n-type FZ silicon (lithium concentration 1.5×10^{15} cm^{-3}) are shown in Figures 3 and 2, respectively. The annealing is first-order, with an activation energy of 0.69 ± 0.02 eV and an effective frequency factor of $(1.5 \pm 0.8) \times 10^7$ sec^{-1}.

3.2. *Electron spin resonance and infrared absorption*

Electron spin resonance and infrared absorption measurements of lithium-diffused silicon are valuable for the insight they give into the effects of lithium on known impurity defect centers such as the Si-B1 (vacancy-oxygen), Si-G7 (divacancy) and Si-G8 (phosphorus-vacancy). These known defect centers have been studied by various investigators and their properties well documented.[14]

3.2A. *Si-B1 center*

Phosphorus-doped (10^{16} cm^{-3}) quartz-crucible silicon samples were diffused to lithium donor densities of 10^{16} cm^{-3} and 4×10^{17} cm^{-3}. Before irradiation, an isotropic resonance line with $g = 1.998$, which we attribute to the LiO donor, was observed in these samples.[15,16] The density of this resonance line was proportional to the lithium concentration. No phosphorus donor signal could be found even though 10^{16} P/cm^3 were present.

The heavily diffused samples (4×10^{17} Li/cm^3) were irradiated with 30-MeV electrons at 295°K to fluences between 1 and 2×10^{17} e/cm^2. Resistivity measurements indicated carrier removal rates of 1.7 ± 0.2 cm^{-1}; the LiO resonance decreased at a rate of 1.5 ± 0.5 cm^{-1}, but the average B1 introduction rate was only 0.025 ± 0.01 cm^{-1}. However, the lightly diffused samples (1×10^{16} Li/cm) irradiated with 5×10^{16} e/cm^2 yielded a B1 introduction rate of 0.14 cm^{-1}, which is comparable to the rate obtained for nondiffused silicon.[17] This indicates the B1 center production is considerably inhibited in samples with large lithium concentrations. Figure 4 shows the isochronal annealing results for the highly diffused samples, and shows simultaneous changes in conductivity, LiO and B1 center concentration. It should be noted that, simultaneous with the B1 center anneal at 325°K, there was a decrease in the electrical conductivity. Above 373°K, the concentration of Si-B1 centers remained undetectable, whereas the electrical conductivity increased. The high conductivity and simultaneous presence of the LiO resonance is evidence that the disappearance of the B1 center is not due to a depopulation of its electrical level.

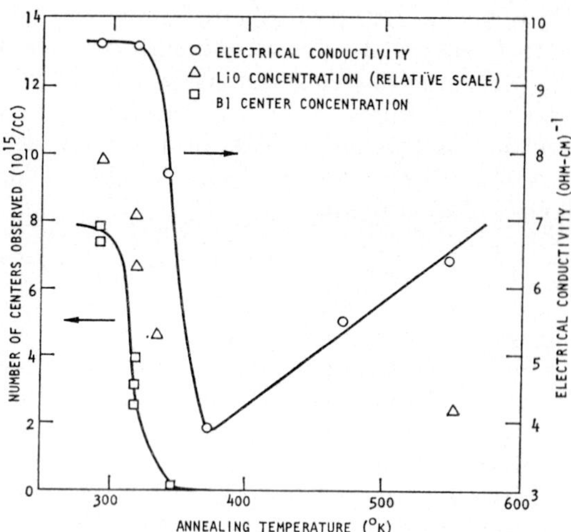

FIG. 4. Isochronal annealing of electrical conductivity, LiO concentration, and B1 center concentration after 30-MeV electron irradiation at 295 °K.

3.2B. *Si-G7 center*

Observation of the divacancy in *n*-type silicon by ESR techniques is complicated by Fermi level requirements of the divacancy being in a paramagnetic charge state. However, infrared absorption bands at 1.8, 3.3 and 3.9 μm have been observed and correlated with the divacancy.[18] Optical absorption samples were fabricated from 10^4 ohm-cm FZ silicon diffused with lithium to 0.15 ohm-cm ($n_0 \simeq 5 \times 10^{16}$ cm³). For comparison, optical samples of the non-diffused 10^4 ohm-cm starting material were also fabricated. These were irradiated with 30-Mev electrons to a fluence of 2.5×10^{17} e/cm². During irradiation, the sample temperature never exceeded 150 °K, and the samples were stored and optically investigated at 77 °K until the annealing study began.

A band at 1.8 μm is attributed to the divacancy, and its introduction rate is the same in both the lithium-diffused and nondiffused material. After annealing at 300 °K for thirty minutes, the 1.8-μm band in the lithium-diffused samples had practically disappeared and new bands near 1.4 and 1.65 μm appeared. These lines have been previously observed by other investigators.[10] These results indicate that divacancies are produced in irradiated lithium-diffused silicon (lithium density of 5×10^{16} cm⁻³) at a rate comparable to that in nondiffused silicon, but that in diffused silicon the divacancy anneals, perhaps to become another defect, responsible for the 1.4- to 1.65-μm bands, near 300 °K

compared with the 500 to 675 °K anneal observed in nondiffused silicon.[18,19]

3.2C. *Si-G8 center*

Phosphorus-doped float-zone silicon samples with initial phosphorus densities of 1×10^{16} cm⁻³ were lithium-diffused to densities of 1×10^{16} and 3.5×10^{16} cm⁻³. ESR measurements were performed on these samples. As in the Si-B1 center studies, the lithium oxide resonance was introduced as a result of the lithium diffusion. These samples were irradiated below 150 °K with 30-MeV electrons to fluences from 1×10^{17} to 2.5×10^{17} cm⁻², respectively. The introduction rate of the Si-G8 center was found to be comparable to nondiffused silicon ($\sim$0.15 cm⁻¹).

The annealing of the Si-G8 center took place in the temperature range centered near 450°K. This annealing of the Si-G8 center in the lithium-diffused silicon occurs over the same temperature range as in nondiffused silicon.[20]

4. DISCUSSION OF RESULTS

4.1. *Introduction of defects*

The lifetime degradation constant due to irradiation with 30-MeV electrons of lithium-diffused float-zone *n*-type silicon depends on the lithium concentration. For lithium densities of about 10^{16} cm⁻³, the degradation constant [$(1.8 \pm 0.3) \times 10^{-7}$ cm²/sec] is twice as large[17] as it is in nondiffused silicon of the same carrier concentration. At lithium concentrations of 10^{14} cm⁻³, the degradation constant is at most 20 per cent larger[17] than for nondiffused silicon of the same donor concentration.

The neutron situation is different in that the degradation constant is the same in lithium-diffused and nondiffused *n*-type silicon[21] of the same resistivity.

The above results are consistent with a defect picture in which the recombination centers in electron-irradiated materials are mainly point defects, while the recombination centers in neutron-irradiated material are clustered in nature. The dependence of the degradation constant on lithium in electron-irradiated material implies that these point defects contain lithium or are affected in their production by lithium. The lithium independence of the neutron-produced recombination centers is consistent with the impurity independence for neutron damage, which is due to the clustered nature of neutron damage.

The introduction rate of the Si-B1 in 30-MeV electron-irradiated lithium-diffused silicon is dependent on the presence of lithium. The Si-B1 center introduction rate is at least a factor of 8 smaller at lithium concentrations of 4×10^{17} cm^{-3} and fluences of 2×10^{17} e/cm^2 and equal to the introduction rate in nondiffused silicon[22] (0.2 cm^{-1}) for lithium concentrations of 10^{16} cm^{-3} and fluences of 5×10^{16} cm^{-3}.

The effect of lithium on the introduction rate of recombination centers and known defect centers is evidence that the presence of lithium determines the nature and quantity of point defects produced by irradiation.

There are two possible causes for the decreased introduction rate of known defect centers in lithium-diffused silicon. If lithium-impurity pairing is possible, then the impurities are tied up by the lithium and not free to combine with the irradiation-produced vacancies. An alternate reason for the decrease in Si-B1 centers may be the greater probability of the irradiation-produced vacancies being captured by lithium rather than by the impurities. The mobility of the irradiation-produced vacancies is much larger than that of the lithium. Therefore, vacancy diffusion governs the nature of the point defects formed, and it is likely that the impurities are competing with lithium for the vacancy. The lithium, due to either its density or large cross section, is more successful in capturing the vacancies. This will decrease the number of known impurity centers and at the same time increase the number of lithium-related defects, which may account for the increase in recombination center production in lithium-diffused material. The independence of the introduction rate of the divacancy on lithium is consistent with the model of the divacancy as an intrinsic defect.

4.2. *Annealing defects*

The isothermal and isochronal annealing of the electron- and neutron-produced recombination centers indicated an annealing process which was first-order. The activation energies for the annealing were 0.80 eV $\pm$ 0.10 eV for samples with lithium concentrations of 2.5×10^{16} cm^{-3}, and 0.75 ± 0.10 eV for lithium concentrations of 4.5×10^{14} cm^{-3}. The effective frequency factors were 1×10^{10} sec^{-1} and 5×10^7 sec^{-1}, respectively. The neutron-produced recombination centers annealed with an activation energy of 0.69 ± 0.02 eV and a frequency factor of 1.5×10^7 sec^{-1}.

The activation energies for all the above annealing processes are not felt to be inconsistent with the 0.66 ± 0.05 eV for the activation energy of lithium diffusion in silicon.[23–25] The effective frequency factors, however, depend on the lithium concentration. These annealing kinetics indicate a process involving long-range migration[26] of lithium to the recombination centers.

The above annealing should be contrasted to the annealing of recombination centers in electron- and neutron-irradiated nonlithium-diffused *n*-type float-zone silicon. The recombination centers produced by 30-MeV electrons anneal in a broad temperature range between 350 and 500 °K,[27] while neutron damage anneals over a temperature range between 300 and 500 °K.[21] From these data, one would expect less than 10 per cent recovery for non-lithium-diffused material subject to the same annealing schedule as used for the lithium-diffused material.

The annealing temperature of impurity-related defect centers is lowered by the presence of lithium. The Si-B1 center in nonlithium-diffused material[28] anneals near 600 °K; when lithium is present in the lattice, this annealing takes place at temperatures near 325 °K. The annealing temperature of the Si-G7 center is similarly lowered in lithium-diffused silicon.

The lowering of the annealing temperature and the similarity with the minority carrier lifetime annealing implies that both the Si-B1 and Si-G7 center are annealed by the long-range diffusion of lithium.

5. SUMMARY

The lifetime degradation constants in lithium-diffused float-zone *n*-type silicon irradiated with 30-MeV electron depends on the lithium concentration and is larger than in equivalent nondiffused silicon. Meanwhile, in neutron-irradiated lithium-diffused *n*-type silicon, the degradation constant is the same as for nondiffused material of the same donor concentration.

These results are consistent with the point defect and clustered nature of damage produced by electrons and neutrons, respectively. In lithium-diffused material, the point defects contain lithium or are affected in their production by lithium, while the clusters are independent of lithium.

The introduction rates of known defect centers (Si-B1) are decreased in lithium-diffused material. This is due to either lithium-impurity pairing or the greater probability of irradiation-produced vacancies combining with lithium than with other impurities.

The annealing of both electron and neutron recombination centers and impurity-related defects (Si-B1, Si-G7) are dependent on lithium. The kinetics of annealing indicate that long-range diffusion of lithium to the defect center is responsible for the annealing.

REFERENCES

1. R. F. Bass and O. L. Curtis, Jr., *IEEE Trans. Nucl. Sci.*, NS-15, 47 (1968).
2. N. Hirata, M. Hirata, H. Saito and J. H. Crawford, Jr., *J. Appl. Phys.*, 38, 2433 (1967).
3. R. V. Tauke and B. J. Faraday, *J. Appl. Phys.*, 37, 5010 (1966).
4. H. J. Stein and F. L. Vook, *Phys. Rev.*, 163, 790 (1967).
5. E. M. Pell, *Symposium on Solid State Physics and Tele-communications*, I, 261 (1960).
6. V. S. Vaivlov, I. V. Smirnova and V. A. Chapnin, *Soviet Phys.–Solid State*, 4, 2469 (1963).
7. J. J. Wysocki, *IEEE Trans. Nucl. Sci.*, NS-14, 103 (1967).
8. J. R. Carter, Jr. and R. G. Downing, *Conference Record of the Sixth Photovoltaics Specialists Conference*, 3, 145 (March 1967).
9. G. J. Brucker, *Phys. Rev.*, 183, 712 (1969).
10. R. C. Young, J. W. Westhead and J. C. Corelli, *J. Appl. Phys.*, 40, 271 (1969).
11. IRE Standards on Measurement of Minority-Carrier Lifetime, *Proc. IRE*, 49, 1293 (1961).
12. B. C. Passenheim and J. A. Naber, *Rad. Effects*, 2, 229 (1970).
13. H. Reiss and C. S. Fuller, *Trans. AIME*, 206, 276 (1956).
14. G. D. Watkins, *Proceedings of Seventh International Conference on the Physics of Semiconductors, Radiation Damage in Semiconductors*, Paris (1965), p. 67.
15. G. Feher, *Phys. Rev.*, 114, 1219 (1959).
16. A. Honig and A. F. Kip, *Phys. Rev.*, 95, 1686 (1954).
17. Unpublished data, Gulf Radiation Technology.
18. L. J. Cheng, J. C. Corelli, J. W. Corbett, and G. D. Watkins, *Phys. Rev.*, 152, 761 (1966).
19. G. D. Watkins and J. W. Corbett, *Phys. Rev.*, 138, A543 (1965).
20. H. Saito, M. Hirata and T. Horiuchi, *Phys. Soc. of Japan*, 18, III, 246 (1963).
21. H. J. Stein, *J. Appl. Phys.*, 37, 3382 (1966).
22. E. G. Wikner and H. Horiye, *IEEE Trans. Nucl. Sci.*, NS-13, 18 (1966).
23. E. M. Pell, *J. Appl. Phys.*, 31, 291 (1960).
24. C. S. Fuller and J. C. Severeins, *Phys. Rev.*, 96, 21 (1960).
25. J. P. Maita, *J. Phys. Chem. Solids*, 4, 68 (1958).
26. A. C. Damask and O. L. Curtis, Jr., *IEEE Trans. Nucl. Sci.*, NS-15, 47 (1968).
27. R. F. Bass and O. L. Curtis, Jr., *IEEE Trans. Nucl. Sci.*, NS-15, 47 (1968).
28. G. D. Watkins and J. W. Corbett, *Phys. Rev.*, 121, 1001 (1961).

DISCUSSION

Comment (CORELLI) We have also found that, in oxygen-rich Si irradiated by $\sim$40 MeV electrons doped with lithium to $\geqslant 10^{16}$ cm^{-3}, the A-center infrared active band at $12\,\mu$ quickly anneals (i.e., disappears) after 15 minutes at $\sim$110 °C with the appearance of a new band at $\sim$10 μ. This implies that the Li atom moves to the A-center at $\sim$110 °C and produces a new defect complex, since the A-center normally requires a heat treatment to $\sim$325–250 °C for its recovery. This result agrees with what you report.

Comment (KIMERLING) Your lifetime degradation data indicates that Li does not interact with the neutron-induced defect cluster. Our carrier removal data on room temperature, neutron-irradiated, Li-doped, float-zoned, Si samples show a strong dependence on Li concentration, tending to always show higher carrier removal rates than observed with conventional substitutional donor impurities. Typical values are 13 cm^{-1} for 7×10^{14} Li/cm^3 to 63 cm^{-1} for 4×10^{15} Li/cm^3.

Comment (CURTIS) The dependence of carrier removal rate and independence of lifetime degradation on Li is consistent with the general observation of the dependence of these properties on impurities. This is because recombination is dominated by disordered regions, while isolated defects are important to carrier removal.

ANNEALING OF γ-RAY IRRADIATED LITHIUM COMPENSATED p-TYPE SILICON

N. B. URLI

Institute 'Rudjer Bošković', Zagreb, Yugoslavia

A floating zone 100 ohm-cm boron-doped silicon, homogeneously compensated with lithium by the drift technique, was aged for several months at room temperature prior to irradiation in the Co^{60}-gamma source at 35 °C. After aging such a material turned to be slightly undercompensated p-type. As a result of irradiation, there was a decrease in the concentration of free lithium ions directly proportional to the gamma fluence. Two deep donor levels, located at $E_v + 0.29$ eV and $E_v + 0.32$ eV were detected immediately after irradiation and attributed to the Li-V or Li-V-O complexes. A relatively fast precipitation of free lithium ions, with a time constant inversely proportional to the gamma fluence, was observed up to 30 to 40 days after irradiation. A short annealing at temperatures from 230 to 275 °C almost restored the initial free lithium ion concentration. Heating at higher temperatures caused lithium outdiffusion to the surface. The present results were compared with those already published and discussed in view of the probabilities for the formation of different defect complexes.

1. INTRODUCTION

In spite of much efforts which have been directed towards better understanding of radiation defects in p-type silicon, the situation is still far from being satisfactory.[1-4] It seems that even a presence of a very small concentration of various impurities strongly affects the formation of a certain defect configuration as well as the post-irradiation behavior of the irradiated p-type silicon.

In this work we decided to study more carefully one aspect of that problem, namely, the behavior of lithium as a compensating impurity in a boron-doped silicon. It was actually an extension of the work reported previously[5] with the initial conditions somewhat changed, however, having as a consequence quite unexpected results. It could be also regarded as a complementary research to more numerous published papers concerning the behavior of irradiated lithium-diffused silicon p-n junctions.[6-8]

Instead of using p-type silicon of very low concentration of boron impurities ($\approx 3 \times 10^{12}$ cm^{-3}) for the preparation of the lithium compensated material,[5] this time the initial material had about 1.5×10^{14} cm^{-3} boron atoms, and, consequently, after the drift process the lithium concentration was greater than before. However, the gamma fluences were approximately the same as in the previous irradiation.[5] In addition, the compensated silicon was aged at room temperature for a period of 6 months prior to irradiation in the Co^{60}-gamma source at 35 °C.

Hall effect and resistivity measurements were used to monitor changes in the electrical properties of such a prepared silicon single crystals before and after irradiation and after the high temperature annealing (up to 900 °C).

2. EXPERIMENTAL

A floating-zone, 100 ohm-cm, boron-doped silicon of low oxygen concentration ($< 10^{16}$ cm^{-3}) was used as a starting material. After evaporation of lithium on a silicon wafer and a short diffusion for half an hour at 350 °C, lithium ions were drifted in the electric field at about 100 °C up to the depth of about 5 mm. Then, in order to achieve a higher degree of compensation, the so-called 'clean-up' method was applied by holding the wafer at the reverse bias of 400 V at room temperature for six days.

The wafer was allowed to age at room temperature for a period of four and a half months, and, then, rectangular samples homogeneously doped with lithium were cut from the middle portion of the wafer.[5] As Hall effect measurements on such samples had shown some peculiar results, a second, control wafer was prepared in such a way that, after cutting, only one half of it was subjected to the clean-up process. Then, the results of measurements on samples from both halves of the wafer were compared.

The samples were irradiated in the Co^{60}-gamma source in the flux of 5.5×10^{14} photons/cm^2/hr. The time integrated flux changed from 2.2×10^{16} up to 6.2×10^{16} photons/cm^2. The temperature during irradiation was maintained at $(35 \pm 0.5)\,°C$. The variation in conductivity of one sample was monitored during irradiation.

Two samples were immersed in the liquid nitrogen immediately after the end of irradiation and stayed there for a couple of days. The idea was to try to prevent any possible modification of defects after irradiation.

The room temperature annealing of the irradiated samples was checked by electrical measurements for several months after irradiation. When the concentration of carriers almost stabilized five months after irradiation, the isochronal and iso-thermal annealing treatments had been performed on the samples in vacuum at higher temperatures.

3. RESULTS

The results of Hall effect measurements for a typical sample before irradiation are shown in Figure 1 (Curve 1). It is important to note that the Hall coefficient is positive at lower temperatures indicating p-type conductivity, that is opposite to what we have observed and published in our previous paper.[5] The second, control wafer has shown that there was no substantial differences between the 'clean-up' samples and the regular ones. Both have shown n-type conductivity as it had been obtained before;[5] however, the measurements were taken immediately after the drift process, not allowing the room temperature aging to take place. By repeating the measurements 3 months later, we obtained again p-type conductivity (for the clean-up samples even at room temperature!)

That was proof that the initially overcompensated n-type silicon changes to slightly under-compensated p-type material after the room temperature aging.

Curve 1 also shows an inversion of the sign of the Hall coefficient somewhat below room temperature. Such an inversion may occur in p-type samples solely due to the fact that the ratio of electron to hole mobilities is greater than 1, and it was observed previously in the case of highly purified p-type silicon,[9] but the inversion occurred at much higher temperatures than in our samples.

Another interesting feature is a high degree of compensation after the room temperature ageing (for example, an excess of only 6×10^8 boron atoms over the donor concentration, as calculated from the low temperature plateau of the R_H vs. $1/T$ curves). Thus, we were able to detect several localized energy levels in the lower half of the

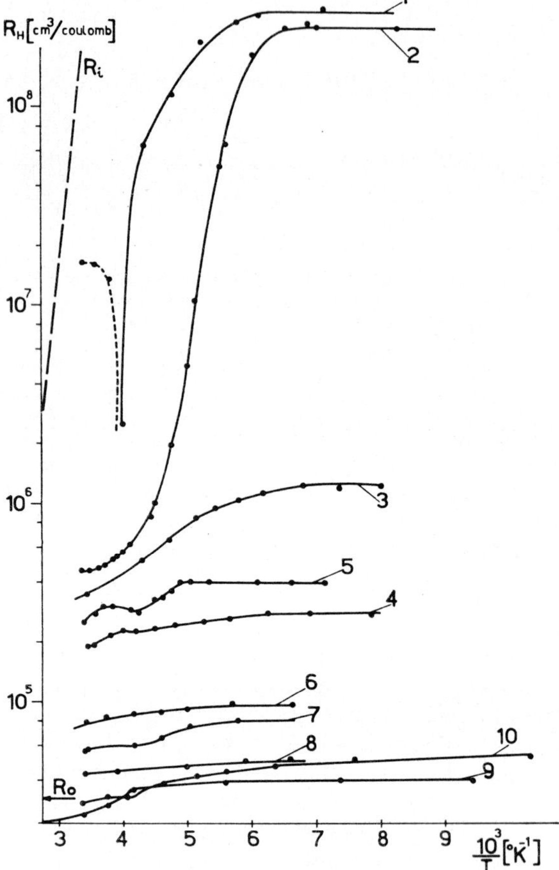

FIG. 1. The temperature dependence of the Hall coefficient of the sample KMA2–2 before and after irradiation ($\phi = 6.2 \times 10^{16}$ photons/cm^2) and after various thermal treatments. 1—before irradiation, 2—one day after irradiation, 3—annealed at room temperature for 6 days, 4—annealed at room temperature for 155 days, 5—after the initial heating at 230 °C. After a half an hour annealing: 6—at 400 °C, 7—at 450 °C, 8—at 500 °C, 9—at 700 °C, 10—at 900 °C. R_i—the calculated intrinsic Hall coefficient. R_0—the Hall coefficient of the sample before the lithium drift process. Full lines designate p-type, dashed lines n-type.

forbidden gap ($E_v + 0.45$ eV, $E_v + 0.36$ eV) even in the concentration as low as 10^9 to 10^{10} cm^{-3}.

It has also been observed that due to the decrease in the donor concentration, the degree of compensation decreased with time, and the Hall coefficient sign inversion shifted to higher temperatures after prolonged aging at room temperature.

From the initial linear decrease in conductivity during irradiation, the removal rate of $(0.8 \pm 0.1) \times 10^{-4}$ cm^{-1} was calculated. It was five times smaller than the value reported previously, but quite similar to the acceptor level introduction rate ($E_v + 0.23$ eV).[5]

It is interesting that the minimum in the conductivity vs. dose curve was reached at a lower gamma fluence (1.5×10^{15} photons/cm²) than in the case of samples with the lower boron and lithium concentrations,[5] indicating that its occurrence depended upon the degree of compensation and not upon the absolute values of the concentration of dopants. After reaching the minimum, the conductivity first increased slowly with fluence, but after about 8×10^{15} photons/cm² the increase was much faster. After 6.2×10^{16} photons/cm² the conductivity of the samples at room temperature was about ten times greater than before irradiation.

Two deep donor levels, located at $E_v + 0.32$ eV (calculated from Figure 1, Curve 2) and $E_v + 0.29$ eV, were detected immediately after irradiation. The latter was observed only in the samples irradiated with lower gamma fluences (2.2×10^{16} photons/cm²); it was quite unstable disappearing in a couple of days at room temperature, or, for longer irradiation times, even during irradiation, giving rise to a more stable defect with the $E_v + 0.32$ eV level. The statistical g-factor for the former was determined by the method using data of several samples[10] and was found to be equal to 1.

The amount of a decrease in the concentration of free lithium ions, calculated from the difference in the positions of the low temperature plateau of the Hall curves before irradiation, and the room temperature value of R_H immediately after irradiation, was directly proportional to the gamma fluence. A decrease in the concentration of the interstitial lithium ions was about 1×10^{13} cm⁻³ for the highest fluence employed, what was only about 10 per cent of the total electrically active lithium concentration. After room temperature annealing of about 5 months (Figure 1, Curve 4) that number increased to 15 per cent, indicating a 'spontaneous damage' process after irradiation.

It was possible to distinguish two different phases in the room temperature annealing of the samples after irradiation, characterized by two exponential processes, which could be represented by an expression

$$N_d/N_d^0 = A \exp\left[-\frac{t}{\tau_1}\right] + B \exp\left[-\frac{t}{\tau_2}\right] \quad (1)$$

where N_d is the total donor concentration at time t, and N_d^0 immediately after irradiation (calculated

from the low temperature plateau of the Hall curves), and τ_1, τ_2 are the time constants.

The first process lasted up to 30 to 40 days after irradiation with a time constant $\tau_1' \approx 193$ days for the highest fluences as shown in Figure 2 (Curves 2 to 4), and $\tau_1'' \approx 546$ days for $\phi = 2.2 \times 10^{16}$ photons/cm² (Figure 2, Curve 1).

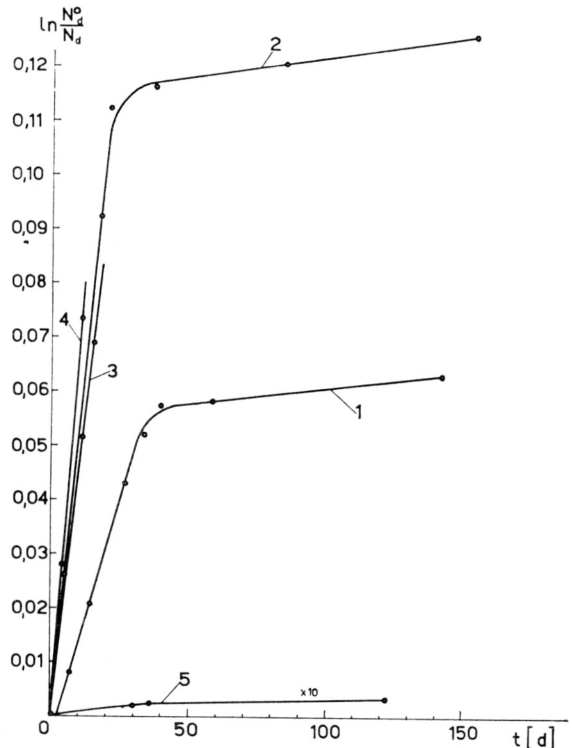

FIG. 2. The room temperature annealing of the samples after irradiation. 1—the sample KMA2-1 ($\phi = 2.2 \times 10^{16}$ photons/cm²), 2—the samples KMA2-2, 3—KMA2-3, 4—KMA3-2 (all with $\phi = 6.2 \times 10^{16}$ photons/cm²), 5—an unirradiated lithium compensated sample (with magnification in the ordinate 10 times).

It is important to note that the time constants were inversely proportional to the gamma fluence, which is different from the results already published for lithium-doped solar cells.[7,8]

It has also been found that the 'spontaneous damage' process was completely stopped during the time the samples were at 77 °K.

The concentration of the $E_v + 0.32$ eV level was found to be dose dependent being greater in the samples irradiated with higher gamma fluences.

The end of the first phase was reflected in the Hall measurements by a disappearance of the

$E_v + 0.32$ eV level in favor of two new levels situated at $E_v + 0.28$ eV and $E_v + 0.35$ eV (Figure 1, Curve 4). The first had a higher concentration in the samples more strongly irradiated, and the second was independent of the dose.

The number of electrically active lithium ions has decreased further at a much slower rate in the second phase, and almost stabilized 5 months after irradiation. A decrease in the degree of compensation was smaller for the samples irradiated with lower gamma fluences (Figure 2, Curve 1). For comparison, the room temperature aging of an unirradiated sample was also included in Figure 2 (Curve 5), with properly adjusted time scale, showing a time constant of about 1.5×10^5 days.

Then, the isochronal and isothermal annealing treatments on the irradiated samples were performed at higher temperatures. This time, the annealing stage centered at about 130 °C, where a type inversion had occurred,[5] was absent.

Partial recovery and an increase in the concentration of free lithium ions (for about 1×10^{13} cm^{-3}) was observed after a 20-minute annealing at temperatures from 230 to 275 °C (as shown in Figure 1, Curve 5). The $E_v + 0.28$ eV and $E_v + 0.35$ eV levels were not affected by that annealing step. After additional heating in the same temperature interval, the active lithium concentration started to decrease again. At this point, we tried to apply the method of the step annealing of one sample,[11] but we found that the process was neither a simple one, nor had it a single activation energy.

Half an hour heating at 350 °C caused the disappearance of the $E_v + 0.35$ eV level. After a 500 °C annealing step (Figure 1, Curve 8) all the introduced levels vanished. Additional heatings at 700 and 900 °C for half an hour have restored the Hall coefficient corresponding to the initial boron concentration (indicated by R_0 in Figure 1). Some steps in the curves due to localized levels were also introduced probably due to in-diffusion of impurities during the high-temperature thermal treatment in vacuum.

By combining the results of the isochronal and isothermal annealing in the temperature interval from 350 to 500 °C, a single activation energy of 0.52 eV was obtained for that process.

4. DISCUSSION

By inspection of the data obtained previously[5] and the present ones, it is obvious that the carrier removal rate is dependent upon the relative concentration of the radiation-introduced vacancies and impurity atoms. It is much higher when there are free lithium ions in excess over the boron impurities, and lower in the opposite, present case. Therefore, the defect introduction rate also varies during the irradiation depending upon the gamma flux strength (by influencing the time required for a certain gamma fluence), and the temperature of irradiation because the defect structures are dependent upon the ability of lithium to diffuse to the defects formed primarily.

After irradiation, in a first room temperature metastable phase, two deep donor levels have been developed which do contain lithium at the expense of a corresponding decrease in a free lithium ion concentration. That decrease was directly proportional to the number of radiation introduced vacancies. Thus, we propose that the $E_v + 0.29$ eV and $E_v + 0.32$ eV levels correspond to defects which consist of the Li-V pairs or the Li-V-O complexes in two configurations.

The donor nature of those levels was deduced from the position of the low temperature plateau of the Hall curves, which was usually higher immediately after irradiation than before irradiation, indicating the higher degree of compensation due to the introduction of deep donor levels.

The first annealing phase was also characterized by a relatively fast spontaneous damage process, which we attributed to a precipitation of free lithium ions. In the second annealing phase, a much slower precipitation process occurred which could be controlled by a disassociation of the Li-O$^+$ pairs.

As a result of a decrease in the shallow donor level concentration, the Fermi level shifted towards the valence band, and two additional levels (at 0.28 eV and 0.35 eV above the valence band) appeared. They were stable throughout the high temperature annealing treatment. The later annealed out at 350 °C in accordance with the previously published data,[2] and as its concentration was independent of the irradiation doses it could have been present even before the irradiation and connected with the presence of oxygen. The former was very stable and annealed out only after the 500 °C annealing step. It could be the same level reported by Vavilov.[1]

It is quite possible that a sudden increase in the concentration of free lithium ions after short heating in the temperature interval from 230 to 275 °C was due to the liberation of trapped lithium atoms from vacancies. The depth of that potential well was calculated to be 2.16 eV.[12] The results of additional heating in the same temperature interval showed that lithium had diffused away and an activation energy of 0.52 eV was obtained at somewhat higher temperatures and attributed to the lithium outdiffusion to the surface or precipitation at dislocation lines. The same value for the activation energy was obtained by combining the results of the isothermal and isochronal measurements on two identical samples, and by the method suggested by Gusev and coworkers[13] for the case of diffusion of the interstitial impurity atoms where the slope of the experimental curves equals half of the activation energy. The value obtained was less than the value quoted usually (0.66 eV),[14] however, it was in an exact agreement with that obtained for the case of radiation-stimulated diffusion in irradiated silicon.[12]

Finally, it is necessary to try to explain substantial differences between the present results and those already published.[5,1,15] In our previous work[5] there was a large disproportion between the number of available lithium ions and radiation introduced vacancies (due to relatively high doses). As a result, there was a strong preference for the formation of a stable lithium-divacancy complex where the presence of the positively charged lithium ion relaxed the Coulomb repulsion which might had otherwise prevented a divacancy formation. Its acceptor nature had changed the type of conductivity of that initially overcompensated silicon. The annealing stage at about 130 °C had caused disassociation of the complex and restoration of the n-type conductivity. This stage was absent in the present work because now there was no deficiency in lithium ions and the relatively unstable Li-V or Li-V-O complexes were favored instead of the Li-divacancy pairs, having also as a consequence a decrease in the shallow donor concentration.

In almost all other published data[6-8,15,1] lithium was present in a significant concentration in respect to the irradiation doses employed (or to the number of vacancies) and a spontaneous recovery in some physical properties was possible at room temperature. After irradiation the Fermi level was usually left in the upper half of the forbidden energy gap in such materials.

The increased stability of damage in lithium diffused solar cells with higher integrated fluxes[7,8,15] may be explained by a lack of lithium available for pairing and neutralization of the action of recombination centers. For even higher integrated fluxes a spontaneous damage should be observed after irradiation.

REFERENCES

1. V. S. Vavilov, *Proc. 7th Intern. Conf. Phys. Semicond.* (Dunod, Paris, 1964), p. 115.
2. E. Sonder and L. C. Templeton, *J. Appl. Phys.*, **36**, 1811 (1965).
3. K. Nakashima and Y. Inuishi, *Radiation Effects in Semiconductors*, Proc. Santa Fe Conf., 1967 (New York, 1968), p. 162.
4. D. Konozenko, A. K. Semenyuk, V. I. Khivrich and G. A. Dobrokhotov, *Fiz. i tekhn. poluprovodnikov*, 3, 155 (1969); *Soviet Physics-Semiconductors*, 3, 130 (1969).
5. N. B. Urli and M. Peršin, *Proc. 9th Intern. Conf. Phys. Semicond.* (Moscow, 1968), Vol. I, p. 118.
6. P. H. Fang and Y. M. Liu, *Appl. Phys. Letters*, **9**, 364 (1966).
7. J. J. Wysocki, *IEEE Trans. on Nucl. Sci.*, **NS-14**, 103 (1967).
8. P. H. Fang, Y. M. Liu, J. R. Carter, Jr. and R. G. Downing, *Appl. Phys. Letters*, **12**, 57 (1968).
9. A. Hoffman, K. Renschel and H. Rupprecht, *J. Phys. Chem. Solids*, **11**, 284 (1959).
10. V. V. Voronkov, G. I. Voronkova and M. I. Iglitzin, *Fiz. i tekhn. poluprovodnikov*, 3, 1720 (1969).
11. F. Bell and R. Sizmann, *phys. stat. solidi*, **15**, 369 (1966).
12. A. E. Kyv and B. L. Oxengendler, *Fiz. i tekhn. poluprovodnikov*, 3, 1178 (1969).
13. V. M. Gusev and V. V. Titov, *Fiz. i tekhn. poluprovodnikov*, 3, 3 (1969).
14. E. M. Pell, *Phys. Rev.*, **119**, 1222 (1960).
15. P. H. Fang and Y. M. Liu, *J. Appl. Phys.*, **38**, 4552 (1967).

LOW TEMPERATURE PHOTOCONDUCTIVITY IN ELECTRON-IRRADIATED p-TYPE Si

P. VAJDA† AND L. J. CHENG‡

Chalk River Nuclear Laboratories, Atomic Energy of Canada Limited, Chalk River, Ontario, Canada

(Presented at the conference by M. SWANSON)

The photoconductivity spectra of p-type silicon irradiated at $\sim15\,°K$ with 1.2 MeV electrons were studied in the wavelength range from 1.2 to 5.5μ at temperatures from 23 to 80 °K. The 3.9μ photoconductivity band appears immediately after irradiation in all crystals already at low temperatures, giving further evidence that it is due to the divacancy formed directly during irradiation by electrons. Three main annealing stages of the photoconductivity have been observed; (a) below 160 °K, (b) 160–250 °K, and (c) 280–360 °K. A radiation-induced deep level at $E_v + (0.12 \pm 0.02\;\mathrm{eV}$ disappears upon annealing at stage b. The annealing behavior of the spectra depends strongly on the measuring temperature. The dependence of the spectra on chopper speed was also investigated.

1. INTRODUCTION

The investigation of photoconductivity (PC) spectra is a powerful tool in the study of radiation induced defects in semiconductors.[1,2] Many studies of photoconductivity spectra have been made on silicon crystals irradiated with gamma rays, 1–3 Mev electrons, and fast neutrons at various temperatures.[3–7] Using polarized light and a uniaxial stress technique, Cheng[8,9] has correlated the $3.9\,\mu$ photoconductivity band (PC band) with the singly positively-charged divacancy in p-type silicon irradiated with fission neutrons at room temperature. Kalma and Corelli,[10] using the same technique, have correlated two defect levels, located at $E_c - 0.39$ eV and $E_c - 0.54$ eV, with the divacancy in electron-irradiated n-type silicon. Another defect level located at $E_v + 0.42$ eV in irradiated p-type silicon has been found to arise from a defect with $\langle 111 \rangle$ electronic symmetry.[9] Recent work of Cherki and Kalma[11] attributes this defect to the interstitial dopant atom (B or Al).

The purpose of the present work was to obtain more information about the defects responsible for the radiation-induced photoconductivity spectra. We, therefore, undertook a detailed annealing study of these spectra for different crystals (pulled and floating-zone refined) and of their temperature dependence. In this paper, we shall report some important results. The detailed results will be published elsewhere.

† Present address: Laboratoire de Chimie Physique, Faculté des Sciences de Paris, 91 Orsay, France.
‡ Present address: Institute of Atomic Energy Research, Lung-Tan, Taiwan, Republic of China.

2. EXPERIMENTAL PROCEDURE

Two types of commercially available p-type crystals were used: floating-zone refined (FZ) silicon of 90 ohm-cm resistivity (1.4×10^{14} boron atoms/cm³, $\lesssim 10^{16}$ oxygen atoms/cm³) and pulled silicon of 43 ohm-cm resistivity (3×10^{14} gallium atoms/cm³, 1×10^{18} oxygen atoms/cm³). The irradiations were performed at the 1.5 MeV Van de Graaff accelerator at Chalk River using 1.2 MeV electrons. The irradiation temperature was $\sim15\,°K$. The infrared light used to measure the photoconductivity was emitted from a SiC globar through a Perkin–Elmer Model 98 Monochromator with a LiF prism and was normally chopped at 17 cps using a symmetric square wave disc, except for a few experiments where the frequency was varied between 5 and 120 cps. The photosignals were amplified by a Princeton Applied Research Model HR8 Lock-in amplifier. The light intensity was determined by a thermocouple infrared detector. The annealing was done *in situ* in the cryostat.

3. EXPERIMENTAL RESULTS AND DISCUSSION

Because of the limitation on the length of the papers to be presented in this Conference, we shall only report some important results from this study.

3.1. *Appearance of the $3.9\,\mu$ PC Band Immediately after Irradiation*

Figure 1 shows the photoconductivity spectra, $\Delta\sigma/I$, of p-type FZ silicon before irradiation, after irradiation, and after several anneals, (similar

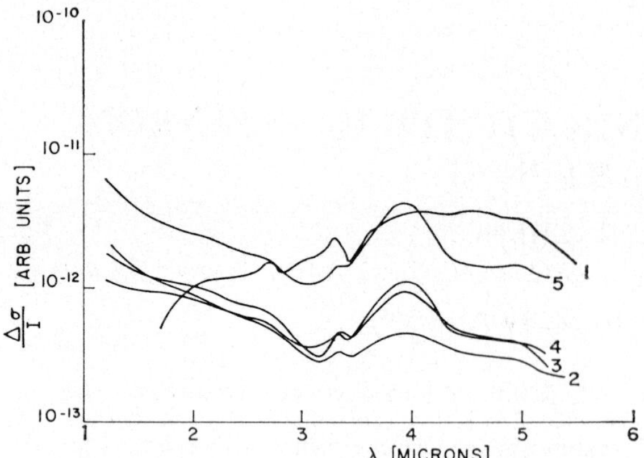

FIG. 1. Photoconductivity spectra, $\Delta\sigma/I$, of FZ silicon irradiated with 6×10^{15} electrons/cm² at $\sim 15\,°$K, $T_m = 23\,°$K.

1. Before irradiation; 2. After irradiation; 3. After $90\,°$K anneal; 4. After $210\,°$K anneal; 5. After $360\,°$K anneal.

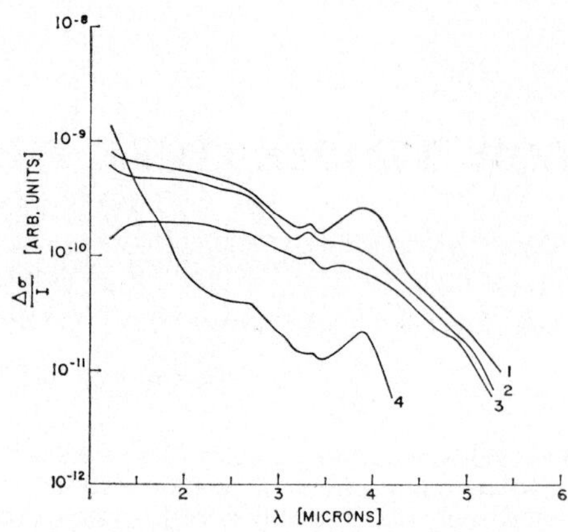

FIG. 2. Photoconductivity spectra, $\Delta\sigma/I$, of FZ silicon irradiated with 6×10^{15} electrons/cm² at $\sim 15\,°$K, $T_m = 41\,°$K.

1. After irradiation; 2. After $90\,°$K anneal; 3. After $210\,°$K anneal; 4. After $360\,°$K anneal.

spectra are also observed for pulled crystals). It is clear that the $3.9\,\mu$ PC band appears immediately after the irradiation, indicating that the defect responsible for the band is formed at low temperatures directly by the primary electrons. Cheng[8,9] has identified the $3.9\,\mu$ PC band as arising from the singly positively-charged divacancy. Though no vacancy motion is expected at these temperatures $(\sim 20\,°$K),[12] the divacancy can be formed directly by particle bombardment through multiple displacement processes[13] and thus gives rise to the $3.9\,\mu$ PC band. This result gives further evidence that the band is due to the divacancy.

Matsui and Baruch[7] observed this band in some of their floating-zone and pulled crystals after $20\,°$K irradiations, while Cherki and Kalma[11] saw it only above $77\,°$K. Neutron-irradiated crystals exhibit the band at temperatures above $30\,°$K.[14] We have found that the appearance of the band depends on measuring temperature and light chopper speed. A comparison between Figure 1 and Figure 2, both measured from the same sample but at different measuring temperatures, indicates clearly the dependence of the appearance of the band on measuring temperature. Figure 2 exhibits the 3.9 PC band after irradiation and warming-up to the measuring temperature, but this band practically disappears after $90\,°$K anneal and reappears only after $360\,°$K. No such phenomenon was observed at $T_m = 23\,°$K. Figure 3 indicates from the dependence on light chopper speed that the defect photoconductivity seems to disappear

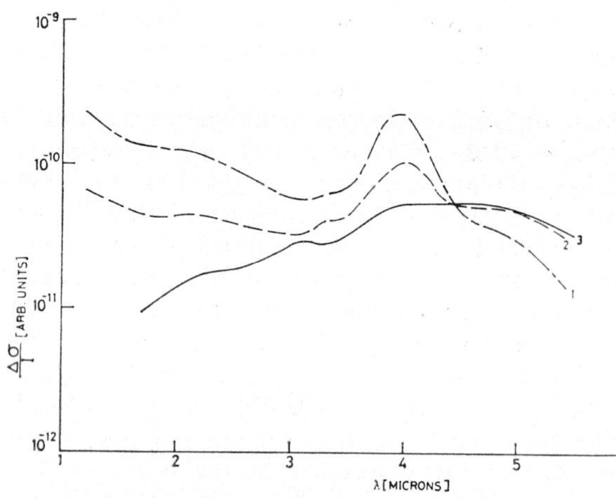

FIG. 3. Dependence of photoconductivity spectra (FZ silicon after irradiation and $400\,°$K anneal) on light chopper speed. $T_m = 23\,°$K.
1. 5 cps; 2. 16 cps; 3. 120 cps.

with high chopper speed (120 cps). These results can be also used to explain why Cherki and Kalma only saw the band above $77\,°$K.

3.2. Temperature Dependence of Defect photoconductivity

In order to investigate the influence of measuring temperature more carefully, we monitored the photoconductivity signal at $3.9\,\mu$ and at $2.1\,\mu$ in the temperature range $23\,°$K $\leqslant T_m \leqslant 80\,°$K. In Figure 4, we have plotted $\Delta\sigma/I$ at $2.1\,\mu$ for a FZ

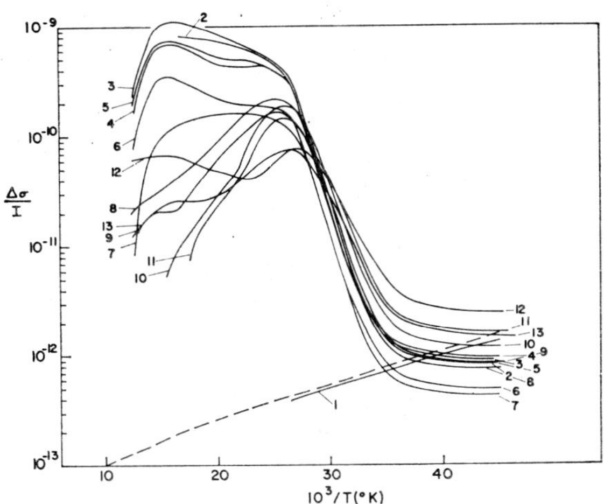

FIG. 4. Photoconductivity, $\Delta\sigma/I$, at $2.1\,\mu$ as a function of measuring temperature (FZ silicon irradiated with 6×10^{15} electrons/cm^2 at $\sim 15\,°$K). The dashed line represents the T_m-dependence of the mobility in *p*-type Si[15].

1. Before irradiation; 2. After irradiation; 3. After $60\,°$K anneal; 4. After $90\,°$K anneal; 5. After $120\,°$K anneal; 6. After $150\,°$K anneal; 7. After $180\,°$K anneal; 8. After $210\,°$K anneal; 9. After $240\,°$K anneal; 10. After $280\,°$K anneal; 11. After $320\,°$K anneal; 12. After $360\,°$K anneal; 13. After $400\,°$K anneal.

sample as a function of reciprocal temperature, with the annealing temperature as parameter. The general features of the temperature dependence of the photoconductivity are the same for pulled and FZ samples measured at $3.9\,\mu$ and $2.1\,\mu$. Two important features appear in the figure.

First, the irradiation did affect the temperature dependence of the photoconductivity drastically. The pre-irradiation curve is found to be similar to the temperature dependence of the mobility in *p*-type silicon,[15] indicating that the number of photo-excited carriers is roughly independent of measuring temperature.

After irradiation, the photo-conductivity $\Delta\sigma/I$, in general, consists of two main parts—a low temperature part where $\Delta\sigma$ decreases exponentially with decreasing T and tends to level out below $27\,°$K, and a high temperature part, where $\Delta\sigma$ decreases with increasing T. The exponential part of the curves in the low temperature range corresponds to an effective activation energy of 0.05 eV. An analysis of the data based on a model of a single trapping level indicates that the temperature dependence in the low temperature part can be explained by the existence of a trapping level. We

have also concluded that the dopant (boron or gallium) acts as the trap.

The second important feature showing in Figure 4 is that all the curves (apart from the pre-irradiation one) could be separated into two groups, one below the $180\,°$K anneal and the other above. This shows clearly that a major defect annealing occurred around 170–$200\,°$K.

The rapid decrease of $\Delta\sigma$ around 60–$70\,°$K before the annealing is very interesting. The decrease can be explained by the existence of a deep trapping level which becomes active around this temperature. The trapping level is located around $E_v + 0.12 \pm 0.02$ eV, estimated from the slope and the temperature where the effect occurs. The defect that disappears upon the annealing is associated with the trapping level. The neutral vacancy existing in *p*-type Si is known to become mobile in the temperature range from 150–$180\,°$K. Therefore, the annealing stage may be associated with migration of the neutral vacancy. However, the recent data of Cheng and Lori[16] obtained from a study of electrical properties of low temperature irradiated *p*-type Si have shown that the annealing stage in this temperature range is very complex and more studies, especially on detailed correlation, are needed in the future.

3.3. *Annealing Stages*

In general the annealing behavior of the photoconductivity is very complex. However, some interesting conclusions can still be made. Because of the strong dependence of the annealing behavior on measuring temperature, a proper and simple presentation of the annealing data was difficult. We have arbitrarily chosen $23\,°$K as a fixed measuring temperature, around which $\Delta\sigma$ was roughly independent of measuring temperature and the $3.9\,\mu$ PC band was always the major portion of $\Delta\sigma$ at $3.9\,\mu$.

Figure 5 shows the annealing behavior of $\Delta\sigma$ at $3.9\,\mu$ and $2.1\,\mu$ for two FZ samples irradiated with different doses. The annealing can be separated into three stages: below $160\,°$K, from $160\,°$K to $250\,°$K and above $280\,°$K.

3.3A. *Below* $160\,°K$

Because none of the known defects in *p*-type Si is mobile in this temperature range, we tend to assume that what we see here is not a real annealing of the defects responsible for the photoconductivity change, but rather a charge redistribution among some shallow levels (acting as photoexcited carrier traps) affecting the carrier lifetime.

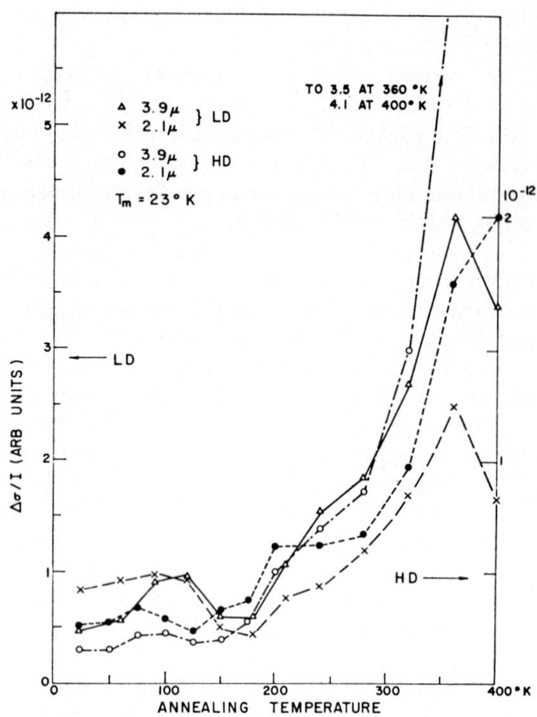

FIG. 5. Annealing of photoconductivity at different wavelengths for FZ silicon irradiated with different doses (HD = 1.9×10^{16} electrons/cm² and LD = 6×10^{15} electrons/cm³) at ∼15 °K. $T_m = 23$ °K.

3.3B. 160–250 °K

This stage has been discussed in the previous section.

3.3C. 280–360 °K

No drastic change in the temperature dependence of $\Delta\sigma$ was observed. However some annealing did occur in this temperature range. For example, in Figure 5, $\Delta\sigma$ at $3.9\,\mu$ and $2.1\,\mu$ in FZ samples increases markedly in this stage. The amount of the increase at $3.9\,\mu$ is much larger than that at $2.1\,\mu$,

suggesting that the number of divacancies might increase in this stage by means of the aggregation of single vacancies after the breaking-up of vacancy-impurity complexes. The same is indicated in pulled specimens. In the latter case, oxygen may play an important role by trapping the available free vacancies and forming A-centers. The results also indicate some dose dependence, especially in the pulled samples. A similar broad annealing stage in carrier concentration was observed in p-type Si irradiated with Co-60 γ-rays at 82 °K.

REFERENCES

1. V. S. Vavilov, *Phys. Stat. Sol.*, **11**, 447 (1965).
2. J. W. Corbett, *Electron Radiation Damage in Semiconductors and Metals* (Academic Press, New York, 1966).
3. V. S. Vavilov, A. F. Plotnikov and V. D. Tkachev, *Sov. Phys.-Solid State*, **4**, 2522 (1963).
4. T. D. Tkachev, A. F. Plotnikov and V. S. Vavilov, *Sov. Phys.-Solid State*, **5**, 1332 (1964).
5. V. D. Tkachev, A. F. Plotnikov and V. S. Vavilov, *Sov. Phys.-Solid State*, 2333 (1964).
6. V. S. Vavilov, S. I. Vintovkin, A. S. Lyutovich, A. F. Plotnikov and A. A. Sodolova, *Sov. Phys.-Solid State*, **7**, 399 (1965).
7. K. Matsui and P. Baruch, *Lattice Defects in Semiconductors*, Ed. R. R. Hasiguti (Univ. of Tokyo Press, 1968), p. 282.
8. L. J. Cheng, *Phys. Letters*, **24A**, 729 (1967).
9. L. J. Cheng, *Radiation Effects in Semiconductors*, Ed. F. L. Vook (Plenum Press, New York, 1968), p. 143.
10. A. H. Kalma and J. C. Corelli, *Phys. Rev.*, **173**, 734 (1968).
11. M. Cherki and A. H. Kalma, *IEEE Annual Conf. on Nuclear and Space Radiation Effects*, University Park, Pa., U.S.A. (July 1969).
12. G. D. Watkins, *Radiation Effects in Semiconductors*, Ed. F. L. Vook (Plenum Press, New York, 1968), p. 67.
13. J. W. Corbett and G. D. Watkins, *Phys. Rev. Letters*, **17**, 314 (1961).
14. L. J. Cheng and M. L. Swanson, *J. Appl. Phys.*, **41**, 2627 (1970).
15. R. A. Logan and A. J. Peters, *J. Appl. Phys.*, **31**, 122 (1960).
16. L. J. Cheng and J. Lori, *Phys. Letters*, to be published.

DISCUSSION

Question (STEIN) Were any significant differences between floating zone and crucible grown Si observed with regard to the photoconductivity spectra?

Answer (SWANSON) I believe the results were generally similar but there did seem to be some dependence for high dose data, especially for the pulled crystals.

Question (GREGORY) The dependences on chopping speed and measurement temperature should be related. Was this investigated?

Answer (SWANSON) I don't believe so. Most of the results were for 17 cps.

RADIATION DEFECTS IN Si OF HIGH PURITY

I. D. KONOZENKO, A. K. SEMENYUK AND V. I. KHIVRICH

Institute of Nuclear Research, Academy of Sciences of Ukrainian SSR, Kiev

Radiation defects created by γ-irradiation of Co^{60} and fast neutrons in high purity p-Si ($\rho = 5 \times 10^3$ to $4 \times 10^4\ \Omega \cdot$cm) and n-Si ($\rho = 4 \times 10^2$ to $5 \times 10^3\ \Omega \cdot$cm) are investigated by measurements of Hall effect, resistivity and minority carrier lifetime. The oxygen concentration in the crystals is in the range of 5×10^{14} to 5×10^{15} cm^{-3}.

It is shown that stable γ-defects at 300 °K are divacancies and complexes of vacancies with donor or acceptor impurities. Divacancies introduced by γ-irradiation are the secondary defects. They become predominant after 'exhaustion' of the dopant. When divacancies become the predominant defects the Fermi level occupies its boundary position $E_v + 0.39$ eV in the gap. At low doses ($\Phi < 10^{16}$ photons/cm^2) vacancy-impurity complexes and at heavy doses ($\Phi > 10^{17}$ photons/cm^2) divacancies play the main role in the recombination process.

In neutron irradiation disordered regions are introduced and the level at $E_v + 0.35$ eV is observed. The Fermi level in both n- and p-Si shifts to the middle of the gap. At the annealing of disordered regions in the interval 200 to 250 °C the level at $E_v + 0.27$ eV appears and Fermi level occupies its boundary position at $E_v + 0.39$ eV. This indicates that divacancies become the predominant defects which can be formed as secondary defects at the destruction of the disordered regions.

1. INTRODUCTION

Electron paramagnetic resonance (EPR), the infrared absorption spectra and electrical measurements have given much information about properties of radiation defects in silicon. At present considerable progress has been achieved in obtaining Si of high purity and resistivity, and with low compensation of electrically active impurities.[1,2] It is interesting to investigate radiation defects in such a material because its properties are more sensitive to irradiation and the concentration of complexes of radiation defects with impurities is limited by the low content of these impurities (10^{11} to 10^{12} cm^{-3}). Radiation defects created in high restivity Si by 1 MeV electrons have been investigated in Ref. (1), by γ-irradiation in Refs. (2) and (3), and in high purity Si created by γ-irradiation in our previous works.[4,5] In this paper the results of investigation of radiation defects created by γ- and neutron irradiation in n- and p-type Si of high purity are given.

2. EXPERIMENTAL CONDITIONS

We have investigated p-Si samples of 5×10^3 to $4 \times 10^4\ \Omega \cdot$cm resistivity and n-Si of 4×10^2 to $5 \times 10^3\ \Omega \cdot$cm resistivity, grown by zone melting material of high purity. The oxygen concentration determined by the radioactivation method was found to be in the range of 5×10^{14} to 5×10^{15} cm^{-3}.

Resistivity and Hall effect measurements were carried out using cross-shaped samples ($1.0 \times 0.3 \times$ 0.1 cm^3) by the standard compensation method. The Sb doped eutectic Au + Si and that doped with Al were used as contacts for n- and p-Si, respectively.[7] Minority carrier lifetime was measured using the microwave absorption method.[8] The measurements of Hall effect, resistivity and minority carrier lifetime were carried out in the 77 to 370 °K temperature range.

γ-irradiation of the samples was performed at room temperature from Co^{60} sources producing fluxes of 4×10^{11} photons $\times$ cm^{-2} and 6×10^{12} photons $\times$ cm^{-2} s^{-1} by doses ranging from 10^{14} photons $\times$ cm^{-2} to 10^{18} photons $\times$ cm^{-2}. Fast neutron irradiation with a flux of $1.2 \times 10^9\ n \times$ cm^{-2} s^{-1} and doses up to $5 \times 10^{13}\ n \times$ cm^{-2} was also performed at room temperature. To remove the effect of thermal neutrons Cd plates were used.

Analysis of the results of Hall effect measurements and minority carrier lifetime, determination of the activation energy of recombination levels and their cross sections are described in our work.[5]

3. RADIATION DEFECTS CREATED BY Co^{60} γ-RAYS IN p- AND n-TYPE Si of HIGH PURITY

3.1. *Experimental results*

The hole concentration in p-Si samples ($\rho_0 = 2.6 \times 10^4$ to $4 \times 10^4\ \Omega \cdot$cm), measured at 300 °K, increases with the doses (Figure 1) and reaches saturation at the value 2.2 to 2.4×10^{12} cm^{-3}.

Temperature dependence of the hole concen-

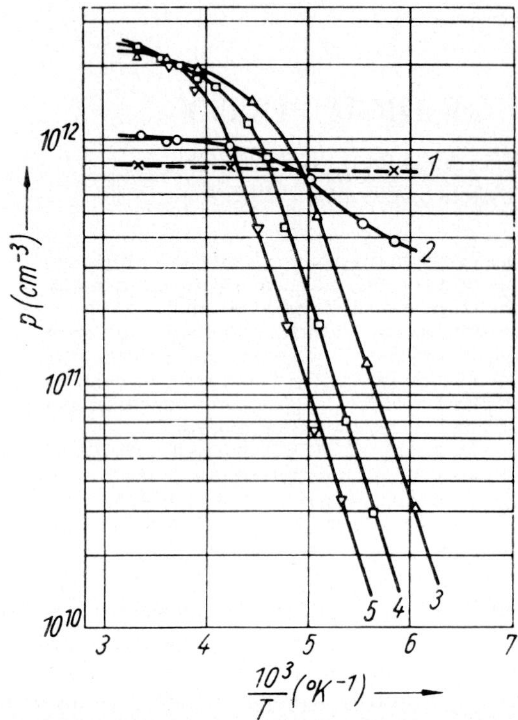

FIG. 1. Temperature dependence of hole concentration in highly pure p-Si ($\rho_0 = 2.6 \times 10^4\ \Omega\cdot$cm) before and after γ-irradiation with different doses: (1) before irradiation, (2) after 2×10^{15} photons $\times$ cm^{-2}, (3) after 2×10^{16} photons $\times$ cm^{-2}, (4) after 1.4×10^{17} photons $\times$ cm^{-2}, (5) after 6.6×10^{17} photons $\times$ cm^{-2}.

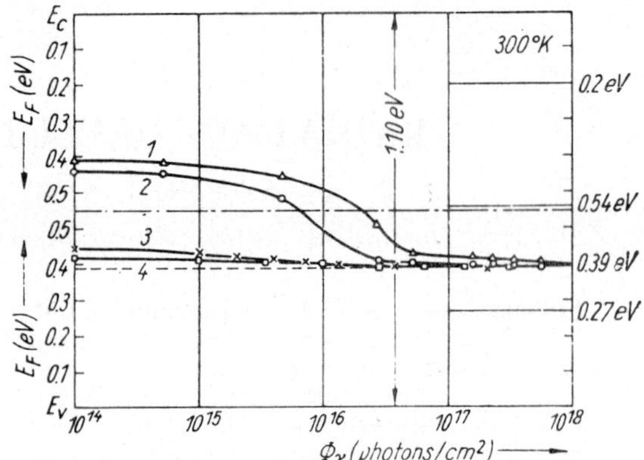

FIG. 2. Fermi level position E_F at 300 °K vs. γ-irradiation dose in n- and p-Si with different initial carrier concentration: (1) n-Si, $\rho_0 = 2 \times 10^3\ \Omega\cdot$cm, $n_0 = 2.4 \times 10^{12}$ cm^{-3}; (2) n-Si, $\rho_0 = 5 \times 10^3\ \Omega\cdot$cm, $n_0 = 9 \times 10^{11}$ cm^{-3}; (3) p-Si, $\rho_0 = 4 \times 10^4\ \Omega\cdot$cm, $p_0 = 4 \times 10^{11}$ cm^{-3}; (4) p-Si, $\rho_0 = 2.6 \times 10^4\ \Omega\cdot$cm, $p_0 = 8 \times 10^{11}$ cm^{-3}. The levels at $E_v + 0.27$ eV, $E_c - 0.54$ eV, $E_c - 0.21$ eV correspond to the divacancy (see the text).

tration gives a level at $E_v + 0.27$ eV the concentration of which increases with increasing γ-irradiation dose (Figure 1, curves 3, 4, 5). The initial rate of introduction of the level at $E_v + 0.27$ eV, determined from curve 2 of Figure 1, gives 3.2×10^{-4} (photons $\times$ cm)$^{-1}$ and decreases with doses.

In n-Si samples the carrier concentration decreases with flux, achieves intrinsic value, the samples then convert into p-type and the hole concentration increases reaching saturation at densities of 2.2 to 2.4×10^{12} cm^{-3}.

These experiments show that the Fermi level after γ-irradiation of high purity Si of both n- and p-type moves to the boundary position in the gap at $E_v + 0.39$ eV (Figure 2). The boundary position of the Fermi level is achieved earlier in samples with lower impurity concentration (curves 2 and 3 of Figure 2).

The carrier mobility at 300 °K in n-Si samples before irradiation was 1200 cm^2 $V^{-1} s^{-1}$ and in p-Si samples it was 400 cm^2 $V^{-1} s^{-1}$, and they have practically not changed under γ-fluxes used. The carrier mobility at 77 °K in n-Si before irradiation was 2.2×10^4 cm^2 $V^{-1} s^{-1}$ and in p-Si samples 1.0 to 1.2×10^4 cm^2 $V^{-1} s^{-1}$. Such high carrier mobility at 77 °K shows that the n- and p-Si used had a low degree of compensation and was homogeneous. The temperature dependence of the carrier mobility observed in the temperature range from 150 to 350 °K for p-Si obeys the law $\mu \sim T^{-2.4}$ and for n-Si $\mu \sim T^{-2.6}$, and practically doesn't change after irradiation.

Minority carrier lifetime of high purity p-Si changes peculiarly after γ-irradiation. The change in $\Delta(1/\tau)$ after successive irradiation is shown in Figure 3. The value of $\Delta(1/\tau)$ is proportional to the recombination centres concentration.[5,9] There are three sections a, b, c on every curve of Figure 3. Section a, a' gives a law of change $\Delta(1/\tau) \sim \Phi^{0.7/0.8}$. After this first part of the dependence follows the saturation part (b, b' for curves 1 and 2, correspondingly), and then again we found an increase of the value $\Delta(1/\tau)$ which for all samples can be expressed by the dependence $\Delta(1/\tau) \sim \Phi^{0.6}$ (parts c and c' for curves 1 and 2, correspondingly).

It is necessary to note that the Fermi level position at the fluxes corresponding to sections b, b' and c, c' in Figure 3 doesn't change and for section a, a' it changes slightly (see Figure 2, curves 3 and 4). We have studied the temperature dependences of the lifetime $\tau = f(10^3/T)$ on isochronal (Figure 4)

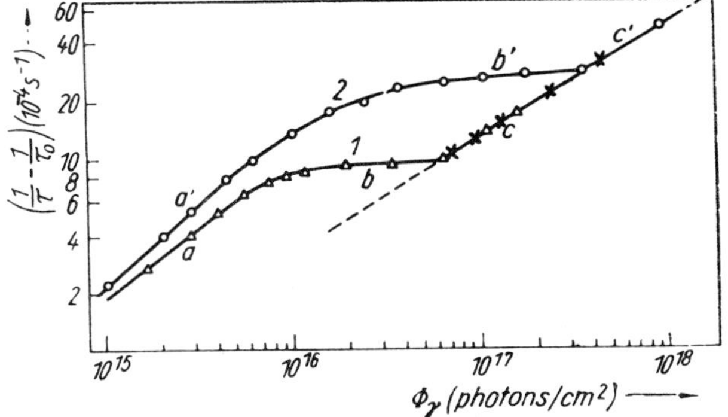

FIG. 3. Change in $\Delta(1/\tau) = 1/\tau - 1/\tau_0$ vs. γ-irradiation doses for high purity p-Si samples: (1) p-Si, $\rho_0 = 2.6 \times 10^4\ \Omega\cdot$cm; (2) p-Si, $\rho_0 = 3.6 \times 10^4\ \Omega\cdot$cm. x—flux 4×10^{11} photons $\times$ cm^{-2}s^{-1}, o, $\triangle$—flux 6×10^{12} photons $\times$ cm^{-2} s^{-1}.

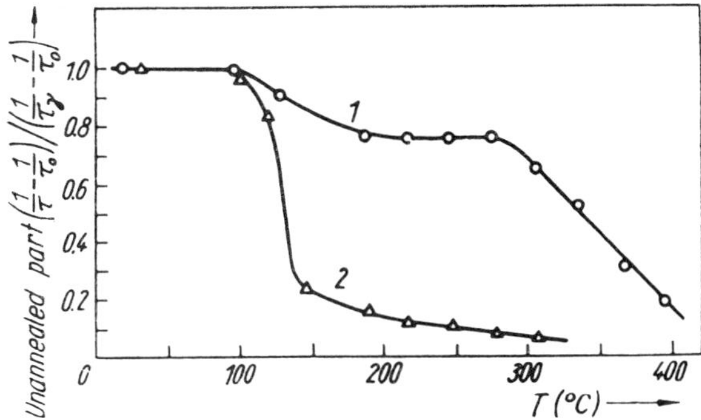

FIG. 4. Isochronal annealing (10 min) for the change of minority carrier lifetime in p-Si ($\rho_0 = 3.6 \times 10^4\ \Omega\cdot$cm) after various γ-irradiation doses: (1) after 8×10^{17} photons $\times$ cm^{-2}; (2) after 5×10^{15} photons $\times$ cm^{-2}.

and isothermal annealing of crystals irradiated with different γ-fluxes to establish the nature of the recombination centers in γ-irradiated Si of high purity.

3.2. Discussion

Generally under irradiation of silicon containing heavy concentration of impurities and oxygen, a decrease of the majority carrier concentration is observed and Fermi level monotonically moves to the middle of the gap.[10–13]

By now it is established that not primary defects but association of primary defects with impurities play the main role. The nature of some of them was established by EPR studies. In n-Si there is the E-center, which represents a phosphorus-vacancy complex and gives an acceptor level at $E_c - 0.4$ eV; in n- and p-Si with heavy concentrations of oxygen (10^{17} to 10^{18} cm^{-3}), there is the A-center, a vacancy-oxygen association, which gives an acceptor level at $E_c - 0.16$ eV. There are also other defects which give deep levels in the energy gap. The nature of these defects is not known and they lead to a compensation of the main doping impurity.

In high purity p-Si a hole concentration increase at $300\,°$K as a result of γ-irradiation is observed and the Fermi level takes the boundary position in the lower half of gap at $E_v + 0.39$ eV for sufficiently heavy doses (Figure 2). High purity n-Si converts into p-type and the Fermi level also moves to the boundary position at $E_v + 0.39$ eV.

The temperature dependence of the carrier concentration in p-Si, $p = f(10^3/T)$ (Figure 1) gives a level at $E_v + 0.27$ eV which is the lowest of the three known divacancy levels.[14,15] It corresponds to the transition from positive charge state (V_2^+) to the neutral one (V_2^0) and, consequently, it is donor level. The two other divacancy levels are located at $E_c - 0.54$ eV[16,17] and $E_c - 0.21$ eV[18] or $E_c - 0.4$ eV[14,16] and correspond to transitions to one (V_2^-) and two $(V_2^=)$ negatively charged states, correspondingly. These levels are acceptors.

Although in the literature two positions in the gap for the upper divacancy acceptor level are given, we found in γ-irradiated n-Si an acceptor level at $E_c - (0.21 \pm 0.01)$ eV. The introduction rate for the level at $E_c - 0.21$ eV was 1.5×10^{-4} (photon $\times$ cm)$^{-1}$, which in close to the introduction rate of the $E_v + 0.27$ eV level (0.7×10^{-4} (photon $\times$ cm)$^{-1}$) under the corresponding integrated γ-fluxes. Annealing of the defects which gives the level at $E_c - 0.21$ eV takes place at temperatures from 300 to 400 °C. That corresponds to the annealing temperature of divacancies in zone floated Si.[14]

The energy level at $E_c - 0.21$ eV also has been assigned to the A-center ($E_c - (0.16$ to $0.17)$ eV).[11] The authors[11] give for the introduction rate of A-centers the value 1×10^{-3} (photons $\times$ cm)$^{-1}$. The A-center annealing temperature interval is also close to that of divacancy annealing. But taking into account the energy location of the level at $E_c - 0.21$ eV in the gap and its introduction rate, which is close to the introduction rate of the level at $E_v + 0.27$ eV and considerably differs from the A-center introduction rate, and taking into account the temperature interval of annealing, it is possible to argue that the level at $E_c - 0.21$ eV belongs to the divacancy. The introduction rate of E-centers, which corresponds to the level at $E_c - 0.4$ eV is 1.8×10^{-5} (photons $\times$ cm)$^{-1}$, is in agreement with the data of Ref. (11).

The results listed above indicate that under γ-irradiation—the main radiation damage centers in Si of high purity are divacancies. The Fermi level boundary position is located between two divacancy levels: a donor level at $E_v + 0.27$ eV and a lower acceptor level.

Using data of Figure 1 it is possible to calculate the concentration of divacancies N_D versus doses of irradiation. Results of our calculations of N_D and also data of other authors[3,12] are presented in Figure 5. One can see that all experimental points well fit to the direct line with the slope 0.6. Thus

$$N_D = K\Phi^{0.6} \qquad (1)$$

where $K = 5 \times 10^2 = $ const. The concentration N_D calculated from formula (1) at doses, for which the Fermi level passes the middle of the gap (Figure 2, curves 1, 2) is equal to the electron concentration in the samples. This indicates that the conversion from n-Si into p-type is due to the lower acceptor level of divacancy.

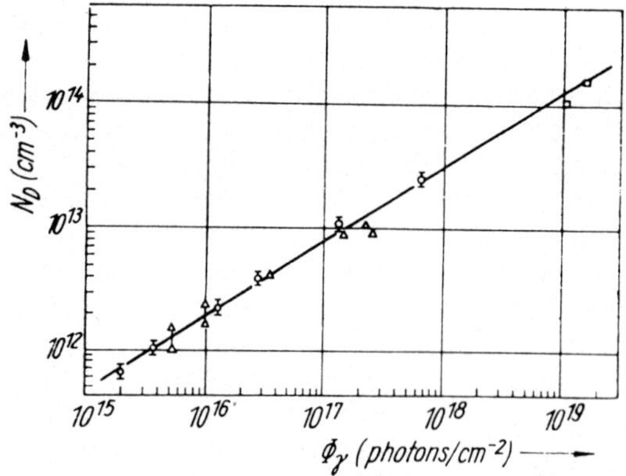

FIG. 5. Divacancy concentration N_D vs. irradiation dose in p-Si of high purity: Φ-data received from our experimental results, $\triangle$—data from (3), $\square$—data from (12).

It is necessary to point out that the maximum defect concentration N_D at the maximum doses used was 3×10^{13} cm^{-3}. This value exceeds by far the doping concentration in the investigated p-Si samples. This is an additional evidence in favour of the fact that the observed defects are not associations of point defects with dopants. The data on the energy level spectrum, data on introduction rate of defects and annealing temperature range indicate that in γ-irradiated Si of high purity the main defects being stable at room temperature are divacancies.

It follows from expression (1) that the intro-

duction rate of divacancies $(dN_D/d\Phi)$ decreases with increasing of dose. The decrease of introduction rate in the range of doses from 2×10^{16} to 2.6×10^{17} photons $\times$ cm^{-2} was observed by Ref. (3) who attempted to describe the dependence $dN_D/d\Phi = f(\Phi)$ with a differential equation. However, although the experimental points of Ref. (3) well fit the dependence (1), the differential equation doesn't describe this dependence over the wide range of γ-doses. The authors[3] point out that the introduction rate of the stable defects strongly depends on the temperature and at a prolonged irradiation at 77 °K it decreases to zero. Taking into account the fact that in p-Si at 77 °K vacancies are immobile[19] one may suppose that the main part of divacancies produced in p-Si under γ-irradiation are not the primary defects but appear as a result of pairing of single vacancies mobile at the irradiation temperature. It should be noted that the disappearance of EPR spectra of the single vacancies and formation of divacancies as secondary defects[14] are observed besides of the formation of primary divacancies under 1.5 MeV electron bombardment.

The introduction rate of the level at $E_c - 0.21$ eV in n-Si is a little different from the introduction rate of the level at $E_v + 0.27$ eV in p-Si at the same irradiation doses. One may suppose that divacancies appear as a result of single vacancy pairing, then the dependence of their formation rate on Fermi level position may be explained by the influence of charge state on the probability of close pairs dissociation[20–22] and on the free vacancy mobility.[19]

The results of the investigation of minority carrier lifetime show that both divacancies and associations of defect-impurities have an influence on the recombination processes in γ-irradiated Si of high purity. At slight doses of irradiation of 10^{15} to 10^{16} photons $\times$ cm^{-2} (section a, a' of Figure 3) probably the main role in recombination is played by radiation defects which include the doping impurity (boron). The presence of the including boron centers is indicated by the lifetime-temperature dependence, which gives the slope 0.43 eV.[5] The level at $E_v + 0.45$ eV in irradiated pure Si was observed in Refs. (1), (4) and (5) and is attributed to the vacancy-boron complex; a recombination level at $E_v + 0.46$ eV was discovered[23] in γ-irradiated p-Si (1000 Ω·cm) with low oxygen density.

Achieving saturation in the dependences 1 and 2 of Figure 3 (section b, b') is connected with exhaustion of boron. This fact allows us to consider the concentration of recombination centers at

$E_v + 0.43$ eV to be equal to the boron concentration. The electron cross section for the level at $E_v + 0.43$ eV was $\sigma_n = 3.2 \times 10^{-14}$ cm².

At doses $\Phi > 10^{17}$ photon $\times$ cm^{-2} the divacancy level at $E_v + 0.27$ eV plays the main role in recombination processes. The law of change of the inverse lifetime on irradiation doses can be expressed for all samples by the dependence $\Delta(1/\tau) \sim \Phi^{0.6}$ (section c, c' of Figure 3), which coincides well with the observed dependence (1). Besides the level at $E_v + 0.43$ eV the temperature dependence of lifetime gives also a level corresponding to the lowest divacancy level.[5]

As one can see in Figure 3 irradiation of crystals with highly different intensities leads to the same dependence $\Delta(1/\tau) \sim \Phi^{0.6}$.

Generally it is considered that the non-linearity of dependence $\Delta(1/\tau) = f(\Phi)$ is connected with the Fermi level shift as a result of irradiation.[24] The results of our work show that the Fermi level position in p-Si for doses $\Phi > 10^{17}$ photons $\times$ cm^{-2} doesn't change and a non-linear dependence $\Delta(1/\tau) = f(\Phi)$ describes the kinetics of recombination centers accumulation, namely divacancies. The electron cross-section of the lowest divacancy level was $\sigma_n = 1.2 \times 10^{-15}$ cm².

Studies of the recombination in n-Si show that before conversion the recombination centers appearing are E-centers; at the conversion n-Si into p-Si measurements are embarrassed.

Investigating the isochronal annealing of recombination centers in p-Si we have observed two stages: a first one at 100 to 150 °C and a second one at 300 to 400 °C (Figure 4). In slightly irradiated samples (γ-doses $\Phi < 10^{16}$ photons $\times$ cm^{-2}), when as we suppose in the recombination the main role is played by vacancy-boron complexes, we found only the first stage of annealing (Figure 4, curve 2). Annealing of the samples irradiated by heavy γ-doses ($\Phi > 10^{17}$ photons $\times$ cm^{-2}) shows the presence of both annealing stages, at the first stage only a small part of recombination centres is annealed and the main part of them is annealed at the second stage. The temperature range of second annealing stage coincides with the annealing temperature range of divacancies.[14] Annealing of the levels at $E_v + 0.27$ eV and $E_c - 0.21$ eV received from Hall effect measurements also takes place in the range from 300 to 400 °C.

Annealing activation energy determined from the minority carrier lifetime isothermal annealing is (0.85 ± 0.1) eV for the first annealing stage and (1.5 ± 0.1) eV for the second one.

4. RADIATION DEFECTS CREATED BY FAST NEUTRONS

4.1. *Experimental results*

As distinguished from γ-irradiation results, the carrier concentration measured at 300 °K monotonically decreases with neutron doses both in n- and p-type Si of high purity. Further the resistivity moves to its intrinsic value $\sim 3 \times 10^5$ $\Omega \cdot$cm at room temperature. The monotonic decrease of carrier concentration with increasing of doses of neutron irradiation of high purity n- and p-Si leads to the shifting of Fermi level to the middle of the gap.

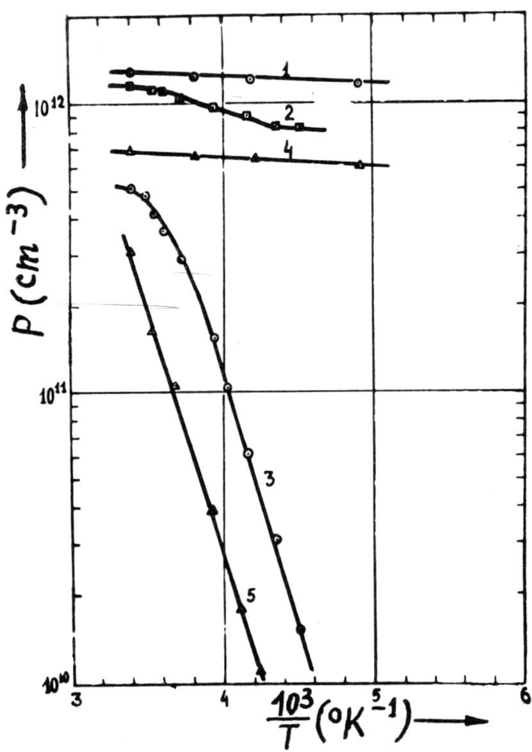

FIG. 6. Temperature dependence of hole concentration in highly pure p-Si irradiated of the fast neutrons: (1) p-Si ($\rho_0 = 2.6 \times 10^4$ $\Omega \cdot$cm) before irradiation; (2) after 9×10^{11} $n \times$ cm^{-2}; (3) after 1.1×10^{13} cm^{-2}; (4) p-Si ($\rho_0 = 4 \times 10^4$ $\Omega \cdot$cm) before irradiation; (5) after 1.1×10^{13} $n \times$ cm^{-2}.

Temperature dependence of the hole density $p = f(10^3/T)$ measured immediately after neutron irradiation of p-Si samples gives the level at $E_v + (0.35 \pm 0.01)$ eV (Figure 6, curve 3.5). The introduction rate of this level was $0.3(n \times$ cm$)^{-1}$. The concentration dependence of centers with the level at $E_v + 0.35$ eV with irradiation doses is linear and is expressed by the equation

$$N_d = K_1 \cdot \Phi, \qquad (2)$$

where $K_1 = 0.3(n \times \mathrm{cm})^{-1}$.

As a result of neutron irradiation the minority carrier lifetime in p- and n-Si of high purity considerably decreases. Judging from dose dependence of the reverse lifetime change $\Delta(1/\tau) = f(\Phi)$, the concentration of recombination centers also linearly increases with integrated neutron flux. From the lifetime temperature dependence the recombination level at $E_v + 0.33$ eV was determined, its cross section for electron being $\sigma_n = 5 \times 10^{-14}$ cm^2.

As distinguished from γ-irradiation, the carrier mobility at low temperatures considerably decreases as the result of fast neutron irradiation. Under annealing of fast neutron irradiated p-Si samples the hole concentration increases to the values 2.2 to 2.4×10^{12} cm^{-3}; that exceeds the hole concentration before irradiation. It is necessary to note that the slow annealing took place even at room temperature. Under annealing of fast

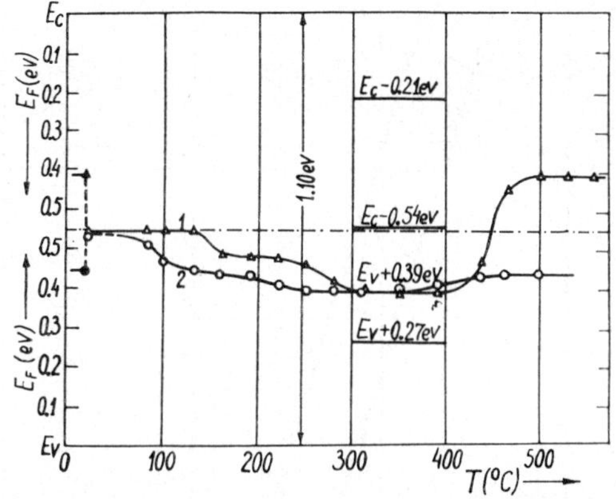

FIG. 7. Fermi level position at 300 °K in the process isochronal annealing after neutron irradiation with the dose 4×10^{13} $n \times$ cm^{-2}: (1) n-Si ($\rho_0 = 2.5 \times 10^3$ $\Omega \cdot$cm); (2) p-Si ($\rho_0 = 2.6 \times 10^4$ $\Omega \cdot$cm; ●, ▲-Fermi level position before irradiation.

neutron irradiated n-Si samples conversion to p-type took place, and the hole concentration increases to the values 2.2 to 2.4×10^{12} cm^{-3} and Fermi level occupies its boundary position at $E_v + 0.39$ eV. The return to the initial concentration occurs in the range of 300 to 500 °C. The Fermi level location at 300 °K in n- and p-Si after isochronal annealing is shown in Figure 7.

4.2. *Discussion*

Irradiation with fast neutron leads to the creation of both simple defects and more complex

defects as divacancies and disordered regions. If the predominant defects were divacancies then we should observe (at the conditions of our experiments) the level at $E_v + 0.27$ eV, conversion n-Si to p-type and also the boundary position of Fermi level $E_v + 0.39$ eV. In this case the hole concentration as the result of irradiation must increase to the value 2.2 to 2.4×10^{12} cm^{-3}, and the mobility doesn't change practically.

The mobility decrease indicates the presence of disordered regions.[25,26]

The fact that the level at $E_v + 0.27$ eV doesn't appear, may be connected with the presence of defects which gives a deeper level at $E_v + 0.35$ eV in Si gap.

Kinetics of the creation of defects with the level at $E_v + 0.35$ eV (2) under neutron irradiation differs from that of divacancy creation under γ-irradiation. The existence of the center at $E_v + 0.35$ eV was established by the authors.[21] Later this level was observed by authors,[12,13] who connected this level with an oxygen containing defect because it is observed under γ- and electron irradiation of oxygen containing silicon. Our data are obtained on the samples of pure silicon with low oxygen concentration ($\sim 10^{15}$ cm^{-3}) under neutron irradiation. Gregory[28] considered that the theoretical derived recombination level at $E_v + 0.35$ eV with cross section 4×10^{-15} cm^2 in neutron irradiated silicon belong to divacancy. Our data on the energy location of the level, its cross section and results of annealing contradict his view. Nor are the defects with the level at $E_v + 0.35$ eV complexes with acceptor impurity atoms, as the concentration achieves a value of 10^{13} cm^{-3} which considerably exceeds the impurity concentration. That's why it is possible to assume that level at $E_v + 0.35$ eV belongs to a complex defect which doesn't include an atom of oxygen or impurity.

The level at $E_v + 0.27$ eV appears under annealing of irradiated samples in the range of temperatures 100 to 200 °C. In this temperature range the conversion n-Si to p-type also takes place (Figure 7) and the hole mobility almost completely recovers. Intensive recombination center annealing took place between 100 to 200 °C. At 300 °C the annealing process slows down and some τ decrease is observed, that tells about recombination centers increase.

The data listed above show that it is likely in the process of annealing the disordered regions are destroyed. The split-off point defects may partly annihilate one another and partly form other defects, for example, divacancies. This process leads to increase of the divacancy concentration in high purity silicon.

When divacancies become the predominant defects the Fermi level occupies its boundary position $E_v + 0.39$ eV. The process of 'ripening' of defects upon annealing neutron irradiated crystals has been observed in a number of works.[29,30] In the case of high pure silicon this process leads to a divacancy concentration increase.

5. SUMMARY

The main defects introduced in high purity Si by Co^{60} γ-irradiation at 300 °K are divacancies and complexes of vacancies with impurity atoms of donor or acceptor type. Divacancies become the predominant defects after exhaustion of the dopant. When divacancies become the dominating defects the Fermi level occupies the boundary position in the Si gap at $E_v + 0.39$ eV and is situated approximately in the middle of the distance between the two divacancy levels at $E_v + 0.27$ eV and $E_c - 0.54$ eV.

The divacancy produced by Co^{60} γ-irradiation are secondary defects.

At γ-irradiation in high purity Si at least two types of recombination centers are introduced. After low doses the main role in recombination process is played by vacancy-impurity complexes, and, after heavy doses, divacancies. Radiation defects annealing in Co^{60} γ-irradiated Si of high purity show two stages, the temperature ranges of which correlate with the annealing ranges of vacancy-boron complexes and divacancies. Upon neutron irradiation of silicon of high purity disordered regions are introduced and the level at $E_v + 0.35$ eV is observed. The 'ripening' of defects in samples is connected with destroying of disordered regions.

Upon warming to 200–250 °C the Fermi level occupies its boundary position at $E_v + 0.39$ eV and the level at $E_v + 0.27$ eV appears. This indicates that divacancies become the predominant defects which can be formed as secondary defects at the destruction of the disordered regions.

REFERENCES

1. V. S. Vavilov, S. I. Vintovkin, A. S. Lutovich, A. F. Plotnikov and A. A. Sokolova, *Fiz. tverd. Tela*, **7**, 502 (1965).
2. S. V. Starodubtzev, V. M. Michaelyan, A. S. Lutovich, V. A. Sinukov and A. N. Suvorov, *Radiatz. Fiz. Nemetal. Kristal*, Izd Naukova Dumka (Kiev, 1967), p. 109.
3. H. Djerassi, J. Merlo-Flores and J. Messier, *J. appl. Phys.*, **37**, 4510 (1966).
4. I. D. Konozenko, A. K. Semenyuk, V. I. Khivrich and G. A. Dobrokhotov, *Fiz. Tekh. Poluprov.*, **3**, 155 (1969).
5. I. D. Konozenko, A. K. Semenyuk and V. I. Khivrich, *Phys. Stat. Sol.*, **35**, 1043 (1969).
6. G. I. Aleksandrova, *Atomnaya Energiya*, **23**, 106 (1967).
7. V. S. Chadrin, *Prib. i Tekh. Eksper.*, **4**, 222 (1965).
8. I. I. Gashka and Y. K. Pogela, *Fiz. tverd. Tela*, **1**, 1431 (1959).
9. J. J. Loferski and P. Rappaport, *J. Appl. Phys.*, **30**, 1181 (1959).
10. K. Lark-Horowitz, Semi-Conducting Materials, Proc. Conf. (London, 1951).
11. E. Sonder and L. C. Templeton, *J. Appl. Phys.*, **34**, 3295 (1963).
12. E. Sonder and L. C. Templeton, *J. Appl. Phys.*, **36**, 1811 (1965).
13. V. D. Tkachev, Ph.D. Thesis (Minsk, 1967), p. 111.
14. G. D. Watkins and J. W. Corbett, *Phys. Rev.*, **138**, A543 (1965).
15. J. W. Corbett and G. D. Watkins, *Phys. Rev.*, **138**, A555 (1965).
16. A. H. Kalma and J. C. Corelli, *Radiation Effects in Semiconductors*, Proc. Santa Fe Conf., 1967 (New York, 1968), p. 153.
17. A. Seeger and K. P. Chik, *Phys. Stat. Sol.*, **29**, 455 (1968).
18. L. J. Cheng, J. C. Corelli, J. W. Corbett and G. D. Watkins, *Phys. Rev.*, **152**, 761 (1966).
19. G. D. Watkins, *J. Phys. Soc. Japan*, **18**, Suppl. II, 25 (1963).
20. J. W. MacKay and E. E. Klontz, *Radiation Damage in Solids*, Vol. 3, 27 (Vienna, 1963).
21. J. W. MacKay and E. E. Klontz, *Radiation Effects in Semiconductors*, Proc. Santa Fe Conf., 1967 (New York, 1968), p. 175.
22. F. L. Vook and H. J. Stein, *Radiation Effects in Semiconductors*, Proc. Santa Fe Conf., 1967 (New York, 1968), p. 99.
23. K. Nakashima and Y. Inuishi, *Radiation Effects in Semiconductors*, Proc. Santa Fe Conf., 1967 (New York, 1968), p. 162.
24. M. Hirata, M. Hirata and H. Saito, *J. Appl. Phys.*, **37**, 1868 (1966).
25. B. R. Gossick, *J. Phys. Soc. Japan*, **18**, Suppl. 3, 226 (1963).
26. M. Bertolotti, *Radiation Effects in Semiconductors*, Proc. Santa Fe Conf., 1967 (New York, 1968), p. 311.
27. V. S. Vavilov, E. N. Lotkova and A. F. Plotnikov, *Proc. II Internat. Conf. on Photoconductivity* (1962), p. 31.
28. B. L. Gregory, *IEEE Trans. on Nuclear Science*, Vol. NS-16, 53 (1969).
29. R. F. Konopleva, S. R. Novikov and E. E. Rubinova, *Lattice Defects in Semiconductors* (Tokyo, 1968), p. 327.
30. G. Swenson, *Arkiv för fys.*, **28**, 67 (1965).

RADIATION INDUCED DEFECTS IN DIAMOND

C. D. CLARK AND E. W. J. MITCHELL†

J. J. Thomson Laboratory, The University of Reading, U.K.

A critical review of radiation damage in diamond is presented, focusing on important specific centres rather than attempting a review of the whole field. The proposed models for these centres are examined and often found wanting. Phenomena observed in ion bombardment of diamond is discussed and found to probably be due to amorphous or graphite surface layers. Throughout a number of suggestions of important experiments are given.

1. INTRODUCTION

The optical effects of defects produced by radiation damage were first systematically studied at Reading by Clark and coworkers[1] in 1956. It was appreciated then that, while some of the effects seemed to be intrinsic, others were influenced by impurities. This was underlined when a centre was found only in Type I diamonds (which from their physical characteristics exhibited impurity effects and were shown by Kaiser and Bond[2] to contain substantial quantities of nitrogen) by heat treatment after irradiation—and not by irradiation alone—and furthermore was shown by Elliott and coworkers[3] 1956 to possess a well defined symmetry axis corresponding to $\langle 110 \rangle$.

Since 1956 many workers in many laboratories have been endeavouring to unravel the intricate interplay of intrinsic defects and impurity effects in diamond. The work on silicon of Watkins and coworkers[4] and Bemski[5] and the continuing work of Watkins[6] has given detailed information about a number of complexes in silicon and this work has been substantially responsible for stimulating an increasing theoretical interest in the electronic structure of defects in covalent lattices. For the *tight-binding approach* used in most theoretical work so far, diamond probably represents in principle the best crystal in which to assess the adequacies, or inadequacies, of the theoretical work. In fact calculations about the structure of defects in diamond were first made by Coulson and Kearsley[7] in 1957 and by Yamaguchi[8] in 1962. However, more recently general ideas gained from the experience in silicon are being incorporated in theories which are being worked out for defects in diamond.

We believe, therefore, it is particularly appropriate to review briefly the experimental situation

† Department of Physics, University of Reading.

in diamond. In presenting this we shall, for completeness, include some relevant data which have appeared in the past[9,10] about radiation effects in diamond as well as the more recent work on which it seems that a constructive comparison is likely to depend.

2. THE CRYSTALS

One of the difficulties about the work on diamond is the extreme technical problem of systematically controlling and modifying a range of specific impurities. Very recent work[11,12] in this field seems more promising than earlier attempts and, hopefully, further development will lead to sufficient control to assist the interpretation of radiation effects.

For the work to be reported, however, the starting materials have had to be drawn from the following:

2.1. *Type Ia*

These are naturally occurring diamond containing as much as $10^{-3} - 5 \times 10^{-1}$ per cent nitrogen, much of it in precipitates on $\{100\}$. It has been pointed out[13,14] that it is extremely improbable that all the nitrogen atoms are in platelet form and recently this has been strongly emphasized by Sobolev and coworkers[15] from measurements of anomalous x-ray spike intensities and optical features. The direct comparison carried out by Sobolev has been questioned by Evans[16] who finds that, from measurements of x-ray spike intensity and direct electron microscopy of platelets, the spike intensity is related not simply to the total amount of precipitated material but is a function also of the size distribution. The relative amounts of nitrogen in different configurations have not unambiguously been established.

Type Ia diamonds usually show the 4150 Å (N3)

absorption and emission (blue) system (probably due[17] to a triangular arrangement of nitrogen atoms on {111}), strong absorption between 3 eV and the intrinsic indirect band to band edge at 5.4 eV,[18] and defect (nitrogen impurity in some form) activated 1-phonon vibrational absorption. Because of the large nitrogen content these diamonds are not particularly suitable for fundamental radiation damage studies. However, they show two interesting effects which could prove to be sensitive tests of theories. The first is the 5032 Å (H3) absorption and emission (green) system which is seen after heat treatment after irradiation. Some Type Ia diamonds contain this centre naturally when it presumably arises from natural geo-radioactivity irradiation plus heat treatment. This centre will be discussed in detail. The second interesting centre in Type Ia diamonds from the point of view of the present paper is the so called ND1 centre (3934 Å) studied by Clark,[19] Dyer and du Preez,[20] Ferdinando,[21] Dean,[22] and Davies and Lightowlers.[23] Although undoubtedly an impurity centre, the observed strength is influenced by radiation damage.

2.2. Type Ib[24]

Most of the Type Ib diamonds are synthetic but some natural samples show some of the features. These crystals contain singly substituted para-magnetic nitrogen centres.[25,26] They also show absorption between 3.0 and 5.4 eV and show defect activated 1-phonon vibrational absorption. They contain immeasurably small numbers of platelets although the synthetic samples frequently show evidence from x-ray studies of metal inclusions characteristic of the metallic solution from which they have been grown. The most important characteristic is the paramagnetic nitrogen centre which has associated with it an optical absorption band centred at 4.5 eV.[24] This is difficult to measure because of the presence of the strong 'Type Ia-like' absorption but there does appear to be some associated fine structure with possible zero-phonon lines at 4.07 eV and 4.58 eV.[24] An interesting feature of these samples is that sub-stitutional nitrogen pairs in nearest neighbour positions have been found and apparently only after irradiation.[27,28] Type Ib crystals may also show the ND1 system referred to in the section on Type Ia.

2.3. Type IIa

The absorption features referred to above are either very weak or absent in these diamonds. Apart from an absorption edge tail of varying strength it is rare for these samples to show any extrinsic features in optical absorption between the intrinsic 2-phonon vibrational absorption and the fundamental absorption edge. Laser excited emission spectra[29] show that at high intensities some of the Type Ia emission spectra may be produced indicating that in small concentrations some of the same centres as in Type Ia exist. Type IIa diamonds are all natural; the nitrogen content is less than 10^{-3} per cent.

2.4. Type IIb[30]

The types referred to above are good insulators. Type IIb diamonds are p-type semiconductors containing typically 5×10^{16} cm^{-3} acceptors (B or Al?),[31,32] and 5×10^{15} cm^{-3} partially compensating donors (believed to be N). Recently[12] this material has been grown on non-conducting seeds by a vapour transport method using BH_3/CH_4 decomposition on to hot diamond seeds under non-equilibrium conditions. Paradoxically natur-ally occurring IIb semiconducting diamonds re-present the purest form of diamond currently available. They have been used for optical and electrical studies of radiation induced defects.[33–35] A carrier removal rate of 0.5 cm was found for 2 MeV electrons at 80 °K.

3. GENERAL PROBLEMS

Before discussing particular centres we offer some general remarks. Without detracting from the considerable amount of theoretical progress which has been made it cannot be overemphasized that all the calculations which have been made represent severe approximations. Just how serious this is, in relation to accounting for experimental data, cannot be assessed since the data have not been uniquely related experimentally to the model whose properties are being calculated. In all theories—in spite of the apparent claims—optical transition energies may be adjusted between about 1 and 10 eV by appropriate choice of undetermined constants. At the present time probably the best assessment of the theoretical methods would be obtained by a calculation of the electronic states of the singly substituted N centre. The ground state is known from EPR and the related optical absorp-tion occurs between 4.0 and 4.6 eV. This calcula-tion unfortunately has not yet been made. How-ever, Every[36] has attempted, with some success, to co-relate the observed EPR spectrum, the optical spectrum and the infra-red vibrational spectrum with a model of the centre.

The most physically satisfying way[37a,b] of determining the wave functions, energy levels and energies of a defect is on the basis of the known Bloch Functions in the conduction and valence bands of the perfect crystal. This has been done for shallow levels in germanium[38] and the reverse process has been carried out for deeper levels in the alkali halides by Pekar.[39] However, in order to know in precisely what form to specify the defect one needs to know the nature of the surrounding distortions of the nominal defect (e.g. vacancy) before starting the calculation based on Bloch Functions. Although in principle some elaborate variational approach should be possible to cope with energy levels and distortions this is so far too demanding a computing problem from which to expect quantitatively meaningful results.

The approach which has been tried extensively is the defect molecule approach. Here the defect and its distortion are effectively decoupled from the remainder of the lattice and many electron molecular orbital theory used to calculate the electronic structure. Since experimentally the vibronic structure is of the medium coupling type it is clear that the limitation of the defect molecule approach (zero coupling) assumes less coupling than is actually the case. On the other hand this seems to be the only way at present capable of yielding predictions which can possibly be tested.[52] The most valuable predictions concern the nature of the electronic states and, in the cases of degenerate states, the distortions (static Jahn–Teller effect) of the nearest neighbours[74] which can be expected to occur. The calculated transition energies are summarized in Table I.

Details of these calculations are being reviewed elsewhere in this conference by Stoneham. Here we wish to emphasize the importance of establishing such characteristic differences—apart from transition energies—as may be calculable between vacancy based centres (vacancy, divacancy, vacancy-impurity complex etc. . . . .) and interstitial based centres.

Some of the obvious possibilities are:
(a) Symmetry of ground state wave function;
(b) Symmetries of the excited states;
(c) The distortion (if any) associated with (a) and (b);
(d) The associated optical spectra and the Huang–Rhys factor;

TABLE I

Summary of theoretical calculations using
Molecular Orbital Method

Reference	Defect	Transition	Energy (eV)
C & K[7]	V^0	$^1E - {}^1T_2$	8.0
		$^3T_1 - {}^3T_1$	9.7
		$^1E - {}^1T_2$	1.8–2.3
Y[8]	V^0	$^1E - {}^1T_2$	8.84
		$^1E - {}^1T_2$	6.46–7.31
H[52]	V^0	$^1E - {}^1T_2$	4.5
	with J–T	$^3T_1 - {}^3E$	3.4
	distortion	$^3T_1 - {}^3T_{1,2}$	3.7
		$^3T_1 - {}^3A_2$	5.1
Y[8]	V^-	$^4A_2 - {}^4T_1$	9.68
		$^4A_2 - {}^4T_1$	3.80–4.72
H[52]	V^-	$^2T_1 - {}^2T_2$	3.5
	with J–T	$^2T_1 - {}^2E$	3.7
	distortion	$^2T_1 - {}^2T_1$	4.5
		$^4A_2 - {}^4T_1$	2.8
Y[73]	I^0	$^1E - {}^1T_2$	1.33–1.72
H[52]	I^-	$^2E - {}^2T$	2.4–3.1
	with J–T	$^4A_2 - {}^4T_1$	2.2
	distortion		
C & L[75]	V_2	$^3A_{2g} - {}^3E_u$	0.75, 1.23
		$^3A_{2g} - {}^3A_{1u}$	3.05, 3.55
		$^3A_{2g} - {}^3A_{1u}$	3.17, 6.73
		$^1A_{1g} - {}^1E_u$	3.59, 3.59

Notes to Table I: All authors use several values of penetration and overlap integrals; where authors state reasons for preferring one value that value is underlined but reference must be made to original papers for details. [Since this paper was prepared two further calculations have been reported: (i) a vacancy calculation, involving whole crystal wavefunctions, by Lenglart at this conference; and (ii) substitutional N and B by Messmer and Watkins at the Boston Semiconductor Conference and for the vacancy at this conference. Both are valuable in relating defect energy levels to the band edges and Lenglart finds a transition energy of 1.8 eV for the neutral vacancy.]

(e) The details of the vibronic spectra;
(f) The stability and energy of motion of the defect.

The theoretical information about 'a–d' is summarized in Table I. Calculations of (e) have been made by Ritter[40] using methods developed by McCombie and coworkers for treating the electron-vibrational defect states as perturbed cases of response functions of the whole crystal. This allows individual structure to appear in the calculated spectrum (which does not appear in other calculations) for comparison with experiment.

In what follows we shall concentrate on such information as exists about symmetries, utilizing results from polarization of luminescence, uniaxial stress splittings of absorption lines and the magnitudes of hydrostatic shifts.

4. THE PRODUCTION AND CHARACTERISTICS OF THE GR1 BAND

The absorption spectra of electron irradiated diamonds all show the prominent features illustrated in Figure 1. Clark, Kemmey and Mitchell[34] found that the rates of production of the various features in Type IIa diamonds during electron irradiation at 290 °K depended upon the energy of the incident electrons. The increase in absorption was nearly linear with electron dose and it is presumed that intrinsic defects are responsible for the effects.

The zero-phonon line at 1.673 eV (see Figure 1) and its associated vibronic structure has been labelled the GR1 band.[1] A fluorescence emission characteristic of the GR1 band has been observed by Norris.[41] A sharp emission line close to 1.673 eV with vibronic structure to lower energies is excited by light in the range 1.67–2.5 eV and above 2.8 eV. It was this emission which was monitored to record the excitation spectrum of Figure 1. It is concluded that the lines labelled GR3 and GR6 (2.887 and 2.957 eV respectively) arise from transitions within the same defect that causes the GR1 band. Moreover, at least part of the UV absorption band at energies greater than 2.8 eV is associated with the GR1 defect.

Confirmation of the above conclusions is provided by the work of Dyer and Ferdinando,[42] where the GR1 and UV absorption bands were not observed in Type IIb semiconducting diamond until the irradiation dose was sufficient to compensate all the impurity acceptor centres in the crystal. Thus the GR1 and UV bands are considered to be associated with irradiation induced

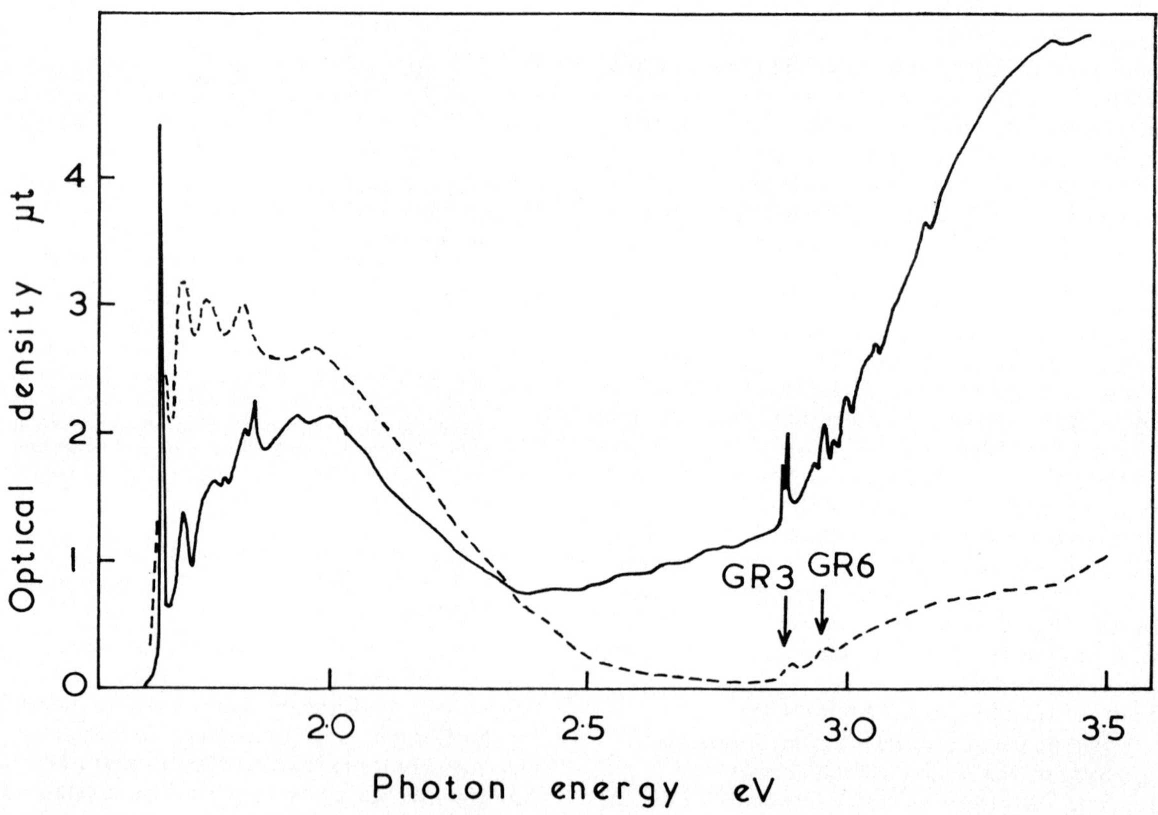

FIG. 1. Absorption spectrum (full line curve) and excitation spectrum (broken curve) of a Type IIa diamond which has been irradiated with 2.0 MeV electrons. Optical measurements were made at 80 °K on instruments of different resolving power.

donor centres. Since it is unlikely that such a donor centre would be positively charged, these results indicate that the defect involved is either neutral or negatively charged.

The rates of production of the GR1 band in Type IIa diamonds at 290 °K during irradiation with electrons of various energies have been compared[34] with calculated curves for the rates of displacement of atoms assuming a range of values for the threshold displacement energy. In the electron energy range 0.3–1.0 MeV the results are consistent with a displacement energy of 80 eV.

The rates of change of the electrical properties[34] of *p*-type semiconducting diamonds can be used to get a quantitative measure of the rate of production of irradiation donor centres. These results are also consistent with a threshold displacement energy of 80 eV.

By comparing the production of the GR1 band in Type IIa and Type IIb diamonds irradiated under the same conditions Ferdinando[21] has calculated the oscillator strength for the GR1 band. It is assumed that each irradiation induced donor centre compensates one of the acceptor states. There was a spread of values for different pairs of IIa/IIb specimens but nevertheless the oscillator strength for the GR1 band was found to lie in the range 0.05–0.15. The strength of the UV band appears to be about 20 times stronger than the GR1 band, but exact estimates are difficult because of the problem of overlapping bands.

Some of the range of oscillator strengths mentioned above probably arises from the changes in occupancy of the electronic ground state of the GR1 defect even in the semiconducting crystals. It is reported[43] that during electron irradiation at 100 °K the strength of the GR1 band gave a sublinear plot against electron dose. Small changes in strength of the band were observed by warming to room temperature, but these were recovered very rapidly during subsequent irradiation at 100 °K.

The detailed shape of the optical absorption spectrum in the 1.6–2.5 eV range can vary from specimen to specimen and probably arises from the presence of several zero-phonon transitions with their accompanying vibronic structure. The presence of these transitions frequently can be established by comparing optical absorption spectra with photo- and cathodo-luminescence spectra. On other occasions the zero-phonon transitions can be recognized by relative changes in strength of the absorption features. For instance, Figure 2 shows some results obtained by Duncan[44] during three successive irradiations on the same crystal. The sample was kept below 100 °K throughout the experiment. It is seen that between (a) and (b) the GR1 band decreased during irradiation and at the same time the 1.678 eV line almost completely disappeared leaving only a faint bump on the absorption curve. Detailed comparison of the curves (a) and (b) showed changes in the vibronic structure. The GR3 (2.887 eV), and GR6 (2.957 eV) peaks also were greatly reduced in strength. Curve (c) of Figure 2 shows the reappearance of the GR1a line (at 1.678 eV) and an increase in the strength of the 'GR1 band' in a succeeding irradiation. These results suggest that there may be a second zero-phonon transition at 1.717 eV (GR1b). In many irradiated crystals the ratio of the absorption coefficients at 1.678 and 1.717 eV is not constant. The results of Figure 2b are presumed to be an extreme case of the relative abundance of the GR1a and GR1b defects.

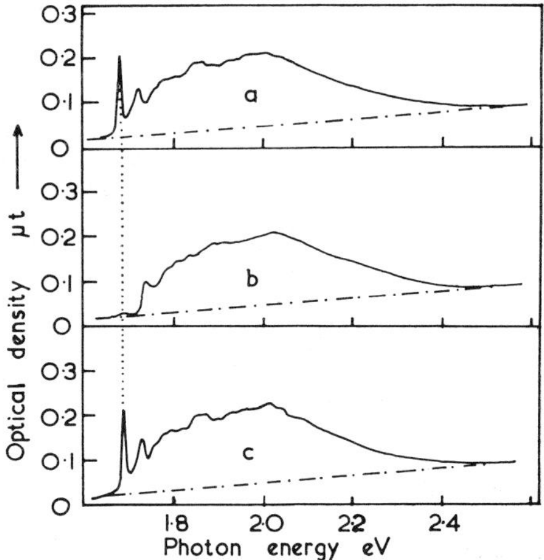

FIG. 2. The GR1 band produced by 1.0 MeV electron irradiation of a Type IIa diamond after doses (a) 4.51×10^{17}, (b) 5.26×10^{17}, and (c) 6.02×10^{17} electrons cm^{-2} at about 100 °K. Optical measurements were recorded at 80 °K. The integrated strengths of the band are

(a) $7.38 \times 10^{-2} \mu t$ eV
(b) $7.05 \times 10^{-2} \mu t$ eV
and (c) $8.20 \times 10^{-2} \mu t$ eV.

There are several reports of 'line splittings' of the GR1a line in the literature.[21,45,46] Runciman[45] found a splitting of about 6 Å in the GR1a line in Type I diamonds and attributed this to internal strains. Ferdinando[21] found that the GR1a line showed a splitting which varied from one Type IIb crystal to another within the range 5–13 Å, and that the relative strengths of the components was not constant. Illumination with light of various wavelengths could be used to alter the splitting of the components. Measurements[21] also showed that some crystals which had a strong birefringence pattern and presumably had considerable internal strain did not show a line splitting whereas other less birefringent crystals did exhibit a splitting. It seems unlikely that local strains can give by themselves an adequate explanation of these results. It is possible that the results are due to the presence of another defect such as substitutional impurity in close proximity to the GR1 defect. A similar explanation has been suggested by Mitchell[47a] to account for the large line width of the GR1a line after neutron irradiation. It was suggested that the GR1a line sharpened during subsequent heat treatment because close vacancy-interstitial pairs annealed to leave only widely separated defects which all absorb at the same energy.

In spite of the above problems of line shape and multiplicity of the GR1a line in unstressed crystals, Runciman[45] and Crowther[48] have measured the absorption spectra in polarized light under uniaxial stresses applied along the three principal directions [100], [110] and [111]. The results of the two authors are not in complete agreement and it is clear that more extensive and detailed experimental work is required. No satisfactory explanation for the above results have been given, though Lannoo and Stoneham[49] discuss the possibility that a dynamic Jahn–Teller distortion within the neutral vacancy may give results similar to those obtained experimentally. In a later publication Stoneham and Lannoo[50] point out that mixed distortions can arise for defects in valence crystals when two electronic states of different symmetry are themselves nearly degenerate. Such effects were suggested to be important in the interpretation of the uniaxial stress data of the GR1a line.

Norris,[41] using the techniques described by Elliott, Matthews and Mitchell[3] and by Clark, Mitchell and Maycraft,[51] has measured the polarization of luminescence of the GR1 band in the temperature range 4.2 to 290 °K, using plane polarized 2.15 and 2.271 eV excitation radiation. Within experimental error the luminescence was found to be unpolarized. This result can be explained[49] in terms of a static tetragonal distortion of the electronic ground state in the neutral vacancy and a trigonal distortion of the excited state. Emission from any trigonally excited state can take place with equal probability to any of the tetragonally distorted ground states to give an unpolarized emission. Under suitable experimental conditions it would seem that such a model would predict that the vacancies in a crystal could be optically pumped to give a preferential orientation to the distortion and thereby give rise to polarized absorption. As far as we are aware this experiment has not been carried out.

Alternatively the absence of polarization of luminescence could be explained by a defect of fairly high symmetry (e.g. T_d) having a non-degenerate ground state, and having excited states which have a number of crystallographically equivalent distortions. If the initial excitation is to an intermediate configurational state from which any of the distorted configurational minima may be reached it is possible to have geometrics for which the net result would be that luminescence from the minima would not be polarized.

Lannoo and Stoneham[49] have estimated that the value of the Huang–Rhys factor for the neutral vacancy is 3.68. This is in good agreement with the experimental value of 3.57 ± 0.06,[44] obtained by measuring the ratio of the intensity of the GR1a zero-phonon line to the integrated strengths of the GR1 band. However, as pointed out earlier the relative strengths of GR1a and 1b do appear to vary (e.g. Figures 2a and b) and while the value of 3.57 at 80 °K is typical of most cases there are inexplicable deviations as for the spectrum of Figure 2b.

Hagston[52] has attempted to calculate the electronic states of vacancies and interstitials allowing coupling to the lattice via the dynamic Jahn–Teller effect. The conclusion of his calculation, contrary to Lannoo and Stoneham, is that the GR1 band is associated with a negatively charged interstitial atom. This centre exhibits several sharp electronic lines, shows photoluminescence and occurs at approximately the correct energy. (The vacancy in Hagston's calculations does not possess these characteristics). In our view it is difficult to maintain this conclusion in the light of other work relevant to interstitial atoms in diamond.

Owen[57] has argued that two of the EPR spectra observed[53–56] are associated with interstitial atoms. One centre has $S = 1$ and an axis along [100] and the other has $S = 1$ and an axis along [994]. He suggests that the two centres arise from different distortions around the interstitial, arising presumably from different charge states.

If both Hagston and Owen are correct then the GR1 optical band should co-relate with the EPR

$\langle 100 \rangle$ centres or the EPR $\langle 994 \rangle$ centres. This has been examined[58] by studying optical and EPR spectra during electron irradiation at various energies in the range 1.0–2.0 MeV. In fact a good co-relation was observed between the production rates of the GR1 band and the $\langle 100 \rangle$ EPR centres. However, in subsequent annealing studies at 450 °C the EPR spectrum was annealed completely[58] but the GR1 band was only partially reduced in strength. These annealing studies were made on different samples and need to be confirmed by simultaneous measurements.

As it stands the annealing result shows that the GR1 band and the EPR $\langle 100 \rangle$ spectrum cannot be associated with the same centre. This does not conflict with the co-relation found for the production of the effects—this requires, for example, two centres being produced in the same proportion during irradiation as would be the case for vacancies and interstitials. If some interstitials were trapped by impurities or other imperfections then mobility of the remaining isolated interstitial could lead to a *decrease* in effects associated with vacancies, but the complete *removal* of effects associated with interstitials.

In silicon and germanium interstitials are mobile at lower temperatures than vacancies, although the mobility does depend on the charge state.[6] Interstitials are mobile in silicon at 4 °K[6] and in germanium in various charge states in the range 4–60 °K; it seems that 730 °K is much too high to correspond to interstitial mobility in diamond, even allowing for the stronger binding. Furthermore Wild[60] has observed an EPR spectrum following electron irradiation below 55 °K which is annealed by heating at 130 °K. The spectrum has not been analysed and it is not recoverable on cooling to 55 °K after heating above 130 °K. The large electron dose involved of 10^{16}–$10^{17}\ e^-$ cm^{-2} makes it unlikely that this could be an ionization effect. Wild[60] proposes, therefore, that this spectrum is associated with the interstitial in some form which is capable of motion at 130 °K. This is an added reason for believing that 730 °K cannot be the interstitial motion temperature. It is more likely that the 730 °K annealing stage involves the thermal release of interstitials from complexes at which they were trapped and the partial removal of the vacancy concentration.

Thus from these annealing studies it is difficult to maintain that the EPR $\langle 100 \rangle$ can be associated with isolated interstitial atoms or vacancies. Equally it is difficult to maintain that the GR1 band which persists to 950 °C could be associated with isolated interstitials. The working kinetic model is that interstitials are mobile at relatively low temperatures and become trapped in several ways. One of these gives the EPR $\langle 100 \rangle$ spectra and at 450 °C detrapping occurs with partial annealing of the vacancies, and a partial removal of the vacancy associated GR1 band.

We conclude that neither Hagston's[52] or Owen's[57] assignments are correct. Further work is required on the low temperature EPR spectra.

We have discussed above a number of experimental aspects of the main radiation induced centres in relation to theoretical calculations. As Lannoo and Stoneham[49] have pointed out it is difficult to distinguish many electronic features of the vacancy and the interstitial. The lattice coupling might be expected to be very different, however, and it is this—through the observed vibronic structure and the presumed higher motion energy for the vacancy compared with any form of isolated interstitial, that leads to the continuing tentative assignment of the GR1 optical band to a vacancy.[47b]

5. THE INTERACTION OF RADIATION INDUCED DEFECTS WITH IMPURITIES

5.1A. *The nitrogen molecule centre*

In this and the following section we discuss two defects which, although their production is greatly influenced by irradiation, do not appear themselves to incorporate either vacant lattice sites or interstitial carbon atoms.

The first is the ionized substitutional nitrogen pair reported by Shcherbakova and coworkers[27] and found in natural Type Ib diamonds. The nitrogen atoms are found to be in substitutional sites in near neighbour positions. Loubser[26] has recently reported that the concentration of these nitrogen pairs can be increased during electron irradiation. It is this result which we wish to discuss in this paper.

Shcherbakova and coworkers[27] have suggested that this effect arises from the ionization of nitrogen pairs already in the crystal in an unobservable charge state. This seems unlikely in view of the high radiation doses required to observe the centres. Indeed one would expect to render the centres observable using UV illumination rather than the extensive high energy electron doses used both by Shcherbakova and by Loubser.

On the other hand, although isolated neutral

substitutional nitrogen is known to be stable in diamond at room temperature[25] it seems unlikely that this defect can provide the necessary atomic mobility for the formation of the substitutional nitrogen pairs. Interstitial nitrogen atoms may provide the necessary mobile species but then one has to consider the formation of that defect. Simple interstitial impurities have been observed[6] in silicon as a result of irradiation at $4\,°K$ in concentrations comparable to the calculated concentration of vacancies and silicon interstitials expected. Since no evidence has been observed for the existence of interstitial silicon atoms after irradiation, Watkins[6] has suggested that the silicon interstitials are extremely mobile at $4.2\,°K$, and diffuse to the sites of substitutional impurities which exchange their substitutional sites for interstitial positions. A similar process in irradiated Type Ib diamonds could produce interstitial nitrogen atoms. Even if this is a possible mechanism, it remains to explain the formation of substitutional nitrogen pairs. One can invoke the existence of nitrogen-vacancy pairs either present before or produced during the irradiation, and then allow these to trap diffusing nitrogen interstitials. This would mean that the vacancies have to be mobile in order to form the nitrogen-vacancy pairs and this is unlikely.

An alternative explanation for the formation of nitrogen pairs could be the degradation of bigger nitrogen complexes by the successive substitution of carbon interstitial atoms for nitrogen atoms in the complex. Such a mechanism would seem to imply that nitrogen pairs themselves would sooner or later finish up being dissociated by migrating carbon interstitials, possibly accompanied by an increase in the isolated paramagnetic substitutional nitrogen concentration and by single interstitial nitrogen atoms. These are tentative mechanisms which are capable of further experimental investigation.

5.1B. *The* ND1 *system*

The defect which has been labelled ND1 by Dyer and du Preez[20] can be produced in Type Ia diamonds by irradiation at room temperature followed by thermal annealing in the range $250–500\,°C$. The optical absorption of the ND1 centre has a zero-phonon transition at 3.149 eV at $80\,°K$ and associated vibronic structure consisting of a series of overlapping broad bands of separation about 0.08 eV. The ND1 absorption band is similar in appearance to the N3 system[19] on which it is often superimposed.[61] No photoluminescence has been found to be associated with the ND1 absorption system.

A defect with almost identical optical absorption characteristics can be produced readily in Type Ib diamonds by electron irradiation at room temperature. Only very feeble ND1 absorption bands have been detected in Type IIa and IIb diamonds and it seems certain that the defect depends upon the presence of nitrogen in the lattice.[23] The different behaviour of the centre in Type Ia and Ib diamonds under irradiation, heat treatment and optical illumination as described by Dyer[62] is perhaps not surprising since the nitrogen is present predominantly in quite different forms.

Davies and Lightowlers[23] have reported measurements of the polarized absorption under uniaxial stress on diamonds containing the ND1 centre. There are two components in the spectrum when the crystal is stressed parallel to either the $\langle 100 \rangle$, $\langle 110 \rangle$, or $\langle 111 \rangle$ directions and the optical absorption is measured in a direction perpendicular to the applied stress. In each case one component is polarized with the electric vector parallel and one with the electric vector perpendicular to the applied stress. The results are consistent with an A–T transition in a centre of tetrahedral symmetry, and were assigned to an interstitial nitrogen atom. The line widths of the stress split components were found to increase with the applied stress and this was attributed to different ND1 centres experiencing different effective stresses due to the inhomogeneous distribution of nitrogen platelets in the crystals investigated. This may be an indication that the ND1 centres are formed peripherally to the nitrogen platelets in Type Ia diamonds.

If the ND1 centre is indeed an interstitial nitrogen atom it raises again the question as to how it is formed during irradiation. A similar mechanism to that referred to above is that interstitial carbon atoms migrate to nitrogen platelets and exchange at the periphery to form interstitial nitrogen.

Since the ND1 centre is stable to at least $500\,°C$[20] and the existence of a mobile interstitial nitrogen cannot be assumed to be one of the stages in the formation of substitutional nitrogen pairs during irradiation or during heat treatments below $500\,°C$, it would seem that measurements of the rates of formation of nitrogen pairs as a function of temperature of irradiation might yield some useful information about ND1 as well as the pairs.

TABLE II

Summary of piezospectroscopic data on H3 and H4 centres as deduced from the paper by Runciman

| Stress direction | Experimental results | | | | Intensity ratios for monoclinic I $I_\parallel : I_\perp$ (Theory) | | | Number of components | | |
| | H3 | | H4 | | | | | | | |
	ΔE (meV)	$I_\parallel : I_\perp$	ΔE (meV)	$I_\parallel : I_\perp$	$Z\langle 110\rangle$	$X\langle 110\rangle$	Expt.	Orthorhombic I (theory)	Monoclinic I (theory)	†(i)
$P\|[100]$	+6.2	2:1	+6.2	2:1	2:1	2:1	2	2	2	(1)
	−9.3	0:1	−7.4	0:1	0:1	0:1				(2)
$P\|[111]$	−3.7	0:4	−2.7	0:6	8:2	0:6				(1)
	+6.2	2:4	+2.5	6:6	0:3	4:1	3	2	3	(2)
	+11.2	6:0	+5.6	10:3	0:3	4:1				(3)
$P\|[110]$	+14.0	1:0	+5.6	3:0	0:2	0:2				(1)
[111] View	0	3:6	0	10:16	3:5⎫	3:5⎫	3	3	4	(2)
					3:5⎭	3:5⎭				(3)
	+21.5	2:0	+9.3	3:0	6:0	6:0				(4)

† i—an integer labelling the centres.

5.1C. *The* H3 (5032 Å) *and* H4 (4965 Å) *systems*

The H3 and H4 centres are produced in large concentrations in Type I diamonds by irradiation and heat treatment.[1,62,63] Thus it seems reasonable to suppose that these centres are due to the association of an irradiation induced defect with nitrogen present as an impurity. Bearing this in mind we now go on to consider what information about the defects may be obtained from measurements of optical absorption under uniaxially applied stresses[45] and hydrostatic stress[64] and from measurements of the polarization of luminescence.[65]

5.2. *Uniaxial and hydrostatic stress experiments*

Piezospectroscopic measurements taken from the paper by Runciman[45] are summarized in Table II. Under the column headings H3 and H4 are given the estimated ratios of the absorption measured in plane polarized light with the electric vector parallel and perpendicular to the stress direction given in column 1. The values ΔE give the energy shift of the measured components under an applied stress of 1.2×10^4 kg cm^{-2} from the energy of the line in the unstressed crystal. The numbers of lines observed experimentally for the different stress directions are compared in the last three columns of Table II with the number of components expected theoretically from defects of orthorhombic I and monoclinic I symmetry showing orientational degeneracy.[45,66] Defects of these symmetries are those for which the predicted number of components best fits the observations. The number i in

brackets in the monoclinic I theory column is the same i used by Kaplyanskii[66] to distinguish between groups of orientationally degenerate centres.

For stress parallel to [111] there are more components observed than are predicted for orthorhombic I, which is a difficult situation to explain. Thus monoclinic I best fits the results and we proceed on the assumption that for $P\|[110]$ the two expected components for which $i = 2$ and 3 are unresolved experimentally.

Kaplyanskii[66] has tabulated expressions for the magnitude of the splitting of the lines under uniaxial stresses parallel to [100], [110] and [111], together with the orientational degeneracy under stress of each line. Comparison of these quantities with the experimental shifts allows the calculation of the components of the piezospectroscopic tensor for the centres, whence the shift of the absorption line under hydrostatic pressure can be calculated.[66]

Figure 3 shows the hydrostatic shift of the H3 and H4 absorption lines as measured by Wedlake.[64] The line shifts are linear with applied stress and are compared in Table III with the hydrostatic shift rate calculated from the uniaxial data of Table II. Considering the inevitable inaccuracies of the data in Table II, the predicted and measured shift rates are in reasonable agreement for stresses applied along [100] and [111] for both H3 and H4. For $P\|[110]$ the predicted shift rate agrees well with the hydrostatic shift rate calculated for $P\|[100]$ and [111] for H4, using the assumption that the components $i = 2$ and 3 overlap. The results for $P\|[110]$ for H3 are not in such good agreement.

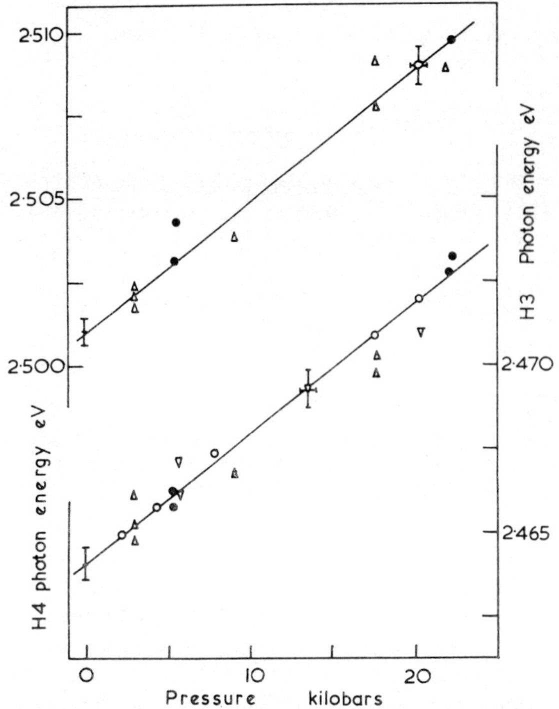

FIG. 3. The energy of the zero-phonon lines H3 and H4 as a function of hydrostatic pressure. Measurements were recorded at about 100 °K. Experimental points represent results on different samples.

TABLE III

Comparison of the hydrostatic shift rate of the H3 and H4 centres

Centre	Hydrostatic shift rate (eV/k bar)	
	Predicted Average of calculated shifts using uniaxial data for $P\|[100]$ and $P\|[111]$	Experimental
H3	$4.5 \pm 1.2 \times 10^{-4}$	$3.9 \pm 0.1 \times 10^{-4}$
H4	$3.2 \pm 1.2 \times 10^{-4}$	$3.9 \pm 0.4 \times 10^{-4}$

The allowed electric dipole transitions in defects of monoclinic I symmetry correspond to a $Z\langle110\rangle$ oscillator orientated parallel to the $\langle110\rangle$ major axis of symmetry of the defect, and a pair of oscillators X and Y mutually at right angles to the $Z\langle110\rangle$ oscillator. The relative orientations and magnitudes of the X and Y oscillators are not specified by symmetry considerations. Under uniaxial stress, the centres which are not orientationally degenerate generally will give different anisotropic

absorption ratios. The absorption anisotropy expected for $X\langle110\rangle$ oscillators in a monoclinic I centre are listed in column 7 of Table II. These may be compared with the experimental values for the H3 and H4 systems in the fourth and sixth columns of the table. Whilst corresponding ratios for $P\|[100]$ are in good agreement the ratios for $P\|[111]$ are far from agreement.

Wedlake[64] has shown that the magnitude of the X (say) oscillator must be much greater than the Y and that with this assumption and with the orientation of an X oscillator along $\langle110\rangle$ very good agreement is obtained with the experimental ratios; the predicted absorption ratios for this are quoted in column eight of Table II. Whilst the anisotropic absorption ratios give reasonable agreement in most cases, nevertheless the ratio of the absorption intensities for components for different values of i are not very satisfactory. A possible explanation for this type of behaviour is that under the effects of applied stress and at sufficiently high temperatures it may be possible for defects to realign themselves preferentially, thereby reducing their total potential energy.

The above results and discussion are reminiscent of the interpretation by Watkins and Corbett[67] for the EPR spectra obtained in conjunction with uniaxial stress on the 'E-centre' in silicon. In order to account for the presence of only three lines in the E-centre spectrum with $P\|[100]$ it was assumed that the lines for which $i = 2$ and 3 were not resolved—a situation analogous to that used earlier to account for the H4 $P\|[100]$ results. The defect model proposed by Watkins and Corbett[67] was a substitutional phosphorus atom in association with a vacant lattice site in a near neighbour position. Such a centre would be trigonal but for the fact that it has a doubly degenerate E ground state. This degeneracy is lifted by a Jahn–Teller distortion to give a centre which behaves as monoclinic I at low temperatures. At higher temperatures a thermally activated reorientation of the distortion or of the defect itself was observed.

The behaviour of the H3 and H4 centres are very similar and a possible model for either of the centres seems to be a nitrogen substitutional atom in a nearest neighbour position to a vacancy. Figure 4 represents the proposed model. In accord with the Watkins and Corbett model[67] for the E-centre, when a nitrogen atom replaces a carbon atom next to a vacancy the two excess electrons of the nitrogen pair off. Any two of the remaining electrons of the three other carbon atoms near the vacancy can

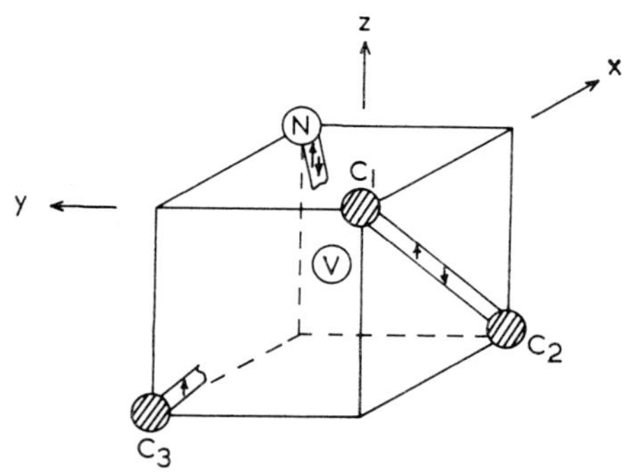

FIG. 4. Proposed model of a nitrogen-vacancy defect. Due to a Jahn–Teller distortion the defect symmetry is monoclinic I.

pair off leaving one unsaturated bond. The paired carbon bond is parallel to $\langle 110 \rangle$. The reflection plane which defines the monoclinic I centre passes through the carbon atom C_2, the nitrogen atom N and the vacancy V. A reorientation of the Jahn–Teller distortion corresponds to the switching of the bond pairing between the three possible $\langle 110 \rangle$ directions associated with any particular nitrogen-vacancy $\langle 111 \rangle$ direction.

The probability that a reorientation will occur depends upon the number of sites to which the distortion is allowed to switch, which in turn depends upon the direction of the applied stress. The probability of occupancy of a particular pair of distortion configurations also depends on the difference in potential energy $\Delta E'$ between the two distorted states. The relative populations of centres which are not orientationally degenerate is then given by the Boltzmann factor $\exp(-\Delta E'/kT)$, where $\Delta E'$ is the energy separation under stress of the absorption components associated with the two distorted configurations of the defect.

In monoclinic I centres $X\langle 110 \rangle$ electric dipole transitions are allowed between A' and A'' states. There are two cases: case (a) in which a Jahn–Teller distortion occurs in the ground state, and case (b) in which distortion occurs in both states, the Jahn–Teller energy being different in the ground and excited states. Wedlake[64] has shown that case (a) does not satisfactorily explain the stress data. It has been shown that the computed changes in strength of the split absorption components under reorientation give an even worse fit to experimental uniaxial results than that obtained assuming no reorientation of the distortion. In principle it seems likely that case (b) may give an adequate explanation but more detailed piezospectroscopic results on the H3 and H4 centres as a function of applied stress and of temperature are needed before much further progress can be made.

5.3. Polarization of luminescence experiments

Some results for the polarization of luminescence of the H3 system obtained by Clark and Norris[65] are reproduced in Figure 5A. It is seen that the polarization of the emitted light for 2.82 eV polarized excitation is higher than that for 3.40 eV polarized excitation.

Figure 6 shows the variation of the polarization p and phase ϵ as a function of wavelength and for a fixed orientation of the electric vector of the exciting radiation. For comparison the excitation spectrum of the H3 system is shown also. It is seen that p and ϵ are nearly independent of the energy of the exciting light until the energy approaches the sharp line at 3.368 eV. The 3.368 eV line also appears in the absorption spectrum (mentioned as H13 by Clark, Ditchburn and Dyer[1]). The evidence is that the 2.464 and 3.368 eV lines arise from transitions within the H3 centre.

The three-level energy scheme which was proposed[65] to account for the above H3 luminescence results is modified somewhat in the explanation which follows.

In the undistorted configurations, $E \to E$ transitions are allowed in trigonal defects. A schematic configuration co-ordinate diagram for the distorted states in a monoclinic I defect is shown in Figure 5B. The ground and excited states are assumed to be distorted similarly so that the three potential wells have approximately the same configuration. The barrier height for reorientation of the distortion is assumed to be greater than kT because it is reported[65] that the polarized luminescence results are not significantly changed in the temperature range 80–300 °K. Considering one centre, it is suggested that an $X[110]$ transition between the two A'' states gives the H3 absorption line while $Z[1\bar{1}0]$ transitions from the A'' ground state to the A' excited state give the H13 absorption line.

Photoluminescence corresponding to $X[110]$ transitions between the A'' states occurs for low energy excitation. After high energy excitation ($A'' - A'$) a non-radiative relaxation of the centre in the excited state to any of the three distorted minima in the excited A'' state will occur with equal probability so that emission ($A'' \to A''$) will be characteristic of $X[110]$, $X[101]$ and $X[0\bar{1}1]$ transitions, all these being perpendicular to the

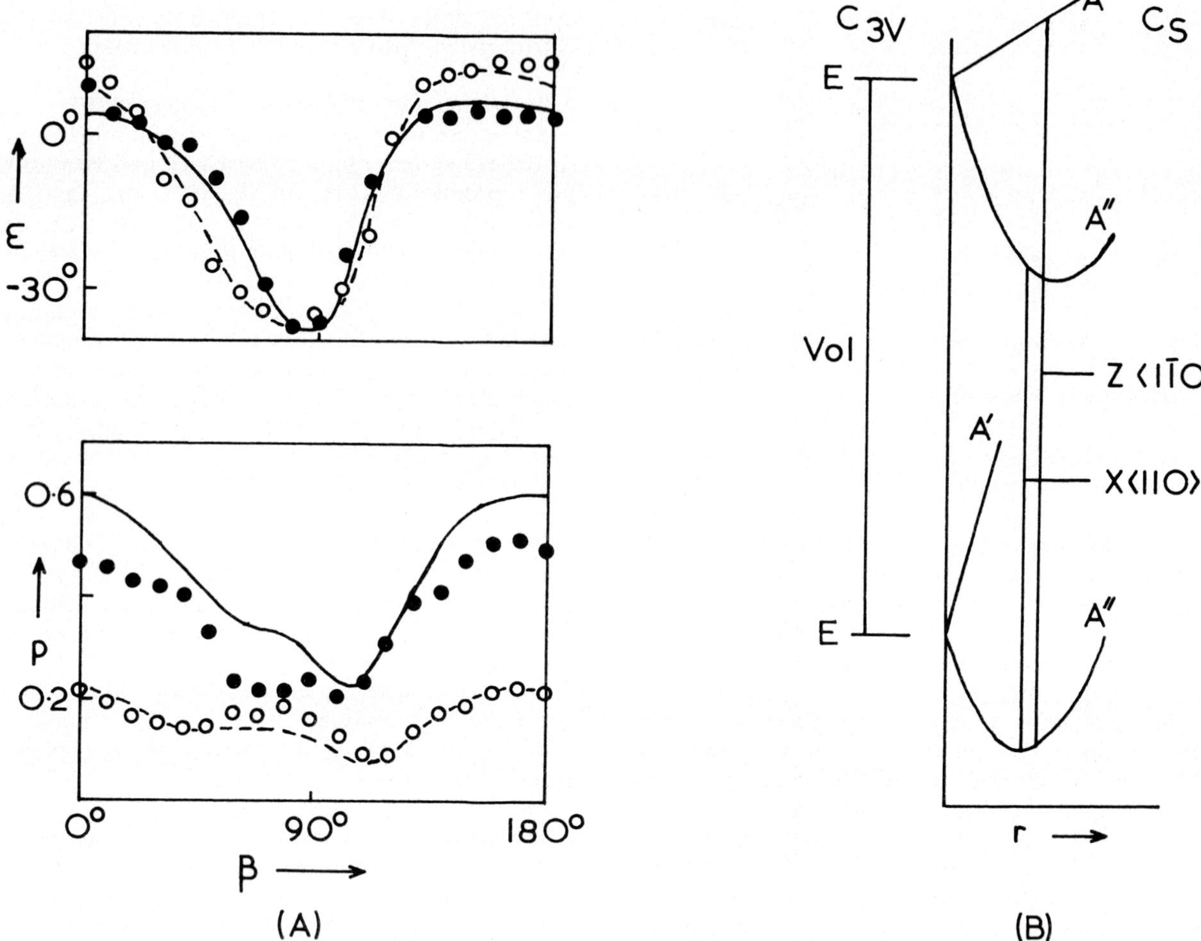

FIG. 5. Polarization p and phase ϵ measurements as a function of the orientation β of the electric vector of the exciting light for the H3 system. Exciting light was incident close to [111] and emitted light observed close to [1̄10]. Full and open circles give results for 2.82 and 3.40 eV excitation respectively. The curves are those predicted by theory as indicated in the text.

[1̄11] direction of one of the nitrogen-vacancy directions in the undistorted trigonal state. Wedlake[64] has shown that the curves for the polarization p and phase ϵ as a function of the angle of the electric vector of the exciting light for the above model are very similar in shape to those obtained experimentally. The predicted curves are indicated in Figure 5A. The full line curves are those calculated for low energy excitation and involve $X\langle110\rangle$ transitions only, while the broken curves are appropriate to transitions in which a reorientation of the distortion may occur in the excited state.

The model discussed above is speculative, but the wide range of data it satisfactorily describes is

sufficient to stimulate measurements of polarized luminescence and piezospectroscopic measurements over a greater temperature range than hitherto, and in more quantitative detail. Also if the model is correct then the centre should show an electron paramagnetic resonance spectrum.

6. ION BOMBARDMENT OF DIAMOND

It is not intended to discuss the electronic excitation conductivity produced by the passage of a charged particle, but the metastable disruptions or modifications of the diamond lattice produced near the surface by ion bombardment. Briefly the observed facts seem to be that conducting surface

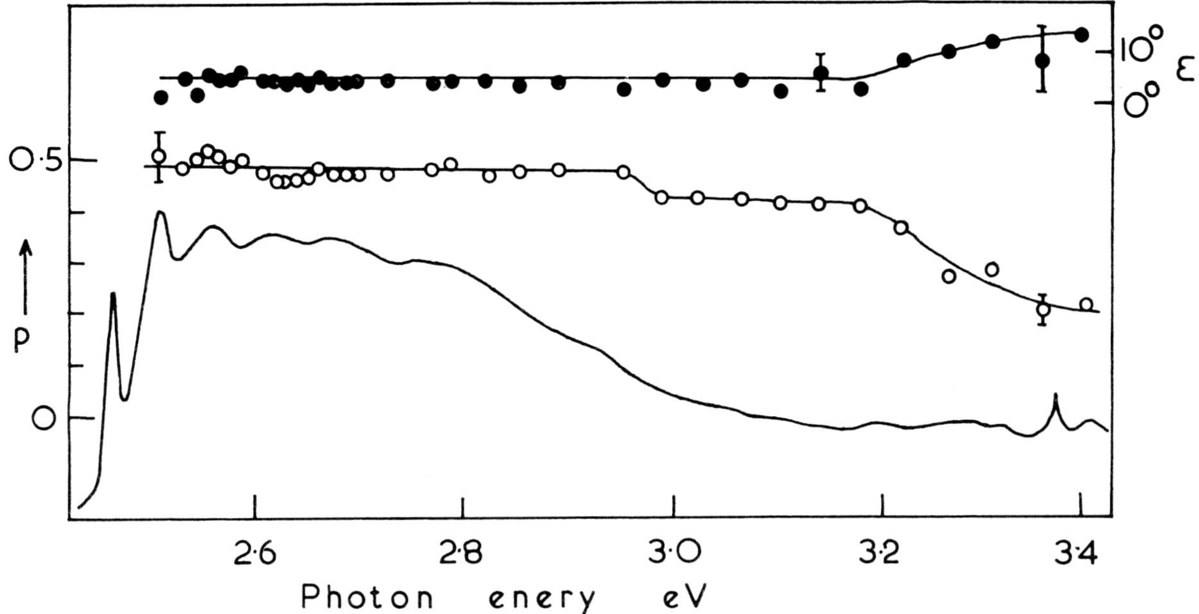

FIG. 6. Measurements of the polarization p (open circles) and phase ϵ (dots) for the light emitted in the range 2.2 to 2.4 eV by the H3 centre, as a function of the energy of the exciting light. The exciting light with electric vector parallel to [11$\bar{2}$] was incident close to [111], and the emitted light was observed close to [$\bar{1}$10]. The lowest curve gives the corresponding 80 °K excitation spectrum plotted on an arbitrary scale.

layers are produced on insulating diamonds by a mixture of nitrogen and argon ions at 2000 V,[68] by lithium ions and by boron ions at 40 keV,[69] by carbon ions,[70] and by phosphorus and boron ions at voltages up to 400 keV.[71,72] There is no doubt that: (i) conducting layers have been observed; (ii) metal–diamond–metal sandwiches show negative temperature coefficients of resistance; (iii) if (ii) is associated with a simple activation energy it is in the range 0.1–0.3 eV; and (iv) elementary thermoelectric power tests indicate n-type layers in some cases and p-type in others.

With one possible exception no Hall effect data have been obtained. Electron diffraction evidence indicates a very diffuse pattern. In the low energy bombardments the effects were partially removed by heating at 400 °C. For annealing at 700 °C Carlson[72] observed changes but the treatment produced a visible film both on the implanted face and the reverse face. For heating above 700 °C resistance increased by temperature coefficients corresponding to 0.18 eV.

The doses used by Carlson were 0.6–3.2 × 10^{15} ions cm^{-2} and Vavilov and coworkers[69] up to 0.9 × 10^{16}. In the former case (100–400 keV), the surface appeared affected to a depth of up to 4200 Å and in the latter case 2000 Å (a range was obtained but this value is typical).

We therefore see that ca. 5 × 10^{15} ions cm^{-2} have lost about 10^5 eV energy in a thickness of about 3000 Å. Quite apart from the 10^{20}–10^{21} ions cm^{-3} which have been incorporated into the surface layers, a substantial number (say 1–10 per ion) lattice displacements would normally have been expected to occur. However, the effective doses here are very high and one cannot expect low dose radiation damage arguments to apply.

Another complicating feature in the diamond case is the dubious role of the heat treatments. As in all ion implant experiments the objective of heat treatment is to incorporate more implanted ions substitutionally and to anneal as much radiation damage as possible. With diamond there is a difficulty. Any heat treatment must be carried out with oxygen partial pressures of less than 10^{-6} mm Hg. If this is done the only graphitization which is observed is the diamond/graphite phase change at 1800 °C. If oxygen is present then at lower temperatures (ca. 1000 °C) surfaces may be oxidized and subjected to non-diamond carbon deposition from the dissociating oxides.

Whether the observed activation energies are

characteristic of the layer conductance or a metal–layer combination is not at all clear and from the work so far reported it would be naïve to associate it with a conventional donor or accepter in a more or less perfect lattice. This is the ion implanter's ideal, but on the basis of the published work we do not believe that anyone has uniquely established such a mechanism for bombarded diamond. Other, in our view, more likely possibilities are:

(i) a radiation induced layer of amorphous carbon containing large amounts of intranetwork and extranetwork impurity.

(ii) a conducting carbon film brought about by irradiation and vacuum heat treatments at $p > 10^{-6}$ mm Hg.

The work reported so far in our view is more related to the phenomena of semiconducting glasses rather than conventional crystalline semiconductor effects.

7. CONCLUSIONS

We have not attempted to review the whole field of radiation damage in diamond but to examine some specific centres. We have tried to put the discussion in the context of what is known about unirradiated diamond.

Substantial advances have been made in theoretical calculations of the electronic structure of defects in diamond. Nevertheless, on the basis of the spectroscopic data alone we cannot yet decide whether the GR1 band is associated with a vacancy or an interstitial. The nature of the vibronic structure of the band indicates a substantial coupling to the lattice, favouring a vacancy rather than an interstitial model. Similarly the stability of GR1 to high temperatures is further evidence in favour of a vacancy model.

We have also discussed the formation of nitrogen molecule centres by irradiation. A mechanism is suggested in which mobile carbon interstitials exchange with nitrogen atoms in higher substitutional complexes, leaving substitutional nitrogen molecule centres as one of the products of break-up of the complex.

The properties of the ND1 system have been reviewed and, as argued by Davies and Lightowlers, the evidence is strongly in favour of interstitial nitrogen. A similar production mechanism is postulated with the break-up of nitrogen platelets by carbon/nitrogen exchange.

In our detailed review of the spectroscopic properties of H3 we conclude that the most likely model is a nitrogen-vacancy complex.

Undoubtedly more work is needed on the piezospectroscopy, the polarization of luminescence and the EPR of the *individual* centres. It is equally clear that the study of the *reactions between centres* should also prove a fruitful study—e.g. are ND1 formed during nitrogen pair formation and vice versa?; does the presence of H3 enhance nitrogen pair formation?

A reasonable scheme now exists comprising the neutral vacancy (GR1), the interstitial nitrogen (ND1), the substitutional nitrogen molecule, the nitrogen-vacancy complex (H3), singly substituted nitrogen and a substituted complex of three nitrogens (N3). The last two have so far not been found to be influenced by irradiation and have, therefore, not been discussed in detail in this paper. One of the problems of working with diamond is that certain centres are best observed in one or other of the diamond types. If suitable crystals can be chosen to study the reactions between the various centres during irradiation further experimental tests of the scheme will be possible.

The occurrence of surface conductivity after ion bombardment and of the nature of its activated process has been briefly referred to. We have suggested that the occurrence may not be a conventional crystalline semiconductor effect, but more probably a consequence of an amorphous or graphitic surface layer on crystalline diamond.

ACKNOWLEDGEMENTS

We wish to thank many colleagues for helpful discussions on the topics in this paper. We would thank particularly Dr. R. J. Wedlake for permission to refer to his recent thesis work.

REFERENCES

1. C. D. Clark, R. W. Ditchburn and H. B. Dyer, *Proc. Roy. Soc.*, **A237**, 75 (1956).
2. W. Kaiser and W. L. Bond, *Phys. Rev.*, **115**, 857 (1959).
3. R. J. Elliott, I. G. Matthews and E. W. J. Mitchell, *Phil. Mag.*, **3**, 360 (1958).
4. G. D. Watkins, J. W. Corbett and R. M. Walker, *J. App. Phys.*, **30**, 1198 (1959).
5. G. Bemski, *J. App. Phys.*, **30**, 1195 (1959).
6. G. D. Watkins, *Proc. 7th Int. Conf. Semi-cond. (Paris)*, **3**, 97 (1964).
7. C. A. Coulson and M. J. Kearsley, *Proc. Roy. Soc.*, **A241**, 433 (1957).
8. T. Yamaguchi, *J. Phys. Soc. Jap.*, **17**, 1359 (1962).
9. E. W. J. Mitchell, *Physical Properties of Diamond*, Ed. R. Berman (The University Press, Oxford, 1965), p. 394.

10. C. D. Clark, Ibid. (1965), p. 295.

11. H. J. Milledge, Samples from G.E.C. (U.S.A.) on display at 'Diamond Exhibition', Geology Museum, Oxford, organized by Dr. J. W. Harris and Dr. J. Milledge during 1970 Diamond Conference (1970).

12. T. J. Dyble, Reported at 1970 Diamond Conference, Oxford (1970).

13. P. J. Kemmey and E. W. J. Mitchell, *Proc. 5th Int. Conf. Semi-cond.* (Prague, 1960).

14. G. Davies, Ph.D. Thesis (University of London, 1970).

15. E. V. Sobolev, V. I. Lisoivan and S. V. Lenskaya (translation of 1967 original), *Sov. Phys. Dokl.*, **12**, 665 (1968).

16. T. Evans, Private communication (1970).

17. E. W. J. Mitchell, *Diamond Research* (Ind. Dia. Inf. Bur., Lond., 1964), **13**.

18. C. D. Clark, *J. Phys. Chem. Solids*, **8**, 481 (1959); C. D. Clark, P. J. Dean and P. V. Harris, *Proc. Roy. Soc.*, **A277**, 312 (1964).

19. C. D. Clark, R. W. Ditchburn and H. B. Dyer, *Proc. Roy. Soc.*, **A234**, 363 (1956).

20. H. B. Dyer and L. du Preez, *J. Chem. Phys.*, **42**, 1898 (1965).

21. P. Ferdinando, Ph.D. Thesis (University of Reading, 1967).

22. P. J. Dean, *Phys. Rev.*, **139**, A588 (1965).

23. G. Davies and E. C. Lightowlers, *J. Phys. C.* (Solid State), In Press (1970).

24. H. B. Dyer, F. A. Raal, L. du Preez and J. H. N. Loubser, *Phil. Mag.*, **11**, 763 (1965).

25. W. V. Smith, P. P. Sorokin, I. L. Gelles and G. J. Lasher, *Phys. Rev.*, **115**, 1546 (1959).

26. J. H. N. Loubser and L. du Preez, *Brit. J. App. Phys.*, **16**, 457 (1965).

27. M. Ya. Shcherbakova, E. V. Sobolev, N. D. Samsonenko and V. K. Aksenov, *Sov. Phys.—Solid State* (transl.), **11**, 1104 (1969).

28. J. H. N. Loubser, Reported at Diamond Conf. (Oxford, 1970).

29. I. Duncan, Reported at Diamond Conf. (Oxford, 1970).

30. J. F. M. Custers, *Physica*, **18**, 489 (1952).

31. E. C. Lightowlers, *Science and Technology of Diamonds* (Ind. Dia. Inf. Bur., London, 1967), p. 27; P. J. Dean, E. C. Lightowlers and D. R. Wright, *Phys. Rev.*, **140**, A352 (1965).

32. A. T. Collins, Reported at Diamond Conf., Oxford. (Measurements excluding A1 in favour of B.) (1970).

33. P. T. Wedepohl, *Proc. Phys. Soc.*, **B, 70**, 177 and 1957, Ph.D. Thesis, University of Reading (1957).

34. C. D. Clark, P. W. Kemmey and E. W. J. Mitchell, *Proc. 5th Int. Conf. Semi-cond.* (Prague, 1960); *Disc. Far. Soc.*, **31**, 96 (1961).

35. S. M. Horszowski and J. A. J. Lourens, *J. Phys. C: Solid St. Phys.*, **3**, 2442 (1970).

36. A. G. Every, Ph.D. Thesis (University of Reading, 1970).

37a. J. Callaway and A. J. Hughes, *Phys. Rev.*, **156**, 860; **164**, 1043 (1967).

37b. K. H. Benneman, *Lattice Defects in Semiconductors* (Tokyo Symposium, Ed. R. R. Hasiguti, 1968), p. 30.

38. W. Kohn *Advances in Solid State Physics*, Ed. F. Seitz and D. Turnbull (Academic Press, 1957), p. 5.

39. S. I. Pekar, V. G. Grachev and M. F. Deigen, *Proc. 9th Conf. Semicond.* (Moscow, 1968), p. 1130.

40. J. T. Ritter, *Solid State Comm.*, **8**, 773 (1970).

41. C. A. Norris, Ph.D. Thesis (University of Reading, 1967).

42. H. B. Dyer and P. Ferdinando, *Brit. J. App. Phys.*, **17**, 419 (1966).

43. C. D. Clark, *Proc. 1st Int. Cong. on Dia. in Industry* (Paris, 1962), p. 253.

44. I. Duncan, Ph.D. Thesis, University of Reading (1963).

45. W. A. Runciman, *Proc. Phys. Soc.*, **86**, 629 (1965).

46. D. M. S. Bagguley, Private communication (1970).

47. E. W. J. Mitchell: a. *J. Phys. Chem. Solids*, **8**, 444 (1959); b. *Science and Technology of Diamonds* (Ind. Inf. Bur., London), **1**, 12 (1967).

48. P. A. Crowther, Private communication quoted in reference 49 (1968).

49. M. Lannoo and A. M. Stoneham, *J. Phys. Chem. Solids*, **29**, 1987 (1968).

50. A. M. Stoneham and M. Lannoo, *J. Phys. Chem. Solids*, **30**, 1769 (1969).

51. C. D. Clark, G. W. Maycroft and E. W. J. Mitchell, *J. App. Phys.*, **33**, 378.

52. W. E. Hagston, *J. Phys. C.* (Solid State), **3**, 791 (1970).

53. J. M. E. Griffiths, J. Owen and I. M. Ward, *Rep. on Conf. on Defects in Crystalline Solids* (London—Phys. Soc., 1955), p. 81.

54. E. A. Faulkner and J. N. Lomer, *Phil. Mag.*, **7**, 1995 (1962).

55. E. A. Faulkner, E. W. J. Mitchell and P. W. Whippey, *Nature*, **198**, 981 (1963).

56. J. A. Baldwin, *Phys. Rev. Letters*, **10**, 220 (1963).

57. J. Owen, *Physical Properties of Diamond*, Ed. R. Berman (The University Press, Oxford, 1965), p. 274.

58. C. D. Clark, I. Duncan, J. N. Lomer and P. W. Whippey, *Proc. Brit. Ceram. Soc.* (*Leamington Conf.*), **1**, 85 (1964).

59. I. Duncan, *Proc. 7th Int. Conf. Semi-cond.* (*Paris*), **3**, 135 (1964); C. D. Clark and D. W. Palmer, *Reactivity of Solids* (Elsevier, Amsterdam, 1960), p. 436; D. W. Palmer, Ph.D. Thesis, University of Reading (1961).

60. A. Wild, Ph.D. Thesis, University of Reading (1970).

61. P. Pringsheim, *Phys. Rev.*, **91**, 551 (1953).

62. H. B. Dyer and L. du Preez, *Science and Technology of Diamonds* (Ind. Dia. Inf. Bur., London), **1**, 23 (1967).

63. N. N. Gerasimenko, V. E. Il'in, L. V. Lezheiko, L. S. Smirnov, E. V. Sobolev and O. P. Yureva, *Soviet Phys.—Solid State* (translation), **9**, 2898 (1968).

64. R. J. Wedlake, Ph.D. Thesis, University of Reading (1970).

65. C. D. Clark and C. A. Norris, *J. Phys. C.* (Solid State), **3**, 651 (1970).

66. A. A. Kaplyanski, *Optics and Spectroscopy*, **16**, 329 (1964).

67. G. D. Watkins and J. W. Corbett, *Phys. Rev.*, **134**, 1359 (1964).

68. R. H. Wentorf and K. A. Darrow, *Phys. Rev.*, **137**, A 1614 (1965).

69. V. S. Vavilov, M. I. Guseva, E. A. Konorova, V. V. Krasnopevtsev, V. F. Sergienko and V. V. Tutov, *Sov. Phys.—Solid State* (translation), **8**, 1560 (1966).

70. V. S. Vavilov, Private communication (1968).

71. C. M. Kellett, W. J. King and F. A. Leith, *Ion Physics Corp. Sci. Report No. 1* (AF19 (628) 4970), quoted in reference 72.

72. R. O. Carlson, *Met. Soc. Conf. of A.I.M.E* (1968), p. 16.

73. T. Yamaguchi, *J. Phys. Soc. Jap.*, **18**, 368 (1963).

74. J. Friedel, M. Lando and G. Leman, *Phys. Rev.*, **164**, 1056 (1967).

75. C. A. Coulson and F. P. Larkins, *J. Phys. Chem. Solid*, **30**, 1963 (1969).

DISCUSSION

Question (VOOK)   Are there any defects produced in diamond by ionization? Is there a threshold displacement energy for the proposed vacancy centre in diamond?

Answer (MITCHELL)   We have no evidence for defect production by ionization processes in diamond. Measurement between 0.3 and 1.0 MeV give a displacement energy of 80 eV both from optical and electrical measurements as reported at the Faraday Discussion in Paris in 1961. In Type IIa diamonds the rate of production of the GR1 band is more or less independent of sample. We have observed ionization effect on the strength of various bands after the defects have been produced.

Question (STONEHAM)   First a comment. You correctly noted that the Runciman and Crowther stress splitting data were not fully in accord. They do, however, agree that the data are not consistent with simple levels (with a tetragonal distortion or a trigonal distortion for either or both levels) in cubic symmetry. The results do not exclude more complex Jahn–Teller effects involving accidental degeneracy in the V^- in silicon. Secondly, on what evidence have you assigned the charge state of the vacancy in your table?

Answer (MITCHELL)   First on your comment, I think that another complicating feature in the case of the GR1 splittings is the effect of internal strains. To answer your question the GR1 band is not produced in semi-conducting diamond until the holes have been compensated. Therefore we do not believe it is associated with V^+. We cannot make any comment between V^0 and V^-.

Question (KOEHLER)   How low a temperature has been used to study damage production and the nature of the resulting defects in diamond?

Answer (MITCHELL)   Damage production has been studied by optical and electrical measurements down to 80 °K. ESR spectra have been looked for during damage production at 10–20 °K.

Comment (WATKINS)   A word of caution is in order concerning the identification of the ND1 band with nitrogen in the interstitial tetrahedral site. Preliminary calculations on our LCAO-MO model (R. P. Messmer and G. D. Watkins, this conference) indicate that nitrogen is not stable in this position.

Question (LARKINS)   Current theoretical calculations suggested that lattice distortion effects are very important in determining electronic properties of simple defects. Are you aware of any experimental work which can give us an estimation of the direction and extent of distortion around for example the isolated vacancy?

Answer (MITCHELL)   I am not aware of any experiments which give directly the amount of distortion around the single vacancy in diamond. I agree that it is important to try to determine their effect experimentally. There are measurements of lattice parameter changes in neutron irradiated diamond but of course these give the integrated effects of the distortions around all the defects induced. For the divacancy in germanium there is some information from neutron scattering.

Question (LARKINS)   Have you a feel for the temperature at which the carbon interstitial becomes mobile in diamond.

Answer (MITCHELL)   There is an ESR spectrum which Wild and Lower (Reading) have found after irradiation at 10–20 °K. The doses required lead us to believe that it is a displacement rather than an ionization effect. It anneals irreversibly at 130 °K. The spectrum has not been analysed but we believe it is a defect motion involved in the anneal. It could be the interstitial but we cannot at the moment be sure.

III–V COMPOUND REVIEW

F. H. EISEN†

North American Rockwell Science Center, Thousand Oaks, California 91360, U.S.A.

Radiation effects in InSb, GaSb, and GaAs are reviewed with emphasis on damage production and comparison of recovery results. The orientation dependence of damage production at electron energies near threshold, the rather low values of the threshold displacement energy and subthreshold damage effects are discussed. Consideration is given to a number of similarities in the recovery data for InSb and GaSb, and some differences between these materials and GaAs are discussed.

There has been a considerable amount of work in radiation damage in III-V compounds over the past years, but the work in these materials has not been nearly as extensive as that in silicon and germanium. At this conference, the program includes several papers on III-V compounds, but more of the work is related to ion implantation than to traditional radiation-damage studies. It is well-known, of course, that there are more different types of intrinsic defect structures possible in these materials than in Si and Ge.[1] A further problem is that there is as yet no identification of any defects in these materials based upon interpretation of microscopic data such as that from electron paramagnetic resonance measurements. This is due, in part, to the fact that in these materials, the measurements would be considerably more difficult than in Si and Ge, or the II-VI compounds because of the high nuclear spins of the Group-III and Group-V elements. In this review, I shall concentrate entirely upon InSb, GaSb, and GaAs, because more detailed studies of radiation effects have been carried out in these materials than in the other III-V compounds. I shall consider damage production, including: (i) the use of the symmetry properties of the III-V (zincblende) structure to determine whether the Group-III or the Group-V element has the lower displacement energy, (ii) the low values of the measured displacement energy, and (iii) the production of changes in InSb by low-energy x-rays, subthreshold electrons, and light. I shall also discuss the recovery and compare the observations in the three different materials, looking for similarities and differences between these materials.

We shall first describe how it is possible in some cases to determine which atom has been displaced

† Visiting scientist at the Institute of Physics, University of Aarhus, Denmark, 1970–71.

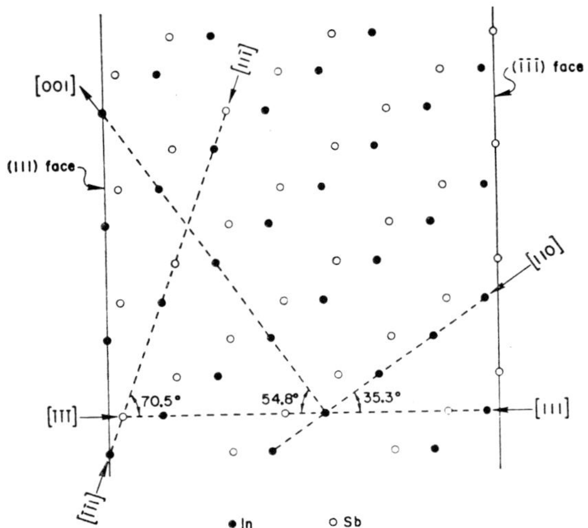

FIG. 1. Plan view of the ($\bar{1}$10) plane in InSb.

in a given irradiation. Figure 1, shows the arrangement of atoms in a zincblende lattice in the ($\bar{1}$10) plane. Note that if one looks at the (111) face traveling into the crystal, the points representing In atoms have empty sites behind them, whereas if one looks at the ($\bar{1}\bar{1}\bar{1}$) face in the opposite direction, the Sb atoms have empty spaces behind them. It should therefore be easier to displace an In atom when the electron beam is perpendicularly incident on the (111) face, and easier to displace an Sb atom with the electron beam incident on the opposite face. Before proceeding to examine the results of such irradiations, it is desirable to examine Figure 2, which shows the isochronal recovery results obtained in electron-irradiated InSb.[2] Note that in *n*-type material, stages I and II‡ are well separated in temperature and that, by annealing to a temperature intermediate between them and to a

‡ Recovery stages are numbered in order of increasing temperature.

"

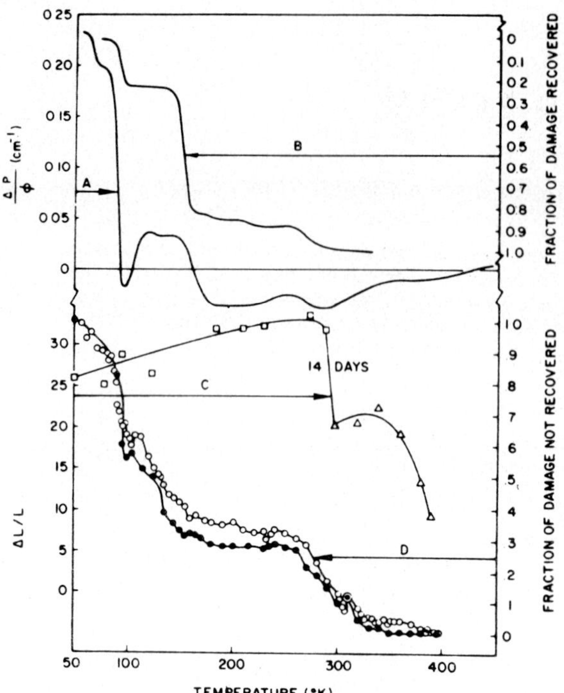

FIG. 2. Isochronal recovery data for InSb. Curves A and B are from data on carrier concentration recovery following 1.0-MeV electron irradiation with n_0 or p_0 about 10^{16} cm^{-3}.[16] Curve C shows the recovery of sample length[27] and curve D shows the recovery of thermal conductivity.[18]

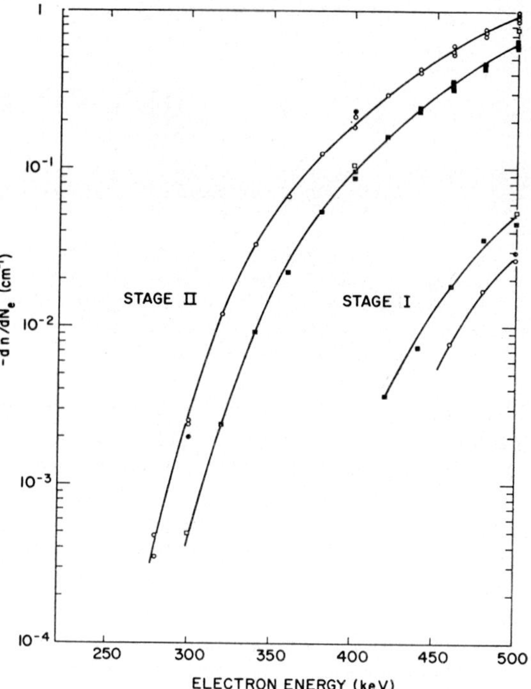

FIG. 3. Carrier removal rate, $-d\bar{n}/dN_e$, for (111)- and ($\bar{1}\bar{1}\bar{1}$)-oriented InSb samples, separated into the portion which recovers in stage I and that which recovers in stage II, versus bombarding electron energy. Circles represent data obtained when irradiating the (111) surface, and squares are for the ($\bar{1}\bar{1}\bar{1}$) surface. Filled symbols are for sample 1 and empty symbols are for sample 2.

temperature slightly above stage II, it should be possible to determine the portion of the damage which anneals in stage I and that which anneals in stage II. Stages III, IV, and V were found to have higher threshold energies and were not produced in sufficient amounts to be of importance in the experiments to be described below.

Figure 3 shows the results of irradiating two {111}-oriented samples with opposite {111} faces exposed to the electron beam.[3] It can be seen that the rate of production of damage which anneals in stage II is higher with the (111) surface exposed to the electron beam, and the rate of production of damage, which anneals in stage I, is higher with the ($\bar{1}\bar{1}\bar{1}$) surface exposed to the beam. The data are interpreted as meaning that the defects which annihilate in stage II are due to the displacement of In atoms, and those which annihilate in stage I are due to the displacement of Sb atoms. Note also that the relative difference in the damage rate between opposite faces for either stage increases as the bombarding electron energy is lowered.

Similar data for GaSb are shown in Figure 4. Here again, it can be seen that the defects which annihilate in stage II have the lowest threshold energy and are due to the displacement of the Group-III element (Ga), whereas those which annihilate in stage II are due to the displacement of antimony. In GaSb, the difference in the Ga and Sb masses is large enough that Ga would be expected to be displaced somewhat lower electron energies than Sb. The fact that the interpretation of the anisotropy of damage production agrees with this expectation may be considered as a confirmation of the interpretation of the anisotropy data. This type of experiment, which makes use of the symmetry of the zincblende lattice, has also been used by Arnold[5] to show that the quenching of photoluminescence in GaAs is primarily due to the displacement of As atoms.

The measured values of the threshold displacement energy in the III-V compounds, i.e. the lowest values of the maximum possible energy transfer to

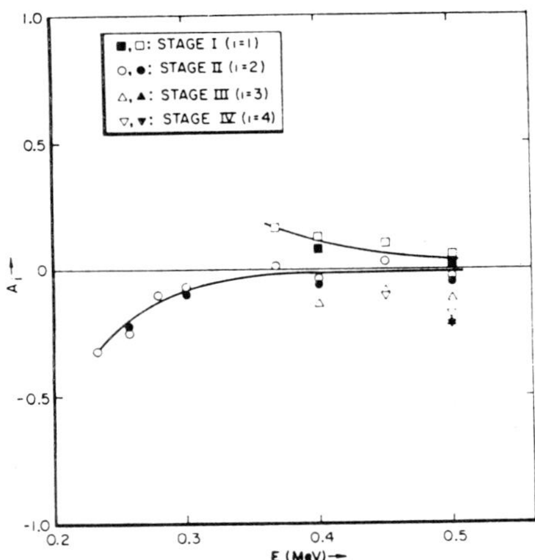

FIG. 4. Anisotropies A_i, $i = 1$–4, versus electron energy for GaSb. $A_i = (\eta_i^+ - \eta_i)/\eta_i^+ + \eta_i^-)$ where η_i^+ and η_i^- are the defect production rates with the beam incident in the [111] direction and in the [$\bar{1}\bar{1}\bar{1}$] direction, respectively, for defects which are annihilated in the ith stage. Full symbols: sample 14; open symbols: sample 16.[4]

a Group-III or Group-V atom for which damage effects can be observed, range generally between 6 and 10 eV.[2–4] There are several possible difficulties connected with these low values of threshold energies. One is that in high-energy irradiations a higher effective threshold energy is generally required in order to obtain good agreement between the calculated number of displacements and the observed number of displacements. A second possible difficulty stems from the suggestion of Banbury,[6] made at the Santa Fe conference, that measured values of the lowest electron energies at which displacements are produced in Si yield a threshold which is somewhat higher than the threshold energy indicated by the anisotropy of the damage production in silicon. He also suggested the possibility that at low energies, the damage is produced by a sub-threshold mechanism.

I would like to briefly consider several possible sub-threshold mechanisms in the III–V compounds. One possibility is that displacements are produced by some ionization-dependent mechanism similar to the Varley[7] mechanism. Such a mechanism would be expected to become more important as the energy of the bombarding electrons is lowered. The fact that in the anisotropy experiments described above, the relative difference in damage production with the two different sample faces exposed to the electron beam increases with decreasing electron energy, indicates that an ionization mechanism cannot play a significant role for displacements in the material. A second possibility is that displacements are produced at some special sites in the crystal such as atoms near a dislocation or a vacancy, or that a light impurity may be displaced which subsequently displaces a Group-III or a Group-V atom. Thommen[4] has examined this problem and concluded that for reasonable assumptions about the density of such special sites, one cannot expect to obtain the number of displacements observed in actual irradiations. Oen[8] has suggested a mechanism which is not readily eliminated. This is the possibility that the electron which produces the displacement also simultaneously ionizes one of the bonds to the displaced atom, therefore lowering the energy required to produce a displacement. However, it does not seem that such a mechanism would lower the threshold displacement energy by more than about 25 per cent. Therefore, even if this mechanism exists, it is probable that the observed lowest energies required to produce displacements, are close to the actual threshold energies. Further, the observed values of the threshold energy are in reasonable agreement with values computed recently by Bailly[9,10] for the binding energy in these materials, as shown in Table I. In fact, the computed binding energies are slightly lower than the observed average threshold energy.

TABLE I

	Mean displacement energy (eV)	Binding energy (eV) Ref. 9
InP	7.7	− 6.6
InAs	7.5	− 6.25
InSb	7.7	− 5.5
GaAs	9.2	− 6.4
GaSb	>6.9	− 5.95

We now consider the production of changes in the electrical and thermal conductivity in InSb by low-energy x-rays, sub-threshold electrons, and light. Some years ago, Arnold and Vook[11] observed that both the electrical and the thermal conductivity of samples converted from n- to p-type by irradiation with 2-MeV electrons, were changed by irradiation with 100-keV X-rays. They also found the same type of changes in samples which were originally p-type. They argued that some form of ionization mechanism must be responsible for these changes since they felt that the extensive irradiations carried out with 2-MeV electrons would saturate all possible ionization effects, and the thermal-conductivity changes would not have been produced by surface effects, Similar effects on the electrical conductivity have been observed in irradiation with low-energy electrons with energies as low as ∼20 keV in *n*- and *p*-type materials.[2,12] Changes in electric conductivity produced by x-ray irradiation have also been observed in the work of the Leningrad group.[13] More recently, Kreutz has reported x-ray results, in which he measured both electrical and thermal conductivity and concluded that at least the electrical conductivity changes were due to surface effects.[14] His conclusions are based upon the dependence of the conductivity change upon the thickness of the sample[15] and upon the effects of various surface treatments on the electrical conductivity changes. Kreutz[14] observed the recovery of the electrical conductivity change at 100 °K in agreement with the earlier work of Vook[11] and Eisen[2,12] and observed the same activation energy for the recovery as was observed in the electron-irradiation work. He did not, however, see a complete annealing of the thermal conductivity even when the samples were annealed to 300 °K and concludes that thermal conductivity results cannot be understood at this time.

The interpretation of the electrical conductivity changes as being due to a surface effect has been somewhat confused by recent work of Vitovskii and coworkers.[13a] They report the observation of changes in the electrical conductivity of *p*-type InSb samples irradiated with light. Figure 5 shows the annealing of their light-irradiated samples compared with the annealing of samples damaged by gamma irradiation. Since one would expect the gamma irradiation to produce damage that would recover in stage II, and since the light-produced conductivity changes also recover at the same temperature as the gamma irradiation, they conclude that they have produced the same type of close pairs with light as those that have been suggested to account for stage-II recovery observed after 1.0-MeV electron irradiation (see below).

In irradiating *p*-type InSb, there is a problem in distinguishing the damage produced by sub-threshold mechanisms from that which ordinarily recovers in stage II. Figure 2 shows that the recovery in stage II in *p*-type material is centered at ∼90 °K. The damage produced by sub-threshold electrons has been found to anneal in the neighbourhood of 100 °K.[11,12] Figure 6 shows annealing

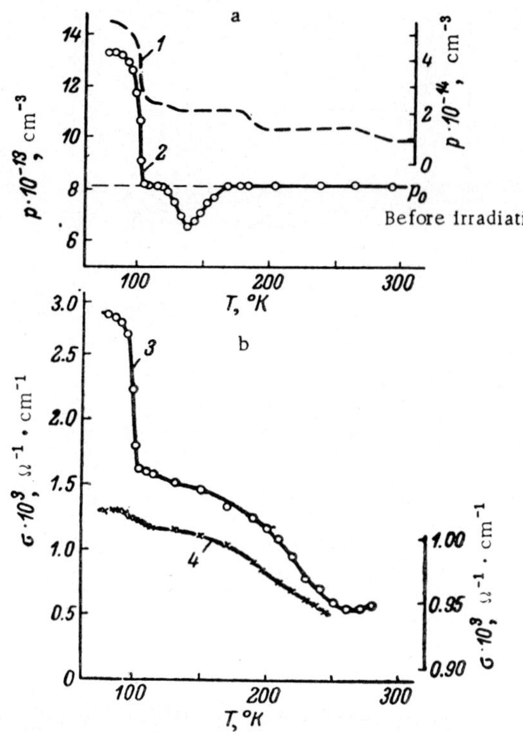

FIG. 5. Changes in the hole density (a) and electrical conductivity (b) during isochronal annealing of *p*-type InSb after irradiation with ^{60}Co γ-rays (1) and x-rays (2) as well as after illumination with light without (3) and with (4) a filter. The ordinate scales on the right apply to curves 1 and 4.[13a]

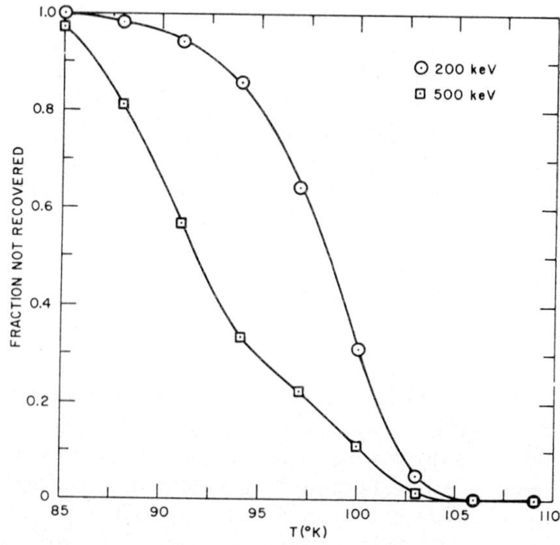

FIG. 6. Isochronal annealing of *p*-type InSb after irradiation with 200 or 500 keV electrons.

results obtained after irradiating p-type InSb ($p_0 \sim 4 \times 10^{15}$/cm³) with 200 and 500 keV electrons. Here one can see that for the 200 keV irradiation, the recovery occurred in the stage at about 100 °K, whereas for the 500-keV irradiation, a distinct difference can be seen between the recovery in stage II and that in the 100 °K stage.

The temperature difference between stage II and the 100 °K stage in the samples used by Vitovskii *et al.*[13a] (p_0 about 1×10^{14}/cm³) would be much smaller than that shown in Fig. 6 (see below). It is possible that this had led to the apparently erroneous conclusion that the defects produced in subthreshold irradiations are the same as those which annihilate in stage II following 1.0-MeV electron irradiation. More careful work therefore seems required to definitely establish whether or not changes produced by low-energy x-rays, low-energy electrons, and light represent a surface effect or a real volume damage effect in InSb.

Let us now consider the results of annealing experiments in these materials. There is no recovery of electrical properties observed in InSb, GaSb, or GaAs below about 70 °K. Figure 2 shows that in InSb, there is a large temperature shift for stage II between p-type and n-type material. This temperature shift has been examined in more detail as a function of the carrier concentration of the sample,[16,17] and the results are shown in Figure 7, which displays both the center temperature of the recovery stage and the activation energy as a function of sample-carrier concentration. This dependence on carrier concentration has been explained by assuming that the two energy levels associated with the defects, one of which lies near the valence band and the other near the conduction band, must both be empty in order for the defects to annihilate. If one then takes into account the effects of sample-carrier concentration on the

percentage of defects which have both these levels empty, the solid curves on Figure 7 result. It can be seen that these calculations agree quite well with the observed dependence of recovery temperature and activation energy on sample-carrier concentration. Observations on highly compensated samples show that the recovery temperature and activation energy are correlated with the net carrier concentration in these samples and not the net impurity concentration. These facts indicate that the dependence of stage-II recovery on carrier concentration is actually produced by changes in the charge-stage of the defects and not by interaction of the defects with impurities in the sample.

The results for the recovery of thermal conductivity obtained by Vook[18] are also represented in Figure 2. Thermal-conductivity measurements require that the irradiation be carried to sufficient doses that the number of displaced atoms introduced should be significantly greater than the number of impurity atoms in the sample. Therefore the recovery would be expected to be due to the annihilation of intrinsic defects rather than to impurity-associated defects. This is a further indication that the effects in stage II are due to intrinsic defects.

Isothermal recovery measurements indicate that stage-II kinetics can be divided into two first-order processes and the recovery of two close, interstitial-vacancy close pairs has been suggested to account for the recovery kinetics.

The isothermal data for the other recovery stages in InSb also cannot be fitted by simple first- or second-order processes. If an attempt is made to fit the data by a chemical rate equation, the order of the reaction obtained for the first four stages is slightly above one, whereas that for the fifth stage is somewhat higher.[12,17] Also, the result for stage V is dependent on the energy of the bombarding electrons, showing an increase in apparent order with increasing electron energy. This energy dependence is not observed for stages I to VI. These results suggest that stage V may involve long-range migration of defects, although the isothermal data have not been analyzed in detail.

Figure 8 shows isochronal recovery data obtained by Thommen[19] for electron-irradiated GaSb. Here, only four recovery stages are seen. Thommen suggests that stage III may actually contain two unresolved stages. His suggestion is based upon the analysis of the activation-energy data and upon the shape of the derivative curve of the isochronal data. Stage I shows first-order recovery of

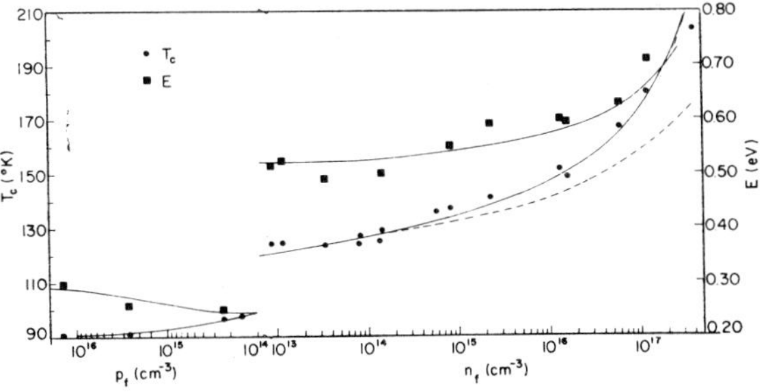

FIG. 7. Recovery temperature T_c and activation energy E as a function of carrier concentration for stage II in electron-irradiated InSb.[16]

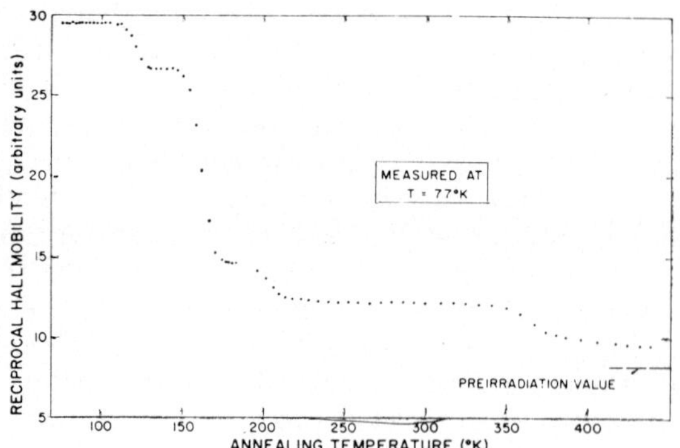

FIG. 8. Isochronal recovery results for reciprocal mobility in electron-irradiated p-type GaSb.[19]

the reciprocal mobility, but the others, again, do not show simple recovery. Stage IV has higher-order apparent kinetics similar to that observed for stage V in InSb. Unpublished data of Thommen indicate that the temperatures for the various recovery stages are higher in tellurium-doped n-type GaSb than they are in p-type GaSb, suggesting that the dependence of defect annihilation on the charge state of the defects may be similar in GaSb and InSb. A further common feature is the previously mentioned fact that stage I in both materials is due to antimony displacements and stage II is due to the displacement of the Group-III atom. In both cases, the lowest threshold energy is observed for the production of the defects which annihilate in stage II.

Isochronal recovery data for GaAs are presented in Figure 9.[20] Here we see the three recovery stages for the electron-irradiation damage. Aukerman[21] has observed a fourth recovery stage at 6–700 °C in neutron-irradiated GaAs. The temperature region in which stage-I and stage-II recovery is observed is roughly the same as that in

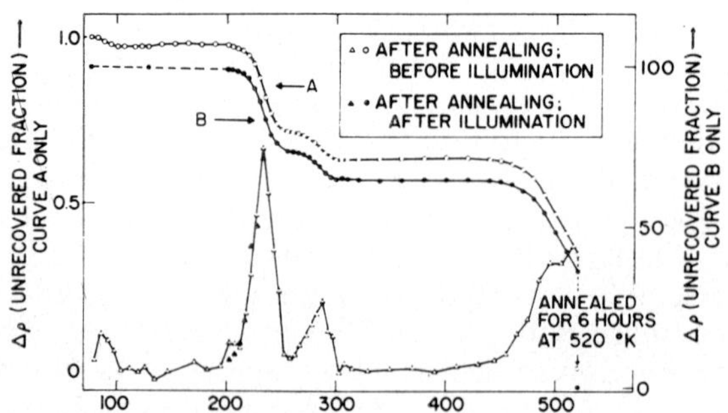

FIG. 9. Isochronal recovery data for electron-irradiated GaAs.[20]

which the recovery of ion-implantation damage was observed by Weisenberger and coworkers.[22] Length change and thermal conductivity had also shown recovery in this temperature range, but previously there had been no indication of recovery of electrical conductivity in the work of Grimshaw and coworkers.[23] Stein[24] has also seen recovery in this temperature range in electrical measurements recently, but he did not resolve stages I and II. A possible reason for the Grimshaw and coworkers missing of the recovery in stages I and II will be discussed below. Thommen has observed first-order kinetics for stages I and II and suggests that the recovery is due to close pairs since long-range migration to traps seems unlikely for reasonable assumptions about the trap density, and the observed frequency factors are in reasonable agreement with those expected for close-pair recovery. Stage III has been studied in some detail by Aukerman,[25] who finds that the recovery can be divided into two first-order substages with the rate constant of the second substage dependent on the sample-carrier concentration. However, in this case, the dependence is the opposite of that observed in InSb, i.e. it is necessary for the defect energy level to be occupied by an electron in order for the recovery to proceed. Aukerman[25] has suggested that this stage may be due to the recovery of close pairs; however, thermal-conductivity data of Vook[26] suggest that at about 325 °K in GaAs, clustering of defects is taking place rather than defect annihilation.

Thommen has also made measurements of the energy-dependence of the production of damage in the three stages that he observed, which are shown in Figure 10. The interesting thing about the results for GaAs is that the stage III now has the lowest threshold energy. Thommen[20] also observed that the damage-production rates in stages I and III were higher at 5 °K than at 77 °K. This is somewhat different from the situation in InSb, where none of the measurements made have revealed any difference in the amount of damage produced in irradiations at about these same two temperatures. Thommen has conjectured that gallium displacements may be responsible for the damage recovery in stage III. This is based on an analogy with GaSb and InSb, where the lowest threshold was observed for the Group-III atom. Also in these materials, the stage having the lowest threshold energy is the one that accounted for a major portion of the recovery as is the case in GaAs. It is possible that the reason that Grimshaw

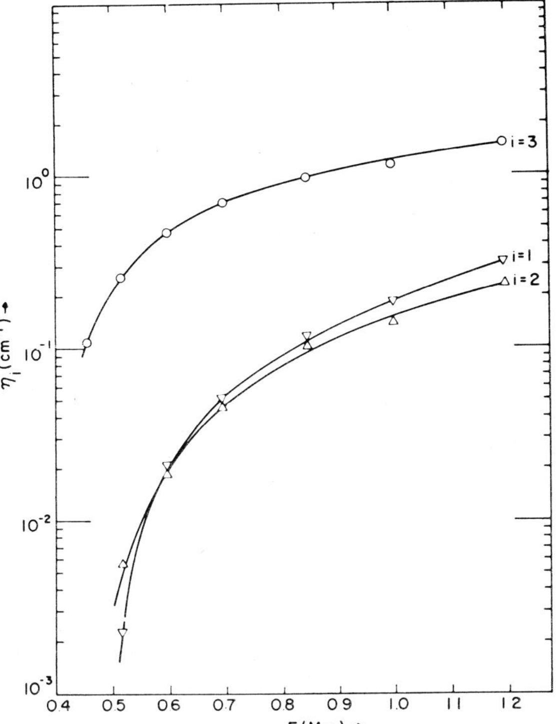

FIG. 10. Carrier removal rate η_i for the ith recovery stage in GaAs as a function of the incident energy of the bombarding electrons.[20]

ind coworkers[23] did not observe recovery in stages I and II is associated with the higher threshold energy for the production of damage recovering in these stages. If their irradiations were performed at low enough energy, these stages might easily have been overlooked. For example at 0.5 MeV, they account for only 1 per cent of the total damage.

We have seen that there is considerable similarity between the recovery in InSb and GaSb and some possible differences between these compounds and GaAs. Further work is necessary to access the significance of these differences. In some cases, models have been suggested for the defects which are responsible for the recovery. These have principally involved a suggestion of interstitial-vacancy pairs, without envoking any interaction of the intrinsic defects with the impurities in the sample. This is partly because there has been no need on the basis of any of the data available so far, to envoke defect-impurity interactions in order to explain the data. Also, no specific studies have really been done to determine whether or not

defect-impurity interactions are of importance. Considering the work in silicon and germanium, where these play a prime role, one must wonder whether or not they are also of considerable importance in the III–V compounds. Perhaps the type of experiment, which should be done, would be one in which controlled doping of a material is undertaken and radiation effects studied in samples with various known impurities included. While this may be a fairly tedious process, it may be experimentally the easiest method of obtaining results on defect-impurity interactions, considering the difficulties involved in electron paramagnetic resonance measurements. However, it seems desirable, if at all possible, to carry out some kind of microscopic measurements in these materials to establish defect identifications with the same type of certainty as is now available for silicon and also, to a certain extent, for germanium.

REFERENCES

1. For comprehensive reviews of the older work in III–V compounds see: J. W. Corbett, *Radiation Damage in Semiconductors and Metals* (Academic Press, New York, 1966); and L. W. Aukerman in *Semiconductor and Semimetals*, Ed. R. K. Willardson and A. C. Beer, **4**, 343 (Academic Press, New York, 1968).
2. F. H. Eisen in *Radiation Damage in Semiconductors* (Dunod Cie, Paris, 1965), p. 163.
3. F. H. Eisen, *Phys. Rev.*, **135**, A1394 (1964).
4. K. Thommen, *Phys. Rev.*, **174**, 938 (1968).
5. G. W. Arnold and G. Gobeli in *Radiation Effects in Semiconductors*, Ed. I. F. Vook (Plenum Press, New York, 1968), p. 435.
6. P. C. Banbury in *Radiation Effects in Semiconductors*, Ed. I. F. Vook (Plenum Press, New York, 1968), p. 280.
7. J. H. O. Varley, *Nature*, **174**, 886 (1954); *J. Nucl. Energy*, **1**, 130 (1954).
8. O. S. Oen in *Radiation Effects in Semiconductors*, Ed. F. L. Vook (Plenum Press, New York, 1968), p. 264.
9. F. Bailly, *J. Phys. Radium*, **27**, 335 (1966).
10. F. J. Bryant and A. F. J. Cox, *J. Phys. Chem.*, **1**, 1734 (1968).
11. G. W. Arnold and F. L. Vook, *Phys. Rev.*, **137**, A1839 (1965).
12. F. H. Eisen (unpublished work).
13. B. Chelustra, R. Yu. Khansevarov, T. V. Mashovets and I. R. K. Kozlova, *Pov. Phys. Solid State*, **9**, 253 (1967).
13a. N. A. Vitovskii, G. A. Vikhlii, V. V. Galavanov, T. V. Mashovets and R. Yu. Khansevarov, *Soviet Physics—Semiconductors*, **3**, 106 (1969).
14. E. W. Kreutz, *Zeits. für ang. Phys.*, **27**, 244 (1969).
15. E. W. Kreutz, H. Pagnia and W. Waidelich, *Phys. Stat. Sol.*, **27**, K111 (1968).
16. F. H. Eisen, *Phys. Rev.*, **148**, 828 (1966).
17. F. H. Eisen, *Phys. Rev.*, **123**, 736 (1961).
18. F. L. Vook, *Phys. Rev.*, **135**, A1750 (1964).

F. H. EISEN

19. K. Thommen, *Phys. Rev.*, **161**, 769 (1967).
20. K. Thommen, *Rad. Effects*, **2**, 201 (1970).
21. L. W. Aukerman, P. W. Davies, R. D. Graft and T. S. Shilladay, *J. Appl. Phys.*, **23**, 3590 (1963).
22. W. H. Weisenberger, S. T. Picraux and F. L. Vook, *Rad. Effects*, this issue.
23. J. A. Grimshaw in *Radiation Damage in Semiconductors* (Dunod Cie, Paris, 1965), p. 377, and J. A. Grimshaw and P. C. Banbury, *Proc. Phys. Soc.*, **84**, 15 (1964).
24. H. J. Stein, *J. Appl. Phys.*, **40**, 5300 (1969).
25. L. W. Aukerman and R. D. Graft, *Phys. Rev.*, **12** 1576 (1962).
26. F. L. Vook, *Phys. Rev.*, **135**, A1742 (1964).
27. F. L. Vook, *J. Phys. Soc. Japan*, **18**, Suppl. II, 19 (1963).

DISCUSSION

Question (KOEHLER)   Have there been any infrared absorption measurements associated with defects?

Answer (EISEN)   Only a few—though some data will be presented later in this session.

Question (FAN)   Would you comment briefly about the nature of the defects revealed by the several stages of annealing?

Answer (EISEN)   The lower temperature stages have usually been interpreted as being due to close interstitial-vacancy pairs, even when the kinetics are not exactly first order.   For stage II in InSb for example, where the kinetics can be represented by the sum of two first order processes, it is possible to construct a model of two slightly different close pairs which will account for th relative amounts and rates of the two first orde processes.   Unfortunately there are no micro scopic data to back up this interpretation of th macroscopic results.

Question (INUISHI)   Is there any clear evidence o donor-acceptor pairs?

Answer (EISEN)   The assumption of the presence o close interstitial-vacancy pairs is based primaril upon the fact that first-order kinetics (sometime involving two first-order processes) are observe for the lowest temperature recovery processes for which the lowest threshold energies are als observed, and that the recovery time observe is in accord with this assumption.

ELECTRON IRRADIATION OF HEAVILY DOPED GaAs:Si AND GaAs:Te

A. KAHAN, L. BOUTHILLETTE AND H. M. DeANGELIS

Air Force Cambridge Research Laboratories, Air Force Systems Command, L. G. Hanscom Field, Bedford, Massachusetts, U.S.A.

Infrared optical reflectivity measurement techniques are applied to determine the carrier removal rate, mobility changes, and annealing characteristics of heavily doped n-type silicon or tellurium-doped gallium arsenide. 1 MeV room temperature electron irradiation reduces both the carrier concentration and the mobility. Removal rates for GaAs: Te and GaAs:Si having initial carrier concentrations of 6×10^{18} cm^{-3} are approximately 3.4 and 3.0 cm^{-1} respectively. The mobility degrades exponentially as a function of fluence. Isochronal anneal experiments show one major recovery stage between 170 and 230°C, and the results are in reasonable agreement with previous data obtained by electric measurements on more lightly doped n-type GaAs.

1. INTRODUCTION

This paper reports on the application of optical techniques to estimate the carrier removal rate, mobility degradation, and annealing behavior of 1 MeV electron irradiated n-type gallium arsenide heavily doped with silicon or tellurium. For these materials the optical experiments reduce to near infrared reflectivity measurements and changes in free carrier absorption are assessed in a routine manner. The optical beam penetrates only to a depth of a few microns, and the advantages associated with this measurement technique, compared to transport property investigations, are discussed by A. Kahan and coworkers.[1] Previous studies[2-6] of electron irradiated n-type GaAs were performed on lightly doped materials (carrier concentrations of 10^{15}–10^{17} cm^{-3}), and these results are compared with data obtained in this investigation on GaAs:Si and GaAs:Te crystals with initial carrier concentrations of 6×10^{18} cm^{-3}.

2. ANALYSIS

The infrared optical reflectivity as a function of frequency v is analyzed in terms of a linear superposition of lattice vibration and free carrier absorption mechanisms. In the classical limit the equations for ϵ_1 and ϵ_2, the real and imaginary parts of the complex dielectric constant, are

$$\epsilon_1 = n^2 - k^2 = \epsilon_\infty + \frac{4\pi\rho(1 - v^2/v_t^2)}{(1 - v^2/v_t^2)^2 + \gamma^2(v/v_t)^2} - \frac{\epsilon_\infty v_{\rm pl}^2}{G^2 + v^2} \tag{1}$$

$$\epsilon_2 = 2nk = \frac{4\pi\rho\gamma(v/v_t)}{(1 - v^2/v_t^2)^2 + \gamma^2(v/v_t)^2} + \frac{G\epsilon_\infty v_{\rm pl}^2}{v(G^2 + v^2)} \tag{2}$$

where ϵ_∞ is the high frequency dielectric constant, and the other terms in ϵ_1 and ϵ_2 are expressions for the classical oscillator model for lattice vibration and for the classical free carrier absorption. The lattice vibration is described by three parameters, $4\pi\rho$, γ, and v_t, the width, strength, and frequency position, respectively. The free carrier absorption is characterized by two parameters, $v_{\rm pl}$ and G, and these are related to carrier concentration N, effective mass m^*, scattering relaxation time τ, and mobility μ, by

$$v_{\rm pl}^2 = \frac{e^2 N}{\pi c^2 m^* \epsilon_\infty}, \quad \tau = \frac{1}{2\pi c G}, \quad \text{and} \quad \mu = \frac{e}{m^*}\tau \tag{2a, b, c}$$

The normal angle of incidence reflectivity R is calculated from the optical constants n and k,

$$R = \frac{(n-1)^2 + k^2}{(n+1)^2 + k^2} \tag{3}$$

The reflectivity variation as a function of lattice vibration, free carrier absorption, and effects of the various parameters have been described by many investigators. In general, the wavelength position of the plasma reflectivity minimum is an indication of carrier concentration, and the plasma edge slope a measure of mobility. For the same mobility the plasma edge shifts with increasing carrier concentration towards the near infrared, and for the same carrier concentration the plasma edge slope decreases with decreasing mobility.

A general computer program, incorporating an

IBM Scientific Subroutine Package subroutine, FMFP—Extremum of Functions, and subsidiary functions of the reflectivity and its derivatives with respect to the lattice and free carrier parameters, is utilized to curve fit and to determine the lattice and free carrier parameters from the best fit to the experimental data. The lattice parameters, $\epsilon_\infty = 11.1$, $4\pi\rho = 1.975$, $\gamma = 0.008$, and $\nu_t = 268.3$ cm^{-1}, are computed from data where free carrier effects are negligible. For the GaAs irradiation and annealing results analysis, 20–30 experimental points are selected from each reflectivity curve, and the previously calculated lattice parameters are held constant in the determination of the free carrier parameters.

3. EXPERIMENTAL DETAILS

The Laue pattern of the bulk GaAs:Si sample, grown at the University of Southern California and designated as USC-WA-12, indicates a mosaic structure. The silicon impurity concentration of this crystal is estimated as 7×10^{19} cm^{-3} and the net n-type carrier concentration, approximated from infrared localized vibrational mode absorption measurements, as 8×10^{18} cm^{-3}. The electrical properties of the bulk GaAs:Te, purchased from Monsanto and designated as C2-1029, are determined by Monsanto as $N = 9.4 \times 10^{18}$ cm^{-3} and $\mu = 1820$ cm^2/volt-sec. Optical reflectivity measurements (assuming $m^* = 0.08$) yield preirradiation values of $N = 6.0 \times 10^{18}$ cm^{-3} and $\mu = 770$ cm^2/volt-sec for GaAs:Si, and $N = 6.25 \times 10^{18}$ cm^{-3} and $\mu = 2060$ cm^2/volt-sec for GaAs:Te.

The sample surfaces are prepared by a mechanical-chemical polishing procedure utilizing a sodium oxychloride solution.[7] This is essential as mechanical polishing results in damaged surfaces. Figure 1 illustrates the reflectivity as a function of wavelength for materials processed by standard mechanical and by mechanical-chemical techniques. The front and back surfaces of the sample polished by mechanical means show substantial differences in reflectivity while the mechanical-chemical process yields identical reflectivity spectra. It is evident that to a depth equivalent to at least the penetration depth of the optical beam, in this particular case 2–3 μm, the damage introduced by mechanical polishing manifests itself in lowered carrier concentration and reduced mobility.

The AFCRL Dynamitron was utilized to irradiate the crystals with 1 MeV electrons at an average flux of 3×10^{14} e/cm^2-sec. The electrons incident

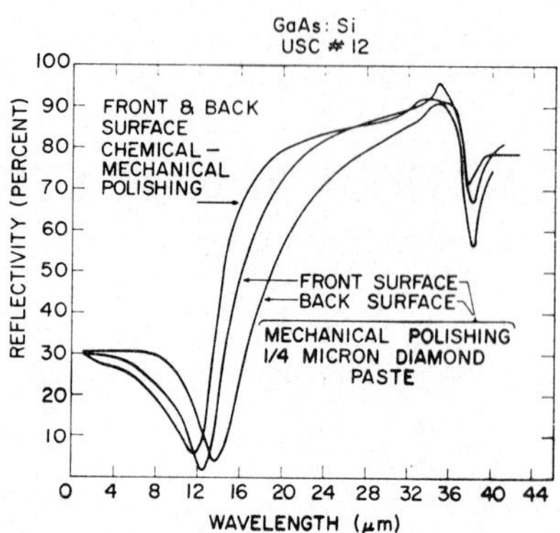

FIG. 1. Effects of surface preparation on optical reflectivity in heavily doped n-type GaAs.

on the sample mount and the sample were collected and monitored with a current integrator and the fluence was estimated from this reading relative to the charge collected in a Faraday cup placed at the sample position prior to each irradiation. The sample temperature during irradiation, with the sample mounted on a liquid nitrogen cooled fixture and in addition sprayed with liquid nitrogen, did not exceed 100 °K.

The optical reflectivity measurements were performed at room temperature in a Perkin–Elmer 112 single-beam double-pass prism spectrometer. A Fourier spectrometer system was employed for some far infrared investigations and, in these cases, studies were extended to 150 μm. The isochronal anneal experiments were performed for 20 minute intervals in an open tube quartz furnace for GaAs:Si while for GaAs:Te the anneals were performed in a nitrogen atmosphere.

4. RESULTS

The experimental room temperature infrared reflectivity as a function of wavelength after each successive fluence of 1 MeV electrons is shown in Figure 2 for the silicon-doped, and, in Figure 3, for the tellurium-doped GaAs. Except for the last two GaAs:Si irradiations, measurements were carried out only to the 30 μm region. The reflectivity curves are for the front surface, the side exposed to the electron beam, of the approximately 1 mm thick GaAs:Si and 0.8 mm thick GaAs:Te samples. The fluence for GaAs:Si varied from 2.5×10^{17} e/cm^2 for the first irradiation to 5×10^{17} e/cm^2 for each successive bombardment, while for GaAs:Te the fluence was sequentially decreased from 5 to 3 to 2×10^{17} e/cm^2. In agreement with

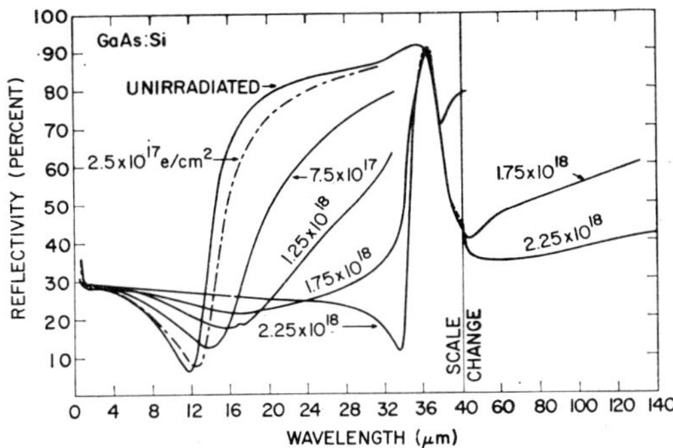

FIG. 2. Reflectivity as a function of wavelength for 1 MeV electron irradiated silicon-doped GaAs.

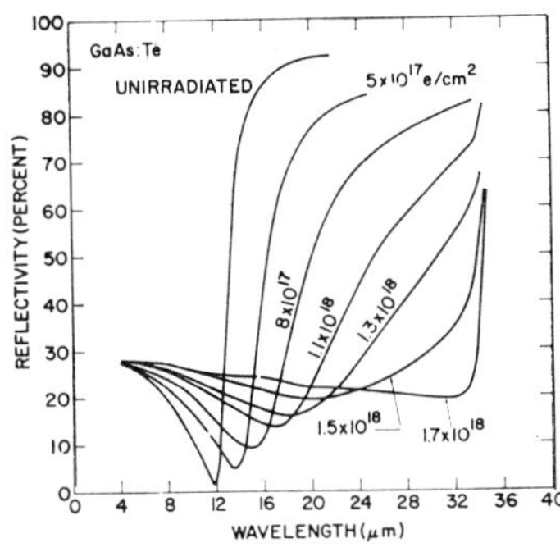

FIG. 3. Reflectivity as a function of wavelength for 1 MeV electron irradiated tellurium-doped GaAs.

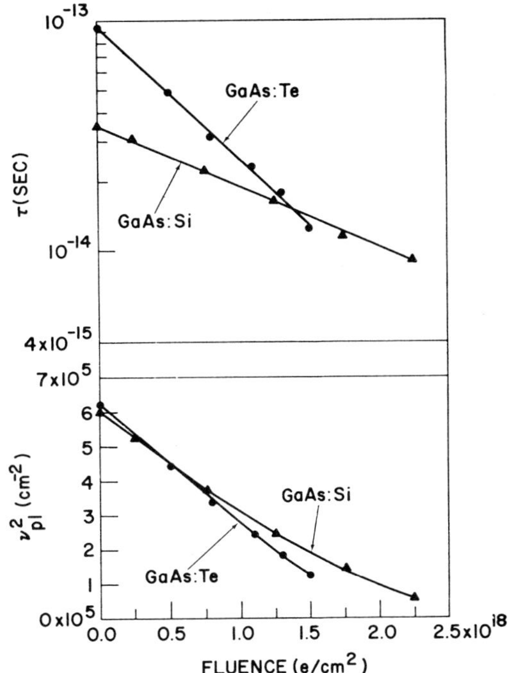

FIG. 4. Scattering relaxation time and plasma frequency as a function of fluence for 1 MeV electron irradiated GaAs:Si and GaAs:Te.

energy degradation relationships for these sample thicknesses no reflectivity changes were observed for the back surface. The general nature of these curves show that with increasing fluence the plasma reflectivity edge shifts towards the far infrared and at the same time its slope is decreasing, indicating a decrease in carrier concentration and mobility.

Figure 4 shows the computed free carrier parameters as a function of fluence. The scattering relaxation time τ, proportional to mobility μ, is plotted in the top part, and the square of the plasma frequency ν_{pl}^2, proportional to carrier concentration N, is depicted in the bottom part of the figure. It is more convenient to relate radiation damage results in terms of changes in N and μ rather than the computer analysis determined parameters N/m^* and μm^*. For discussion purposes, one then neglects the effective mass variation as a function of carrier concentration, assumes for n-type GaAs a constant effective mass $m^* = 0.08$, and with this assumption then Eqs. (2a) and (2c) reduce to

$N = (10^{13})\nu_{\text{pl}}^2$ cm^{-3} and $\mu = (2.2 \times 10^{16})\tau$ cm^2/volt-sec, respectively.

Figure 3 also shows GaAs: Te data for a 1.7×10^{18} e/cm^2 irradiation. The reflectivity features, below the lattice vibration resonance, for the resulting carrier concentration and mobility are indistinct. For accurate free carrier parameter determination measurements need to be extended to the far infrared, as in the case of GaAs:Si. Consequently, the computed parameters of this irradiation are omitted from Figure 4.

Both GaAs: Si and GaAs: Te show a major recovery stage between 170 and 230 °C. Figure 5 shows, for GaAs: Te, the free carrier parameters as a function of anneal temperature. At 230 °C the carrier concentration has recovered to approximately 93 per cent and the mobility to 90 per cent of their preirradiation values. No additional changes in either parameter are observed after a 260 °C isochronal anneal. The GaAs: Si recovery stage is similar to GaAs: Te, but at 225 °C both parameters have recovered only to 80 per cent of their initial values with no additional changes even at a 325 °C anneal. Figure 6 shows a comparison of $1/\tau$ as a function of ν_{pl}^2 for both irradiation and

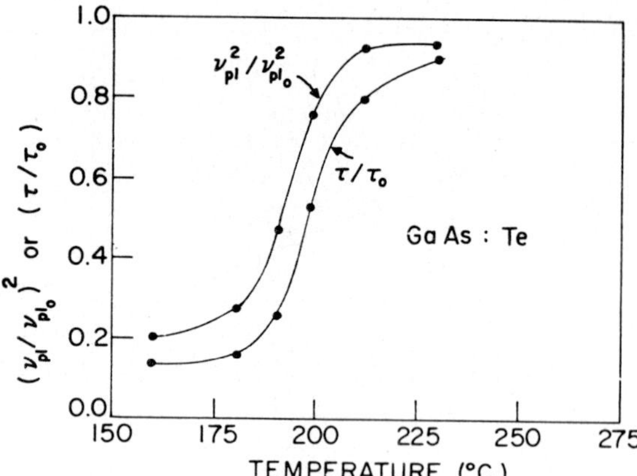

FIG. 5. Isochronal anneal of GaAs:Te. The free carrier parameters are normalized to their unir-radiated values.

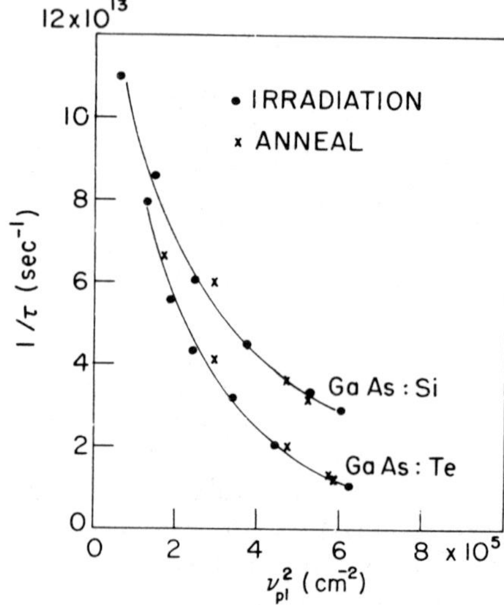

FIG. 6. Reciprocal scattering relaxation time as a function of plasma frequency squared for GaAs:Si and GaAs:Te for both the irradiation (closed circles) and the anneal (crosses) cycle.

anneal experiments indicating that the recovery of the free carrier parameters are associated with the annealing of the irradiation-induced defects.

5. DISCUSSION

For GaAs:Te the carrier concentration decreases linearly as a function of fluence, and over the same range GaAs:Si shows non-linear behavior, Figure

4. The carrier removal rates for carrier concentrations between 6×10^{18} and 3×10^{18} cm^{-3} are approximately 3.4 cm^{-1} for GaAs:Te and 3.0 cm^{-1} for GaAs:Si. For lightly doped GaAs, 10^{15}–10^{17} cm^{-3}, room temperature carrier removal rates determined from electric transport measurements are reported[2-6] to be between 0.5 and 2.5 cm^{-1}.

For both GaAs:Si and GaAs:Te the dominant radiation effect on scattering relaxation time is its exponential decrease as a function of fluence, with GaAs:Te degrading at a faster rate. This behavior was also evident in the exponential decrease in electrical conductivity of a GaAs:Te sample, initial carrier concentration 3×10^{18} cm^{-3}, irradiated sequentially to 1.2×10^{17} e/cm^2. The carrier removal rate for GaAs:Si diminishes with continued irradiation, but the mobility degrades with the same rate constant even at the higher fluences.

Optical reflectivity techniques locate the center of the annealing stage near 200 °C. Also, the annealing behavior of the GaAs:Te 3×10^{18} cm^{-3} sample, as determined from conductivity investigations, parallels the results obtained from optical measurements. This stage was also observed by electric transport measurements by Auckerman and Graft[2] near 225 °C with complete recovery of the electrical conductivity, by Thommen[6] near 250 °C with complete recovery in conductivity, and by Stein[5] near 250 °C with almost complete recovery in carrier concentration but only 90 per cent recovery in mobility. The electric transport measurements of Stein and Thommen were carried out at 80 °K.

For the silicon-doped material utilized in this investigation, an electron irradiation induced absorption band is observed[8] at 369 cm^{-1}. This band is not found in either GaAs:Te or in GaAs:Zn, and may be connected with the amphoteric nature of the silicon dopant. It is possible that the defects associated with the 369 cm^{-1} band are responsible for the lower free carrier parameter recovery observed in GaAs:Si.

Several difficulties arise in interpreting the data in a consistent manner. While a linear decrease in carrier concentration as a function of fluence is reasonable, the exponential decrease in mobility is not understood. In spite of the different physical characteristics of the two samples, as evidenced from X-ray analysis and from the large impurity concentration in GaAs:Si, their initial carrier removal rates are nearly identical. Based on this observation it would be tempting to suggest that the introduction of defects which reduce the **free** carrier concentration is not dependent on material quality or dopant species. However, the differences noted for these samples in the other irradiation and anneal experiments preclude this.

REFERENCES

1. A. Kahan, L. Bouthillette and W. G. Spitzer, *J. Appl. Phys.*, **40**, 2678 (1969).
2. L. W. Auckerman and R. D. Graft, *Phys. Rev.*, **127**, 1576 (1962).
3. L. W. Auckerman, P. W. Davis, R. D. Graft and T. S. Shilliday, *J. Appl. Phys.*, **34**, 3540 (1963).
4. J. A. Grimshaw and P. C. Banbury, *Proc. Phys. Soc.*, **84**, 151 (1964).
5. H. J. Stein, *J. Appl. Phys.*, **40**, 5300 (1969).
6. K. Thommen, *Rad. Effects*, **2**, 201 (1970).
7. The samples were polished at Semiconductor Processing Co., Hingham, Massachusetts, and details of a similar process are disclosed in A. Reisman and R. L. Rohr, *J. Electroch. Soc.* **111**, 1425 (1964).
8. W. G. Spitzer, A. Kahan and L. Bouthillette, *J. Appl. Phys.*, **40**, 3398 (1969).

DISCUSSION

Question (STRNISA)   Please describe your chemical-mechanical polishing technique.

Answer (KAHAN)   The details of this procedure are given in Reference 7.

Question (STEIN)   Have you performed similar measurements on neutron-irradiated GaAs?

Answer (KAHAN)   Not at the present time, but the experiment is planned for the near future.

Question (PICRAUX)   Do you take the variations in results you mentioned for GaAs materials from different manufacturers as further indication of the importance of impurities in GaAs radiation effects?

Answer (KAHAN)   Preliminary experiments on a Monsanto GaAs: Si sample with a lower initial carrier concentration and higher mobility indicates a significantly higher initial carrier removal rate and a faster decrease in mobility as a function of fluence, than either the GaAs: Si or GaAs: Te sample shown in Figure 4. Some other material property, perhaps oxygen concentration, is influencing the radiation behavior. This points out the importance of performing irradiation experiments only on well-characterized crystals.

Question (INUISHI)   Did you observe concentration dependence of the annealing temperature?

Answer (KAHAN)   For the samples studied, recovery occurs in the same temperature range, which is lower than observed by other investigators on lightly doped samples.

ANNEALING BEHAVIOR OF BULK *n*-GaAs IRRADIATED BY ELECTRONS AT 77 °K

M. U. JEONG, J. SHIRAFUJI AND Y. INUISHI

Faculty of Engineering, Osaka University, Suita, Osaka, Japan

Annealing behavior of electrical properties and photoluminescence spectra both at 77 °K in electron-irradiated melt-grown *n*-GaAs were investigated. Defects electrically active in the Hall mobility and carrier removal anneal through two stages centered at 250 ° and 460 °K. From the temperature dependence of carrier concentration the existence of a defect level located near 0.15 eV below the conduction band is supposed. Several emission bands are resolved at 1.51, 1.47, 1.415, 1.305 and ∼1.2 eV in photoluminescence experiments. Electron irradiation (1.5 – 2.0 MeV) causes a remarkable decrease in emission intensity of 1.51 and ∼1.2 eV bands. Recovery of emission intensity occurs remarkably when samples are annealed to 520 °K which would correspond to the 460 °K annealing stage for carrier concentration and Hall mobility. The 250 °K annealing stage is not observed in photoluminescence experiments. The 1.415 eV peak appears clearly after irradiation and grows remarkably with the 520 °K annealing, especially in Si-doped samples, resulting in large reverse annealing. This band is tentatively speculated to be a complex of Si on As site with As vacancy. Moreover, in samples doped with Te a new emission band at 1.305 eV (9500 Å) is observed after 470° – 620 °K annealing.

1. INTRODUCTION

Gallium arsenide is now one of the most promising semiconducting materials for bulk effect and optoelectronic devices. Because of little understanding of lattice defects and poor controllability of crystal perfection, however, satisfactory performance, in particular in reproducibility and life, of such devices has not been attained yet.

Investigations of lattice defects in GaAs have been mainly carried out by means of annealing experiments of electrical properties in Cu-diffused *p*-type[1–3] or as-grown semi-insulating *n*-type crystals.[4,5] For studies on lattice defects related to radiative transitions, photoluminescence measurement is effective. Williams has measured photoluminescence spectra of *n*-type crystals doped with various donor impurities and assigned an emission peak near 1.2 eV to be donor-Ga vacancy complex.[6–9]

Kressel et al. have studied photoluminescence spectra of *n*- and *p*-type GaAs grown epitaxially from gallium solution containing a small amount of Si or Ge.[10–12] They observed in Si-doped crystals, in addition to the emission band due to Si acceptors, the 1.40 eV emission band which was speculated to be a complex of Si on As sites with either vacancies or donors.[10] The corresponding band in Ge-doped crystals was at 1.44 eV.[12] On the other hand, Rosztoczy and coworkers[13] have assigned the 1.44 eV band in Ge-doped *p*-GaAs to be an LO phonon replica of the 1.48 eV band due

to transitions between the conduction band and isolated Ge acceptors. Kressel and coworkers have found a 1.34 eV peak in Ge-compensated *n*-GaAs, speculating that it is a Ge-As vacancy complex.

Radiation effects on electrical properties of GaAs have been studied by Aukerman and coworkers[14] and by Stein.[15] Loferski and coworkers have made a preliminary photoluminescence experiment of *n*- and *p*-type GaAs irradiated by electrons,[16] followed by a more extensive study by Arnold on electron-irradiated Cd-doped *p*-GaAs.[17] Saji and Inuishi[18] have investigated γ-ray irradiation effects on spectra of GaAs laser diodes, concluding that besides the decrease of quantum efficiency the lasing peak wavelength at 8414 Å before irradiation shifted slightly to shorter wavelength with increasing γ-ray irradiation dose. Although the laser emission intensity recovered after 200 °C annealing, the wavelength of the peak did not.

In order to get more detailed informations about lattice defects we investigated the annealing behavior of photoluminescence spectra in electron-irradiated *n*-type GaAs single crystals doped with various impurities and compared them to annealing effects on electrical properties such as carrier concentration and Hall mobility.

2. EXPERIMENTAL PROCEDURES

Samples used were cut from melt-grown *n*-type GaAs single crystals doped with various impurities such as Sn, Si and Te having carrier concentration of 5×10^{15}–7×10^{17} cm^{-3}. Electrical properties of samples at room temperature are tabulated in Table I. Sample faces were oriented to $\langle 111 \rangle$

TABLE I
Electrical properties of samples at room temperature

Sample No.	Dopant	$\sigma(\Omega^{-1}\,\mathrm{cm}^{-1})$	$n(\mathrm{cm}^{-3})$	$\mu(\mathrm{cm}^2\,\mathrm{v}^{-1}\,\mathrm{sec}^{-1})$
F2†	Sn	1.0×10	8.9×10^{15}	7.2×10^3
F4†	Sn	1.0×10	6.1×10^{15}	1.2×10^3
HB-160	Sn	1.9×10^2	6.8×10^{17}	1.8×10^3
UN-1	Te	1.9×10^2	3.4×10^{17}	3.4×10^3
SIB	Si	1.7×10^2	4.0×10^{17}	2.7×10^3

† Measured at 77 °K.

direction and etched in a mixture with composition of H_2SO_4, 1: H_2O_2, 1: H_2O, 1.

Irradiations were made at a sample temperature of 77 °K with fast electrons of 1.5–2.0 MeV from a Van de Graaff accelerator. Total incident flux was varied from 5×10^{15} to 2×10^{17} electrons/cm².

Electrical measurements were carried out by the ordinary d.c. method in the dark and under illumination of visible light. Optical excitation for photoluminescence experiments was made by using a 500-W Xe arc lamp with a filter of saturated aqueous solution of $CuSO_4$ to cut the infrared emission. The emitted light from the front surface of the sample immersed in liquid nitrogen was scanned by a grating monochromater with dispersion of about 30 Å/mm at the energy gap of GaAs. A cooled ($\sim$170 °K) photomultiplier of S-1 response connected with a lock-in amplifier was used to detect the emission sensitively.

Isochronal annealing experiments were done on a cold finger with a heater which can heat the sample to temperatures as high as 400 °K. After each annealing for 30 minutes the sample was quenched by immersing in liquid nitrogen. Annealing above 400 °K was made in a quartz tube with flowing nitrogen gas.

3. EXPERIMENTAL RESULTS

3.1. *Electrical measurements*

Figure 1 shows the total carrier removal Δn as a function of total electron fluence Φ_e. The effects of 30 min isochronal annealing on carrier concentration and Hall mobility in Sn-doped sample are shown in Figure 2, where the fraction not annealed in carrier concentration $f(n)$ and in reciprocal

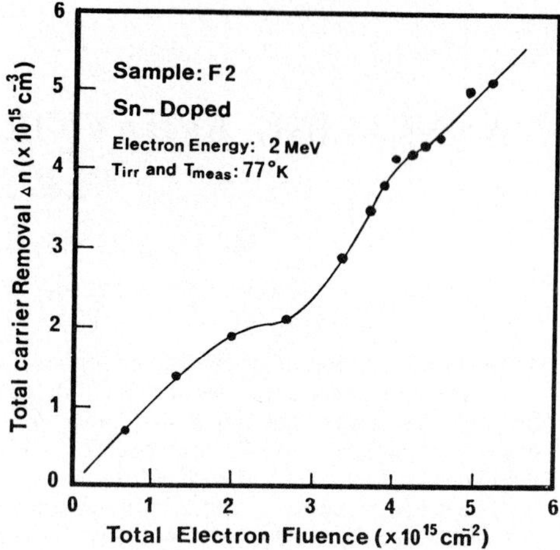

FIG. 1. Total carrier removal Δn as a function of total electron fluence Φ_e in Sn-doped sample.

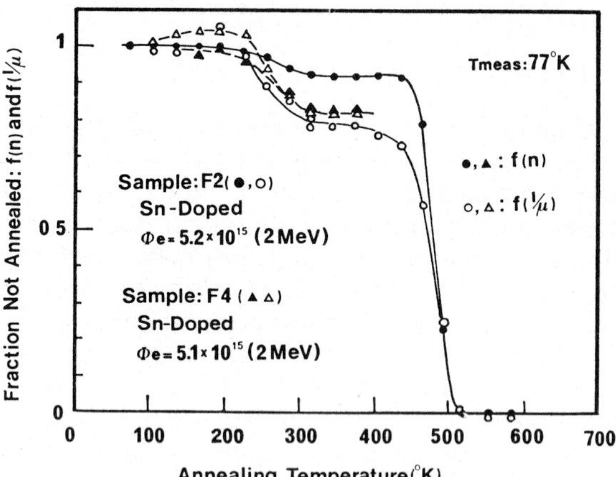

FIG. 2. Effects of 30 min isochronal annealing on carrier concentration and Hall mobility in Sn-doped samples. In the vertical coordinate the fraction not annealed in carrier concentration $f(n)$ and reciprocal mobility $f(1/\mu)$ are plotted.

mobility $f(1/\mu)$ are plotted against annealing temperature. $f(n)$ and $f(1/\mu)$ are defined as

$$f(n) = \frac{n_0 - n_t}{n_0 - n_r}$$

$$f(1/\mu) = \frac{1/\mu_0 - 1/\mu_t}{1/\mu_0 - 1/\mu_r}$$

where the suffixes, 0, r and t, represent pre-irradiation, post-irradiation and after-annealing, respectively. Noticeable difference between annealing in the dark and under illumination was not observed. The defects introduced by electron irradiation anneal through two stages centered at about 250 and 460 °K. These results are in good agreement with those observed by Stein.[15] Some sample showed an anomalous reverse annealing in Hall mobility at temperatures between 150° and 300 °K. It seems to be probably due to either deep traps such as Cu or localized space-charge regions having a large scattering cross-section as suggested by Weisberg.[19] From the temperature dependence of Hall coefficient, a defect level located near 0.15 eV below the conduction band, being in good accord with previous results,[14,15] was supposed to exist in Sn-doped sample after 280 °K annealing.

3.2. *Photoluminescence measurements*

Figure 3 shows variation of photoluminescence spectra in Sn-doped sample with electron irradiation and subsequent annealing. Four emission bands (tentatively named as A, B, C and D) are re-

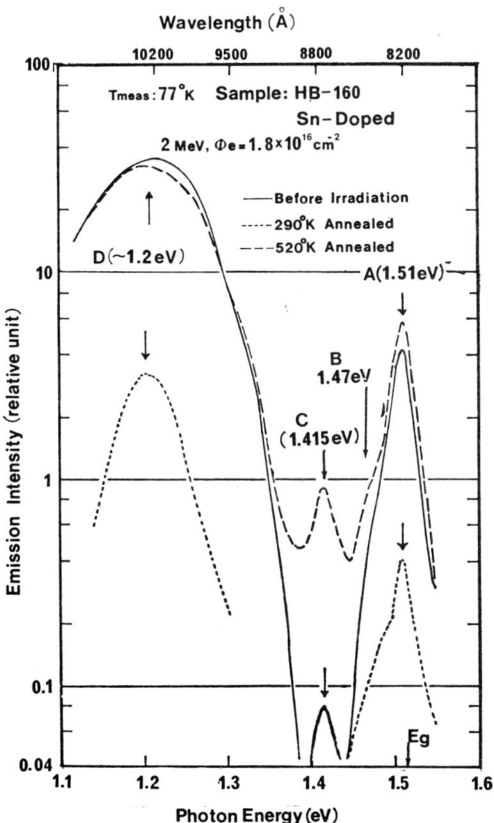

FIG. 3. Annealing behaviors of photoluminescence spectra at 77 °K in Sn-doped sample.

solved. They occur at photon energies of 1.51, 1.47, 1.415 and ∼1.2 eV respectively. The emission intensities of the dominant bands A and D decrease remarkably with similar ratio after electron irradiation, but they recover almost completely by 520 °K annealing. On the other hand, the band C does not decrease due to irradiation and grows over the initial intensity after 520 °K annealing, showing a remarkable reverse annealing. This situation is also shown in Figure 4 in which

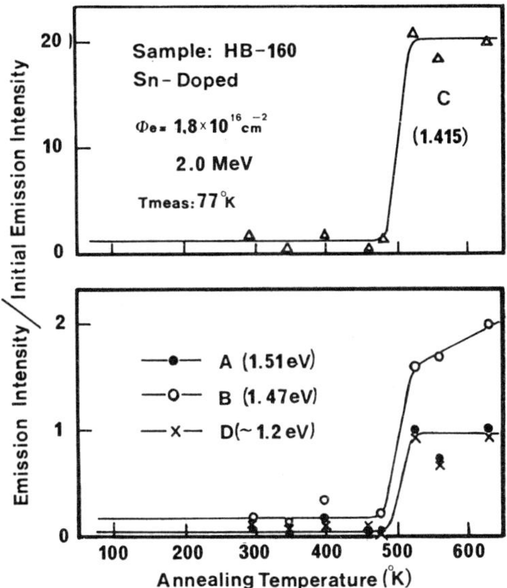

FIG. 4. Changes in emission intensities with annealing temperature in Sn-doped sample.

annealing behavior of emission intensities for four bands are plotted as a function of annealing temperature. Drastic change in emission intensity after 520 °K annealing is clearly demonstrated. It should be emphasized that the emission intensity of C band does not anneal even upon annealing at 620 °K, showing only a monotonic reverse annealing. Qualitatively similar behavior of annealing in photoluminescence spectra were also observed in all other samples doped with different impurities. The location of C band does not seem to depend on the kind of impurity.

Figure 5 shows the annealing behavior of a Si-doped sample. The characteristic difference in Si-doped samples from Sn-doped samples is that the band grows to a comparable level with or in excess of the A band after 520 °K annealing. The growth of C band became larger with higher total electron fluence. The weak emission band at

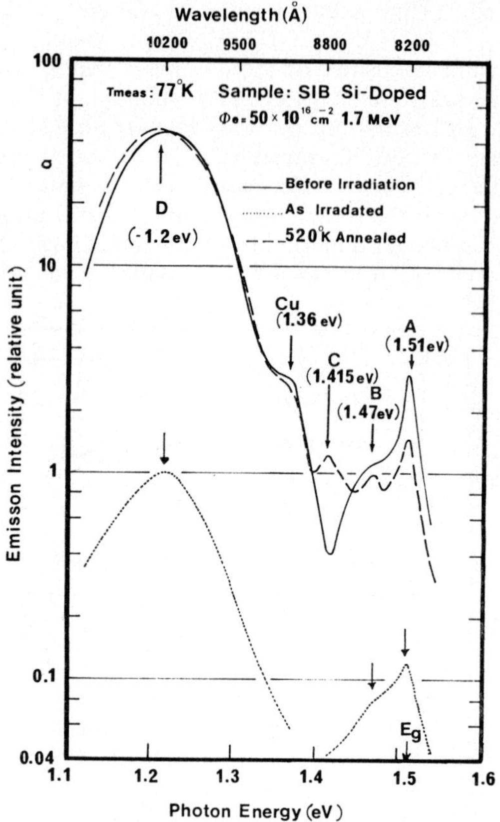

FIG. 5. Annealing behavior of photoluminescence spectra at 77 °K in Si-doped sample.

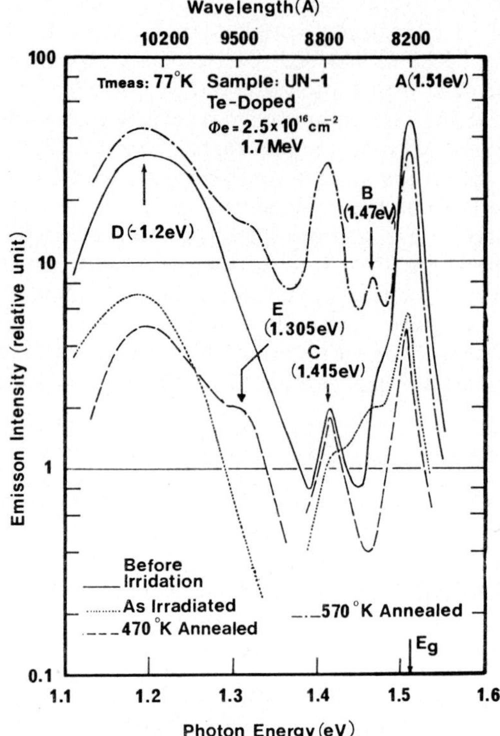

FIG. 6. Annealing behavior of photoluminescence spectra at 77 °K in Te-doped sample.

1.36 eV seen in Figure 5 is probably due to residual Cu acceptors.[23]

Te-doped samples behaved in somewhat different manner from Sn- and Si-doped samples, when samples were annealed at temperatures above 470 °K. Figure 6 shows the annealing behavior of emission spectra in the Te-doped sample. After 470 °K annealing, a new broad band E centered at 1.305 eV appears. This band becomes more intense with higher temperature annealing. In another Te-doped sample the E band appeared upon 620 °K annealing. The band E was not detected in virgin samples and in unirradiated samples subjected to the similar heat treatment as annealing.

4. DISCUSSION

Electrically active defects anneal through two stages centered at 250 and 460 °K as seen in Figure 2. This has already been observed by Stein.[15] Annealing of defects related to luminescent transitions occurs only near 520 °K. The lack of the 250 °K annealing stage in photoluminescence spectra has also been observed in luminescence studies on Cd-doped p-GaAs.[17] Stein has supposed a conductivity-type dependence for the defect annealing at 250 °K, since there is no 250 °K annealing stage in photoluminescence spectra in p-type samples.[11] Our findings, however, indicate that the lack of the 250 °K annealing stage in photoluminescence spectra also occurs in n-type GaAs.

The reduction of the emission intensities of A and D bands in an equal ratio due to irradiation may be due to an introduction of some killer centers which compete with luminescent centers.

The band A is the usual near-band-gap emission observed in n-type GaAs and involves either electron transitions from the conduction band tail states to free holes[12] or free exciton recombination.[20] The band B seems to be caused by transitions between states within the tail below the conduction band and shallow acceptors.[10,12] Although the origin of shallow acceptors is not identified, Si on As site, particularly in Si-doped samples, is suspected.

Electron irradiation followed by annealing enhances the band C in all samples examined in the present experiments. The band C grows drastically by 520 °K annealing, especially in Si-doped samples. These facts suggest that C band should be related to lattice defects and impurities, possibly Si. Queisser has observed an intense emission band at 1.42 eV in Si-compensated melt-grown *n*-GaAs following high temperature heat treatment.[21] He supposed as an origin of the band that the heat treatment causes in-diffusion of As vacancies from the sample surface and their subsequent replacement by Si atoms on Ga sites. Queisser's assumption, however, is in conflict with the reasonable conclusion deduced in Si- and Ge-doped liquid-epitaxially-grown GaAs by Kressel and coworkers.[10–12] and by Rosztoczy and coworkers[13] that transitions from the conduction band to Si or Ge acceptors give a band near 1.48 eV. Kressel and coworkers[10] have also found the emission band near 1.40 eV in Si-compensated solution-grown GaAs, corresponding to the band C in the present experiments. They assumed that the 1.40 eV band is due to a formation of a complex of Si on As sites with either vacancies or donors. Our findings seem to support the possibility of a complex between Si on As sites and vacancies. Furthermore, if the 520 °K annealing stage is due to the movement of As vacancies as suggested by Potts and Pearson,[22] the band C would be due to a complex involving Si on As site and As vacancy.

In Te-doped samples, a new emission band E centered at 1.305 eV was revealed after annealing at temperatures above 470 °K. This band is very close to the 1.32 eV band which has been observed in Cu-diffused GaAs and assigned to a complex involving Te and Cu on Ga site.[23] Kressel and coworkers have observed the 1.34 eV band in Ge-doped *n*-GaAs[12] and argued that it is not due to a contaminant such as Cu. They speculated the band to be resulted from a complex of Ge with As vacancies.

Nonradiative centers will have an important influence on luminescence spectra. A large decrease in emission intensity after electron irradiation would be attributed to an introduction of non-radiative centers which dominate the radiative centers. However, the introduction of deeper centers which cause not-observed transitions in infrared region should be taken into consideration.

5. SUMMARY

Annealing experiments of electrical properties and photoluminescence spectra in electron-irradiated *n*-type GaAs were performed. Carrier concentration and Hall mobility anneal through two annealing stages centered at 250 and 460 °K. The irradiation by electrons reduces remarkably the emission intensities of the predominant bands A (1.51 eV) and D (~1.2 eV). These bands recover to the initial values by the 520 °K annealing. The 250 °K annealing stage of electrical properties does not appear in photoluminescence spectra. The growth and negative annealing of two bands, C (1.415 eV) and E (1.305 eV), being possibly related to vacancies, are found after irradiation and subsequent annealing. The 1.415 eV band is tentatively speculated to be a complex of Si on As site with As vacancy.

ACKNOWLEDGEMENTS

The authors are greatly indebted to Takasaki Radiation Research Establishment, Japan Atomic Energy Research Institute for the kind convenience of electron irradiation. We are also grateful to Dr. T. Suzuki of Research and Development Laboratories, Sumitomo Electric Industries, Ltd. and to Dr. O. Ryuzan of Semiconductor Research Department, Fujitsu Laboratories, Ltd. for kindly supplying GaAs crystals.

REFERENCES

1. C. S. Fuller and K. B. Wolfstirn, *J. Phys. Chem. Solids*, **27**, 1889 (1966).
2. C. S Fuller, K. B. Wolfstirn and H. W. Allison, *J. Appl. Phys.*, **38**, 2873 (1967).
3. C. S. Fuller, K. B. Wolfstirn and H. W. Allison, *J. Appl. Phys.*, **38**, 4339 (1967).
4. J. Blanc, R. H. Bube and L. R. Weisberg, *Phys. Rev. Letters*, **9**, 252 (1962).
5. J. Blanc, R. H. Bube and L. R. Weisberg, *J. Phys. Chem. Solids*, **25**, 225 (1964).
6. E. W. Williams and D. M. Blacknall, *Trans. AIME*, **239**, 387 (1967).
7. E. W. Williams, *Phys. Rev.*, **168**, 922 (1968).
8. M. B. Panish, H. J. Queisser and L. Derick, *Solid-State Electronics*, **9**, 311 (1966).
9. A. R. Goodwin, J. Gordon and C. D. Dobson, *Brit. J. Appl. Phys.*, **19**, 115 (1968).
10. H. Kressel, J. U. Dunse, H. Nelson and F. Z. Hawrylo, *J. Appl. Phys.*, **39**, 2006 (1968).
11. H. Kressel and H. Nelson, *J. Appl. Phys.*, **40**, 3720 (1969).
12. H. Kressel, F. Z. Hawrylo and P. LeFur, *J. Appl. Phys.*, **39**, 4059 (1968).
13. F. E. Rosztoczy, F. Ermanis, I. Hayashi and B. Schwartz, *J. Appl. Phys.*, **41**, 264 (1970).

14. L. W. Aukerman, P. W. Davis, R. D. Graft and T. S. Shilliday, *J. Appl. Phys.*, **34**, 3590 (1963).
15. H. J. Stein, *J. Appl. Phys.*, **40**, 5300 (1969).
16. J. J. Loferski, H. Flicker, R. M. Esposto and M. H. Wu, *Lattice Defects in Semiconductors*, Ed. R. R. Hashiguchi (University of Tokyo Press, Tokyo, 1968), p. 355.
17. G. W. Arnold, *Phys. Rev.*, **183**, 777 (1969).
18. M. Saji and Y. Inuishi, *Japan. J. Appl. Phys.*, **4**, 830 (1965).
19. L. R. Weisberg, *J. Appl. Phys.*, **33**, 1817 (1962).
20. M. I. Nathan and G. Burns, *Phys. Rev.*, **129**, 125 (1963).
21. H. J. Queisser, *J. Appl. Phys.*, **37**, 2909 (1966).
22. H. R. Potts and G. L. Pearson, *J. Appl. Phys.*, **37**, 2098 (1966).
23. H. J. Queisser and C. S. Fuller, *J. Appl. Phys.*, **37**, 4895 (1966).

DISCUSSION

Question (URLI)   Could you make any comment on the steps in the carrier removal versus fluence curve?

Answer (INUISHI)   I don't want to place much significance on these steps.

Question (BARNES)   Did you heat an unirradiated photoluminescence sample to approximately 550 °K in order to eliminate the possibility that the increase in the 'C' band following anneal is merely due to sample heating and, therefore, not associated with any radiation effect?

Answer (INUISHI)   Yes we did.   No remarkable change was found in unirradiated samples.   However, we have to be careful about the change of surface.

INFRARED ABSORPTION STUDIES IN NEUTRON- AND ELECTRON-IRRADIATED GaAs

K. V. VAIDYANATHAN AND L. A. K. WATT

Department of Electrical Engineering, University of Waterloo, Waterloo, Ontario, Canada

Mechanically and chemically polished samples of GaAs were irradiated with fission neutrons to fluences varying from 2×10^{16} to 1.4×10^{18} neutrons/cm^2 and with 1.5 MeV electrons with doses varying from 10^{17} to 10^{18} el/cm^2. Infrared absorption in irradiated crystals was measured at 300°K and 85°K.

In neutron-irradiated samples absorption bands centered at 0.47 eV and at 0.25 eV are observable. It is suggested that the 0.47 eV band is connected with transitions from a level at $(E_c - 0.5)$ eV reported earlier from electrical measurements. The absorption coefficient was found to vary linearly with the square of the photon energy over a limited energy range.

For electron-irradiated GaAs, the absorption in the band gap had an exponential variation with energy. The possible reasons for such a behaviour are discussed. The absorption spectrum at 85°K in electron-irradiated samples shows the presence of a band at 1.0 eV. Isothermal and isochronal annealing behaviour of this absorption band was studied. An analysis of data suggests that the recovery obeys first order kinetics with an activation energy of 0.88 ± 0.07 eV.

1. INTRODUCTION

It is known that the optical properties of semiconductors are modified when defects are introduced by electron- and neutron-bombardment. Several infrared absorption bands caused by such defects have been observed and identified in the case of silicon.[1]

The earliest attempt to study infrared absorption in irradiated GaAs was carried out by Aukerman and coworkers.[2] However, no definite structure was observed either in electron or in neutron-irradiated samples. Conflicting results have been reported about the presence of discrete levels at 0.5 eV and at 0.25 eV in neutron-irradiated GaAs.

In this paper, we report the results of infrared absorption measured at room temperature and at lower temperatures for electron- and neutron-irradiated GaAs.

2. EXPERIMENTAL TECHNIQUES

Rectangular shaped samples (1.0 cm × 0.25 cm) were cut from single crystal ingots of GaAs. The crystals were mechanically polished with various grades of diamond abrasive and finally polished with 0.05 micron alumina to provide a smooth mirror like surface. Some samples were also chemically polished in a solution of 3:1:1 of H_2SO_4:H_2O_2: H_2O.
Infrared absorption measurements in the wavelength range $0.82\,\mu$ to $7.5\,\mu$ were carried out using a Perkin Elmer Model 137 G double beam grating spectrophotometer. For measurements at liquid nitrogen temperature, a small glass cryostat with quartz or sodium chloride windows was used.

A 2 MeV Van de Graaff accelerator at Chalk River Nuclear Laboratories was used for electron-irradiations. The beam energy was kept constant at 1.5 MeV. The samples were mounted on a water-cooled copper sample holder. The dose received was computed from the current used and the time of irradiation. The temperature during irradiation was ~ 50 °C.

The neutron irradiations were performed in the E-3 experimental hole of the NRX reactor at Chalk River. The fission neutron flux at this position was 5×10^{12} neutrons/cm^2/sec. The samples were wrapped in cadmium foils to minimize transmutation effects due to thermal neutrons. The temperature during irradiation was ~ 100 °C.

Isothermal and isochronal anneals were performed either in vacuum or in an inert gas atmosphere using a large thermal capacity oil bath. The temperature during anneals was controlled within ± 0.5°C. Infrared absorption measurements were carried out at 85 °K after each anneal.

3. EXPERIMENTAL RESULTS

3.1. *Data reduction*

In all the cases, the absorption coefficient α was

computed from transmission measurements using the relation

$$T = \frac{(1 - R)^2 \, e^{-\alpha x}}{1 - R^2 \, e^{-2\alpha x}} \qquad (1)$$

where T and R are the transmission and reflection coefficient respectively and x, the thickness of the sample. In the case of electron-irradiated samples, all measurements were normalized at 0.6 eV while for neutron-irradiated samples the normalization was carried out at 0.17 eV. By normalizing, we mean that the light intensity with and without the sample in the path of the beam was measured at that photon energy. A value of 0.30 for R was used in our analysis of electron-irradiated samples. In the case of neutron-irradiated samples, values of R were obtained from V. C. Burkig and coworkers.[6]

3.2. *Neutron-irradiation damage*

Figure 1 shows the variation of the absorption coefficient as a function of photon energy for neutron doses varying from 2×10^{16} neutrons/cm² to 1.4×10^{18} neutrons/cm². In the case of samples irradiated to low neutron fluences, measurements

indicate that the band edge is not significantly altered. As seen from Figure 1, the absorption increases rapidly with the dose at low photon energies. For the most heavily irradiated sample (1.8×10^{18} neutrons/cm²), no measurements beyond 0.4 eV could be taken even with an extremely thin sample. The data show the presence of an absorption band centered at 0.47 eV. In the heavily irradiated sample an additional band at ~ 0.25 eV is just observable.

3.3. *Electron-irradiation damage*

Figure 2 shows the variation of the absorption coefficient as a function of the photon energy, E, at room temperature in electron-irradiated samples. The figure shows the characteristic exponential variation in absorption below the band gap. No structure in absorption can be observed and this is in agreement with the results of Aukerman and coworkers.[2]

Figure 3 shows the absorption spectrum at 85 °K. The data show the presence of a band at 1.0 eV. This band grows in intensity with the electron dose.

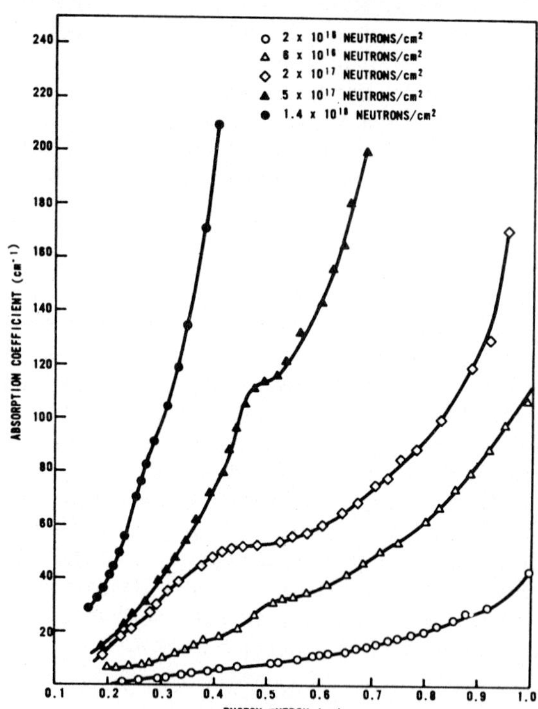

FIG. 1. Variation of absorption coefficient at room temperature as a function of photon energy in neutron-irradiated GaAs.

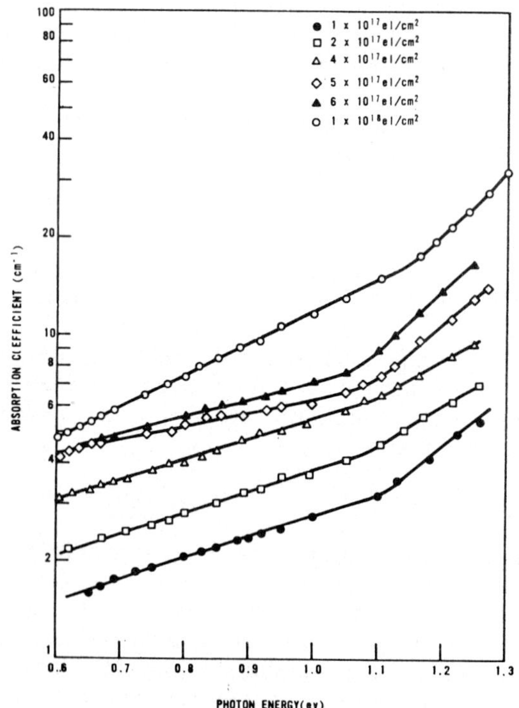

FIG. 2. Variation of absorption coefficient at room temperature as a function of photon energy in electron-irradiated GaAs.

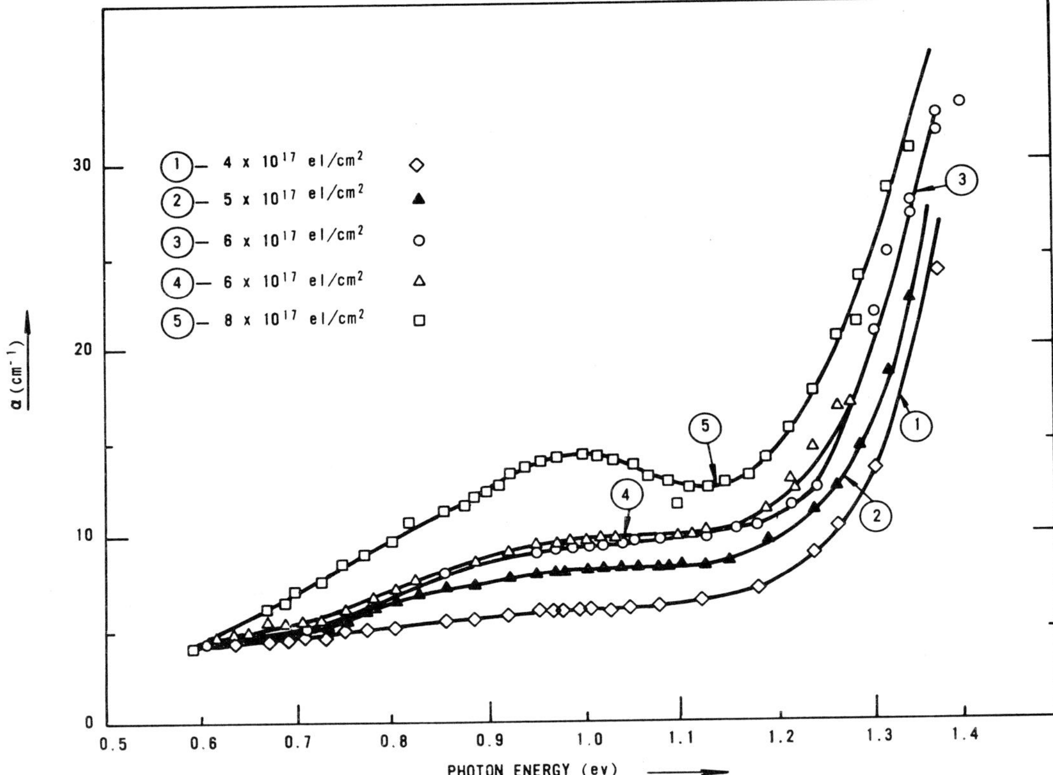

FIG. 3. Variation of absorption coefficient at 85 °K as a function of photon energy in electron-irradiated GaAs.

This increase of the intensity of absorption with dose is plotted in Figure 4.

The 1.0 eV absorption band was found to be very broad, with a half width of 0.35 eV at 85 °K. The shape of the band was asymmetric with the low energy tail smeared out. We also found that the spectrum at 11 °K had the same shape although the half width was slightly reduced. No other fine structure in the band was observable.

The results of isothermal and isochronal anneals are shown in Figures 5 and 6. The fraction of defects remaining after an anneal is computed from the area of the absorption band. The recovery of the fundamental absorption edge is also shown in Figure 5.

In an attempt to study the effects of impurities, we irradiated three different types of samples. The n-type undoped material had $\sim 9 \times 10^{15}$ carriers/cc, the n-type Sn-doped one had 7×10^{16}/cc while the Zn-doped p-type had 1.3×10^{17} carriers/cc. The dislocation density of the samples also differed considerably. In all these cases the 1.0 eV absorption band was still observable. The production rate in all these cases was almost identical.

Preliminary results indicate that no photoconductivity is associated with the 1.0 eV absorption band.

4. DISCUSSION

4.1. *Neutron-irradiation damage*

Aukerman and coworkers[2] were the first to study changes in optical properties of neutron-irradiated GaAs. They did not observe any structure either at room temperature or at 60 °K. Vodopyanov and Kurdiani[3] reported that while no discrete levels were observable at room temperature, two levels, one at 0.5 eV and another at 0.15 eV, were clearly observable at 80 °K. Both authors[2,3] reported that the absorption coefficient varied as the square of the photon energy over a wide range of photon energy.

McNichols and Ginell,[4] using Aukerman's data, have shown that the observed variation of α with E^2

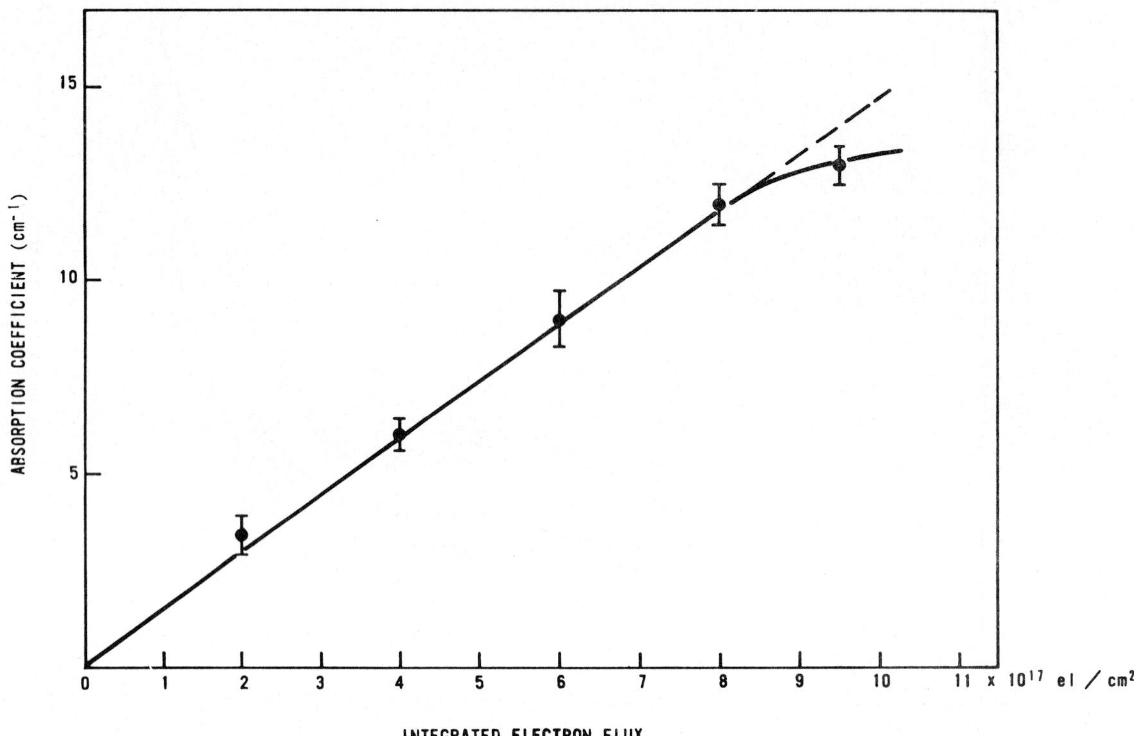

FIG. 4. Variation of absorption coefficient at 1.0 eV at 85 °K in electron-irradiated GaAs as a function of electron dose.

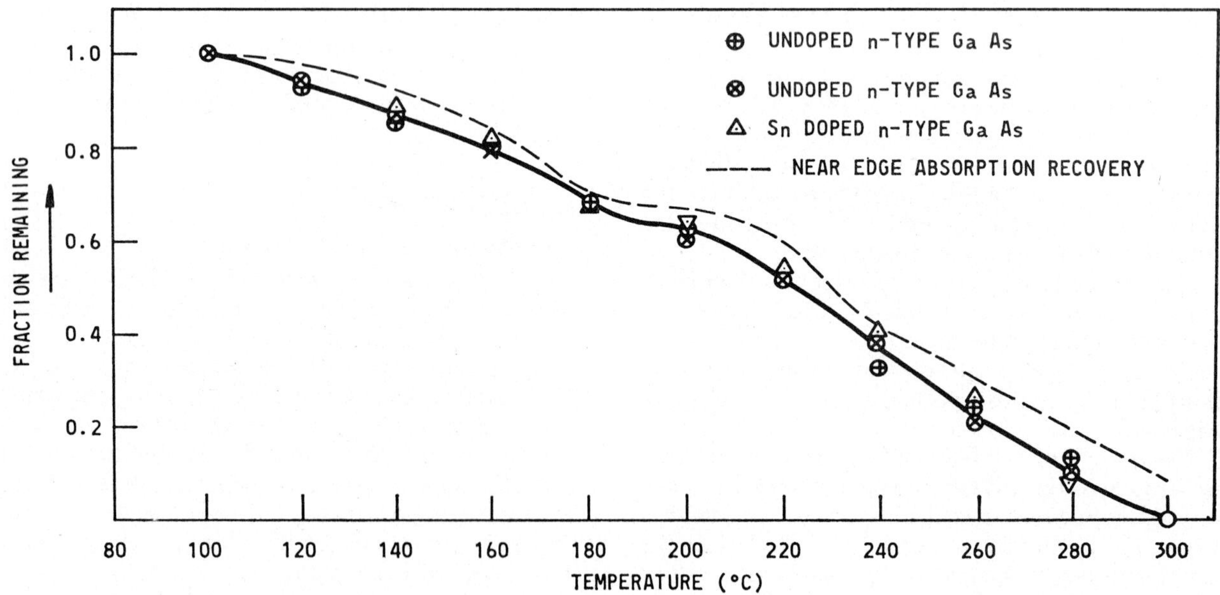

FIG. 5. Isochronal recovery of the 1.0 eV absorption band after 30 minutes anneal at each temperature in electron-irradiated GaAs.

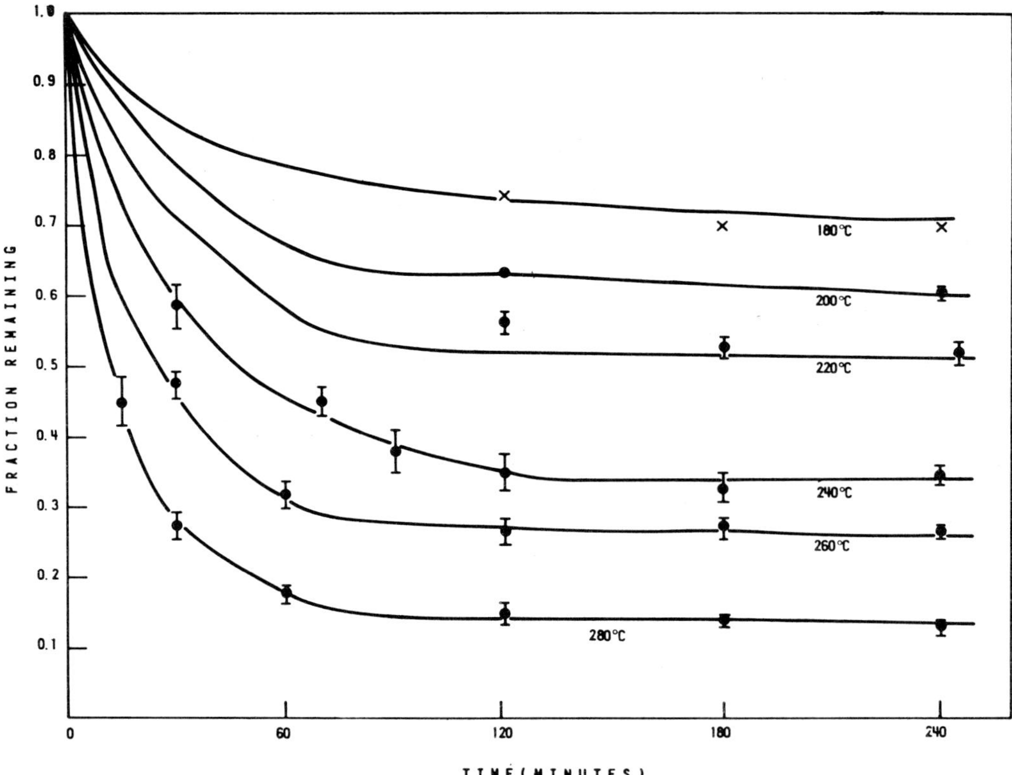

FIG. 6. Isothermal recovery of the 1.0 eV absorption band in electron-irradiated GaAs.

can be explained if small metallic inhomogeneities of $\sim 100\,\text{Å}$ in radius are assumed to be present in the otherwise normal matrix. They argued that such regions could be due to the thermal spike pheonomenon. The absorption in the presence of such heavily damaged regions according to the above authors would be given by

$$\alpha(E) = \alpha_0(E) + \Lambda\,E^2 \qquad (2)$$

where

$$\Lambda = 6.9 \times 10^{22} \int \frac{\epsilon_1 n_1}{\sigma^2} P(a)\,V(a)\,\mathrm{d}a \qquad (3)$$

ϵ_1 and n_1 are the permittivity and refractive index respectively of the host lattice and σ the static conductivity of the metallic zone. They estimate that radius of such zones would be $\sim 100\,\text{Å}$ which should be observable by transmission electron microscopy. The presence of such regions, as far as we know, has not been definitely established in neutron-irradiated III-V compounds. The only evidence for the presence of such regions is the excellent work of Gonser and Okkerse[5] in deuteron-irradiated GaSb. However, these spike regions

annealed out at $\sim 160\,^\circ$K. Thus, the spike regions which are stable at room temperature as proposed by McNichols and Ginell would be of a different nature. Attempts are being made to study this problem in detail.

Burkig, McNichols and Ginell[6] reported that no discrete levels were observable at $100\,^\circ$K. Their samples were irradiated to doses varying from $10^{15}\,n/\text{cm}^2$ to $1.5 \times 10^{18}\,n/\text{cm}^2$.

In our data we find strong evidence for the presence of two absorption bands, one at 0.47 eV and the other at 0.25 eV. We find that over 50 data points are necessary in the energy range 0.8 to 0.2 eV to see these bands. We also find that the linear variation of α with E^2 can be fitted only over a limited energy range (0.2 to 0.8 eV). These results are in qualitative agreement with the recent data of Pankey and Davey.[7]

It is interesting to note that Aukerman observed an energy level at $(E_c - 0.5)$ eV from his electrical measurements on neutron-irradiated GaAs.[2] We feel that the 0.47 eV absorption band is probably caused by a transition from this defect level to the conduction band.

Spitzer and Whelan[8] have observed that at low photon energies, the absorption coefficient is proportional to the carrier concentration. This offers a method of evaluating carrier removal rate from optical data.[6] The change in carrier concentration is related to the change in absorption coefficient by

$$\frac{\Delta N}{N_0} = \frac{\Delta \alpha}{\alpha_0} \tag{4}$$

Using the above equation, we found that for the 6×10^{16} carriers/cc, n-type material, the carrier removal rates were 7/cm and ~ 5.5/cm for fluences of 10^{16} and 2×10^{16} neutrons/cm² respectively at 0.3 eV. These figures compare favourably with the removal rate of 8/cm reported by Aukerman.[9]

4.2. *Electron-irradiation damage*

It has been suggested by Fischer[10] that exponential variations in the absorption coefficient at photon energies below the band gap in irradiated semiconductors may be due to the presence of electric fields caused by the defects (as in Figure 2). He has suggested that the internal Franz-Keldysh effect may explain the behaviour. Redfield and Afromowitz[11] have applied the same kind of idea with some success to the case of heavily doped GaAs. Tharmalingam[12] has shown that the absorption coefficient in the presence of a field F would be given by

$$\alpha(E) = \frac{e^3 \mu F C_0^2}{2n\, m^2\, h^3\, c\, \omega(\omega_1 - \omega)} \exp$$
$$\left[-\frac{4(\omega_1 - \omega)^{3/2}(2\mu h)^{1/2}}{3\, e\, F} \right] \tag{5}$$

where e and m are the charge and rest mass of an electron respectively, ω the angular frequency corresponding to the photon energy, ω_1 that corresponding to the band gap, n the refractive index and μ the reduced effective mass of the host lattice, c the velocity of light and C_0 the matrix element for direct transitions having units of momentum. The above relationship is valid only for energies very low compared to the band gap.

Our attempts to find a fit between our data and Eq. (5) have not been successful. This is probably due to the fact that Eq. (5) assumes a uniform field F. In an actual situation, the fields would be non-uniform and highly localized.

Franz[13] in a recent paper has considered the case of impurity absorption in the presence of an electric field and has shown that the low energy tail of the absorption band would be greatly smeared out. This agrees qualitatively with our observations (Figure 3).

In heavily doped crystals, Halperin and Lax[14] have calculated the density of electronic states in the impurity band and have found an exponential tail which seems to correlate with the low energy tails in impurity absorption bands. Assuming a screened Coulomb potential between impurities, they find that the density of states, $\rho(E)$, is given by

$$\rho(E) \cong \exp[-\beta(E)] \tag{6}$$

where $\beta(E)$ varies from $E^{1/2}$ to E^2 depending on several material parameters. We feel that this type of an approach would be more applicable to our present case. More work in this area is required before any definite conclusions can be drawn.

Figure 5 shows the isochronal recovery of defects in electron-irradiated samples. The annealing was carried out in vacuum for 30 minute intervals in steps of 20 °C from 100 °C to 300 °C. The absorption was measured at the end of each anneal at 85 °K. The solid curve represents the recovery of the 1.0 eV band while the dotted curve represents the recovery of the absorption near the band edge. The two curves follow each other quite closely.

The recovery seems to take place in two stages. One near 140 °C and the second and dominant one at ~ 240 °C. The fact that the absorption band changed shape during the anneal led us to believe that the band is made up of more than one peak. However, our measurements at 11 °K do not support this view. At present we can not offer any explanation for this annealing behaviour.

Figure 6 shows the isothermal recovery of the 1.0 eV band in electron-irradiated samples. By using the cross-cut method[15] of analyzing annealing curves, we obtained an activation energy of 0.88 ± 0.07 eV at 260 °C. We also found that the recovery obeyed first order kinetics.

Aukerman and coworkers[16] had studied the annealing behaviour in electron-irradiated GaAs. They found that the recovery was complete by ~ 260 °C. It was also observed that there were two unresolved substages with activation energy of $(1.10 \pm 0.05$ eV) and 1.55 ± 0.05 eV). Our activation energy is lower and the annealing temperature higher than their data. Our results (Figure 5) are in qualitative agreement with the recent work of Thommen,[17] where about 45 per cent to 50 per cent of the defects still remain even after an anneal at ~ 250 °C. He did not follow the recovery beyond this temperature due to experimental difficulties.

An activation energy of 0.88 eV suggests that the annihilation is due to long range migration of the defects to fixed sinks. This mechanism has been suggested by earlier workers also.[16,17]

The carrier removal rate calculated from Eq. (4) at 0.6 eV for electron-irradiated samples is 1.1/cm and 0.85/cm for electron fluences of 10^{16} and 2×10^{16} el/cm^2 respectively. This is in agreement with the removal rate of 1.5/cm reported by Aukerman.[9]

Regarding the nature of the transition causing the 1.0 eV band, we note that the absence of photoconductivity associated with this band suggests that the absorption is due to transitions from the ground state of the defect to its excited state in the band gap.

5. SUMMARY

Absorption bands at 0.47 eV and 0.25 eV have been observed in neutron-irradiated samples at room temperature. One of these bands (0.47 eV) is probably caused by a transition from a defect level to the conduction band.

In neutron-irradiated samples the absorption coefficient varies linearly with the square of the photon energy over a limited energy range.

In electron-irradiated samples an exponential absorption is observed below the band gap at room temperature.

An absorption band at 1.0 eV at 85 °K has been observed in electron-irradiated sample. This band anneals out at 300 °C with an apparent activation energy of (0.88 ± 0.07 eV).

ACKNOWLEDGEMENTS

We would like to thank M. L. Swanson for the many fruitful discussions and helpful suggestions during the course of this work. One of us (K.V.V.) would also like to thank the Atomic Energy of Canada Limited for the use of various facilities without which this work would not have been possible.

REFERENCES

1. J. W. Corbett, Electron Radiation Damage in Semiconductors, *Solid State Physics*, suppl. 7 Academic Press (1966).
2. L. W. Aukerman, P. W. Davis, R. D. Graft and T. S. Shilliday, *J. Appl. Phys.*, **34**, 3590 (1963).
3. L. K. Vodopyanov and N. I. Kurdiani, *Sov. Phys. Solid State*, **8**, 204 (1966).
4. J. L. McNichols and W. S. Ginell, *J. Appl. Phys.*, **38**, 656 (1967).
5. U. Gonser and B. Okkerse, *Phys. Rev.*, **105**, 757 (1957); **109**, 663 (1958).
6. V. C. Burkig, J. L. McNichols and W. S. Ginell, *J. Appl. Phys.*, **40**, 3268 (1969).
7. T. Pankey and J. E. Davey, *J. Appl. Phys.*, **41**, 697 (1970).
8. W. G. Spitzer and J. M. Whelan, *Phys. Rev.*, **114**, 59 (1959).
9. L. W. Aukerman, *Semiconductors and Semimetals*, Vol. 4 (Academic Press, 1969).
10. J. E. Fischer, *Phys. Rev.*, **181**, 1368 (1969).
11. D. Redfield and M. H. Afromovitz, *Appl. Phys. Letters*, **11**, 138 (1967).
12. K. Tharmalingam, *Phys. Rev.*, **130**, 2204 (1963).
13. W. Franz, *Tunneling Phenomenon in Solids*, (Plenum Press, New York, 1969), p. 207.
14. B. I. Halperin and M. Lax, *Phys. Rev.*, **148**, 722 (1966).
15. A. C. Damask and G. J. Dienes, *Point Defects in Metals*, (Gordon and Breach, 1963), p. 146.
16. L. W. Aukerman and R. D. Graft, *Phys. Rev.*, **127**, 1576 (1962).
17. K. Thommen, *Rad. Effects*, **2**, 201 (1970).

DISCUSSION

Question (KAHAN)   Have you investigated corresponding changes in reflectivity as a function of electron and neutron irradiation, and are the changes large enough to affect the absorption coefficient calculations?

Answer (VAIDYANATHAN)   No, we have not measured reflectivity changes in electron-irradiated samples. For our analysis of the neutron-irradiated samples, we have used the reflectivity data from reference 6 of the text. Changes in reflectivity we feel are too small to affect our absorption coefficient calculations.

Question (MITCHELL)   Are you aware of the work of Pegler in our laboratory who observed this band in electron-irradiated GaAs about seven years ago and also of the work of Norris of our laboratory who reported at Tokyo on neutron-irradiated GaAs and who obtained very similar results.

Answer (VAIDYANATHAN)   No I hadn't seen this paper.

Question (BORDERS)   Could Dr. Mitchell comment on the difference between the neutron results described by himself and Norris and those given here?

Answer (MITCHELL)   I understand that Dr. Vaidyanathan has not examined his neutron data on a log/linear plot. In any case over the limited energy range his function might be similar to ours. The best thing will be to plot the data together.

Question (FAN) Professor Mitchell. Referring to your comment, does the result of the previous measurements made at Reading agree with the results reported here? I have in mind particularly the absorption peak.

Answer (MITCHELL) Yes.

Comment (VAIDYANATHAN) We are not aware of the results of the work at Reading. However, the results as pointed out by Professor Mitchell are in good agreement.

IRRADIATION EFFECTS IN II–VI COMPOUNDS

G. D. WATKINS

GE Research and Development Center, Schenectady, New York, U.S.A.

A summary of published results on electron energy threshold measurements in II-VI compounds, and their interpretation in terms of damage on the metal and chalcogenide sublattices, is given. EPR results in irradiated II-VI compounds are also reviewed. These include the F^+ center in ZnS; ZnO, and BeO and new results on the zinc vacancy (V) in ZnSe. For the zinc vacancy, optical absorption bands at 4680Å and 8850Å are identified with V^- and a band at 5000Å is tentatively assigned to $V^=$. The activation energy for anneal of the vacancy is measured to be $1.26 \pm .06$ eV, and this is tentatively identified as the activation energy for vacancy motion. An EPR spectrum produced in a 20.4 °K irradiation is tentatively identified as a zinc vacancy-interstitial pair. Complete annealing occurs for this defect in the range 60–100 °K.

1. INTRODUCTION

Before considering radiation effect studies in II–VI compounds, let us briefly consider what is known about intrinsic lattice defects from other types of studies.[1] It has been known for a long time, for instance, that stoichiometry is very important in determining the physical properties of these materials. By cooking in the various component vapors, substantial changes occur in the materials, indicating that intrinsic lattice defects, i.e., interstitials and vacancies, are easily formed and play an important role in determining their properties.

For many years the role of these defects in the luminescence properties of these materials has been an active field of study and controversy. More recently, self-diffusion and electrical property studies under the component vapors have also been performed. There is, therefore, quite a bit of experimental information that bears rather directly on intrinsic lattice defects in II–VI compounds. Ultimately of course, all of this information will fit together and a coherent self consistent picture must emerge. However, at present, it is certainly not unfair to say that this is not the case. Very little is actually known with any certainty. In fact, much of the experimental results often appear contradictory and defy simple interpretation.

Part of the problem is, of course, that a compound semiconductor has two sublattices. There are two kinds of vacancies, two kinds of interstitial atoms, each with several choices for interstitial positions, and we have also the possibility of atom A on a B site, etc. Add to these simple intrinsic point defects, defect interactions—with each other, and with impurities—and the situation very quickly gets out of hand. And this brings up another very real problem in II–VI materials: impurity concentrations are high, and largely undetermined, and may well be completely dominating the chemistry of the defects.

This is the picture then—a very complex situation, poorly understood, but with a great deal of experimental data already available that ultimately must be explained.

Radiation damage experiments offer some unique advantages over these more conventional thermochemical approaches that, with luck, should help in unraveling the puzzle. One of the most important is the existence of a sharp displacement threshold energy for damage which can be different for the two component atoms. Using high energy electrons, these thresholds can be explored and under favorable circumstances defects produced on the two sublattices can be distinguished. Another trick exists for separating the sublattices for cadmium containing compounds. For these, the high thermal neutron cross section for the Cd^{113} (n, γ) Cd^{114} reaction allows preferential damage to the cadmium lattice via the γ-recoil energy of the Cd^{114} nucleus.

Another distinguishing feature of radiation damage is that defects are produced in a *cold* lattice and highly mobile defects, normally not retained after high temperature cooking experiments, can be frozen in for study. This is, of course, a mixed blessing because new defect structures are being added to the problem at the same time. However, coupled with a microscopic tool such as EPR, that can partially sort out the defects, this distinguishing feature can be put to advantage.

In this short review, I will not attempt to

catalogue all of the radiation damage results in the literature or the many defect models that have been suggested in these studies. Rather I will limit myself to those few experiments from which I feel positive results have emerged. This includes a brief review of threshold measurements, and then a review of the few EPR studies that have led to definite defect models.

2. DAMAGE THRESHOLDS

In experiments with high energy electrons, electron energy thresholds for damage have been observed in all II–VI compounds that have been studied so far. An example is shown in Figure 1 from the work of Kulp and Detweiler.[2] Here the rate of damage as monitored by the growth of fluorescent bands at 5460 Å and 5850 Å in ZnSe is plotted versus electron bombardment energy at 85°K. This sharp threshold for damage, typical of all II–VI compounds, means that ionization is not an effective mechanism for damage and that damage occurs only by elastic Rutherford scattering of the atomic nuclei by the incoming electrons.

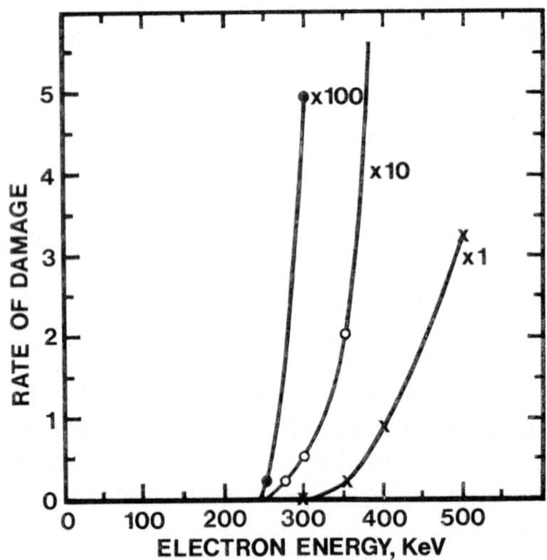

FIG. 1. Damage in ZnSe as monitored by fluorescence as a function of bombarding electron energy at 85°K (from Kulp and Detweiler[2]).

Table I contains a summary of the threshold experiments in II–VI compounds. In the table the maximum kinetic energy that can be transmitted to each atom is given for each electron threshold energy. Only one of these corresponds to the true atomic displacement threshold in each case.

The choice between the two is straightforward when the masses of the two atoms differ greatly, the lower electron energy threshold being assigned to the lighter atom. In making this assignment, the displacement thresholds for the two atoms turn out to be nearly equal, as expected, but with the metal atom slightly smaller. Using this as a guide, it has been the general practice to assign the lower atomic displacement threshold to the metal ion when the masses of the two atoms are comparable. These choices are shown in bold type.

Table I demonstrates graphically the power of threshold studies in separating out damage associated with the two sublattices. Care must still be exercised, however, in interpreting such results with detailed models based purely on this type of experiment. There are still many possible complexities on a single lattice. Our experience in silicon should serve as a warning here!

Table I also illustrates another point of interest. In almost all cases where low temperature studies have been made, low temperature annealing stages are observed. Figure 2 illustrates the results of Detweiler and Kulp[3] for 10°K irradiations of ZnSe. At this lower temperature a second threshold at 195 keV was also observed. Above the 195 keV threshold but below that at 240 keV (Figure 1), where presumably only zinc atoms are being displaced (see Table I), complete annealing takes place in a single step at 90°K. Above the 240 keV threshold, annealing stages emerge at 60°K and 135°K as well, the latter stage being present only in 'less pure' materials. Again the apparent complexity and the hint of high mobility defects should serve as a warning against over simplified models.

3. THERMAL NEUTRON DAMAGE

Cd^{113} has a large thermal neutron cross section and the resulting Cd^{113} (n, γ) Cd^{114} reaction gives recoil energies estimated to be $\sim$50 eV to the Cd^{114} nucleus.[4] This is considerably more than the displacement threshold (Table I) and cadmium ion displacements will occur. The principal uncertainty in interpreting such experiments is whether this recoil energy is large enough to cause disruptions in turn to the other atoms and thereby destroy the sublattice separation,[5] or not.[4] Two papers at this Conference deal with such experiments in CdS[6,7] and CdTe[7] and we will leave this question to them.

TABLE I

Summary of threshold experiments in II–VI compounds

Mat'l (MX)	Electron threshold (keV)	T (°K)	Method†	Anneal	Max recoil energy (eV) M	X
ZnO	~600	80[a]	A (~4100 Å)	150 °K, 250 °C	~30	120
ZnS	185	300[b]	TL (195 °K)		7.3	**15.0**
	240	300[b,c]	TL (235 °K)		**9.9**	20.2
ZnSe	195	10[d]	L (6100 Å)	90 °K	7.6	6.2
	240	85[e]	L (5460 Å, 5850 Å)	135 °K	10.0	**8.2**
		10[d]	L (5850 Å, 6400 Å)	60 °K, 135 °K		
CdS	115	300[f]	L (5140 Å, 7200 Å)		2.4	**8.7**
	125	77[g]	L (10,200 Å)		2.7	**9.6**
	285	77[h]	L (6050 Å, 10,300 Å)		7.3	25.6
CdSe	250	77[i]	L (12,500 Å)		6.0	**8.6**
	320	5[j]	L (11,200 Å)	62 °K	**8.1**	11.6
CdTe	235	15[k]	L (8050 Å)		5.6	5.0
	~250	110[l]	E			
	340	77[m]	L (11,300 Å)	110–140 °K	8.9	**7.8**
		15[k]	L (11,300 Å)	110–140 °K		

† Key to abbreviations: A, optical absorption; L, luminescence; TL, thermoluminescence; E, electrical properties. The value in parenthesis denotes the wavelength of the optical band, or in the case of thermoluminescence, the temperature of the luminescent burst.

References for Table I

(a) W. E. Vehse, W. A. Sibley, F. J. Keller and Y. Chen, *Phys. Rev.*, **167**, 828 (1968).
(b) F. J. Bryant and S. A. Hamid, *Phys. Rev. Letters*, **23**, 304 (1969).
(c) F. J. Bryant and A. F. J. Cox, *Physics Letters*, **20**, 108 (1966).
(d) R. M. Detweiler and B. A. Kulp, *Phys. Rev.*, **146**, 513 (1966).
(e) B. A. Kulp and R. M. Detweiler, *Phys. Rev.*, **129**, 2422 (1963).
(f) B. A. Kulp and R. H. Kelley, *J. Appl. Phys.*, **31**, 1057 (1960).
(g) F. J. Bryant and A. F. J. Cox, *Proc. Roy. Soc.*, **A310**, 319 (1968).
(h) B. A. Kulp, *Phys. Rev.*, **125**, 1865 (1962).
(i) B. A. Kulp, *J. Appl. Phys.*, **37**, 4936 (1966).
(j) H. J. Schulz and B. A. Kulp, *Phys. Rev.*, **159**, 603 (1967).
(k) F. J. Bryant, A. F. J. Cox and E. Webster, *J. Phys. C.* (*Proc. Phys. Soc.*) [2], **1**, 1737 (1968).
(l) K. Matsuura, N. Itoh and T. Suita, *J. Phys. Soc. Japan*, **22**, 1118 (1967).
(m) F. J. Bryant and E. Webster, *Phys. Stat. Sol.*, **21**, 315 (1967).

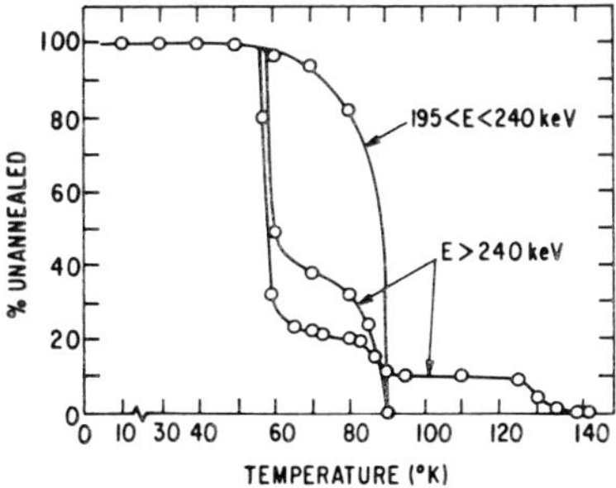

FIG. 2. Isochronal annealing of radiation produced fluorescence in ZnSe, bombarded by electrons of various energies at liquid helium temperatures. The stage at 135 °K for $E > 240$ keV is present in 'less pure' materials (after Detweiler and Kulp[3]).

4. EPR STUDIES

So far, definitive EPR identification of intrinsic lattice defects in II–VI compounds, produced by irradiation or otherwise, has been sparse. Recently spectra identified as arising from the isolated chalcogenide vacancy in ZnS,[8] ZnO[9] and BeO[10] have been reported. In this paper, we will describe new results on the isolated zinc vacancy in ZnSe.[11] These plus the earlier work on the zinc vacancy-donor pair spectra[12] sometimes present in *unirradiated* ZnS and ZnSe represent for the most part the total that is known about intrinsic defects from EPR studies in these materials. In the next section we will review briefly the results on the chalcogenide vacancy. In the following section we will present in somewhat more detail the results on the zinc vacancy in ZnSe and attempt a rough correlation with other studies.

4.1. *The Chalcogenide Vacancy*

In ZnS, Schneider and Rauber[8] have observed an EPR center which they identify as a single electron trapped at an isolated sulfur vacancy. The spectrum is isotropic and resolved hyperfine interactions with 4.1 per cent Zn^{67} reveals that the wavefunction is spread equally over the four zinc nearest neighbors. The center is thus highly similar to the F center in alkali halides, or the F^+ center in MgO.

This F^+ center can be produced by irradiation with fast neutrons. These authors report, however, that Co^{60} γ-rays or 1 MeV electrons did not produce the center. In view of the 185 keV threshold assignment in Table I for sulfur displacement, this result is surprising, and should probably be checked further. The centers are also produced by additive coloration (cooking in liquid zinc at about 1100 °C) in crystals of sufficient purity. This reveals the important additional information that the sulfur vacancy is not very mobile since it can be introduced in its isolated state by this high temperature process.

Leutwein, Rauber and Schneider[13] also demonstrated a strong correlation between optical absorption bands at 4300 Å and 5450 Å, previously studied by Yoshida and coworkers[14] in neutron irradiated ZnS, and the EPR center. These bands were shown to have the same optical excitation and bleaching spectra as that for the generation and bleaching of the EPR center and they concluded tentatively that the bands represent optical transitions directly associated with the F^+ center.

Recently Smith and Vehse[9] have reported the EPR of a similar F^+ center in 2 MeV electron irradiated ZnO. They indicate that annealing and production rate studies for the EPR center correlate with those of an optical absorption band at $\sim$4100 Å. Again there appears some inconsistency with the threshold studies (Table I). In these earlier studies by Vehse and coworkers,[15] the apparent high threshold for the 4100 Å band had suggested that displacement on the zinc lattice was occurring.

The F^+ center in neutron-irradiated BeO has also been reported by DuVarney, Garrison and Thorland.[10] In an EPR and ENDOR study, they were able to base their identification on hyperfine interactions observed with the four $Be^{(9)}$ neighbors

surrounding the oxygen vacancy. No optical absorption bands have been identified with the center as yet.

To complete the information on the chalcogenide vacancy, we mention briefly two other pieces of work. Dieleman and coworkers[16] have reported an EPR center in unirradiated Cu-, Ag- and Au-doped ZnS which they identified as a sulfur vacancy next to the impurity ion. As pointed out by Schneider,[17] there may be some uncertainty in this identification, however, because the observed Zn^{67} hyperfine interactions are much smaller than those that have subsequently been found to be associated with the sulfur vacancy. Secondly, Morigaki and Hoshina[18] have attributed a center in CdS to the sulfur vacancy. No hyperfine interactions are observed so that this identification, although quite possibly correct, cannot be substantiated at this point.

4.2. *The zinc vacancy in ZnSe*

Irradiation of *n*-type ZnSe by 1.5 MeV electrons at room temperature produces a strong well-resolved anisotropic spectrum that has been identified as arising from the isolated zinc vacancy.[10] The spectrum is axially symmetric along a $\langle 111 \rangle$ axis in the crystal and Se^{77} hyperfine interactions reveal that the unpaired electron wavefunction is localized primarily on a single selenium atom, which by symmetry must lie on this axis.

A simple molecular orbital model for the defect, constructed from the 'dangling orbitals' on the four neighbor selenium atoms, is shown in Figure 3. As shown, the EPR arises from the single negatively charged state of the vacancy and the $\langle 111 \rangle$ symmetry axis results from a trigonal Jahn–Teller distortion. (The distortion can also be easily

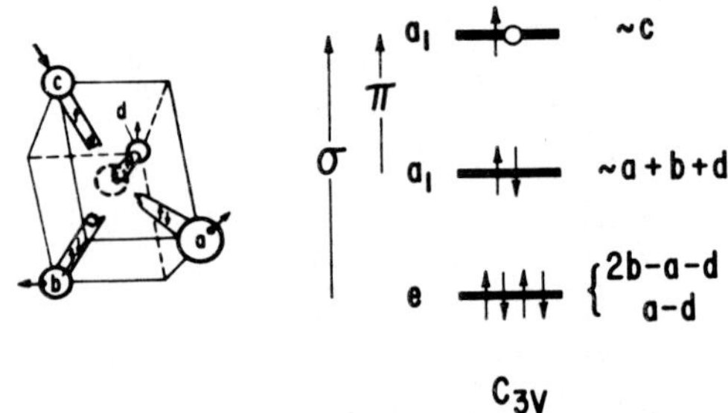

FIG. 3. A simple molecular orbital model for the single minus charge state of the zinc vacancy in ZnSe, after a trigonal Jahn–Teller distortion. Shown also are allowed optical transitions predicted by the model.

visualized on an ionic model. A hole on one selenium ion means that the corresponding Se^- ion is repelled outwards less than the three $Se^=$ ones.) Shown also in the figure are allowed optical transitions that such a molecular orbital model predicts. We shall return to these later.

That this anisotropy does indeed result from a Jahn–Teller distortion (rather than, say, from the presence of a nearby defect) can be demonstrated conclusively by experiment. The application of uniaxial stress to the crystal produces a preferential alignment of the defects which can be monitored directly in the EPR. This means that each defect in the lattice can distort equally well in each of the four $\langle 111 \rangle$ directions and is free to reorient under the applied stress. The distortion therefore must be an intrinsic property of the defect which otherwise has the full tetrahedral (T_d) symmetry of a vacant lattice site. The kinetics for reorientation are temperature dependent and at $2.0\,°K$, for instance, the characteristic reorientation time is 20 seconds.

If the sample is cooled in the dark, no EPR signals are observed. They are generated by light centered around 5000 Å. A model consistent with this is shown in Figure 4. The vacancy appears to be a double acceptor which is normally doubly charged ($V^=$) in the n-type material. To generate the EPR spectrum, 5000 Å light photo-ionizes the center and the conduction electron produced is trapped at an original donor site. The spectrum can be quenched with infrared light centered over a broad region $\sim$10,000 Å.

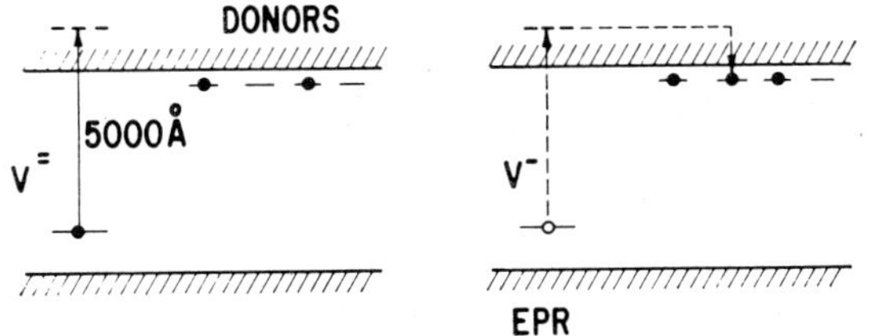

FIG. 4. Model of the electrical level structure of the zinc vacancy in ZnSe which is consistent with the optical generation and bleaching properties of the EPR and optical spectra.

Optical absorption bands can also be identified with the vacancy. This is illustrated in Figure 5, where the optical absorption at $2.0\,°K$ is shown before and after irradiation. The dominant effect of irradiation has been to introduce strong additional absorption near the band edge. This does not appear to be associated with the vacancy and its origin is not understood. On the other hand, a weak band at 5000 Å is revealed by its removal after 5000 Å bleach. It can be regenerated by $\sim$10,000 Å light and can be cycled back and forth in direct correlation to the generation and quenching of the EPR signal. Consistent with Figure 4, we therefore tentatively identify this 5000 Å band with the $V^=$ state of the vacancy.

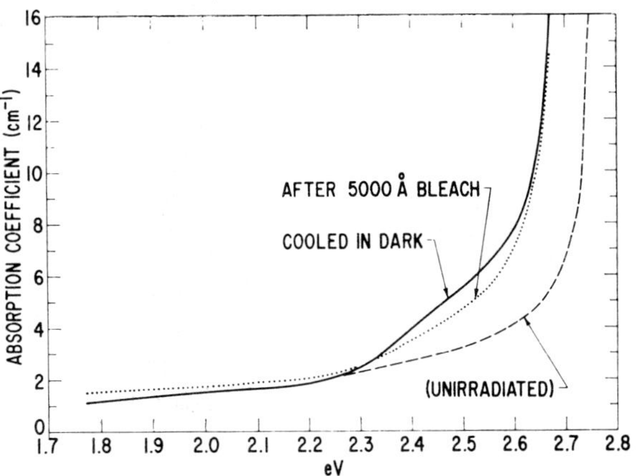

FIG. 5. Optical absorption in ZnSe at $2.0\,°K$ before and after 1.5 MeV electron irradiation at room temperature. The material was originally n-type ($\sim 10^{17}$ donors/cm³) and the irradiation dose was $\sim 2.10^{18}$ e/cm². Bleaching at 5000 Å reveals a band tentatively identified with the double minus charge state of the zinc vacancy.

After bleaching with 5000 Å light, an additional pair of weak optical absorption bands can be identified with the V^- charge state. They are not apparent in Figure 5 because they are so weak. They can be extracted, however, by taking advantage of the known stress alignment properties of the V^- center. Using polarized light, the change of absorption versus applied uniaxial stress, shown in Figure 6, reveals the two bands, which have different polarization properties. The time constant for the polarization process was measured directly to be 20 seconds at $2.0\,°K$ proving conclusively that the bands arise from the V^- center. The analysis of the stress results shows the 1.4 eV band (8850 Å) to have its transition dipole moment parallel to the trigonal axis of the defect (π polarization) and the 2.65 eV band (4680 Å), perpendicular (σ). In Figure 3, we see that it is possible to assign two transitions within the molecular orbital model which may account for these bands.

Luminescence has also been studied. Figure 7 shows the luminescence spectra at $4.2\,°K$ before and

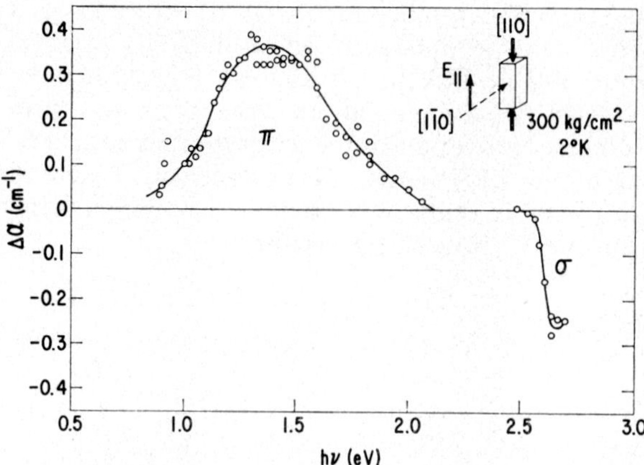

FIG. 6. Bands associated with the single minus charge state of the zinc vacancy in ZnSe, as revealed by stress-induced dichroism at 2.0 °K.

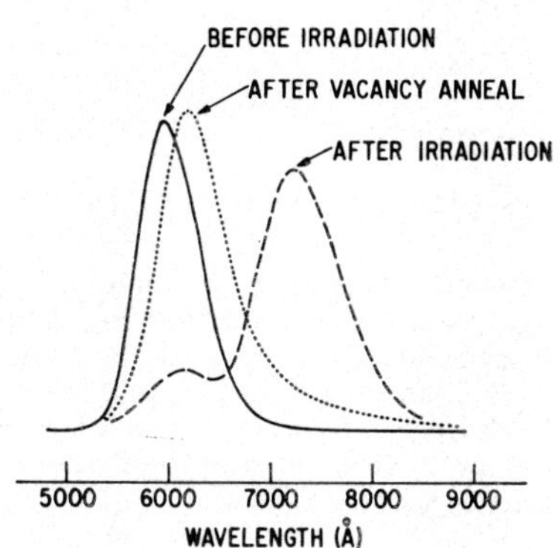

FIG. 7. Luminescence at 4.2 °K in ZnSe before and after 1.5 MeV electron irradiation at room temperature, and after an anneal in which the vacancy has been removed.

after electron irradiation and after an anneal in which the vacancy has been removed. The 7200 Å emission band appears to grow and anneal in rough 1:1 correlation with the production and annealing of the vacancy EPR spectrum. However, attempts to observe polarized effects either in the excitation or emission spectrum for this band under conditions of uniaxial stress at 4.2 °K where the V^- centers are known to be oriented have failed. Similarly, there appears to be no correlation between a thermoluminescence burst of 7200 Å light observed on warming the crystal and the temperature at which the V^- center recaptures an electron as evidenced by the loss of the EPR signal. We tentatively conclude therefore that the 7200 Å band is not directly associated with the vacancy and that its close correlation in annealing may be accidental.

The results of annealing studies on the vacancy using EPR are shown in Figure 8. At a given temperature, the vacancy disappearance follows approximately a simple exponential decay, characteristic of first-order reaction kinetics. The plot in Figure 8 of the characteristic decay time versus $1/T$ gives an activation energy of 1.26 ± 0.06 eV. Studies in materials of two different origins are shown. The same activation energy is obtained for both, although the pre-exponential factors differ by an order of magnitude.

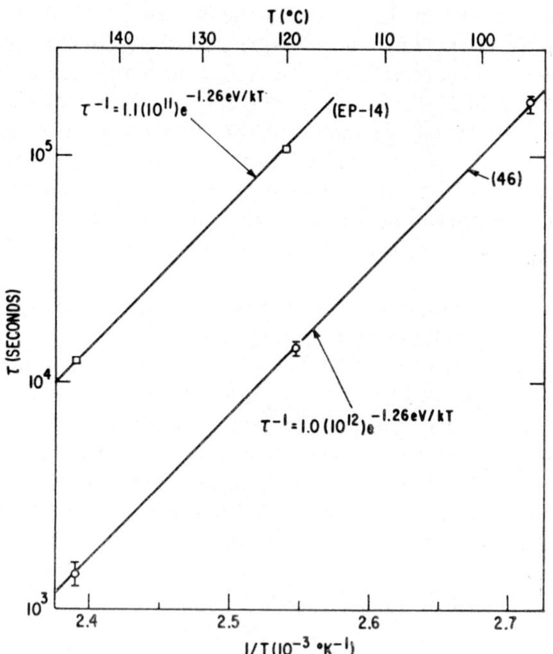

FIG. 8. Characteristic time constant for isothermal anneal of the zinc vacancy in ZnSe versus annealing temperature. Results for materials of two different origins are included.

As the vacancy disappears, new EPR spectra emerge which can be identified as vacancy-defect pairs, the dominant one being different for the two materials. This suggests diffusion of the vacancy until trapped by the corresponding defect in the lattice. Consistent with this, sample EP-14 (Figure 8) was chlorine-doped in the growth and, for it, the dominant center that emerges is one previously studied by Holton and coworkers[19] and identified as a vacancy paired with a chlorine donor substituting for one of the four nearest selenium neighbors (A-center). This sample was also the sample used for the optical studies, and again consistent with this, the 6200 Å luminescence which emerges after anneal (Figure 7) has been identified by Holton and coworkers[19] as arising from this center.

The dominant center that emerges in sample 46 has not been identified, however. This sample was not intentionally doped and was believed to be of high chemical purity. The defect involved in this material, and its apparent high concentration if this vacancy diffusion model is correct, poses an interesting mystery.

We therefore tentatively identify the 1.26 ± 0.06 eV as the activation energy for motion of the zinc vacancy. It is interesting to note that this activation energy agrees closely with an estimate by Swank and coworkers[22] of 1.2 eV for an acceptor which appears to diffuse in rapidly from the surface of ZnSe, when cooked in the absence of zinc vapor. They postulated that the diffusing entity was the zinc vacancy and our results suggest that their educated guess may be right!

Finally, let me mention EPR studies of low temperature irradiations. Irradiating ZnSe with 1.5 MeV electrons at 20.4 °K produces the V^- spectrum, but, in addition, another spectrum of comparable intensity is also observed. It is highly similar to the V^- spectrum suggesting that it also is associated with a zinc vacancy. An important clue to its identity comes from uniaxial stress experiments. Under applied stress, it does not align. We conclude, therefore, that it is a vacancy that is perturbed by the strain field of a nearby defect. This strongly suggests a nearby interstitial atom, presumably the one formed in the vacancy production.

The annealing kinetics of this center have been explored to see if a correlation exists with the low temperature annealing studies of Detweiler and Kulp,[3] Figure 2. Our preliminary results are shown in Figure 9. Curve A shows the annealing in the dark. That is, after first generating the signal with 5000 Å light, the sample receives no further light and the signal amplitude, monitored at 20.4 °K, is measured after isochronal anneals at each temperature. A single annealing stage at ∼80–90 °K is indicated. On the other hand, curve B shows the result if ambient room light is allowed to reach the sample at 20.4 °K after each anneal (performed in the dark). The presence of light has apparently introduced an extra annealing stage at ∼60 °K. Particularly interesting about this stage is the fact that after the 60 °K anneal but before light is admitted, there is no evidence of loss of the centers. Apparently the 60 °K anneal has somehow altered things so that annealing can

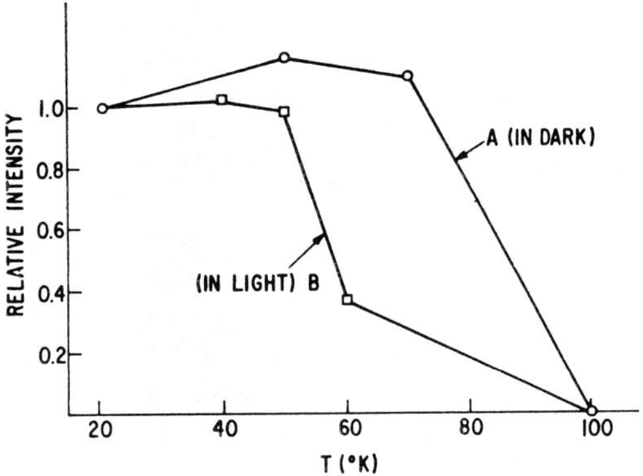

FIG. 9.. Low temperature isochronal annealing of the EPR spectrum identified as a 'perturbed' zinc vacancy in ZnSe, after 20.4 °K irradiation.

subsequently be 'triggered' by light. (Before the 60 °K anneal, say after the 50 °K anneal, light has no effect.)

The annealing of this center is, therefore, actually quite a complex process. It is tempting to conclude that the annealing stages of this center are related to the 60 °K and/or 90 °K stages seen by Detweiler and Kulp,[3] Figure 2. However, in view of the complex character of the EPR center annealing, and the lack of certainty as regards the role of stray light or the electron irradiation used to excite the fluorescence, etc., in the study of Detweiler and Kulp, a detailed correlation should await further study.

Still it is interesting to point out that the defect studied in the EPR anneal is a perturbed *zinc vacancy* and as such is associated with the zinc lattice. Detweiler and Kulp,[3] on the other hand, assigned the 60 °K anneal to the selenium lattice and the 90 °K one to the zinc lattice (Table I). A detailed study by EPR through this region gives promise, therefore, of a critical evaluation and comparison to these threshold assignments.

5. SUMMARY

Radiation damage, as a means of introducing intrinsic point defects into II–VI compounds for study, has a number of unique features which can often give information that other types of experiments cannot. Particularly important in this regard, are threshold measurements with high energy electrons which, under favorable conditions, can serve to separate the defects on the two sublattices. Threshold measurements have been completed now for most of the II–VI compounds and the broad features have been fairly well mapped out. The high thermal neutron cross section for the Cd^{113} (n, γ) Cd^{114} reaction serves as another convenient experimental trick for producing damage preferentially on the metal sublattice in cadmium containing compounds.

Irradiation also produces damage in the cold lattice and primary defects, perhaps not present in other experimental approaches which involve high temperature treatments, can be frozen in for study. However, it is still almost impossible to make much detailed headway without a microscopic tool such as EPR.

Three years ago, at the time of our last radiation effects conference, the sulfur vacancy in irradiated ZnS had just been identified[8] by EPR. In the intervening three years, the oxygen vacancy in ZnO^9 and BeO^{10}, and the zinc vacancy in $ZnSe^{11}$ have been added to the list. That's about it.[21] Expressed this way, the progress has not been impressive.

On the encouraging side, however, these few successful studies have given us insights that we didn't have before. We are beginning to get a 'feel' for vacancies, at least, and what their properties are. And we have found that EPR can work in these materials—sometimes, at least.

How do we best proceed from here? It's clear that EPR still remains the best hope. Optical and electrical studies can begin now to attempt correlation with the existing EPR results. In this correlation, uniaxial stress and polarized light greatly expand the power of the optical studies. In fact, it might be wise to make an educated guess that Jahn–Teller distortions are characteristic of metal vacancies and use stress-induced dichroism techniques as a routine tool—even in the absence of detailed EPR results. Low temperature irradiations are important. Angular dependence of damage near threshold has not been reported and might turn out to be interesting.[22] In addition, a lot of work, better ideas than I have suggested here, and a lot of luck will be required.

ACKNOWLEDGEMENT

The author is grateful to H. H. Woodbury for several helpful discussions about the mysteries of defects in II–VI compounds.

REFERENCES

1. A recent review of the state of knowledge in II–VI compounds is found in *Physics and Chemistry of II–VI Compounds*, Eds. M. Aven and J. S. Prener (North-Holland Pub. Co., Amsterdam, 1967). See also the many review and contributed papers in *II–VI Semiconducting Compounds: 1967 International Conference*, Ed. D. G. Thomas (W. A. Benjamin, Inc., New York, 1967).
2. B. A. Kulp and R. M. Detweiler, *Phys. Rev.*, **129**, 2422 (1963).
3. R. M. Detweiler and B. A. Kulp, *Phys. Rev.*, **146**, 513 (1966).
4. C. Barnes and C. Kikuchi, *Nuclear Sci. and Eng.*, **31**, 513 (1968).
5. R. O. Chester, *J. Appl. Phys.*, **38**, 1745 (1967).
6. C. Kikuchi, this conference.
7. A. A. Abramov, V. S. Vavilov and L. K. Vodopianov, this conference.
8. J. Schneider and A. Rauber, *Solid State Comm.*, **5**, 779 (1967).
9. J. M. Smith and W. H. Vehse, *Physics Letters*, **31A**, 147 (1970).
10. R. C. DuVarney, A. K. Garrison and R. H. Thorland, *Phys. Rev.*, **188**, 657 (1969).
11. G. D. Watkins, to be published. Preliminary reports of this work have been presented in *B. Am. Phys. Soc. II*, **14**, 312 (1969); **15**, 290 (1970).
12. See J. Schneider, *II–VI Semiconducting Compounds 1967 International Conference*, Ed. D. G. Thomas (W. A. Benjamin, Inc., New York, 1967), p. 40 for a review of this work.
13. K. Leutwein, A. Räuber and J. Schneider, *Solid State Comm.*, **5**, 783 (1967).
14. T. Yoshida, T. Seiyama, Y. Shono and M. Kitagawa, *Appl. Phys. Letters*, **9**, 26 (1966).
15. W. E. Vehse, W. A. Sibley, F. J. Keller and Y. Chen, *Phys. Rev.*, **167**, 828 (1968).
16. J. Dieleman, S. H. deBruin, C. Z. vanDoorn and J. H. Haanstra, *Phillips Res. Reports*, **19**, 311 (1964).
17. J. Schneider, *II–VI Semiconducting Compounds: 1967 International Conference*, Ed. D. G. Thomas (W. A. Benjamin, Inc., New York, 1967), p. 40.
18. K. Morigaki and T. Hoshina, *J. Phys. Soc. Japan*, **24**, 120 (1968).
19. W. C. Holton, M. deWit and T. L. Estle, *International Symposium on Luminescence, Munich 1965*, Eds. N. Riehl and H. Kallmann (Verlag Karl Thiemig KG, Munchen, 1966), p. 454.
20. R. K. Swank, M. Aven and J. Z. Devine, *J. Appl. Phys.*, **40**, 89 (1969).
21. Note added in proof: Hervé and Maffeo have very recently reported a spectrum which they identify as arising from the isolated beryllium vacancy (V^-) in BeO. [A. Hervé and B. Maffeo, *Physics Letters*, **32A**, 247 (1970).] Dr. Hervé informs me at this conference that his group has seen a similar center in ZnO and also an $S = 1$ center that they attribute to the neutral vacancy in ZnO (to be published).
22. In the II–VI cubic lattice, the [111] and [$\bar{1}\bar{1}\bar{1}$] directions are not equivalent. Experiments comparing irradiations along these two directions, as first demonstrated by Eisen [F. H. Eisen, *Phys. Rev.*, **135**, A1394 (1964)] for III–V compounds, serve to assign the sublattice damage less ambiguously when the two masses are comparable. They could also turn up interesting surprises of their own.

DISCUSSION

Question (SEEGER) In one of the experiments on the disappearance of the Zn vacancy, the pre-exponential of the annealing rate was 10^{12} sec^{-1}. This means that the concentration of the trapping impurity must be of the order of magnitude of 1 per cent. This may be high enough for detecting these impurities by chemical or neutron-activation means.

Answer (WATKINS) I agree. These high pre-exponential factors certainly indicate high concentrations of the vacancy traps. Sample EP-14 (see Figure 8) was originally chlorine-doped by adding 10^{-3} mole fraction $SrCl_2$ to the charge before vapor growth, and high concentrations of chlorine are therefore not unreasonable. Consistent with this, the chlorine-vacancy A-center is the dominant center to emerge upon vacancy anneal. (Electrical measurements indicate, however, strong compensation with only $\sim 10^{17}$ cm^3 *net* chlorine donors before irradiation.)

The case of sample 46 presents more of a mystery. This sample was not intentionally doped and was believed to be of high chemical purity. Mass spectrographic analysis revealed no impurity greater than a few parts per million with the exception of carbon and oxygen which were of the order of a few hundred parts per million. This is still quite low, suggesting that the dominant trapping defect in this material may be an intrinsic one.

Question (GIBSON) Recent work by Henry and coworkers has shown that optical luminescent centers in CdS which have long been ascribed to Cd vacancies are actually due to Li and Na impurities—presumably sitting on a Cd site. These centers show characteristic and sharp annealing behavior as might be expected for an intrinsic defect. Could you comment on the characterization of defects in CdS by EPR and also on the possible role of alkali impurities in the other II–VI materials?

Answer (WATKINS) In CdS, Morigaki and Hoshina have attributed an EPR center to the sulfur vacancy. As is described in the text of my talk this identification may be correct but cannot be directly demonstrated because of the lack of resolved hyperfine interactions. I believe this is the only defect center in CdS that has been described as being associated with an intrinsic center. I am not aware of any EPR work that bears on the alkali impurities.

ELECTRON RADIATION DAMAGE AND THE EDGE EMISSION OF CADMIUM TELLURIDE

F. J. BRYANT AND D. H. J. TOTTERDELL

Department of Physics, University of Hull, Hull, England

The cathodoluminescence edge emission of cadmium telluride at 4.2 °K has been observed, (*a*) before and after displacement of cadmium atoms or displacement of cadmium and tellurium atoms by monoenergetic electrons, (*b*) firing in vacuum or an over-pressure of the cation.

The intensities of emission bands occurring at 1.555 eV, 1.572 eV and 1.588 eV are enhanced by cadmium displacement and the defect centres responsible for these emissions are believed to involve the first charged state of the cadmium vacancy, the neutral cadmium vacancy and an exciton bound to the neutral cadmium vacancy respectively. The intensity of the emission band at 1.553 eV is enhanced by tellurium displacement or by firing the material in a high cadmium overpressure and is believed to be the double acceptor centre already reported and involving a tellurium vacancy-halogen complex.

1. INTRODUCTION

In a binary compound the controlled production of point defects using radiation damage produced by monoenergetic electrons can give rise to the displacement of one or both species of the host atoms. For the irradiated material the point defects produced modify considerably the physical properties (such as cathodoluminescence) by which the damage may be observed. Firing the material in an atmosphere of either the cation or the anion can also introduce intrinsic defects, and, by comparison of the results of such firing studies with those of radiation damage studies, a positive identification of the species of displaced atom may be made, together with the determination of the position of the defect level within the forbidden energy gap.

This paper reports the results of such damage and heat treatment studies performed on cadmium telluride in an attempt to understand the edge emission observed under electron excitation at 4.2 °K. Identification of such shallow levels as those observed in the edge emission of cadmium telluride is difficult, but a combination of controlled radiation damage and heat treatment results enables emission bands occurring at similar wavelength positions to be separated and leads to an assignment for the energies corresponding to the double acceptor centre (tellurium vacancy and halogen complex), the first charged state of the cadmium vacancy, the neutrally charged state of the cadmium vacancy and an exciton bound to the neutrally charged state of the cadmium vacancy.

2. EXPERIMENTAL PROCEDURE

Each of the cadmium telluride samples investigated was ground from single crystal material and mounted in a thermally conducting silver paste on a copper substrate which was in turn mounted on one of the cold faces of a multisample liquid helium cryostat. Radiation damage was achieved by a beam of monoenergetic electrons from a High Voltage Engineering AS 400 Van de Graaff accelerator, the energy of which could be controlled to ± 2 keV at any energy in the range 75 keV to 400 keV. The vacuum common to both accelerator drift tube and cryostat was better than 10^{-6} torr. Cathodoluminescence at electron energies below the atomic displacement threshold was used to monitor the behaviour of the samples before and after damage at energies capable of displacing cadmium atoms only or both cadmium and tellurium atoms. Threshold energies of 235 keV for cadmium and 340 keV for tellurium corresponding to displacement energies of 5.6 eV and 7.8 eV respectively have been established[2]. Cathodoluminescence was excited by 100 keV electrons at a beam current of 1.0 μA incident normally on the sample and observed at 45° to this direction.

The emission was mechanically chopped at 800 Hz, analysed by a Bausch and Lomb monochromator with a 500 mm grating blazed at 1.0 μm and detected by a red sensitive CVP 150 photomultiplier which fed a phase sensitive detection system and potentiometric recorder.

Throughout the experiment the sample temperature was monitored continuously by a copper-constantan thermocouple and a gold-iron vs. chromel thermocouple. The cold junctions of the thermocouples were cemented to the cold faces of the cryostat by thermally conducting silver paste.

3. EXPERIMENTAL RESULTS

3.1. *Radiation damage*

The cathodoluminescence spectra of cadmium telluride at 4.2 °K are shown in Figures 1 and 2. The spectrum of a typical cadmium grown sample (Figure 1) is seen to have emission peaks at

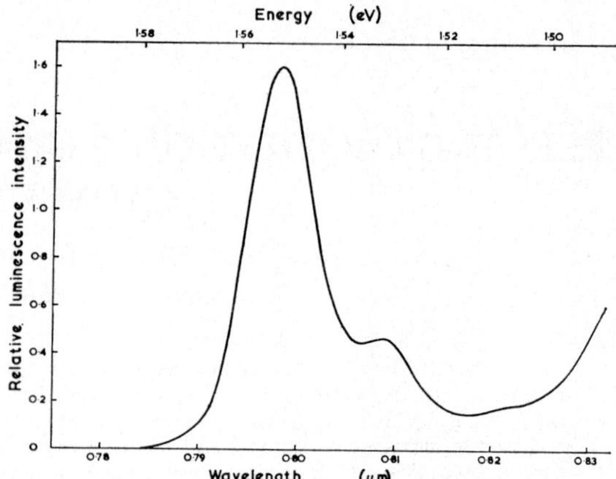

FIG. 3. Cathodoluminescence spectrum at 4.2 °K excited by 100 keV electrons of a cadmium grown cadmium telluride sample fired in vacuo at 600 °C for one hour and subsequently fired in a cadmium overpressure at 1000 °C.

1.587 eV (0.7805 μm), 1.542 eV (0.8030 μm) and 1.527 eV (0.8110 μm). With the exception of the 1.587 eV emission these emissions occur at lower energies than the corresponding emissions in the excess tellurium grown sample which has emission peaks at 1.587 eV (0.7805 μm), 1.572 eV (0.7880 μm), 1.554 eV (0.7975 μm) and 1.532 eV (0.8085 μm). The chief difference in the two spectra is the 1.572 eV emission which appears in the cadmium deficient sample.

Radiation damage of cadmium telluride at an energy above the energy threshold producing cadmium displacements only, increases the emission intensities at 1.587 eV and 1.572 eV as shown in Figure 4. The emission at 1.554 eV is also increased by cadmium displacement in some samples and the 1.532 eV emission is also observed to increase slightly in intensity with damage. Figure 8 illustrates the behaviour of these emission intensities with dose of electrons at 260 keV. It is observed that the 1.572 eV emission and the 1.587 eV emission behave in a similar manner and that both tend towards saturation.

With cadmium displacement the emission at 1.587 eV is observed to increase in intensity rapidly with electron dose and observed to shift to lower energy. A shift of 11.1 meV after a dose of 8.94×10^{16} electrons cm^{-2} has been observed. Similarly the emission at 1.572 eV shifts to lower energy with electron dose, in this case a shift of 5 meV after 8.94×10^{16} electrons cm^{-2} has been observed, however the rate of intensity increase of

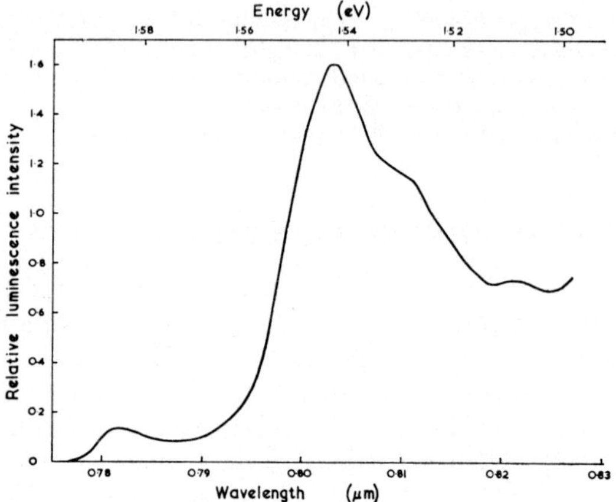

FIG. 1. Cathodoluminescence spectrum at 4.2 °K of a typical cadmium grown cadmium telluride sample excited by 100 keV electrons.

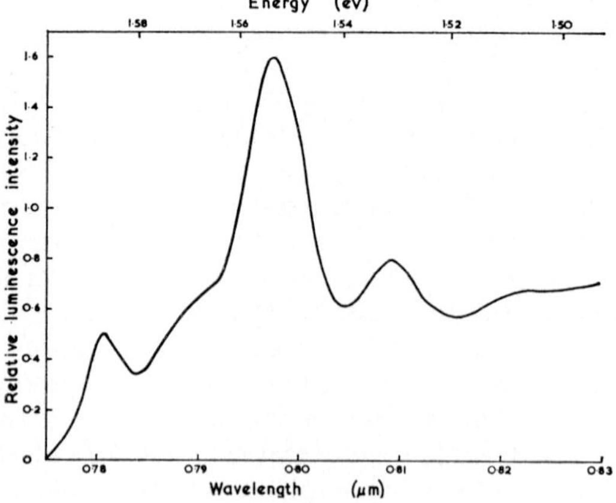

FIG. 2. Cathodoluminescence spectrum at 4.2 °K of a typical tellurium grown cadmium telluride sample excited by 100 keV electrons.

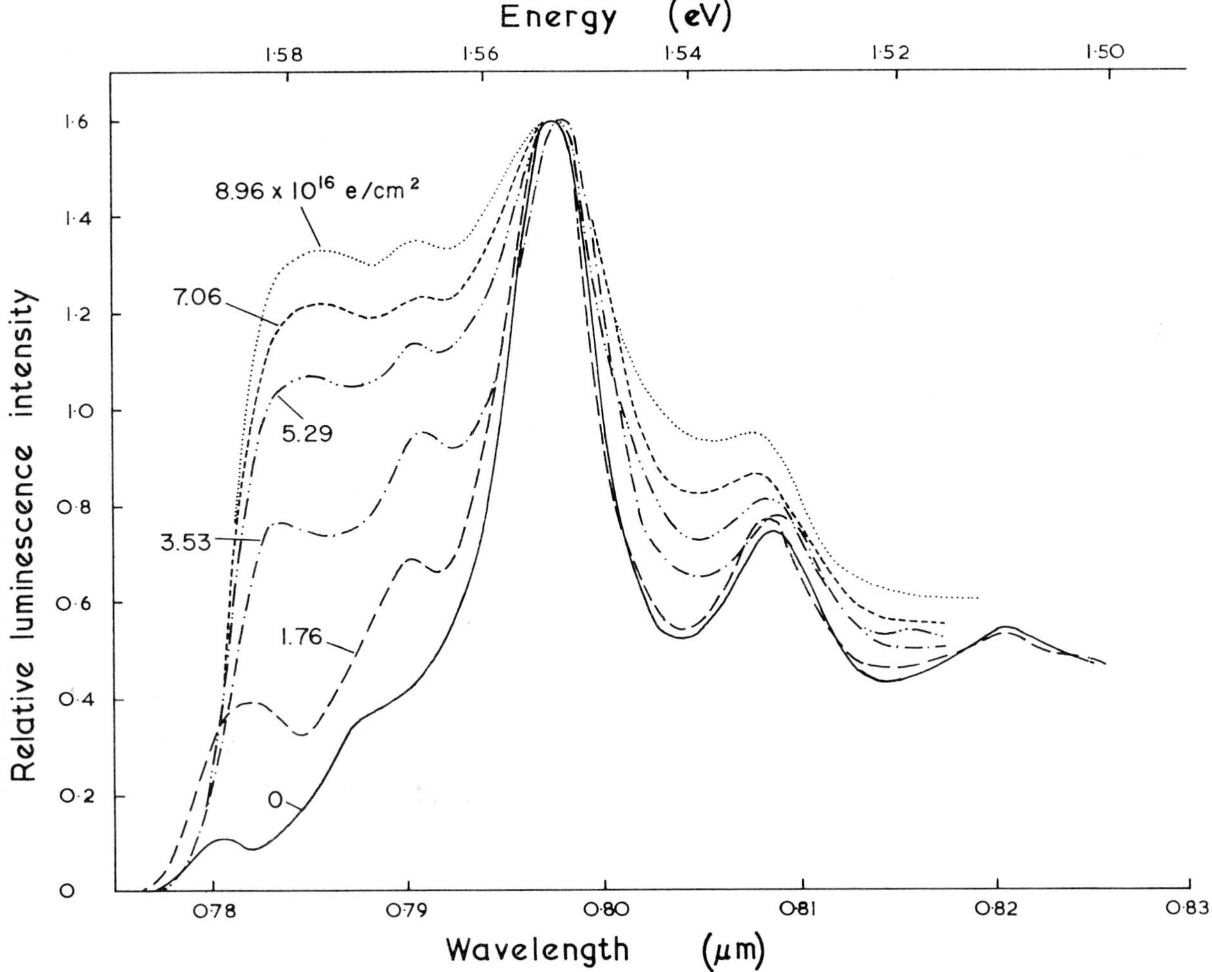

FIG. 4. Cathodoluminescence spectrum of cadmium telluride at 4.2 °K excited by 100 keV electrons; before irradiation at higher energies – 0; after 1.76×10^{16} electrons cm⁻² at 260 keV; after 3.53×10^{16} electrons cm⁻² at 260 keV; after 5.29×10^{16} electrons cm⁻² at 260 keV; after 7.06×10^{16} electrons cm⁻² at 260 keV; after 8.96×10^{16} electrons cm⁻² at 260 keV.

the 1.572 eV emission is not as large as that of the 1.587 eV emission.

When both cadmium atoms and tellurium atoms are displaced (see Figures 5 and 6) the emission peak at 1.554 eV is observed to behave in a different manner than when only cadmium atoms are displaced. Displacement of cadmium atoms only causes the 1.554 eV emission to increase in intensity and to shift to higher energy. Whereas when both cadmium and tellurium atoms are displaced the peak of the emission energy remains unaltered.

It is also observed that for a given dose of electrons producing cadmium displacements only the increase in intensity of the 1.554 eV emission is comparable to the increase observed for the same electron dose producing both cadmium and tellurium displacements, whereas when both atoms are displaced the increase in intensity of the 1.532 eV emission is much larger than when cadmium atoms only are displaced.

Figure 7 shows that variation of the energy of the exciting electrons affects the relative intensities of the emission bands in the edge emission of cadmium telluride. The emission at 1.544 eV increases in intensity with higher energy electrons whereas the higher energy emission intensity varies little.

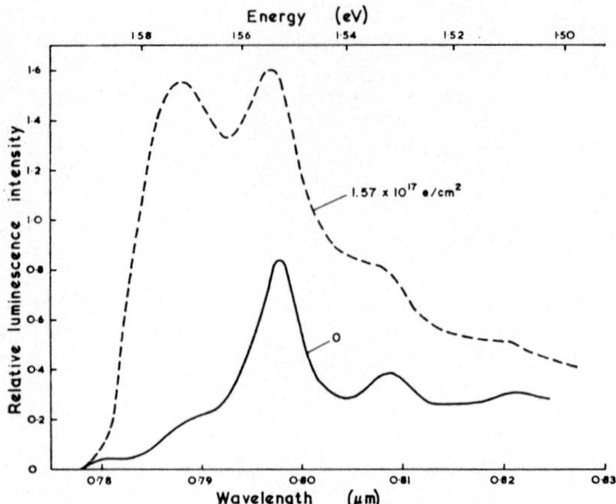

FIG. 5. Cathodoluminescence spectrum of cadmium telluride at 4.2 °K excited by 100 keV electrons; before irradiation at higher energies – 0; after 1.57×10^{17} electrons cm^{-2} at 260 keV.

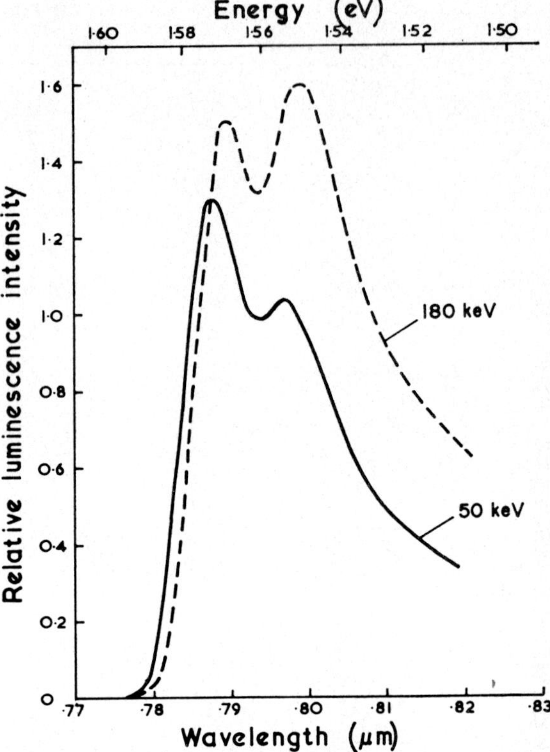

FIG. 7. Cathodoluminescence spectrum of cadmium telluride at 4.2 °K; excited by 50 keV electrons; excited by 180 keV electrons.

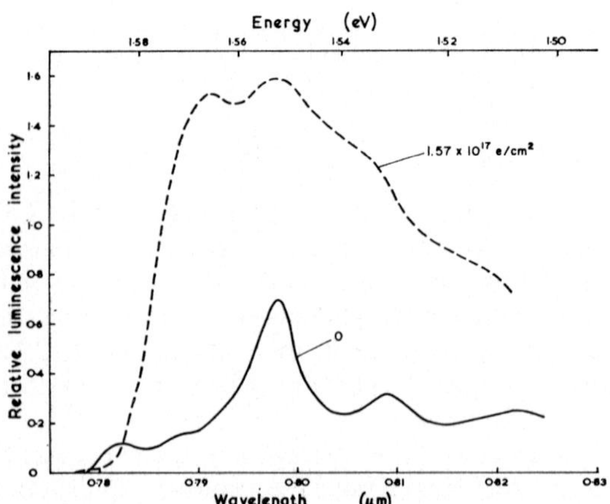

FIG. 6. Cathodoluminescence spectrum of cadmium telluride at 4.2 °K excited by 100 keV electrons; before irradiation at higher energies – 0; after 1.57×10^{17} electrons cm^{-2} at 400 keV.

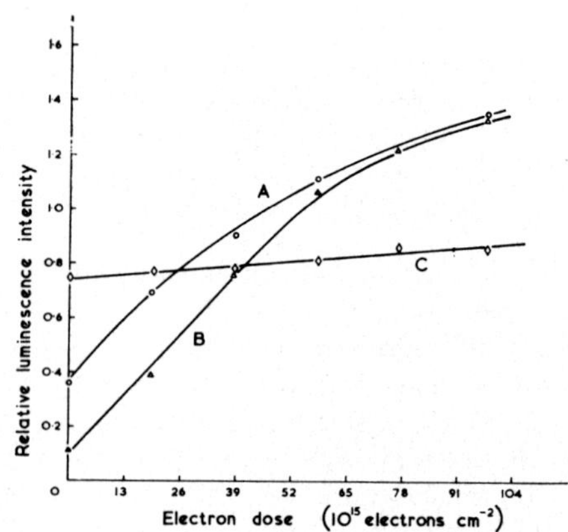

FIG. 8. Emission intensity dependence of cathodoluminescence of cadmium telluride at 4.2 °K (excited by 100 keV electrons) with dose of 260 keV electrons capable of producing cadmium displacements; A, 1.572 eV emission; B, 1.587 eV emission; C, 1.532 eV emission.

3.2. *Heat treatment*

Firing cadmium telluride in vacuo at 600 °C for one hour greatly reduces or even destroys the edge emission. Upon refiring this vacuum annealed sample in an overpressure of cadmium, an emission is reproduced at 1.552 eV (0.7980 μm) as shown in Figure 3. Successive firing under a cadmium overpressure causes this emission to shift to lower energies. The emissions at 1.587 eV and 1.572 eV are always removed or greatly reduced by any firing treatment.

4. DISCUSSION

The behaviour of the dominant emission (1.554 eV) in the edge emission of the as grown samples may be explained by assuming it to consist of two emissions resulting from different defect centres.

Radiation damage producing displacement of cadmium atoms only results in a shift of the 1.554 eV emission to higher energy (see Figure 5) and an increase in the emission intensity with electron dose which is dependent in magnitude upon the emission intensity in the original sample. However when the as grown samples are annealed in vacuo the edge emission is destroyed, and refiring of these annealed samples reproduces an emission in this region. Therefore an emission can be produced in this wavelength region either by cadmium displacement or by firing the material in a cadmium overpressure. In these two processes cadmium vacancies and tellurium vacancies are formed respectively. However since no evidence of a bound-to-bound type of transition has been observed, this peak is believed to be composed of two separate emission bands. One of these bands is due to a transition from the conduction band into the first charged state of the cadmium vacancy and the other band is due to a transition from a level involving a tellurium vacancy to the valance band. Evidence of such an acceptor level at $E_v + 0.05$ eV has been reported,[5] and has been assigned to the first charged state of the cadmium vacancy. The second emission is attributed to the double acceptor centre[6] and believed to involve a tellurium vacancy and halogen complex.[7] Confirmation of this separation into two emission bands is given when both cadmium and tellurium atoms are displaced (see Figure 6). No shift in the peak energy is observed, a result which would be expected if this emission involves the two defects.

The peak energy of this composite emission varies from sample to sample, as would be expected since different concentrations of each defect would contribute a varying amount to the total emission intensity. The peak energy of an emission in this region has also been shown to be very dependent on the preparation and firing conditions of the material.[4]

The emission at 1.572 eV is also observed to increase with cadmium displacement. A marked feature of the damage of this emission is the shift to lower energy with electron dose (see Figure 4). Previously this emission has been attributed to involve a neutral cadmium vacancy[2] and our results confirm this assignment. The observed shift with dose can be explained on the basis of the neutral defects interacting with each other because they have large orbits.

The damage sensitive emission at 1.587 eV is believed to be an exciton bound to a neutral cadmium vacancy. This emission is observed to increase in intensity with cadmium displacement. From theoretical considerations[1] it may be shown that such a bound exciton would have an emission energy slightly less than that of the free exciton, as is observed here. The rate of increase in intensity with electron dose is much greater than for any other observed emission in most samples. This would be expected because with cadmium displacement not only the cadmium vacancy concentration will increase but also the exciton capture cross section will increase and the emission intensity is dependent upon both these quantities. The observed large shift to lower energy with electron dose of the 1.587 eV emission is due to overlapping of the wavefunctions of the defects and broadening of the exciton band.

Variation of the penetration depth of the exciting electrons may be achieved by varying the electron energy. An increase in the penetration depth of the electrons is seen to cause a major increase in the intensity of the 1.554 eV emission as shown in Figure 7, whereas the exciton emission (1.587 eV) increases to a lesser extent. Considering these results it is believed that the 1.554 eV emission is a result of defects occurring in the bulk of the crystal whereas the exciton emission occurs mainly in the surface region.

ACKNOWLEDGEMENT

One of us (D.H.J.T.) wishes to thank the Science Research Council for the award of a research studentship.

REFERENCES

1. E. H. Bogardus and H. B. Bebb, *Phys. Rev.*, **176**, 993–1002 (1968).

2. F. J. Bryant, A. F. J. Cox and E. Webster, *J. Phys. C.* (*Proc. Phys. Soc.*), ser. 2, vol. 1, 1737–1745 (1968).

3. F. J. Bryant and C. J. Radford, *Cryogenics*, 10, 329–331 (1970).

4. R. E. Halsted, M. R. Lorenz and B. Segall, *J. Phys. Chem. Sol.*, **22**, 109–116 (1961).

5. M. R. Lorenz and B. Segall, *Phys. Letters*, **7**, 18–20 (1963).

6. M. R. Lorenz, B. Segall and H. H. Woodbury, *Phys. Rev.*, **134**, 751–760 (1964).

7. H. H. Woodbury and M. Aven, *Gen. Elec. Res. Repts.*, No. 64-RL-36599 (1964).

DISCUSSION

Question (RANDOLPH)   What is the effect of room temperature irradiation?

Answer (TOTTERDELL)   All the observed damage to the edge emission reported in this paper involving cadmium displacements is seen to have annealed at 77 °K. Room temperature irradiations do not produce defects stable at that temperature which are involved in the edge emission.

THERMAL NEUTRON DAMAGE IN CdS AND CdTe†

C. KIKUCHI‡

Institute for Chemical Research, Kyoto University, Kyoto, Japan

Photoluminescence and electrical resistivity changes in CdS and CdTe produced by thermal neutrons are discussed. The damage is produced principally by the neutron capture reaction ^{113}Cd (n,γ) ^{114}Cd. Since the reaction product ^{114}Cd is stable, complications arising from impurity introduction is minimal. The cumulative recoil nuclear recoil energy is about 143 eV, but is not the recoil energy at the time atomic displacement occurs. Thermal and fast neutrons enhance the CdS luminescence band at 7200A in the ratio of 28:1, but the resistivity changes are in the ratio of 40:1. Cd interstitial is suggested as the luminescence center. Hall measurements on *n*-type CdTe suggest that only Cd defects are produced for low thermal neutron doses. The acceptor introduction rate is about 1.0 to 0.6, compared to 0.098 for CdS. These are in good agreement with the values reported by R. O. Chester. The fast neutron effects in high resistivity CdS reported by Johnson indicate the need for further measurement.

The purpose of this paper is to discuss some of the pronounced photoluminescence and electrical resistivity changes in CdS and CdTe produced by thermal neutron irradiation.[1-3] Atomic displacement by thermal neutrons perhaps better known as the Szilard–Chalmers effect, has been a subject of intensive investigation by hot atom chemists but has received relatively little attention by physicists working in the field of radiation damage. Consequently a search for the effect was made, and looked for in Cd compounds, because the thermal neutron capture cross section is large (2450 b). This large cross section is due primarily to the 12.26 per cent abundant ^{113}Cd, whose cross section is known to be over 20,000 b. Of the possible Cd compounds, CdS was selected as the first material to study because its photoluminescence is known to be sensitive to defect structure.

The mechanism for the Szilard–Chalmers effect is that nuclei recoil when capture gamma rays are emitted. For Cd, as an example, the neutron capture process ^{113}Cd (n, γ) ^{114}Cd results in the release of about 9.044 MeV of gamma energy. Measurements indicate that about 4.1 gammas are emitted in the transition of the highly excited state to the ground state of the stable ^{114}Cd nucleus. The successive emission of γ-rays results in the cumulative recoil energy of about 143 eV, which is obtained by taking the weighted average of the capture gamma spectrum.[4] This amount, however, is not necessarily the energy with which the Cd atom is displaced from its site.

The theory of gamma decay from highly excited nuclear states[5] suggests that the first one or two gammas are energetic, and these provide the necessary energy for displacement. Subsequently the moving ^{114}Cd atom intermittently radiates the remaining energy as it makes its zig-zag interstitial flights. An artist's concept of what may be taking place is shown in Figure 1.

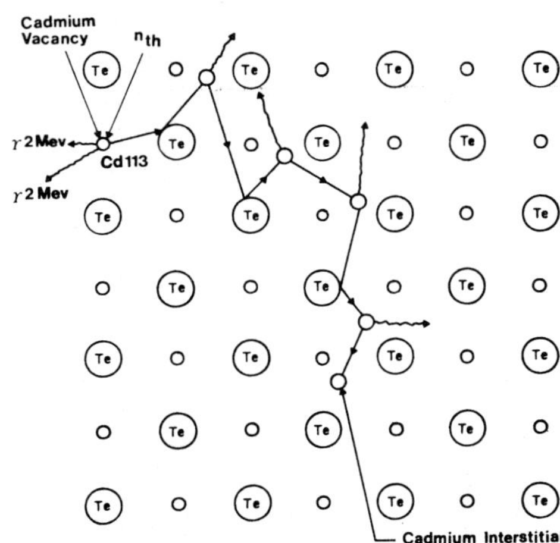

FIG. 1. Possible course of recoiling ^{114}Cd as it collides with lattice atoms and emits gamma rays.

The neutron irradiations were carried out in the University of Michigan swimming-pool type reactor. During irradiation, the samples were placed near the reactor core, where the temperature varied from about 50 to 90 °C. At these temperatures, defect annealing is important, as has been found for CdTe[3] and CdS.[6]

The thermal to fast neutron flux ratio is about

† Supported in part by National Science Foundation U.S.-Japan Cooperative Science Program.
‡ On Sabbatical Leave from the University of Michigan during the 1969–70 Academic Year.

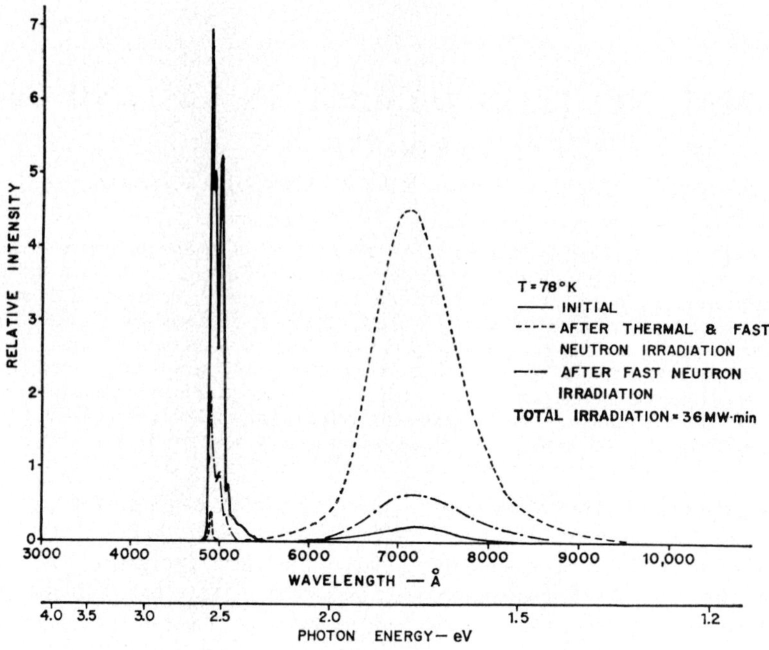

FIG. 2. Red luminescence induced by thermal neutrons ($\sim 10^{16}/cm^2$).

10, and this value leads to 18:1 for the estimated ratio of thermal and fast neutron damage in CdS. We need, however, to be careful in applying this ratio to a particular experiment, because thermal neutrons are expected to produce predominantly Cd defects whereas fast neutrons will produce both Cd and CdS defect in nearly equal concentrations. For example for photoluminescence, which involves only one kind of defect, say Cd, the ratio of the changes could be as high as 36:1. In resistivity measurements, on the other, the Cd and S defects are likely to have opposite effects upon the electron concentration so that the ratio can be expected to be even larger.

Figure 2 shows the effect of thermal neutron irradiation on photoluminescence. There is a noticeable decrease in the intensities of the edge and blue luminescence and a marked increase in the red luminescence at 7200A. If the sample is shielded from thermal neutrons, the increase in the luminescence intensity is reduced by about 28. This ratio of 28:1 is in good agreement with the estimated upper limit of 36:1. Measurement of the resistivity changes shows that the ratio is 40:1, again consistent with expectations. These results suggest that either the Cd interstitial or vacancy is the luminescence center.

Turning to CdTe, we find that under moderate thermal neutron fluence, Te defects are not pro-duced. The Hall plots given in Figure 3 indicate a level at about E_c–0.012 eV, which the GE workers[7] have attributed to chemical donors. The sample, which Dr. Halsted gave us is n-type, with initial carrier concentration of the order of $10^{15}/cm^3$. The evidence for a level at E_c–0.06 eV, which has been attributed to a Te vacancy appears only after

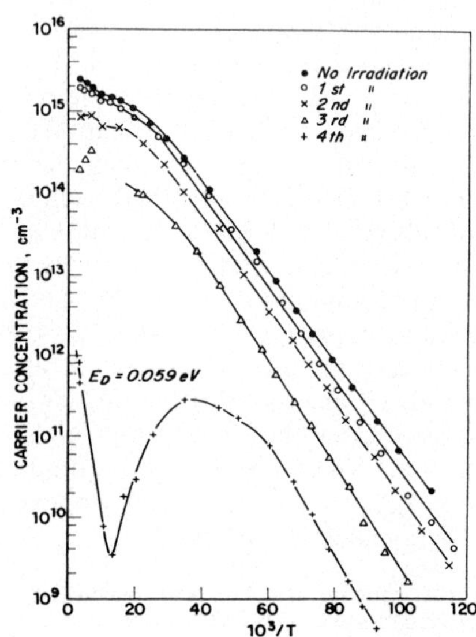

FIG. 3. Hall plot for neutron irradiated n-type CdTe sample. Sample presented by R. E. Halsted.

prolonged irradiation. One reason for this may be that the threshold for Te displacement (7.8 eV) is appreciably larger than that for Cd (5.6 eV). Another possibility is annealing because of the relatively high irradiation temperature. The carrier removal rates for the two *n*-type samples are 1.0 and 0.6 in good agreement with the value 0.6 reported by R. O. Chester.[8] For purposes of comparison, the Hall plot and carrier removal curve for CdS are given in Figures 4 and 5. The

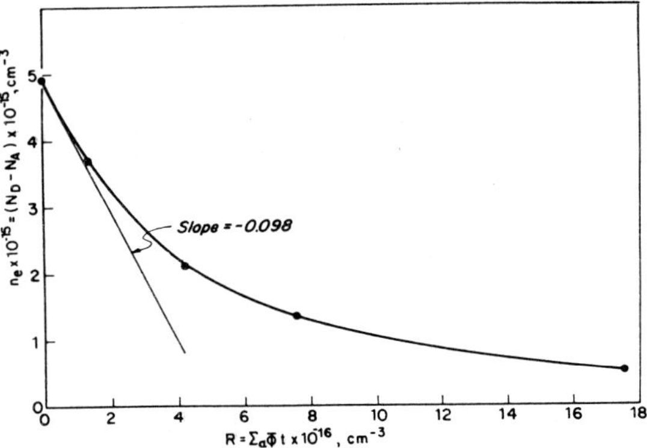

FIG. 5. Carrier removal curve for CdS crystal.

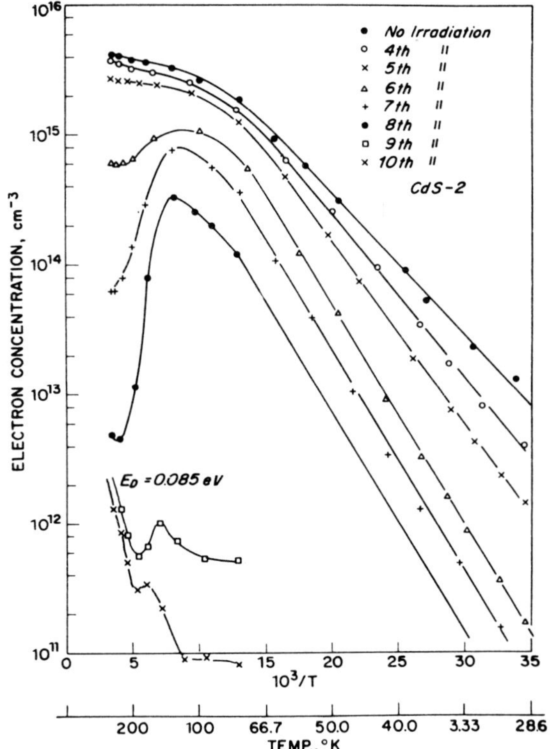

FIG. 4. Hall plot for neutron irradiated CdS crystal (Fluence for last irradiation $\sim 10^{17}/cm^2$).

value 0.085 eV suggests the double acceptor level reported by R. A. Anselmo and H. H. Woodbury.[9]

Comparable measurements have been made on *p*-type CdTe, purchased from Metal Hydrides having initial carrier concentrations of the order of $10^{16}/cm^3$. The carrier removal rate is again about 1.0. Upon prolonged irradiation the hole concentration dropped to about $10^9/cm^3$, but there was no evidence of change to *n*-type, as has been reported by others.[10,11] The reason for this is not clear at present, but may be due to sample preparation procedures. For our measurements, the CdTe samples were enclosed in glass tubes, flushed with

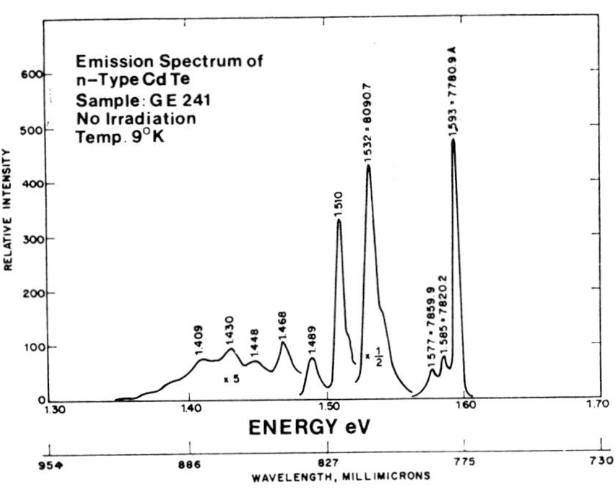

(a) No Irradiation

(b) After First Irradiation

FIG. 6. Photoluminescence spectra of *n*-type CdTe before and after neutron irradiation.

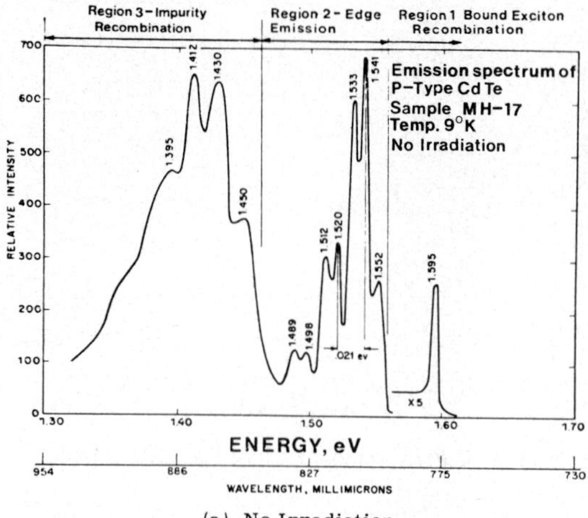

(a) No Irradiation

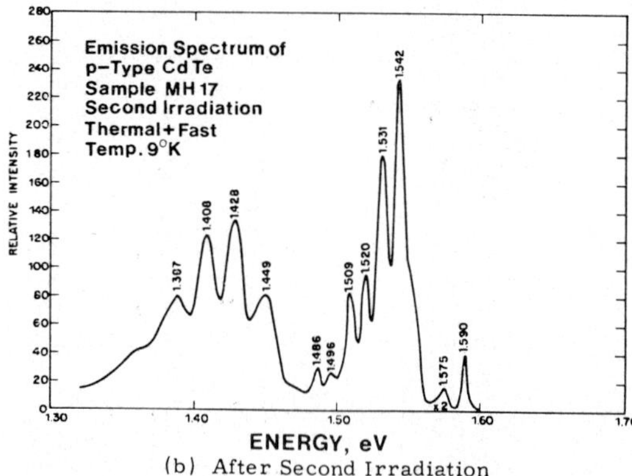

(b) After Second Irradiation

FIG. 7. Photoluminescence spectra of *p*-type CdTe before and after neutron irradiation.

argon, and sealed in about 2/3 atm argon pressure during irradiation.

Changes in photoluminescence for *n*- and *p*-type CdTe are shown in Figures 6 and 7. For the *p*-type sample (Figure 7) a weak emission line at 1.575 eV appears after irradiation. This may possibly be identified with the 7896A line, which has been attributed to Cd vacancy.[12]

Our results combined with the luminescent measurements on plastically bent CdS suggest that the Cd interstitial is the center for the 7200A luminescence band. According to the results by Mitsuhashi, Chikawa and Nakayama[13] the luminescence bands at 6350 and 7400A are strong in samples bent to produce excess Cd rich dislocations.[14] Furthermore they noted that the 7400A band intensity is enhanced relative to the 6350A

band upon iodine doping. These results, if the reversal of the *c*-axis direction is taken into account, suggest that the 7400A band arises from a Cd interstitial coupled to an iodine. Consequently, it appears reasonable to assign the 7200A band to a Cd interstitial bonded possibly to a different halogen, say a chlorine.

R. T. Johnson[15] has reported that the sign of the carrier removal rates depends upon the initial carrier concentration. In particular for high resistivity CdS ($\sim 10^8$ ohm cm) the resistivity decreases upon irradiation.[16] In an earlier experiment, S. Tanaka and T. Tanaka[17] bombarded high resistivity crystals (10^{10} ohm cm) with 2 MeV deuterons, found the resistivity to decrease, and attributed it to a *S* donor level at about 0.4 eV below the conduction band. The nature of this center appears to have been confirmed by K. Morigaki and T. Hoshina's[18] ESR measurements. Donor depth of 0.26 eV was estimated from the *g*-values ($g_\parallel = 1.783$, $g_\perp = 1.74$).

ACKNOWLEDGEMENTS

I wish to thank Professor Mitsuhashi and Dr. Chikawa for helpful discussions, Drs. C. E. Barnes (Sandia) and R. B. Oswald (Harry Diamond Laboratories) for suggestions and permissions to present unpublished results, and Professor Sakae Shimizu of the Kyoto University Radio-Isotope Laboratory for identifying the various radiation effects programs in Japanese laboratories.

REFERENCES

1. R. B. Oswald and C. Kikuchi, *Nuc. Sci. and Eng.*, **23** 354 (1965).
2. C. E. Barnes and C. Kikuchi, *Nuc. Sci. and Eng.*, **31**, 513 (1968).
3. C. E. Barnes and C. Kikuchi, *Rad. Effects*, **2**, 243 (1970).
4. L. V. Groshev, A. M. Demidov, V. N. Lutsenko and V. I. Pelekhov (Transl. by J. B. Sykes), *Atlas of γ-Ray Spectra From Radiative Capture of Thermal Neutrons* (Pergamon Press, 1959).
5. J. M. Blatt and V. F. Weisskopf, *Theoretical Nuclear Physics* (John Wiley and Sons, 1952), p. 647.
6. L. P. Randolph and R. B. Oswald, *Bull. Am. Phys. Soc.*, **15**, 398 (1970).
7. B. Segall, M. R. Lorenz and R. E. Halsted, *Phys. Rev.*, **129**, 2471 (1968).
8. R. O. Chester, *J. Appl. Phys.*, **38**, 1745 (1967).
9. R. A. Anselmo and H. H. Woodbury, *Bull. Am. Phys. Soc.*, **9**, 248 (1964).
10. N. B. Urli, *J. Phys. Soc. Japan*, **21**, Suppl. 259 (1966).
11. Quoted in Billington and Crawford *Radiation Damage in Solids*.

12. F. J. Bryant A. F. J. Cox and E. Webster, *J. Phys. C. (Proc. Phys. Soc.)*, **1**, 1737 (1968).
13. H. Mitsuhashi, J. Chikawa and T. Nakayama, *App. Phys. Letters*, **10**, 339 (1967).
14. E. P. Warekois, M. C. Lavine, A. N. Mariano and H. C. Gatos, *J. Appl. Phys.*, **37**, 2203 (1966).
15. R. T. Johnson, *J. Appl. Phys.*, **39**, 3517 (1968).
16. M. Kitagawa has reported similar results for electron bombarded CdS. See, for example, *Ann. Rept. of Radiation Center of Osaka Prefecture*, **5**, 73 (1964).
17. S. Tanaka and T. Tanaka, *J. Phys. Soc. Japan*, **14**, 113 (1959).
18. K. Morigaki and T. Hoshina, *J. Phys. Soc. Japan*, **24**, 120 (1968).

DISCUSSION

Question (MEESE) Is the sharp line emission at higher energies associated with the decay of an exciton bound to a defect?

Answer (KIKUCHI) Yes, the Cd vacancy.

RADIATION EFFECTS IN CdTe INDUCED BY THERMAL NEUTRONS AND 1 MeV ELECTRONS

A. A. ABRAMOV, V. S. VAVILOV AND L. K. VODOPIANOV

P. N. Lebedev Institute, Moscow, U.S.S.R.

Neutron and electron bombardments of *n*- and *p*-type CdTe were carried out. Carrier concentration, Hall mobility and photoconductivity were measured. Energy levels at $E_c - 0.06$ eV and $E_v + 0.17$ eV were found and are attributed to complicated centers, including the Cd vacancy.

The electrical and optical properties of CdTe not specially doped during the growing are determined by the defects of the structure.[1,2] It is convenient to use a radiation method for studying the influence of structural defects on the properties of the crystals. By choosing different types of irradiation (slow neutrons and electrons) it is possible to create the defects mainly in sublattice of Cd (the neutron absorption cross-section for Cd is very large compared to that to Te) or in the case of electrons in both sublattices with the same probability (the displacement threshold for Cd is near to that of Te).

In the present work we have investigated the structural defects created in CdTe by neutron and electron bombardment. The experimental technique and methods of preparation and irradiation of samples were described elsewhere.[3] The neutron irradiation experiments which involved cadmium thermal-neutrons shields showed that in CdTe the main effect is due to thermal neutrons. The samples of *n*-CdTe having the initial concentration of $n \simeq 1 \times 10^{15}$ cm^{-3} after irradiation with a flux of slow neutrons at doses of an order of 1×10^{16} cm^{-2} showed an inversion of the type of conductivity. The rate of defect creation decreased with increasing of initial concentration of current carriers in *n*-CdTe. In the samples having the initial concentration of electrons of 5×10^{16} cm^{-3} and more we did not observe an inversion of the type of conductivity up to doses of 5×10^{18} $n \cdot$cm^{-2}. These samples showed only decreasing of mobility and increasing of specific resistance.

In the crystals suffering an inversion of the type of conductivity after irradiation with slow neutrons, the concentration of acceptors did not exceed 3×10^{15} cm^{-3}, and mobility of the holes was not higher than 50 cm$^2 \cdot$v$^{-1} \cdot$sec^{-1}.

Further irradiation did not change the properties of these crystals.

Irradiation of the crystals of *p*-type in the range of doses from 5×10^{15} up to 5×10^{18} $n \cdot$cm^{-2} showed that they are less sensitive to neutrons compared to *n*-type. In the samples of *p*-type irradiated with thermal neutrons and also in the crystals suffering an inversion we discovered one acceptor level $E_v + 0.17$ eV by measuring the temperature dependence of the Hall constant and of the conductivity.

In the crystals of *n*-CdTe after neutron irradiation with doses not enough for an inversion a large increasing of photosensitivity was observed. The dark conductivity of such samples in the temperature range 150–110 °K decreased very rapidly (Figure 1, curve B). It was not possible to determine at these temperatures the energy of ionisation of the centre causing the dark conductivity due to the fact that the stationary value of the Hall constant took a few hours to be established. This means that electronic equilibrium is established very slowly. When such a sample at liquid nitrogen temperature was irradiated with white-light a sharp increase in the concentration of electrons in the conduction band was noted. This higher concentration of carriers after the light was turned off gradually decreased over the period of half an hour. Under light illumination the stable concentration of carriers was established rather rapidly and we could measure the electron concentration over the range of temperatures from 20 to 200 °K (Figure 1, curve B′).

As one can see from Figure 2 the dark mobility of electrons in the irradiated samples is much less than that before irradiation (see curves B and A respectively). With illumination the mobility of electrons was increased nearly up to the initial values. It is possible to explain all this totality of experimental data by assuming the creation under irradiation of doubly charged acceptor centres.

To determine the activation energy of the radia-

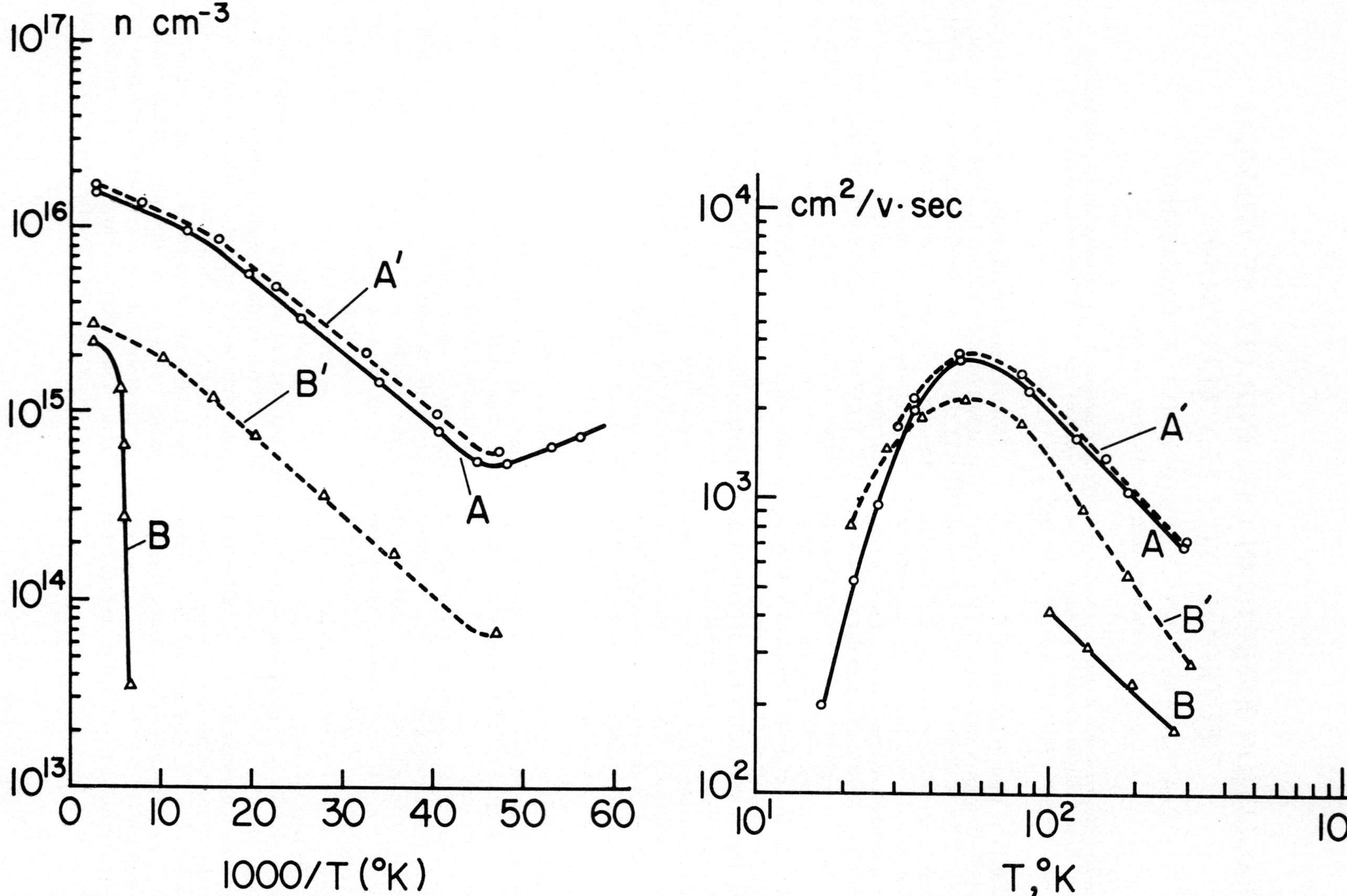

FIG. 1. Electron concentration of n-CdTe as a function of temperature before and after irradiation with thermal-neutrons (curves A and A'—before irradiation without and with light respectively; curves B and B'—the same, but after irradiation).

FIG. 2. Electron Hall mobilities for samples shown in Figure 1 (the same notations as in Figure 1).

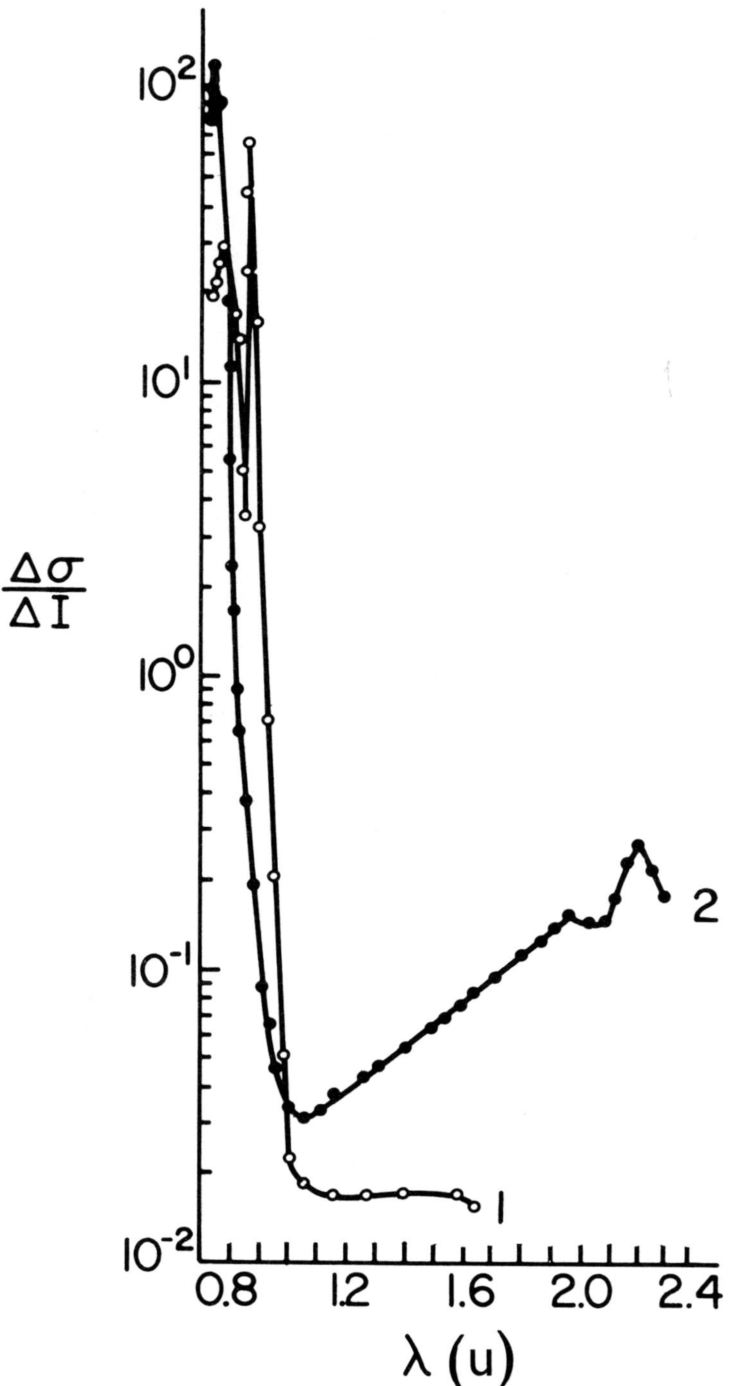

FIG. 3. Spectra of photoconductivity of n-CdTe irradiated with thermal-neutrons at 300 °K (curve 1) and at 100 °K (curve 2).

tion defects induced by thermal neutrons and electrons the spectra of photoconductivity were measured over the wavelength range from 0.7 to $3.0\,\mu$ (Figure 3). The spectra of photoconductivity measured at 300°K indicate the existence of pronounced peak at $\sim 0.9\,\mu$ (Figure 3, curve 1). This peak is due to the radiation defects corresponding to a level about 0.06 eV below the conduction band. At the temperature of 300°K the quanta of light having an energy ~ 1.45 eV excites the electrons from the valence band to the level $E_c - 0.06$ eV (the Fermi level at 300°C is situated about 0.1–0.15 eV below the conduction band). At low temperatures this level is occupied by electrons and probability of the transition from the valence band to the $E_c - 0.06$ eV level becoming very small. This corresponds to the fact that the peak of photoconductivity disappears at 100°K (Figure 3, curve 2).

The level $E_c - 0.06$ eV was also discovered from spectra of photoconductivity in samples of n-CdTe with initial concentration of the electrons of 5×10^{15} cm^{-3} after electron irradiation starting with a dose of 1×10^{17} elec. cm^{-2}. Measurement of the Hall constant and conductivity showed that defects created by electron irradiation have an acceptor nature. We consider that an acceptor level $E_c - 0.06$ eV—is a second charged state of a centre, the nature of which is unknown. Thus, an inspection of electrical and photoelectrical measurements shows that thermal-neutron and electron irradiations of n-CdTe resulted in the introduction of acceptor levels at $E_c - 0.06$ eV and $E_v + 0.17$ eV. We assume that radiation defects responsible for these levels are complicated centres, including the Cd vacancy.

REFERENCES

1. D. Nobel, *Phil. Res. Rep.*, **14**, 430 (1959).
2. S. Yamada, *J. Phys. Soc. Japan*, **17**, 645 (1962).
3. L. Vodopianov and A. Abramov, Collection: "Cadmium Telluride," Moscow, 1968.
4. M. Lorentz, B. Segall and H. Woodbury, *Phys. Rev.*, **134**, A751 (1964).

LUMINESCENCE OF IRRADIATED β-SiC

I. I. GEICZY, A. A. NESTEROV AND L. S. SMIRNOV

Institute of Semiconductor Physics, Academy of Sciences of the U.S.S.R., Siberian Branch, Novosibirsk, U.S.S.R.

The influence of electron irradiation with energy of 3.5 MeV on the luminescence of cubic Silicon Carbide (β-SiC) was investigated. It was shown that irradiation results in the diminution of edge emission and appearance of the red band (2.0 – 1.3 eV). Obtained results give an opportunity to consider the red band as transitions from excited states. The data on temperature stability of irradiated damage is given. The probable nature of irradiated defects are discussed.

1. INTRODUCTION

Despite certain difficulties connected with growing silicon carbide single crystals of fairly large size, purity and structural perfection the interest in this material connected with prospects of its use is continuously increasing.[1]

At present optical and electrophysical properties of α-SiC (hexagonal modification) are relatively well investigated, but very little information is available concerning the similar properties of β-SiC.[2-6] At the same time it is known that β-SiC has some advantages in comparison with α-SiC (higher electron mobility, brightness of the electroluminescence.[7-9] The irradiation effect by fast particles and hard radiation on α- and β-SiC properties has been poorly studied. There are only a few articles in this field.[10-12]

In this connection the purpose of the present paper was to investigate the changes of luminescence properties of β-SiC upon irradiating by fast electrons. The preference to β-SiC (zincblend structure) simplifies the interpretation of obtained results due to the absence of non-equivalent sites in the lattice that occur in other modifications.[13]

2. EXPERIMENTAL PROCEDURE

Tabular single crystals of β-SiC grown by the gasphase method, n-type conductivity $\rho = 0.1$–1.0 ohm·cm have been investigated.[14] The surface (111) of samples was ground and polished. Before measuring the samples were etched in the boiling mixture $4HF + HNO_3$. The irradiation was at room temperature with 3.5 MeV electrons. The luminescence was excited by an electron beam with the energy of 30 KeV and was registered by cooled photomultipliers.[11] The annealing was performed

in the air with 10-min annealing times at every temperature.

3. EXPERIMENTAL RESULTS

Figure 1–1 shows the typical spectrum of cathodoluminescence of the initial samples. The structure I_1, I_2, I_3 is apparently caused by annihilation of free excitons with simultaneous emission of TA, LA and TO phonons. The main band is accompanied by phonon-assisted components too (LO_Γ and TO_Γ). Figure 1–a shows the dependence of intensity in maximum of the band on the density of excitation.

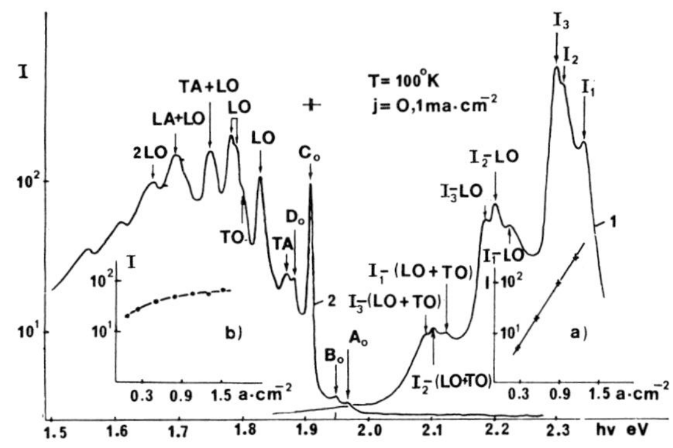

FIG. 1. Cathodoluminescence spectra β-SiC. 1—before irradiation; a—emission intensity in edge band as a function of excitation density; 2—after irradiation by electrons 3.5 MeV (dose ~10^{17} e/cm²); b—emission intensity in red band as a function of excitation density.

The irradiation of the β-SiC samples by fast electrons results in a decrease in intensity of the edge emission (2.3 eV) and the appearance of a wide red band (2–1.3 eV). At 100 °K this band has a well developed fine structure, mainly defined by the zero-phonon line C_0 (1.91 eV) and its phonon-assisted transitions (Table 1). Besides series C zero-phonon lines A_0 (1.97 eV), B_0 (1.95 eV) and D_0 (1.89 eV) are visible (Figure 1–2).

TABLE I

Identification of fine structure of series A, C, D.

	λ(Å)	Δ(meV)	Phonons
A-series	6295 ± 4	0	—
	6406 ± 10	35 ± 4	TA
	6525 ± 5	69 ± 2	LA
	6622 ± 5	98 ± 2	TO
	$(6670, 6686) \pm 3$	$(110, 116) \pm 3$	LO
C-series	6490 ± 4	0	—
	6600 ± 10	33 ± 4	TA
	6755 ± 5	77 ± 2	LA
	6835 ± 5	97 ± 2	TO
	$(6872, 6903) \pm 8$	$(107, 112) \pm 3$	LO
	7015 ± 10	145 ± 3	TA + LO
	7187 ± 10	186 ± 3	LA + LO
D-series	6565 ± 5	0	—
	6680 ± 10	33 ± 4	TA
	$(6807, 6832) \pm 5$	$(67, 74) \pm 2$	LA
	6925 ± 5	99 ± 2	TO
	$(6980, 7000) \pm 8$	$(110, 116) \pm 3$	LO

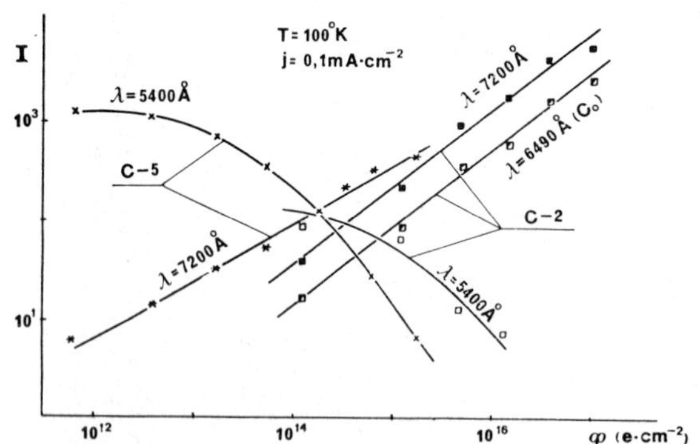

FIG. 2. Dependence of emission intensity on irradiation dose.

Figure 2 shows the dependence of the edge emission intensity 5400 Å (2.3 eV), red band 7200 Å (1.72 eV) and line C_0 6490 Å (1.91 eV) as a function of irradiation dose. No significant changes in the band shape were observed. The basic results of the investigation of the red band can be presented as follows:

—the half-width of the zero-phonon line C_0 at 100 °K equals ~ 6 meV;

—the line intensities decrease with increasing temperature and they 'fall' into the continuous spectrum; .

—the fine structure practically disappears at 180 °K though the intensity of continuous spectrum changes only slightly;

—the intensity of the band sharply quenches at $T > 300$ °K ($E_{\text{act}} \simeq 0.48$ eV);

—the intensity reduction with increasing temperature is accompanied by a decrease in after-glow time;

—changing the excitation intensity over a wide range $(0.1{-}1 \times 10^3)$ mA/cm² does not result in an appreciable shift of fine structure or a change in the spectrum's form; the similar result has been obtained upon recording the spectrum after a delay, i.e. at various times after the excitation impulse;

—intensity dependence of C_0 and of the long-wave edge of the red band as a function of the excitation density have the same (sublinear) character (Figure 1–b).

The above facts show that the recombination of localized charges takes place and the continuous spectrum and fine structure present the same mechanism of luminescence. The difference in the shape of the red band at various temperatures is determined by whether the zero-phonon transition occurs or not. In the first case the fine structure appears to be a zero-phonon line and its phonon-assisted transitions. With increasing temperature the probability of zero-phonon transitions sharply decreases (optical analogue of the Mössbauer's effect)[15] and radiative transitions occur at the interaction with phonons, both at the boundary of Brillouin's zones and the parts of phonon branches as well.

The results of the experiment demonstrate that the fine structure cannot be explained by the donor-acceptor pairs mechanism of radiation recombination. We suppose that the radiative transitions of the excited states take place here.

To determine the temperature stability of the radiation damage responsible for the red band of luminescence, isochronal annealing was performed. Figure 3 shows the intensity changes of zero-phonon lines upon annealing. The characteristics are as follows:

—up to 600 °C series C is dominant in the spectrum and then it disappears in the range of 600–630 °C;

—from 650 to 900 °C series C is not observed in

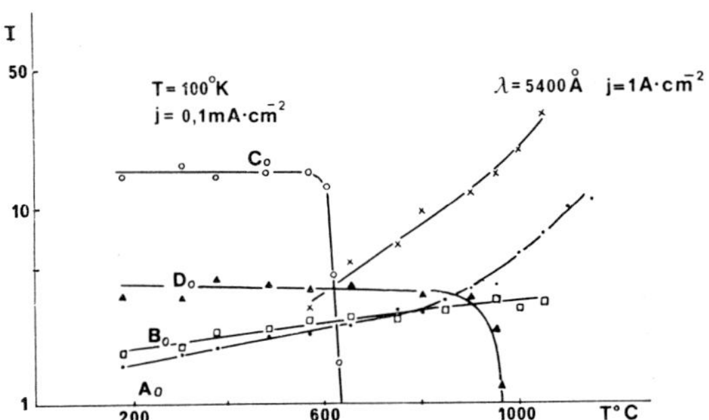

FIG. 3. Behaviour of line intensities of fine structure in irradiated β-SiC at isochronal annealing.

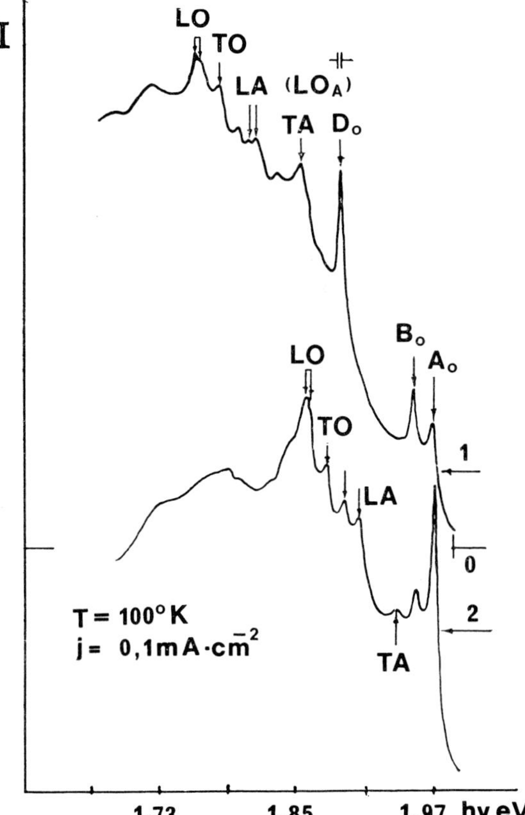

FIG. 4. Cathodoluminescence spectra of irradiated β-SiC after isochronal annealing (1—700 °C; 2—1100 °C).

the spectrum and lines A_0, B_0, D_0 are clearly visible (Figure 4–1);

—after the annealing at $T > 1000\,°C$ series A is dominant in the spectrum and increases with increasing temperature;

—the intensity of the edge emission increases upon annealing;

—the annealing of non-irradiated samples at $T \simeq 1100\,°C$ results in only series A, but its emission intensity is 1.5 to 2 orders of magnitude weaker.

To determine the activation energy of series C, isothermal annealing at 610 and 625 °C was performed, the activation energy calculated by section method equals 2.2 eV.[16]

Repeated irradiation of annealed crystals resulted in the preferential increase of series C and decrease of series A. Nevertheless, the intensity of series A is higher than after the first irradiation. The treatment of spectra with well-developed series A, B, D (after the annealing of series C) permitted us to identify the fine structure beyond peak D_0 as its phonon replicas (Table 1). Series D disappears in the range of 950–1000 °C. This series is controlled by LO-phonon of line D_0 because D_0 coincides with one of the A_0 replicas connected with localized vibration of the lattice[17] or with LA-phonon for point L of the Brillouin zone.[5] After the annealing at 1200 °C series A is dominant (Figure 4–2, Table 1). Qualitative coincidence of luminescence characteristics of series A, B, C and D means that the mechanism of the radiative transitions responsible for these series is similar.

As a result of the irradiation by electrons with various energies significant differences in the luminescence spectra have been obtained (Figure 5).

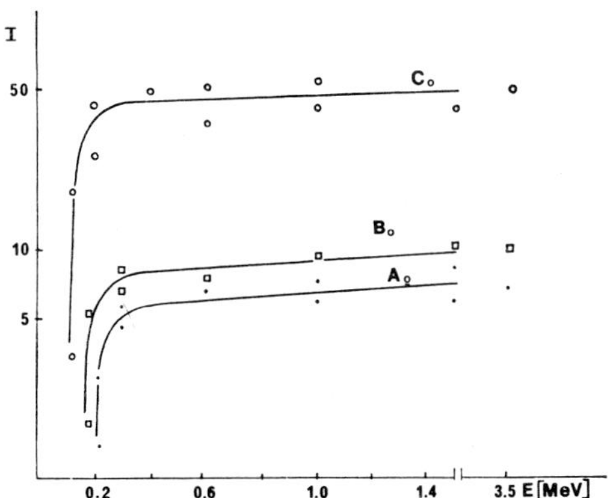

FIG. 5. Intensity dependence of lines A_0, B_0 C_0, on fast electron energy (dose $\sim 4.10^{16}$ e/cm²).

These are preliminary results as yet but we can state that with decreasing energy of fast electrons series A, B, D and C disappear successively, i.e. the energy of defect ‘C’ formation is the very low.

4. DISCUSSION

The experimental results permit us to draw some conclusions about the properties and nature of radiative centres which appear in β-SiC after the irradiation. Many data point out the independence of luminescence channels responsible for series A, B, C, D from each other and from the channel of edge emission. The annealing of centres 'C' did not result in the changes of intensities A, B and D lines which on the one hand shows the absence of competition between centres on capture of non-equilibrium carriers and on the other hand that the simplest defects released at destruction of centres 'C' do not participate in centres 'A', 'B', 'D' formation.

The character of edge emission whose intensity is connected with electrons and holes concentration in free zones[5-7] makes us to suppose that after the irradiation radiative and non-radiative recombination channels appear. The intensity of edge emission depends upon the efficiency of these channels. This is proved by the fact that the irradiation by electrons with low energies (< 200 KeV) effects the intensity of edge emission in a relatively smaller degree than the irradiation with energy 3.5 MeV (at equal doses).

Emission independence in various series allows us to compare the relative concentrations of damage centers by the intensity of lines corresponding to them. But we have to understand fully that this situation is possible only in case of the excitation of inside-centre transitions.

Series A behaviour has a peculiar character. Series A appears at irradiation and heating at 1200 °C as well. But if the irradiated sample is heated, series A grows more intensely, i.e. the stimulation of the formation process of these centres by irradiation takes place. However repeated irradiation 'destroys' some centres of A type the concentration of which tends to quasi-equilibrium magnitude. We think that the defect of the lattice on silicon is included in centre 'A'. On the one hand it is known that the high-temperature heating ($T > 1100$ °C) results in the silicon evaporation.[18] (It is necessary to stress that in the specific case the question is about concentrations of 10^{12}–10^{14} cm^{-3} order). On the other hand since there are two thresholds for atom displacement in SiC, we expect a decrease of the number of defects on the silicon lattice with decreasing electron energy.[12] Our results with the irradiation by electrons of energy less than 200 KeV are quite consistent with the above-stated supposition. The defect on silicon may be to some extent characteristic of centres 'B' and 'D'.

Since the relative intensity of series C at the irradiation by electrons with energy less than 200 KeV decreases we may suppose that the defect on carbon is dominant because the displacement energy for carbon atoms is less than for silicon atoms. It is indirectly shown by the low energy of activation at annealing of series C.

ACKNOWLEDGEMENTS

The authors are acknowledged to S. N. Gorin for crystals and Yu. M. Limasov, V. A. Patrenin, V. I. Abramenko and S. A. Sokolov for the irradiation of samples.

REFERENCES

1. Proc. Int. Conf. on Silicon Carbide, University Park, Pennsylvania, October 1968 (Pergamon Press, New York, 1969).
2. H. R. Philipp, *Phys. Rev.*, **111**, 440 (1958).
3. W. J. Choyke, D. R. Hamilton and L. Patrick, *Phys. Rev.*, **133**, A1163 (1964).
4. M. L. Belle, N. K. Prokof'eva and M. B. Reifman, *Fiz. Tekh. Poluprov.*, **1**, 383 (1967).
5. D. S. Nedzvetskii, V. V. Novikov, N. K. Prokof'eva and M. B. Reifman, *Fiz. Tekh. Poluprov*, **2**, 1029 (1968).
6. G. Zanmarchi, *J. Phys. Chem. Sol.*, **29**, 1727 (1968).
7. L. Patrick, *J. Appl. Phys.*, **37**, 4911 (1966).
8. A. Rosengreen, Proc. Int. Conf. on Silicon Carbide, University Park, Pennsylvania, October 1968 (Pergamon Press, New York, 1969), p. S355.
9. Yu. M. Altaiskii, Abstract Ph.D. Thesis, Polytekh. Inst., Kiev (1969).
10. V. V. Makarov, *Fiz. Tverd. Tela*, **9**, 597 (1967).
11. I. I. Geiczy, A. A. Nesterov and L. S. Smirnov, *Fiz. Tekh. Poluprov.* **4**, 897 (1970).
12. J. H. N. Loubser, J. A. de Sousa Balona and W. P. van Ryneveld, Proc. Int. Conf. on Silicon Carbide, University Park, Pennsylvania, October 1968 (Pergamon Press, New York, 1969), p. S249.
13. L. Patrick, *Phys. Rev.*, **127**, 1878 (1962).
14. S. N. Gorin and A. A. Pletyushkin, *Izv. AN SSSR, ser. fiz.*, **28**, 1310 (1964).
15. E. F. Gross, B. S. Razbirin and S. A. Permogorov, *Dokl. AN SSSR*, **147**, 338 (1962).
16. A. C. Damask and G. J. Dienes, Point Defects in Metals, Izd. 'Mir', Moscow (1966), p. 149. (Eng. edition published by Gordon and Breach, New York.)
17. D. R. Hamilton, W. J. Choyke and L. Patrick, *Phys. Rev.*, **131**, 131 (1963).
18. F. A. Halden, Silicon Carbide-A High Temperature Semiconductor, (Pergamon Press, London, 1960), p. 115.

THE EFFECT OF NEUTRON AND ALPHA-PARTICLE IRRADIATION UPON THE PHYSICAL PROPERTIES OF α-SiC

Y. V. BARINOV, Y. V. BULGAKOV, M. I. IGLITZYN, M. A. ILYIN, M. G. KOSSAGANOVA,
N. M. PAVLOV, M. B. REIFMANN, B. A. SAKHAROV, AND V. N. SOLOMATIN

Giredmet, Moscow, U.S.S.R.

Alpha-particle (25 MeV) and neutron irradiation of sublimation-grown α-SiC were carried out. A new EPR spectrum was observed but no model yet established for the defect responsible for this spectrum. IR absorption (0.4–25 mkm.) was also studied; no distinct bands were observed beyond a broad near-edge non-selective absorption.

1. INTRODUCTION

In the present work the silicon carbide single crystals grown out of a gas phase by the sublimation technique were investigated. Prior to the study the crystals had been thoroughly checked for their resistivity homogeneity at room temperature both on the surface and into the depth.

The samples were irradiated with 25 MeV α-particles in a cyclotron having a 120 cm polar diameter. The temperature therein did not exceed 50 °C.

The irradiation of the samples with fast neutrons was carried out in the channel (130 mm diameter) of a water-cooled and water-moderated reactor. The fast neutron energy spectrum and thermal neutron flux distribution are discussed in an article by Mr. Zvonov and coworkers.[1]

The emphasis in the present work was laid upon the study of the electron-paramagnetic resonance (EPR) and that of an irradiated SiC optical transmission in the $\lambda = 0.4$ to 25 mkm wavelength range.

2. THE ELECTRON-PARAMAGNETIC RESONANCE

The EPR of an irradiated n-type α-SiC containing nitrogen as a predominant impurity was investigated.

In the SiC crystals the EPR was observed at a frequency of 9235 MHz and a temperature of 100 °K. The EPR spectrum of a nitrogen-doped silicon carbide consisted of three lines at a nitrogen concentration of about 10^{17} cm^{-3} and less[2] and after irradiating in a $3.6 \cdot 10^{13}$ α/cm^2 flux of α-particles it did not alter. By increasing the flux by approximately one order of magnitude the nitrogen spectrum was found to vanish or be significantly weakened. The latter was observed in the case when the sample thickness was larger than the depth of α-particles penetration. A completely new spectrum emerged. It consisted of one intense line with an isotropic g-factor equal to 2.0033. This line was of a somewhat asymmetric shape dependent on the magnetic field orientation. In the samples with a free carrier concentration of $1.5 \cdot 10^{17}$ cm^{-3} or less at room temperature there was a series of weaker lines added to the intense one. These were first of all twelve anisotropic lines 1.7 oersted broad, similar to those observed in an electron-irradiated SiC,[3,4] and a big number of considerably weaker anisotropic lines. Exactly the same alterations in the EPR spectrum were observed as a result of irradiating the SiC samples with fast neutrons.

The 12-line spectrum was studied in detail in three samples irradiated with α-particles: a 6H polytype with a free carrier concentration of $4.4 \cdot 10^{16}$ cm^{-3} at room temperature, irradiated in an integrated $3.6 \cdot 10^{14}$ α/cm^2 flux; a 4H polytype (free carrier concentration of 1.10^{17} cm^{-3}, an integrated flux of $2.2 \cdot 10^{14}$ α/cm^2); a 15R polytype (free carrier concentration of $0.8 \cdot 10^{16}$ cm^{-3}, an integrated flux of $2.2 \cdot 10^{14}$ α/cm^2 and one more 6H polytype sample with a free carrier concentration of $2.9 \cdot 10^{16}$ cm^{-3}, irradiated to an integrated $6.9 \cdot 10^{15}$ n/cm^2 neutron flux.

This spectrum is well described by a spin-Hamiltonian:

$$\mathscr{H} = \beta \bar{\mathrm{H}} \cdot \bar{g} \cdot \bar{S} + D[S_z^2 - \tfrac{1}{3}S(S+1)] + E(S_x^2 - S_y^2)$$
with $S = 1$.

The main values for g-tensor as well as D- and E-values for this principal defect are presented in the Table II. It follows from the angular dependence of the line positions that there exist six directions of defect axes which are located in pairs in the perpendicular to the a-axes planes of the crystal. The angles between the defect axes in each of these planes which are alike at various forms of irradiation (for the 6H polytype), slightly differ for various polytypes. These angles values are given in the Table I. The angles between the defect axes axes and 'c'-axis of the crystal are near 45°. The possible precise values of these angles for various polytypes, calculated on the basis of experimental data, are presented in the Table I. Nevertheless, it is not excluded that these angle values, in particular for the 6H polytype, for both directions are alike and equal to $\beta/2$.

TABLE I

The angles β between the defect axes and possible values of the angles α between 'c'-axis of the crystal and defect axes for various polytypes of α-SiC.

Polytype	β	α_1	α_2
4H	91.6°	47.5°	44.1°
15	91.2°	46.5°	44.7°
6H	90.8°	46.3°	44.5°

Out of a large number of weaker lines (one part of which is obviously due to a superfine interaction between the lattice atoms and Si^{29} and C^{13} nuclei) for the 6H and 15R polytypes samples (the weaker lines spectrum of the 4H polytype sample being left without analysis), we succeeded in singling out two groups of lines which may be expressed by means of the spin-Hamiltonians of the same kind. The parameters of the spin-Hamiltonian for these centres, which will be referred to further on as 'a'- and 'b'-defects respectively and which we were able to evaluate from the angular dependence of the line positions are presented in the Table II. The amplitudes of the 'a'-defect spectrum lines relative to those of the main defect spectrum lines accounted

for about 8 per cent with the breadth of the lines. This defect was observed in both the 6H and 15R polytype samples. The relative amplitude of the 'b'-defect lines was as low as 2 per cent while the breadth of the lines was almost twice as large as that of the main defect lines. This defect was revealed only in the 6H polytype sample (perhaps it is due to the fact that the 15R polytype sample was considerably smaller and its spectrum was respectively weaker). These defects axes too had two directions lying in the same planes as the main defect, the angles for these directions having been found to be constant with the respective main defect axes. These angle values are also presented in the Table II. It is noteworthy that whereas for all three kinds of the defects the g- and D-values differ little, the E-value for the main defect and 'b'-defect differ by an order of magnitude and for the 'a'-defect the E-value changes its sign.

It should be noted that the weaker line spectrum is not yet fully decoded; it must obviously contain the lines of other defects of the same type the parameters of which are not yet determined with a sufficient degree of reliability.

Similar lines were observed in all the samples near the vicinity of 1600 oersted. We suppose that these lines are due to the quasi-forbidden $\Delta M = 2$ junctions which is corroborated by the fact that these lines cannot be observed at a 35000 MHz frequency.

Due to the fact that for all the defects the main values of the g-tensor differ little and approach the value of the g-factor of a free electron and due to the large D constant, we may assume that the spectrum fine structure is accounted for by the magnetic dipole-dipole interaction of two localized spins. In this case:

$D = \frac{3}{2}g^2\,\beta^2\,r^{-3}$, where r is the distance between the two localized spins.

The distances r calculated with the above formula are also presented in the Table II.

TABLE II

The defect parameters†

Defects	g_x	g_y	g_z	$\lvert D \rvert$ (cm^{-1})	E (cm^{-1})‡	Angles with 'c' axis	$r(A)°$
Main defect	2.0016	2.0014	2.0010	0.0560	$+6.6\cdot10^{-4}$	α	3.59
'a'-defect	2.002	2.001	2.001	0.054	-9.10^{-4}	$\alpha - 9.4°$	3.6
'b'-defect	2.004	2.004	2.007	0.033	$+65.10^{-4}$	$\alpha + 8.0°$	4.3

† The values of α are given in the Table I.
‡ The sign of the E value is related to that of the D value.

We deem it premature to put forward any conclusions as for the nature of these centres. It is highly probable that a nitrogen atom participates in their formation to which testifies the above-mentioned fact of a simultaneous observation of the defects and nitrogen spectra in one sample when the sample thickness exceeded the depth of alpha-particles penetration. We may hope that further analysis of the data available and new investigations will make it possible to disclose the nature of this defect.

3. THE OPTICAL TRANSMISSION

The optical transmission was investigated of the 6H polytype p-type boron-doped α-SiC from 0.4 to 25 mkm.

For irradiation with fast neutrons the samples were divided into three groups; each of them had respectively the following integrated irradiation dose: $2.3 \cdot 10^{15}\,n/cm^2$; $3.9 \cdot 10^{15}\,n/cm^2$; $6.7 \cdot 10^{15}\,n/cm^2$.

The electrical parameters of the samples before and after irradiating them are presented in the Table III along with the designation of the incident particles used and irradiation doses.

The typical transmission spectra in a wavelength 0.4 to 1 mkm range for the samples before and after irradiation with α-particles and neutrons are shown in the Figures 1 and 2. The transmission spectra for the samples before and after irradiation with fast neutrons in a 2 to 25 mkm wavelength range are presented in the Figure 3. As it is known, the strong bands of a phonon absorption[5,6] are observed in α-SiC in the 4.2–4.6 mkm and 6–15 mkm spectral ranges. We exclude these spectral regions from our study. In all these spectra there may be observed the well known peak of lattice absorption[5,6] in the 6H polytype α-SiC in a 4.8 mkm wavelength.

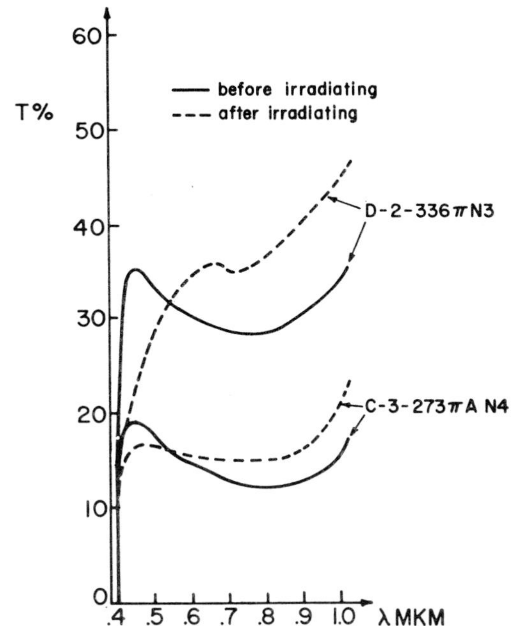

FIG. 1. The transmission spectra of the α-SiC samples with the concentration of free carriers about $4.10^{14}\ cm^{-3}$ before and after irradiating them with α-particles.

It follows from the analysis of the transmission spectra for the samples before and after their irradiation in the wavelength region being studied:

1. There are no new peaks appearing in the transmission spectra of the samples irradiated and measured at a room temperature as compared to the spectra of the non-irradiated samples.

2. After irradiating with both α-particles and neutrons in all the samples the absorption coefficient K in the spectral region near the absorption edge proper increased significantly (near-edge non-selective absorption, Figures 1 and 2).

3. The intensification of the neutron integrated

TABLE III

Electrical parameters of the samples before and after an irradiation

| Sample | Before an irradiation | | | After an irradiation | | | Characteristics and dose of the irradiation |
	p om cm	h cm^{-3}	m cm²/v sec	p om cm	h cm^{-3}	m cm²/v sec	partions/cm²
D-2-336 N3	324	$4.4 \cdot 10^{14}$	44	3420	$3.95 \cdot 10^{13}$	46	$2.2 \cdot 10^{14}\ \alpha/cm^2$
C-3-273ΠA N4	290	$4.0 \cdot 10^{14}$	53.6	642	$2.36 \cdot 10^{14}$	41.2	$3.0 \cdot 10^{13}\ \alpha/CM^2$
D-2-340ΠA N2	93	$2.5 \cdot 10^{15}$	26.7	1000	$1.1 \cdot 10^{15}$	5.0	$6.7 \cdot 10^{15}\ h/cm^2$
D-2-341ΠA N6	72.5	$2.9 \cdot 10^{15}$	30.0	461	$1.43 \cdot 10^{15}$	9.5	$3.9 \cdot 10^{15}\ n/cm^2$
D-2-336ΠA N2	262	$5.2 \cdot 10^{14}$	46.0	11800	$7.0 \cdot 10^{13}$	5.0	$2.3 \cdot 10^{15}\ n/cm^2$
D-2-353Π N1	629	$2.1 \cdot 10^{14}$	46.5	10^7	—	—	$3.9 \cdot 10^{15}\ n/cm^2$

                                    Y. V. BARINOV ET AL.

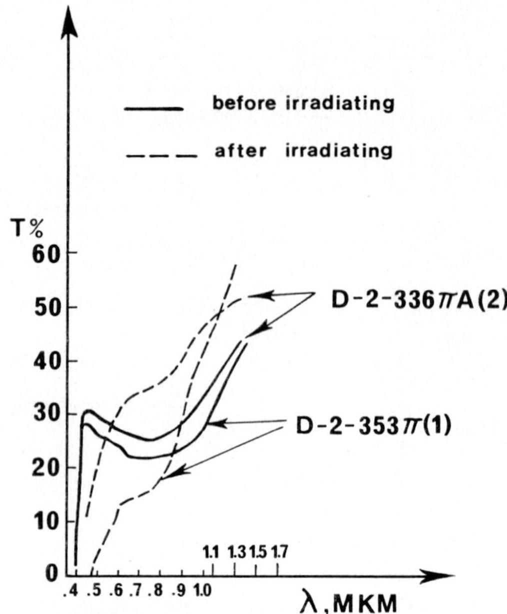

FIG. 2. The transmission spectra of the α-SiC samples before and after irradiating them with various doses of neutrons.

flux results in increasing the K value near the edge as well as in widening the spectral region where the near-edge nonselective absorption occurs. Thus, with an irradiation dose of $2.3 \cdot 10^{15}$ n/cm² and $\lambda = 0.52$–0.55 mkm, the T values are equal before and after irradiating, while with a $3.9 \cdot 10^{15}$ n/cm² dose the parity of the transmission coefficients before and after irradiating is observed at considerably greater wavelengths about 0.85 mkm (Figure 2).

4. The absorption broad band (0.48 mkm–0.87 mkm) is usually attributed to the presence of boron impurity in α-SiC.[7] After irradiating with both a-particles and neutrons the intensity of this band decreases. This effect is noticeable only with such irradiation doses when the nonselective absorption is small in this spectral range.

5. In the 2–6 mkm spectral range increased transmission in all the samples is observed. Increasing the irradiation dose results in a sharp increase of the degree of improved transmission.

6. In the wavelength range of $\lambda = 15$–25 mkm the transmission of the samples after irradiating alters little (Figure 3).

The following models may be suggested to interpret the nonselective absorption phenomenon observed near the absorption edge proper.

The increase of absorption is connected with the formation after irradiation of a number of donor-type defects giving rise to a series of energy levels near the edge of the forbidden gap. Two alternatives are both possible: either the formation of shallow ionized donors or the emergence of very deep neutral donors. In the former case the light absorption is accompanied with the transition of electrons from the valence band to the ionized donor levels, in the latter—transition of electrons from the neutral donor levels to the conduction band. The measurements of the electrical parameters of the samples at a room temperature after irradiation (Table III), the decrease of spin concentration as well as the initial results of the study of photoconductivity of the samples irradiated (the electrical conductivity of the crystals increases sharply when the crystals are exposed to an indirect lighting)—all this is a persuasive argument in favour of the first assumption. The appearance of a structure in the spectral region investigated corresponding to the level-band transitions may be expected in low temperature measurements.

Another alternative for interpreting the near-edge nonselective absorption phenomenon may be proposed. It might be brought about by the emergence of the disordered regions as a result of irradiating the samples, as it had been observed in some studies.[8–11]

The decrease of the impurity absorption band 0.48–0.87 mkm, which may be observed after irradiating α-SiC (this band, as it is shown,[7] is due to the transitions of the electrons from the valence band to the boron atoms levels, the atoms being ionized at a room temperature), may be due to the following reasons. First, it may be due to the boron atoms becoming optically inactive after being irradiated (e.g. as a result of their being displaced from the lattice sites). Next, as it had been said, bombarding with heavy particles results in an intensive formation of the radiation donor-type defects making up the material. In the crystal there remain less neutral boron atoms which are needed to bring about an absorption band.

In the 2–6 mkm region the absorption in the non-irradiated samples increases as the boron concentration (dopant impurity) increases. There was some improved transmission observed in all the neutron-irradiated samples in the 2–6 mkm wavelength region. If we assume that the absorption in this region is due to the impurity, then its decrease after irradiating may be also caused by the boron atoms passing into an optically inactive state. Our

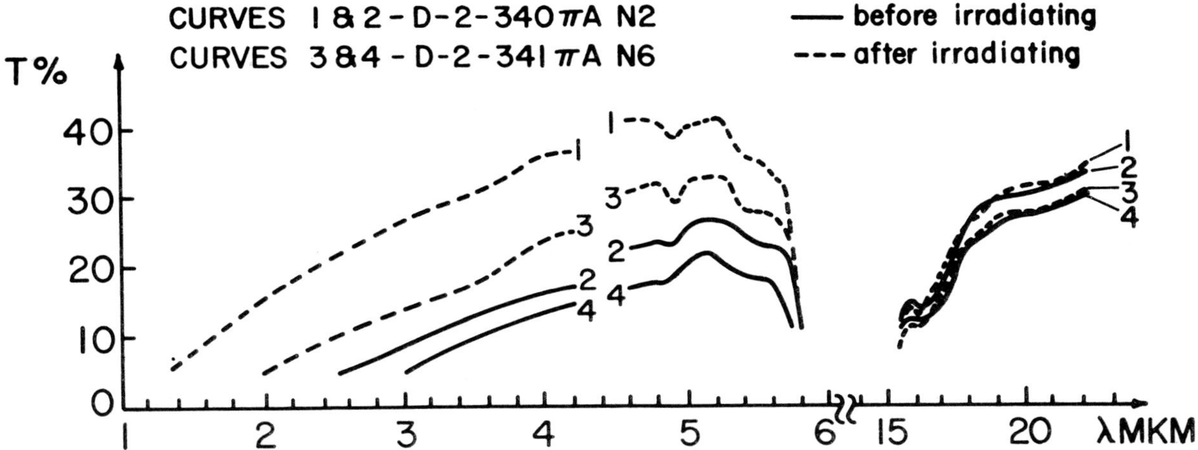

FIG. 3. The transmission spectra of the α-SiC samples before and after irradiating them with various doses of neutrons.

second model for the intense formation of the radiation donor-type defects would also explain the improved transmission phenomenon after irradiating the samples in the 2–6 mkm region since it is liable to result in the absorption decrease both when it is connected with the availability of free holes and when it is, by its nature, due to an impurity presence (the making up of the main dopant impurity, the decrease of the free holes concentration). Further clarification in the 2–6 mkm spectral region as the neutron irradiation dose increases may be easily explained by means of both models proposed. As a matter of fact, the greater the irradiation dose, the more defects are formed in the material, either because of the transition of the boron atoms into an optically inactive state or the emergence of radiation donor-type defects.

It is to be noted that even at a room temperature the radiation defects, emerging while irradiating α-SiC with fast neutrons and α-particles, are partially annealed, since the EPR signal and electrical and optical measurements show decreases.

ACKNOWLEDGEMENTS

The authors of the present paper express their gratitude to Dr. V. N. Mordkovitch and Dr. V. G. Fomin for their having taken part in the discussion of the results.

REFERENCES

1. N. V. Zvonov, A. I. Nisskavitch, I. V. Rogoshkin, V. I. Tereschenko, T. I. Turkov, V. P. Utkin, *Atomic energy*, N 2, II (1962).
2. H. H. Woodbury and G. W. Ludwig, *Phys. Rev.*, **124**, 1083 (1961).
3. J. H. N. Loubser, J. A. de Sousa Balona and W. P. van Ryneveld, *Mat. Res. Bull.* **4**, S249 (1969).
4. W. P. van Ryneveld and J. H. N. Loubser, *Brit. J. Appl. Phys.*, **17**, 1277 (1966).
5. W. J. Choyke, *Phys. Rev.*, **123**, 813 (1961).
6. L. Patrick, *Phys. Rev.*, **127**, 1868 (1962).
7. G. N. Violina, 'The study of the electrical and optical properties of SiC with a view of its use in the resistors and computers', a doctorate thesis, Leningrad (1966).
8. I. L. McNichols and W. S. Ginell, *J. Appl. Phys.*, **38**, 656 (1967).
9. R. Truell, *Phys. Rev.*, **116**, 890 (1959).
10. L. W. Aukerman, *J. Appl. Phys.*, **34**, 3590 (1963).
11. M. T. Lappo and V. D. Tkachev, in press.

LOW TEMPERATURE ELECTRON IRRADIATION OF TELLURIUM

E. GMELIN, R. STAPF, P. KLEMT, G. LANDWEHR, W. LICHTENBERG AND
A. PRZYBYLSKI

Physikalisches Institut der Universität Würzburg, D–87 Würzburg, Germany

Tellurium single crystal samples with a hole concentration of $3 \times 10^{14}/cm^3$ were irradiated at 10 K with electrons with an energy of 0.6 and 1 MeV. In the range investigated resistivity and Hall-coefficient R_H both decreased linearly with the integrated electron flux. The hole generation rate was 0.09 cm^{-1} and 0.47 cm^{-1} for 0.6 and 1 MeV electrons, respectively. The Hall-mobility R_H/ρ increased with irradiation.

Annealing of the radiation damage by raising the temperature clearly revealed three recovery stages in the resistivity- and Hall-data. At 180 K ρ and R_H returned to their pre-irradiation values. The original Hall-mobility was already restored close to 90 K.

A more detailed study of the first recovery stage, which occurs at about 50 K, revealed an activation energy of 170 ± 40 meV. It is most likely, that the observed lattice defects are Frenkel-defects. There are indications, that the point defects interact with dislocations.

1. INTRODUCTION

There are indications, that the acceptors in pure tellurium with a low temperature hole concentration of the order $10^{14}/cm^3$ are lattice defects. The energy levels of the acceptors must be located within the valence band, because they cannot be depopulated even by reducing the temperature to 0.1 K. Nor can they be 'frozen out' by strong magnetic fields, as one would expect for hydrogen-like impurities. At present practically nothing is known of the microscopic nature of the acceptors in high purity tellurium. Because irradiation with electrons with energies in the 1 MeV range should produce isolated interstitial-vacancy pairs, the study of which might give insight into the acceptor problem, a systematic study of the recovery behavior of low temperature irradiated tellurium was initiated.

No extensive investigation of the effects of electron irradiation in tellurium has been published so far. Van Lint and Roth[1] irradiated tellurium single crystals at 80 K with 20 MeV electrons and measured the radiation induced changes in conductivity and Hall-coefficient. However, no systematic recovery studies were performed.

2. EXPERIMENTAL

Our experimental setup allows to irradiate samples at helium temperatures between 7 and 10 K with electrons which are accelerated by an electrostatic generator. The electron energy could be varied from 0.4 to 1.2 MeV. The properties we investigate are resistivity and Hall-effect, which are measured with conventional d.c.-techniques. In the cryostat constant temperatures between 4 and 300 K can be established with an accuracy of 0.2 K or better. The temperature is measured with a platinum resistance thermometer. The tellurium sample which is studied is attached at one end to a copper block, so that no strain due to differential contraction can develop. The specimen is located at the center of a nitrogen cooled copper solenoid which gives a magnetic field of about 1000 G. The two samples investigated had a hole concentration of $3 \times 10^{14}/cm^3$ at 4.2 K and a hole mobility of about 3100 cm$^2/Vs$ at this temperature. They were cut from a Czochralski-grown single crystal with their long axes perpendicular to the trigonal axis of the lattice. In order to avoid lattice damage, the specimens were cut with an acid saw and with great care lapped down and finally etched to a thickness of 0.125 and 0.196 mm. After irradiation the recovery of the lattice damage was studied by raising the temperature in steps from 5 to 10 K and watching resistivity and Hall-effect for 10 minutes. Whenever no isothermal annealing could be detected, the temperature was increased further. The first recovery stage was studied in more detail.

3. RESULTS

In Figure 1 the resistivity ρ of sample II with a thickness of 0.125 mm is shown as a function of temperature before and after irradiation at about 10 K. A dose of 6×10^{14} electrons/cm^2 with an

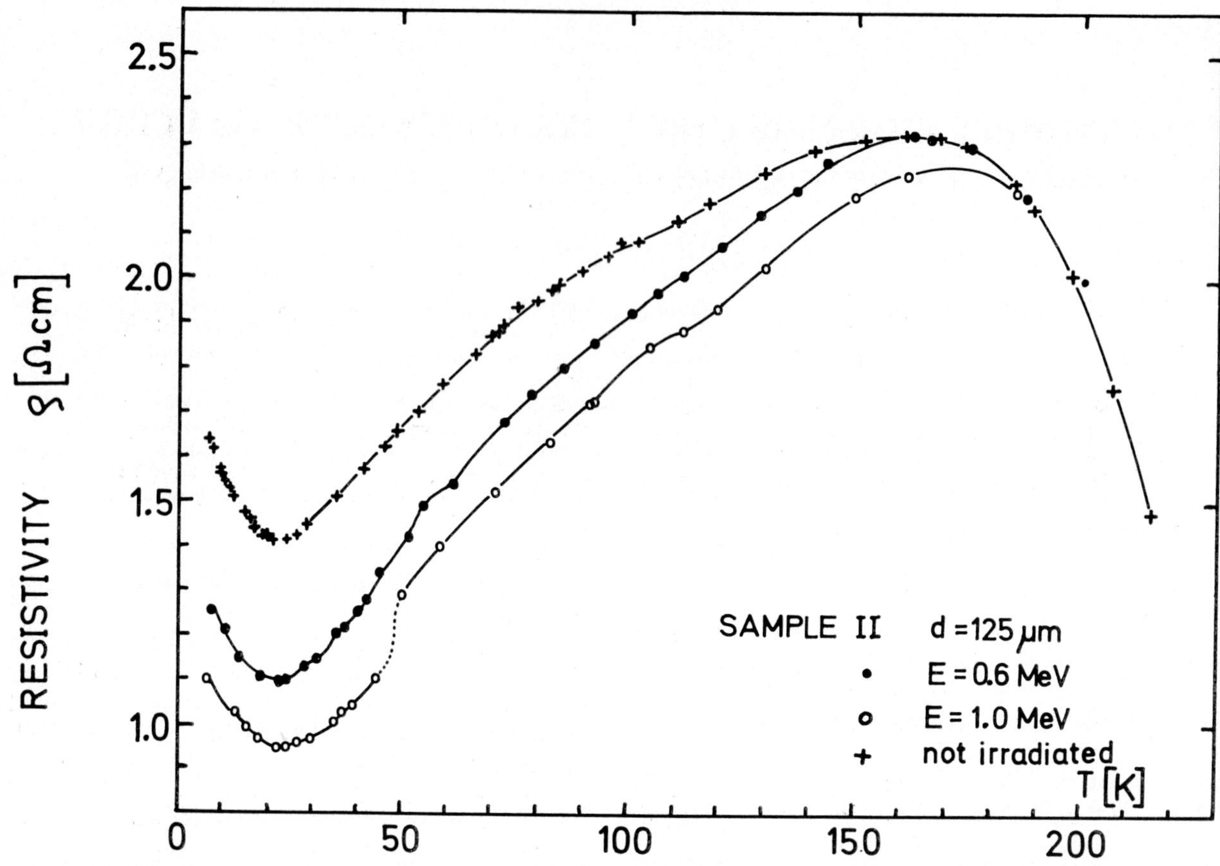

FIG. 1. Electrical resistivity ρ as a function of temperature before and after irradiation with 0.6 MeV and 1 MeV electrons; integrated flux ϕ (0.6 MeV) $= 6 \times 10^{14}$ electrons/cm², ϕ (1 MeV) $= 1.7 \times 10^{14}$ electrons/cm².

energy of 0.6 MeV resulted in a resistivity decrease of ~ 24 per cent, and $1.7 \cdot 10^{14}$ 1 MeV electrons/ cm² lowered ρ by ~ 33 per cent. In the lower curve one recognizes immediately two recovery stages around 50 and 100 K, and the disappearing of the radiation damage at about 180 K. A plot of the unannealed fraction of the radiation induced change in resistivity reveals three or possibly 4 recovery stages (Figure 2). It may be seen, that the thicker sample I shows essentially the same behaviour as sample II. Because the same holds for the Hall-coefficient, only data for sample II are shown in the following. Moreover, the radiation damage in this specimen should be more uniform than in the thicker one. (The transmission rate for sample II is about 85 per cent for 1 MeV electrons.)

The Hall-coefficient R_H decreased after electron bombardment, 17 per cent and 22 per cent for 0.6 MeV and 1 MeV, respectively. A plot of the unannealed fraction $\Delta R_H / R_H$ as a function of T

indicates 4 annealing stages in the temperature ranges 30–45 K, 60–80 K, 90–110 K and above 120 K. The second stage around 70 K is not visible in the resistivity. The Hall-mobility R_H / ρ increases after irradiation with 0.6 and 1 MeV electrons by 11 and 13 per cent, respectively. About 75 per cent of the mobility change disappears between 30–50 K, and the rest between 70 and 90 K, as may be seen from Figure 4. Only the first recovery stage was studied in some detail by isochronal annealing. The reference temperature was 23 K, where the resistivity minimum occurs. The result is plotted in Figure 5. It is evident, that the recovery step is rather sharp.

4. ANALYSIS OF THE DATA AND DISCUSSION

The Hall-data clearly demonstrate, that additional acceptors are introduced by electron irradi-

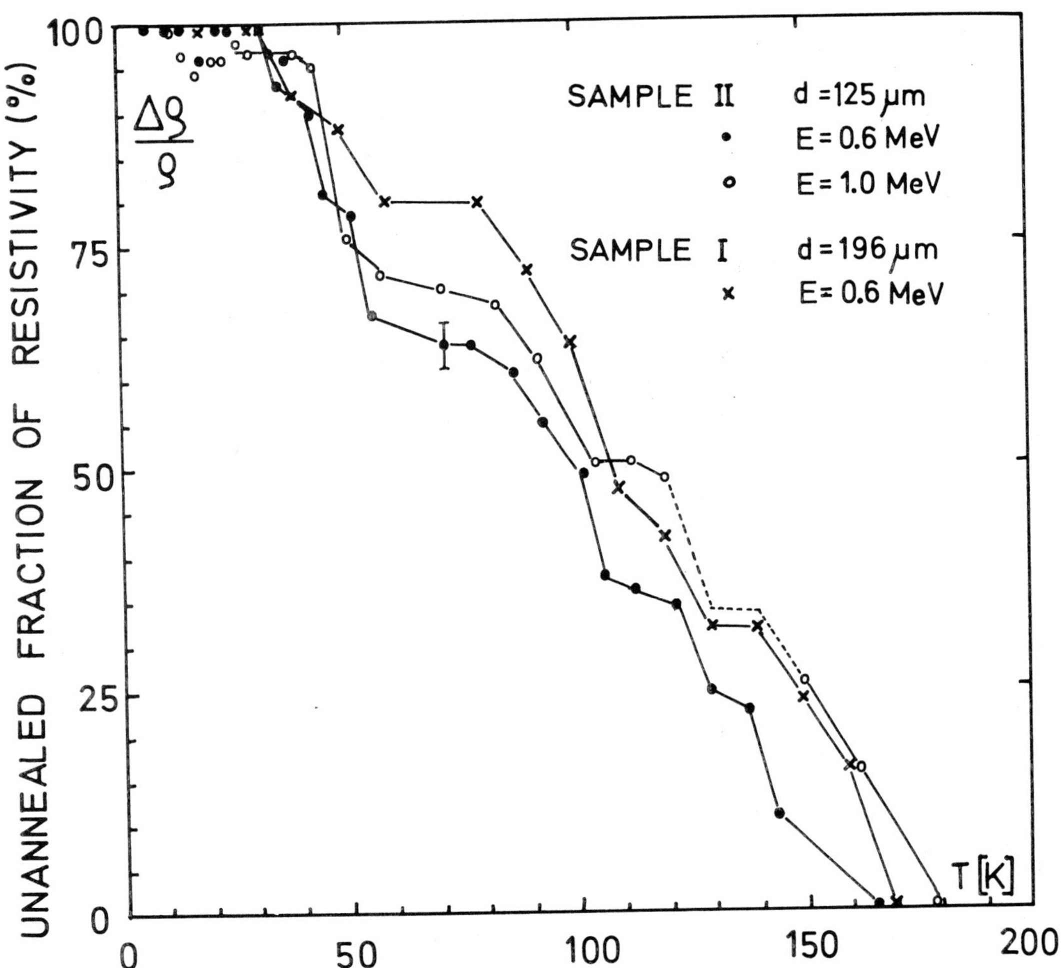

FIG. 2. Fractional recovery of radiation damage in resistivity caused by 0.6 MeV and 1 MeV electron irradiation as a function of temperature. Sample thickness $d = 125\,\mu$m and $d = 196\,\mu$m. Integrated electron flux ϕ (0.6 MeV) = 6×10^{14} electrons/cm². ϕ (1 MeV) = 1.7×10^{14} electrons/cm².

ation. They must have a very small activation energy or be located within the valence band, because they are ionized at 4.2 K. The hole generation rate was 0.09 cm⁻¹ and 0.47 cm⁻¹ for 0.6 and 1 MeV electrons, respectively. Van Lint et al.[1] obtained a value of 4.4/cm⁻¹ at 20 MeV. The number of defects per electron flux was calculated to be 1.1/cm and 4.4/cm for the two energies. This leads to the result, that 0.1 holes are generated per defect by 0.6 MeV electrons. The number for 1 MeV electrons is 0.13 holes/defect. It is interesting to note, that the same result can be deduced from the data of Van Lint and coworkers,[1] when a displacement cross section[2] of 400 barns is assumed for 20 MeV electrons. Because it is likely, that the defect associated with stage I at about 50 K is an

isolated Frenkel defect we have performed a few isothermal annealing experiments, in which the change of resistance was measured as a function of time for various temperatures in the vicinity of the recovery step. Assuming a first order annealing rate according to a relation

$$-\frac{1}{p}\frac{\mathrm{d}p}{\mathrm{d}t} = K \exp\left(-E_a/kT\right)$$

(p = excess hole concentration)

and neglecting the change in mobility, we obtain a reasonably good fit with an activation energy of $E_a = 190 \pm 40$ meV. The error is rather large because the main part of the annealing occurs in a very small temperature range (see Figure 4). The

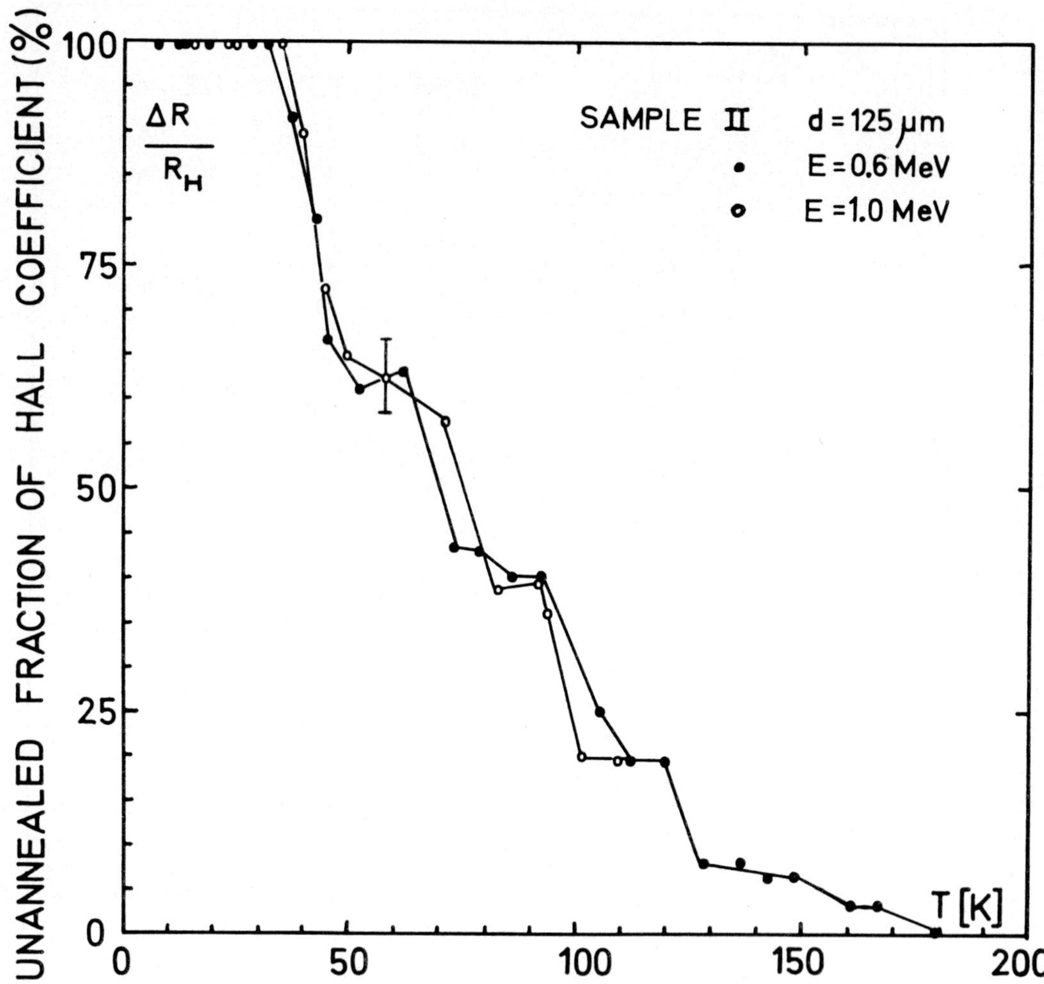

FIG. 3. Fractional recovery of radiation damage in Hall-coefficient R_H, caused by 0.6 MeV and 1 MeV electron irradiation as a function of temperature.

thermal activation energy has also been calculated with the method used by Overhauser,[3] yielding a value of $E_a = 160 \pm 30$ meV. It is obvious, that the defects which anneal at about 50 and 70 K, could not have been observed by Van Lint and co-workers,[1] disregarding for the moment that 20 MeV electrons most likely will create clusters, and not single point defects. It is also evident, that the lattice defects which we have studied are not identical with those, which are generated by thermal quenching.[4] For these defects hours are needed to anneal at room temperature.

Although it is probable, that the defects which were created by the low temperature irradiation of tellurium with 1 MeV electrons are vacancy-interstitial pairs, nothing can be said at present about the microscopic nature of the radiation damage. Much more has to be learned about lattice defects in tellurium, before one can even make educated guesses. There are indications, that interactions with dislocations are playing a role. After several runs with sample I one of the contacts broke loose. During the repair the sample was inadvertently bent and plastically deformed, which resulted in an increase of resistivity at 4.2 K of about 60 per cent. The hole density was increased by ~15 per cent. Subsequent irradiation with 0.6 MeV electrons decreased the carrier density, instead of increasing it. During warm-up the second recovery stage in the resistivity around

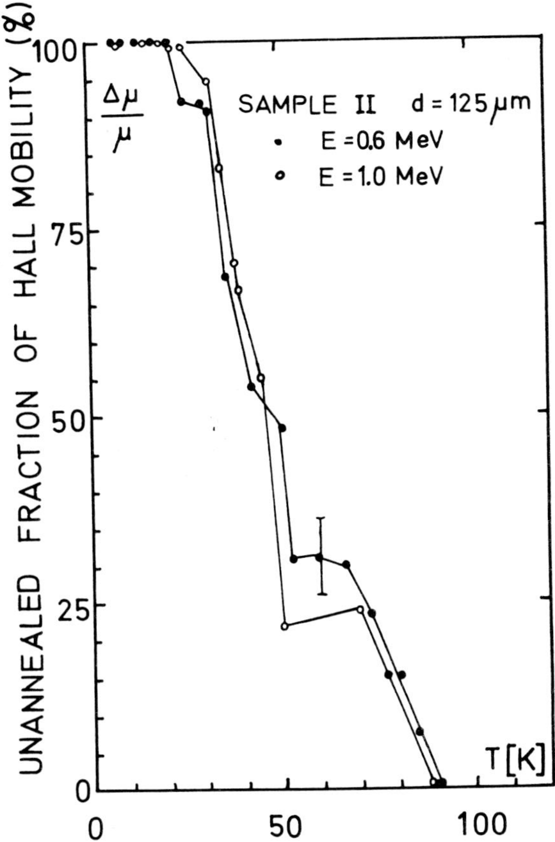

FIG. 4. Fractional recovery of change in Hall-mobility μ, caused by 0.6 MeV and 1 MeV electron irradiation as a function of temperature.

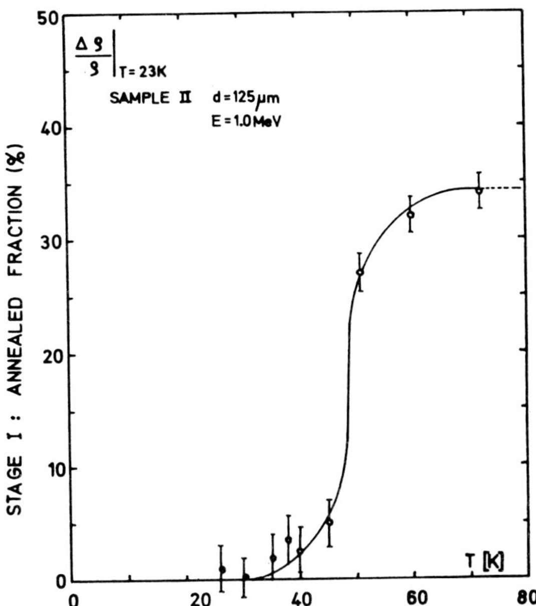

FIG. 5. Temperature dependence of isochronal annealing of stage I for sample II, irradiated with 1.7×10^{14} electrons/cm² of an energy of 1 MeV; reference temperature is the resistivity minimum at 23 K.

100 K was clearly missing. Because tellurium is extremely susceptible to lattice damage, this indicates again that extreme care is necessary during sample preparation.

A puzzling aspect of our data is the increase of the hole mobility after irradiation. The effect has been observed before by Van Lint and coworkers[1] and by Hermann and coworkers[4] on quenched samples. The increase can certainly not be explained by the peculiar band structure of tellurium. According to the presently accepted model, the constant energy surfaces consist of 4 ellipsoids of revolution close to point H of the Brillouin zone at hole energies characteristic for our samples. A much larger irradiation, leading to a carrier concentration of about 10^{17}/cm³ would be necessary to get the sample into a range, where the Fermi surface is changing its shape, and could give rise to different scattering. In addition, the number of introduced acceptors is not large enough, to change the amount if ionized impurity scattering appreciably. Moreover, additional scattering by charge centers should decrease, but not increase the mobility. Thus one is inclined to believe, that the interaction of radiation induced defects with other imperfections of the lattice is responsible for the increase in the hole mobility. It is interesting to note that only the defects associated with the first two recovery stages affect the mobility.

Our program to investigate radiation damage in tellurium has just been initiated, more extensive experiments are being planned. We intend to study the recovery steps in more detail and to investigate the damage caused by irradiation perpendicular to the trigonal axis and the influence of doping.

ACKNOWLEDGEMENTS

The authors would like to thank the 'Bundesministerium für Bildung und Wissenschaft' for sponsoring the project and the 'Deutsche Forschungsgemeinschaft' for travel support. We had interesting discussions with Dr. D. Harder.

REFERENCES

1. V. A. J. van Lint and H. Roth, *J. Appl. Phys.*, **30,** 1235 (1959). V. A. J. van Lint, E. G. Winkler and P. H. Miller, Jr., Proc. Int. Conf. Semicond. Phys., Prague (1960), p. 306.
2. O. S. Oen, Report ORNL—3813 (May 1968).
3. A. W. Overhauser, *Phys. Rev.*, **90,** 393 (1953).
4. K. H. Hermann, R. Link and W. D. Rentsch, *Phys. Stat. Sol.*, **8,** 719 (1965).

DISCUSSION

Question (ALBANY)   The variation of electrical resistivity ρ with the temperature of the sample before irradiation appears to exhibit the three temperature intervals. Could you comment on the similar observed behavior of ρ after irradiation?

Answer (GMELIN)   This variation is normal for tellurium and the dependence is observed both before and after irradiation.

THEORY OF AMORPHOUS SEMICONDUCTORS[†‡]

M. H. COHEN

James Franck Institute and Department of Physics, University of Chicago, Chicago, Illinois 60637, U.S.A.

The basic concepts underlying the present theory of amorphous semiconductors are reviewed. Emphasis is put on simple band models.

Amorphous semiconductors are of great current interest. They display novel and fascinating physical phenomena. There is relatively little well defined experimental data. The field is potentially very large with respect to the variety of materials and phenomena and the range of parameters available. The materials are technologically promising. The field as a whole is of fundamental theoretical interest.

We are concerned here with this theoretical interest. Amorphous semiconductors are examples of disordered materials. Ordered materials (crystals) have universal features to their electronic structures. Disordered materials similarly should. Experiments on amorphous semiconductors probe just those aspects of the electronic structure differing most from crystals. Theory and experiment on amorphous semiconductors are therefore very illuminating in relation to the larger subject of the electronic structures of disordered materials.

The most important elements in amorphous semiconductors are Si and Ge in Group IV; P, As, Sb, and Bi in Group V, and S, Se, and Te in Group VI. Other elements appear usually as minority constituents. The materials form covalent amorphous structures with excellent short-range and no long range order. They may be elemental, e.g., bulk S, Se, or Te or films of Si or Ge, compounds, e.g., bulk Ag_2Te_3 or multicomponent alloys. The structures are linear when Group VI elements dominate, planar for Group V, and three-dimensional networks for Group IV. The structures appear to approach that of an ideal covalent glass, excellent short-range order, a perfect network, no dangling bonds, every atom has its valence requirements satisfied.

The electrical conductivity σ is primarily intrinsic even for alloys of varying valences over broad ranges of composition. The optical absorption shows evidence for a band gap closely related to the activation energy in σ. These facts can be understood qualitatively in terms of the ideal glass concept and simple bonding considerations. Detailed understanding requires a band model.

Mott in 1967 synthesized our understanding of the electronic structures of disordered materials into the basic band model (BBM). There are bands of extended states with only short-range order in their phases. There are tails of localized states. There are sharp transitions at particular energies from localized to extended states.

The simplest extension of the BBM for amorphous semiconductors is a two band model having a valence band with localized states above E_v and a conduction band with localized states below E_c. In a crystal, intrinsic behavior of σ follows from the sharp band edges. Here, to get intrinsic behavior, we must assume in addition a sharp drop in the mobility at E_c and E_v for the localized states. E_v and E_c are mobility edges, and E_c-E_v is the mobility gap. It is the mobility gap which enters σ. Recent experiments of Donovan, Spicer and colleagues and of Spears can be interpreted in terms of the concept of an ideal glassy structure and the simple band model.

Non-ideal glasses can have well defined imperfections which introduce nonmonotonicity into the density of states in the tails. Increased randomness as in alloys; films deposited rapidly, or on a cold substrate, or evaporated at too low a temperature; or liquids can lead to overlap of tails.

† A lecture covering the same material was delivered to the Tenth Annual International Conference on the Physics of Semiconductors, the text of which will appear in the Proceedings of that conference. What appears here is an abbreviated version. For details and references, see Proc. Tenth Ann. Int. Conf. Phys. Semiconductors.

‡ This work was supported in part by the Office of Naval Research and benefited from general support of Materials Science at the University of Chicago by ARPA.

This pins the Fermi level, gives rise to hopping conduction and charged traps.

Nine independent assumptions can be identified in these band models. Of these the existence of bands of extended states with tails of localized states may be regarded as secure. The proof of the existence of sharp energies of transition from localized states which in addition are mobility edges is much more difficult. It rests on Anderson's 1958 demonstration of the absence of diffusion in certain random lattices. This difficult paper has been clarified by Ziman in 1969, Thouless in 1970, and most recently Economou and Cohen have arrived at a nearly complete proof of the existence of mobility edges. There is little room for doubt, and now the theory is progressing to approximate calculation of mobility edge positions.

The next question is the energy dependence of the mobility above the mobility edge. Fluctuation theory, classical percolation theory, tunnelling arguments, analogies with phase transitions, and other arguments suggest the possibility that $\mu(E) \propto (E - E_c)^s, s < 1, E > E_c$ while $\mu(E) = 0, E < E_c$ for $T = 0$. However, this point is controversial.

Finally, the most immediate outstanding problems relate to actual calculation of σ and similar quantities such as optical absorption and to more accurate characterization of the localized states.

DISCUSSION

Question (STONEHAM)   In your talk you commented that the conductivity of amorphous materials is insensitive to irradiation. Is this a prediction, or is this well founded experimentally?

Answer (COHEN)   There is ample experimental evidence on this effect—studies carried out under neutron and pulse gamma radiation conditions.

Question (NAGELS)   Some authors have described the electrical properties of amorphous semiconductors in terms of a simple model of a highly compensated semiconductor. In this way they claimed to be able to explain some particular characteristics of amorphous semiconductors, e.g., the low Hall mobility and the sign anomaly. What do you think of the validity of such a simplified model?

Answer (COHEN)   Such models can explain the low Hall mobility and the sign anomaly only in a limited temperature range without highly artificial use of parameters. With natural parameters, they would find a Hall reversal and ordinary size Hall mobility, with strong temperature dependence, in contrast to the observations.

Question (BISHAY)   As you may know, some multi-component glasses which contain very high concentrations of transition elements, e.g., Fe, show semiconducting properties. Can you apply your theory to these glasses?

Answer (COHEN)   The present theory would have to be augmented to include the effects of local moments on the transition metal ions before it would be directly applicable to such glasses.

14 MeV-NEUTRON IRRADIATION OF *p*-TYPE GERMANIUM AT 77 °K

D. WOLF

Institut für Reine und Angewandte Kernphysik der Universität, Kiel, 23 Kiel, Olshausenstrasse 40–60, Germany

P-type germanium has been irradiated at 77 °K by monoenergetic 14 MeV-neutrons, and the change of resistivity and Hall-coefficient has been measured. Subsequently isochronal annealing experiments have been performed. The results can be explained by the assumption that by energetic neutrons disordered regions are created, which decay by annealing between 77 °K and 230 °K, probably forming well-conducting zones. This decay is superposed by the annealing behaviour of a type of defect, which is observed as well after electron irradiation.

1. INTRODUCTION

In several neutron irradiated semiconductors disordered regions have been observed.[1-3] These are highly damaged zones, surrounded by space charge regions. In *n*-type germanium the mean radius of a damaged zone is $r_1 \approx 100 \,\text{Å}$, and the mean radius of the surrounding space charge is some 1000 Å. Generally it is assumed that the damaged zones have the same properties as bulk material, which has been irradiated by very high neutron fluences. So silicon becomes intrinsic and germanium becomes *p*-type at room temperature. Since the space charge region is usually regarded as an essential feature of the disordered regions, at room temperature no such regions are expected to be found in *p*-type germanium.

By exposures at 77 °K, however, *n*- and *p*-type germanium becomes intrinsic at high fluences.[4] So it is suggested that at this temperature disordered regions are created. Intrinsic means that the Fermi-level has shifted to near the middle of the bandgap. (According to investigations of Cleland, Crawford, and Pigg,[5] the Fermi-level moves by irradiation independently of temperature to a limiting level, which lies 0.12 eV above the valence band. Their experiments have been performed only down to dry ice temperature. As will be shown, nearly all annealing has taken place below this temperature. So their observations are not in contradiction to the above statement.) Therefore the height of the potential barrier between the damaged zone and the undisturbed lattice will be comparable in either types of germanium. As this potential difference ψ is directly proportional to the volume V of a disordered region, the damage rate, e.g. the measured carrier removal rate should amount to the same order of magnitude in *n*- and *p*-type germanium. This is different from observations in electron-irradiated material (3.5 MeV-electrons), where the carrier removal rate of *n*-Ge is about 20 times that of *p*-Ge.[6,7]

Pulse annealing experiments on germanium bombarded with reactor neutrons have been reported by several authors.[4,5,8] At 77 °K the carrier concentration of *p*-type germanium decreases during irradiation. Pulse annealing leads to a further decrease up to 150 °K Between 150 °K and 230 °K the hole concentration increases, and overshoots a little the pre-irradiation value. More recent investigations[9,11] indicate a similar behaviour of electron irradiated material. Therefore it is suggested that the reverse annealing stage in reactor-neutron irradiated material is due to small defects as can be created as well by energetic electrons.

Further informations should be obtained by irradiations with monoenergetic 14 MeV-neutrons, because these are supposed to produce mainly disordered regions, whereas all kinds of defects will be produced by reactor-neutrons because of their broad energy spectrum.

2. EXPERIMENT AND RESULTS

Samples listed in Table I were irradiated at 77 °K by monoenergetic 14 MeV-neutrons. The neutron flux was $\sim 10^8 \,\text{cm}^{-2}\,\text{s}^{-1}$. The samples remained in the dark for the whole time. The changes of the 77 °K-resistivity and Hall-constant were recorded. For comparison the data of an *n*-type sample are also given in Table I. The observed electron removal rate of $\Delta n / \Delta \phi = 8 \,\text{cm}^{-1}$ has also been reported by Curtis and Cleland.[12] In agreement

D. WOLF

TABLE I

The investigated samples, radiation induced changes, and dates concerning the disordered region model

Sample	Type	Carrier concentration [cm^{-3}]	Neutron fluence [nvt]	$\Delta p/p$ [%]	$\Delta p/\phi$ [cm^{-1}]	f_ρ [%]	f_{AH} [%]	r_2 [Å]	ψ [V]
P I	p	2.2×10^{14}	2×10^{12}	4.8	4	4.6	6.3	3800	0.4
			3×10^{12}	4.3	3	4.9	5.7	3200	0.3
VK7/395	p	1.7×10^{14}	2×10^{12}	2.9	3	3.8	3.8	3300	0.2
P III	p	1.4×10^{14}	2×10^{12}	3.1	3	5.3	4.2	3500	0.2
N V	n	2.7×10^{14}	2×10^{12}	6.7	8	9.1	8.9	4400	0.7

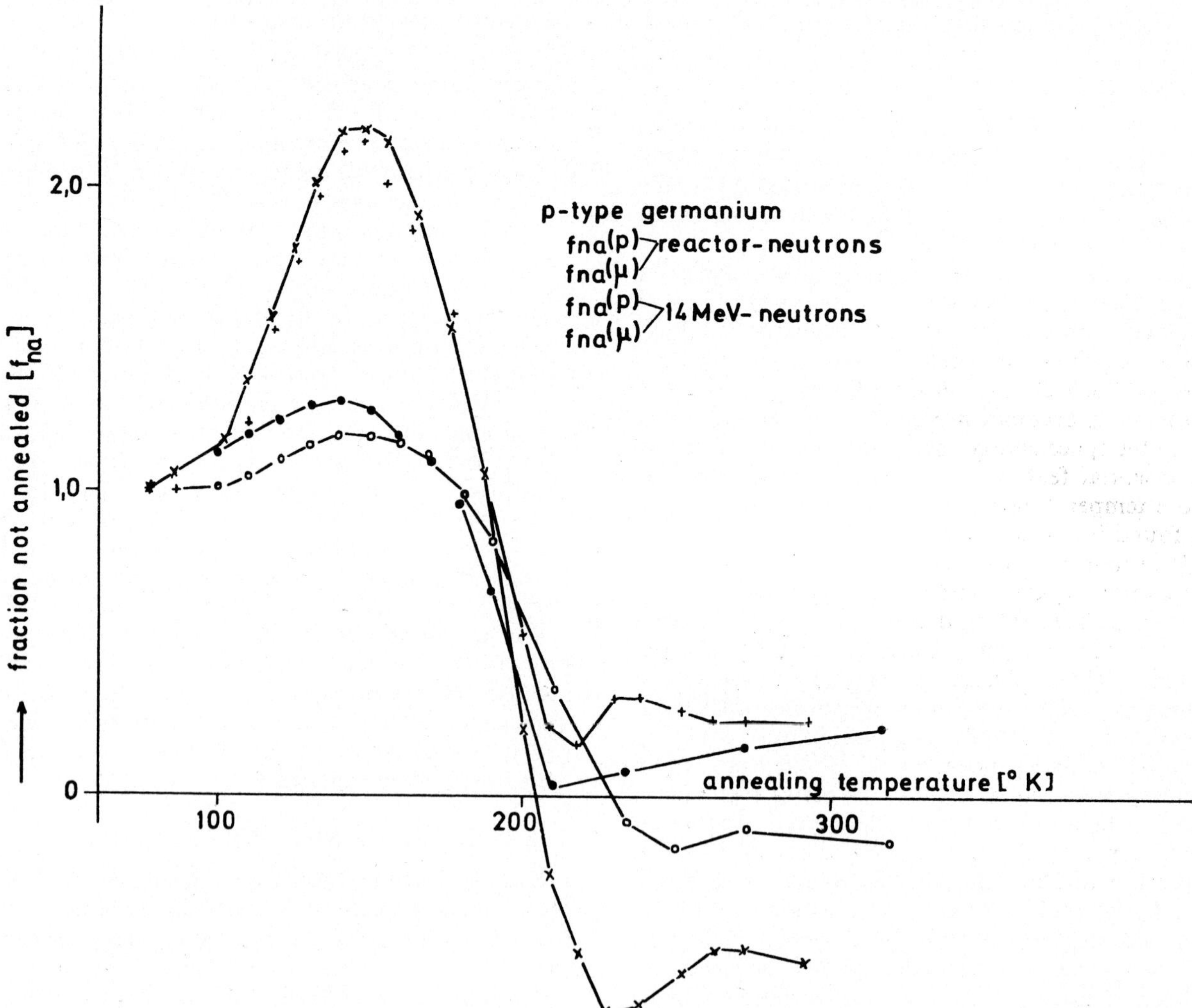

FIG. 1. 10-minutes isochronal anneal of hole concentration p and mobility μ of neutron-irradiated p-type germanium.

with the above considerations the hole removal rate is nearly one half of the electron removal rate.

Assuming that disordered regions do exist, the fraction f of the total volume occupied by the regions has been determined in the usual way by measurement of resistivity before and after irradiation using the relation[13]

$$f_\rho = \frac{\rho_{\text{after}} - \rho_{\text{before}}}{\rho_{\text{after}} + \rho_{\text{before}}/2}.$$

Furthermore it has been computed from the change of the Hall-constant, using

$$f_{AH} = \frac{A_{\text{after}} - A_{\text{before}}}{A_{\text{after}} - A_{\text{before}}/4}.$$

These two independently obtained values agree on the average within 15 per cent. This could be regarded as an additional support of the disordered region hypothesis.

The values of the potential difference ψ have been obtained using the equations[1,2]

$$f = V \cdot \Sigma \cdot \phi \quad \text{and} \quad V = \frac{4\pi}{3} r_2^3 = \frac{4\pi\epsilon\psi}{qN_A} r_1$$

where V is the volume of a disordered region; $\Sigma = 0.13\ \text{cm}^{-1}$ is the macroscopic scattering cross-section for fast neutrons in germanium, ϵ the dielectric constant of Ge, q the electron charge, and N_A is the acceptor concentration of p-Ge. It has been assumed $r_1 = 100\ \text{Å}$ in accordance with electron microscope observations.[3]

The result that $\psi(n\text{-Ge}) \geqslant 2\psi(p\text{-Ge})$ indicates that the damaged zone is not exactly intrinsic, but is still weakly p-conducting. The hole concentration at 77°K corresponding to a Fermi-level 0.2 eV above the valence band is only $\sim 6 \times 10^4\ \text{cm}^{-3}$. It may be, however, that this weak p-conductivity is responsible for the lack of the strong photo-conductivity which is observed in neutron irradiated silicon[14] and n-type germanium,[12] and which is a phenomenon of the disordered regions.[14,15]

After irradiation the samples had been subjected to 10 minutes isochronal anneals. The results are

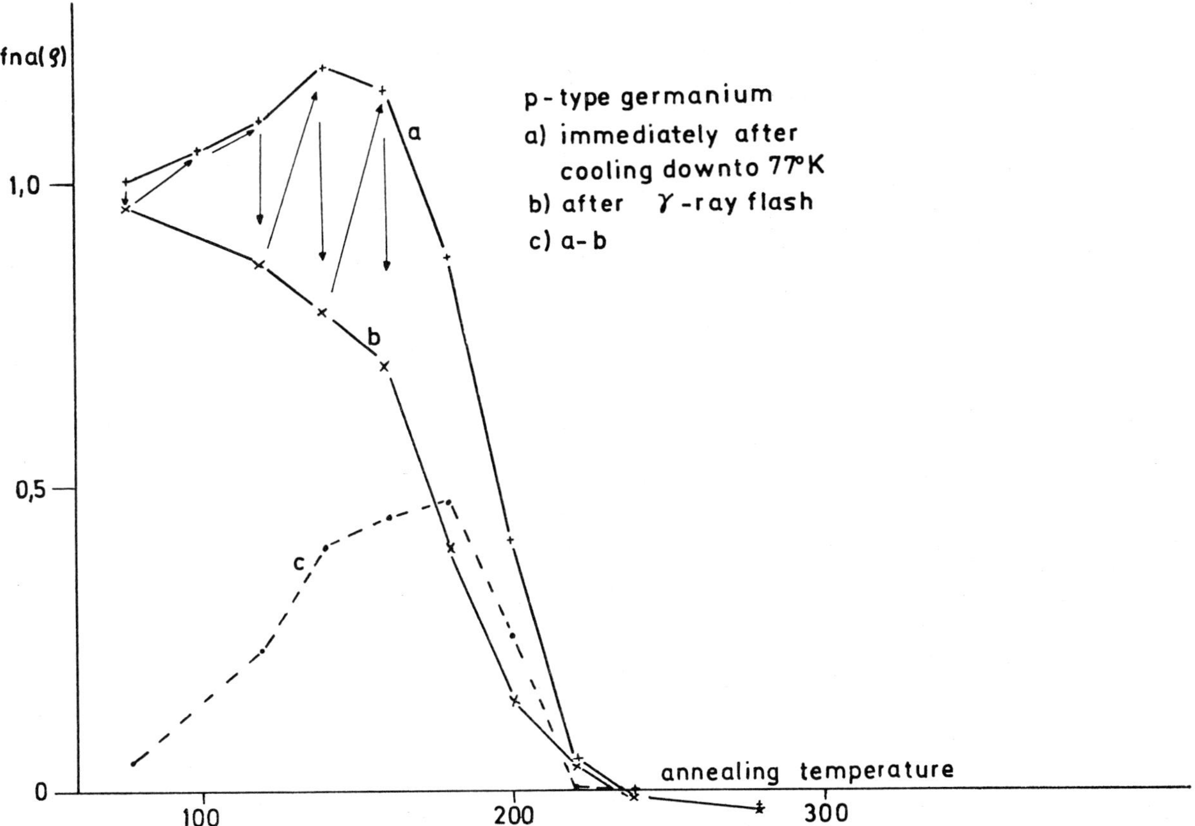

FIG. 2. Isochronal annealing. After measurement of each point of curve *a* the sample was exposed to an X-ray flash. The arrows indicate chronological sequence of measurement.

presented in Figure 1. Here the fraction not annealed of hole concentration $f_{\mathrm{na}}(p)$ and of mobility $f_{\mathrm{na}}(\mu)$ is plotted against annealing temperature. For comparison values obtained after reactor-neutron irradiation[8] are inserted. The curves have the same characteristics, only the amplitudes are higher after reactor-neutrons.

This behaviour is in agreement with the conclusion that the reverse annealing is due to small defects. If reactor-neutrons produce all kinds of defects, and monoenergetic 14 MeV-neutrons pro-duce mainly disordered regions, the reverse annealing should be higher after neutron irradiation, and this in fact can be seen in Figure 1.

The increase and the decrease of the $f_{\mathrm{na}}(p)$-curve are governed by first order processes[8] with activation energies $E_{\mathrm{incr}} = 0.1$ eV and $E_{\mathrm{decr}} = 0.2$ eV. In electron-irradiated p-type germanium the same order of kinetics has been observed. Stas' and Smirnov[11] determined activation energies $E_{\mathrm{incr}} = 0.19$ eV and $E_{\mathrm{decr}} = 0.58$ eV, whereas Flanagan and Klontz found $E_{\mathrm{incr}} = 0.1$ eV and $E_{\mathrm{decr}} = 0.4$ eV.

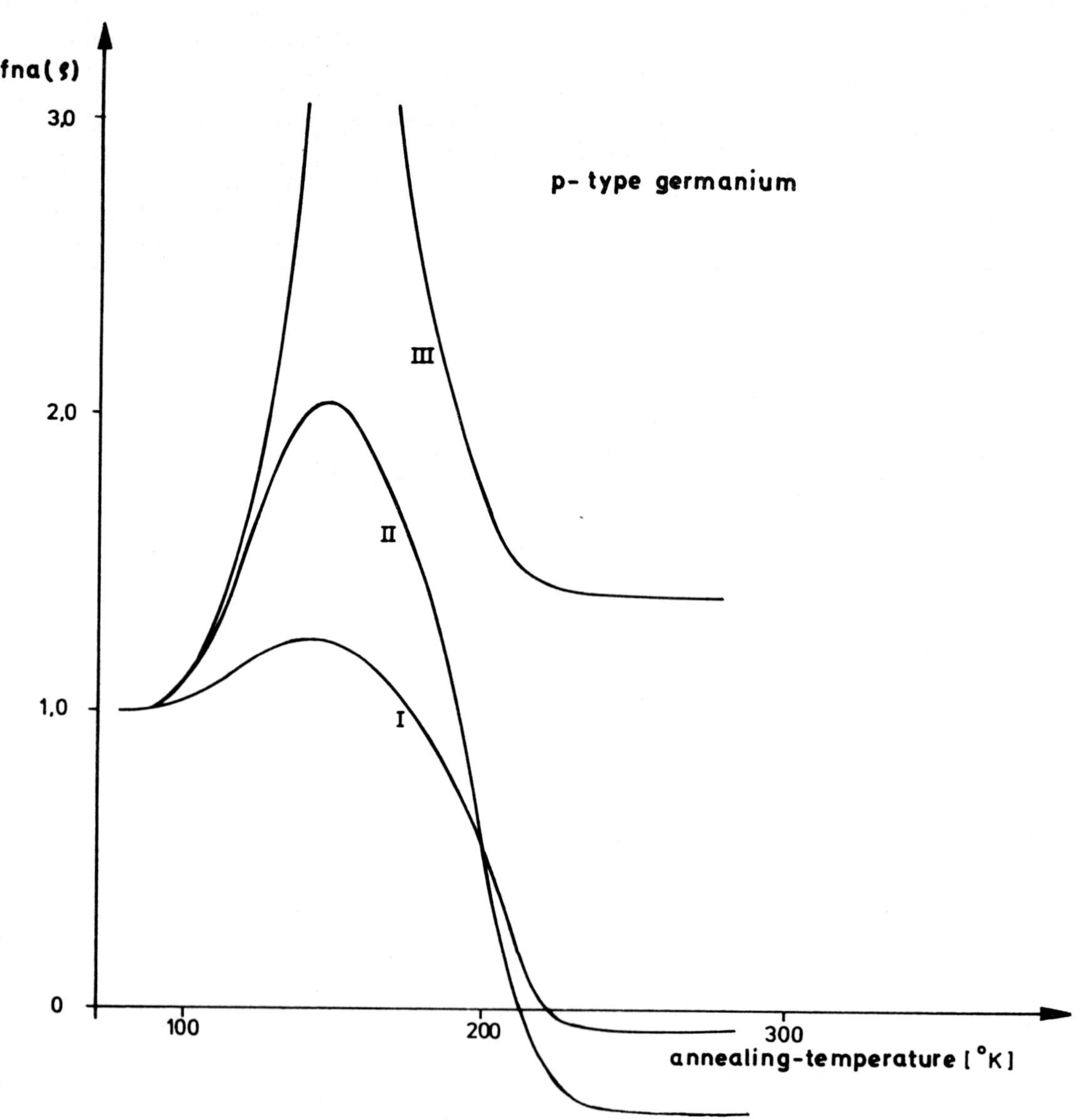

FIG. 3. Schematic graph of annealing curves after (I) 14 MeV-neutrons, (II) reactor-neutrons, (III) ~3 MeV-electrons.

The difference of the values may arise from the different methods, how they have been determined. Unfortunately these methods have not been described in the publications. In electron-irradiated material, the defect responsible for the reverse annealing peak changes its load by ionizing radiation and by thermal stimulation. Details are described elsewhere.[10] In order to demonstrate that the same type of defect is present after neutron bombardment the following procedure was performed: after each annealing pulse the sample was exposed to an X-ray flash (Febetron X-ray flash system, model 705). Thereby the resistivity sank to a certain lower value and did not sink by further flashes. (That means one flash was strong enough to lead to saturation.) When the sample was reheated, the previous value reappeared. This was the expected behaviour due to the defect in question.

In Figure 2 the values before and after flash are graphed. Curve (a) shows the values before flash; it is identical with an annealing curve, which would be obtained if no ionizing radiation would be applied to the sample. Curve (b) shows the values after flash. The difference between these two values are drawn in curve (c), which therefore indicates the magnitude of the defect.

The annealing curve of neutron-bombarded *p*-type germanium shows, however, one marked difference to the annealing curve of electron-irradiated material, as is schematically visualized in Figure 3. After the annealing of the electron-irradiated material, the resistivity is still higher than it was after exposure before annealing, whereas the resistivity of the neutron-bombarded samples is even less, than it was before exposure. Therefore an additional process must be superposed in the latter. This process is probably the decay of the disordered regions. It must not simply be seen as a decomposition of the damaged zones into small clusters or single defects. In all probability the primarily intrinsic zones convert into well-conducting *p*-type clusters by thermally activated rearrangement of the defects. This will give a possible explanation for the overshooting of the carrier concentration over the pre-irradiation value.

REFERENCES

1. J. H. Crawford and J. W. Cleland, *J. Appl. Phys.*, **30**, 1204 (1959).
2. B. R. Gossick *J. Appl. Phys.*, **30**, 1214 (1959).
3. M. Bertolotti, in *Radiation Effects in Semiconductors*, Ed. F. L. Vook (New York, 1968), p. 311.
4. R. F. Konopleva and S. R. Novikov, *Soviet Physics—Solid State*, **6**, 819 (1964).
5. J. W. Cleland, J. H. Crawford and J. C. Pigg, *Phys. Rev.*, **99**, 1170 (1955).
6. J. Kortright, Ph.D. thesis, Purdue University, 1963 (unpublished, cited by T. A. Callcott and J. W. MacKay, *Phys. Rev.*, **161**, 698 (1967)).
7. M. Persin, *Period. Math.—Phys. Astron.* (Yugoslavia), **20**, 277 (1965).
8. G. Lautz and D. Wolf, *Atomkernenergie*, **11**, 457 (1966).
9. H. Saito, N. Fukuoka, H. Hattori and J. H. Crawford, in *Radiation Effects in Semiconductors*, p. 232.
10. T. M. Flanagan and E. E. Klontz, *Phys. Rev.*, **167**, 789 (1968).
11. V. F. Stas' and L. S. Smirnov, *Sov. Ph.—Semicond.*, **2**, 1147 (1969).
12. O. L. Curtis and J. W. Cleland, *J. Appl. Phys.*, **31**, 423 (1960).
13. H. J. Juretschke, R. Landauer and J. A. Swanson, *J. Appl. Phys.*, **27**, 838 (1956).
14. H. J. Stein, *IEEE Trans. Nucl. Sci.*, **NS-15**, No. 6, p. 69 (1968).
15. B. L. Gregory, *Appl. Phys. Letters*, **16**, 67 (1970).

DISCUSSION

Comment (MITCHELL) Clark, Stewart and I have examined the long wavelength (6–16 Å) neutron scattering from germanium reactor irradiated to fast neutron dose of 1.1×10^{20} n/cm² measured on a Ni scale. We observe a peak near 10 Å in the defect scattering cross-section. This peak is characteristic of a defect together with relaxation of surrounding atoms. The only fit we have been able to obtain for various strains around interstitial, vacancies, divacancies, pentavacancies and higher complexes, is for the *divacancy*. Even then we have to make *critical* assumptions about the relaxation—

(1) that it is large—nearest neighbors (N.N.) relax inwards by ~20 per cent;
(2) that the N.N.N. from the ends of the divacancy relax in, but those lying along the length relax out;
(3) that the third N.N. from the mid-point of the divacancy relax in; and
(4) that the fit to the experimental results is not improved by going beyond third N.N.

The long wavelength neutrons originate from a liquid H_2/D_2 moderator of 5 liter placed alongside the core of the 5 MW reactor.

PRODUCING DISORDERED REGIONS IN GERMANIUM BY 28 MeV ELECTRONS

N. A. UKHIN AND A. K. ABIEV

Kurchatov Institute of Atomic Energy, Moscow, Russia

SUMMARY

Electrons with energy more 20 MeV are capable of transferring to the germanium atoms sufficient energy to produce the complicated defects like disordered regions in neutron irradiated semiconductors.

Irradiation of n-type germanium samples with different dopants and different resistivity was carried out with 28 MeV electrons from the Kurchatov Institute Linear Accelerator.

The Hall effect and conductivity were the parameters measured prior and during irradiation at 88 °K and after each annealing step.

It was observed that not only conductivity but Hall mobility of majority carriers decreased greatly during irradiation at low temperature and after annealing at room temperature. The initial rate of decrease of the Hall mobility does not depend on dopant, but it depends on resistivity of the samples.

The initial rate of decrease of the relative Hall mobility varied inversely to the majority carrier concentration corresponding to the Gossick model of disordered regions. This permitted calculation by computer of the most probable range of average threshold energy for producing disordered regions in germanium. It was from 15 keV to 19 keV.

Special calculations show that the low and high limits of the average threshold energy range depends very weakly on the average acceptor concentration inside disordered regions.

Annealing from 88 °K to the room temperature shows that the fraction of the total volume occupied by disordered regions decreases but the total concentration of point defects increases.

In particular the concentration of energy levels $E_c - 0.2$ eV increases also. It does seem reasonable to draw on the following conclusions.

The p-type disordered regions are produced in germanium in the low temperature 28 MeV electron irradiation. The disordered regions are one of the main causes of the resistivity and Hall mobility change. The increase in the total concentration of the acceptor type point defects during annealing to room temperature is induced by the decay of periphery layers of the disordered regions. If one assumes the acceptor type defects inside disordered region are not annihilated during annealing one can estimate the concentration this defects inside disordered regions. The concentration found was 3×10^{17} cm^{-3} at threshold energy equal 15 keV and 2×10^{16} cm^{-3} at threshold energy equal 19 keV.†

† This paper was accepted but not presented at the conference.

THE PROCESSES OF FORMATION AND ANNEALING OF RADIATION DEFECTS IN GERMANIUM, IRRADIATED BY HIGH ENERGY PROTONS

R. F. KONOPLEVA, S. R. NOVIKOV AND E. E. RUBINOVA

A. F. Ioffe Physico-Technical Institute Academy of Science of the U.S.S.R., Leningrad

The rate of damage formation, the spectrum of energy levels and the annealing results on n- and p-type Ge irradiated by 660 MeV protons were studied and compared to similar results on γ-irradiated samples. Reverse annealing (and defect creation) is found upon the decay of disordered regions. The formation upon annealing of complexes with impurities is also deduced.

1. INTRODUCTION

The rate of damage formation, the spectrum of energy levels and the annealing results on n- and p-type Ge irradiated by 660 MeV protons were studied and compared to similar results on γ-irradiated samples. Reverse annealing (and defect creation) is found upon the decay of disordered regions. The formation upon annealing of complexes with impurities is also deduced.

2. EXPERIMENTAL DATA

For this study use was made of n-Ge specimens with the initial concentration of Sb being 2.10^{15} cm^{-3} and p-Ge with 4.10^{14} cm^{-3} Ga.

The specimens were irradiated in the accelerator by 660 MeV protons with various integral fluxes (from 1.10^{11} up to 1.10^{14} p/cm^2). The reaction ^{27}Al $(p, 3pn)$ ^{24}Na was used to measure the dose on the specimen.

2.1. *The rate of defect formation*

The initial rate of removal (or of introduction) of free current carriers was taken as the rate of defect formation. The rate of carrier removal was determined from the temperature dependence of the Hall constant for different integral proton fluxes. The rate of current carrier removal (introduction) determined for a number of n- and p-Ge specimens proved to be 10^2 cm^{-1}.

The conductivity was found to decrease in n-material and to grow in p-material. It enabled us to suggest that in the course of Germanium being irradiated by high energy protons acceptor levels are mostly introduced.

We tried to compare the experimental value of the defect formation rate with the number of displaced atoms, obtained by calculation in order to estimate the contribution to the defect formation due to the Rutherford scattering and nuclear elastic and inelastic interaction in the case of Germanium irradiated by 660 MeV protons. The calculation of the total number of displaced atoms was carried out according to the Kinchin–Pease displacement theory.[1] For 660 MeV protons, the data calculated for the total number of displaced atoms in the case of the Coulomb scattering are presented in Table I.

TABLE I

Data calculated for the number of displaced atoms in Ge, irradiated by 660 MeV protons.

Parametres	Values of parameters	References
σ_n	$0.9 \cdot 10^{-24}$ cm^2	
$E(E_{ne})$	120 KeV	
σ_{ne}	$0.6 \cdot 10^{-24}$ cm^2	(4)
N_k	12 cm^{-1}	
N_{ne}	88 cm^{-1}	(1)
N_{ne}	70 cm^{-1}	(4)
σ_{ni}	$0.75 \cdot 10^{-24}$ cm^2	(5)
$E(E_{ni})$	10^3 KeV	(4)
N_{ni}	10^3 cm^{-1}	(1)
σ_k	$90 \cdot 10^{-24}$ cm^2	—
$E(E_k)$	225 eV	—

In the case of heavy charge particles with energy exceeding 100 MeV, nuclear interactions begin to play the deciding part in the formation of displaced atoms.[2,3] Then the total number of displaced atoms is determined by two components: the Coulomb scattering and nuclear interaction.

$$N_{\text{compl.}} = N_k + N_n = \frac{N_0 \varphi t}{2E_d} [\sigma_k \bar{E}(\bar{E}_k) + \sigma_n \bar{E}(\bar{E}_n)]$$

where N_0 is the number of atoms per unit crystal volume, φ—the density of bombarding particles flux, t—irradiation time, σ_k, σ_n are effective cross-sections due to Coulomb scattering and nuclear interaction, $\bar{E}(\bar{E}_k)$ and $\bar{E}(\bar{E}_n)$ are average energies, transmitted by the recoil nucleus to the surrounding atoms due to Coulomb scattering and nuclear interaction. The value N_n in its turn is also determined by two components, associated with nuclear elastic and inelastic interaction

$$N_n = N_{ne} + N_{ni}$$
$$= \frac{N_0 \varphi t}{2E_d}[\sigma_{ne}\bar{E}(\bar{E}_{ne}) + \sigma_{ni}\bar{E}(\bar{E}_{ni})] \qquad (2)$$

The value σ_{ne} and $\bar{E}(\bar{E}_{ne})$ was taken from reference (4).

In the case of nuclear inelastic interaction the value $\bar{E}(\bar{E}_{ni})$ can be also determined from reference (4), since it is known that the average energy of recoil nuclei due to different secondary particles, emitted during the typical fission, is 10 MeV.[2] Unfortunately, at present there are no available data on cross-sections of inelastic interaction between 660 MeV protons and Ge nuclei. However, in a first approximation, in order to estimate σ_{ni} we made use of the value for copper which is equal to $0.75 \cdot 10^{-24}$ cm².[5] In this case the N_{ni} value is $1 \cdot 10^3$ cm⁻¹. The values σ_{ne}, $\bar{E}(\bar{E}_{ne})$, σ_{ni}, $\bar{E}(\bar{E}_{ni})$, N_k, N_{ne}, N_{ni} for Germanium, irradiated by 660 MeV protons, are given in Table I.

From the Table and the expressions (1) and (2) it can be seen that the total number of displaced atoms is mainly determined by nuclear interactions. Indeed, for the values σ_k, σ_{ni} and σ_{ne} of the order of 1 B the value $\bar{E}(\bar{E}_k)$ is by three orders less than $\bar{E}(\bar{E}_{ne})$ and by four orders less than $\bar{E}(\bar{E}_{ni})$. Therefore, the value N_k, associated with the Rutherford scattering, compared with N_n due to nuclear interaction, is negligible.

Thus, the value of the number of displaced atoms, obtained by calculation, is by an order larger than the experimental value. The discrepancy observed can be apparently explained, as well as is the case of Germanium irradiated by fast neutrons, as due to the formation of disordered regions, which screen the defect carrier capture within these regions.

2.2. The energy spectrum of defect levels

The energy spectrum of defect levels was determined from the temperature dependence of the Hall constant and spectral curves of impurity photo-conductivity. From the Hall constant curves six local levels were found in the forbidden gap for a number of specimens of n-Ge, irradiated by different integral proton fluxes. The levels $E_c - 0.075$ and $E_c - 0.20$ eV were detected in n-Ge, and $E_v + 0.26$, $E_v + 0.20$, $E_v + 0.13$ and $E_v + 0.065$ eV in the Germanium, converted into p-type. The complete spectrum of levels, determined from the Hall constant temperature dependence, is shown in Figure 1(a).

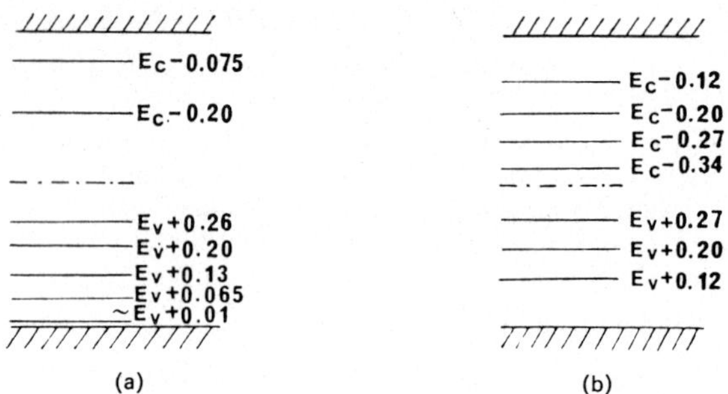

FIG. 1. Spectrum of energy defect levels in Ge, irradiated by protons according to the data from the Hall's measurements (a) and photoconductivity (b).

We also tried to estimate the rate of the introduction of the levels observed. Figure 2 schematically represents the relative efficiency of the introduction of the levels, which was found from the Hall measurements. As seen from Figure 2, the levels situated close to bands are introduced at considerably higher rates than the deep levels.

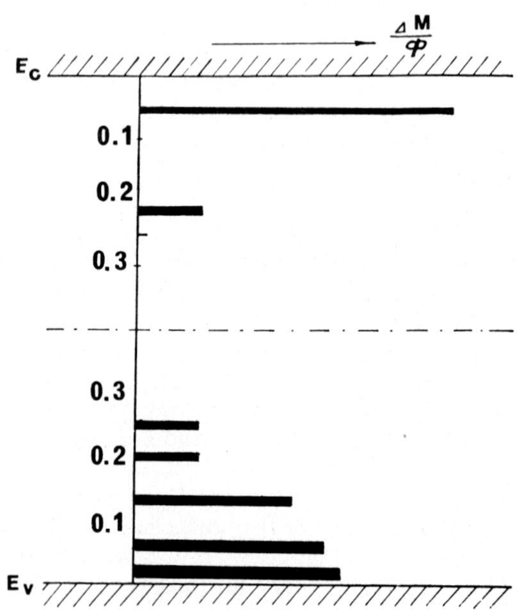

FIG. 2. Relative introduction level efficiency, found from the Hall's measurements.

The energy spectrum of defect levels was also determined from spectral curves of impurity photoconductivity. The complete diagram of the levels, found from spectral curves of impurity photoconductivity is presented in Figure 1(b).

The comparison between the activation energy values for impurity levels obtained from the measurements of impurity photoconductivity and the data of the Hall's measurements shows a sufficiently good coincidence of the position of levels.

2.3. *The high temperature annealing of defects*

The Ge specimens investigated were subjected to high temperature isochronal and isothermal annealing. The annealing was performed in air within the temperature range 100–400 °C. After each annealing stage the measurements were made of conductivity, the Hall constant temperature dependence and spectral distribution of impurity photoconductivity. The activation energy at each annealing stage was determined by the angle coefficients method.[6] Further, it was supposed, that the kinetics of radiation defects could be described by the equation

$$\frac{\mathrm{d}n}{\mathrm{d}t} = K_1 n^2 + K_2 n \qquad (3)$$

where K_1 and K_2 are the averaged coefficients of bi-polar and monopolar annealing processes and defect formation in the case of heating the crystal. The coefficients K_1 and K_2 were determined by the least-square variance method using the electronic computer.

Figure 3 (a, b) show the curves of the change in conductivity and mobility of specimens of n-type and p-Ge converted irradiation in the process of isochronal annealing.† From the curves presented in Figure 3 (a, b) it can be seen, that in n-Ge two annealing stages are observed, and in the case of converted p-Ge three annealing stages. The activation energy of the first stage (100–180 °C) is ∼0.7 eV for p-type‡ according to the data obtained by the angle coefficients method and it is equal to ∼0.62 eV according to the data obtained from the electronic computer for monopolar and bi-polar processes.

The second annealing stage (180–250 °C) is

---

† The reverse conversion temperature lies within the range of 250–300 °C, being directly dependent on the concentration of defects introduced.

‡ We failed to determine the value of the activation energy for n-type within this temperature region.

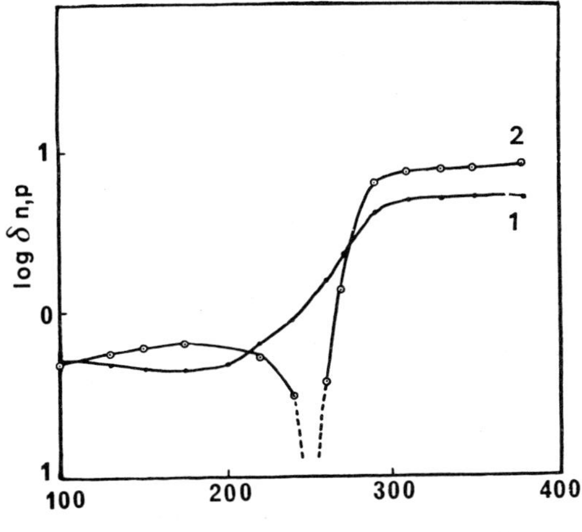

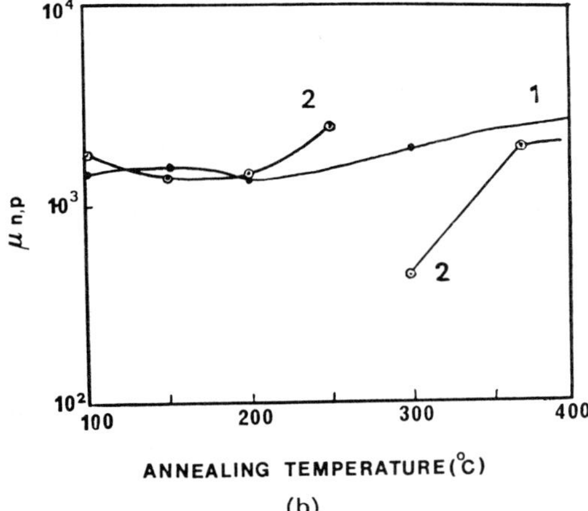

FIG. 3. Changes of conductivity (a) and the Hall's mobility (b) of Ge, irradiated by 660 MeV protons in the course of isochrone annealing (measured at 300 °C)
1. n-type,
2. p-type converted by irradiation.

characterized by the activation energy ∼1.3 eV for p-type and ∼1.5 eV for n-type.

The third annealing stage lies within the temperature range 250–400 °C and is characterized by the activation energy ∼2.4 eV.

Table II shows the spectrum of levels, revealed in the process of annealing, which was obtained from the study of impurity photoconductivity and the Hall's constant temperature dependence. It can be seen from the Table, that within the temperature range investigated all the spectrum of radiation defect levels, produced in Ge by 660 MeV

TABLE II

Activation energy and level spectrum of radiation defects

Type of conductivity after irradiation	Annealing stage	Annealing temperature °C	Activation energy eV n-type	Activation energy eV p-type	Level spectrum, observed at different stages of annealing and conversion — Levels eV	Time and temperature °C of annealing	Level spectrum of radiation defects, introduced into n-Ge, irradiated by 660 MeV protons
n, p	1	100–180	—	0.66 0.77	$E_c - 0.12$ $E_c - 0.28$	After annealing	$E_c - 0.12$
					$E_c - 0.12$ $E_c - 0.2$ $E_c - 0.28$	60 min, 100	$E_c - 0.20$ $E_c - 0.27$
					$E_c - 0.12$ $E_c - 0.28$ and the self-damping of impurity photo-conductivity with infra-red boundary 0.3 eV	60 min, 150	$E_c - 0.35$ $E_v + 0.35$ $E_v + 0.31$
n	2	180–400	1.28 1.58	1.13	$E_c - 0.12$ $E_c - 0.2$ $E_c - 0.28$	60 min, 200	$E_v + 0.27$ $E_v + 0.20$
p		180–250	1.34 1.78	1.2 1.49	$E_c - 0.12$ $E_c - 0.28$	60 min, 250	$E_v + 0.12$
p	3	250–400	—	2.37 2.29 2.52 2.38 2.13	$E_c - 0.27$ $E_c - 0.35$ $E_v + 0.32$ $E_v + 0.20$ $E_v + 0.12$	60 min, 300 60 min, 350	
					$E_c - 0.27$ $E_v + 0.28$ $E_v + 0.18$ $E_v + 0.12$	60 min, 350	

protons, becomes apparent,[7] these levels being thermostable up to $\sim$350 °C.

Figure 4 presents a part of unannealed defects arising in the process of isochronal annealing. It evidently follows from the figure, that at the first annealing stage (100–180 °C), the fraction increases whereas in the subsequent annealing it decreases. The main part of defects ($\sim$80 per cent) is annealed at temperature higher than 250 °C. At 400 °C the defects are practically annealed completely.

According to the data on isothermal annealing obtained from using Eq. (3) the coefficients K_1 and K_2 were determined. These coefficients were also found for the case of n-Ge irradiated by γ-rays and fast neutrons in accordance with the data from Ishino and coworkers[8] and Konopleva and coworkers,[9] (Table III).

The analysis of the values and signs obtained for K_1 and K_2 show, that in the case of proton irradiation as the annealing temperature grows K_1 at the first stage increases (100–150 °C), and as the temperature continues to rise, it decreases, all the while being larger than K_2, but of the same order. Since with the growth of temperature the number of remaining defects decreases, and the term $K_1 n^2$ falls faster than $K_2 n$, one can say that at the first annealing stage defect formation is prevalent and at high temperature defect annealing competes with it.

The positive value of K_1 and its being of the same

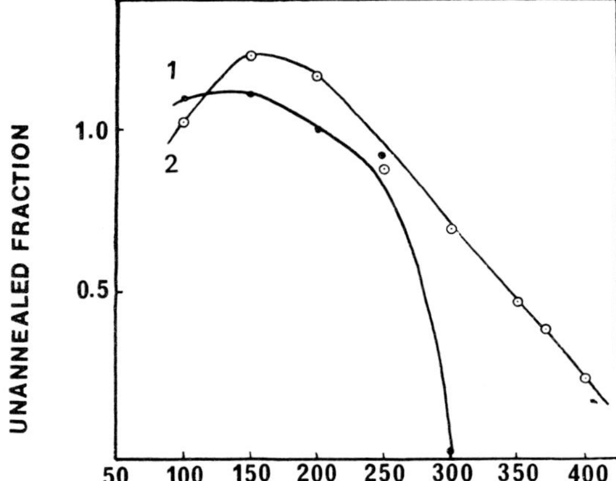

FIG. 4. Contribution of unannealed defects to Ge. irradiated by 660 MeV protons in the course of isochronal annealing (measured at 77 °K)
1. *n*-type,
2. *p*-type, converted by irradiation.

TABLE III

The coefficients for bipolar and monopolar components according to the data, obtained from the annealing of *n*-Ge, irradiated by *P*; γ-rays Co^{60} [8] and by fast neutrons of the reactor[9] (see formula (3)).

Annealing temp. °C	P		n_f		γ-rays	
	K_1	K_2	K_1	K_2	K_1	K_2
100	$+0.54$	-0.397	—	—	—	—
150	$+5.5$	-4.14	-0.04	$+0.6$	$+1.56$	-15.3
200	—	—	-1.31	$+13.5$	$+1.28$	-12.8
350	$+3$	-1.31	—	—	—	—

order as the value of K_2 over all the temperature range investigated give evidence of the fact, that at any annealing stage, even when a strong restoration of crystal properties takes place, defect formation occurs.

In the case of *n*-Ge, irradiated by γ-rays of Co^{60} at 20 °C,[8] the annealing proceeds through three stages, which are essentially dependent on the type of impurity. However irrespectively of it at 160 °C, 50 per cent of introduced defects were annealed.

As follows from Table III, in the case of Co^{60} irradiated by γ-rays the coefficient K_2 in the monopolar process is an order higher than K_1 and is negative, which indicates that at given temperatures the annealing of point defects or of their complexes with impurities takes place.

The study of isochronal annealing in Ge,

irradiated by fast neutrons from the reactor[10] showed that the radiation defect annealing goes through two stages with the centres at 150 and 400 °C.

The main part of defects (70 per cent) was annealed at the first annealing stage. The annealing stage at 150 °C was considered by the authors as due to the decay of disordered regions and migration of point defects to sinks. Indeed, as seen from the Table III in the case of neutron irradiation K_2 is an order higher than K_1 in the temperature range considered and positive, which points to the bi-polar mechanism of radiation defects annealing within this temperature range and the prevalence of defect formation over this process. According to the data from Konopleva and coworkers,[9] a share of unannealed defects is larger than a unit. Apparently in this case the defect growth is conditioned by point defect formation due to the decay of disordered regions, which is confirmed by the data from Baldwin[10] and Baldwin and coworkers.[11]

All which was said above enables one to make a suggestion, that the first annealing stage (100–180 °C) in the case of Ge, irradiated by protons is due to the decay of disordered regions. The annealing of Ge, irradiated by γ-rays in the investigated temperature range is connected with the annealing of point defects or their complexes with impurities. The comparison of the signs for K_1 and K_2 in the case of the annealing of defects produced by neutrons and protons enabled one to make a conclusion, that in the second case the decay of disordered regions is followed by the formation of complexes, composed of simple defects of one type, which form disordered regions. Such complexes are apparently divacancies. Then in the case of the annealing of defects, formed by high energy protons, the second annealing stage (180–250 °C) is connected with the disappearance of divacancies, of those formed directly during irradiation as well as of those formed at the first annealing stage as the result of the decay of disordered regions. The vacancies which are formed under such conditions may go to sinks, form complexes with atoms of impurities and with each other.

The third annealing stage (250–400 °C) is apparently connected with the decay of complicated vacancy complexes between each other and atoms of impurities.

Thus, the comparison of the result of the experiments or defect annealing, produced in *n*-Ge by high energy protons, fast neutrons and γ-rays, reveals the clearly pronounced qualitative difference in the behavior of defects. This difference can be explained on the basis of the notions of the different relative contribution to the change in properties of the material irradiated by different types of radiation.

ACKNOWLEDGEMENTS

The authors wish to express their gratitude to L. N. Nikituk and V. N. Pocrovskiy for their assistance in irradiating the specimens, and also V. P. Sadikov and N. A. Uhin for the calculation of the annealing coefficients.

REFERENCES

1. G. Kinchin and R. Pease, *Prog. in Phys.*, **18** (1955).
2. C. F. Powell, P. H. Fowler and D. H. Perkins, *The study of elementary particles by the photographic method*, London, New York, Paris, Los Angeles (1959).
3. Hans A. Bethe and P. Morrison, *Elementary nuclear theory*, New York, London (1956).
4. F. Gambon and D. Daspet, *C. R. Acad. Sci., Paris*, **261**, 4709 (1965).
5. N. A. Perfilov, O. V. Loshkin and I. O. Ostroumov, *The nuclear reactions under act of particles of high energy*, M.-L., AN USSR (1962).
6. A. C. Damask and G. J. Dienes, *Point defects in metals*, New York, London (1963).
7. R. F. Konopleva, S. R. Novikov, E. E. Rubinova, U. A. Zaporoshenko, V. N. Pocrovskiy and L. N. Nikituk, *F.T.P.*, **3**, 1119 (1969).
8. J. Ishino et al., *J. Phys. Chem. Solids*, **24**, 1033 (1963).
9. R. F. Konopleva, S. R. Novikov and S. M. Ryvkin, *FTT*, **6**, 426 (1964).
10. T. O. Baldwin, *Phys. Rev. Letters*, **21**, 901 (1968).
11. T. O. Baldwin and J. E. Tomas, *J. Appl. Phys.*, **39**, 4391, (1968).

DAMAGED REGIONS IN NEUTRON-IRRADIATED AND ION-BOMBARDED Ge AND Si

M. L. SWANSON, J. R. PARSONS AND C. W. HOELKE

Chalk River Nuclear Laboratories, Atomic Energy of Canada Limited, Chalk River, Ontario, Canada

The formation of amorphous or crystalline damaged regions by ion bombardment or neutron irradiation of Ge and Si is discussed. Several experiments have demonstrated that Ge and Si become amorphous when bombarded to high fluences with ions of 20–200 keV energy, but remain crystalline after low ion fluences ($<10^{14}$/cm²) or after fission neutron irradiations up to very high fluences. The present results indicate that a fission neutron fluence of 5×10^{20} n/cm² at ~50 °C and at a flux of 6×10^{13} n/cm²/sec produces no amorphism in Ge or Si, but that micro-crystals appear after annealing at 450 °C.

It is concluded that the individual damaged regions produced by high-energy ions or by primary knock-on atoms from fission neutron collisions are crystalline. When these damaged regions are created rapidly enough so that overlap occurs before appreciable annealing of point defects takes place, the damaged regions become amorphous. The critical defect concentration required for spontaneous transformation to amorphism is estimated to be 0.02.

1. INTRODUCTION

Although it is now well-known that high fluence ion bombardment of crystalline Ge and Si produces an amorphous damaged layer,[1-3] the nature of the damage produced by fission neutrons is not so clear. Early X-ray experiments,[4] as well as more recent data,[5,6] suggest that neutron-irradiated Ge and Si remain crystalline. This difference is rather unexpected, since the energy of primary knock-on collisions E_p for 20–200 keV ion bombardments and for fission neutron irradiations are similar. However, there are many other factors besides E_p which may influence the crystallinity of bombarded material, such as the irradiation flux, integrated flux (fluence), and temperature, as well as the interatomic forces of the material. We shall consider the nature of the individual damaged regions, and describe a critical experiment showing that they remain crystalline in Ge and Si after high fluence fission neutron irradiation.

2. PRESENT EXPERIMENTAL RESULTS

We have examined neutron-irradiated Ge and Si by transmission electron microscopy. Single crystal samples of 2 ohm cm n-type Ge (10^{15} Sb atoms/cm³) and 1 ohm cm p-type Si (10^{16} B atoms/cm³), with surfaces perpendicular to the $\langle 111 \rangle$ direction, were irradiated at ~50 °C in a fission neutron flux of 5×10^{12} n/cm²/sec to fluences of 5×10^{16}, 5×10^{17} and 2.5×10^{18} n/cm², and in a fission flux of 6×10^{13} n/cm²/sec to fluences of 8×10^{19} and 5×10^{20} n/cm². The samples were thinned by a standard jet polishing technique. Individual damaged regions were observed in bright field images as spots ~50 Å in diameter for Ge and a similar size for Si. The concentration of spots for the fluence of 5×10^{17} n/cm² corresponded to one visible damaged region per primary knock-on, assuming a neutron scattering cross-section of 3 barns. For this fluence, Figure 1 shows a typical bright field image of Ge under two beam dynamical conditions [(220) reflection]. Several spots have a black-white contrast, which indicates a strain field associated with the damaged regions.

For our maximum fluence of 5×10^{20} n/cm², the average distance between primary knock-on collisions was only 24 Å, thus ensuring considerable overlap of damaged regions. Bright field images of these high fluence samples showed a low concentration of spots against a rather diffuse background, as expected because of the large overlap. Even for this high fluence (as well as for lower fluences), no amorphous rings were observed in the electron diffraction patterns of Ge or Si. The usual crystalline spot pattern was observed. We conclude that the damaged regions in neutron-irradiated Ge and Si were still crystalline.

The high fluence samples were also examined after annealing for 1 hr at 450 °C. Figures 2a and 2b show the same area of a Ge sample in bright field and in dark field. Figure 2b is a high resolution dark field image taken with the objective aperture displaced to a position on the (220) ring

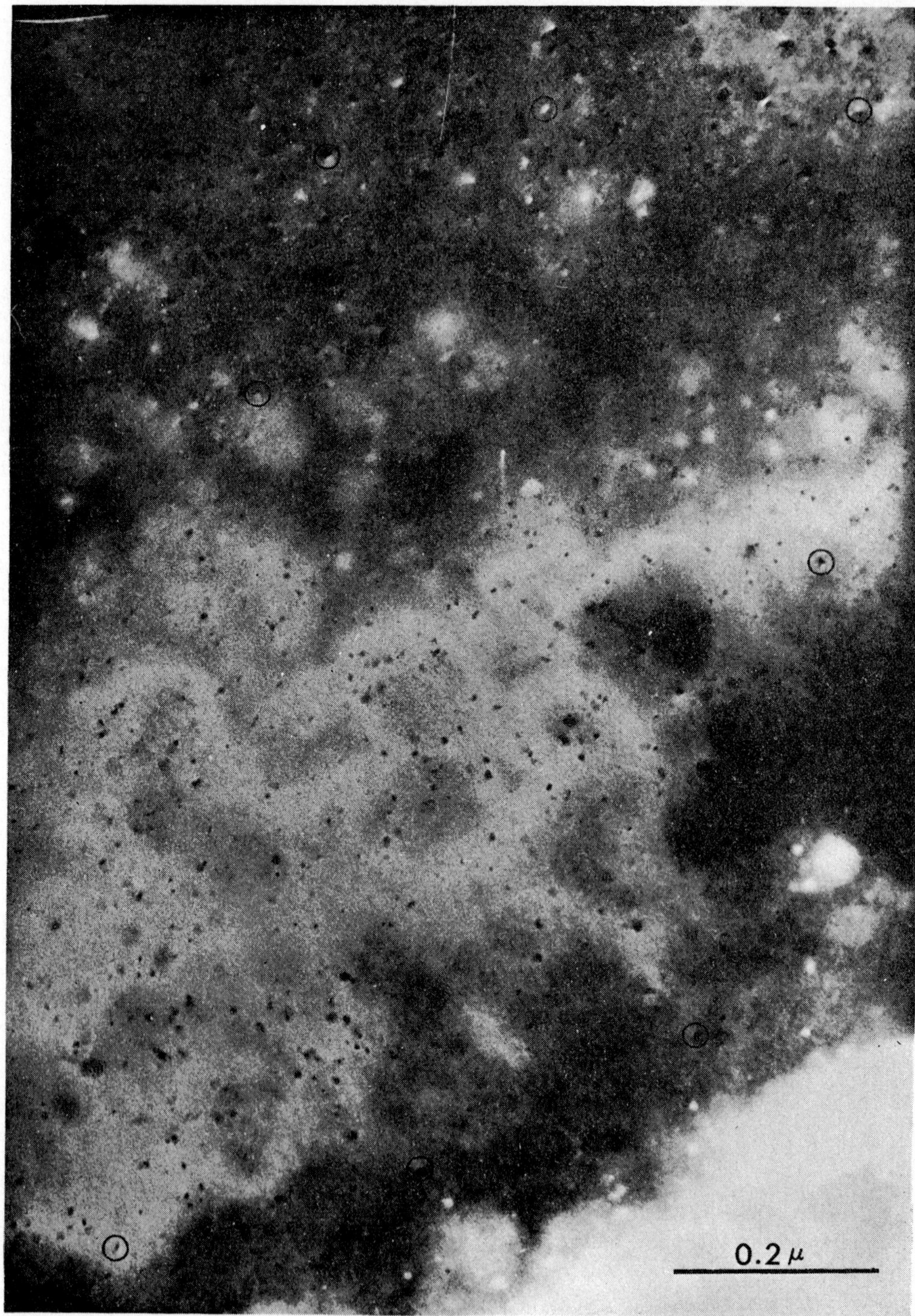

FIG. 1. Bright field image of Ge irradiated with 5×10^{17} n/cm², recorded under two beam dynamical conditions [(220) reflection]. The black-white contrast of the circled damage regions indicates the presence of a localized strain field.

which did not include a matrix reflection. The correspondence between diffracting regions in Figures 2a and 2b indicates that the regions are microcrystals with an orientation different from that of the matrix. A similar experiment performed on a low dose specimen did not indicate any microcrystalline regions.

3. CRYSTALLINE OR AMORPHOUS DAMAGED REGIONS?

A number of other experiments have indicated that Ge and Si become amorphous after bombardment with high ion fluences, but remain crystalline after lower ion fluences ($<10^{14}$/cm²) or after neutron irradiation up to very high fluences at $\sim 50\,°$C.

3.1. *Evidence for amorphism*

(a) For Ge bombarded with $>10^{12}$ 100 keV 0⁻ ions/cm², an amorphous ring pattern was superimposed on the normal crystalline spot pattern seen by transmission electron diffraction.[1] The ring pattern gradually replaced the spot pattern for higher doses, until after a fluence of 10^{15}/cm² only the ring pattern was seen for film thicknesses $<600\,$Å. By using an aperture which transmitted only electrons scattered into the second diffuse diffraction ring, individual damaged regions appeared white on the resulting dark field photograph. This is the most direct evidence which has been obtained to show that a *single* damaged region is amorphous.

(b) After bombardment with various heavy ions to fluences of about 10^{14}/cm² at $300\,°$K, the surfaces of both Ge and Si appeared milky, and the electron diffraction patterns indicated that the surface layer was amorphous.[2,3]

(c) The examination of neutron-irradiated Ge by two beam phase contrast electron microscopy revealed that the (111) atomic planes did not extend through a damaged region which intersected both surfaces of the foil. Although this result could be caused by the damaged region being amorphous,[7] it could also be caused by a different diffraction within the damaged region.

3.2. *Evidence for crystallinity*

(a) In Ge and Si which were irradiated at $\sim 50\,°$C to neutron fluences of $>10^{20}$/cm², X-ray diffraction measurements indicated that the material was still crystalline,[4,5] as do our transmission electron diffraction patterns.

(b) Cheng[6] found from infrared absorption measurements that ~ 40 divacancies were produced for each primary knock-on collision in neutron-irradiated Si. This represents a considerable fraction (perhaps 40 per cent) of the total number of defects expected to remain after room temperature irradiation.

(c) The damage produced by low ion fluences annealed out at a much lower temperature than that produced by high ion fluences.[2] As pointed out by Vook and Stein,[8] the lower temperature annealing stage of Si correlates well with simple defect (divacancy) annealing.

(d) The number of displaced atoms produced in Si by 20–225 keV Bi ions (as measured by backscattering of 1 MeV helium ions) increased linearly with ion energy for low fluences.[9] This result is explained better by a simple atomic displacement theory than by amorphous damaged regions (whose volume would not increase linearly with ion energy).

(e) The damaged regions produced by 600 MeV proton irradiation of Ge and Si are about 2000 Å in diameter, have a defect density of 10^{19}/cm³, and retain their crystallinity (as determined by electrical measurements).[10]

4. FACTORS AFFECTING CRYSTALLINITY OF DAMAGED REGIONS

From the previously mentioned results it can be concluded that individual damaged regions in Ge and Si remain crystalline, both for ion bombardment and neutron irradiation, but that in the former case overlap of damaged regions renders them amorphous. It is to be noted, however, that high temperature ion bombardment ($>300\,°$C) does not produce amorphism, even for high fluences.[1–3] Since the neutron irradiations are normally carried out at fluxes which correspond to about 10^{-5} of the energy deposition rate of ion bombardments, it seems plausible that the flux and temperature are critical factors in determining amorphism.[8] In addition, other factors, such as energy dissipation by ionization, and the effect of implanted ions should be considered. The distribution of damaged regions, and of defects within the damaged regions, may also be important. Some evidence for different defect distributions in ion-bombarded and neutron-irradiated Si has been provided by measurement of the temperature dependence of the damage rate. In low dose Sb-ion bombarded Si, a pronounced temperature dependence was observed from 50–200 °C[11], but no such

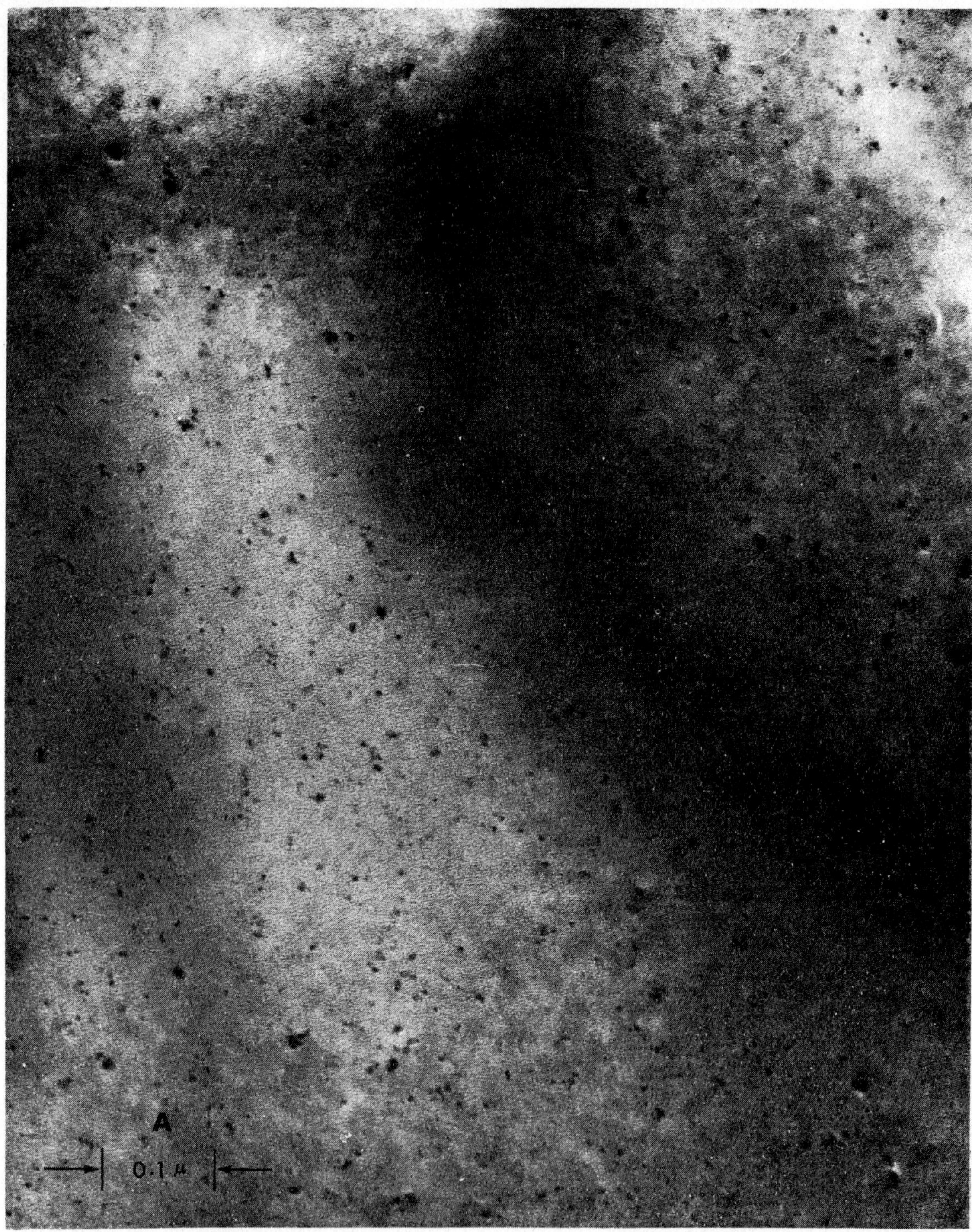

FIG. 2. Images of the same area of a Ge sample irradiated with 5×10^{20} n/cm^2 and annealed 1 hour at 450 °C. (a) Bright field image.

FIG. 2. (b) Dark field image with a 10 μm objective aperture located at a non-matrix position on the (220) polycrystalline diffraction ring. Note the correspondence between diffracting regions in (a) and (b).

dependence was seen for neutron-irradiated Si[12].

The transformation from crystallinity to amorphism, and also the phenomenon of melting, have not yet been satisfactorily treated in a theoretical manner. Since the lattice vibrations in a material near its melting point influence the fusion phenomenon profoundly, it is expected that the athermal vibrations which accompany a displacement spike will greatly affect the transformation to an amorphous region. For a damaged region containing about 10,000 atoms, which is created by a knocked-on atom having an energy of 10 keV, there is an energy of 1 eV/atom dissipated in the form of atomic vibrations (as the total energy of the 100 or more stable defects remaining is negligible). This is comparable to the energy of fusion, which is $\frac{1}{3}$ eV/atom for Ge and $\frac{1}{2}$ eV/atom for Si. The average defect concentration in such a damaged region is about 0.01 atomic fraction. It is expected that 0.02 is approximately the concentration required to transform the region into an amorphous volume, under the influence of the atomic vibrations. (The critical concentration which Vook and Stein[8] assume, about 0.5, is much too large, since it is even greater that the total concentration of defects produced in the 10^{14} ion/cm^2 bombardments, before athermal and thermal annealing is considered.)

The critical defect concentration at which a damaged region might spontaneously transform to a stable or metastable amorphous region can be estimated by assuming that it occurs when the free energy of the amorphous region equals that of the defect-rich damaged region. These energies are difficult to calculate from the available experimental data, but rough estimates can be made. It appears that the structure of Ge which is rendered amorphous by ion bombardment[3] is very similar to that of evaporated Ge.[13,14] The model of Richter and Breitling[13,14] will be adopted for amorphous Ge: a layer structure exists, in which the tetrahedral bonding is retained but one bond is stretched by 0.27 Å. Then, if the other misorientation effects are neglected, the energy of amorphous Ge can be calculated simply as the energy of these stretched bonds. Using the generalized Morse potential previously derived for vacancy energy calculations, the bond energy,[15]

$$E = D\left\{\frac{\alpha_2}{\alpha_1 - \alpha_2} \exp[-\alpha_1(r - r_0)] - \frac{\alpha_1}{\alpha_1 - \alpha_2} \exp[-\alpha_2(r - r_0)]\right\} \qquad (1)$$

where D = dissociation energy, r_0 = equilibrium separation of atoms, and α_1 and α_2 are constants determined from compressibility data. For Ge, the increase in bond energy,

$$\Delta E = E + D = D\{1 + 0.244 \exp[-2.625(r - r_0)]$$
$$- 1.244 \exp[-0.515(r - r_0)]\} \qquad (2)$$

If $r - r_0 = 0.27$ Å, $\Delta E = 0.23$ eV. Since this energy is shared by two atoms, the increase in energy per atom is 0.11 eV. This energy corresponds to a defect concentration of 0.04, if each defect has an energy of 3 eV.[15] When entropy changes and the probable deviation from perfect tetrahedral bonding are considered, it is concluded that a crystal containing a defect concentration of 0.02–0.04 would be unstable with respect to the amorphous state.

When overlap of damaged regions occurs, the cascade damage process will be altered by the large defect concentration already present. Energy will be transported less readily in a damaged region because of reduction in collision chain length, so that the damage will be more concentrated. Thus the critical defect concentration required for amorphism could be easily exceeded in overlapped damaged regions. An experiment to test this idea would be to irradiate a sample with electrons to a very high dose at a low temperature. Then the whole lattice would be 'sensitized' by defects, and *every* ion (or neutron) might produce an amorphous region. This should result in amorphous rings in a diffraction pattern at low fluences, an increased slope of number of displaced atoms versus fluence, and a saturation at lower fluence (for the back-scattering type of experiment).

5. ANNEALING IN DAMAGED REGIONS DURING NEUTRON IRRADIATION

The high fluence neutron irradiations of Ge and Si have been carried out over periods of several months at $\sim 50\,°$C. Under these conditions, many of the irradiation defects will have annealed out before overlap of damaged region occurs (whereas for typical ion bombardments which last only minutes, this annealing is much less). However, it is impossible at the present time to calculate the defect concentration, type and distribution after a given irradiation. A discussion of the effects of correlated recovery and saturation fluences has been given by Corbett.[16]

Vook and Stein[8] have attempted to express the

critical defect concentration for amorphism by a simple rate equation describing divacancy migration. If a similar approach is used for neutron irradiations, it is easy to show that divacancy annealing at 50 °C is too slow to affect the transition to amorphism in Si. Let us assume that the whole crystal has been neutron-irradiated to an approximately uniform defect density of 0.01 (which takes 6 days at a flux of 6×10^{13} n/cm²/sec, if 100 defects are produced per neutron collision). Then in order for the defect concentration to increase beyond that value, the rate of defect creation must exceed the rate of defect reduction by annealing: $40\,n_p > 0.004\,C_s\,N_0\nu\,\exp(-E^M/kT)$ if only divacancies are considered. Here n_p is the number of primary knock-on collisions/cm³/sec ($= 10^{13}$/cm³/ sec), and it is assumed that 40 out of every 100 defects are divacancies. The effective sink concentration, C_s is taken as 0.01 from the results of Cheng and Lori,[6] $E^M = 1.25$ eV for divacancies, N_0 is the concentration of atoms, and $\nu \simeq 10^{13}$/sec. The inequality becomes $4 \times 10^{14} > 6 \times 10^{11}$/cm³/sec at 320 °K. Although this estimate is crude, it does seem to exclude the possibility that divacancies determine the transition to amorphism. Of course, athermal effects have been neglected; they might well dominate in this case. In particular, γ-rays and low energy neutron collisions might affect recovery considerably.

We conclude that the absence of amorphism in Ge and Si which were irradiated to a fission neutron fluence of 5×10^{20} n/cm² at 50 °C can be attributed to thermal migration of a more mobile defect than the divacancy, or to athermal annealing during the irradiation. It is expected that high fluence neutron irradiations at lower temperatures will create amorphous damaged regions in Ge and Si, similarly to results for 200 keV boron ion bombardment of Si.[17]

REFERENCES

1. J. R. Parsons, *Phil. Mag.*, **12**, 1159 (1965).
2. J. A. Davies, J. Denhartog, L. Eriksson and J. W. Mayer, *Can. J. Phys.*, **45**, 4053 (1967).
3. D. J. Mazey, R. S. Nelson and R. S. Barnes, *Phil. Mag.*, **17**, 1145 (1968).
4. M. C. Wittels, *J. Appl. Phys.*, **28**, 921 (1957).
5. M. C. Wittels, *J. Appl. Phys.*, **40**, 2909 (1969).
6. L. J. Cheng and J. Lori, *Phys. Rev.*, **171**, 856 (1968).
7. J. R. Parsons, M. Rainville and C. W. Hoelke, *Phil. Mag.*, **21**, 1105 (1970).
8. F. L. Vook and H. J. Stein, *Rad. Effects*, **2**, 23 (1969).
9. D. A. Marsden, G. R. Bellavance, J. A. Davies, M. Martini and P. Sigmund, *Phys. Stat. Sol.*, **35**, 269 (1969).
10. S. R. Novikov and R. F. Konopleva, private communication.
11. S. T. Picraux, J. E. Westmoreland, J. W. Mayer, R. R. Hart and O. J. Marsh, *Appl. Phys. Letters*, **14**, 7 (1969).
12. L. J. Cheng and J. Lori, *Appl. Phys. Letters*, **16**, 324 (1970).
13. H. Richter and G. Breitling, *Z. Naturforsch*, **13a**, 988 (1958).
14. G. Breitling, *J. Vac. Sci. Technol.*, **6**, 628 (1969).
15. A. Seeger and M. L. Swanson, *Lattice Defects in Semiconductors*, Ed. R. R. Hasiguti (University of Tokyo Press, 1968), p. 93.
16. J. W. Corbett, *Rad. Effects*, **1**, 85 (1969).
17. J. E. Westmoreland, J. W. Mayer, F. H. Eisen and B. Welch, *Appl. Phys. Letters*, **15**, 308 (1969).

DISCUSSION

Question (FISCHER)   How can overlapping crystalline regions produce an amorphous structure?

Answer (SWANSON)   As stated in the text, overlap of defect-rich regions creates regions with continuously increasing defect concentration. When this defect concentration exceeds a critical value, the crystalline state becomes energetically unstable, and could transform to the amorphous state. This is analogous to the transformation of a supersaturated metallic solid solution to different stable or metastable (martensitic) phases.

Comment (STEIN)   Low-fluence ion implantation, like neutron irradiation, produces crystal lattice defects such as divacancies and A centers in Si. Divacancy quenching occurs at fluences less than that to form an amorphous layer, and the results are consistent with divacancy self-quenching at a concentration of $\sim 7 \times 10^{19}$/cm³.

Recent measurements of oxygen associated localized modes in Si have shown that these defects increase in concentration with oxygen ion fluence to fluences within a factor of two of that required to produce a discoloration of the surface. Such discoloration is another indication of amorphous layer formation. These results indicate that most of the implanted layer behaves as disordered crystalline material to quite high fluences. The amorphous transition is then assumed to occur when the crystal lattice defect concentration within a region exceeds a critical value. The critical concentration being built up by an overlapping of damaged crystalline regions.

Comment (VOOK) Silicon divacancy annealing at 50 °C is *not* too slow over the long irradiation time of 3 months $\approx 10^7$ sec to affect the crystalline to amorphous transition. If the irradiation time and temperature are compared with divacancy annealing rate[8] it can be seen that neutron induced divacancy annealing will occur during the irradiation. Therefore, for these conditions it is not necessary to invoke a more mobile defect than the divacancy (or athermal annealing) to explain the absence of amorphism in silicon following the irradiation.

Reply (SWANSON) The question of whether thermal annealing of divacancies during 50 °C reactor irradiations prevents the transformation to amorphism depends on the critical defect concentration for the amorphous transformation. It does appear that a critical concentration of 0.02 indicates that thermal divacancy annealing is too slow at 50 °C to affect the transformation, but this is not an important point, since athermal annealing might well be important. In addition, the annealing of other defects should be considered. The effect of the implanted ions may also be significant in transformation of ion-implanted samples, as these impurity concentrations are the order of 10^{19}–10^{20}/cm^3.

DISCUSSION

Remark (SEEGER) I should like to remark that the energy difference between amorphous and crystalline germanium calculated by you appears to compare quite well with the calorimetric data of H. S. Chen and D. Turnbull (*J. Appl. Phys.*, **40**, 4214, 1969), who have found that the energy release in the transition from the amorphous to the crystalline state is 2.75 cal/*g* atom.

ION IMPLANTATION—LATTICE DISORDER

J. W. MAYER

California Institute of Technology, Pasadena, California 91109, U.S.A.

This review is directed toward some of the recent developments in the analysis of lattice disorder in ion-implanted silicon. Although the disorder picture has not been entirely clarified in silicon, many of the salient features have been identified. IR absorption and EPR measurements have been used to identify specific defects and correlation was found with anneal behavior in implanted and fast-neutron irradiated samples. Analysis of low temperature implants suggested that release of point defects (vacancies) played an important role in the growth of divacancies and the anneal of disorder. Depth distribution of defects have been investigated and compared with ion ranges and energy lost in atomic processes. Various models have been proposed for the nature of the disordered region around the ion track and the formation of an amorphous layer at high doses. The data indicates that the individual disordered regions are not primarily amorphous. The nature of ion implanted layers in compound semiconductors have not yet been as thoroughly investigated as those in silicon.

1. INTRODUCTION

The purpose of this review paper is to consider some of the recent developments in the analysis of disorder in ion-implanted semiconductors.[1,2] It is more in the nature of a progress report, since our knowledge of implantation processes is going through a strong transition period. The work up to the latter part of 1969 was primarily directed toward obtaining a consistent picture of the amount of disorder in implanted layers and the influence of implantation parameters such as substrate temperature, ion species, energy, and dose. At present, more sophisticated questions are being asked concerning the nature of the disordered region around the track of particle, the depth distribution of the disorder and the inventory of defects in the implanted layer. Only comparatively recently have techniques more common to defect identification in radiation damage studies, such as electron paramagnetic resonance (EPR) and infrared absorption, been applied to analysis of implanted layers. The picture is far from complete.

It should be noted that the situation in compound semiconductors has still not been resolved. There are major questions concerning the anneal behavior, the influence of implantation temperature, dose dependence, the nature of defects in the implanted layer, etc. While these problems are being intensively investigated by a number of groups,[2-4] no coherent pattern has emerged. Consequently, this review will be primarily concerned with the nature of the disorder in ion implanted silicon—the material in which the majority of the investigations have been made. Even here, one can only touch on limited aspects of the work. The examples used in this review were chosen to indicate the type of study being made rather than to cover all the investigations in the past months.

2. BACKGROUND†

As an implanted ion slows down and comes to rest in a crystal it makes many violent collisions with lattice atoms which results in the production of a highly disordered region around the path of each ion. At sufficiently high doses, a non-crystalline or amorphous layer is formed. Figure 1 shows how

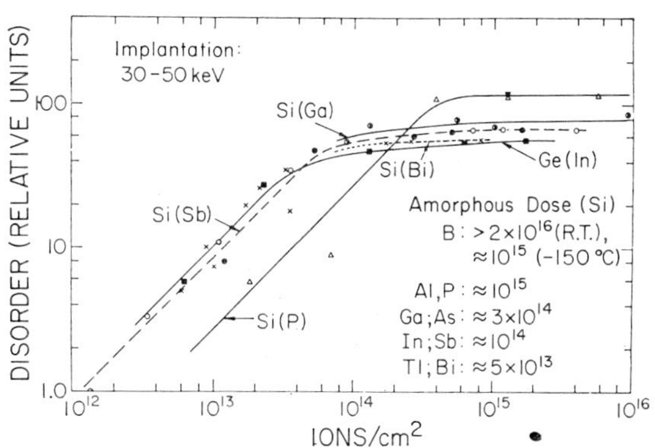

FIG. 1. Channeling-effect measurements of the amount of disorder produced in 30–50 keV room temperature implant atoms in silicon and germanium. The dose required to form an amorphous layer in Si is indicated. (Data from Davies and coworkers[5] and Mayer and coworkers[6].)

† In the review of background material available up to early 1970, only a portion of available references were cited. For a more complete list, see references (1), (2).

lattice disorder varies with dose for a number of different ion species implanted at room temperature.[5,6] It can be seen that the disorder increases linearly at first (the region where the individual disordered regions well-separated) but reaches a saturation value at higher doses. In channeling effect measurements such as these, this saturation has usually been associated with the formation of an amorphous layer. Indeed for high dose implants, electron transmission microscopy and electron diffraction studies have shown that this layer does not possess long-range order.[1] EPR[7] and optical reflectivity measurements[8] also show a correlation between the characteristics of amorphous sputtered films and 'saturated-dose' implanted layers.†

We turn to the question of disorder anneal of the implanted layer. For low-dose implantations there is a strong decrease in the amount of disorder at about 200 °C.[1] The anneal curve for disorder as measured by Stein and coworkers[9] for infrared (divacancy) absorption and that for channeling effect measurements[5,6] agree rather closely with each other.[10] Similar results have been found for the Si P–3 center.[11,12] The anneal behavior is also similar to that obtained by Cheng and Lori on the anneal of fast-neutron irradiated silicon.[13,14] Such correlation might be expected because the disordered regions are created in each case by silicon recoils.

The amorphous layer anneals at higher temperatures (between 550 and 660 °C) by reordering 'epitaxially' on the underlying crystalline substrate.[1,6] As shown in Figure 2 there are also indications of a correlation between implantation and high-fluence anneal behavior.

The electrical characteristics of implanted layers have been investigated in detail.[1,2] In Hall-effect measurements, the presence of an amorphous layer has a pronounced effect on the anneal characteristics.[1,2,15,16] As shown in Figure 3a there is a strong increase in the number of carriers per cm² in the temperature region around 550 to 600 °C associated with the reordering of the amorphous layer. This same correlation with disorder anneal

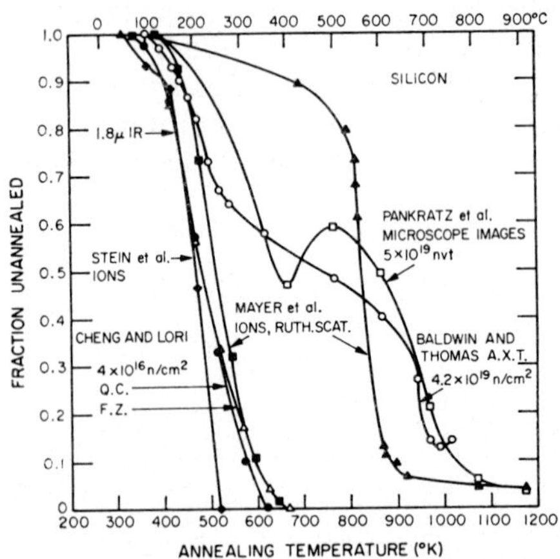

FIG. 2. Comparison of the isochronal annealing of neutron and ion produced defects in Si: $1.8\,\mu$ IR absorption band (1.75×10^{14} 400 keV oxygen ions and 4×10^{16} n/cm²), anomalous x-ray transmission (A.X.T.) (4.2×10^{19} n/cm²), electron microscope cluster images (5×10^{19} nvt), and channeling effect measurements of low dose ($\approx 10^{13}$ ions/cm²) and above saturation dose ($\approx 3 \times 10^{14}$ ions/cm²) R.T. implantations at 40 keV. (Taken from Vook and Stein[10].)

† For the most recent account of the difficulties of characterizing disorder measured by different techniques, see reference (2).

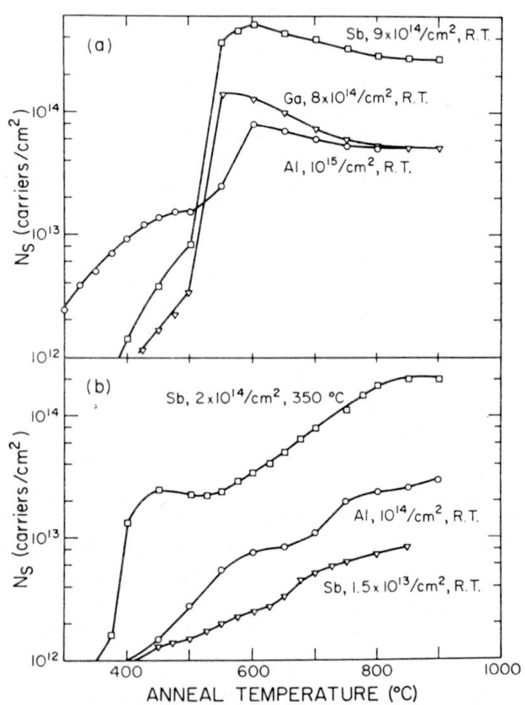

FIG. 3. Number of carriers/cm² measured in implanted layers in silicon versus anneal temperature: (a) anneal behavior for samples implanted at room temperature to a dose of $\sim 10^{15}$ ions/cm² such that an amorphous layer was formed. (b) anneal behavior of a hot-substrate (350 °C) Sb implantation and of lower-dose room temperature Sb and Al implantations under conditions such that an amorphous layer was not found. (From Baron and coworkers[16] and Johansson and Mayer[18].)

is not found on either hot substrate or low dose implantations where an amorphous layer is not formed (Figure 3b). In this case, annealing of electrically active, compensating defects can apparently occur up to temperatures near 800 °C, viz. well above any identified defect anneal range.[17,18] This requirement of high (~800 °C) anneal temperature is found for all ion species and is not understood at present. This is one of the areas where further work is obviously required.

Rather more attention has been paid to boron implanted layers and the explanation for electrical behavior has been slowly unfolding. There were many aspects that seemed anomalous: reverse annealing behavior,[19] a requirement for $\gtrsim 900$ °C temperatures to achieve ≈ 100 per cent electrical activity[20] and the 550 °C anneal stage found in liquid N_2 temperature implants ($\gtrsim 10^{15}$ B/cm²).[21] Channeling effect measurements indicated that marked annealing of disorder occurred during R.T. implantation[22,23] and that a significant amount of boron was substitutional prior to any anneal treatment.[24,25] Upon anneal (Figure 4) the

substitutional content decreased (accounting for reverse anneal behavior) and temperatures of 900 °C or larger were required for >75 per cent of the boron to move back to substitutional sites (accounting for the high temperature anneal observed in the electrical measurements). In contrast, implantation at Liq. N_2 temperature to doses of $\gtrsim 10^{15}$/cm² resulted in the formation of an amorphous layer (accounting for the 550 °C anneal stage). These phenomena are being examined in greater detail and it appears that much of the electrical behavior can be correlated with disorder and the lattice site of the boron atoms.[26]

The above should not be construed to imply that there exists a complete and consistent empirical description of all the characteristics of the implanted layers in silicon for the different ion species. However, there is enough data available to make reasonable estimates of the disorder and electrical characteristics over a wide range of implantation parameters. It happens that in areas dealing with subjects such as enhanced diffusion[27] and carrier lifetime[28,29] few measurements exist on which to base similar extensive empirical predictions.

3. DISORDER MEASUREMENT TECHNIQUES

Disorder in ion implanted layers has been studied by a number of techniques: visual observation, optical reflectivity, optical absorption, channeling effect, EPR, x-ray topography, electron microscopy, copper decoration, Hall effect, and carrier lifetime measurements. In general, each of these techniques yields information about different aspects of disorder.

For example, optical absorption (1.8 μ, divacancy) and EPR (for example, Si–P3 centers) measurements give microscopic information about specific defects that must be located in essentially crystalline regions for identification. Optical reflectivity and EPR (isotropic center) can be used to evaluate the formation of the amorphous layer. Electron microscopy can give information on the presence of disordered regions and the growth of dislocations. Channeling effect measurements detect disorder and its depth distribution but do not supply information about the microscopic nature of the defects.

While all these methods are continually being refined there have been recent developments in channeling effect measurements which should be mentioned. In this technique,[1] the samples are aligned so that the beam is incident along low-index

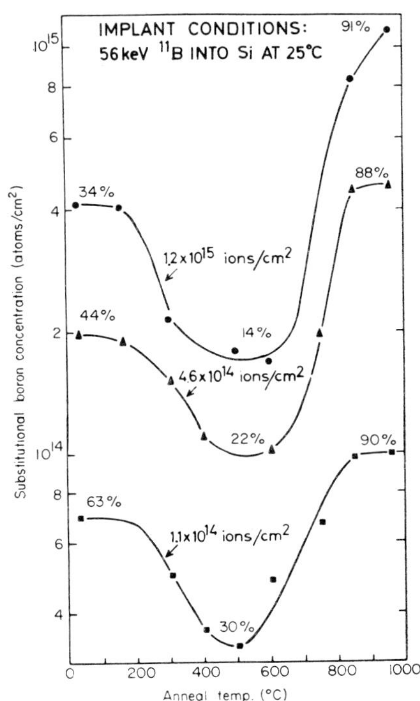

FIG. 4. Channeling effect measurements of the anneal behavior of the number of boron atoms on substitutional sites for three different dose, 56 keV boron implants in Si. The percentage of substitutional boron atoms are indicated at several places. (From Fladda and coworkers[25].)

axial directions defined by the underlying un-damaged lattice. The number of scattering centers contributing to the backscattering yield in the aligned direction is considered to be the number of lattice atoms located more than 0.1–0.2 Å off lattice sites. It was pointed out earlier[1,30] in disorder studies that the number of 'off-lattice-site' atoms was measured to be about a factor of 2 to 3 greater than the calculated number of atoms displaced by atomic processes. The current view is that this discrepancy may be due to strain and distortions of the lattice in the implanted layer. In-vestigations are under way to test this.

A newly revealed aspect of channeling is 'flux-peaking' which, for certain beam-to-substrate orientations (such as the $\langle 110 \rangle$) results in an enhanced backscattering yield from atoms located in well defined interstitial sites over that for random orientation.[31,32] This is a critical effect in deter-mining the percentage of implanted atoms on tetra-hedral interstitial sites. It may not be as critical in disorder studies because the magnitude of the effect would be decreased by the increase in angular spread in the beam due to multiple scattering by defects.

It should be remarked that now that the major features of disorder have been described, progress in establishing an inventory of defects and in elucidating some of the physical mechanisms in-volved will probably require use of more than one analysis technique. In this regard one should be careful to compare data on samples implanted under the same conditions since factors such as substrate temperature, ion-channeling, dose, and dose rate play a large role.

4. TEMPERATURE DEPENDENCE OF DISORDER

Implantation at substrate temperatures well above R.T. results in less lattice disorder than that for room temperature implantation. The annealing occurring during room temperature implantation was not generally appreciated, however, until the electrical behavior[21] of boron implanted layers required explanation. It was found[22,23] that the level of disorder can be a factor of 20 to 30 lower in R.T. implants than in those implanted at, for ex-ample, –150 °C.

Earlier this was thought to be characteristic of only light ion species. More recently Picraux and coworkers[33] have shown that the disorder pro-duced in 200 keV Sb implantations has a similar

but not quite as pronounced implantation tem-perature dependence (Figure 5). There is a 3-to-4-fold reduction in disorder in R.T. implants as compared to those at –180 °C. Further, the anneal characteristics of low temperature B and Sb im-plants are nearly identical (Figure 6). It should be

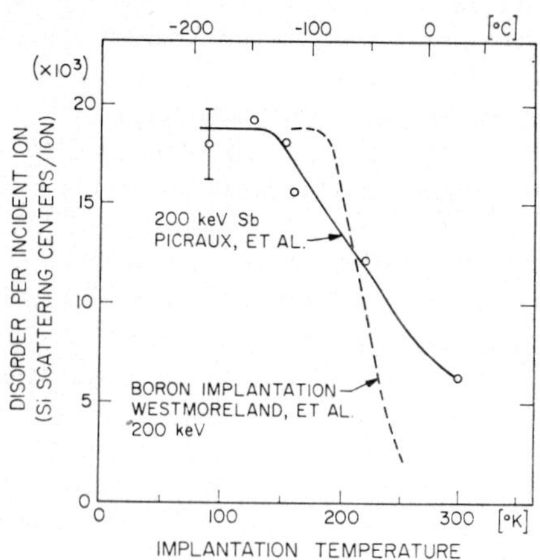

FIG. 5. Comparison of channeling effect measure-ments of the implantation temperature dependence of lattice disorder per incident ion for 200 keV Sb with that for 200 keV B. The B results have been normalized to 1.9×10^4 Si centers/ion at $T = 158$ °K for comparison purposes. (From Picraux and co-workers[33].)

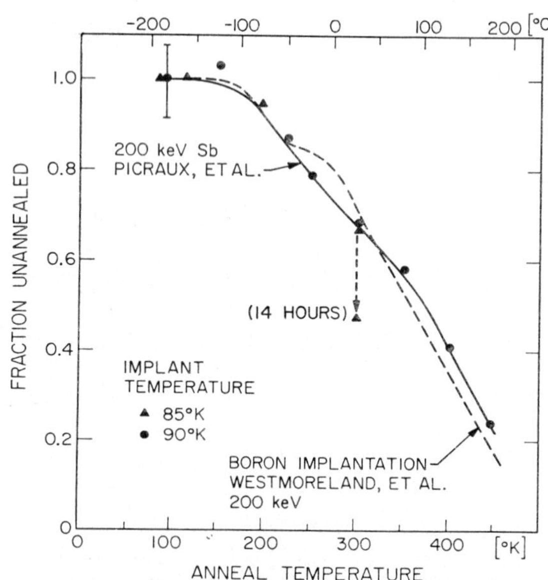

FIG. 6. Channeling effect measurements of disorder anneal behavior for low temperature 200 keV Sb and B implants in silicon. (From Picraux and coworkers[33].)

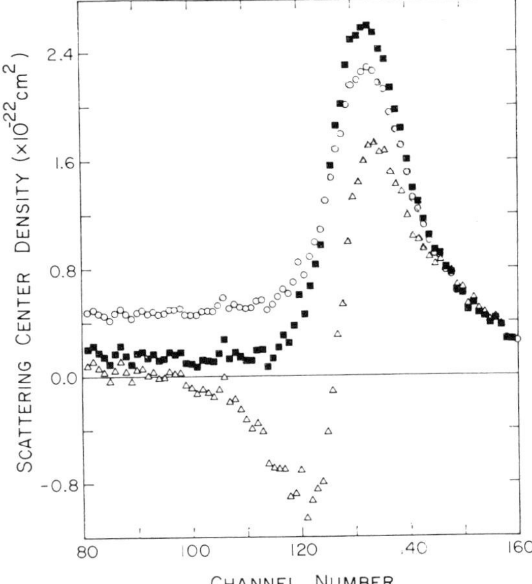

FIG. 7. Scattering center densities in Si implanted at $-150\,°C$ with 200 keV B $(5 \times 10^{14}$ ions/cm$^2)$ calculated from channeling effect data obtained with 1.8 MeV He$^+$. The densities were calculated from (1) squares-plural scattering, (2) circles-single scattering and (3) triangles-multiple scattering. (From Westmoreland and coworkers[40].)

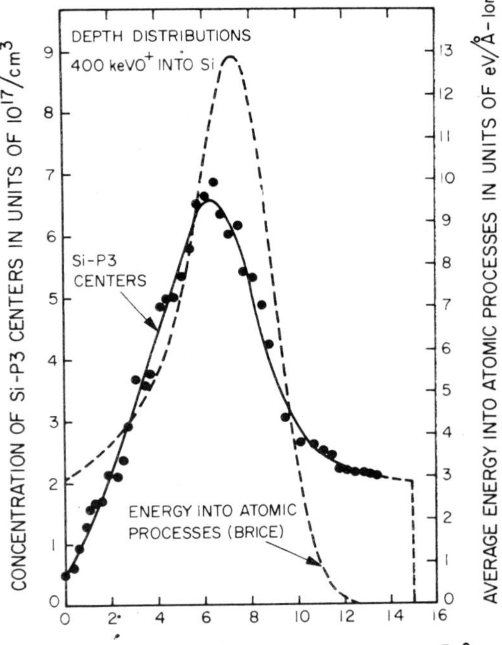

FIG. 8. Depth distribution of Si–P3 centers (solid line) determined from layer removal studies and EPR measurements and calculated average energy into atomic processes (dashed line). The sample was implanted at R.T. with 3×10^{13} O ions/cm^2 at 400 keV. (From Brower and coworkers[42].)

noted that a decrease of disorder occurs in the temperature range up to 300 °K in channeling effect measurements whereas there is a marked increase in the number of divacancies[34] (cf. Figure 9).

The fact that the disorder dependence on anneal and implantation temperatures are relatively independent of ion mass† indicates (as pointed out by Picraux and coworkers) that the disorder produced depends primarily on the properties of the target material. The existence of low temperature anneal stages in Si strongly points up the possibility of low temperature stages in Ge, GaP or other compound semiconductors. Such behavior has been found in GaAs.[4]

5. DEPTH DEPENDENCE OF DISORDER

Calculations of the depth dependence of the amount of disorder have been made by a number of authors[36–38] (see also Mayer and coworkers[1]). These may be tested against two basic types of

† This may not be strictly true at lower ion energies. There is some indication that for 30–50 keV implantations of heavy ions such as Sb, there may not be such pronounced annealing effects.[35] This may be due to the proximity of the disorder to the surface and should be checked further.

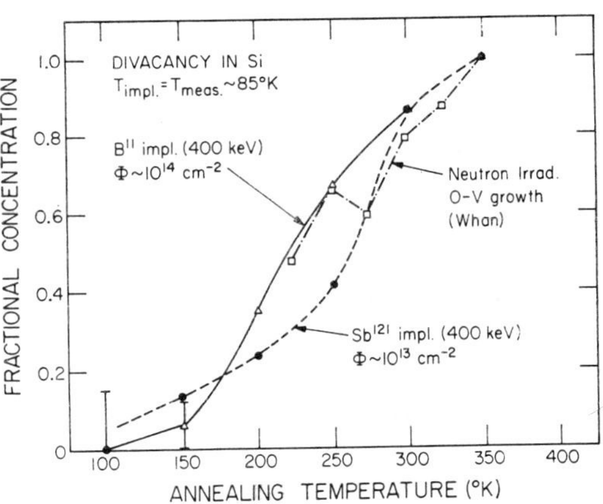

FIG. 9. Fraction of divacancies versus annealing temperature for 400 keV-B and Sb ion implantation into Si at 85 °K. The annealing periods were 20 minutes. Also shown is the vacancy-oxygen (A center) growth following neutron irradiation. (From Vook and Stein[34].)

experimental investigations: the one, channeling effect measurements and the other, layer removal coupled with EPR or IR absorption, for instance.

Channeling effect measurements of the depth distribution of disorder have primarily been concerned with light ions (B, C, Ne) where the penetration depth is large.[39,40] The major problem is in accounting correctly for the dechanneling of the incident aligned beam. Aside from lattice strain effects, the dominant scattering mode producing dechanneling must be identified. Figure 7 shows disorder distributions derived for 3 different scattering calculations (single, multiple, plural). In this case the plural treatment gave the most consistent results.

The results of such channeling measurements[39,40] indicate that the peak in the disorder distribution is 10 to 20 per cent less than the calculated range of the ion (for B, C, and Ne implants). This is roughly in accord with theoretical predictions. Disorder distributions determined from results of layer removal studies also are consistent with calculated distributions. Measurements[41] of the $1.8\,\mu$ IR absorption in 400 keV-oxygen implants as a function of depth indicated that the divacancy depth distribution was similar to that calculated by Brice.[37] Figure 8 shows the differential distribution of Si-P3 center for a 400 keV 0^+ implant into Si and the distribution of energy into atomic processes.[42]

The IBM Group have made extensive investigations[15,43,44] to compare the depth distribution of the number of electrically active centers with the disorder profile. The ratio of the location of the damage peak to that of the peak in the concentration of electrically active centers was about 0.7 for P and 0.8 for As.[43] These values are in reasonable agreement (although somewhat lower) than calculated values. For low doses (10^{11} to 10^{13} ions/cm²), the location of the damage peak was determined by changes in the electrical properties of thin uniformly doped Si layers as a function of depth. For higher doses, the growth of a continuous amorphous layer was studied with EPR, optical transmission and visual observation of disorder.

A further interesting aspect of this work was that annealing of the radiation damage centers was not uniform throughout the implanted layer.[15,43] For example that portion of a phosphorus implanted layer in which an amorphous condition was formed becomes essentially completely electrically active and uncompensated after a 600 °C 30-minute anneal.

Beyond the boundaries of this region, i.e. in material which was heavily damaged but not completely amorphous, a 600 °C anneal was not sufficient to eliminate completely the compensation effects, i.e. radiation damage persisted in some measure.

6. NATURE OF THE DISORDERED REGION

One of the fundamental problems is the nature of the disordered region around the track of the particle. One deceptively simple question to ask is whether the region is primarily amorphous (heavily disordered) or is primarily crystalline with a large defect concentration. Since specific defects such as the divacancy and Si-P3 center are only defined (and detected) in a locally crystalline environment, a measurement of their number might be used to analyze the character of the disordered region. However, most divacancies are not formed directly but only as a result of vacancy migration, i.e. a large fraction of the divacancy population may be formed outside the initially disordered region.

Divacancies[45] are found in R.T. implanted silicon. However, the number of divacancies for a 200 keV Sb implant is approximately a factor of 600 (60 from comparison of initial production rates) lower than the number of scattering centers determined in channeling effect measurements. This is too large a discrepancy to be explained in terms of lattice distortion around each divacancy and we must conclude that the measured divacancy population accounts for only a small fraction of total number of displaced Si atoms.

In low temperature implants ($\sim$80–90 °K) the formation of divacancies is inhibited by the low mobility of the single vacancies.[34] Figure 9 shows the growth of the number of divacancies as a function of anneal temperature for B and Sb implants at 85 °K. Relatively few divacancies are observed immediately after 85 °K implants and growth of divacancies occurs between 150 and 300 °K. At 300 °K the divacancy density is almost equal to that for the same ion fluence at 300 °K implant temperature. It can also be noted that the growth of the fractional concentration of divacancies is similar to the growth of the O-V (oxygen-vacancy) center in neutron irradiated material. From this and other data, Vook and Stein[34] conclude that neutral vacancy motion and trapping control the divacancy formation upon annealing.

Another approach to the analysis of the disordered region is provided by transmission electron microscopy studies which can image disorder zones rather than identify the specific defects.[46,47]

Chadderton and Eisen[47] have used channeling-effect measurements, electron microscopy and infrared absorption to examine the damage produced in room and liquid nitrogen temperature boron implantations. They concluded that in R.T. implants, fast moving point defects nucleate at clusters somewhat homogeneously while at liquid nitrogen temperatures the motion of the defects is arrested and that damage is 'frozen in'. Electron transmission micrographs indicated that the disordered region around the track could not be described as completely amorphous. Even in low temperature implants only a small portion of the disorder regions are amorphous and these amorphous zones do not disappear until temperatures slightly in excess of 450 °C are attained—temperatures well above that where the majority of the disorder (as measured by IR absorption and by channeling effect) has been annealed out, i.e. the greater part of the disorder signal seems not to be associated with the amorphous zones.

Note that there are major differences in the nature of the disorder generated in room and low temperature implants—in the dose and dose rate dependence of the disorder,[48] the nature of the dislocations formed upon anneal and in the presence of clearly identified defects.[47] This again implies that low-temperature implantation conditions should not be overlooked in evaluating disorder effects in other materials.

7. CRYSTALLINE TO AMORPHOUS TRANSITION

We are now led to ask how a continuously amorphous layer is formed if the individual disordered regions are not amorphous. The formation of the amorphous zones in low temperature implants probably results from the overlap of individual disordered regions. In R.T. implants, it could be due to an overlapping of disordered regions and/or coalescing of defects at pre-existing clusters and amorphous zones.[47]

Stein and coworkers[45] have shown that the divacancy production rate saturates and then decreases at increasingly heavy ion doses in R.T. implants (Figure 10). The absence of saturation for boron implants was attributed to diffusion of vacancy-associated defects. Comparisons of calculations of the total energy spent in atomic pro-

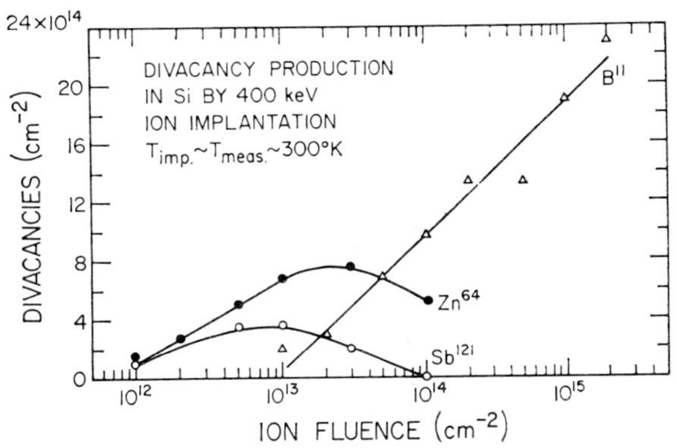

FIG. 10. Divacancy density versus dose for 400 keV Sb, Zn and B ions implanted at room temperature in silicon. (From Stein and coworkers[45].)

cesses with data for heavy ion (Zn and Sb) implants suggests that there is a maximum divacancy concentration the lattice can support which is related to a critical energy E spent in atomic processes per unit volume. The maximum concentration ($\approx 7 \times 10^{19}/cm^3$) is reached when $E \approx 10^{20}$ keV/cm³ and an amorphous layer is formed when $E \approx 10^{21}$ keV/cm³. Picraux and coworkers[33] suggest for low temperature implantations that a value of 5×10^{20} keV/cm³ can be used to estimate the dose necessary to reach the transition to an amorphous layer.

The annealing growth of divacancies following low temperature ion implantation has been attributed to vacancy liberation from defect clusters (Vook and Stein[34]). Similarly, they conclude that the crystalline-to-amorphous transition is limited by the annealing of neutral vacancies which move into relatively undamaged crystalline Si from isolated ion-produced defect clusters.

Morehead and Crowder[49] have considered a model where vacancies escape via thermal diffusion from the disordered core surrounding the ion track. The formation of the amorphous layer is due to the overlap of the disordered regions remaining around the track. They find reasonable agreement between calculations and their data on the ion species and temperature dependence of the critical dose required to form an amorphous layer.

This does not imply that the mechanisms responsible for the crystalline-to-amorphous transition have been resolved. There are somewhat conflicting views in the different models. However, with the variety of experimental techniques available, one can anticipate that the situation will be clarified shortly.

8. SUMMARY

There is a vigorous attack directed toward the analysis of disorder effects in ion implantation processes—the diversity of techniques and approaches that have been tried is impressive. The fact that many of the techniques used in investigations of electron and neutron irradiated semiconductors can be applied to implanted layers has provided the means for a direct comparison between implantation and irradiation effects. Other techniques, such as channeling effect measurements, have not only provided insight into implantation phenomena but also their usefulness and limitations have been better defined as a result of these investigations.

Much of the thrust behind these studies comes from the requirements imposed by the use of implanted layers in device applications. In many cases apparent anomalies in electrical behavior suggested investigations that exposed new facets of implantation phenomena—boron in silicon provides an example of this.

As a result of these investigations, the situation in regard to silicon is improving. The identification of specific defect species in implanted layers has led to indications of a strong correlation between anneal behavior in fast-neutron irradiated and implanted samples. The increase of the number of divacancies as an implanted sample is warmed to 300 °K is attributed to the motion of vacancies—a phenomena also attributed to the dependence of disorder on implantation temperature.

The depth distribution of disorder in implanted layers seems to be in reasonable agreement with theoretical predictions; at least no major conflict has developed. It has also been shown that the disorder distribution in relation to that of the implanted dopants can have a major influence on the depth distribution of electrically active centers following anneal cycles. The reordering of the amorphous layer at about 550–600 °C has a key influence on the electrical behavior of the implanted region.

There are, of course, major areas that require further clarification. The necessity for high anneal temperatures for near-complete electrical activity when an amorphous layer has not been formed requires explanation. The nature of the disordered region around the particle track is not completely settled although the evidence indicates that these regions are not primarily amorphous. This also bears on the question of the formation of the amorphous layer in high dose implants. There are also areas where the problems have not been sufficiently defined such as in lifetime studies or in the influence upon electrical characteristics of dislocations or of enhanced diffusion.

Implantation phenomena in compound semiconductors should receive increased attention due to both device potential and insight gained from silicon. An approach would be to investigate the behavior of samples implanted at low temperatures to determine if an amorphous layer could be formed that would exhibit well defined anneal behavior.

ACKNOWLEDGEMENTS

This review was difficult to write in a concise form because of the explosion of data and the diversity of approaches. I wish to thank those who sent preprints of their data and listened to my attempts to summarize the situation: Lew Chadderton, Bill Crowder, John Davies, Fred Eisen, Al Mac Rae, Ogden Marsh, Tom Picraux, and Herman Stein. Special thanks are due to Ian Mitchell who disposed of the rambling, first draft of this review and forced the many revisions. His comments clarified many of the issues.

REFERENCES

1. J. W. Mayer, L. Eriksson and J. A. Davies, *Ion Implantation in Semiconductors* (Academic Press, New York, 1970).
2. Proceedings of International Conference on Ion Implantation in Semiconductors. Thousand Oaks, California, May 1970. Published in *Rad. Effects*.
3. R. R. Hart, H. L. Dunlap and O. J. Marsh, this conference.
4. W. H. Weisenberger, S. T. Picraux and F. L. Vook, this conference.
5. J. A. Davies, J. Denhartog, L. Eriksson and J. W. Mayer, *Can. J. Phys.*, **45**, 4053 (1967).
6. J. W. Mayer, L. Eriksson, S. T. Picraux and J. A. Davies, *Can. J. Phys.*, **46**, 663 (1968).
7. B. L. Crowder, R. S. Title, M. H. Brodsky and G. D. Pettit, *Appl. Phys. Letters*, **16**, 205 (1970).
8. T. C. McGill, S. L. Kurtin and G. A. Shifrin, *J. Appl. Phys.*, **41**, 246 (1970).
9. H. J. Stein, F. L. Vook and J. A. Borders, *Appl. Phys. Letters*, **14**, 328 (1970).
10. F. L. Vook and H. J. Stein, *Rad. Effects*, **2**, 23 (1969).
11. D. F. Daly and K. A. Pickar, *Appl. Phys. Letters*, **15**, 267 (1969).
12. K. L. Brower, F. L. Vook and J. A. Borders, *Appl. Phys. Letters*, **15**, 208 (1969).
13. L. J. Cheng and J. Lori, *Phys. Rev.*, **171**, 856 (1968).
14. J. W. Mayer, *IEEE*, **NS15** (No. 6), 10 (1968).
15. B. L. Crowder, *J. Electrochem. Soc.*, **117**, 671 (1970).
16. R. Baron, G. A. Shifrin, O. J. Marsh and J. W. Mayer, *J. Appl. Phys.*, **40**, 3702 (1969).
17. H. J. Stein in reference (2).

18. N. G. E. Johansson and J. W. Mayer, *Solid State Electron.*, **13**, 123 (1970).
19. A. H. Clark and K. E. Manchester, *Trans. AIME*, **242**, 1173 (1968).
20. T. E. Seidel and A. U. Mac Rae, *Trans. AIME*, **245**, 491 (1969).
21. D. E. Davies, *Appl. Phys. Letters*, **14**, 227 (1969).
22. R. R. Hart and O. J. Marsh, *Appl. Phys. Letters*, **15**, 206 (1969).
23. J. E. Westmoreland, J. W. Mayer, F. H. Eisen and B. Welch, *Appl. Phys. Letters*, **15**, 308 (1969).
24. J. C. North and W. M. Gibson, *Appl. Phys. Letters*, **16**, 125 (1970).
25. G. Fladda, K. Bjorkqvist, L. Eriksson and D. Sigurd, **16**, 313 (1970).
26. Papers contained in reference (2); see for example J. C. North and W. M. Gibson as well as T. E. Seidel and A. U. Mac Rae, therein.
27. O. Meyer and J. W. Mayer, *J. Appl. Phys.*, **41**, 4166 (1970).
28. D. E. Davies and S. A. Roosild, *Appl. Phys. Letters*, **17**, 107 (1970).
29. K. A. Pickar and J. V. Dalton in reference (2).
30. P. Sigmund, *Appl. Phys. Letters*, **14**, 114 (1969).
31. J. U. Andersen, O. Andreasen, J. A. Davies and E. Uggerhoj in reference (2).
32. N. G. E. Johansson, G. Fladda and B. Domeij in reference (2).
33. S. T. Picraux, W. H. Weisenberger and F. L. Vook, in reference (2).
34. F. L. Vook and H. J. Stein in reference (2).
35. S. T. Picraux, J. E. Westmoreland, J. W. Mayer, R. R. Hart and O. J. Marsh, *Appl. Phys. Letters*, **14**, 7 (1969).
36. P. V. Pavlov, D. I. Tetel'baum, E. I. Zorin and V. I. Alekseev, *Soviet Phys.—Solid State*, **8**, 2141 (1967).
37. D. K. Brice, *Appl. Phys. Letters*, **16**, 103 (1970) and in reference (2).
38. J. E. Westmoreland and P. Sigmund in reference (2).
39. L. C. Feldman and J. W. Rogers, *J. Appl. Phys.*, **41**, 3776 (1970).
40. J. E. Westmoreland, J. W. Mayer, F. H. Eisen and B. Welch in reference (2).
41. H. J. Stein, F. L. Vook and J. A. Borders, *Appl. Phys. Letters*, **16**, 106 (1970).
42. K. L. Brower, F. L. Vook and J. A. Borders, *Appl. Phys. Letters*, **16**, 108 (1970).
43. B. L. Crowder and R. S. Title in reference (2).
44. B. J. Masters, J. M. Fairfield and B. L. Crowder in reference (2).
45. H. J. Stein, F. L. Vook, D. K. Brice, J. A. Borders and S. T. Picraux in reference (2).
46. R. W. Bicknell and R. M. Allen in reference (2).
47. L. T. Chadderton and F. H. Eisen in reference (2).
48. F. H. Eisen and B. Welch in reference (2).
49. F. F. Morehead and B. L. Crowder in reference (2).

DISCUSSION

Question (DONOVAN) What structural information is available on the amorphous character of the implanted layers?

Answer (MAYER) I have made no such measurements myself but Parsons and others have made electron-diffraction studies; it looks just like genuine amorphous Si (or Ge).

Comments from Bill Crowder *added in proof*.

Basically, the IBM group and the Sandia group are studying different aspects of the same problem; hence, the different simple models advanced by the two groups. I do not feel that the two models are at odds, but that they are rather simple approximations to reality which stress different aspects of the problem. The Sandia group is interested in the behavior of simple defects, which has lead them to a model emphasizing a 'homogeneous' nucleation of such point defects in which migration and recombination play an important role. We are interested in amorphous clusters, which we strongly feel are produced by a 'heterogeneous' nucleation process under almost all experimental conditions. The change of ESR signal (isotropic) with dose can be characterized by a relatively linear increase up to a saturation dose (even for B implantations at room temperature for fixed dose rate!). This result implies a 'heterogeneous' nucleation. This does not imply that all lattice disorder is correctly described by a heterogeneous model—the amount of lattice disorder present as amorphous clusters is not necessarily a large fraction of the total disorder (e.g. B in Si at room temperature).

LOW TEMPERATURE CHANNELING MEASUREMENTS OF ION IMPLANTATION LATTICE DISORDER IN GaAs†

W. H. WEISENBERGER,‡ S. T. PICRAUX, AND F. L. VOOK

Sandia Laboratories, Albuquerque, New Mexico 87115, U.S.A.

Lattice disorder resulting from 140 keV Zn and 151 keV Xe implantations below 100 °K was studied by channeling effect analysis using 400 keV protons. Implantations and analyses were performed in the same system without warmup. Isochronal anneal curves show significant annealing of disorder below room temperature for low temperature, low fluence implants. The annealing is similar for both Zn and Xe implants and also is similar to previous measurements of the annealing of thermal conductivity following low temperature electron irradiation. For 298 °K implants, the disorder production is strongly dose rate dependent and increases significantly with increasing dose rate. For dose rates used in these measurements, the disorder production at 298 °K was greatly reduced from that for the low temperature implantations.

1. INTRODUCTION

Although low temperature irradiation and annealing experiments have been reported for electron[1-3] and neutron[1,4] irradiations in GaAs, no previous measurements have been reported for ion implantation into GaAs below room temperature. Previous implantations into GaAs have been made at or above room temperature.[5-8] Channeling effect measurements of lattice disorder for room temperature implants indicate annealing begins above 350 °K for low fluence implants.[5,6] Saturation disorder implantations require higher anneal temperatures near 900 °K.[5,6,8] Electrical measurements have been made primarily for saturation disorder implantations and have required anneal temperatures greater than 900 °K to obtain significant electrical activity.[7]

Electrical measurements[1] of neutron damage introduced into GaAs at liquid nitrogen temperature show little annealing below 600 °K; but disorder introduced by electron damage in GaAs near liquid nitrogen temperature shows considerable annealing below room temperature for electrical mobility,[1,2] carrier concentration,[1,2] and thermal conductivity measurements.[3] Because of the strong relationship of radiation effects to ion implantation, channeling effect measurements were made to investigate the production and annealing of lattice disorder introduced by low-temperature ion implantation in GaAs. It is also of interest to see if the ion species implanted into GaAs strongly affects the disorder or its annealing. In order to investigate such a dependence on ion species, implants were made with Zn; a standard dopant atom for GaAs, and Xe a heavier inert gas ion.

2. EXPERIMENTAL TECHNIQUE

Single crystal GaAs samples were aligned with a 400 keV proton beam to $\leqslant 0.1°$ of the $\langle 111 \rangle$ crystal axis using the channeling effect technique.[9,10] Analysis of each sample was made prior to implantation in order to assure good crystalline quality. Implants were made into the 'B' (As) face.[11] A preselected orientation, 5° away from the $\langle 111 \rangle$ axis, was used for all implantations and for the non-channeled spectra. All analyses were performed along the $\langle 111 \rangle$ direction with a 400 keV beam near 95 °K.

The implanted ions were mass separated by a 15° analyzing magnet, quadrupole focused, electrostatically steered and uniformly swept over the implanted area. A beam profile monitor and the current from a conducting ring surrounding the sample were used to verify the location and the uniform scanning of the beam. Figure 1 shows the experimental arrangement.

The channeling analysis was performed with a monoenergetic proton beam collimated for a full angular divergence of $\sim 0.08°$. Protons scattered at 145° were detected by a cooled surface barrier detector. Typical energy resolution was 8 keV (FWHM) which corresponds to a depth resolution of about 400 Å near the surface. The projected

---

† This work supported by U.S. Atomic Energy Commission.

‡ Associated Western Universities AEC-DNET Graduate Fellow from U. of Wyoming.

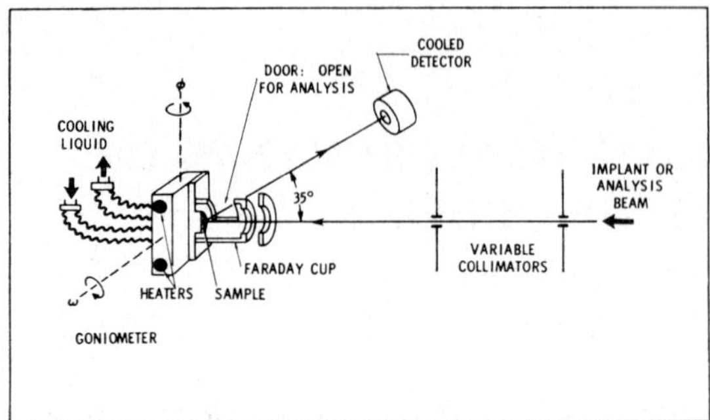

FIG. 1. Schematic of experimental arrangement.

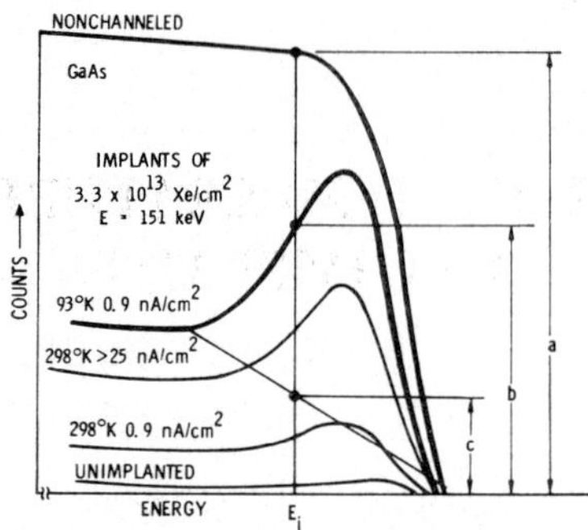

FIG. 2. Backscattered spectra for $\langle 111 \rangle$ channeled 400 KeV protons in GaAs for an unimplanted sample and for three 151 KeV Xe implants. The implants were made to the same fluence level ($3.3 \times 10^{13}/\text{cm}^2$) under different temperatures and dose rate conditions. A non-channeled spectrum also is shown.

range of 140 keV Zn is 540 Å.[12] Relative ion energies used in these experiments were measured to 1 per cent accuracy. However, absolute values have only been calibrated to approximately $\pm$ 10 keV.

The crystals were oriented by means of a variable temperature goniometer. The sample was tilted about three mutually perpendicular axes by means of Datex Encoderdyne motors which have direct angular readout of shaft position corresponding to 0.01°. Temperatures between 90 and 500 °K could be maintained by means of heaters between the cold block and the sample (Figure 1). Bellows from the liquid nitrogen lines to the cold block allowed free movement of the temperature stage in the goniometer. To reduce annealing before or during analyses, measurements were always performed at low temperatures, typically 95 °K. During implantations, the samples were held in a Faraday cup (Figure 1) with a backscattering collection solid angle of 98.5 per cent of 2π. A door on the side of the Faraday cup was opened for backscattering analysis.

3. ANALYSIS

Examples of proton backscattering spectra are shown in Figure 2 for unimplanted GaAs and for 151 keV Xe implantations to 3.3×10^{13} Xe/cm^2. The spectra shown are for one implantation at 93 °K and for two implantations with different dose rates at 298 °K. The minimum yield (χ min) for the unimplanted samples (the ratio of the $\langle 111 \rangle$ channeled yield to the nonchanneled yield) was typically 0.025 or less. It can be seen from the relative sizes of the surface peaks that the disorder is greatly reduced for the 298 °K implants.

The increased yield in the disorder peak corresponds to scattering from lattice atoms $\gtrsim 0.10$ Å from their normal lattice position along the $\langle 111 \rangle$ row. The area of the peak after subtraction for

background dechanneling gives a quantitative measure of the disorder. The data can be converted into the total number of Ga and As scattering centers/cm^2 by comparison to the scattering yield for the non-channeled orientation using the known atomic density for GaAs and the stopping power of protons.

The method used[13] for channeling effect analysis[14–16] of backscattered proton spectra will be summarized here with respect to the representative experimental data of Figure 2. At a given energy (E_i) corresponding to a mean depth t into the crystal, the dechanneled fraction of the analyzing beam is estimated to be $c/a = \chi_r(E_i)$ using a straight line dechanneling background subtraction. Thus, $1 - \chi_r(E_i)$ is the fraction of the probing beam sensitive to atoms off lattice sites. The fractional scattering yield at this energy resulting from atoms displaced from regular lattice sites is $(b - c)/a = \chi_2(E_i) - \chi_r(E_i)$. If a channeled beam were traversing a completely disordered region which is thick compared to the resolution of the system and if the dechanneled level at E_i were c, the backscattered level would be at the nonchanneled level, and the fractional yield corresponding to this disordered region would be $(a - c)/a = 1 - \chi_r$. Thus the density of scattering centers at a channel of energy E_i can be estimated by

$$n'(E_i) = \left[\frac{\chi_2(E_i) - \chi_r(E_i)}{1 - \chi_r(E_i)} \right] n_0,$$

where n_0 is the atomic density of GaAs.

The total number of scattering centers in the disordered region is found by summing the disorder of each channel. This gives

$$D = \sum_{i=1}^{m} \left[\frac{\chi_2(E_i) - \chi_r(E_i)}{1 - \chi_r(E_i)} \right] n_0 \Delta t_{\mathrm{ch}},$$

where D is the total disorder in scattering centers/cm². The depth corresponding to one channel Δt_{ch} was assumed to be constant over the small depth of the implanted region and is given by

$$\Delta t_{\mathrm{ch}} = \Delta E_{\mathrm{ch}}/(\Delta E/\Delta x),$$

where $\Delta E_{\mathrm{ch}} = E_{i+1} - E_i$, and ΔE is the average net incremental energy loss per backscattered particle over the total additional path for a depth Δx into the crystal. A constant was used for the non-channeled level, a, in these calculations. Experimental stopping power results[17] were used together with a geometry of 145° backscattering to obtain $\Delta E/\Delta x$. Uncertainties caused by detector resolution, the level of dechanneling and the lack of specific knowledge of dechanneling mechanisms give rise to an uncertainty in the analyzed experimental results of at least 10 per cent.

4. RESULTS

The dependence of GaAs lattice disorder on implantation fluence for 140 keV Zn and 151 keV Xe implants is shown in Figure 3. The disorder production for a given fluence is much less for the 298 °K implantations than for the low temperature implants. Thus a significant amount of annealing can occur during implantation at room temperature for GaAs. The production of disorder per incident ion has been taken from the initial slope of the low temperature disorder production curves and is seen to be very similar for Xe and Zn. For 151 keV Xe

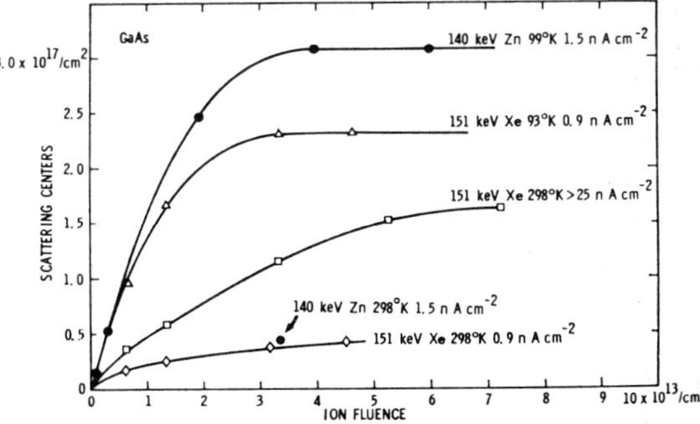

FIG. 3. Dependence of GaAs lattice disorder versus implantation fluence for 140 KeV Zn (1.5 nA/cm² at 93 °K and 298 °K) (●) also for 151 KeV Xe [93 °K (△) and at 298 °K at two different dose rates (□, ◇).] All analyses were performed below 100 °K.

we observe 14,300 scattering centers per incident ion, and for 140 keV Zn, 14,200 scattering centers per incident ion.

From Figure 3 the saturation levels for cold implantations of 140 keV Zn and 151 keV Xe are 3.1×10^{17} and 2.3×10^{17} scattering centers/cm² respectively. These levels are expected to differ in proportion to the implanted layer depth. For a completely disordered region all the atoms would act as scattering centers. Assuming these saturation levels correspond to completely disordered regions, these levels together with the density of GaAs give estimates of the total disorder depths of 700 Å for 140 keV Zn and 520 Å for 151 keV Xe. These compare reasonably well with estimates of the total depths of the heavily disordered regions using the ion range[12] plus half the mean spread in range to give 693 Å for 140 keV Zn and 455 Å for 151 keV Xe.

The room temperature Xe implants show a strong dose rate effect. The disorder at the same fluence level was approximately three times greater for an increase in the dose rate from 0.9 nA/cn², to levels between 25 and 85 nA/cm². This is a further indication that annealing is occurring during implantation at room temperature. During all other Xe implants, a constant dose rate of 0.9 na/cm² was maintained.

In order to compare implantation dose rates into GaAs for Xe and Zn ions i.e., different mass and energy ions, it is desirable to maintain the same energy density into atomic processes per second (energy density rate).[13,18] From Brice's calculation[12] the energy into atomic processes is 107 keV for 151 keV Xe. Again we use the projected range plus half of the mean spread in range as our calculated estimate of the depth of the disordered layer. The average energy density rate for a 0.9 nA/cm² Xe beam with 107 keV into atomic processes into a depth of 455 Å is thus 1.3×10^{20} eV/cm³-sec. For 140 keV Zn implants the energy into atomic processes is 94 keV and the calculated depth of the disordered layer is 693 Å. To maintain the same energy density rate, the beam current for 140 keV Zn is 1.5 nA/cm². This beam current was used for all our Zn implants in these experiments as shown in Figure 3.

The results for 15 minute isochronal annealing of disorder introduced into GaAs at low temperatures are shown in Figure 4. All analyses were performed at ≈95 °K. The low fluence Zn (circles) and Xe (triangles) implants show similar strong anneal behavior below 350 °K. A cold Xe implant to

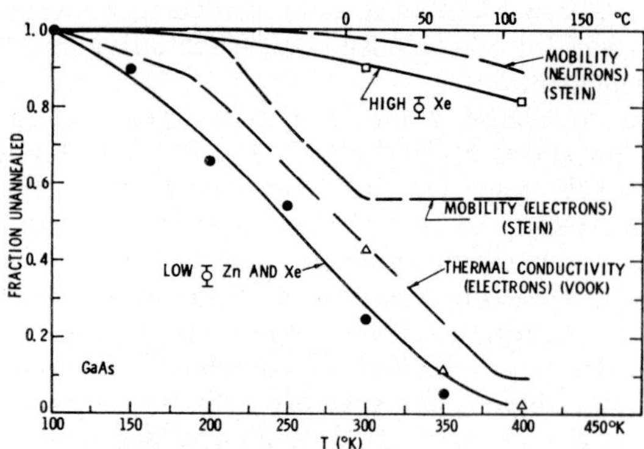

FIG. 4. Anneal behavior of low fluence Zn implant (●), low fluence Xe implant (△) and saturation level Xe implant (□). Anneal behavior of electron mobility for neutron and electron irradiation[1] and of thermal conductivity[3] are also shown for comparison (dashed lines).

saturation disorder (squares) does not show significant annealing below 400 °K. Anneal behavior of electron mobility for low temperature electron and neutron irradiation by Stein[1] and the anneal behavior of thermal conductivity for electron irradiation by Vook[3] are also shown in Figure 4 for comparison (dashed lines). The neutron irradiation fluence was far below that necessary for a saturation disorder level since the total average energy density into atomic processes transferred to the lattice was about a factor of 10^5 lower than for the low fluence ion implantations.

5. DISCUSSION

Based on the nature of energy transfer to the lattice and on previous results for silicon[18] it was expected that ion implantation damage should be more like neutron than electron damage in GaAs. Therefore, the large recovery of GaAs lattice disorder was very unexpected since previous measurements of luminescence,[4] carrier mobility,[1] and carrier concentration[1] following low temperature neutron irradiation showed little recovery below 600 °K.

The anneals of the disorder for both Zn and Xe low dose ion implantations as shown in Figure 4 very closely resemble the anneal of the thermal conductivity after heavy 2 MeV electron irradiation.[3] This would suggest that the same types of defects are controlling these anneals. Furthermore, the carrier mobility measurements show partial recovery in this same temperature range for electron irradiation but not for neutron irradiation.[1] All recovery of the mobility from electron damage does

not occur, however, until 550 °K where another annealing stage occurs.[1] The thermal conductivity and channeling effect measurements are more linear measures of the total amount of disorder at high defect concentrations than are Hall mobility measurements which regain their sensitivity only when the concentration of defects recovers to approximately the original carrier concentration. Therefore, we may expect better agreement of these channeling results with the thermal conductivity than with the mobility results. However, with such a strong anneal stage in the below-room-temperature ion implantation disorder, it is surprising that no corresponding anneal stages have been observed for neutron damage in GaAs.

Of particular importance is the fact that both the Zn and Xe cold low-dose implants showed similar anneal characteristics, indicating that the disorder does not have a strong ion species dependence. Similarities in the low temperature anneal behavior of lattice disorder have also been observed in silicon for quite different implanted ions and interpreted in terms of the basic properties of the target material.[13]

The saturation fluence Xe cold implant did not anneal appreciably below 400 °K. This is consistent with low temperature implantations[13] into Si for which a saturation disorder level implant did not anneal appreciably below room temperature even though a low fluence implantation showed a large amount of annealing. For room temperature implantations into Si and GaAs, a difference between the anneal temperatures of low fluence and saturation fluence implantations is also observed.[5,6,19] Room temperature GaAs implants have been observed to anneal around 500 °K for low fluence implants,[5,6] but temperatures near 900 °K are required to anneal the lattice disorder for saturation levels.[5,6,8]

A consequence of an annealing stage near room temperature is a lower disorder production per ion for the room temperature implants. This annealing is also responsible for the large dose rate effect observed. By increasing the dose rate from 0.9 nA/cm² to between 25 and 85 nA/cm², the disorder introduction per ion is increased by about a factor of 3. This clearly indicates the importance of specifying the dose rate and temperature for GaAs implants near room temperature. One 298 °K Zn implant was made at the same energy density rate as the 0.9 nA/cm² Xe implant. Figure 3 shows that the disorder for this Zn implant agrees with that for the Xe implant.

The energy into atomic processes per scattering center can be determined from the slope of the production curve and the energy per ion into atomic processes. For 140 keV Zn at 99 °K this value is 94,000 eV/ion per 14,200 scattering center/ion = 7.5 eV/scattering center. For 151 keV Xe at 93 °K, this value is 6.6 eV/scattering center. This rough agreement again indicates no strong dependence on ion mass or species for lattice disorder production at low temperature. These values are lower than the threshold values of 9.1 to 9.8 eV/defect measured by Bäuerlein[20] for electron produced lattice damage in GaAs. Since the energy per displaced atom would be expected to be at least twice the threshold value,[21] we may suggest that the defects produce strained lattice regions which result in several scattering centers per defect.

The similarity of the annealing results for low temperature electron irradiation and ion implantation suggests that the same defect or defects are responsible for the observed results in thermal conductivity, carrier mobility, and channeling experiments. Such a defect must cause enough lattice strain to scatter phonons, and atoms must be displaced sufficiently to serve as scattering centers for channeling measurements.

Arsenic vacancy motion has been tentatively identified with an annealing stage with an activation energy of 1.0 eV by the As overpressure experiments of Potts and Pearson.[22] They correlated As vacancy motion with the annealing stage near 500 °K observed by Aukerman and Graft.[23] It is logical that vacancy motion would have a much higher activation energy for compound semiconductors since motion is confined to the sublattice of that atom species. Arnold[24] has tentatively correlated the annealing of the Ga vacancy with an annealing stage near 425 °K. Further support for a high anneal temperature for a single vacancy in a compound semiconductor is the positive identification by EPR and infrared absorption of the Zn vacancy in ZnSe and its motion with an activation energy of 1.25 eV.[25] The annealing below room temperature in GaAs is, therefore, not believed to be associated with monovacancy annealing.

The defects responsible for our observed results are probably simple defects of the type which can be produced by electron irradiation. They must have the right configuration to scatter phonons and produce scattering centers for channeling measurements, and they are more mobile than isolated vacancies. Possibilities for these defects are interstitials, vacancy interstitial pairs, or AB divacancies.

ACKNOWLEDGEMENT

The authors wish to thank R. G. Swier and J. H. Smalley for their valuable technical assistance and D. K. Brice, H. J. Stein and J. A. Borders for stimulating discussions.

REFERENCES

1. H. J. Stein, *J. Appl. Phys.*, **40**, 5300 (1969).
2. K. Thommen, *Rad. Effects*, **2**, 201 (1970).
3. F. L. Vook, *Phys. Rev.*, **135**, A1742 (1964).
4. C. E. Barnes, *IEEC Trans. on Nuc. Sci.*, NS–17 (1970).
5. G. Carter, W. A. Grant, J. D. Haskell and G. A. Stevens, *Rad. Effects*, **6**, 277 (1970).
6. J. S. Harris and F. H. Eisen, *Rad. Effects*, **7**, 123 (1971).
7. R. G. Hunsperger and O. J. Marsh, *J. Electrochem. Soc.*, **116**, 488 (1969); *Rad. Effects*, **6**, 263 (1970).
8. J. L. Whitton and G. Carter, 'Atomic Collision Phenomena in Solids' (North-Holland, Amsterdam, 1970), p. 615.
9. J. U. Andersen, J. A. Davies, K. O. Nielsen and S. L. Andersen, *Nucl. Inst. and Meth.*, **38**, 210 (1965).
10. J. A. Borders and S. T. Picraux, *Rev. Sci. Inst.*, **41**, 1230 (1970).
11. For particles incident on the 'B' face along the [$\bar{1}\bar{1}\bar{1}$] direction, the order of the atoms is Ga-As-space-space-Ga-As-space-space-etc. The 'B' face can be distinguished by the shiny side and the A face by etch pits after a bromine-methanol etch.
12. D. K. Brice, private communication.
13. S. T. Picraux, W. H. Weisenberger and F. L. Vook, *Rad. Effects*, **7**, 101 (1971).
14. E. Bøgh, *Can. J. Phys.*, **46**, 653 (1968).
15. F. H. Eisen, B. Welch, J. E. Westmoreland and J. W. Mayer, 'Atomic Collision Phenomena in Solids' (North-Holland, Amsterdam, 1970), p. 111.
16. L. C. Feldman and J. W. Rodgers, *J. Appl. Phys.*, **41**, 3776 (1970).
17. S. Gorodetzky, A. Pape, E. L. Cooperman, A. Chevallier, J. C. Sens and R. Armbruster, *Nuc. Inst. Mcthc.*, **70**, 11, (1969)
18. F. L. Vook and H. J. Stein, *Rad. Effects*, **2**, 23 (1969); *Rad. Effects*, **6**, 11 (1970).
19. J. A. Davies, J. Denhartog, L. Eriksson and J. W. Mayer, *Can. J. Phys.*, **45**, 4053 (1967).
20. R. Bauerlein, *Z. Naturforsch.*, **14a**, 1069 (1959).
21. P. Sigmund, *Appl. Phys. Letters*, **14**, 114 (1969).
22. H. R. Potts and G. L. Pearson, *J. Appl. Phys.*, **37**, 2098 (1966).
23. L. W. Aukerman and R. D. Graft, *Phys. Rev.*, **127**, 1576 (1965).
24. G. W. Arnold, *Phys. Rev.*, **183**, 777 (1969).
25. G. D. Watkins, *Bull. Am. Phys. Soc.*, **14**, 312 (1969); **15**, 290 (1970).

ION IMPLANTATION OF SULPHUR INTO GaAs, GaP AND Ge MONOCRYSTALS

J. L. WHITTON AND G. R. BELLAVANCE

Solid State Science Branch, Atomic Energy of Canada Limited, Chalk River, Ontario, Canada

The possibility of using ion implantation to form high concentration junctions in semiconductors has been explored for the specific case of sulphur in GaAs, GaP and Ge. The effects of ion dose, ion energy, crystal orientation and target temperature have been investigated by means of radiotracers and sectioning techniques.

It is shown that high concentration junctions can be formed using an incident ion having high electronic stopping cross-section and implanted along the $\langle 110 \rangle$ channeling directions of the crystals. A large increase in junction concentration may be obtained when the GaAs and GaP crystals are maintained at 150 °C during the implantation process, but this is not the case with Ge. Rutherford back-scattering of 1 MeV He$^+$ ions has been used to measure the ion-bombardment induced damage in the crystals and to show how this damage can be annealed by heating the crystal during the implantation. The annealing, at temperatures up to 150 °C, is most effective in GaAs and least effective in Ge.

The introduction of donors and acceptors into III–V compound semiconductor monocrystals by thermal diffusion is difficult because of (a) preferential evaporation of the Group V element at normal diffusing temperatures, (b) chemical combination of the dopant with the compound surface and (c) the different rates of diffusion along the separate sub-lattices.

One attractive alternative method is that of ion implantation, which promises not only easily controlled conditions at ambient or low temperatures, but also, using the channeling effect, offers the possibility of forming high concentration, relatively undamaged, junctions at various depths within a crystalline target. A case in point is the possible formation of a well defined junction of oxygen or sulphur in a zinc-doped GaP luminescent device.

The formation of junctions by ion implantation has, to date, been done in crystals deliberately tilted from channeling directions in order to avoid the unwanted effect of deep penetration of the incident ions. However, it has been shown[1,2] that complete suppression of channeling is not possible simply by tilting a crystal off-axis, and therefore implantation into so-called random directions of a crystal can never result in an abrupt junction. It appears possible to form such a junction in a semiconductor crystal by making use of the channeling phenomenon and it is this possibility that will be considered here.

Before ion implanted junctions can be formed, with a reasonable degree of control, an appreciation of the processes involved in the slowing down of energetic ions in crystalline solids is necessary. Therefore, a knowledge is required of the effect on the implantation profile of the parameters: crystal orientation, incident ion energy, ion type, ion dose and target temperature.

Heavy ions implanted into solids lose energy by two mechanisms, nuclear and electronic stopping, the latter dominating at high energies. The channeling phenomenon, however, discriminates against nuclear collisions. The two mechanisms may be separated as follows: (a) the ions that enter the channels (electronic stopping) and (b) those that are prevented from doing so by large angle collisions (nuclear stopping). However, an overlap of the two is always present at the energies used here.

Recent work, both experimental and theoretical,[3–7] has shown that the electronic stopping cross-section (S_e) varies in a periodic fashion with the atomic number of the incident ion, and that these oscillations are related to the electronic shell structure of the ions. Those ions with closed outer shells have minimum S_e and those with 2/3 filled outer shells have maximum S_e values.

The significance of the oscillating S_e in the ion implantation of the III–V compound semiconductors is that the Group II acceptors have minimum S_e while the Group VI donors have quite high S_e. Since the slowing down of channeled ions depends on the S_e, it is expected that the Group VI donors should be slowed down rapidly and will reach their maximum range (R_{max}) at a shallow depth within the crystal. Conversely, the low S_e of the Group II acceptors will lead to less efficient slowing down, hence much deeper R_{max}. Thus is has been shown for 40 keV S ions (high S_e) in GaAs, $R_{max} = 1.5\,\mu$. On the other hand, for 40 keV ions of minimum S_e, the R_{max} of Zn in GaAs is $>4\,\mu$, of Na in Ge is $>7\,\mu$[1] and of Cu in Si is $>4.5\,\mu$[2]. Since the likelihood of dechanneling increases with the distance travelled by the ion, the possibility of forming a high concentration shallow junction should be more favourable for the donors S, Se and Te than for the acceptors Zn and Cd.

These considerations have been applied in the present study, specifically to obtain high concentration sulphur junctions in GaAs, GaP and Ge. The radiotracer ^{35}S was implanted into carefully prepared crystal surfaces and the concentration profiles were measured by a sequence of radioactivity measurements combined with a precise sectioning technique capable of removing layers as thin as 50 Å.[8] The residual activity measured at different depths allows the construction of integral depth distribution curves.

The effect of crystal orientation is shown in Figure 1, where 20 keV ^{35}S has been implanted into the three low index directions of GaAs. Penetration of the ions increases as the channel size, i.e. in the order $\langle 111 \rangle$, $\langle 100 \rangle$, $\langle 110 \rangle$ and, more important, a well defined R_{max} is seen only in the

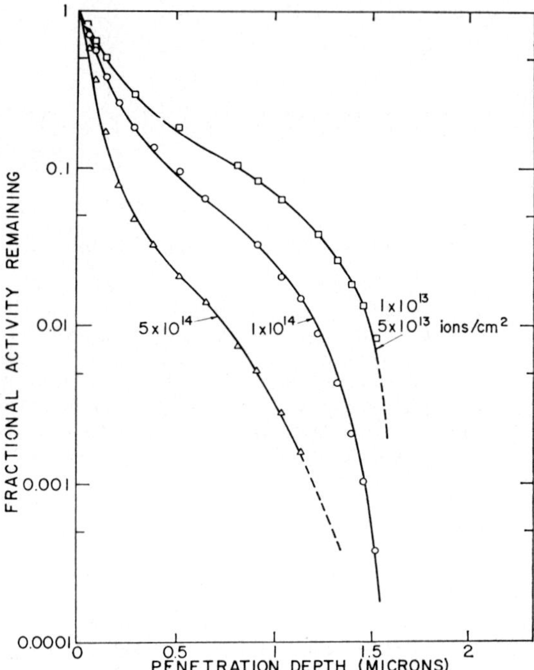

FIG. 2. Effect of increase in dose on the channeled fraction of incident 40 keV ^{35}S in $\langle 110 \rangle$ GaAs.

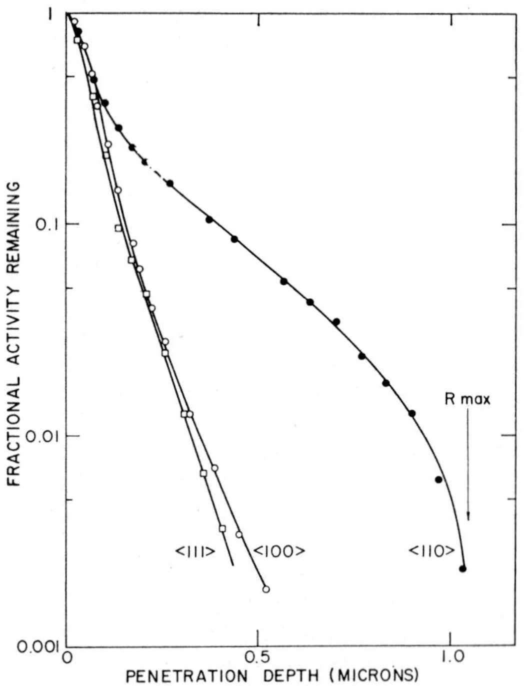

FIG. 1. Effect of crystal orientation on the implantation profiles of 20 keV ^{35}S in GaAs.

$\langle 110 \rangle$ case. This makes clear that for electronic stopping to be the dominant energy loss mechanism, the rate of dechanneling, due to large angle collisions in the channels, must be kept low. For this reason, all crystals used in further study were aligned to $\pm 1°$ or less from the $\langle 110 \rangle$ direction.

Ion bombardment induced radiation damage occurs at relatively low doses in semiconductor crystals and is the major obstacle to the introduction of high impurity concentrations to a layer deep within the crystal. Gross disorder in the crystal surface reduces the fraction of incident ions that can become channeled and this causes a reduction in ion range. Figure 2 shows range profiles for 40 keV ^{35}S ions in $\langle 110 \rangle$ GaAs and demonstrates that the fraction of channeled ions decreases with increasing dose. Little effect is seen up to doses of 5×10^{13} ions/cm² but a dramatic change occurs as the dose is increased to 5×10^{14} ions/cm². The maximum concentration of impurity atoms in the deepest 3000 Å layer is $\sim 7 \times 10^{16}$/cm³. Any subsequent implantation results in an increased concentration of ions only near the surface, i.e. the induced radiation damage has the effect of disordering the crystal lattice to the point that channels are no longer available for the deep penetration of ions.

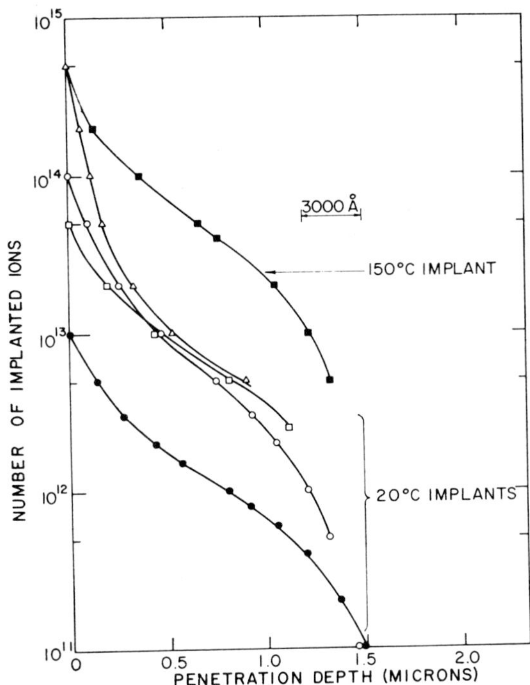

FIG. 3. Effect of increase in dose on the number of incident 40 keV ^{35}S ions channeled in $\langle 110 \rangle$ GaAs —upper curve shows effect of implanting at 150 °C.

This is shown more clearly in Figure 3 which has the data from Figure 2 replotted to show the number (rather than the fraction) of impurity atoms at any depth. The number of channeled atoms increases with increase in dose from 1×10^{13} to 5×10^{13} ions/cm² but a further increase of factor 10 in dose results in no increase in the number of channeled ions.

This radiation damage effect would seem to set an upper limit on the number of channeled ions. However, annealing studies of ion implanted semiconductors indicate that much of the damage may be removed by heating the target during implantation.[1,9,10] The three crystal types were therefore implanted to high (5×10^{14} 40 keV S ions/cm²) 'damage' doses at 20 °C and 150 °C. The net effect in GaAs is seen in Figure 3 (upper curve). This 150 °C implant has allowed an increase in concentration of S ions at the deepest 3000 Å layer to 4×10^{17}/cm³, a factor of ~ 6 increase over the 20 °C implant.

Similar concentration levels have been measured for 40 keV S ions in GaP and Ge and show an increased junction concentration of a factor 2 to 3 in GaP but little or no increase in Ge. The maxi-

mum ranges, or junction depths, are at $1.5\,\mu$ in GaAs and Ge and at $0.7\,\mu$ in GaP.

An assessment of the annealing effect in the three crystal types has been made by Rutherford backscattering studies of 1 MeV He$^+$ in crystals implanted at temperatures from 20 °C to 150 °C. These data are presented in Figures 4, 5 and 6 and serve to confirm the results obtained by the radiotracer depth distribution measurements. Both the GaAs and GaP show significant annealing of surface damage when implanted at 150 °C and, in the GaAs case, the crystal lattice is almost completely reordered by the annealing treatment. The effect is not so marked in GaP but significant annealing has taken place, unlike Ge where implantation at 150 °C shows little or no change from room temperature implantation.

These annealing data in GaAs are in good agreement with the electron microscope observations of Mazey and Nelson[10] who observed dense arrays of dislocations but no indication of amorphicity, as evidenced by electron diffraction patterns, when they implanted 1×10^{16} 80 keV Ne^{20} ions into GaAs held at temperatures between 50 °C and 200 °C.

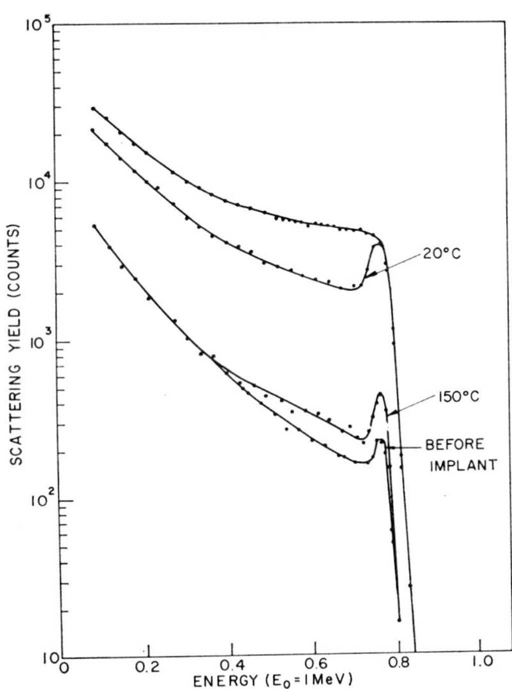

FIG. 4. Effect of substrate temperature on disorder in ion-implanted [(5×10^{14} 40 keV S/cm²) $\langle 110 \rangle$ GaAs] as shown by the spectra of Rutherford back-scattered 1 MeV He$^+$ ions.

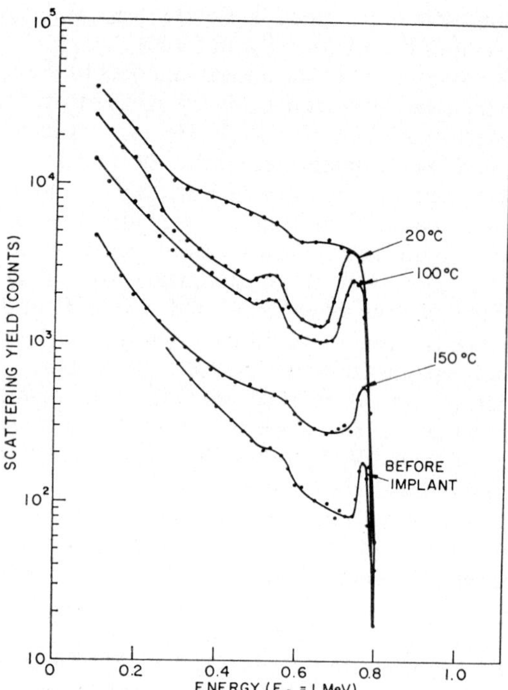

FIG. 5. Effect of substrate temperature on disorder in [(5 × 10¹⁴ 40 keV S/cm²) ⟨110⟩ GaP] as shown by the spectra of Rutherford back-scattered 1 MeV He⁺ ions.

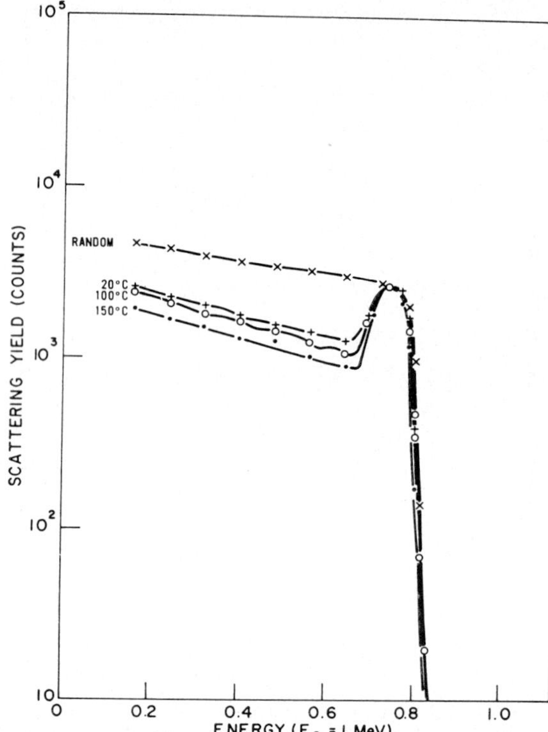

FIG. 6. Effect of substrate temperature on disorder in [(5 × 10¹⁴ 40 keV S/cm²) ⟨110⟩ Ge] as shown by the spectra of Rutherford backscattered 1 MeV He⁺ ions.

It has been previously established that the temperature required for the annealing of disorder during implantation is less than that required for post-implant annealing.[1,11] Post-annealing of heavy ion bombarded GaAs required a temperature of 300 °C,[1] a factor 2 higher than the 150 °C required here. This factor of approximately 2 has been observed previously for heavy ion implantation of Si.[11]

These results show that channeling is a useful method of forming deeply embedded concentration junctions and that annealing at the moderately low temperature of 150 °C can increase significantly the junction impurity concentration.

An added advantage of this relatively low annealing temperature lies in the fact that the thermal vibrational amplitude of the lattice atoms at 150 °C is scarcely changed from room temperature. An increase in atomic thermal vibrational amplitude could lead to a decrease in channeling, because of reduced effective channel size. This reduction in channeled fraction has been demonstrated by Dearnaley and coworkers[12] on implanting phosphorus into ⟨110⟩ silicon at 400 °C.

The junction concentration of 4×10^{17} S ions/cm³

obtained here may be regarded as a lower limit, since more precise crystal orientation (to better than the $\pm 1°$ used here) will allow an increase in channeled ions as will an increase in incident energy; higher energies favour electronic stopping. The shift in R_{max} (junction depth) with increase in incident energy is small and can be readily forecast, as R_{max} has an $E^{0.5}$ dependence.[1,4]

ACKNOWLEDGEMENTS

The authors are pleased to acknowledge the help of D. C. Santry, O. Westcott and C. Sitter in the ion bombardments. Some of the work reported here was done during a stay by one of us (J.L.W.) at the Electrical Engineering Department, University of Salford, England.

REFERENCES

1. J. L. Whitton and G. Carter, Inter. Conf. on Atomic Collision Phenomena in Solids, Sussex, England (North-Holland Publishing Co.) (1970), p. 615.
2. G. Dearnaley, M. A. Wilkins, P. D. Goode, J. H. Freeman and G. A. Gard, ibid., p. 633.

3. B. Fastrup, A. Borup and P. Hvelplund, *Can. J. Phys.*, **46**, 489 (1968) (and references therein).
4. L. Eriksson, J. A. Davies and P. Jespergard, *Phys. Rev.*, **161**, 219 (1967).
5. F. H. Eisen, *Can. J. Phys.*, **46**, 561 (1968).
6. J. Bøttiger and F. Bazon, *Rad. Effects*, **2**, 105 (1969).
7. I. M. Cheshire and J. M. Poate, Inter. Conf. on Atomic Collision Phenomena in Solids, Sussex, England (North-Holland Publishing Co.) (1970), p. 351 (and references therein).
8. J. L. Whitton, *J. Appl. Phys.*, **36**, 12 (1965).
9. S. T. Picraux, J. E. Westmoreland and J. W. Mayer, *Appl. Phys. Letters*, **14**, 1 (1969).
10. D. J. Mazey and R. S. Nelson, *Rad. Effects*, **1**, 229 (1969).
11. J. A. Davies, J. Denhartog, L. Eriksson and J. W Mayer, *Can. J. Phys.*, **45**, 4053 (1967).
12. G. Dearnaley, J. H. Freeman, G. A. Gard and M. A. Wilkins, *Can. J. Phys.*, **46**, 587 (1968).

DISCUSSION

Question (GIBSON)   Is it possible that the high penetration observed for Na implantation might be due to interstitial diffusion?

Answer (WHITTON)   Interstitial diffusion could possibly account for some enhanced penetration, but the fact that the maximum range of 40 keV Na in Ge is not observed at a depth of 7μ indicates that the very low electronic stopping cross-section of Na is the factor mainly responsible for high penetration.   High penetration has also been seen for 40 keV Zn in GaAs and for 40 keV Cu in Si.   In the latter case Dearnaley and coworkers, [Inter. Conf. on Atomic Collision Phenomena in Solids, Sussex, England p. 615 (North-Holland Publishing Co.) 1970] have shown by transmission of Na through thin crystals of Si that the penetration is instantaneous, therefore not due to diffusion.

Question (GLOTIN)   What is your beam divergence as compared to the critical angle?

Answer (WHITTON)   The combination of crystal misalignment and beam divergence is less than half the critical angle.

LUMINESCENCE FROM IMPLANTED Zn IONS IN GaAs†

G. W. ARNOLD, R. E. WHAN, J. K. MAURIN AND J. A. BORDERS

Sandia Laboratories, Albuquerque, New Mexico 87115, U.S.A.

Abrupt changes in the near-band-edge luminescence of n-type undoped GaAs after implantation with 400 keV Zn ions and vacuum annealing at 580 °C are reported. The good agreement of the spectral position, half-width and temperature dependence of the emission obtained after implantation and annealing with that of melt-doped GaAs:Zn indicates that implanted Zn ions have been incorporated at Ga lattice sites. The larger number of Zn substitutions obtained when bombardment is made on the Ga face than for an equivalent fluence on the As face demonstrates the existence of a polar implant effect.

1. INTRODUCTION

The ion implantation process creates lattice damage and strain in semiconducting materials which degrade device performance through their effect on charge carrier mobility and lifetime. The characterization and elimination of these lattice imperfections is one of the major problems in implantation technology. Most commonly, the defect concentration is either minimized by maintaining the material at elevated temperatures during bombardment, or is reduced after implantation by annealing under the appropriate conditions. Thermal annealing dissociates lattice damage defect complexes and allows the implanted ion to attain a minimum energy configuration at a lattice site. In such positions an implanted dopant ion should impart the same characteristics to the material as if incorporated in the lattice from the melt or by diffusion.

The near-band-edge luminescence of GaAs can be excited with relatively high quantum efficiency and is a particularly useful monitor of radiation damage brought about by neutron,[1] electron,[2-4] γ-ray[5] or ion bombardment.[6] The material has a high damage sensitivity as evidenced by the rapid luminescence degradation brought about by charge-carrier trapping at non-radiative recombination quenching centers produced by the irradiation. The annealing of such defects in the host lattice can also be followed by monitoring the luminescence as a function of annealing time and temperature. In those regions of the bombarded sample where the concentration of implanted ions is large, thermal annealing may lead to a substitutional dopant concentration sufficiently great so that luminescence

brought about by recombination involving the dopant center dominates the near-band-edge spectrum. A great deal of information exists concerning the luminescent process involving dopant centers in GaAs, particularly for Zn. Thus, the observation of recombination radiation involving the Zn acceptor level gives, in itself, the useful information that the Zn ion has been incorporated on a Ga lattice site.

Electroluminescence from ion-bombarded GaAs has been observed after Zn ion implantation by Rougham and Manchester[7] and by Hunsperger and Marsh.[8] Electroluminescence peaks from GaAs diodes, however, shift with current density and can occur either from emission in the n-type region or from the p-type side, and it is not clear that the observed spectra involve the Zn level. The present paper utilizes front-surface photoluminescence as a monitor of the damage and recovery of the host lattice brought about by Zn-ion bombardment and annealing, and compares the known spectral position, half-width and temperature dependence of emission involving Zn in as-grown material with that brought about by Zn-ion bombardment and subsequent annealing.

2. EXPERIMENTAL

Room temperature 400 keV Zn-ion implantations were made on both the A(Ga) and B(As) (111) faces of boat-grown n-type GaAs of relatively high purity ($n \cong 10^{15} - 10^{16}/\mathrm{cm}^3$). Channeling effects were minimized by rotating the sample about 5° with respect to the ion beam. The B faces were chemically polished and the A faces were mechanically lapped. Front-surface photoluminescence measurements were made at either liquid He temperature or in the boil-off He gas from a storage

† This work supported by the U.S. Atomic Energy Commission.

dewar. Excitation of the luminescence was made with a He-Ne 5 mW laser. The luminescence was scanned and detected by means of a Perkin-Elmer 210 G grating spectrophotometer and an RCA 7102 cooled photomultiplier. Incremental removal of surface layers was by means of a Syntron vibratory polisher. Annealing was done in vacuum ($\sim$10^{-6} Torr) in a tube furnace with temperatures held to $\pm 2\,°$C. Surface characteristics were monitored by means of a model JSM2 JEOLCO scanning microscope and a Siemens electron-diffraction camera.

3. RESULTS AND DISCUSSION

The 5 °K near-band-edge luminescence from boat-grown n-type GaAs of relatively high purity is as shown in Figure 1. The peak at 1.513 eV is that due to free exciton recombination. The transition at 1.491 eV and its LO phonon replica at 1.454 eV is believed to be due to recombination at the Si acceptor level.[9] Silicon is the major impurity[10] consistently found in unintentionally doped boat-grown GaAs and acts amphoterically to form shallow donor and acceptor levels. The luminescence exhibited by this sample was virtually quenched by a 400 keV Zn-ion fluence of 10^{14} Zn ions/cm^2. Irradiation to such fluence levels results in plastic deformation of the implanted material[6,11] and a consequent loss of crystalline order. The enhanced reactivity of ion-bombarded GaAs (111) surfaces[12] precludes thermal annealing in gaseous atmospheres or bombardment through passivating SiO$_2$ coatings. Such treatment usually results in further luminescence degradation. Host-lattice luminescence was substantially recovered by annealing in vacuum at 580 °C for 75 minutes. The surface depth deterioration due to As loss at 580 °C for the annealing time used is expected to be of the order of a monolayer.[13]

After recovery of the host-lattice luminescence, vibratory polishing was employed to strip off surface layers in order to allow excitation of larger concentrations of Zn ions. The projected range of 400 keV Zn ions in GaAs is about 1590 Å.[14] At a depth between 1300–1900 Å below the original implanted surface (B face) the host-lattice luminescence changed abruptly to that shown in Figure 2. Comparisons of this new luminescence with that obtained from GaAs grown with Zn added to the melt show good agreement with regard to spectral position and half-width ($\sim$0.016 eV). The divergence between the two curves on the low-energy

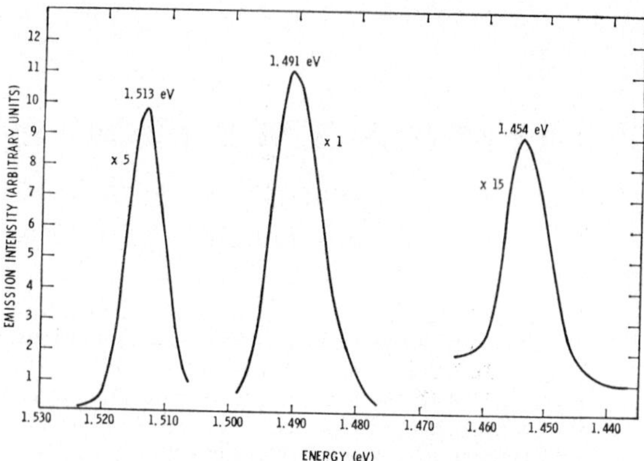

FIG. 1. Near-band-edge emission intensity vs photon energy for n-type boat-grown GaAs of relatively high purity. Measurement made at 5 °K.

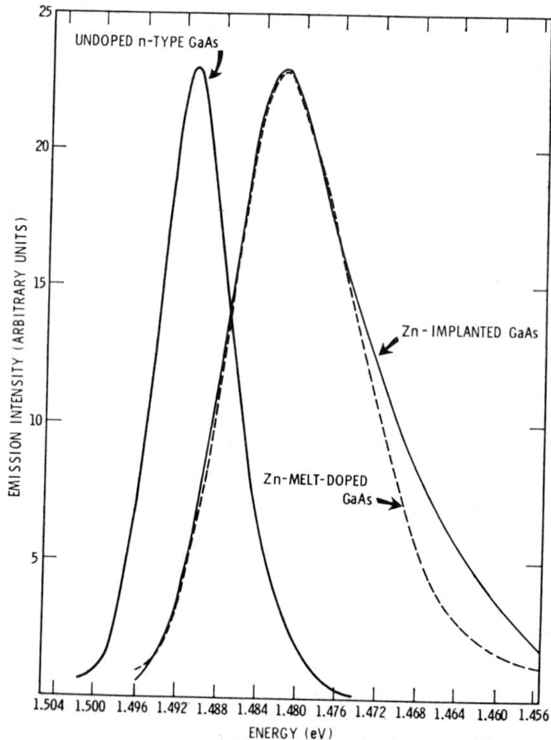

FIG. 2. Emission intensity vs photon energy for undoped GaAs before implantation, after implantation on the As face with 10^{14} 400 KeV-Zn ions/cm^2 followed by annealing and polishing, and for a melt-doped GaAs:Zn sample. All intensities normalized to the same value. Measurements made at 5 °K.

side of the peak is probably due to increased phonon emission caused by strain in the implanted sample.

The shift with temperature of the spectral position of the near-band-edge luminescence in relatively pure GaAs and in melt-doped GaAs:Zn is quite different. The population of higher excited states in Zn-doped material as the temperature is increased causes substantial shifts to higher energies.[15] The energy shifts with temperature of the near-band-edge luminescence in undoped GaAs are much smaller in magnitude. In Figure 3, the spectral position of the luminescence maximum as a function of temperature is shown for undoped material and for the Zn-implanted and Zn-melt doped GaAs. The excellent agreement of the shifts for both the GaAs:Zn samples in contrast to the much smaller shift of the host-lattice luminescence, offers convincing evidence that the luminescence from the implanted GaAs:Zn sample is the same as in melt-doped GaAs:Zn and that the annealing treatment has incorporated Zn at Ga lattice sites.

Zn-ion implantations have also been made on the Ga face of the (111) oriented samples. A comparison is made in Figure 4 of the luminescence obtained from a Ga-implanted unpolished face with that from an As-implanted and polished face. Of particular importance is the fact that even without stripping by vibratory polishing, the characteristic Zn luminescence is seen in the annealed Ga face implant. The differences in half-width and spectral position are due to the competitive host-lattice luminescence which is still present in the sample irradiated on the Ga surface. It has been assumed that the polar nature of the zinc blende lattice along the $\langle 111 \rangle$ axis will cause a higher displacement rate for A atoms when the irradiation is made on the A face than when an equivalent fluence of bombarding particles (usually electrons) are incident on the B face.[16] For the Zn-ion implant made on the Ga face, therefore, it might be expected that a larger number of acceptor-level Ga lattice sites would be produced for the implanted Zn ions to occupy, and this effect is apparently confirmed by the higher Zn acceptor concentration evidenced by the data. Although previous experiments[3,16] have been analysed under the assumption that a particular imbalance of atom displacements can be produced in III-V compounds by proper electron beam orientation with respect to the polar axis, it is believed that the present experiment provides the first direct evidence for the identity of the majority species and the occurrence of this effect for ion implantation. This polar

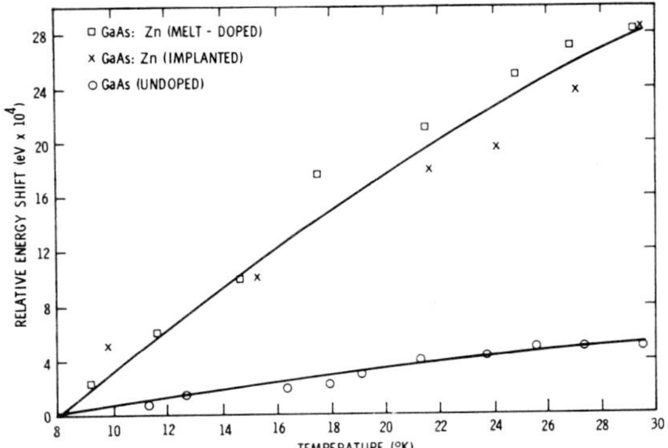

FIG. 3. Shift of luminescence maximum vs temperature for undoped GaAs and for melt-doped and implanted GaAs:Zn.

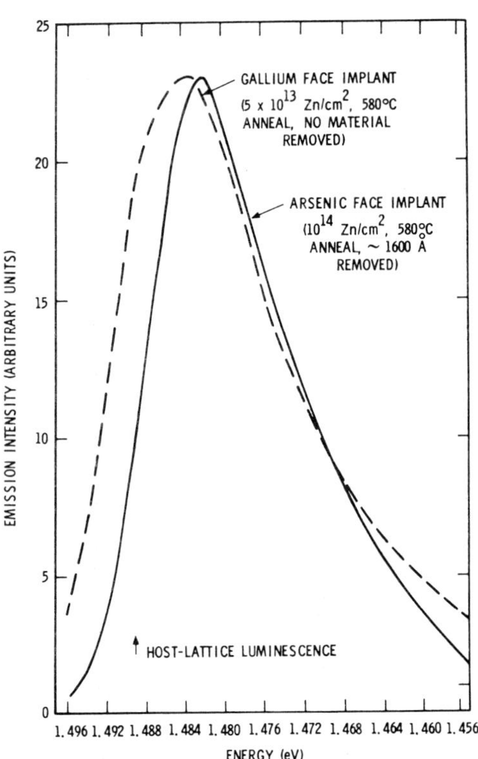

FIG. 4. Emission intensity vs photon energy for Zn-implanted and annealed GaAs samples with the Zn-ions incident on both the Ga and the As faces. Both intensities normalized to the same value. Measurements made at 5 °K.

implant effect should be of considerable importance and usefulness in the fabrication of ion-implanted III-V devices.

REFERENCES

1. M. C. Petree, *Appl. Phys. Letters*, **3**, 67 (1963).
2. G. W. Arnold, *Phys. Rev.*, **149**, 679 (1966).
3. G. W. Arnold and G. W. Gobeli in *Proceedings of the Santa Fe Conference on Radiation Effects in Semiconductors*, Ed. F. L. Vook (Plenum Press, New York, 1968), p. 435.
4. G. W. Arnold, *Phys. Rev.*, **183**, 777 (1969).
5. C. E. Barnes, *Phys. Rev.*, **1**, 4735 (1970).
6. G. W. Arnold and R. E. Whan, *Rad. Effects* **7**, 109 (1971).
7. R. E. Rougham and K. E. Manchester, *J. Electrochem. Soc.*, **116**, 278 (1969).
8. R. G. Hunsperger and O. J. Marsh, *Metallurgical Trans.*, **1**, 603 (1970).
9. E. W. Williams and D. M. Blacknall, *Trans. Metall. Soc. AIME*, **239**, 387 (1967).
10. K. B. Wolfstirn, *J. Phys. Chem. Solids*, **31**, 601 (1970).
11. R. E. Whan and G. W. Arnold, *Appl. Phys. Letters*, **17**, 378 (1970).
12. R. E. Whan, G. W. Arnold, W. E. Chambers and J. K. Maurin (to be published).
13. A. Y. Cho, *Surface Science*, **17**, 494 (1969).
14. W. S. Johnson and J. F. Gibbons, *Projected Range Statistics in Semiconductors* (Stanford University Press, Stanford, California, 1970).
15. G. W. Arnold and D. K. Brice, *Appl. Phys. Letters*, **13**, 51 (1968).
16. F. H. Eisen, *Phys. Rev.*, **135**, A1394 (1964).

DISCUSSION

Question (WHITTON)   Since the implantation dose of 1×10^{14} 400 keV Zn into GaAs is high enough to make the surface at least partly amorphous, is it reasonable to use the crystal polarity effect to explain the increased vacancy concentration in the 'A' side of the crystal.

Answer (ARNOLD)   The polar implant effect is important before the amorphous state is reached.

Question (UNKNOWN)   Did you make any electrical measurements?

Answer (ARNOLD)   No.

MEASUREMENT OF LATTICE DAMAGE CAUSED BY ION-IMPLANTATION DOPING OF SEMICONDUCTORS†

R. G. HUNSPERGER, E. D. WOLF, G. A. SHIFRIN, O. J. MARSH AND D. M. JAMBA

Hughes Research Laboratories, 3011 Malibu Canyon Road, Malibu, California 90265, U.S.A.

Two new techniques have been used to measure the lattice damage produced in gallium arsenide by the implantation of 60 keV cadmium ions. In one of these methods, optical reflection spectra of the ion-implanted samples were measured in the wavelength range from 2000 to 4600 Å. The decrease in reflectivity resulting from ion-implantation was used to determine the relative amount of lattice damage as a function of ion dose. The second technique employed the scanning electron microscope. Patterns very similar in appearance to Kikuchi electron diffraction patterns are obtained when the secondary and/or backscattered electron intensity is displayed in the scanning electron microscope as a function of the angle of incidence of the electron beam on a single crystal surface. The degradation of these 'Coates–Kikuchi' patterns resulting from ion implantation was used to obtain a quantitative measure of the lattice damage caused by the implantation process. The results of measurements made by both of the methods described have been compared with each other, and with data obtained by the more established method of measuring lattice damage by Rutherford scattering of 1 MeV helium ions.

During the process of doping a semiconductor by ion implantation some lattice damage is inevitably produced. This damage most often has been measured by observing the Rutherford backscattering of He^+ ions,[1] which is an extremely useful technique for establishing the relative amounts of lattice disorder present in samples and for identifying the location of implanted ions (substitutional or interstitial). However, a high energy accelerator is required, and the method involves time consuming critical alignment of the sample with the analysis beam. This paper describes two new techniques to measure lattice damage that are more simple to employ, using either a scanning electron microscope or a spectrophotometer. Measurements of lattice damage produced in gallium arsenide by implantation of 60 keV Cd^+ ions and its anneal behavior are used as the basis for comparison of the two newer methods with the more established backscattering technique.

The samples were prepared from GaAs $\langle 111 \rangle$ wafers, with the 'B' face etch-polished in a solution of methyl alcohol and bromine to remove damage from cutting and lapping. The substrates ($0.2\,\Omega$ cm, n-type) were implanted at room temperature with a 60-keV, mass separated Cd^+ ion beam. Doses ranged from 1×10^{12} to $1 \times 10^{15}/cm^2$ in different samples, while Cd^+ ion beam current was $\sim 10^{-7}$ a/cm². Samples implanted at low dose

† This work has been supported in part by NASA Electronic Research Center, Cambridge, Mass.

levels were tilted $\sim 8°$ off axis to minimize channeling effects.

The scanning electron microscope can be used to measure lattice disorder in the following manner. Patterns very similar to Kikuchi electron diffraction patterns are obtained when the secondary and/or backscattered electron intensity is displayed in the scanning-electron-microscope as a function of the angle of incidence of the electron beam on a single crystal surface.[2,3] The quality of these 'Coates–Kikuchi' patterns is sensitive to any chemical or physical process which tends to destroy or shift the periodicity of the first few hundred angstroms of a single crystal surface. In particular the lattice damage resulting from ion-implantation doping causes pattern degradation as shown in Figure 1. Along with the photographs that illustrate qualitative changes in pattern resolution are shown line-scan plots of the angular dependence of backscattered electron intensity, corresponding to the horizontal white lines in the photographs. To obtain a parameter which is a measure of pattern quality the fractional change in intensity ($\Delta I/\Delta I_{\text{ref}}$) across the $(0\bar{4}4)$ band edge (line) was determined by dividing the change observed for an ion-implanted sample (ΔI) by the intensity change for the reference crystalline sample (ΔI_{ref}). This fractional intensity change ($\Delta I/\Delta I_{\text{ref}}$) has been defined[4] as the normalized pattern quality I^*. A plot of the parameter $(1 - I^*)$ in Figure 2 shows Coates–Kikuchi pattern degradation as a function of implanted ion dose for electron beam acceleration plotted here so that total disorder coincides with complete loss of Coates–Kikuchi pattern detail; i.e., where $I^* = 0$ and $(1 - I^*)$, the complement of the normalized pattern quality, is equal to unity.

The two most significant features of Figure 2 are that all curves show a break point or saturation at about 5×10^{13} Cd$^+$/cm^2 dose, and the lattice damage for a given dose caused a greater degradation of the Coates–Kikuchi pattern quality at lower primary electron beam acceleration voltages. It is not yet clear just how the parameter $(1 - I^*)$ is related to the lattice disorder, but it is apparent that there is good agreement between the variation of $(1 - I^*)$ with implanted ion dose and the relative lattice disorder as determined by the standard He$^+$ backscattering technique. We do know from earlier experiments (see Figure 3[4]) that the backscattered electron intensity level for the completely amorphous condition (i.e., no pattern) was approximately the midpoint between all maxima and minima. That is, as lattice disorder destroys the identity of a direction of high penetration (i.e., originally a direction of minimum backscattered signal), the backscattered electron intensity increases. Conversely, the backscattered electron signal decreases from its maximum when the beam is aligned along a low index direction, e.g., within the (220) band, as the 'string effect' is destroyed by the lattice disorder.

A further important point evident in Figure 2 is that the range of measurement can be increased by using different electron energies. If the electron beam acceleration voltage is lowered, the low-dose regime $< 10^{12}$ Cd$^+$/cm^2 can be explored, and if the voltage is increased, the high-dose regime $> 10^{14}$Cd$^+$/cm^2 can be examined. However, one must be careful in interpreting these results in terms of an 'amorphous' layer thickness. The thickness of the implantation-caused damage layer in these samples was always much less than the maximum electron range even at 5.6 kV, and it did not enter the measured value of I^* merely as a weighted average of the electron penetration depth. For example, if the damage is extensive enough (dose $> 4 \times 10^{14}$) over the thickness of the theoretical 60-keV Cd$^+$ ion penetration depth (about 400 Å), then there is only random electron scattering and no pattern appears even for the 30 kV electron beam acceleration voltage.

Optical reflectivity measurements in the near ultraviolet portion of the spectrum have also been used to examine damage in these ion implanted layers and they indicate essentially the same

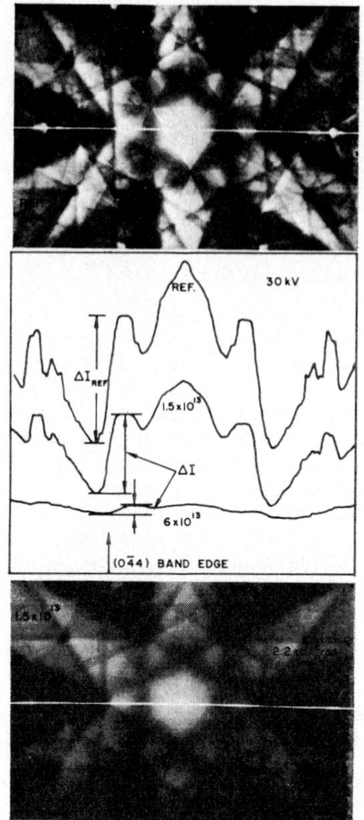

FIG. 1. Coates–Kikuchi patterns of the reference standard (top) and 1.5×10^{13} Cd$^+$/cm^2 ion dose (bottom). The line scans (center) of detector signal versus incident angle of the 30 keV electron beam were taken at the location indicated by the bright horizontal lines (top and bottom).

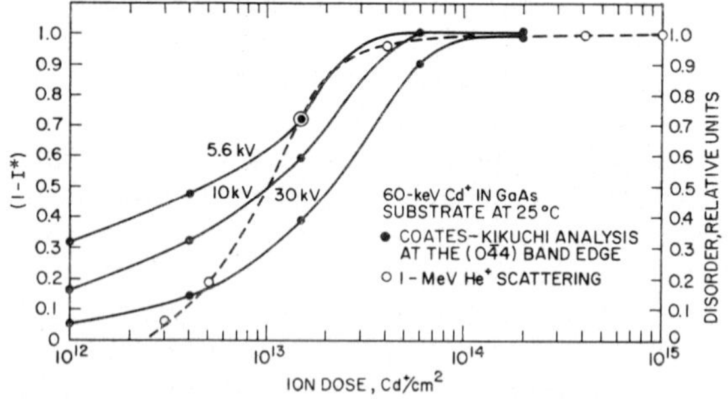

FIG. 2. Pattern degradation as a function of Cd$^+$ ion dose for 5.6, 10, and 30 keV electron beam accelerating voltages.

voltages of 5.6, 10, and 30 kV. Also included are the data obtained from Rutherford backscattering of 1 MeV He$^+$ ions.[5] The ordinate on the right in Figure 2 is a measure of lattice disorder in relative units as determined by the Rutherford backscattering measurements, and the data are dependence of damage on implanted ion dose as was observed with the scanning electron microscope. Reflectance measurements were performed on a Cary Model 14 double beam spectrophotometer with a specular reflectance attachment. (Details of this measurement technique have been previously published.[6,7,8]) Spectra of reflectivity as a function of photon energy for typical samples of Cd+ ion implanted GaAs are shown in Figure 3.

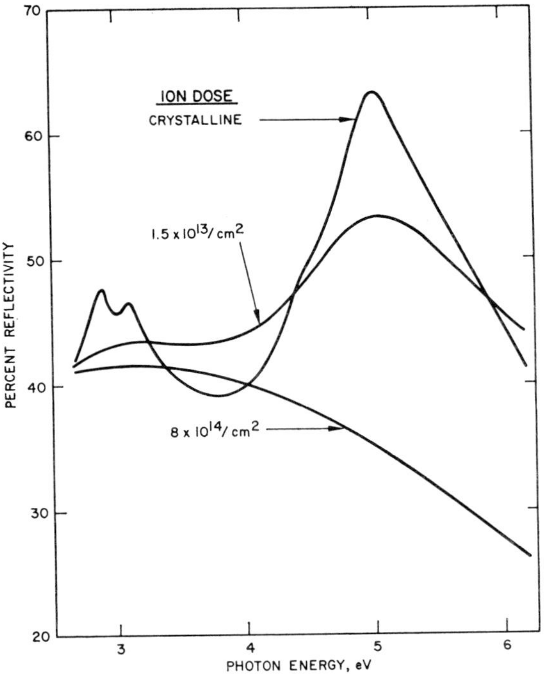

FIG. 3. Optical reflectivity spectra of 60 keV Cd+ implanted layers in GaAs (in the 3–6 eV range of photon energy).

The spectrum of unimplanted crystalline GaAs obtained on the same apparatus is shown for reference; prominent peaks occur at approximately 5.1 eV and 3 eV. The decrease in reflectivity at these peaks with increasing ion dose can be used as a measure of relative lattice damage in much the same way that the normalized pattern quality I^* was used in the scanning electron microscope measurements. Figure 4 shows the fractional change in reflectivity $[(R_0 - R)/R_0]$ at the 5.1 eV (2450 Å) peak as a function of ion dose, where R is the reflectivity of an ion implanted sample and R_0 is the reflectivity of a non-implanted crystalline sample (approx. 62 per cent). Note that a dose of 10^{12} ions/cm² only slightly damaged the GaAs, while doses greater than approximately 10^{14} ions/

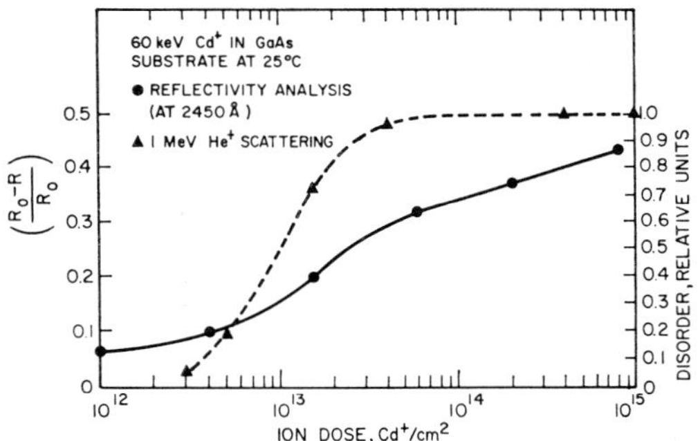

FIG. 4. Fractional change in reflectivity caused by implantation of 60 keV Cd+ ions. (R_0 = reflectivity of non-implanted crystalline sample = 62 per cent; R = reflectivity of ion implanted samples.)

cm² caused much damage. 'Saturation' of the damage versus dose curve begins at a dose of about 5×10^{13} ions/cm². For ion doses in this 'saturation' range the characteristic maxima and minima of the reflectivity spectra are eliminated, and spectra show a marked decrease in the rate of change of reflectivity with increasing dose. However, for the maximum dose used, a true saturation effect (i.e., no further spectral change) was not reached. Comparison of the reflectivity data with the relative lattice disorder determined by backscattering measurements[5] shows that both methods indicate 'saturation' of lattice disorder for ion doses $\gtrsim 5 \times 10^{13}$/cm², corresponding to formation of an essentially amorphous layer.

To verify that the observed changes in reflectivity and Coates–Kikuchi pattern quality were the result of lattice disorder and not merely caused by roughening of the sample surface (with resultant diffuse scattering of light[8]) or by formation of a surface oxide or contaminant layer, post-implantation annealing of the Cd+ ion-bombarded samples was performed in a series of isochronal (15 minute) cycles at temperatures ranging from 100 to 450 °C in 50 °C steps. (Annealing at 500 or 600 °C was also performed in some cases.) The annealing oven was slightly pressurized with a flowing nitrogen atmosphere to avoid contamination. As lattice damage was removed by annealing, the reflectivity and the Coates–Kikuchi patterns of the samples were restored.

The reflectivity spectra for annealed samples tended to progress toward that of crystalline GaAs as the anneal temperature was increased; characteristic peaks and valleys began to emerge. The effect of annealing on reflectivity at a photon energy of 5.1 eV is indicated in Figure 5 for three

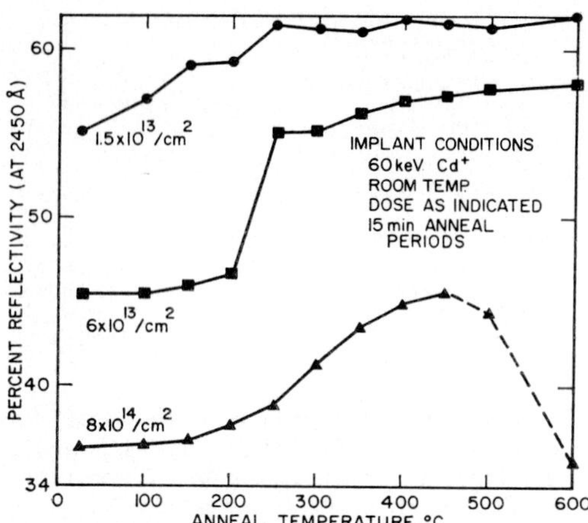

FIG. 5. Anneal behavior of reflectivity (at a photon energy of 5.1 eV).

different ion doses. At 1.5×10^{13} ions/cm² (less than saturation) the implantation did not reduce the reflectivity much below the crystalline value. With annealing, the reflectivity tended to return toward, and almost reach, the crystalline value, but after about 300 °C there was relatively little change up to 600 °C. At 6×10^{13} ions/cm² (just above saturation) the implantation initially reduced the reflectivity substantially below the crystalline value. However, annealing between 200 and 250 °C produced a rapid increase in reflectivity and (above this range) a gradual increase up to the maximum annealing temperature of 600 °C. The value of reflectivity even after the 600 °C anneal was significantly less than the crystalline value and was not much greater than that for the more lightly implanted layer (1.5×10^{13} ions/cm²) before annealing. The layer implanted with the largest ion dose (8×10^{14}/cm²) showed yet a different characteristic. The reflectivity before annealing was quite low. Annealing increased the reflectivity, but even after 450 °C it had only just reached the value which the intermediate sample had prior to annealing. However, upon further annealing at higher temperature this sample showed a marked decrease in reflectivity. This decrease is believed to be associated with decomposition of the crystal (perhaps through the loss of arsenic) when annealed above 400 °C, after having received an implanted dose sufficiently high to produce much lattice disorder. It is possible that the decomposition might have been prevented or reduced if the sample had been protected with an oxide layer before being annealed at this temperature; thin layers of SiO_2 ($\cong 2000$ Å thick) have been

proven effective in limiting decomposition during annealing to the extent that certain electrical properties of ion-implanted GaAs layers[9] are not measurably affected.

The anneal behavior of samples examined with the scanning electron microscope was essentially the same as that observed by reflectivity measurements. As lattice damage was removed by annealing, the Coates–Kikuchi patterns of the samples were restored. A quantitative measure of restored pattern quality is shown in Figure 6 where I^* is plotted as a function of anneal temperature for a number of samples. Comparison of the data of Figures 5 and 6 shows that both techniques of measurement indicate that substantial damage annealing occurred at temperatures below 250 °C for samples implanted with 1.5×10^{13} Cd⁺/cm². An anneal stage at about 200 to 250 °C was observed in samples implanted with 6×10^{13} Cd⁺/cm², while the samples implanted with a dose of 8×10^{14} Cd⁺/cm² exhibited only slight restoration after annealing even at 300 °C. (Mazey and Nelson have observed an anneal stage between 270 and 300 °C in Ne bombarded GaAs, using transmission electron microscopy.[10])

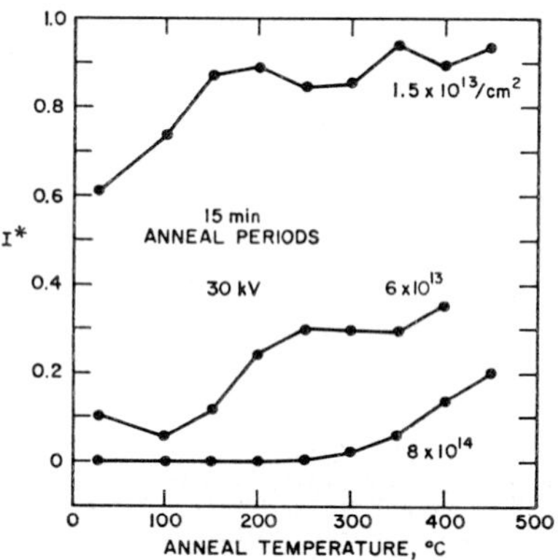

FIG. 6. Normalized Coates–Kikuchi pattern quality as a function of anneal temperature. (30 keV electron energy.)

Rutherford scattering measurements of lattice disorder[5] following annealing confirm this pattern. As shown in Figure 7, the lattice disorder (in terms of the number of scattering centers/cm² for an aligned He⁺ beam) was reduced almost to the background value ($\cong 1 \times 10^{16}$/cm²) in a sample implanted with 5×10^{12} Cd⁺/cm² after annealing at 300 °C. A sample implanted with 4×10^{13} Cd⁺/cm² exhibited gradual annealing at temperatures about 200 °C, while restoration of crystallinity was

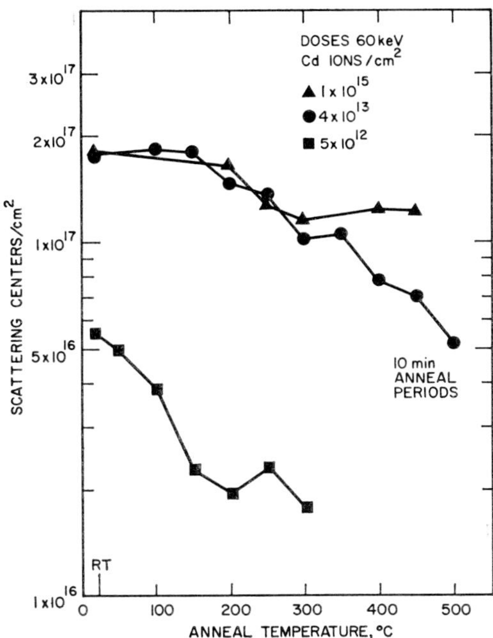

FIG. 7. Effect of annealing on the scattering centers (lattice disorder) produced in GaAs by 60 keV Cd⁺ ions implanted at room temperature.

relatively slight in the sample implanted with 1×10^{15} ions/cm² even after annealing at 450 °C. Surface decomposition of the samples implanted with the largest Cd ion dose was not observed; in Rutherford backscattering measurements, surface decomposition appears as increased crystallinity and hence is difficult to distinguish from annealing of damage.[11]

The ion dose dependence and anneal behavior of lattice damage, as measured using the spectrophotometer or the scanning electron microscope, agree well with results of similar measurements made by the standard Rutherford scattering technique. This agreement suggests that the newer methods of measurement of spectral reflectivity or Coates–Kikuchi pattern quality provide a convenient way to investigate lattice damage produced by ion implantation. The theory and technology of these methods have not yet been developed to the extent that determinations of damage versus depth profiles or implanted-ion lattice locations can be made, as they can by the Rutherford scattering method. However, the two newer methods have advantages (in addition to convenience) such as the fact that 30 keV electrons and 5 eV photons do not produce additional lattice disorder during the measurement process, while 1 MeV He⁺ ions do. Also, the spectral reflectivity method gives a clear indication of surface decomposition which cannot be mistaken for annealing of lattice damage.

REFERENCES

1. J. A. Davies, J. Denhartog, L. Eriksson and J. W. Mayer, *Can. J. Phys.*, **45**, 4053 (1967).
2. D. G. Coates, *Phil. Mag.*, **16**, 1179 (1967).
3. E. D. Wolf and R. G. Hunsperger, *Appl. Phys. Letters*, **16**, 526 (1970).
4. E. D. Wolf, M. Braunstein and A. I. Braunstein, *Appl. Phys. Letters*, **15**, 389 (1969).
5. E. Westmoreland, O. Marsh and R. Hunsperger, *Rad. Effects*, **5**, 245 (1970).
6. G. A. Shifrin and R. G. Hunsperger, *Appl. Phys. Letters* **17**, 274 (1970).
7. S. Kurtin, G. Shifrin and T. McGill, *Appl. Phys. Letters*, **14**, 223 (1969).
8. T. McGill, S. Kurtin and G. Shifrin, *J. Appl. Phys.*, **41**, 246 (1970).
9. R. G. Hunsperger and O. J. Marsh, *J. Electrochemical Soc.*, **116**, 489 (1969).
10. D. J. Mazey and R. S. Nelson, *Rad. Effects*, **1**, 229 (1969).
11. R. R. Hart, Hughes Research Laboratories, Malibu, California, private communication.

DISCUSSION

Question (KAHAN) Have you checked the back surface of the sample in the spectral reflectivity measurements? Did the 450 °C anneal, or decomposition, also occur on the back surface? We have observed the introduction of *p*-type absorption at 450 °C in annealing electron-induced damage. At this temperature, an *n*-type sample with initial carrier concentration of 6×10^{18} cm⁻³ is reduced to 5×10^{18} cm⁻³.

Answer (HUNSPERGER) No, we did not check the back surface because the chemical etch polish we use polishes only the 'B' face and leaves the 'A' surface pitted. Thus we were not able to separate the effect of decomposition from that of damage annealing, and the points at 450 °C and above are not corrected to account for decomposition. [We have also observed *p*-type electrically compensating centers in samples implanted and/or annealed above 450 °C—R. G. Hunsperger and O. J. Marsh, *Metallurgical Trans.*, **1**, 603 (1970).]

Comment (FISCHER) Modulation techniques

(electro-reflectance, wavelength modulation, etc.) are more sensitive to long-range order than ordinary reflectance and would thus probably suit your purpose better.

Reply (HUNSPERGER) Modulation techniques might provide better sensitivity; however we have experienced no difficulty resulting from lack of sensitivity, and one of the advantages of the reflectivity measurement technique that we used is its simplicity in that a commonly available spectrophotometer is employed.

Question (FISCHER) What is the minimum micro-crystallite size that would be consistent with the disappearance of structure in your scanning-electron microscope measurement?

Answer (HUNSPERGER) It would be on the order of 100 microns, so one could not say that a truly amorphous layer was formed solely on the basis of the disappearance of structure in the Ccates–Kikuchi pattern.

Question (DONOVAN) Have you considered surface roughness as a possible explanation for your observed reflectance behavior. Surface roughness would be most important at high photon energies where you observe the greatest decrease in reflectance with dose.

Answer (HUNSPERGER) Yes, we have considered roughness, and we found that it did not affect our reflectance measurements. Based on the recovery of the reflectivity spectra following annealing (as described in this paper) one could conclude that surface roughness played no important role in this case. In addition, conventional scanning electron microscope observation of the surface showed no increase in roughness as a result of implantation. (More extensive measurements on ion-implanted Si samples have also shown no increase in surface roughness.[2])

Comment (BORDERS) I have made measurements of the ion-implantation induced optical extinction on the low energy side of the band gap in GaAs and observed the same annealing stage that you see at $\sim$225 °C. It is also of interest to note that both Stein and Aukerman have observed an annealing stage at this temperature using electrical measurements on electron- and neutron-damaged GaAs.

PHYSICAL PROPERTIES OF RADIATION INDUCED AND GROWTH DEFECTS IN II–VI SEMICONDUCTORS

P. M. WILLIAMS AND A. D. YOFFE

Surface Physics, Cavendish Laboratory, University of Cambridge, Cambridge, England.

Defects in wurtzite single crystals of ZnS, ZnSe and CdS have been observed by means of both scanning and transmission electron microscopy. Application of the scanning technique to the observation of cathodoluminescence in these materials has demonstrated localized effects on the emission close to twin boundaries and stacking faults in the as-grown crystals, which may be interpreted in terms of changes in band gap at the defect. This method has also been applied to heavy ion bombarded crystals; in addition, transmission electron microscope studies have been made of these irradiated samples and the annealing behaviour of the radiation damage has been observed. Finally, in-situ optical absorption experiments have been carried out on samples irradiated at liquid nitrogen temperature, and effects on the characteristic exciton absorption have been noted.

1. INTRODUCTION

The existence of defects in semiconducting materials may be demonstrated by many techniques such as electron spin resonance, but, in general, their effects on the physical properties of the host lattice can be inferred only from indirect measurements. The aim of the present investigation, therefore, has been to observe directly the behaviour of defects in both irradiated and as-grown single crystals of II–VI semiconductors by means of electron microscopy, using both scanning and transmission instruments. In the scanning technique, luminescence is detected from thin platelet crystals, as opposed to secondary electrons in conventional use, and by correlating these observations with transmission measurements on the same crystals, the effect of defects on the electronic centres responsible for radiative recombination may be deduced. The transmission electron microscope has also been used to observe the formation and subsequent annealing of defects in heavy ion bombarded materials; in addition, in-situ optical absorption experiments have been carried out on thin platelets subjected to similar bombardment in order to assess the effects on their optical absorption properties produced by the radiation damage levels observed in the transmission microscope specimens.

2. SCANNING ELECTRON MICROSCOPE OBSERVATIONS

The technique of detecting luminescence, as opposed to secondary or scattered primary electrons, in the scanning electron microscope has been described in detail elsewhere.[1–3] The essential feature of the system is that a quartz light pipe, close to the specimen so as to subtend as large an angle as possible, is substituted for the conventional electron collector/scintillator arrangement; a luminescent 'map' of the surface of a specimen may then be built up using the normal display system of the microscope, enabling localized changes in emission close to defects to be observed. In addition, by means of a lens in the wall of the vacuum system, an image of the specimen may be focused on the entrance slit of a monochromator, so that the emission spectrum from any desired region may be observed, in a manner analogous to the X-ray microanalyser. The specimen is cooled during observations to 80 °K on a moveable liquid nitrogen cold finger.

The application of this technique to II–VI semiconductors has yielded to date much interesting information on the role played by extended defects, such as stacking faults and twin boundaries, in the luminescent properties of these materials. Thin, wurtzite platelets in $(11\bar{2}0)$ orientation, i.e. with the c axis in the growth plane, of ZnS, ZnSe and CdS have been examined. These crystals are known to contain basal plane stacking faults,[4] which may be envisaged as a thin intergrowth of the zincblende modification, three or four atomic layers thick, within the wurtzite lattice, as well as broader intergrowths of both zincblende and polytype structures. Figure 1(a) shows a typical faulted intergrowth about 2500 Å thick, in a wurtzite ZnS crystal recorded at 100 keV in transmission. The width of the faulted region is small so that in this case electron diffraction analysis of the region was not possible; in other specimens, broad homogeneous regions ($\sim 10\,\mu$m wide) of 4H, 6H, 8H and 3C polytypes have been observed in an otherwise 2H lattice. The effect produced by this fault band on the luminescence is clearly demonstrated in Figure 1(b), a luminescence micrograph recorded in the scanning microscope with a 30 keV electron beam of 5.10^{-9} amps, at 85 °K (the change in magnification between (a) and (b) should be noted). The faults, then, appear as a line of intense emission, and as can be seen from Figure 1(c) and (d) where emission spectra have been recorded from areas of pure 2H material (c) and 2H with the fault, (d), this emission is at a longer wavelength (i.e. lower energy) than that from the pure 2H material. The fact that the band gap of these semiconductors decreases as the structure changes from 2H (wurtzite) to 3C (zincblende)—in the case of ZnSe difference isthe 50 meV—suggests that recombination within the faulted band takes place in a lattice of lower band gap than the 2H surroundings. The wavelength of emission (3270 Å) is close to that observed for the 6H polytype (3280 Å) under similar conditions. This suggestion in turn leads to a model which can explain the anomalous increase in emission intensity at the intergrowth in terms of the lower band gap of the material within the latter. This decrease in gap produces an effective potential well into which carriers drift before recombination, in a manner analogous to their behaviour at other heterojunctions, such as a copper sulphide/zinc sulphide interface. A high concentration of carriers results, giving emission of greater intensity than in the surrounding lattice.

Figure 1(e) is a luminescence micrograph of a twin boundary in CdS and shows a pronounced absence of emission at the boundary itself. In this case, however, the few atomic layers comprising the boundary cannot be thought of as an intergrowth of lower band gap. The self-energy of the defect may, in fact, result in a distortion of the bands locally to give an increase in gap, so that carriers will tend to drift away from the boundary, producing a decrease in emission intensity. This phenomena should, however, be an entirely local one, extending over a few atomic planes (~ 100 Å), so that the width of the depleted region in Figure 1(e) (~ 2500 Å) suggests that another mechanism may also be operating. A twin boundary will inevitably act as a sink for diffusing defects during growth, so that it is likely that there is a high concentration of trapping sites responsible for non-radiative transitions close to the boundary; this is, however, a tentative hypothesis and will be difficult to test as a result of the comparatively low defect concentration required to produce the predicted effects.

The application of the technique to irradiated samples is illustrated in Figure 1(f). This shows a crystal of ZnS irradiated with 5.10^{13}, 100 keV Zn^+ ions cm^{-2} (c.f. Figure 3(c)) and demonstrates the marked decrease in luminescence efficiency usually associated with irradiated materials without subsequent annealing. Transmission microscope observations of ion-bombarded ZnS show that on annealing, large dislocation loops (up to 250 Å diameter) may be formed and it is hoped to observe effects due to these by improving the presently available resolution of the instrument (~ 1000 Å in the luminescence mode). This can be effected by increasing the signal to noise ratio at given lens currents by cooling the specimen to helium or hydrogen temperatures.

3. TRANSMISSION ELECTRON MICROSCOPE OBSERVATIONS

Thin wurtzite crystals of ZnS, ZnSe and CdS have been irradiated with heavy ions such as Zn^+ and Mn^+ in order to investigate the formation and annealing of the radiation damage which is inevitably associated with introduction of impurities for doping purposes by ion implantation. The latter has been suggested as a possible low-temperature method for producing type conversion in these materials which avoids auto compensation effects associated with high temperature diffusion, and it is important therefore to discover whether or not it is possible to overcome damage effects in the implanted layer.

Thin specimens (~ 1000 Å thick) supported on carbon film on a copper electron microscope grid were irradiated at room temperature with beams of Zn^+ or Mn^+ at 100 keV at a current density of approximately $0.1\,\mu$A cm^{-2} to total doses of 5.10^{12}, 5.10^{13} and 5.10^{14} ions cm^{-2}. Figures 2(a), 2(c) and 2(e) are transmission electron micrographs of wurtzite ZnSe irradiated to these three total doses respectively, with 100 keV Zn^+ ions. At the lowest dose (5.10^{12} ions cm^{-2}), Figure 2(a), isolated defect clusters, 10–30 Å in diameter are visible, and on raising the total dose to 5.10^{13} ions cm^{-2}, Figure 2(c), the concentration of these clusters which are formed by the diffusion of isolated point defects increases considerably. Their mean separation has become small (~ 50 Å) and when the total dose reaches 5.10^{14} ions cm^{-2}, Figure 2(e), the discrete

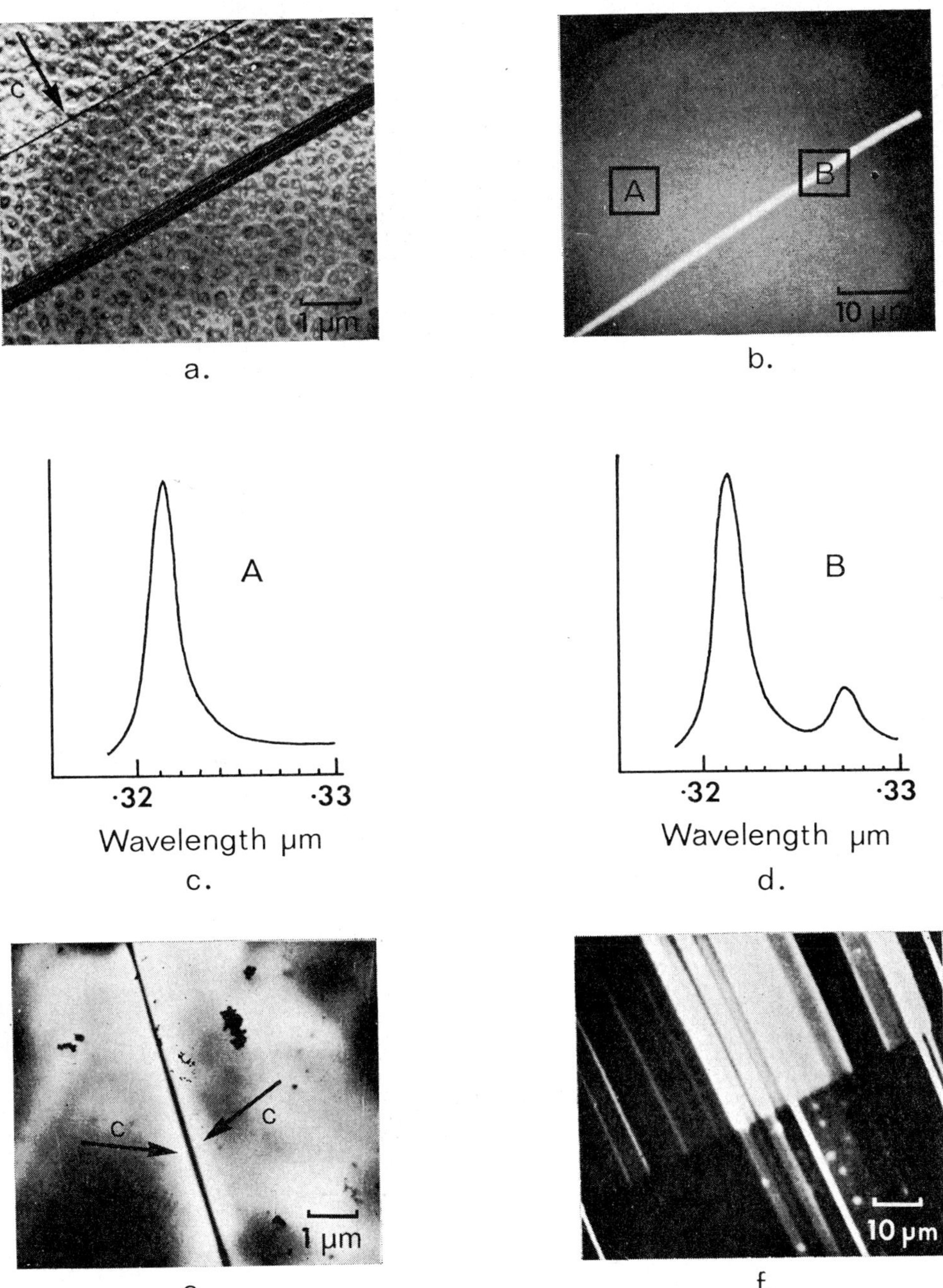

FIG. 1. (a) Transmission electron micrograph of group of stacking faults in ZnS single crystal (spotting is hydrocarbon contamination). (b) Scanning luminescence micrograph of same group of faults as in (a), recorded at 85 °K. (c) and (d) Emission spectra from areas A and B delineated in micrograph (b) respectively. (e) Scanning luminescence micrograph of twin boundary in CdS, recorded at 85 °K. (f) Scanning luminescence micrograph of ZnS single crystal, lower half irradiated with 5.10^{13} 100 keV Zn^+ ions cm^{-2}.

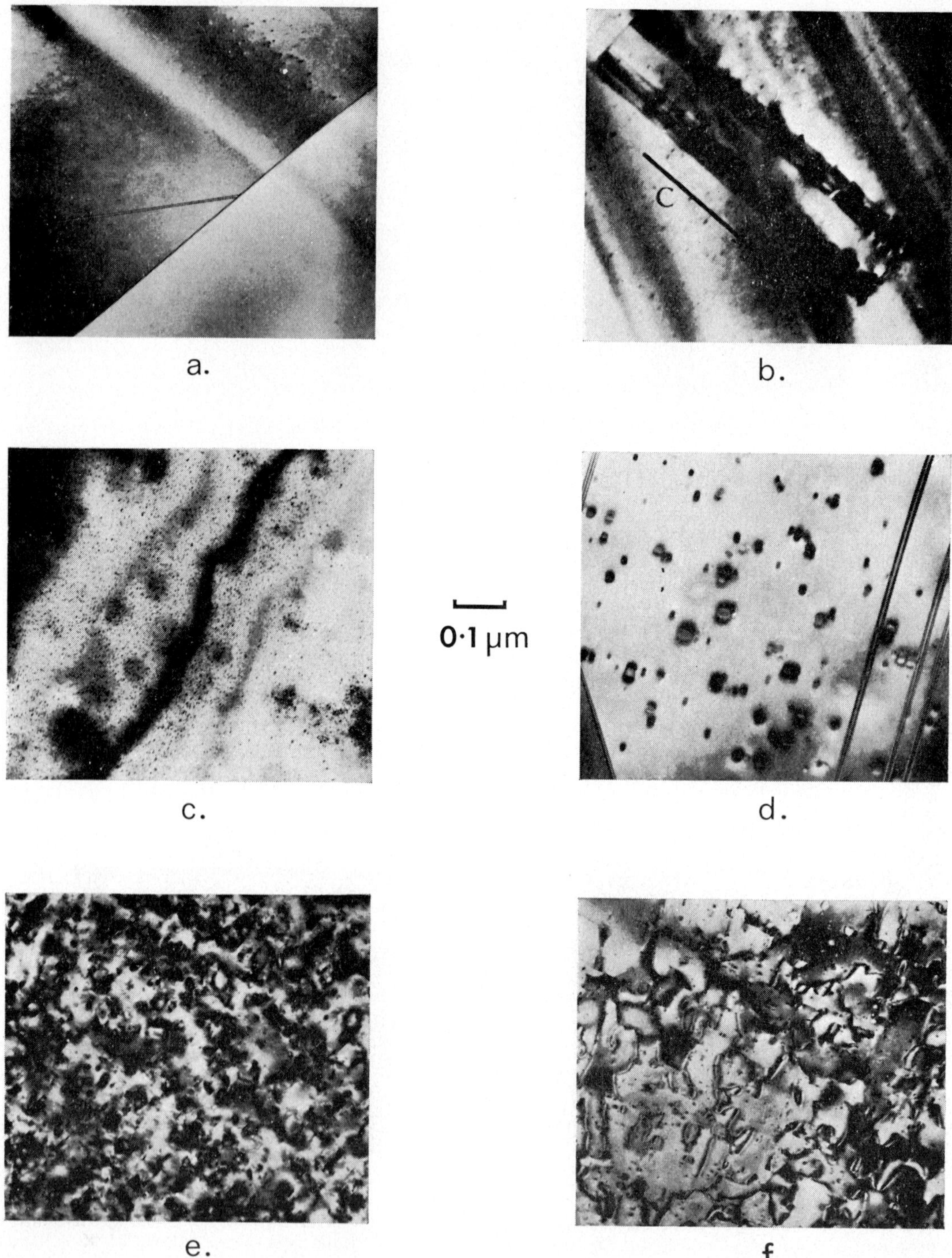

FIG. 2. (a) ZnSe irradiated with 5.10^{12} Zn$^+$ ions cm^{-2} at 100 keV. (b) Removal of prismatic plane stacking fault in ZnSe by irradiation with 5.10^{13} Zn$^+$ ions cm^{-2} at 100 keV. (c) ZnSe irradiated with 5.10^{13} Zn$^+$ ions cm^{-2} at 100 keV. (d) Crystal in (c) annealed for 2 hrs in vacuum at 500 °C, showing dislocation loops on prismatic planes. (e) ZnSe irradiated with 5.10^{14} Zn$^+$ ions cm^{-2} at 100 keV. (f) Crystal in (e) annealed for 2 hrs in vacuum at 500 °C, showing dislocation lines and loops on prismatic planes.

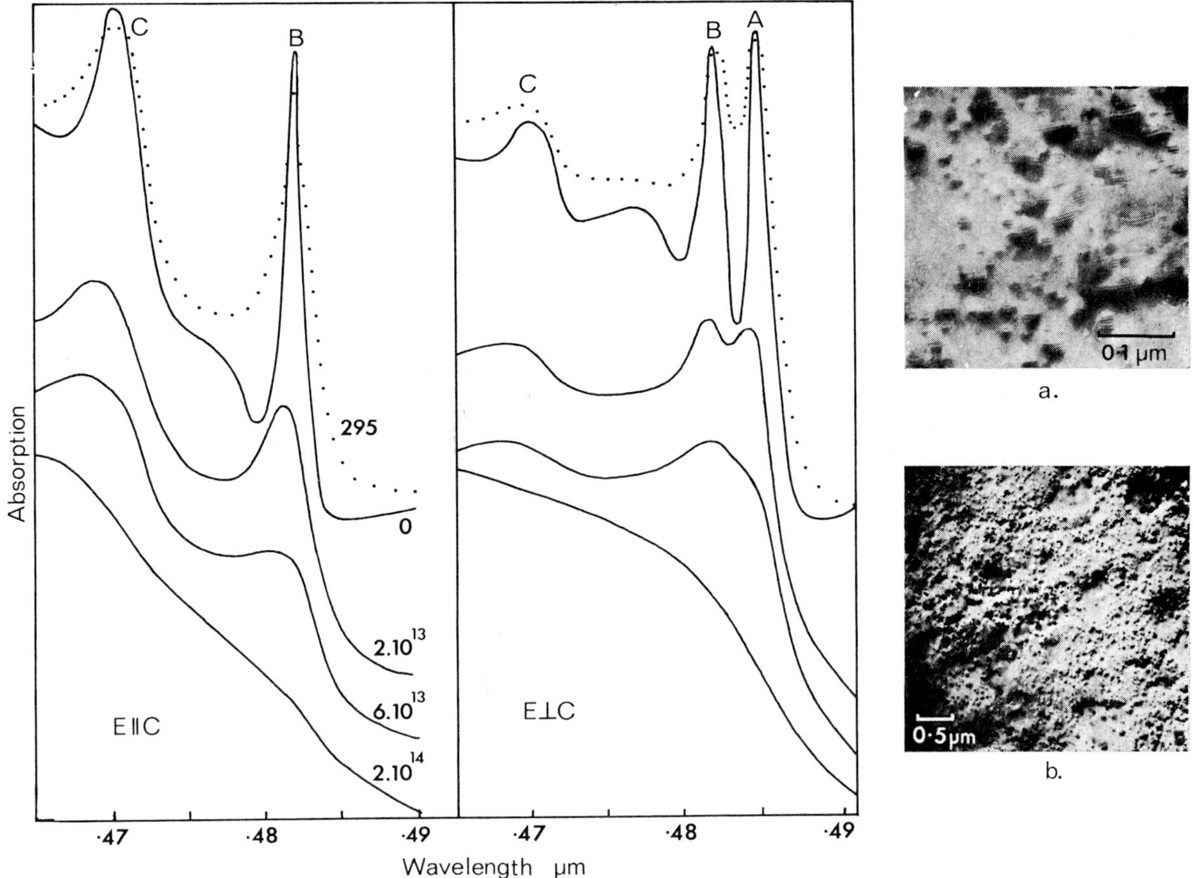

FIG. 3. Optical absorption spectra of CdS single crystal at 77 °K for polarisations $E\|c$ and $E \perp c$, recorded before irradiation with 100 keV protons, at doses of 2.10^{13} ions cm^{-2}, 6.10^{13} ions cm^{-2}, 2.10^{14} ions cm^{-2}, and finally after annealing for 12 hrs at 295 °K. Micrographs (a) and (b) show damage structure following annealing.

clusters overlap to give a complex tangle of dislocation lines and loop-like features.

The annealing behaviour of the damage has been investigated by vacuum anneals at 500 °C and appears to be strongly dependent on initial post irradiation conditions in ZnS and ZnSe. Figure 2(d) shows the effect of a 2 hr anneal at 500 °C on the crystal in Figure 2(c); dislocation loops up to 250 Å in diameter are formed on planes which have been shown by dark field microscopy to be prismatic planes in the wurtzite structure. This conclusion is further supported by the effect of irradiation on prismatic growth faults in the structure; previous investigations[5] have shown that such faults may be removed by intense electron bombardment causing interstitial or vacancy type defects to diffuse and remove either intrinsic or extrinsic faults respectively. In this case, Figure 2(b), it can be seen that the defects produced by a

total dose of 5.10^{13} 100 keV Zn$^+$ ions are sufficiently mobile to partially remove the fault, even without subsequent annealing. It should be noted that the right hand side of the fault in Figure 2(b) is almost unaffected, having been shielded from bombardment by a copper mask. Further annealing, however, beyond the stage at which dislocation loops are formed appears to nucleate at the defect sites a phase transformation from wurtzite to zincblende, since the former is only metastable at the annealing temperatures in question (zincblende is energetically the more favourable structure below approximately 700–800 °C), and it is hoped to publish these findings shortly in greater detail.

Annealing of the crystal in Figure 2(e) for 2 hrs at 500 °C produces the defect pattern shown in Figure 2(f). Some dislocation loops are formed, as with the lower dose case, but in general the complex of dislocation lines persists—in fact anneals up to 6 hrs have produced little change in this behaviour and the presence of such a high concentration of dislocations appears to inhibit the phase transformation discussed previously.

4. OPTICAL ABSORPTION EXPERIMENTS

In order to assess the effects of such damage levels on the macroscopically observable properties of II–VI semiconductors, experiments have been carried out in situ on irradiated single crystals to measure the change in their optical absorption characteristics, in addition to the scanning electron microscope observations described previously. Thin crystals of CdS were mounted on a specially designed sapphire cold finger at 77 °K, so that the beam current density at the specimen could be measured, and were then irradiated with 100 keV protons up to a total dose of 2.10^{14} ions cm^{-2}. Empirically, it appears from electron microscope observations that protons at 100 keV produce a density of radiation damage in the thin crystals used in these experiments between 10 and 20 times less than Zn^+ ions at the same energy, so that this dose is small by comparison with those discussed previously. The average projected range of 100 keV protons in CdS, however, is expected to be considerably greater than the crystal thickness (~ 500–1000 Å) so that few remain in the lattice following bombardment and the effects observed can be attributed entirely to radiation damage.

The exciton absorption properties of CdS are well known and Figure 3 shows spectra recorded for polarisations $E \parallel c$ and $E \perp c$ at 77 °K. The A, B and C excitons are sharp and well resolved prior to irradiation, but a comparatively low dose of 2.10^{14} ions cm^{-2} is sufficient to remove all but traces of the strong exciton absorption, although there is still considerable dichroism. On annealing at room temperature for 12 hrs, however, and cooling once more to 77 °K in order to record spectra, the lattice appears to recover the exciton absorption to a great extent. This is interpreted as being due to diffusion of radiation induced defects between 77 °K and 295 °K; at the lower temperature, assuming these defects to be largely

'frozen' in, their mean separation will be less than the radius of the ground state exciton in CdS (approximately 29 Å) so that the creation of such quasi particles will be inhibited, considerably reducing the absorption of photons at such energies. On annealing at 295 °K, however, the defects aggregate into larger clusters whose mean separation may be seen from micrograph 3(b) to be greater than 29 Å and hence the absorption reappears (micrographs 3(a) and 3(a) were recorded at room temperature on the same crystal used in the absorption experiments). The increase in half width of the peaks compared with the unirradiated case is a result of lifetime broadening caused by scattering of excitons during diffusion by the defect clusters. The origin of the Moiré-like fringing observed in the defect clusters in Figure 3(a) is not clear; if the defects are loops on prismatic planes inclined at 60° to the growth plane, then it is possible that the fringes are due to stacking fault contrast, as is observed at inclined faults,[4] but in general the features bear a stronger resemblance to Moiré fringes.

The optical absorption method has also been applied to materials such as As_2S_3 and MoS_2 in order to study the changes which take place during the gradual transition from crystalline to amorphous structure and these results will be published shortly.[6]

ACKNOWLEDGEMENTS

We wish to thank Dr. R. V. Hesketh of the C.E.G.B. Laboratories, Berkeley, Gloucestershire, for providing accelerator facilities; the Science Research Council and Ministry of Technology for grants to the laboratory.

REFERENCES

1. P. M. Williams and A. D. Yoffe, *Phil. Mag.*, **18**, 555 (1968).
2. P. M. Williams and A. D. Yoffe, *Rad. Effects*, **1**, 61 (1969).
3. P. M. Williams and A. D. Yoffe, *Nature*, **221**, 952 (1969).
4. L. T. Chadderton, A. G. Fitzgerald and A. D. Yoffe, *Phil. Mag.*, **8**, 167 (1963).
5. L. T. Chadderton, A. G. Fitzgerald and A. D. Yoffe, *J. Appl. Phys.*, **35**, 1582 (1964).
6. J. A. Olley and A. D. Yoffe, unpublished.

DISCUSSION

Question (STREETMAN) What was the source material? Were these vapor-grown platelets?

Answer (YOFFE) Yes.

Question (RANDOLPH) What is the characteristic annealing temperature for a reintroduction of bound excitons in CdS which were quenched following Zn ion implantation?

Answer (YOFFE) The features seen here are free excitons, and these could be seen again after a room temperature anneal.

DISORDER PRODUCED IN SiC BY ION BOMBARDMENT†

R. R. HART, H. L. DUNLAP AND O. J. MARSH

Hughes Research Laboratories, 3011 Malibu Canyon Road, Malibu, California 90265, U.S.A.

We have measured the amount of residual lattice disorder produced in α-SiC after implantation of 40-keV Sb^+ and 30-keV N^+. The implantations were performed at 23°C and, to avoid channeling effects, at a preselected random-equivalent orientation of the target with respect to the ion beam. The disorder, defined as the number of lattice atoms displaced greater than ~ 0.2Å from lattice sites, was determined by backscattered energy analysis of 280-keV He^{++} and 140-keV H^+ incident along the $\langle 0001 \rangle$ channel and in a random direction. The amount of disorder increased linearly with dose of the heavy ion (Sb^+) to a saturation condition, corresponding to the formation of an amorphous layer at a dose of $\sim 10^{14}$ Sb^+/cm^2. The disorder produced by the light ion (N^+) at first increased as the square root of N^+ dose but at $\sim 10^{14}$ N^+/cm^2 increased approximately linearly with dose to a saturation value at $\sim 10^{15}$ N^+/cm^2. Similar behavior for these two ions has been observed in Si. The number of displaced atoms of Si and C produced by an Sb ion is ≈ 2400. Comparison of this result with Sb ion displacement in Si indicates a displacement energy E_d for SiC of 17- to 30-eV. A significant annealing of lightly disordered regions in the Sb^+ and N^+ implanted layers is observed after 500°C. More heavily damaged layers anneal at 750°C and above. Little disorder remains after a 1200°C anneal. Annealing at 1650°C or greater produces evidence of decomposition, although there are indications that the implanted species inhibits further decomposition. There is no evidence of outdiffusion of Tl after 1650°C anneal and only slight evidence of outdiffusion of Sb after 1700°C anneal. Channeling analysis of Sb^+ implants after a 1500°C anneal indicate that ~ 50 per cent of the Sb ions are located along the $\langle 0001 \rangle$ atomic rows.

1. INTRODUCTION

The advent of ion implantation as a doping process for semiconductors has required investigations of the formation and annealing of disorder in order to intelligently apply the process to device fabrication. The difficulties experienced in controllably doping SiC by other processes, such as diffusion, and the ease with which ion implantation has been used to dope other semiconductors, suggested the application of ion implantation to the doping of SiC. It was the objective of this investigation to determine the parameters important both to the formation and annealing of disorder and to the location of implanted impurities in the lattice of SiC.

2. EXPERIMENTAL

The production and subsequent anneal of disorder produced in α-SiC by 40-keV Sb^+ and 30-keV N^+ implantations were investigated by measurements of the backscattered energy spectra of 280-keV He^{++}, incident in $\langle 0001 \rangle$ and random-equivalent orientations. The implantations and backscattering analysis were performed in the same system. The beam was magnetically mass separated and, to provide dose uniformity, was electrically swept

† This work was supported in part by NASA/ERC.

over the implanted area during implantations. Secondary electrons were suppressed with a -270 V shield which surrounded the target. The target chamber pressure was less than 2×10^{-7} Torr, and a cryowall near the target effectively prevented hydrocarbon contamination of the target surface.

The target was mounted on a goniometer which permitted the selection of random or $\langle 0001 \rangle$ aligned orientations to better than 0.1° by utilizing the orientation dependence of backscattering. All implantations were made in preselected random-equivalent orientations. The angular divergence of the He^{++} analysis beam was less than 0.06°, and the backscattered particles were detected at an angle of 160° to the incident direction by a cooled, surface-barrier detector (FWHM $\sim$ 7 keV).

The SiC target was grown by the Lely process and consisted of a platelet ~ 6 mm in diameter by ~ 1 mm thick, with the $\langle 0001 \rangle$ axis approximately normal to the surface. Coates–Kikuchi patterns,[1] obtained with a scanning electron microscope, suggested that the SiC crystal was of the 15R polytype. Backscattering indicated that the crystal was of good quality, since χ_{min}, the ratio of the minimum $\langle 0001 \rangle$ aligned yield near the surface to the random yield, was <0.05.

3. RESULTS AND DISCUSSION

The $\langle 0001 \rangle$ channel critical angle, $\psi_c \simeq 1.35° \pm 0.1°$, was determined from the width of the characteristic dip in yield of 280-keV He^{++}, backscattered near the surface. Comparison of this result with the critical angle measurement of Matzke and Koniger[2] ($\psi_c \simeq 1.15°$ for 500-keV He$^+$) indicates that, within experimental error, the expected $E^{-1/2}$ dependence of the critical angle on incident energy[3] is obeyed.

Backscattered energy spectra for both random and aligned orientation after implantation to various doses of 40-keV Sb$^+$ are shown in Figure 1a. The pronounced peaks near the Si surface edge at

$\sim$160 keV are caused primarily by backscattering of the aligned He^{++} from disordered Si atoms, i.e. Si atoms displaced greater than approximately the Thomas–Fermi radius, $a \sim 0.2 \, \text{Å}$, from normal lattice sites.[4,5] It can be seen that the peak height increases with increasing dose and coincides with the random level at a dose of $\sim 9 \times 10^{13}$ Sb$^+$/cm^2, indicating a near-saturation level of disorder as measured by backscattering. The peak near 240-keV represents He^{++} backscattered from the implanted Sb atoms after a dose of 9×10^{13} Sb$^+$/cm^2 and is seen to be well separated from the Si edge.

Similar measurements after 30-keV N$^+$ implantations are presented in Figure 1b. In this case, the depth distribution of disordered Si atoms is approximately three times deeper and twice as wide as that obtained with the Sb$^+$ implantations. Further, the major part of the disorder is clearly submerged below the target surface. The disorder peak coincides with the random level after a dose of about 9×10^{14} N$^+$/cm^2; however, some crystallinity remains near the target surface since the aligned yield is lower than the random yield near the surface. The small peak at the target surface in both the unimplanted and lower-dose aligned spectra is due to displaced Si atoms in the surface oxide.

The disorder was calculated for the Sb$^+$ implantations from the areas of the disorder peaks in Figure 1a according to the procedure discussed in Reference (5) and is shown in Figure 2. The disorder increases linearly with dose to the near-saturation level at 9×10^{13} Sb$^+$/cm^2; this behavior is similar to the production of disorder by 40-keV Sb$^+$ implantations of Si.[4,5] From the slope of the curve the number of displaced Si atoms per incident

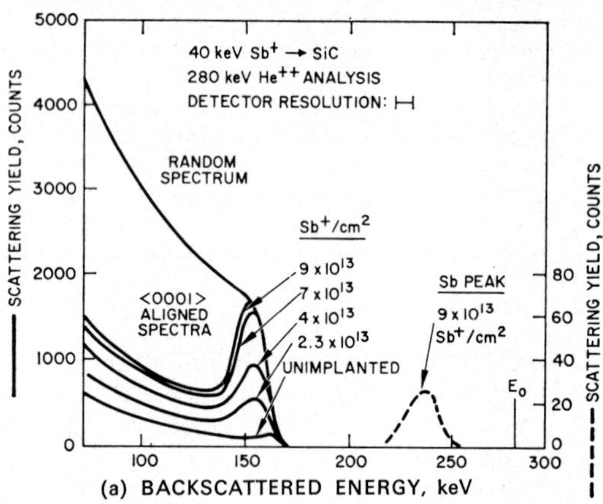

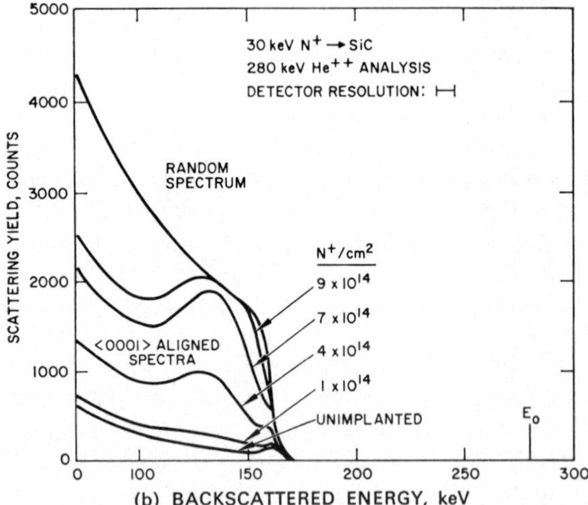

FIG. 1. Backscattered energy spectra as a function of dose for $E_0 = 280$-keV He^{++} incident on α-SiC after implantation of (a) 40-keV Sb$^+$ and (b) 30-keV N$^+$. A random–equivalent spectrum and an aligned spectrum from unimplanted crystal are included for comparison.

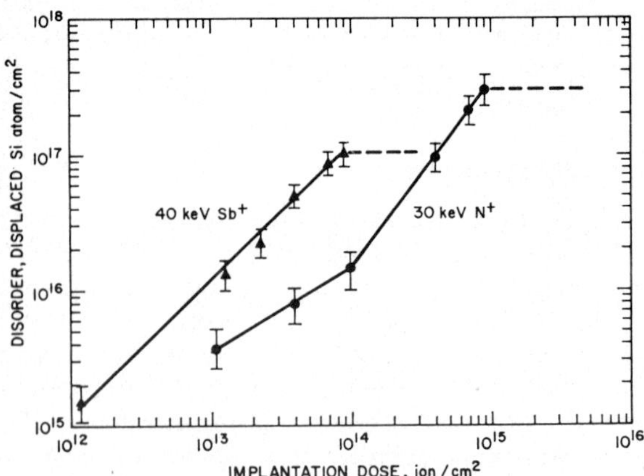

FIG. 2. The disorder produced in α-SiC by implantation of 40-keV Sb$^+$ and 30-keV N$^+$.

ion is $\sim$1200. As will be discussed later, the ratio of displaced Si atoms to displaced C atoms is approximately one. Therefore, $\sim$ 2400 Si and C atoms are displaced per incident ion. This result is lower than that measured after 40-keV Sb⁺ implantations of Si, i.e. $\sim$ 3000 to 5000 displaced Si atoms per incident ion.[4,5] Consequently, the displacement threshold energy E_d of SiC may be 1.2 to 2.1 times greater than that of Si, which is $\sim$14 eV.[6] Thus, E_d in SiC may be 17 to 30 eV.

Dechanneling of the aligned He⁺⁺ beam by disorder, i.e. scattering of the aligned particles out of the $\langle 0001 \rangle$ channel by the displaced Si and C atoms, is sufficiently large after the N⁺ implantations to preclude accurate determination of the disorder from the areas under the disorder peaks in Figure 1b. Dechanneling is manifested by the strong increase in the aligned yields in the undamaged crystal behind the disorder as the disorder increases. However, as discussed in Reference (5) on the basis of the single-scattering theory of dechanneling, the total disorder may be calculated from the yields behind the disorder peak.

D (displaced atoms/cm²)

$$\propto \ln \{ [y_r(t) - y_a(t)] / [y_r(t) - y'_a(t)] \}$$

where $y_r(t) \equiv$ random yield at depth t behind the disorder, $y_a(t) \equiv$ aligned yield from the unimplanted crystal and $y'_a(t) \equiv$ aligned yield from the implanted crystal. The results, normalized by the use of the area technique of disorder measurement calculated after the 9×10^{14} N⁺/cm² implantation, in which the dechanneling correction is small, are also presented in Figure 2. It should be noted that within experimental error the same normalization constant was determined by comparing the results of area and dechanneling calculations for the Sb⁺ implantations.

The disorder produced by the 30-keV N⁺ implantations shows two distinct regimes. At doses less than $\sim 10^{14}$ N⁺/cm², the disorder increases approximately as the square root of the dose; whereas at higher doses, the disorder increases approximately linearly with dose, although the slope may be somewhat greater than one. Since in the low dose region the slope is less than one, considerable annealing occurs during implantation; this annealing may be associated with the diffusion and partial annihilation of point defects. Above 10^{14} N⁺/cm², the amount of disorder is apparently large enough to inhibit the diffusion and annihilation of the point defects. We have observed a

similar behavior of disorder formation produced by 40-keV B⁺ implantations of Si, as have Eisen and Welch,[7] who studied 200-keV B⁺ implantations of Si. These authors also found a significant dose-rate dependence of disorder in the high-dose region, so that similar results may be expected for SiC. The dose-rate of the 30-keV N⁺ implantations of SiC in the high-dose region was $\sim 0.2 \, \mu\text{A/cm}^2$.

Also it should be noted from Figure 2, that the dose required to reach near-saturation of the disorder is about a factor of 10 higher for the N⁺ as compared to the Sb⁺. Again, we have noted a similar ratio for N⁺ and Sb⁺ implantations of Si. Further, the saturation levels of disorder have a ratio of about three which is consistent with the ratio of the disorder depths in Figure 1a and 1b.

To determine the ratio between the number of displaced Si atoms to the number of displaced C atoms, backscattering measurements were also performed with 140-keV H⁺. Backscattered energy spectra for both random and aligned orientations after implantation with 9×10^{13}, 40-keV Sb⁺/cm are shown in Figure 3. The peak at channel number 87 in the aligned spectrum is the Si disorder peak, whereas, the smaller peak at channel number 72 is caused by backscattering from displaced C atoms. The ratio between the two areas, when corrected by the Z_2^2 dependence of the Rutherford scattering cross section, is 1.0 ± 0.1, which indicates that approximately equal numbers of Si and C atoms

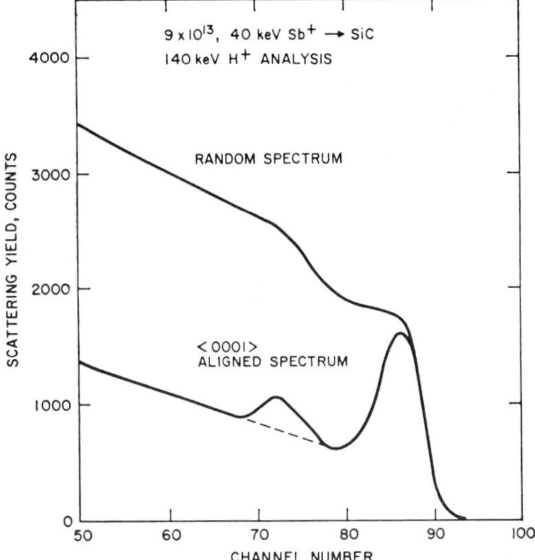

FIG. 3. Backscattered energy spectra for random-equivalent and aligned orientation of SiC after implantation of 9×10^{13} Sb⁺/cm² at 40 keV. Analysis beam is 140-keV H⁺.

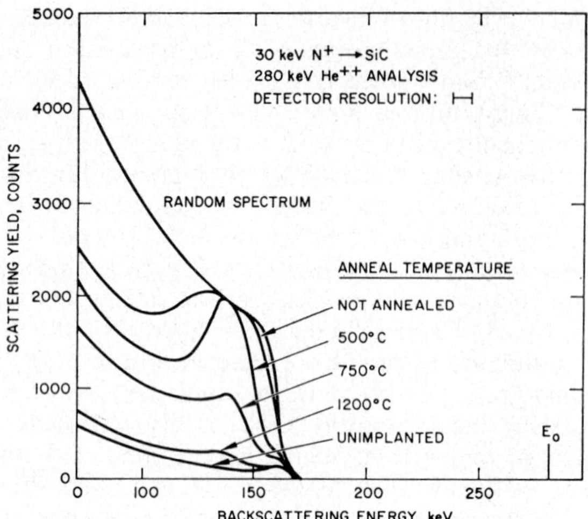

FIG. 4. Backscattered energy spectra of SiC as a function of anneal temperature for the sample described in Figure 1b after the $9 \times 10^{14} \, N^+/cm^2$ implant.

are displaced. The same result was measured after lower dose Sb^+ implantations, although the experimental error was larger. The step in the random spectrum at channel number 75 is caused by backscattering from the C atoms in the SiC substrate. The Si disorder peak in the aligned spectrum does not quite reach the random level as it did for the He^{++} spectrum because the detector resolution corresponds to a depth slightly greater than the width of the disorder.

After the $9 \times 10^{13} \, Sb^+/cm^2$ and $9 \times 10^{14} \, N^+/cm^2$ implantations the SiC target was subsequently annealed for 15 min periods in a N_2 atmosphere for temperatures up to 1300 °C and for 2 min periods in vacuum ($\approx 10^{-6}$ Torr) at 1500 °C and 1700 °C. After each anneal the amount of residual disorder was measured at 23 °C by the backscattering technique. Before each measurement the target was washed in HF to remove any thick surface oxide which may have been formed during annealing. In addition, any C deposits, formed by surface decomposition of the SiC during the two high temperature vacuum anneals was removed by oxidation at 600 °C. No measurable oxidation of SiC takes place at this temperature.[8]

Backscattered energy spectra after several anneals of the N^+-produced disorder are shown in Figure 4. Regrowth of the disorder first proceeds from both the underlying substrate and the target surface, as observed after the 500 °C anneal. This result suggests that lightly disordered SiC anneals signi-

ficantly by 500 °C in agreement with Canepa who observed measurable annealing of proton produced damage after 200 °C annealing.[9] The disorder peak then decreases in height at higher anneal temperatures. Although not presented, the backscattered spectra for the $9 \times 10^{13} \, Sb^+/cm^2$ implanted layer shows regrowth first from only the underlying substrate and then a disorder peak decreasing with increasing anneal temperature.

To obtain quantitative anneal data, the disorder was analyzed after each anneal in the same manner as discussed earlier. Figure 5 gives the relative disorder remaining for both implants as a function of anneal temperature. Both implants behave similarly with an annealing stage centered around $\sim$750 °C and then leveling off to a small amount of residual disorder even after the highest temperature anneals studied here. The dashed line indicates surface decomposition which will be discussed later. The annealing behavior is somewhat less abrupt than that measured after 40-keV Sb^+ implantation into Si at a dose sufficient to produce a saturation level of disorder.[4] In addition, the SiC anneal stage is $\sim$200 °C higher than the Si anneal stage at $\sim$550 °C. Other investigators have also reported an anneal stage, as measured by electrical properties, at approximately 800 °C in SiC damaged by electrons,[10] protons,[9] and neutrons.[11] It is also

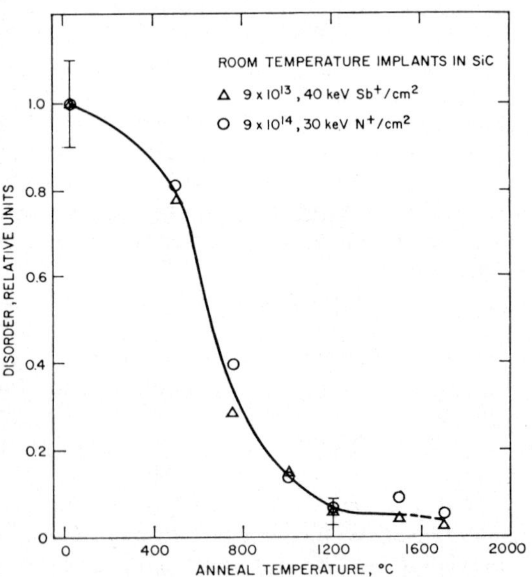

FIG. 5. The annealing of disorder in SiC introduced by $9 \times 10^{13} \, Sb^+/cm^2$ or $9 \times 10^{14} \, N^+/cm^2$ implantations. For both cases the relative disorder was normalized to 1.0 after 23 °C anneal.

interesting to note that the first strong indications of a *p-n* junction formed by nitrogen implants into *p*-type SiC are found after 750 °C anneal.[12] Although there appears to be little measurable annealing taking place at temperatures above 1200 °C, as shown in Figure 5, the measured electron Hall mobility of nitrogen implanted layers in SiC[12,13] continues to increase significantly with annealing to 1700 °C, indicating a strong reduction in the density of electron scattering centers.

In order to determine whether implanted heavy ions might significantly diffuse during high temperature anneals, as well as to determine the decomposition temperature, the depth distribution of implanted Tl and Sb ions was also measured by backscattering after each anneal. Energy spectra of He^{++}, backscattered from Tl atoms implanted at 200-keV to a dose of 8×10^{14} ions/cm^2 and after 1450 °C and 1650 °C vacuum anneals, are shown in Figure 6a. The spectra closely represent the depth distribution of the Tl atoms except at the Tl surface edge where the detector resolution must be considered. The Tl surface edge, representing the backscattering energy for Tl atoms on the target surface, was calculated from elastic collision theory, based on the energy of the Si surface edge. The depth scale was obtained from the projected range calculations for Tl in SiC of Johnson and Gibbons.[14] The Tl distribution did not significantly change up to an anneal temperature of 1450 °C; however, after 1650 °C ~ 250 Å of the surface was decomposed. Evidence for this was a visible discoloration of the surface caused by the carbon decomposition product. In addition, there was a significant energy loss of the incident He^{++} before backscattering from Si atoms as shown by a shift to lower energy at the Si surface edge. Removal of the carbon by a 500 °C oxidation step eliminated the discoloration and returned the Si surface edge to its normal value. The Tl edge had now moved to the surface with little change in the shape of the distribution. Also there was little loss of Tl atoms since the areas under the two peaks in Figure 6a are approximately equal. This indicates that the decomposition halted near the leading edge of the implant.

Somewhat similar results were obtained for the 40-keV 9×10^{13} Sb$^+$/cm^2 distribution as are shown in Figure 6b after 1500 °C and 1700 °C anneals. Again, no significant change in the Sb distribution was observed up to 1500 °C, indicating that little diffusion occured. After the 1700 °C anneal, approximately 75 Å of the surface decomposed,

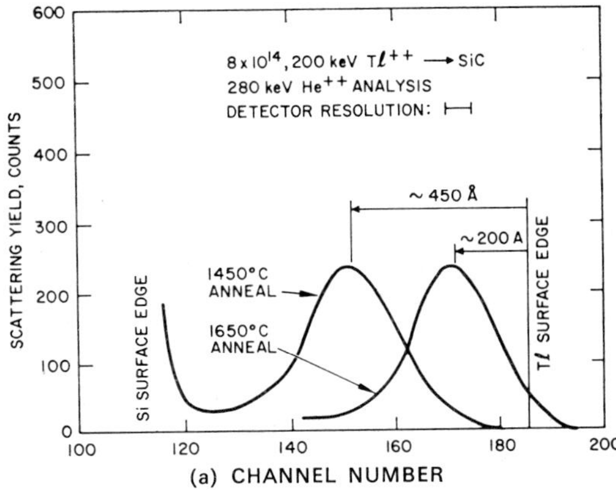

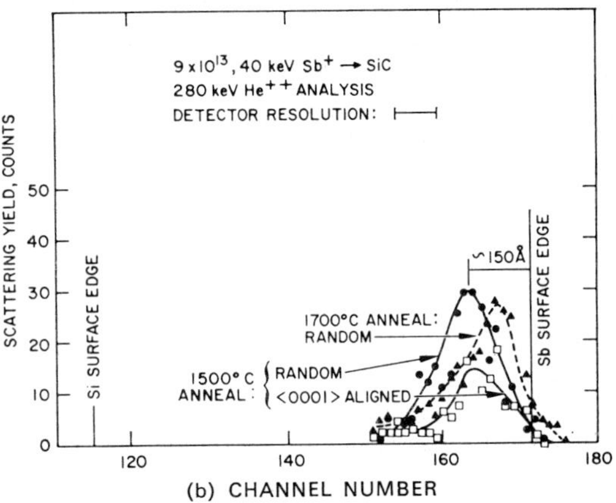

FIG. 6. Spectra of 280-keV He^{++} ions backscattered from impurity atoms in SiC: (a) from Tl$^+$ implanted at 200-keV to a dose of 8×10^{14} Tl$^+$/cm^2 and after anneal at 1450 °C and 1650 °C; (b) from Sb$^+$ implanted at 40-keV to a dose of 9×10^{13} Sb$^+$/cm^2 and after anneal at 1500 °C and 1700 °C.

but the decomposition again halted near the leading edge of the implant. Since the shape and area of the Sb peak appears to have changed, some diffusion and loss of Sb may have occurred after the 1700 °C anneal. However, the counting statistics for this low dose implant preclude a definitive statement.

Also included in Figure 6b is an aligned spectrum for the Sb peak after 1500 °C anneal. Since the area of the aligned peak is less than the area of the random peak by about a factor of $\frac{1}{2}$, approximately 50 per cent of the Sb atoms are along the $\langle 0001 \rangle$ atomic rows and appear to be substitutional; this is suggested by a similar reduction in the Sb peak along a channeling direction other than the $\langle 0001 \rangle$.[4] Previous studies of the electrical behavior of Sb implanted layers in SiC have indicated that an n-type layer is formed.[12,13] The channeling studies tend to confirm that the n-type behavior is due to donor behavior of Sb in a normal substitutional lattice site.

Analysis of the electrical behavior of layers in SiC implanted with acceptor type dopants (Column III of the Periodic Table) after annealing at $\approx 1700\,°C$ have not indicated p-type conduction.[12,13] A possible explanation is that the dopants do not occupy a substitutional site. The Tl implants of Figure 6a after 1650 °C anneal showed no change of the Tl spectrum with aligned and random orientation indicating that no measurable substitutional Tl was present. Although this result tends to confirm the above hypothesis, the dose implanted may substantially exceed the solid solubility for Tl in SiC.

4. CONCLUSIONS

The production of disorder in α-SiC with implantation of 40-keV Sb^+ or 30-keV N^+ at room temperature, as measured by the backscattered energy spectra of 280-keV He^{++}, is similar to that observed for these ions implanted into Si. The disorder produced by a heavy ion, Sb^+, increases linearly with dose to a saturation level at $\sim 10^{14}\,Sb^+/cm^2$. The number of Si and C atoms displaced per incident Sb^+ is ~ 2400 and the ratio $Si/C = 1 \pm 0.1$ (from 140-keV H^+ spectra). A comparison with Si[4,5] of the number of displaced atoms per incident Sb^+ suggests that the displacement threshold energy E_d of SiC is 17 to 30 eV. The disorder produced by a light ion, N^+, increases as the square root of dose at low dose values and approximately linearly with dose at higher values to a saturation condition at $\sim 10^{15}\,N^+/cm^2$.

Considerable reordering is observed in the lightly disordered regions of the N^+ and Sb^+ implants after a 500 °C anneal. The more heavily disordered regions show significant annealing at 750 °C. Other investigations of radiation damage in SiC have observed an anneal stage at approximately this temperature.[9,10,11] Two-minute anneals of heavy ion (Sb^+ and Tl^+) implanted layers in vacuum at temperatures of 1650–1700 °C result in measurable

decomposition of the SiC surface. The decomposition appears to be inhibited by the presence of the implanted impurity. No outdiffusion of Tl is observed after anneal at 1650 °C and there are only slight indications of outdiffusion for Sb after a 1700 °C anneal.

A comparison of 280-keV He^{++} channeled and aligned spectra of a 40-keV implant of $\sim 10^{14}\,Sb^+/cm^2$ after a 1500 °C anneal indicates that ~ 50 per cent of the Sb atoms are located along the $\langle 0001 \rangle$ atomic rows. This result suggests that the n-type behavior observed in annealed Sb^+ implanted layers in SiC[12,13] is due to a normal substitutional donor behavior of Sb.

ACKNOWLEDGEMENT

We wish to thank R. B. Campbell of the Westinghouse Astronuclear Laboratories for the SiC used in this work and E. D. Wolf for his analysis of crystal structure.

REFERENCES

1. E. D. Wolf and R. G. Hunsperger, *Appl. Phys. Letters*, **16**, 526 (1970).
2. H. Matzke and M. Koniger, *Phys. Stat. Sol.* (a) **1**, 469 (1970).
3. J. Lindhard, *Mat. Fys. Medd. Dan. Vid. Selsk.*, **34** (1965).
4. J. A. Davies, J. Denhartog, L. Eriksson and J. W. Mayer, *Can. Journ. Phys.*, **45**, 4053 (1967).
5. R. R. Hart, *Rad. Effects*, **6**, 51 (1970).
6. P. Sigmund, *Appl. Phys. Letters*, **14**, 114 (1969).
7. F. H. Eisen and B. Welch, *Proc. Conf. on Ion Implantation*, Reading, Englans, 1970.
8. Guy Ervin, Jr., *Jour. Am. Ceram. Soc.*, **41**, 347 (1958).
9. P. C. Canepa, P. Malinaric, R. B. Campbell and J. Ostroski, *IEEE Tran. on Nuclear Science*, **NS-11**, 262 (1964).
10. E. W. J. Mitchell and M. J. Moore, *Radiation Damage in Semiconductors* (Academic Press, New York, 1964), p. 235.
11. P. Nagels and M. Denayer, 225, ibid.
12. O. J. Marsh and H. L. Dunlap, *Rad. Effects*, **6**, 301 (1970).
13. H. L. Dunlap and O. J. Marsh, *Appl. Phys. Letters*, **15**, 311 (1969).
14. W. S. Johnson and J. F. Gibbons, *Projected Range Statistics in Semiconductors*, distributed by Stanford University Bookstore (1969).

SUMMARY OF THE CONFERENCE
(Radiation Effects in Semiconductors—1970)

H. Y. FAN

Department of Physics, Purdue University, Lafayette, Indiana, U.S.A.

This paper summarizes the work of the conference and in the field, emphasizing important remaining problems and giving a general overall view of progress in the field.

1. INTRODUCTION

I wish to take this opportunity to express on behalf of the participants our deep appreciation to Professor Corbett and all the people who were responsible for the local organization of this successful Conference. The convenience and comfort provided by the thoughtful arrangements made it easy for us to devote our attentions to the technical matters. For the highly interesting program we owe a debt to the Program Committee. Several excellent reviews were contained in the program, covering important aspects of the field. Some important problems of the field were pointed out in the keynote address. In the following brief talk I shall be mainly concerned with the new contributions presented at the Conference and with a general overall view of the progress in the field.

2. POINT DEFECTS

2.1. *Theory*

So far, only simple defects—vacancy, divacancy and interstitial—have been considered by theoretical treatments. Two different approaches have been used. In one approach, the problem is treated as a perturbation in the band theory of the crystal. The second approach considers a limited number of atoms using a molecular orbital treatment. The problem involves the inherent complication of lattice relaxation, i.e. shifts of the surrounding atoms from their regular positions. Band theory calculations are complicated even when the atomic positions are prescribed and not to be determined. With a clearly simplified basis, the defect molecule approach may lead to more tractable calculations and has been used to investigate effects of lattice relaxation, particularly the Jahn–Teller effect of lattice distortion.

A treatment has been presented which considers a molecule made up of a large number (34) of atoms.[1] Similar to some treatments of defects in metals, the approach has the advantage that the number of atoms is finite but should cover those atoms of significance for the defect. Computer calculations were made for diamond without taking into account electron-electron interactions explicitly. For a molecule without the center atom, a level in the energy gap was obtained which was assigned to a neutral vacancy. Atomic relaxation was considered. An outward symmetric relaxation of the neighboring atoms was calculated to be about 10 per cent, and a Jahn–Teller energy of ~ 0.5 eV was indicated by some investigation of trigonal and tetragonal distortions. Lattice relaxation was also included in the calculation presented for a vacancy in various charge states in silicon.[2] The usual model of a defect molecule made up of the neighbouring atoms was used and electron-electron interaction was taken into account. The symmetric relaxation was inward and it lowered the various electronic levels by different amounts in the range of 3 to 5 eV. For the centers considered—V^0, V^-, and $V^=$—a first order treatment of the Jahn–Teller effect predicted very small splittings, less than 0.1 eV, for most of the electronic levels, with no change in the order of the lowest levels. It is interesting that the Jahn–Teller effect predicted is much smaller than it was thought to be and than the indication mentioned above for diamond. The paper pointed out that the first order treatment may need improvement in view of the large symmetric distortion. Also, a careful calculation of the Jahn–Teller distortion has yet to be made in the treatment for diamond.

2.2. *Germanium and silicon*

2.2.A. *Defect identification.* An important step made in the understanding of radiation damage is the realization that a large variety of defects may be involved, the transformation of which affect the properties of the material. Characteristic effects which exhibit sufficient details, e.g., EPR and infrared absorption, are effective means for distinguishing different types of defects. The identification consists of establishing the structure or the model for defects of a distinct type. For silicon, a large number of defects have been isolated. Models have been fairly well established for a number of them, including single and double vacancies, associations of vacancies with oxygen and with various group III and group V impurities, and interstitial group III atoms. The situation is less advanced for germanium. A number of defects containing oxygen have been recognized, one of which was identified as an oxygen-vacancy complex similar to the A-center in silicon. Some defects are recognized by more or less clear characteristics. Among these are the defects associated respectively with the 65 °K and the 35 °K annealing stages of n-type germanium, and the defects connected with the ~ 60 °K annealing stage and the 'two-level defect' in p-type germanium. The 65 °K defects in n-type germanium are believed to be vacancy-interstitial pairs and the 35 °K defects apparently involve an impurity. Indications of the existence of other defects have been provided by various studies, and models have been suggested. It seems that, except for the oxygen-vacancy complex, clear identification is needed in most cases.

The following contributions of this Conference may be noted. Consider first silicon. EPR measurements on neutron- and electron-irradiated silicon[3] revealed a new type of defect center; from the symmetries of g-tensor and hyperfine interaction, a model consisting of an oxygen atom and several vacancies was deduced. Luminescence measurements showed that electron irradiation introduced into silicon two sets of zero phonon and phonon assisted lines;[4] they were found to be associated with divacancies and K-centers, respectively. In Li-doped silicon, lines were observed which were attributed to Li-containing defect complexes. Structures due to two different kinds of Li-containing defects were found also in EPR spectra after electron irradiation.[5] Studies of Mössbauer effect have been made on Co-doped silicon;[6] a broad spectrum which was produced by neutron irradiation and decayed with time was assumed to be given by mobile Co interstitials in various charge states. In C-doped silicon infrared lattice absorption revealed various defect complexes which involved the impurity.[7,8]

In the case of germanium, electron irradiation of P-doped material was found to introduce a new EPR spectrum,[9] from the g-tensor symmetry and the effect of applied stress, it was deduced that the defects responsible were consistent with P interstitials which had undergone [100] displacement due to the Jahn–Teller effect. A luminescence spectrum introduced by electron or gamma ray irradiation was observed;[10] the set of lines was believed to be associated with a type of defect center containing some impurity.

2.2.B. *Defect migration.* Information about defect migration is often deduced from indications of defect transformations. The stability and evolution of defect centers are known to depend on the charge state of the centers or their components. The effect is evident in illumination-induced and radiation-induced annealings of n-type germanium. It has been reported that the 50–65 °K defects in p-type germanium can also be induced to anneal at liquid helium temperatures by white light as well as by electron irradiation.[11] We have to bear in mind also that the evolution of defect complexes involves not only the motion but also the association or dissociation of the components. The activation energy or the characteristic temperature deduced pertains to the process rather than the migration itself.

We have the following information regarding the mobilities of vacancies and interstitials. The charge state of the defect can be seen to have a strong effect. For silicon,[12] the migration energy E is ~ 0.18 eV for $V^=$, ~ 0.33 eV for V^0 and ~ 1.3 eV for V_2, the divacancy. The temperature T at which there is detectable motion of interstitials is 4.2 °K in p-type and probably ~ 140 °K in n-type silicon. For charged vacancies in germanium, $T \sim 65$ °K was deduced from the development of local mode absorption and an activation energy of $E_a \sim 0.2$ eV was estimated for the process.[13] The 65 °K defect in electron-irradiated germanium was found to anneal with an activation energy of $E_a \sim 0.15$ eV[14] which has been assumed to be associated with the motion of interstitials.

Additional information is provided by the present Conference. Studies of local mode absorption associated with carbon indicate that interstitials in p-type silicon can migrate at 100 °K.[8] In n-type germanium irradiated by electrons, annealing of the 65 °K defect as induced by 0.5 MeV electrons indicates a migration energy of 0.004 eV for interstitials whose charge state is altered under the radiation which induces the annealing.[15] As to the motion of impurities, annealing studies of germanium under illumination suggest that motion of singly charged Sb interstitials takes place at 27 °K and motion of doubly charged Sb occurs at 4.5 °K.[16] It is proposed from studies of the 220–270 °K annealing of p-type germanium[17] that the motion of In interstitials occurs with an activation energy of $E_a = 0.7$ eV while for Ga interstitials $E_a = 0.4$ eV. In silicon kept at room temperature, changes of the EPR spectra attributed to Li-containing complexes are interpreted as resulting from the diffusion of Li.[5] The room temperature decay of the Mössbauer spectrum attributed to Co interstitials is assumed to be the result of the migration and precipitation of Co.[6]

2.2.C. *Defect evolution.*

Most of the defect centers contain some impurity. Impurities may also affect the evolution of defects which do not contain them as constituents. Several papers show that the formation or annihilation of defect centers depends upon which of the group III acceptors or which of the group V donors is involved. Thermal as well as radiation-induced annealing of the 65 °K defects in n-type germanium proceeds more rapidly in As-doped than in Sb-doped material.[18] For p-type germanium, the rate of defect introduction by electron-irradiation at ~10 °K was found to be higher in Ga-doped than in In-doped samples.[11] Meanwhile, from annealing studies in the temperature range 80–170 °K it was concluded that E-center like vacancy-impurity association is favored more by In as compared with Ga, suggesting that the larger sized impurity is more favorable for the formation.[17] In silicon containing 0.6 per cent germanium atoms,[8] studies of the 12 micron absorption band of A centers indicated that Ge traps vacancies to form centers which are stable up to 220 °K.

Concerning Li impurity, the introduction rate of $(O–V)$ centers in silicon as shown by EPR measurements was greatly reduced in the presence of Li and even more so regarding the $(P–V)$ centers.[19] The observations indicate that Li captures vacancies in effective competition with the two other impurities and that it also reduces the amount of available P by pairing. Indications of a similar effect of Li were given also by studies on germanium.[20] The introduction rate of a defect acceptor level was found to be much lower in samples containing Li; the effect was apparently caused by an association of Li with the impurity needed to form the defect. Also, a reduction in the annealing rate was observed for the introduced

donor levels in the upper half of the energy gap, and the explanation was suggested that the presence of Li reduces the concentration of free vacancies which produce the anneal.

One of the few clear cut effects known about interstitials is that they produce Al interstitials in silicon by exchanging places with the substitutional impurity. It has also been suggested from studies of annealing by electrical measurement that a similar effect takes place with Sb impurity in germanium.[21] Studies reported here[8] on local mode absorption of C-containing defects in p-type silicon indicate that the defects were produced by a similar effect of interstitials which were mobile at ~100 °K; mobile vacancies were ruled out as being responsible by the absence of A centers. The measurements of Mössbauer effect in Co-doped silicon[6] showed that the +0.06 cm/sec line of substitutional Co was weakened by neutron irradiation, and the exchange of interstitials with substitutional Co is thought to be the cause of this effect.

2.3. *Compound semiconductors*

Radiation damage studies have emphasized germanium and silicon, the two typical semiconductors with relatively simple crystal structures and best understood properties. It is to be expected that more attention will be given to other semiconductors as progress is made. Different and larger varieties of defects may be found in materials with different structures. The production and evolution of the defects as well as effects on the properties of the material may involve new considerations.

Beside the review papers, the present Conference contained studies on Te, SiC, GaAs and CdTe. In the case of tellurium, irradiation by low energy (620 kV) electrons at 15 °K produced the interesting observation that both the hole concentration and the Hall mobility were increased.[22] Since the introduction of a net amount of ionized acceptors normally tends to reduce the mobility, verification and explanation of the result may provide some insight about the defects introduced.

For GaAs, photoluminescence revealed a type of defect center introduced by electron irradiation.[23] The defect is probably As vacancy. Optical measurements at room temperature had shown that electron or neutron irradiation introduced only a structureless absorption which increased smoothly with frequency toward the absorption edge. Two distinct absorption steps had been found, however, in measurements on neutron irradiated samples at $\sim 80\,°$K. The measurements reported here[23a] reveal a peak at ~ 1.25 microns in electron irradiated samples at liquid nitrogen temperature. Annealing of the band with changes of shape took place between 150 and 300 °C and was rather broad, indicating that the band might consist of overlapping peaks which anneal differently. Other studies[24] show strong anneals of carrier concentration and mobility within the annealing temperature range of this band.

Electrons of various energies were used for the irradiation of CdTe.[25] The fact that the threshold electron energy is different for the displacement of Cd and Te was utilized to identify the defects corresponding to various observed lines of cathodoluminescence. Considering also the observed effect of heat treatments in Cd vapor, several lines were attributed to Cd vacancies and one of the lines was thought to be given by defects containing a Te vacancy. For CdTe and CdS, defect production by recoil Cd of (n, γ) reaction becomes important in view of the large cross section of Cd for neutron capture, and the effect of thermal neutrons in a reactor is shown to be many times larger than that of the fast neutrons.[26]

For various polytypes of α-SiC irradiated with neutrons or α-particles, different types of defects have been characterized by EPR and optical transmission measurements.[27]

3. DAMAGE REGIONS

3.1. *Particle irradiation*

In irradiations with heavy or high energy particles, an atom displaced with a large recoil energy produces many defects in a small region. Such damage regions may be referred to as disordered regions or defect clusters in case their structure and properties may be considered in terms of the same kind of crystal with a large concentration of defects. In general, it is conceivable to have regions of entirely different structures, including other crystalline phases, the amorphous phase, and precipitates of some constituent. The existence of damage regions has been established for various kinds of irradiations: neutron, deutron and high energy electrons, and the regions are usually thought of as disordered regions.

Ion bombardment has gained increased attention in recent years and there is evidence that regions of amorphous material are produced. Since the energy of primary displaced atoms under particle

irradiations may be comparable to the energy used in ion bombardments, similar types of damage might be expected for the two cases. Studies were presented which concern this question.[28] Electron transmission microscopy measurements showed that silicon and germanium remained crystalline after irradiation at ~ 50 °C with large fluences $(5 \times 10^{20}\ cm^{-2})$ of neutrons; a similar conclusion was reached by X-ray diffraction measurements. On the other hand, germanium bombarded with 100 keV O^- ions $(\sim 10^{12}$ ions/cm$^2)$ showed an amorphous ring pattern superimposed on the normal spot pattern of electron transmission diffraction; after a dose of 10^{15} ions/cm^2 only the ring pattern was seen. It was suggested that individual damage regions remain crystalline in both cases. In ion bombardments, overlap of damage regions renders them amorphous since the cascade damage process can be altered by the resulting large defect concentration. Energy transport may be decreased because of a reduction in collision chain length and the damage will be more concentrated. In the neutron irradiation, the samples were at ~ 50 °C for several months and many of the introduced defects annealed out before an overlap of damage regions occurred. The consideration leads one to expect that amorphous regions can be produced by neutron irradiation at low temperatures.

Consider the papers on particle irradiations. Annealing studies on neutron irradiated p-type germanium were made from 77 °K to room temperature.[29] In contrast to the case of irradiation by 3 MeV electrons, the resistivity leveled at a value lower than that before irradiation. It was suggested that neutron irradiation introduces p-type, low resistivity regions as well as disordered regions with space charge.

In neutron-irradiated silicon, infrared absorption measurements provided evidence for the liberation of vacancies as a result of the decay of disordered regions.[30] Following an irradiation at 76 °K, the 1.8 micron absorption band of divacancies increased with annealing at ~ 140 °K, similar to the recovery of minority carrier lifetime. In EPR measurements of silicon irradiated by neutrons at room temperature, it was possible to identify a number of centers similar to those produced by electron irradiation.[31] In particular, the G7 center of the divacancy was observed together with the spectra of the neutral P impurity and also vacancy-oxygen centers in samples with oxygen. The observation indicates that the Fermi level was not uniform with respect to the energy band such that the divacancies were paramagnetic in the disordered regions with low Fermi level, and the neutral P and $(V-O)$ centers were detectable outside where the Fermi level was close to the conduction band.

3.2. Ion bombardment

Ion bombardment concerns directly the problem of moving atoms in a solid. The channeling effect is especially interesting and it is useful for the detection of lattice disorder. Estimates of displacements for Ge and for Si produced by α-particle bombardment of a Si-Ge alloy were obtained from measurements of α-particle backscattering.[32] In the papers presented on ion implantation, increase of backscattering for channeled beams and reduction in the number of channeled ions were used as a means for estimating lattice disorder,[33] and the energy spectrum of backscattered ions served to indicate the depth of scattering centers.[34] Channeling has an important effect on the ease of implantation. The depth of implanted ions was observed to depend on the crystal orientation, increasing with the size of channels.[33] Of the two mechanisms for energy loss of the ions being implanted, nuclear collisions dominate at high energies whereas electronic stopping is important for channeled ions. For III-V semiconductors, the depth of channeled ions was found to be generally larger for group II acceptors than for group VI donor ions, due to the difference in electronic stopping.

The lattice damage produced by ion bombardment plays an important role in limiting the depth of implantation. This is easily understandable in the case of channeled implantation since atomic displacement evidently reduces the passage along the channels. The depth was shown to decrease with continued implantation and additional ions concentrated closer and closer to the surface.[33] The near-band-edge luminescence of GaAs was quenched by Zn ion implantation due to the damage introduced.[35] Annealing at 480 °C built up the luminescence. As surface layers were removed, the luminescence assumed the form of a peak characteristic of Zn-doped GaAs, at a depth of 1280 to 1930 Å below the surface, indicating substitution of Zn for Ga. For the three materials: GaAs, GaP and Ge, bombardment with 20 keV S yielded a much larger concentration of implanted ions at 150 °C than at room temperature.[33] Also, channeling measurement of backscattering showed that implantation of 140 keV Zn in GaAs produced less scattering centers at 165 °K than at 100 °K, and

annealing at 75 °C caused nearly complete reduction of scattering centers.[34] These results indicate that the disorder anneals at rather low temperatures, similar in this sense to the damage introduced by electron irradiation.

With continued implantation of 100 keV Sb^+ or 200 keV Tl^{++}, into α-SiC backscattering, as well as the electrical properties of the implanted layer, was shown to approach a saturation which indicates the formation of an amorphous material.[36] A decrease of optical reflectivity and a degradation of Coates-Kikuchi pattern in electron diffraction were utilized to study surface damage. The studies also indicated a saturation with dose which can be attributed to the formation of an amorphous layer.[37] The damage produced in II-VI compounds by 100 keV Zn^+ bombardment was investigated by means of transmission electron spectroscopy.[38] Bombardment at room temperature with a dose of 5×10^{14} ions caused the lattice to saturate with defects resembling dislocation loops and tangles which were not removed by high temperature annealing. With low doses, $\sim 5 \times 10^{13}$ ions, defect clusters of ~ 10 Å in diameter formed large loops upon annealing at 500 °C. Continued annealing at this temperature initiated a phase change from wurtzite to sphalerite as a result of the motion of dislocations.

4. SUMMARY

The following are some comments about the present state and future prospects of the field.

The identification of several additional types of defects has been reported at this Conference. Also, such effects as Mössbauer spectroscopy and hopping conductivity[39] have been shown to be useful techniques for this purpose. Reviewing the situation, we note that germanium is still behind silicon with regard to the number of defects clearly identified. Secondly, it is desirable to have more information about the interstitials. A full account of the displaced atoms remains a question.

Additional information has been reported concerning the motion of an interstitial in p-type silicon and the motion of an interstitial in some charge state in germanium. We should have more detailed information about interstitial migration in each material and vacancy migration in germanium, for various charge states of the defect in question. Effects of the migration of Li and Co in silicon, and of Sb, In and Ga in germanium are indicated in some of the papers. Definitive information is needed about the migration of various impurities.

Interesting effects of various impurities on the formation and evolution of defect centers are shown in several papers. Some results seem to indicate that an impurity of larger atomic size is more effective in associating with a vacancy. Such empirical notions may serve as useful guides for the understanding of interactions between lattice imperfections.

A natural development is to extend careful studies to various other semiconductors. One obvious attraction of compound semiconductors is that a composition of different atoms presents a larger variety of defects. There is also the possibility of producing displacements of different kinds of atoms discriminately by the choice of the direction or energy of the incident radiation. This possibility can help to separate different types of defects as it is exploited in some of the papers presented.

There is a great deal to be learned about the structures and properties of damage regions. The present practice is to think broadly in terms of two categories, disordered crystalline regions and amorphous regions. Sometimes it is assumed that the properties of a disordered region can be predicted by extrapolation from the properties of the crystal with small concentrations of point defects. The properties of the amorphous region have not yet been seriously considered. It may be necessary to know some specific details about the structure of such a region. Finally, there are damage regions not covered by the two categories mentioned. The dislocation tangles and regions of a different crystalline phase which have been reported for the II–VI compounds are such examples.

Regarding theory, even the simplest defects are difficult to treat. At this Conference, interesting contributions have been presented on the treatment of a vacancy. The promise shown by the computer calculation for a large assembly of atoms may encourage other attempts along this line. Perhaps, the interstitial problem will also be considered. Treatments capable of including lattice relaxation may help to solve the problem of migration. The observed strong dependence of migration on the charge state of the defect is an interesting effect to explain. For the time being, however, semiempirical treatments may be the practical approach for most problems in radiation damage.

Radiation damage concerns lattice defects which constitute a major area of solid state physics. It is particularly important for semiconductors, the properties of which are strongly sensitive to lattice imperfections. After more than two decades of research, the field has achieved some degree of maturity. We may expect significant knowledge to be obtained ever more effectively from studies in this field of large scope.

REFERENCES

1. R. P. Messmer and G. D. Watkins, this conference.
2. F. P. Larkins, this conference.
3. K. L. Brower, this conference.
4. C. E. Jones and W. D. Compton, this conference; E. S. Johnson and W. D. Compton, this conference.
5. B. Goldstein, this conference.
6. K. Matsui, R. R. Hasiguti and H. Onodera, this conference.
7. R. C. Newman and A. R. Bean, this conference. See also reference (12).
8. A. Brelot and J. Charlemagne, this conference.
9. A. Hiraki, this conference.
10. R. J. Spry and J. D. Henes, this conference.
11. R. A. Matula and E. E. Klontz, this conference.
12. G. D. Watkins, *Radiation Effects in Semiconductors*, Ed. F. L. Vook (Plenum Press, New York, 1968), p. 67; G. D. Watkins and J. W. Corbett, *Phys. Rev.*, **138**, A543 (1965).
13. R. E. Whan, *Phys. Rev.*, **140**, A690 (1965).
14. J. Zizine, *Radiation Effects in Semiconductors*, Ed. F. L. Vook (Plenum Press, New York, 1968), p. 186.
15. W. D. Hyatt and J. S. Koehler, this conference.
16. J. Bourgoin and F. Mollot, this conference.
17. H. Saito, N. Fukuoka and Y. Tatsumi, this conference.
18. J. M. Meese and J. W. MacKay, this conference.
19. J. A. Naber, H. Horiye and B. C. Passenheim, this conference.
20. V. S. Vavilov, A. V. Spitsyn and M. V. Tchukichev, this conference.
21. A. Hiraki, J. W. Cleland and J. H. Crawford, Jr., *Radiation Effects in Semiconductors*, Ed. F. L. Vook (Plenum Press, New York, 1968), p. 224.
22. E. Gmelin, R. Stapf, P. Klemt, G. Landwehr, W. Lichtenberg and A. Przybylski, this conference.
23. M. U. Jeong and Y. Inuishi, this conference.
23a. K. V. Vaidyanathan and L. A. K. Watt, this conference.
24. A. Kahan, L. Bouthilette and H. M. DeAngelis, this conference.
25. F. J. Bryant and D. H. J. Totterdell, this conference.
26. C. Kikuchi, this conference.
27. J. V. Barinov, J. V. Bulgakov, M. I. Iglitzin, M. A. Iljin, M. G. Kosaganova, N. M. Pavlov, M. B. Reifman, B. A. Sakharov, V. N. Solomatin, this conference.
28. M. L. Swanson, J. R. Parsons, C. W. Hoelke, this conference.
29. D. Wolf, this conference.
30. C. E. Barnes, this conference.
31. D. F. Daly and H. E. Noffke, this conference.
32. P. Baruch, F. Abel, C. Cohen and M. Bruneaux, this conference.
33. J. L. Whitton, this conference.
34. W. H. Weisenberger, S. T. Picraux and F. L. Vook, this conference.

35. G. W. Arnold, R. E. Whan, J. K. Maurin, J. A. Borders, this conference.

36. R. R. Hart, H. L. Dunlap and O. J. Marsh, this conference.

37. R. G. Hunsperger, E. D. Wolf, G. A. Shifrin, O. J. Marsh, D. M. Jamba, this conference.

38. P. M. Williams and A. D. Yoffe, this conference.

39. R. E. McKeighen and J. S. Koehler, this conference.

List of Participants

Albany, H. J.
Commissariat a l'Energie Atomique
SEP-Centre d'Etudes Nucleaires de Saclay
B.P. no. 2 Gif-Sur Yvette (91) FRANCE

Ammerlaan, C. A. J.
Physical Laboratory
University of Amsterdam
Valckenierstraat 65
Amsterdam, NETHERLANDS

Arnold, G. W.
Sandia Laboratories
Division 5111, P.O. Box 5800
Albuquerque, New Mexico 87115

Arizumi, Tetsuya
Nagoya University
Faculty of Science
Chikusa-ku
Nagoya, JAPAN

Barnes, Charles
Sandia Laboratories
Division 5112, Box 5800
Albuquerque, New Mexico 87115

Baroody, Eugene M.
Battelle Memorial Institute
505 King Avenue
Columbus, Ohio 43210

Baruch, P.
Groupe de Physique des Solides
de'l Ecole Normale Superieure
Tour 23–9, quai Saint-Bernard
Paris 5e, FRANCE

Becker, J.
Physics Department
Rensselaer Polytechnic Institute
Troy, New York

Beezhold, Wendland
Sandia Laboratories
Division 5112, Box 5800
Albuquerque, New Mexico 87115

Berry, Brian S.
IBM Research Center
P.O. Box 218
Yorktown Heights, New York 10598

Bielle-Daspet, Danielle
C.E.S.R.
B.P. 4057
31 Toulouse, FRANCE

Bishay, Adli
The American University in Cairo
113 Sharia Kasr El Aini
Cairo, EGYPT

Blankenship, James L.
Oak Ridge National Laboratory
P.O. Box X
Oak Ridge, Tennessee

Borders, J. A.
Sandia Laboratories
Division 5111, P. O. Box 5800
Albuquerque, New Mexico 87115

Bourgoin, Jacques C.
Physics Department
SUNY/Albany
Albany, N.Y. 12203

Brelot, A.
Groupe de Physique des Solides
de'l Ecole Normale Superieure
Tour 23–9, quai Saint-Bernard
Paris 5e, FRANCE

Brower, K. L.
Sandia Laboratories
Division 5111, P.O. Box 5800
Albuquerque, New Mexico 87115

Brownridge, James D.
SUNY/Binghamton
R.D. 2
Binghamton, New York 13903

Brucker, George J.
RCA Astro Electronics Division
Princeton, New Jersey

Butler, Dwain K.
B-319, U.S. Naval Ordnance Laboratory
White Oak, Maryland 20740

Castner, Theodore G., Jr.
Department of Physics and Astronomy
University of Rochester
Rochester, New York

Causey, Charles
Isotopes Power Systems Branch
SNS-USAEC
Mail Stop F 309
Washington, D.C. 20545

Charlemagne, J.
Groupe de Physique des Solides
de'l Ecole Normale Superieure
Tour 23–9, quai Saint-Bernard
Paris 5e, FRANCE

Chen, C. S.
Physics Department
Rensselar Polytechnic Institute
Troy, New York

Choyke, W. J.
Westinghouse Research Laboratories
Churchill Boro
Pittsburgh, Pennsylvania 15235

Cleland, Jown W.
Oak Ridge National Laboratories
Bldg. 2000 X-10 ORNL
Oak Ridge, Tennessee 37830

Cohen, Morrel H.
The University of Chicago
James Franck Institute
5640 S. Ellis Avenue
Chicago, Illinois 60637

Collins, Alan T.
Physics Department
King's College
Stand W.C.2
London, ENGLAND

Comas, James
Naval Research Laboratory
Code 5212
Washington, D.C. 20390

Compton, W. Dale
Ford Motor Company
Scientific Research Staff
P.O. Box 2053
Dearborn, Michigan 48121

Cooper, Larry R.
Office of Naval Research, Code 422
800 Quincey Street
Arlington, Virginia 22217

Corelli, John C.
Physics Department
Rensselaer Polytechnic Institute
Troy, New York

Crawford, James H., Jr.
Physics Department, Phillips Hall
University of North Carolina
Chapel Hill, North Carolina 27514

Crossman, Leon D.
Dow Corning Corporation
Semiconductor Division
Geddes Road
Hemlock, Michigan 48626

Curtis, Orlie, Jr.
Northrop Corporation Laboratories
3401 West Broadway
Hawthorne, California

Daly, Daniel F.
Bell Telephone Laboratories (2B-138)
Whippany, New Jersey 07981

Davies, Gordon
University of London, King's College
Physics Department
Strand, London, W.C.2, ENGLAND

DeAngelis, Henry M.
AFCRL (CRWH) L. G. Hanscom Field
Air Force Cambridge Research Laboratories
Solid State Sciences Laboratory
Radiation Effects Branch
Bedford, Massachusetts 01730

Donovan, Terence
Michelson Laboratory
China Lake, California 93555

Dresselhaus, M. S.
Electrical Eng. Department, 13–3026
Cambridge, Mass. 02139

Drevinsky, Peter J.
AFCRL (CRWH) L. G. Hanscom Field
Air Force Cambridge Research Laboratories
Solid State Sciences Laboratory
Radiation Effects Branch
Bedford, Massachusetts, 01730

Eisen, Fred H.
Science Center
North American Rockwell
1049 Camino Dos Rios
Thousand Oaks, California 91360

Euler, Ferdinand
Air Force Cambridge Research Laboratories
CRWD
L. G. Hanscom Field
Bedford, Massachusetts 01730

Fan, H. Y.
Department of Physics
Purdue University
Lafayette, Indiana 47907

Faraday, Bruce J.
Code 6065, Naval Research Laboratory
Washington, D.C. 20390

Fischer, John E.
Code 6019, Michelson Laboratory
China Lake, California 93555

Gibson, Walter M.
Bell Telephone Laboratories
Murray Hill, New Jersey 07974

Glotin, Philippe M.
Bell Telephone Laboratory
Mountain Avenue
Murray Hill, New Jersey 07974

Gmelin, Eberhard
Physikalisches Institut
Universitat
8700 Wurzburg, WEST GERMANY

Goben, Charles A.
University of Missouri
Graduate Center for Materials Research
Rolla, Missouri 65401

Goldstein, Bernard
RCA Laboratories
Princeton, New Jersey 08540

Gregory, Bob Lee
Sandia Laboratories
Division 2653
Albuquerque, New Mexico 87115

Grosmann, Michel
University of Strasbourg
9 Rue Curie
67 Strasbourg 3, FRANCE

Hale, Edward B.
University of Missouri
Physics Department
Rolla, Missouri 65401

Hart, Ron R.
Hughes Research Laboratories
3011 Malibu Canyon Road
Malibu, California 90265

Hasiguti, Ryukiti R.
Department of Metallurgy and Materials Science
Faculty of Engineering
University of Tokyo
Bunkyo-ku, Tokyo, JAPAN

Herve, Alain
Laboratoire de Resonance Magnetique
CENG
28, Cours de la Liberation
Grenoble 38, FRANCE

Hiraki, Akio
c/o Professor J. W. Mayer
Department of Electrical Engineering
California Institute of Technology
Pasadena, California 91109

Hirata, Mitsuji
Physics Department
Boston College
Boston, Massachusetts

Hunsperger, Robert G.
Hughes Research Laboratories
3011 Malibu Canyon Road
Malibu, California 90265

Hyatt, William D.
Physics Department
University of Illinois
Urbana, Illinois 61801

Inuishi, Yoshio
Faculty of Engineering
Osaka University
Yamada-Kami, Suita
Osaka, JAPAN

Johnson, Eric Shanks
Coordinated Science Laboratory
University of Illinois
Urbana, Illinois 61801

Jones, Colin E.
Room D316, Eng. Quad.
Princeton University
Princeton, New Jersey 08540

Kahan, Alfred
Air Force Cambridge Research Laboratories
CRWD
L. G. Hanscom Field
Bedford, Massachusetts 01730

Kao, K. C.
Electrical Engineering Department
University of Manitoba
Winnipeg 19
Manitoba, CANADA

Kelly, Thomas M.
Research Laboratories B-59
Eastman Kodak Company
Rochester, New York 14650

Kikuchi, Chihiro
Department of Nuclear Engineering
University of Michigan
Ann Arbor, Michigan 48105

Kimerling, Lionel C.
Air Force Cambridge Research Laboratories
CRWH, L. G. Hanscom Field
Bedford, Massachusetts 01730

Klontz, Everett E.
Physics Department
Purdue University
Lafayette, Indiana 47907

Koehler, J. S.
Physics Department
University of Illinois
Urbana, Illinois 61801

Komatsubara, Kiichi F.
Hitachi Central Research Laboratory
Kokubunji, Tokyo, JAPAN

Komiya, Hiroyoshi
Mitsubishi Electric Corporation
Central Research Laboratory
Solid State Physics Section
Minamishimizu, Amagasaki, JAPAN

Landwehr, Gottfried
Physikalisches Institut
Universitat Wurzburg
Rontgenring 8,
87 Wurzburg, GERMANY

Larkins, F. P.
Theoretical Physics Division
Atomic Energy Research Establishment
Harwell, Berks., ENGLAND

Leadon, Roland
Gulf General Atomic
P.O. Box 608
San Diego, California 92112

Lee, Chul-Chu
Department of Physics
Yonsei University
Seoul, KOREA

Lenglart, Pierre
Institut Superieur d'Electronique du Nord 3
rue F. Baes 59
Lille, FRANCE

MacKay, John W.
Physics Department
Purdue University
Lafayette, Indiana 47907

Marsden, D. A.
Solid State Science Branch
Atomic Energy of Canada Ltd.
Chalk River, Ontario, CANADA

Mashovets, Mrs. T. V.
Laboratory of Professor S. M. Ryvkin
IOFFE Physico-Technical Institute
Leningrad, USSR

Massarani, Bassain
Groupe de Physique des Solides
10 rue Ch. Peguy
93 Stains, FRANCE

Masters, B. J.
Bldg. 340–58, Department 171
IBM Components Division
Hopewell Junction, New York 12533

Matula, Richard A.
Purdue University
Physics Department
Lafayette, Indiana 47907

Mayer, James
Electrical Engineering Department
California Institute of Technology
Pasadena, California 91109

McKeighen, Ronald E.
University of Illinois
Urbana, Illinois 61801

Meese, Jon M.
Physics Department
University of Dayton
Dayton, Ohio

Merz, James L.
Bell Telephone Laboratories
Mountain Avenue
Murray Hill, New Jersey 07974

Messmer, Richard P.
G. E. Research and Development Center
P.O. Box 8
Schenectady, N.Y. 12301

Mitchell, E. W. J.
J. J. Thomson Physical Laboratory
Whiteknights Park
Reading, ENGLAND

Mollot, Francis
Groupe de Physique des Solides
de'l Ecole Normale Superieure
Tour 23–9, quai Saint-Bernard
Paris 5e, FRANCE

Moorehead, Frederick F., Jr.
IBM Research
Box 218
Yorktown Heights, New York 10598

Muir, Arthur H., Jr.
Science Center
North American Rockwell Corporation
1049 Camino Dos Rios
Thousand Oaks, California 91360

Naber, James A.
Gulf General Atomic
P.O. Box 608
San Diego, California 92112

Nagels, Piet
Solid State Physics Department
S.C.K./C.E.N.
B-2400 MOL, BELGIUM

Nelson, William D.
International Business Machines Corporation
Route 17C
Owego, New York

Newman, R. C.
J. J. Thomson Physical Laboratory
Whiteknights Park
Reading, Berks., ENGLAND

North, James C.
Bell Telephone Laboratories
Murray Hill, New Jersey

Oldham, William G.
Department of Electrical Engineering
University of California
Berkeley, California 94720

Padgett, Doran W.
Office of Naval Research
Arlington, Virginia 22217

Palkuti, Leslie J.
Battelle Memorial Institute
Columbus Laboratories
505 King Avenue
Columbus, Ohio 43201

Palmer, D. W.
School of Mathematical and Physical Sciences
University of Sussex
Brighton, BN1 9QH, UNITED KINGDOM

Pearson, Gerald L.
Stanford Electronics Laboratory
Stanford, California 94305

Picraux, S. T.
Sandia Laboratories
Division 5111, P.O. Box 5800
Albuquerque, New Mexico 87115

Poirier, Raymond
Laboratoire de Recherches de la SESCOSEM
Domaine de Corbeville 91
Orsay, FRANCE

Ramm, Wolfgang J.
U.S. Army Electronics Command
Institute for Exploratory Research
AMSEL-XL-G
Fort Monmouth, New Jersey 07703

Randolph, Lynwood P.
Harry Diamond Laboratories
Army Material Command
Washington D.C. 20438

Robison, Charles H.
Qtrs. 6301 B
U. S. Air Force Academy
Colorado 80840

Roizes, Alain
ONERA/DERTS
51 Rue Caranan
31 Toulouse, FRANCE

Roosild, Sven A.
Air Force Cambridge Research Laboratories
L. G. Hanscom Field
Bedford, Massachusetts 01821

Saito, Haruo
Department of Physics
College of General Education
Osaka University
Toyonaka, Osaka, JAPAN

Saris, Frans
FDM Institute for Atomic and Molecular Physics
Kruislaan 407, Amsterdam, HOLLAND

Seeger, Alfred K.
Max-Planck-Institut fur Metallforschung
Institut fur Physik
7000 Stuttgart 1,
Azenbergstr. 12, GERMANY

Shulman, H.
Teledyne-Isotopes
50 Van Buren Avenue
Westwood, New Jersey 07675

Smith, Thomas C.
VEECO Instruments, Inc.
Terminal Drive
Plainview, New York 11803

Spitsyn, Alexei V.
P.N. Lebedev Physics Institute
Academy of Sciences of the U.S.S.R.
Leninsky Prospect 53
Moscow, USSR

Spry, Robert James
Air Force Materials Laboratory, M.A.Y.E.
Wright-Patterson A.F.B.
Ohio 45433

Srour, Joseph R.
Northrop Corporate Laboratories
3401 West Broadway
Hawthorne, California 90250

Stein, Herman J.
Sandia Laboratories
Division 5111, P.O. Box 5800
Albuquerque, New Mexico 87115

Stoneham, A. M.
Theoretical Physics, B.8.9, A.E.R.E.
Harwell, Didcot
Berks, UNITED KINGDOM

Streetman, B. G.
Coordinated Science Laboratory
University of Illinois
Urbana, Illinois 61801

Swanson, M. L.
Atomic Energy of Canada Ltd.
Solid State Science Branch, St. 82
Chalk River, Ontario, CANADA

Tan, Swie-In
IBM, T. J. Watson Research Center
P.O. Box 218
Yorktown Heights, New York 10598

Thiessen, Klaus
German Academy of Sciences
108 Berlin, Mohrenstr. 40141 DDR, GERMANY

Thompson, M. W.
Physics Department
University of Sussex
Brighton, BN1 9QH
Sussex, ENGLAND

Totterdell, D. H. J.
Physics Department
University of Hull
HU6 7RX, ENGLAND

Ure, Roland W., Jr.
University of Utah
Materials Science Division
Salt Lake City, Utah 84112

Urli, Natko B.
Institute " RUDJER BOSKOVIC "
Gajeva 57, Zabreb, YUGOSLAVIA

Vaidyanathan, K. V.
Solid State Science Branch
Chalk River Nuclear Laboratories
Waterloo, Ontario, CANADA

Van der Weg, W. F.
Philips Research Laboratories
Oosterringdyk 18
Amsterdam, NETHERLANDS

Vook, F. L.
Sandia Laboratories
Division 5111, Box 5800
Albuquerque, New Mexico 87115

Walker, David M.
Georgia Tech. Nuclear Research Center
900 Atlantic Drive
Atlanta, Georgia 30318

Walker, James W.
University of Illinois
202 EERL Dept. of Electrical Engineering
Urbana, Illinois 61801

Watkins, George
G. E. Research and Development Center
Schenectady, New York 12301

Weisenberger, Wesley
Sandia Laboratories
Division 5111, Box 5800
Albuquerque, New Mexico 87115

Whitehouse, John E.
J. J. Thomson Physical Laboratory
Whiteknight Park
Reading, ENGLAND

Whitton, J. L.
Atomic Energy of Canada Ltd.
Solid State Science Branch
Chalk River, Ontario, CANADA

Wilsey, Neal D.
Code 6460
U.S. Naval Research Laboratory
Washington, D.C. 20390

Wilson, Donald K.
Bell Telephone Laboratories
Whippany, New Jersey

Witteles, A. A.
Singer-General Precision, Inc.
1225 McBride Avenue
Little Falls, New Jersey 07424

Wolf, Dieter
Institut fur Reine u Angewandte Kernphysik
23 Kiel
Olshausenstr. 40–60, GERMANY

Yoffe, A. D.
University of Cambridge
Cavendish Laboratory
Cambridge, ENGLAND

Young, R. C.
Physics Department
Rensselaer Polytechnic Institute
Troy, New York

Ziegler, James F.
IBM Research
Yorktown Heights, New York 10598

Author Index